Throughout his life Lewis Fry Richardson made many inspired contributions to various disciplines. He is best known for his wealth of important work on meteorology, and his ground-breaking application of mathematics to the causes of war. But his field of interest was in no way limited to these, and various aspects of psychology and mathematical approximation also benefited from his novel modes of thought.

Richardson had a rare determination to trust his own ideas, even when they were not well received. These two volumes show that much of his thinking has long been under-rated, and that much of his work was ahead of its time.

Collected Papers of Lewis Fry Richardson

VOLUME 2

LEWIS FRY RICHARDSON, D.Sc., FRS

Reproduced with permission of Godfrey Argent

Collected Papers of
LEWIS FRY RICHARDSON

Edited by

Oliver M. Ashford
H. Charnock
P. G. Drazin
J. C. R. Hunt
P. Smoker
Ian Sutherland

VOLUME 2 **Quantitative psychology and studies of conflict**

General Editor: Ian Sutherland

CAMBRIDGE UNIVERSITY PRESS
Cambridge, New York, Melbourne, Madrid, Cape Town, Singapore, São Paulo, Delhi

Cambridge University Press
The Edinburgh Building, Cambridge CB2 8RU, UK

Published in the United States of America by Cambridge University Press, New York

www.cambridge.org
Information on this title: www.cambridge.org/9780521115353

First published 1993
This digitally printed version 2009

A catalogue record for this publication is available from the British Library

Library of Congress Cataloguing in Publication data
Richardson, Lewis Fry, 1881–1953.
Quantitative psychology and studies of conflict/edited by Oliver M.
Ashford ... [et al.]: general editor, Ian Sutherland.
p. cm. – (Collected papers of Lewis Fry Richardson ; v. 2)
Includes bibliographical references.
ISBN 0 521 38298 X
1. Psychometrics. 2. Senses and sensation – Statistical methods.
3. Conflict (Psychology) – Statistical methods. 4. War – Psychological
aspects – Statistical methods. I. Ashford, Oliver M. II. Sutherland,
Ian, 1921– . III. Title. IV. Series: Richardson, Lewis Fry, 1881–1953.
Essays: v. 2.
QC861.2.R4725 1993 vol. 2

[BF39]
551.5 s–dc20
[150'.1'51] 92-6788 CIP

ISBN 978-0-521-38297-7 hardback (volume one)
ISBN 978-0-521-13591-7 paperback (set of two parts)
ISBN 978-0-521-13592-4 paperback (part one of volume one)
ISBN 978-0-521-13593-1 paperback (part two of volume one)
ISBN 978-0-521-38298-4 hardback (volume two)
ISBN 978-0-521-11535-3 paperback (volume two)

Contents

Acknowledgements

The editors are grateful to Anthony and Joan Rampton and to the Royal Meteorological Society for their encouragement and generous financial support.

The editors are grateful to the following copyright holders, who have generously given permission for the republication of Richardson's work, as indicated.

The American Meteorological Society (1948:3)

The British Association (1915, 1938:2)

British Journal of Sociology courtesy of Routledge (1952:1)

The British Psychological Society (1928:3, 1929:4, 1930:3, 1933:4, 1937:1, 1950:1, 1952:4)

The Friend (1926:2)

The Galton Institute (1913:1, 1950:2)

Heldref Publications (1929:3, 1930:2)

The Institute of Physics (1908:1, 1911, 1919:5, 1920:5, 1923:2, 1924:2, 1928:2, 1928:5, 1929:2, 1932:4, 1933:2, 1933:5, 1935:1)

International Thomson Publishing Services Ltd (1952:1)

The International Union of Geodesy and Geophysics (1928:1)

The London Mathematical Society (1921:3)

The Mathematical Association (1925:4)

Nature courtesy of Macmillan Magazines (1919:6, 1932:2, 1933:3, 1935:2, 1935:3, 1938:3, 1941:1, 1941:2, 1944:1, 1945, 1946:2, 1947:1, 1951:4, 1951:5)

Psychometrika (1948:2)

The Richardson Family (1919:1, 1936, 1951:3)

The Royal Astronomical Society (1926:6, 1933:1)

The Royal Dublin Society (1908:2)

The Royal Meteorological Society (1919:2, 1920:4, 1922:3, 1924:1, 1925:1, 1925:2, 1949:1, 1952:3)

The Royal Society (1910, 1919:3, 1919:4, 1920:2, 1920:3, 1923:3, 1926:1, 1927:1, 1937:2, 1946:3, 1950:3, 1952:5)

The Royal Statistical Society (1944:2, 1946:1, 1952:2)

Sankhyā (1953:3)

Philosophical Magazine courtesy of Taylor & Francis Ltd (1925:3, 1928:4, 1953:2)

Friedrich Vieweg & Sohn (1929:1)

Foreword

by Cedric A. B. Smith
Chairman, Conflict Research Society

I have been honoured to be asked to write a foreword to this volume of Lewis Fry Richardson's work. But the best advice to the reader is to go direct to Richardson's writings. Richardson sets forth so well his explanation of what he is trying to do, and why, that any further explanation would be superfluous, and probably not be so clear as that of the author himself.

But it seems useful to think about what Richardson has achieved in scientific and political terms, how it bears on today's world, and how it may influence the future.

During the First World War Lewis Richardson felt that as a meteorologist he was leading a rather comfortable life while so much destruction and suffering was going on at the front. Being a Quaker, he would not be a combatant; but he joined the Friends' Ambulance Unit in order to help the wounded (which, incidentally, says something of his physical and moral courage). This brought him into close contact with the misery, suffering and death involved in the war. In his spare moments he began to consider how to investigate further the causes of war, thus hopefully leading ultimately to its prevention.

The view is often put forward that our code of morals lags far behind our technical abilities. Compare the total elimination of smallpox, once a very common disease, with the widespread misery produced by wars, famines, and other disasters. That view may have some validity, but it might be more profitable to contrast scientific and political attitudes and ways of behaviour. Ideally a scientist regards facts as tests of the correctness of his theories. Most people, when thinking politically, regard facts, whatever they may be, as confirming their previous opinions. The result can be catastrophic.

Richardson's outstanding contribution was to look upon political matters as ones which could be investigated by scientific method. He introduced numerical calculation and mathematical and statistical relations. For example, one can argue verbally that armaments deter wars, by frightening possible aggressors. Or one can argue verbally that armaments promote wars, by stimulating a build-up of armaments by the other side. One can argue verbally until the cows come home. But suppose one collects data about how particular nations build up their armaments, and how often they are involved in wars. Then if one should find that the most heavily armed nations rarely fight, that would throw doubt on the second theory, whereas if the opposite was true, the first theory would seem less tenable. Of course, as every statistician is taught, correlation does not necessarily imply causation, and such data have to be handled with care. But they can be a useful corrective to seemingly plausible but misleading theories.

Since Lewis Richardson was trained as a physical scientist, he naturally chose the tools which had been so successful in physical science, such as differential equations. He argued that although individual decisions may be unpredictable, in the mass they could average out in a way which behaves regularly and predictably. He found that the arms race before the First World War agreed remarkably well with his postulated differential equations. He also made use of new statistical methods which were rapidly being developed during the period 1920–50. Although there was a precedent in the mathematical and statistical development of economics, Richardson's application of mathematical and statistical methods to politics was an original and heterodox approach, not always welcomed even today by international relations experts. Like many other ideas, this seems obvious once it has been put forward. But it needed Richardson to think of it.

One recommendation which is often very sound, when one is approaching a new subject, is to go to the original papers. Those who originate a new way of thought are often very conscious of the difficulties of the subject, and of the limitations of their approach, and set them out clearly. So it is with Richardson. As one reads his papers, one has almost a feeling of sitting with him, chatting with him as he gradually develops his ideas and his techniques, and explores one possibility after another. Beside the traditional Occam's razor, that hypotheses are not to be multiplied beyond necessity, he had a stylistic one, that complications of explanation are only to be brought in when unavoidable. His prose is lucid, and his mathematical methods are as simple and straightforward as are possible, compatible with his assumptions and the facts. From time to time, there are simple but illuminating remarks, such as 'it would appear that the partial removal of a grievance may stimulate efforts for its total abolition'. Some of his findings may surprise the reader. It would astonish the ordinary good-humoured Briton to know that Britain has been easily the most belligerent nation during the past few centuries.

One may wonder how long it will be before Richardson's work and its successors will have an appreciable influence on political matters. He himself tried to bring it to the attention of political and military decision makers, but with little success. I have heard that he felt isolated, with almost no evidence of outside interest in what he had been doing. He first published his ideas in 1919. Around 40 years later, from 1959 onwards, there began to be a steady growth in centres of peace and conflict research. As so often occurs, there was a considerable delay between sowing the seed and the vigorous growth of the plant. In the meantime, new tools have been developed. Surprising though it may be, although life is full of decisions, there seems to be have been no thorough mathematical approach to the problem of how to make decisions until Wald published his ideas in 1950, and their full significance only became clear with the work of Savage in 1954. In 1944 von Neumann and Morgenstern in their book *Theory of Games and Economic Behaviour* tackled systematically, for the first time, the problems of interaction between the decisions of two or more people, which are still not entirely understood. It has also been found that under certain conditions very small changes in circumstances at one moment can result in very large divergences in the future, making prediction hazardous. These new developments would have been of great interest to Richardson, if he had lived to see them.

Unfortunately, 70 years after Richardson's 1919 essay, there remains a big gap between the discoveries of the workers studying conflict and cooperation and the still largely traditional attitudes and actions of the political and military decision makers. This is not a matter of minor importance; in addition to numerous devastating conventional wars, the world has already several times been uncomfortably near the brink of unintended nuclear war. Recently tension between the superpowers has greatly diminished. But conventional ways of thinking still persist, and the military are reluctant to diminish their armaments, preferring to invent even more devastating ones.

There is a race between the establishment of a more rational social and political order, on the one hand, and a war which would obliterate civilization on the other. But the gap between theory and practice is closing, even if slowly, as is shown by the creation of centres of conflict and peace research in many countries, and the growing interest in such ventures. The re-publication of Richardson's papers may be a small step towards the encouragement of rationality, and eventually a peaceful and prosperous world order which he would have so much desired.

A general introduction to the life and work of L. F. Richardson†

by J. C. R. Hunt

The discoveries of many great scientists and mathematicians, including Lewis Fry Richardson's, are directly related to the political, economic and other human concerns of their period. He expressed himself clearly at the outset of his career (1908:1)‡, 'The root of the matter is that the greatest stimulus of scientific discovery are its practical applications'.

During the period of his life from 1881 to 1953 science and its applications developed rapidly and in quite new directions. It is well known how the lives of those involved in the developments of modern physics are woven into the history of this century, such as Einstein and Bohr, to name but two. But pioneers of other branches of mathematical sciences have also contributed to the great changes in the world by their discoveries; and as with the physicists, the researches and the lives of many of these scientists and mathematicians were also strongly influenced by the history of this period, especially the two world wars and their peacetime repercussions. Richardson was one of these pioneers, and was outstanding in making important contributions to several different fields. One can add his name to the inventors of computational mathematics (such as von Neumann, Courant, Turing); of modern meteorology and fluid mechanics (V. Bjerkens, Taylor, Prandtl); of quantitative techniques in psychology and social sciences (W. James); and of analysis and modelling of complex systems (Wiener). In all these fields his work is still cited in current research papers. In one citation index there were over 200 references between 1980 and 1984.

Knowing something about the lives and beliefs of creative people usually helps one understand and appreciate their work; this is especially so for Richardson, much of whose scientific work changed and evolved as a direct result of the political and technological changes during his life, and his reactions to them as a Quaker. One of the reasons for collecting his papers in these two volumes is to show how a mathematical scientist can respond to the problems of the world around him, and, one might add, not necessarily in accordance with the ways currently favoured or promoted by the established organizations that direct and finance science.

Richardson's special contribution to all these fields was to apply quantitative and mathematical thinking to problems that were considered outside the scope of mathematics at that time, and to have been so effective in his thinking that his formulae and his methods are still used daily by working scientists and mathematicians. His

† Much of this introduction is based on the biographical accounts by Ashford (1985) and Gold (1954). It is also based on recollections by various members of Lewis and Dorothy Richardson's family, including myself.

‡ In this book Richardson's own publications are cited in this way. An explanation is given on page 47 in the 'Note on the arrangement and presentation of the papers'.

personal stamp on the work is unusual in that many of his results are still referred to by his name. However, he was himself especially conscious of the dangers of applying mathematical ideas and techniques to complex human behaviour and natural phenomena as he aptly summarized in *Mathematical Psychology of War* (1919:1, p. 2).

> Mathematical expressions have, however, their special tendencies to pervert thought: the definiteness may be spurious, existing in the equations but not in the phenomena to be described; and the brevity may be due to the omission of the more important things, simply because they cannot be mathematized. Against these faults we must constantly be on our guard. It will probably be impossible to avoid them entirely, and so they ought to be realized and admitted.

Early life

Born on 11 October, 1881, of Quaker parents, David and Catherine Richardson in Newcastle upon Tyne, Lewis Fry Richardson was the youngest of seven children. David Richardson, who ran a prosperous tanning and leather business, had been trained in chemistry. His technical abilities enabled him to design new machinery and some new production methods. Lewis showed early on an independent mind and an empirical approach, as when he tested at the age of five the proposition learnt from his elder sister that 'money grows in the bank'. He buried some money in the garden and was disappointed to find it did not grow! After a period at Newcastle Preparatory School, where his chief enjoyment was Euclid 'as taught by Mr. Wilkinson', he was sent, aged twelve, to a Quaker boarding school, Bootham, at York. There he had excellent teaching, especially in science, which stimulated a great interest in natural history. He kept a diary of birds, insects, flowers and weather. He collected 167 species of insects and made detailed studies of plants (he returned to collecting statistics, of wars, in the last phase of his research). Another important part of his character was developed by a teacher who 'left me with the conviction that science only has to be subordinate to morals'.

His higher education began in 1898 with two years at Newcastle University (formerly, Durham College of Science) with courses in mathematical physics, chemistry, botany and zoology. He proceeded in 1900 to King's College, at the University of Cambridge, where he was taught Physics in the Natural Science Tripos by (among others) Professor J. J. Thomson (the discoverer of the ratio of the charge to the mass of an electron, e/m) and graduated with a first class degree in 1903.

In Table 1 the main appointments in his career are listed.

Early career 1903–13

In the ten years (1903–13) after graduating L. F. Richardson took a series of short research posts, a career not unfamiliar to today's scientists, first in a government research laboratory (the National Physical Laboratory), then in physics departments (at University College of Wales, Aberystwyth and at Manchester University) and in industry. Some of the research required in these posts clearly did not interest him much (such as in metallurgy and metrology). However, it was as a *chemist* with the National

Table 1*

1903–4	Student Assistant at the National Physical Laboratory (Metallurgy Department)
1905–6	Junior Demonstrator in Physics, University College, Aberystwyth
1906–7	Chemist to National Peat Industries Ltd, 'whose managing director stole a large sum and fled abroad'
1907–9	Assistant at the National Physical Laboratory (Metrology Department)
1909–12	In charge of the chemical and physical laboratory of the Sunbeam Lamp Co., Gateshead (where he organized a troop of scouts)
1912–13	Demonstrator and Lecturer in Physics, Manchester College of Technology
1913–16	Meteorological Office, Superintendent of Eskdalemuir Observatory
1916–19	In the Friends' Ambulance Unit in a motor convoy (SSA 13) attached to the 16th French Infantry Division
1919–20	Meteorological Office, nominally in charge of experiments in the computation of the sequence of weather by numerical processes. He worked at Benson, Oxfordshire, where W. H. Dines was in charge of 'Upper Air Investigation'. (Richardson revived the village troop of scouts.)
1920–29	In charge of the physics department at Westminster Training College
1929–40	Principal of Paisley Technical College and School of Art
1940–53	Retired to do research on wars. Research also in eddy-diffusion

* From Gold (1954)

Peat Industries Limited from 1906 to 1907 that Richardson began his first pioneering research, although it was in mathematics and not chemistry. He was faced with the problem 'Given the annual rainfall, how must the drains (i.e. channels in the peat moss) be cut in order to remove just the right amount of water?' He found that the percolation of water through the peat could be described by the well-known (eighteenth century) equation of Laplace ($\partial^2\phi/\partial x^2 + \partial^2\phi/\partial y^2 = 0$), but that the boundaries around the region where the equation had to be solved did not usually have the nice shapes, such as circles and rectangles, that had been studied by mathematicians and for which solutions were known. Although he realized that exact mathematical methods could often be found for more complicated regions, these methods were difficult to derive and were not general for any shape; so faster and more general, if less accurate, methods were necessary. His prescient remarks on this dilemma are as relevant today as in 1908 – though most of today's problems require the solution of more complicated equations.

> Further than this, the method of solution must be easier to become skilled In than the usual methods [i.e. analytical solutions]. Few have time to spend in learning their mysteries. And the results must be easy to verify – much easier than is the case with a complicated piece of algebra. Moreover, the time required to arrive at the desired result by analytical methods cannot be foreseen with any certainty. It may come out in a morning, it may be unfinished at the end of a month. It is no wonder that the practical engineer is shy of anything so risky.

Richardson (1908:1) showed first that a broad brush solution for the peat flow could be obtained by drawing lines of the flow free-hand according to certain rules (which would require a good eraser, a soft pencil and some patience) until the lines satisfied these rules – an approach that was still taught to engineering students in the 1960s. But a more accurate and systematic method for obtaining approximate solutions was to convert the differential equation, defining the smooth continuous changes of the variable (say y, the height of a curve), into an approximate equation relating the small changes in the variable (δy) to the small distances (δx) (or steps) over which they occur.

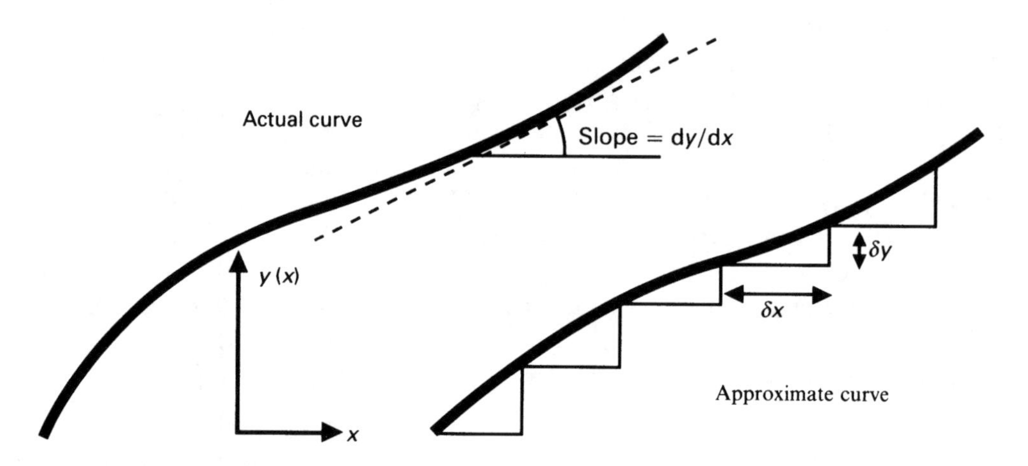

Fig. 1. A sketch to illustrate how a smooth curve is approximated by a series of simple curves between points separated by δx and δy. This is the basis of Richardson's methods for computing the solutions to differential equations.

Then the rules of arithmetic (and algebra) can be used for the sequence of steps rather than the special techniques of differential calculus for continuous functions (in which there are infinitely many steps on the staircase, of infinitesimal size). Richardson (1927:1) later explained that, paradoxically, this approach was an historical regression to a time before the invention of calculus!

Richardson's (1910) paper on this method made use of recent research into finite differences by Sheppard (1899). He also commented that the approximate but more formal methods introduced in Germany by Runge (1895), Kutta and Ganz, for solving differential equations were not suitable, at that time, for solving the kind of practical problem that interested him, such as the percolation of water, the stresses in masonry dams, and, later, meteorology. Not surprisingly to anyone familiar with the scientific and academic worlds this new approach was at first too new for the referees who reviewed the paper for the *Philosophical Transactions of the Royal Society*. It was only after much deliberation and correspondence that it was published (in 1910).

More importantly for Richardson's career this approach to the approximate solution of differential equations was also too new for King's College, Cambridge, where Richardson submitted this work, as a dissertation, in the competition for a Fellowship (i.e. a research and teaching post). Apparently (I am indebted to Professor Huppert for this information from the files of King's College, Cambridge) opinions were sought from the mathematicians at Trinity College, who said this was approximate mathematics and they were not impressed. So Richardson never returned to Cambridge and for the rest of his career he did not work in any of the main centres of academic research. At the time he did not regret this, commenting (apropos of Manchester) that he liked to work 'somewhere where there are fewer people buzzing around'. But this isolation probably affected the development of his research and hindered its appreciation in the scientific community. Perhaps the lack of collaboration with colleagues explained why the presentation of his research was often idiosyncratic, and probably also meant that he did not receive suggestions from other researchers as to how a line of research might profitably develop. It may explain why he moved from one subject to another suddenly and quite often. But one must also recognize that the lack of the 'guiding' influences of colleagues may have been a factor in the great originality and diversity of his research; as G. I. Taylor (1959) commented, Richardson was 'a very interesting and original character who seldom thought on the same lines as his contemporaries and often was not understood by them'.

These early papers helped pioneer the development of 'numerical' methods for the solution of differential equations (reviewed by L. Fox, in Volume 1), a subject he returned to later when it was gaining greatly in importance with the growing use of calculating machines and, later, the arrival of electronic computers (1927:1, 1950:3). At the time Richardson saw the main importance of this work as laying the basis for solving the equations required for predicting weather (reviewed by Charnock, in Volume 1).

Despite the progress he had made in this field of research during this early stage of his career, there were no funds to continue it; indeed, as mentioned earlier, he had to move between short-lived research and teaching posts. No clear direction for his future was emerging. Nevertheless two of these brief projects did reveal his intention of eventually moving away from the physical to the behavioural sciences. He had said to himself as an undergraduate that 'I would like to spend the first half of my life under the strict discipline of physics, and afterwards to apply that training to researches on living things'. According to his own autobiographical notes, 'I kept this programme a secret'.

In 1907 he sold his physics books and briefly went to work as an assistant 'to learn about statistical proof' under Professor Karl Pearson at University College, London, an authority on mathematics, genetics and the philosophy of science. In fact Richardson worked on stresses on masonry dams (1910) and helped prepare an index of the journal *Biometrika* founded for the statistical study of biological problems. He had to leave within a few months because there were no funds for his proposed research into quantitative studies of heredity. This subject together with eugenics were topics he took up between 1912 and 1913 when he was a lecturer at the Municipal School of Technology in Manchester (now the University of Manchester Institute of Science and Technology (UMIST)). He published the first of his papers on the quantitative analysis of 'living things', entitled 'On the measurement of mental "nature" and the study of adopted children' in *Eugenics Review* in 1913:1. It built on some earlier research of Pearson.

During this period of his life L. F. Richardson used to have holidays in a resort on the south coast of England, at Seaview, Isle of Wight, with his friend Stuart Garnett, whom he had met in Cambridge and who, like Richardson, had a wide range of interests; he worked as a barrister, pioneering the new developments in child law, and lived in the East End of London where he helped found the Sea Scouts for the local boys; he also dabbled in Conservative politics and wrote an early book on steam turbines. They stayed with Stuart's parents, William and Rebecca Garnett, who had a substantial seaside house with a boathouse underneath it. Much of the Garnetts' family life was centred on education, boating and religion. William had worked under Professor James Clerk Maxwell as a demonstrator in the Cavendish Laboratory at the University of Cambridge in the 1870s, and later became an administrator and exponent of adult education (after whom Garnett College in London is named). His sons Maxwell, Stuart and Kenneth all excelled in their mathematical studies at Trinity College, Cambridge. Together with their sisters and friends (including also G. I. Taylor from Cambridge) they all joined enthusiastically in sailing and rowing and, on Alpine holidays, in mountaineering.

There, in the Isle of Wight, Richardson met Stuart's sister Dorothy. The family story is that he was attracted by her great sense of fun shown by throwing water over her brothers! They were married in 1909 in London. Dorothy's religion changed from the patriotic, but ascetic, Church of England Christianity of her Garnett family to the pacific Quaker religion of the Richardsons. Nevertheless Dorothy always retained an energetic, optimistic and often headstrong approach to life that was quite a contrast to Richardson's persistently enquiring attitude and his deep and quiet philosophy. Despite the differences in their personalities, their joint sense of fun and adventure led to a happy marriage. Characteristically for that period, Dorothy devoted her life to being his wife and carer of the home and family. However, it turned out that their blood groups were incompatible, so she could not have children. Their three children, Olaf (b. 1917), Stephen (b. 1920) and Elaine (b. 1927) were adopted. Later in the 1930s Dorothy was an energetic voluntary worker and proselytizer for the peace movements, including the League of Nations Union, whose British secretary was her brother Maxwell.

While Richardson and Dorothy were in the Isle of Wight in the Easter holiday of 1912, the large passenger liner *Titanic* collided with an iceberg in the Atlantic, off Newfoundland in foggy weather, and sank with great loss of life. Richardson immediately had the idea that this kind of accident could be avoided if ships sent out a focused beam of sound and then measured the delay time for any echo to be detected by a sensitive receiver. He tested this idea in Seagrove Bay, in a dinghy rowed by Dorothy (as she told the story) to different distances from Seaview Pier (now destroyed). He blew a penny whistle and by measuring the time for the return of the echo, using an umbrella held over his shoulder to amplify the echo, he calculated the distance. He found that the method worked well; so he filed a patent in October 1912, the importance of which led the writers of a text book on fluid mechanics in 1936 (Drysdale *et al.*) to hope that Richardson's 'method will ultimately eliminate the last of the serious dangers of navigation'! This impromptu but highly original experiment was typical of many that Richardson undertook with the minimum of expense, but which usually led to important new ideas. (See also Charnock (1981).)

The sinking of the passenger liner *Titanic* led to a scientific expedition to study the atmosphere and ocean off Newfoundland; the meteorologist was Richardson's great contemporary G. I. Taylor, whose early ideas on turbulence stemmed from his observations there. Richardson's and G. I. Taylor's interest touched again in the next phase of Richardson's career.

1913–16; Meteorology – the first phase

On the recommendation of Richardson's former colleagues at the National Physical Laboratory, he applied for and was appointed to the position of Superintendent of the Eskdalemuir Observatory in Southern Scotland. The Observatory had been set up in a remote location primarily to record magnetic fields and seismic vibration, but also to record meteorological measurements. Richardson had no previous professional experience in meteorology, but, as explained to him in a letter from Napier Shaw, the chairman of the appointments committee (and one of the founders of meteorology as a quantitative science, in Britain), he had been appointed to bring a more theoretical approach to the understanding of meteorology, and a critical examination of the methods of measurement.

This was an attractive position for Richardson, as it came with a house and considerable freedom to pursue his major research interest, that of devising a method for calculating the weather a few hours or days ahead, using the relevant theoretical equations which describe the behaviour of the atmosphere. Although most of the equations were known, there were a number of new aspects of the physics that had to be estimated: also there was the new task of transforming these equations into the appropriate form for some step-by-step method. 'In the bleak and humid solitude of Eskdalemuir', as he described it, he completed the first draft of his book entitled, then, *Weather Prediction by Arithmetical Finite Differences*.

He also participated in some developments of seismography and new techniques for the detection of thunderstorms from their electrical and magnetic fields, using telephone lines. At the same time he was performing the administrative duties of superintendent, which became more onerous after the outbreak of the First World War. Correspondence between Richardson and Napier Shaw (Ashford 1985, p. 50) shows that Richardson was uneasy about using his own and the observatory's scientific knowledge and equipment for military purposes, such as measuring the vibration in the ground caused by distant artillery fire.

Whether or not this was the cause, Richardson resigned from the Meteorological Office on 16 May 1916 to join the Friends' Ambulance Unit in France. He had asked earlier to be released for work with the 'Red Cross Unit of the Ambulance Corps' at the outbreak of war, but had been refused permission. Clearly his parting with the Meteorological Office was amicable, because he was allowed to rejoin in 1919 on his return from France.

1916–19; France and the psychology of war

The Society of Friends always urged its members (Quakers) not to take part in war, and although in the 1914–18 war a few Quakers of military age joined the armed forces, many did not, and either did humanitarian work connected with the war or refused to have anything to do with war activities. Many of the latter were imprisoned, though usually with other conscientious objectors (such as Bertrand Russell).

Richardson followed the pacifist course of action and in 1916 joined the Friends' Ambulance Unit which was financially supported by the Society of Friends. It was set up in 1914, following an initiative by Philip Noel Baker.

At this time many families were divided in their response to the war. Richardson's Cambridge friend and brother-in-law Stuart Garnett was killed while test-flying an aircraft in 1916 in the Royal Flying Corps, and Stuart's younger brother Kenneth, aged 25, died slowly during 1916–17 after being injured with a shrapnel wound in the Battle of the Somme. Richardson was exceptional amongst scientists (in Britain or Germany) in deliberately ceasing to do scientific research financed by the government during the war; other scientists not only continued with their research, but also gave advice to armed forces that was used to great effect, particularly in aerodynamics (such as G. I. Taylor, at Cambridge, L. Prandtl, at Göttingen), ballistics (J. E. Littlewood, at Cambridge), and the chemistry of explosives and gases (C. Weizmann at Manchester).

After initial training Richardson was attached as a driver to the Section Sanitaire Anglaise (SSA 13), a group of fifty-six men with twenty-two ambulances working with the 16th French infantry division. They worked alongside the French military ambulance unit 'Section Sanitaire'. Richardson was, in the words of the ambulance maintenance crew, 'a careful and conscientious driver and managed to avoid careless driving through shell-holes'! For periods of two weeks at a time he was engaged in transporting wounded soldiers, often under shell fire. Then he would have a few days of respite. Occasionally he returned to England for brief periods of leave to see his wife and family.

In his spare time in France Richardson set up and designed various simple meteorological instruments and took readings. Also he had brought along the first draft of his book on numerical weather prediction, and during a six-week period he worked on a specific calculation ('on a heap of hay in a cold wet rest billet') to show how the numerical forecasting system might be used in practice. At one stage the manuscript was lost during the battle of Champagne in April 1917, but fortunately was rediscovered some months later under a heap of coal. The work was finally published in 1922, when he had taken up meteorological research again. In France he also did some 'laboratory' experiments on the motion of water in a vessel on a rotating gramophone turntable; the apparatus was too crude to test the effects of thermal convection and rotation that he was investigating.

The main effect of the war on his research and thinking was to direct him towards studying the causes of wars and how they may be prevented. As usual he began with a quite new approach that was also mathematical. He believed that conflict between nations resulted from a number of factors, chief amongst which were misunderstandings, the general admiration of warlike attitudes, unstable patterns of mutual hostility, the interest of certain industries in war and the absence of effective international organizations for settling differences between governments.

In an unpublished manuscript 'The condition of a lasting peace in Europe' in 1915, he wrote that misunderstandings between peoples and governments could be greatly reduced if there was an international language. He learnt Esperanto, the best known of the international languages at that time, and was a strong advocate of it. (He tried it out on German prisoners of war, but neither they nor any non-Esperanto speaker could or would understand it!) He also wrote about many other psychological and cultural causes of wars, including the factor, so alien to Quakers but common to most societies, of the admiration of military courage for its own sake.

From this beginning he began to develop a mathematical model for how the animosity between two mutually suspicious and well-armed nations might develop over time. He suggested that the animosity of each side (the *Entente Cordiale*, i.e. Britain, France, on one side and Germany on the other) could be expressed in terms of numbers derived from measurements. This was therefore a mathematical quantity and could be repesented in equations by the symbols A_E and A_G respectively, which showed how A_E depends on A_G and vice versa. Although mathematics was already in use in various social sciences, it had not previously been used for modelling war *behaviour*. However it had begun to be used for modelling *tactics* of war especially on the basis of the fundamental, and still used, equation of F. W. Lanchester (1916) – another unorthodox Englishman who is equally well known for his pioneering research on aerodynamics, and innovations in automobile engineering.

Richardson wrote up these studies in a fifty-page essay of plain language and equations, *Mathematical Psychology of War* (1919:1) which he dedicated to his 'comrades of the motor ambulance convoy'. They and subsequent readers found the algebra and calculus early on in the text so off-putting that they could not reach the end, where there are understandable and novel conclusions about the circumstances in which the relations between countries can become 'unstable' and wars might ensue.

[Nowadays equations similar to these are taught to students in biology as predator–prey equations. They are not easy, and have to be explained with the use of diagrams and graphs which were missing from Richardson's (1919:1) essay, but were included in his later publication (1939:1).]

The introductory section contains the famous 'apology for the use of mathematics', the first part of which extols the virtues of a mathematical approach in the analysis of complex problems, including those in the social sciences.

> To have to translate one's verbal statements into mathematical formulae compels one carefully to scrutinize the ideas therein expressed. Next the possession of formulae makes it much easier to deduce the consequences. In this way absurd implications, which might have passed unnoticed in a verbal statement, are brought clearly into view and stimulate one to amend the formula.

But, as mentioned earlier, he was also well aware of the dangers of this approach (see Sutherland, below). In arranging to have this essay published. Richardson first approached a commercial publisher, for whom Bertrand Russell reviewed it favourably as an academic book, but Russell also reckoned that it would not sell many copies. The latter point was more important to the publishers, so Richardson had to publish the work privately; three hundred copies cost him £35. But the ideas in this book were certainly not lost; they were the basis of the later mathematical studies of conflict, including Richardson's in the period 1935 to 1953 and much modern research (see Sutherland and Nicholson, below).

Richardson's other early studies of conflict were on how an international assembly, which would have to be created after the war, could function most effectively. Once again he suggested a new measurable quantity – 'internationality' – based in part on international trade, and that this should define the voting strength of a nation in the assembly. Apparently this was considered by a committee of the British Foreign Office; but they preferred the more obvious principle of one vote per state.

1919–26 – Meteorology again; Benson and Westminster College

From France Richardson wrote to Napier Shaw of the Meteorological Office asking whether he could return there, with the specific aim of working on upper-air sounding experiments with a view to making 'weather predictions by a numerical process ... a practical system'. In his letter Richardson floated the idea that, with a large technical staff of twelve to fifteen, he could make rapid progress in overcoming some of the outstanding observational and mathematical problems. If this was not possible he suggested an alternative proposal of a modest research position for himself with an assistant. Shaw replied that the large scheme might be possible technically and even financially, but that other forecasting schemes requiring funds had also been proposed, such as V. Bjerknes' graphical scheme based on the analysis of fronts that had been developed at Bergen during the 1914–18 war, and which subsequently became the basis for understanding weather systems and qualitative weather forecasting throughout the world. Interestingly Bjerknes, who later collaborated with Richardson, had also hinted at the possibility of numerical weather forecasting in 1904, but did not pursue it himself. The outcome was that Richardson was appointed by the Meteorological Office to W. H. Dines' upper-atmosphere research group at Benson, near Oxford.

With a view to providing the data that were a necessary input for the numerical model, and also to test the model, Richardson developed three different kinds of instrument for atmospheric measurements up to several kilometres above the ground; they were all characteristically original, but in fact none were continued after he left Benson (see Charnock, in Volume 1). He developed a complicated method for measuring the upper wind speed by shooting metal spheres of various diameters upwards at small angles to the vertical (1923:1, 1924:1, 1924:2, 1924:4). A nice Richardson touch to the experiments was the arrangement whereby, to protect themselves and record the point of impact, the shooters stood in a shelter underneath a large metal sheet and fired the gun upwards through a central hole (see 1924:4, Fig. 2).

A major task in developing an underlying theory for the numerical method of forecasting was to improve the theory of turbulence and turbulent mixing in the lowest 2 km of the atmosphere. The turbulent eddies determine how rapidly the upper-level atmospheric winds can slip over the earth's surface and how much heat and moisture can be carried up from or down to the earth's surface. The earlier research by Schmidt in Austria, Boussinesq in France, and G. I. Taylor in England had shown that these eddies behave in *some respects* like molecules in a gas so that it was meaningful to define an 'eddy' viscosity or 'eddy' conductivity K (1922:1, p. 67). However, whereas the viscosity or conductivity of a gas is independent of its velocity or kind of motion and independent of the scale over which there are variations in velocity or temperature, and has the same value everywhere (provided the gas is at the same temperature and pressure), none of these properties are true of a turbulent flow. Richardson's research between 1919 and 1926 led to great advances in understanding these special properties of turbulent eddies and in providing a novel kind of physical explanation.

At Benson he followed up some experiments which he had begun in France to study how material randomly disperses in the turbulent eddies in the atmosphere, by measuring the widths of smoke plumes and the distances between floating objects (from seeds to balloons) released into the wind. Some of these experiments provided more data that largely confirmed previous results, such as how K increases with height above the ground in the atmosphere and with wind speed (1920:2). Later he conducted further experiments with balloons which travelled for several hours from their release point (1926:3). Some of these experiments were in the characteristic informal Richardson style; in some cases the balloons were released at publicity events in Hyde Park, or in a 'balloon competition' on Brighton beach. Many of them travelled over France, Belgium and Holland. Labels attached to the balloons were returned by people finding the balloons when they landed, from which their trajectories were plotted. The results showed that their variability or dispersion was greater the further they travelled. As is not unusual in science, once he had his own data he found earlier data by other scientists that was consistent with the pattern he had begun to observe. The most novel experiments (1929:1) involved releasing seeds (with Dorothy and Stephen on Hindhead Common) and small balloons *simultaneously* at different distances apart. Both these sets of experiments were used by Richardson (1926:1) to derive a general law that the rate of increase of the square of the separation (i.e. the rate of diffusion) between objects diffusing in a turbulent stream increases in proportion to the separation raised to the power 4/3 – the famous 4/3 law. This showed conclusively that turbulence is not quite like the molecules of a gas and contains eddies with many length scales, and that different methods of analysis are necessary. Many of the questions he raised in 1926 are still not satisfactorily resolved, though his insights are still instructive.

Jonathan Swift, author of *Gulliver's Travels*, had contemptuously described poets' use of each other's work in a well-known satirical verse that compared them to fleas. Richardson (1922:1, p. 66) adapted the verse to describe his conception of turbulence by casting the fleas as eddies. 'Big whirls have little whirls that feed on their velocity, and little whirls have lesser whirls and so on to viscosity – in the molecular sense'.

This rhyme is still used to summarize the dynamics of turbulent eddies both in text-books and in lectures (see Ashford, p. 85). Although nowadays rhyme is an unusual 'teaching aid', in the Victorian era humorous limericks were widely used for this purpose. Richardson would have known about Maxwell's rhymes on dynamics, 'Gin a body meet a body/Flying through the air/...',; and on topology 'My soul's an amphicheiral knot/Upon a liquid vortex wrought/...', which were collected in the biography by Lewis Campbell and William Garnett (Richardson's father-in-law) (1882).

Richardson's next important insight into the nature of turbulence also arose from experiments at Benson. He observed how the fluctuations in wind speed increased or decreased depending on both the difference between the wind speeds at two heights (1 m and 26 m in his case) and the difference between the temperatures at these two heights; so that in the evening as the ground temperature fell and the temperature difference increased, the fluctuation decreased. In his analysis he derived estimates for the relative contributions to the energy of the turbulence from, on the one hand, the buoyancy forces caused by eddies moving between levels of higher and lower temperatures (which might reduce or amplify the turbulence), and from, on the other hand, the accelerations of the eddies moving between levels of high and low wind speed (based on the earlier statistical analysis of Osborne Reynolds at Manchester thirty years earlier, and his and G. I. Taylor's experiments on turbulent fluctuations). The ratio of these two contributions is now called the Richardson Number (*Ri*) (usually said as 'R I'), the name given by H. Schlichting (1935). Richardson showed that if *Ri* was greater than one the turbulence was suppressed. Dimensionless numbers like theorems, commemorate physicists; but the Richardson Number is probably unique in that it takes positive and negative values, depending on whether the atmosphere is stable or unstable!

This insight into the occurrence and strength of turbulence was acknowledged in the scientific world within a few years, probably because it crystallized ideas that had already been suggested by G. I. Taylor and others. In fact Prandtl incorporated Richardson's criterion into his text book *Fluid Dynamics* as early as 1931. It took somewhat longer for Richardson's concepts of eddy structure and of eddy diffusion of clouds of material to gain comparable recognition. Unfortunately Richardson could not at that time have been aware of how strongly they influenced Kolmogorov (1941) and Oboukhov (1941) in Moscow in their development of the fundamental theory of turbulence. (Although in his discussion of turbulence Kolmogorov (1941) essentially paraphrases Richardson (1922:1, p. 66), Richardson's work was not explicitly referenced by Kolmogorov himself until 1962.) However Richardson's contributions were certainly recognized by G. K. Batchelor (1952) in Cambridge in his exposition and extensions of diffusion theory.

Richardson's wife Dorothy used to say how unfortunate it was that the research of Richardson, who was a pacifist, had been used in the military-sponsored research on the dispersion of gases and vapours, such as that led by O. G. Sutton at Porton during the 1920s and 1930s. This probably explains why Richardson replied rather discouragingly when Sutton wrote to ask his advice on this research. Sutton well understood Richardson's position, but he disagreed with it because 'it is never possible to ensure that any original work will not be used by others in ways that the originator would condemn' (Ashford 1985, p. 128). This point, ironically, was demonstrated by Richardson himself, who complained in 1941 that, during the Second World War, the UK government's instructions to civil defence about the dangers of gas warfare did not refer to the importance of assessing atmospheric windspeed and turbulence before advising civilians what to do (Ashford 1985, p. 169). In other words he seems to have been blaming them for ignoring his research! Nowadays regulations covering the safety of chemical plant and emission of pollutants into the atmosphere have been carefully devised to take into account recent research on atmospheric turbulence, much of which has been built upon Richardson's and Sutton's original work.

The third main element in Richardson's programme of research on improved methods of forecasting was his study of radiation and the thermodynamics in the atmosphere. At Benson he focused on water in clouds (1919:4), and the thermodynamics of moving and radiating 'parcels' of air (1919:3). Later, between 1927 and 1928, studying the reflection of radiation from the earth's surface, he was able to mount his instruments on a small aircraft of the De Havilland Aircraft Company, which measured the reflectivity (or albedo) of woods, fields and suburbs between London and St Albans (1930:1). With G. C. Simpson, Richardson (1928:1) recognized earlier than other meteorologists how reflection has an important but 'as yet imperfectly understood effect on local climate'.

These studies were part of an international programme, organized at the Madrid meeting of the International Union of Geodesy and Geophysics (IUGG) in 1924, which involved comparative measurements of the albedo in different countries using the *same* photometer instruments. This was one of the instruments that Richardson had developed while in France with the Friends' Ambulance Unit. (The cost in 1927 was £14.10s.)

While at Benson in 1919 Richardson sent the completed manuscript for his book on 'Weather Prediction by Numerical Process' to Cambridge University Press for publishing, but it did not appear in print until 1922. In the book the results of Richardson's scientific and mathematical studies of atmospheric processes are collected and it is shown how these formed the basis of a numerical method for weather prediction. But the book is best known for the great *failure* of the method when it came to be applied in a particular calculation of the meteorology over a six-hour period over Germany in 1910, using the data from the maps of V. Bjerknes. (This was the calculation he had performed in France.) So it was a surprising and brave decision to include this calculation. (The reasons for the error, which became clear about twenty years later, are given by Charnock, in Volume 1).

In the final chapter he briefly discusses the practical organization of such forecasting, using only humans computing with slide rules and calculating machines. He imagined a large hall, similar to a concert hall, with the chief forecaster acting like a conductor organizing the information flowing in to him from all parts of the hall. But his estimate for the number of computers required to 'race the weather for the whole globe' was 64,000, which meant the hall would be more like a football stadium! (At that time he did not imagine that the calculating machines could eventually do the whole job unaided.)

He saw substantial practical and economic advantages flowing from improved forecasting, mainly to agriculture. He did not mention the great advantage to other activities, especially aviation, which has been the primary reason for the financial support of meteorological research from that time on.

The protracted period between 1919 and the final publishing of the book in 1922 involved the publisher in extra costs. So they asked the author for a subsidy of £50, and a further subsidy of £50 came from the Meteorological Office! 750 copies were printed and sold at £1.5s. The reviewers of the book (quoted extensively by Ashford (1985), pp. 94–9) were impressed by its 'originality, its coordinated treatment of the dynamical processes', and agreed that it provided the basis for systematic quantitative forecasting. But both they and the forecasters within meteorological organizations did not believe that this provided a practical approach at that time. In his review of the book in the *Quarterly Journal of the Royal Meteorological Society*, S. Chapman suggested that its more immediate impact might be in calling attention to the kind of observations and measurements that are required for theoretical studies (see Charnock, Browning, in Volume 1). Richardson's ideas and methods later formed part of the programme led by von Neumann and Charney to introduce numerical weather forecasting in the USA, which became practical in the 1940s and 1950s with the arrival of electronic computers (see Chapman 1965). Richardson's book continues to be referenced, and his vision is often described in the popular press around the world when they attempt to explain the mysteries of numerical weather forecasting!

Richardson's time at Benson came painfully to an end in 1920, as a result of his pacifist convictions. The Meteorological Office had been financed as a scientific organization, with the Royal Society appointing its governing Council. During the war of 1914–18 the armed forces found that meteorology was so important that they set up their own separate meteorological services. The Government decided in 1918 that these should be brought into one organization: its first recommendation was that this should be a government scientific organization, analogous to the National Physical Laboratory, but then it decided that the Meteorological Office should indeed be incorporated into the Air Ministry (which controlled the Royal Air Force). Richardson followed these developments with great anxiety. The final decision came in July 1919, and he eventually resigned with effect from 1 September 1920. There were many expressions of regret by his colleagues in the Meteorological Office, with whom he continued to collaborate on scientific matters for the rest of his life.

Because of the outcome of the Government decision on the Meteorological Office, in May 1920 Richardson had applied for a job as a lecturer at Westminster Training College (then near Westminster Abbey), teaching physics and mathematics to prospective school teachers up to level of a bachelor's degree. There had been other distinguished scientists before him at this college and he was told by the principal, H. B. Workman, that he would have a reasonable amount of time for research work. He became a very conscientious teacher; he believed in knowing exactly what his students would have to face in examinations, so he took some of them himself, including pass and honours B.Sc. degrees in mathematics and psychology in 1925 and 1929; this was his first university qualification in mathematics, and psychology was the subject in which he became steadily more interested during the 1920s.

During this period he and his family lived in the North London suburb of Golders Green, near Dorothy's family and other relatives in Hampstead where they could walk over the wild and beautiful heath. Dorothy enjoyed the activities of London including the theatre, while Richardson spent more of his spare time on study and research. However, they both enjoyed meeting a wide circle of friends including many scientists from other countries. Richardson's papers refer to conversations with his brother-in-law Maxwell Garnett on psychology and international relations, and with Richardson's nephew, the actor Ralph Richardson – later Sir Ralph.

At Westminster, from 1920 to about 1925, Richardson's main research interests continued to be in meteorology. He was secretary of the Royal Meteorological Society from 1920 to 1924, attending meetings and helping instigate the new series for reporting meteorology and research, *Memoirs of the Royal Meteorological Society*. He contributed the first paper to the first issue (1926:3), on atmospheric dispersion. He attended two major international symposia, at Bergen in 1921 and Leipzig in 1927, which helped establish many of the internationally agreed new methods of forecasting by analysing the movements of fronts and by developing new methods of meteorological research. V. Bjerknes and others expressed surprise that a scientist of Richardson's stature did not have a secure position to continue his meteorological research at a high level (Ashford 1985, p. 157). Richardson was elected to a Fellowship of the Royal Society of London in 1926; the citation read 'Distinguished for his knowledge of physics and eminent in the application of mathematics to physical problems of the atmosphere and other structures. Author of "Weather Prediction by Numerical Process" (Camb. Univ. Press), and of numerous papers of great originality both in scientific idea and experimental method . . .'.

In about 1926 Richardson changed his field of research to psychology, with the main objective of applying the ideas and methods of mathematics and physics to this field. He had read McDougall's (1905) book on *Physiological Psychology* and he began his research by attempting to express its results in mathematical form. The main themes of his work over the next eight years can be summarized as:

(a) establishing that many different sensations are quantifiable (a commonplace idea now, but then highly controversial);

(b) finding *methods* of measuring and quantifying these sensations and relating them to external physical conditions or stimuli; which is nowadays the field of applied psychology. (In the introduction to his paper of 1929(:4) Richardson states 'Psychology will never be an exact science unless psychic intensities can be measured. Some authorities say that such measurement is impossible'.)

(c) modelling these sensations with mathematical equations or in terms of analogous processes in physics (the aspect emphasized here; Poulton discusses his major contributions to psychology below).

He collaborated in these studies with two colleagues on the staff of Westminster College, J. S. Ross and R. S. Maxwell.

To quantify the *sensation* of touch, he asked the subject to estimate the distance apart of two pin pricks (the stimulus) touching his or her arm – a standard neurological test. He devised the kind of graph that is now standard in experimental and applied psychology by plotting a quantitative measure of sensation (in this case millimetres) against a physical stimulus (also measured in millimetres). Although the graphs differed between the subjects (his colleague Ross and his wife Dorothy), both had a similar shape that curves upwards and then levels out (1928:3).

By recording the intensity of his own mental image when thinking of a word (or the opposite of a given word), timing the decay of the images and then plotting the result, he found that many kinds of mental images also produced a set of curves that were similar to each other (1929:3). These results, which were only published with difficulty because of the scepticism of academic referees, were the basis of Richardson's more general ideas about how thoughts arise then slowly decay, and then sometimes repeat themselves. He also developed general ideas about how the mind solves problems.

To answer a fundamental question raised by the Harvard psychologist and philosopher William James (1890) (brother of the novelist Henry James), he studied whether certain colours, such as pink, are perceived as a mixture of primary colours or at being intrinsic. In 1929:4 Richardson introduced the use of what is now called in psychological research, a 'semantic differential': i.e. the subject's response is a mark on a line (usually 10 cm long) drawn between two words with opposite meaning, e.g. white and scarlet. By asking subjects to look at spinning discs covered with different zones of white and scarlet and to assess the shade of pink (by marking the line somewhere between white and scarlet), Richardson and R. S. Maxwell verified the mixture hypothesis. Despite criticism of the experiment, Richardson concluded that 'we have cleared the way towards a more scientific study of quantitativeness in sensations of various kinds'.

This controversy continued, even into the 1930s, with some scientists believing that there could be no systematic measure of sensation.

In another important study of sensation, his subjects, who included the college organist, had to estimate the loudness produced in a pair of head phones by a measured electric current, the stimulus (1930:2). He established for the first time that the ratio of the loudness of the signal to that of a standard signal varied logarithmically with the stimulus, a result that helped lay the basis for the quantitative science of acoustics and for acoustical engineering.

For the fifth sensation, *pain*, Richardson used himself as the subject (following an operation on his thumb), as he did with mental images. In this case too he invented a quantitative 'pain' scale, A to E, for measuring the sensation. He saw then that such a scale could be used generally by patients to indicate quantitatively how the sensation of pain is changing. (This is now standard clinical practice.)

As a postscript to this work it is interesting to note the recognition since the 1940s of the importance of the quantitative study of sensation (e.g. Poulton 1989) not only for clinical psychology and the cure of suffering and disabilities, but also for the design and control of environments where people live and work, ranging from the provision of correct light and sound levels in ships and offices to the planning of humanly acceptable wind environments of cities where both meteorology and psychology are important (cf. Hunt *et al.* 1976).

In each of these cases, after finding suitable measurable responses that define sensations, and providing quantitative scales for the measurements, Richardson described the results as mathematical functions or simple differential equations. But these did not *explain* or model any of the physiological processes. Later he returned (1930:4) to consider some of the mental processes involved in sensations, especially how the mental processes are initiated and vary with time.

It had been noted by several authors (including H. Poincaré, the eminent French mathematical physicist) that the solution of difficult mental problems came suddenly, and frequently after a period of relaxation. Richardson developed the idea of McDougall (1901) that there are similarities between the sudden passage of current in gases, e.g. sparks and mental images, by studying the flickering currents of neon lamps; he suggested that the physical and mental processes are dormant until some pulse of charge, or the physiological equivalent, is applied to bring the system above a threshold level. The physical analogy for mental images of the fluorescent tube has not subsequently been pursued by other researchers, but there are some similarities between this model and the actual physical process of sharp electrical impulses along nerves (see Poulton, below).

These studies show clearly that Richardson did not assume that mental and other biological processes are analogous to simple and completely predictable physical or mathematical systems, such as a pendulum or an oscillating atom, but rather that they were analogous to systems with *irregular* behaviour. The mathematics of irregular behaviour was only just beginning at that time, an important step being the discovery by van der Pol (1926) of nonlinear differential equations whose solutions show oscillations at one frequency which then slowly or suddenly change to another frequency. In 1937:1 Richardson wrote in an excited vein that these kinds of differential equations might be good models for psychological and biological processes. This idea is now well established, but Richardson also saw a wider cultural dimension to this discovery. He suggested that, since mental processes are not quite periodic, this might explain the popularity of syncopated music. Although Richardson's musical tastes inclined to classical chamber music, and certainly not to popular concerts, he was presumably making an elliptical reference to jazz, which had invaded Europe in the 1920s and 1930s!

1929–43 Paisley and the studies of conflict

In 1929 Richardson moved again, when Westminster Training College ceased to provide scientific education up to the level of a bachelor's degree; its teaching was then restricted to one-year courses in education. Richardson applied for the post of Principal at Paisley Technical College, and after an interview was selected from a shortlist of four (out of 38 candidates). Paisley is an industrial town near Glasgow, which used to be famous for its textiles and the Paisley pattern. It was not then (or now) usual for a Fellow of the Royal Society to take on such a post, but at that time there were few academic positions available in British universities for someone wishing to pursue research in meteorology, let alone a new kind of quantitative psychology! (He had earlier turned down the offer of a post in New Zealand.) His salary was £1,000 per year (comparable to that of a university professor) and a house was also provided for the Principal in a residential area at Castlehead; this salary was a substantial increase over the £650 he earned at Westminster with no house.

During these years he and his wife Dorothy were active in the Society of Friends and in peace movements, as well as providing a warm and stimulating home for their children and the friends and relations who visited them. Although Richardson sought refuge in his study at the earliest opportunity, the Richardsons' parties, with the many exciting scientific and energetic games are still remembered by those who were there, or (like myself) heard about them twenty years later!

The Principal at Paisley was often expected to teach for many hours each week; the previous Principal had taught for 38 hours. Richardson found that after teaching and doing administration, such as preparing budgets all day, he still had to give sixteen lectures a week in the evenings! He reported to the Governors that this amount of work was 'unusually much, personally disagreeable, and not for the good of the College!' Despite then being relieved from teaching some elementary classes, he was still teaching bachelor's level mathematics and physics to evening classes. He was generally a patient teacher, but in the class, as in his home, he sometimes lost his temper!

It is not surprising that he could only work on his research at weekends and during the holidays. Nevertheless he was able to continue the research on quantitative psychology that he had begun at Westminster College (and that has already been described). During the 1930s he regularly attended meetings of the British Association where there were several controversial discussions about the nature of psychological research. Until 1933 Richardson sat as the representative of the Royal Meteorological Society on the British National Committee of the International Union of Geodesy and Geophysics, although he made few contributions to the official business. Despite being in Paisley, away from the main centres of scientific activity, he continued an extensive correspondence with other research workers all over the world, including his old colleagues in meteorology, notably V. Bjerknes in Norway, with authors of the publications he was reading, and with contacts (including those in his and his wife's large family) who could help to provide him with data, particularly those on wars.

In 1935 Richardson's research returned to the study of the causes and the prevention of wars and other conflicts. This was the last turning point in his scientific career; just as he had largely given up meteorology in 1926, he now ceased further work in psychology. In the 'comparatively tranquil' years from 1919 to 1935 that followed the First World War he did not pursue his earlier research on conflict or attempt to publish his earlier work. He and others had hoped that the League of Nations in Geneva would be effective in preventing further wars.

However, the League's Disarmament Conference in Geneva from 1932 to 1934, was not successful, and rearmament in many countries followed. This led Richardson 'to reconsider and republish' his earlier ideas (see Sutherland, below).

The first of the major themes of Richardson's research in this field was the development of a suitable mathematical model for the tendencies of nations to prepare for war. He worked out the implications of the model for previous times using historical data and then made predictions for 1935 onwards using the model and recent economic and defence expenditure data. The model equations for the variable x, which he now called 'preparedness for war', were based on those he had conceived in the First World War and privately published in 1919 without any supporting data, or without prior presentation and discussion with other scientists. In 1935(:2,:3) Richardson first presented the simplified forms of the equations in a letter to *Nature*, pointing out that the equations were consistent with the rapid, exponential growth of armaments by all sides before 1914. But it worried him that the equations also showed that the unilateral disarmament of Germany after 1918, enforced by the Allied Powers, combined with the persistent level of armaments of the victor countries would lead to the level of Germany's armaments growing again. In other words the post 1918 situation was not stable. From the model he concluded that great statesmanship involving changes of policy would be needed to prevent an unstable situation developing. He noted that if such changes of policies materialized there would have to be new terms in the equations! Subsequently he published a full account of the theory together with the data in a monograph *Generalized Foreign Politics* (1939:1), which he later regretted not giving the more obvious title of *The Theory of Arms Races*.

This book is not only a mathematical study, but also invites the reader to consider a new approach to the politics of international affairs. Richardson was well aware of the general belief that there is an unbridgeable gulf between mathematics and politics, so he opens the book with an imaginary dialogue between a sceptical critic and the author.

> *Critic* '...How can anyone possibly make scientific statements about foreign politics? These are questions of right, of loyalty, of power, of the dignity of free choice. They touch a little on law, but are far beyond the reach of any science'.
>
> *Author* 'I admit that the discussion of free choice is better left to the dramatists. But nowadays science does usefully treat many phenomena that are only in part deterministic, witness the many social applications of statistics and the astounding progress of theories as to the probable positions of an electron'.
>
> .. *Critic* 'Can you predict the date at which the next war will break out?'
>
> *Author* 'No, of course not, ... The process described by the ensuing equations is not to be thought of as inevitable. It is what *would occur if instinct and tradition were allowed to act uncontrolled* ...'.

Richardson later explained that the idea implicit in the 'author's' second reply was that the model was describing events averaged over a certain time scale, or even averaged over many different conflicts. Richardson drew on his research in turbulence to explain that his mathematical models of 'averaged' foreign politics were analogous to statistical models of the average effects of turbulent flows, whereas most politicians and journalists are concerned with the individual 'eddies' of politics – the text of the dispatch, the movements of warships or the facial expressions of ambassadors!

In support of his approach Richardson quotes the judgement of the British Foreign

Secretary of 1914, Sir Edward Grey, as to the underlying cause of the First World War. This was very similar to the main conclusion from the model equations and the data on armaments expenditure, namely that the preparedness for war of each side increased in proportion to the preparedness of the other side. Grey's explanation was disputed by other politicians mainly on the naive grounds that a deterministic explanation based on the general trend of events and the prevailing attitudes could not actually explain the sequence of particular events at the time (such as the shooting of Archduke Francis Ferdinand at Sarajevo, and so on). Richardson's deterministic model was also criticized for similar reasons, both at the British Association meeting in 1938 and afterwards in newspapers. It was also criticized by many colleagues, friends and relations, including his brother-in-law Dr Maxwell Garnett, secretary of the British League of Nations Union. (However, as Rapoport (1968) pointed out in a discussion of Richardson's work, this debate has a long history with protagonists as outstanding as Tolstoy and Clausewitz.)

A balanced review of Richardson's book was given in *Nature* in October 1939 by a professor of mathematics specializing in differential equations, H. T. H. Piaggio, who pointed out that Richardson had made some important assumptions in deriving the form of the differential equations governing the variables, such as preparedness for war. He then suggested that there may be a flaw in this approach since it is well known that a small change in the form of such equations 'can sometimes produce large changes in the solution'. His other concern was that the theory did not include the effect of intelligent aggression planned by a leader that might be as deliberate as the moves in a game of chess – a reference to Fascist aggression at that time. (Richardson attempted to answer this point in his microfilm, published posthumously in 1960(:2), p. 231.) On the other hand another reviewer was impressed by the remarkably close agreement (even by the standards of physics) between the theory and the data for the 1909–14 arms race, which partially answered Piaggio's first point.

Richardson returned to this mathematical model of arms races in 1951 after the Second World War, when a new nuclear arms race was beginning between the USA and the USSR. He began another of his remarkable letters to *Nature*: 'There have only been three great arms-races. The first two of them ended in wars in 1914 and 1939; the third is still going on.' He considered whether there were any new factors that ought to be considered that could possibly lead to an end of the arms race without fighting.

Although an arms race leads to continual growth of armaments, the differences between the level of armaments in rival countries that may be small initially also tend to be amplified. These differences may become so great as to make the 'losing' nation 'submissive' and stop increasing its arms. Richardson introduced this effect, as a 'submissiveness' factor, into the equations; an approximate mathematical analysis suggests that the arms race could then cease. He asked 'Could events really happen thus? As far as I know, they never yet have done so.' (One might ask whether after 1989–90 there is a possibility that the answer is now yes, even if the reasons do not exactly correspond to those in Richardson's model.) However, following a more complete analysis of the equations in 1953, he withdrew his earlier conclusions (see Sutherland, Nicholson, below).

As well as working on his researches in Paisley, largely on his own but with the help of extensive correspondence, Richardson made a few forays to England and beyond. In August 1938 he attended the British Association in Cambridge and presented an outline of his monograph on *Generalized Foreign Politics* to the psychological Section. At the time of this meeting there was a distinct possibility of war breaking out over Czechoslovakia; although this was averted by the Munich agreement, there followed a rapid increase in expenditure on armaments by all the countries involved. These facts were included in an appendix of his monograph (1939: 1) to substantiate his thesis about the instability of the 'traditional' and 'uncontrolled' behaviour of countries in warlike situations.

In August 1939, Richardson paid a brief visit to the Baltic coastal town of Danzig, which lay in the 'corridor' between Germany proper and the Prussian province to the east of Danzig. One of the reasons for the German invasion of Poland in September 1939 was to reunite these two parts of Germany. On his return he wrote articles in the *Paisley Daily Express* and *Northern Echo* about the diametrically different views of the Germans and the Poles about the 'justice' of their case. This was a vivid example of the kinds of 'geographical' and cultural factors that influence the likelihood of wars, and that Richardson was to study in the last phase of his research on wars.

On his return to Paisley Richardson decided to spend more time on his studies of 'instability of peace' and so tendered his resignation to the Board of Governors of Paisley Technical College with effect from 16 May 1940 'when I will have finished the evening classes which I teach personally". He was 59 years old. The College kindly allowed him and his family, to continue to live in the Principal's house, which they did until 1943. Their children left home while they were at Paisley; Olaf became an automobile engineer, Stephen began his career in the British Merchant Service before settling in the USA in the 1940s and beginning a new career in social sciences. Elaine became a teacher.

Despite the intense bombing of Glasgow and the Clyde, there was no great danger in Paisley. Nevertheless Lewis and Dorothy were active in civil defence against the effects of possible air-raids. Also Richardson instructed groups of people about the possible hazards that might result from any enemy use of poison gas (which did not occur).

In fact during his retirement first at Paisley and later in Kilmun, Richardson did not extend much further his studies of 'instability of peace' or causes of wars. Perhaps it was because of the inevitability of a Second World War occurring in his lifetime that, around 1940, he changed the direction of his research on wars to the study of their statistics – when and where they occurred, which kinds of people were involved, and which kinds of geographical factors made them more or less prevalent.

As in his meteorological and psychological research, he began by collecting novel empirical data and then constructing quite new theoretical approaches, drawing on an even broader range of disciplines, in particular psychoanalysis, geography, history and politics, as well as deepening his use of mathematics.

But data are rarely gathered without some idea as to how they will be used. In his study of wars Richardson investigated the 'Freudian thesis' that 'from a psychological point of view a war, a riot, and a murder, though differing in many important aspects, social, legal and ethical, have at least this in common that they are all manifestations of the instinct of aggressiveness' (1948: 4). This was the 'justification for looking to see whether there is any statistical connection between war, riot and murder'. (See also Freud 1933). So starting in about 1939, Richardson began the massive task of accumulating these data, from history books, encyclopaedias, newspaper files, and in some cases by correspondence. He made a card index in which each card contained a symbolic summary of the key features of these 'deadly quarrels' (in his terminology). Although the summaries were almost hieroglyphic and could not be used by the general historian, Richardson's data are now held on computer data-bases at academic institutions in the USA, and elsewhere, specializing in conflict research.

After Richardson had begun this work, another systematic study of data on wars was published in 1942 by Quincy Wright in the USA. Wright's choice of data and method of analysis was based on the legal criteria usually followed by historians (so the data were largely confined to wars between states, etc). Although the data-set differed from Richardson's, which had been chosen on psychological and statistical grounds, he found that Wright's data on the larger 'quarrels' were consistent with his own. On discovering Wright's *Study of War*, Richardson's wrote to him to welcome his 'great book', and to comment on some of Wright's remarks about his own papers. Thereafter they had a warm and fruitful correspondence; Wright helped with the eventual publication of Richardson's writings on peace research after his death.

Richardson's best known finding from his own data, which he first reported (1941:1) in another of his topical letters to *Nature*, is that the number of quarrels (after 1820) decreased in frequency directly in relation to their *magnitude*, defined as the logarithm of the numbers killed on *all sides* of the quarrel. His data ranged from smaller scale conflicts such as banditry in Manchuria in 1935 and gang fights in Chicago (where the usual numbers were less than 10 and their logarithm less than 1) to world wars where the numbers were about 10 million (with logarithm of 7). He found that the same formula was also applicable within restricted ranges of data, and argued that therefore there is some general 'law' for all kinds of conflict.

The data also showed that the average time between wars of different magnitudes (within this period) was random, but did have a particular (Poisson) statistical distribution (1944:2). So for example this means that you can estimate the most likely number of years before the next war or deadly quarrel of a given magnitude will occur (always assuming the future is similar to the past!) – analogous to the time of waiting to cross a road. Recently, in a popular science book '*Cosmos*' Professor Carl Sagan (1980) of Cornell University extrapolated (perhaps tendentiously) from Richardson's result to predict that it will probably be about 1,000 years before there is a conflict so large (with logarithm of magnitude 10) that the entire world population (10,000 million) will be annihilated.

This statistical study and his earlier mathematical model for the causes of war, had a more general significance for Richardson because they were important examples of where new insights into history could come from a mathematical approach. Richardson (1960:1) wondered whether the gulf between history and mathematics could only be bridged by new institutions and whether existing institutions could adapt. (There is some evidence, particularly in the use of statistics in the study of economic history and politics, that they have done so, e.g. Floud (1973).)

Deadly quarrels like other complex phenomena occur randomly, but nevertheless are determined by different factors which can be investigated and quantified; in Richardson's words 'chaos restricted by geography and modified by infectiousness' (1946:1, p. 155).

His data had shown that in different countries and in different populations there were different tendencies for quarrels to occur.

Other investigators might have sought historical, geographical or sociological explanations but, as with his work on psychology, Richardson sought 'explanations' of his empirical and statistical discoveries by developing models drawn from physics. He attempted to explain the occurrence of wars in terms of how populations are distributed in concentrated groups within regions, and also how the shape, length and contiguities of the regions affect the propensity to war of populations within these regions.

Where quarrels involve groups of people in different regions, they must cross common boundaries of intervening regions. So an empirical study of the topographical aspect of geographical factors requires a close study of the geometry of the regions and their boundaries. As ever, Richardson took nothing for granted and began his own systematic study (1951:2) of how to quantify the relevant geometrical aspects of these boundaries. One aspect was simply counting the number of vertices and edges of boundaries and the number of regions, a mathematical problem first considered by Euler in the eighteenth century when he analysed the streets and bridges of Königsberg. By then subdividing the boundaries into different classes of edges, such as those adjacent to another nation or to neutral regions such as deserts or seas, he found that the population of those countries with many external frontiers, such as Great Britain and its Empire, were more prone to participating in wars.

Since many deadly quarrels involve people *within* nations (such as civil wars and riots) he also considered the shapes of the areas in which the population is distributed. He divided countries up into separate polygonal areas of equal population and found by trial and error that these polygons were generally shaped with equal diameters in different directions, and were usually hexagons. The same pattern emerged whether the populations in the areas were small (10,000) or large (10 million). These geometrical studies led Richardson to the interesting general conclusion (1952:2) that there was a greater incidence of low-level (i.e. where the magnitudes were less than 4.5 so that the number *killed* was less than about 30,000) wars between nations than of civil wars and riots of the same magnitude involving people having a common government. Richardson argued, in another of his dialogues, that this supported the need for federal or even world government.

One aspect of this topographical analysis did lead to a lasting contribution to mathematics in the study of curves that are highly irregular and wiggly. Richardson's discovery was not fundamentally a new result, but his empirical approach continues to illuminate research into this aspect of mathematics and its application in many fields. (For a detailed comment see Drazin, in Volume 1). In dividing up countries and regions into hexagonal boxes of different shapes for constructing models of population distribution, he needed to calculate the lengths of the frontiers from the sum of the lengths of the sides of the boxes on the frontier. But there was a problem. An embarrassing doubt arose as to whether actual frontiers were so intricate that their length could not be approximated by the perimeters of simple geometrical figures circumscribing the frontier. 'A special investigation was made to settle this question!' So he made 'measurements... by walking a pair of dividers along a map of the frontier so as to count the number of equal sides of a polygon, the corners of which lie on the frontier'. By summing the steps, an approximation of the total measured length of the frontier was calculated for each value of the step length between the points of the divider. Richardson pointed out that Archimedes used the same method in his approximate calculation of the circumference of the circle.

For a straight line or a smooth curve, it is found that the perimeter is independent of the 'step length' between the dividers. But for wiggly convoluted coast lines, as the step length decreases, more and more of the irregularities are included in the measurement of total length. So the measured perimeter is greater the smaller the step length. For a very wiggly coast line, such as the west coast of Britain, this variation is significant whereas for the rounded coast of South Africa it is not. Thus the relation between the perimeter and the step length produces a mathematical measure of 'wiggliness'. He used this study to classify and correct the simple relation between the length of a frontier and the number of 'hexagonal boxes' of a given area in the country in which the population was counted.

Richardson seemed to regard this mathematical result as an interesting curiosity; unusually for him he said he had 'no theory'! He remarked that 'the relation (relating the total length to the step length) is in marked contrast with the ordinary behaviour of smooth curves (whose) property is used in the "deferred approach to the limit"' (Richardson developed the description of his finite difference method with Gaunt (1927:1)).

Since then there have been many other studies that have shown how the natural (as well as the mathematical) world is full of curves and surfaces that cannot be represented by smooth mathematical functions, and Richardson's pioneering discovery is now as widely known as any of his other work largely because of the full account in Benoit Mandelbrot's very popular book *The Fractal Geometry of Nature* (1982). Ashford (1985, p. 260) describes how Mandelbrot, who is a specialist in this field of mathematics, accidentally discovered Richardson's work on 'fractals' when he was clearing out old papers and happened to read the *General Systems Yearbook* of 1961, where this work was posthumously published.

Retirement and his final researches at Kilmun 1943–53

In 1943 Richardson and his wife moved to their last home, at Hillside House, Kilmun, situated amongst a small group of houses on the shore of Holy Loch below high moorland, about 25 miles from Glasgow. Although the journey usually involved a train to Gourock, a ferry crossing over the Clyde and a short walk from one of the piers, they could still make occasional trips to Glasgow for Quaker meetings or library visits, and their children and friends could visit them.

Richardson had by now analysed all the statistics of the wars and quarrels that he had earlier assembled, and wrote up the main results of these studies in two large book-sized typescripts, which he microfilmed (and kept in a large safe in his home to protect it from possible fire; the safe was unlocked to protect it against burglars using explosives!), and which were deposited in copyright libraries in his lifetime. Steadily over this period, starting in 1941, he also wrote short articles in learned journals describing some of the results, especially the more mathematical ones. However the mathematical analysis of the spatial distribution of population, and the 'fractal' nature of 'wiggly' or irregular curves were only published posthumously.

He spent much time in anxious correspondence about publishing these books but with no success during his life. Fortunately for posterity, his American friends did succeed; *Arms and Insecurity; a Mathematical Study of the Causes of War*, edited by N. Rashevsky and E. Trucco, and *Statistics of Deadly Quarrels*, edited by Quincy

Wright and C. C. Lienau, were both published by Boxwood and Quadrangle Presses in Pittsburgh and Chicago. Although publishers were not surprisingly fearful about their possible losses on such unusual books, in fact they were a success! It is sad that Richardson did not live to see it.

Whilst working during the Second World War on these statistics of wars and quarrels, he felt the need to keep abreast of all the news. He tried to cross-reference *The Times* but gave up after three days (as he told me). So he used *Keesings' Archives* and he took his own brief notes from the newspapers and the radio.

He did not forget his other research interest in his retirement. A visit by H. Stommel, an American oceanographer from Woods Hole Oceanographic Institution on Cape Cod, Massachusetts, was the opportunity to return to the question of how pairs, or a cloud, of particles separate in a turbulent flow, such as he saw everyday from his house on the waters of Holy Loch. Stommel (1985) has recently described his own reminiscences of this experiment. Richardson was certainly very pleased with this research which he explained to my brother and me whilst we were visiting him in 1952.

He also returned to his research on the numerical solutions of differential equations and the associated study of the solution of sets of linear equations. He made use of important developments in the latter field, as he commented (1950:3), 'my two previous accounts of the method can now be much improved, by alliance with the great science of algebra; a coalescence suggested to me in 1948 by Arnold Lubin' (see Fox, in Volume1).

It appears that by the 1940s Richardson was aware that it would be possible to make use of electrical circuits and electronic valves either to model differential equations (the analogue approach) or to act as logical devices to perform arithmetical operations. This would enable differential equations, such as those needed for weather predictions, to be calculated automatically without thousands of human 'computers', as they were then called. In fact, as Dr David Eversley (1988) has recorded, Richardson began making an analogue computer in his 'laboratory' at his home at Kilmun using variable resistances and on/off electronic valves to represent different inputs like wind strength, barometric pressure, temperature at different levels in the atmosphere, and so on. During visits there Eversley helped him with soldering the circuits. None of this was mentioned in his papers of the 1940s, in which he continued to use the word 'computer' to refer to a human, and not a machine.

However at this time J. von Neumann and J. Charney at Princeton in the USA were beginning to use an electronic digital computer ENIAC to calculate the weather, using equations and methods quite similar to those set out by Richardson in his 1922 book on *Weather Prediction by Numerical Process*. Charney (1950) sent Richardson a copy of the paper published in *Tellus* in November 1950 containing the first results. Richardson made a quick empirical test of how good the computations were. The essential results were contained in diagrams of the initial pressure distribution, and the predicted and measured pressure 24 hours later. Richardson asked his wife which of the first two diagrams had greater resemblance to the measurements 24 hours later. In fact the difference to an untutored eye is barely noticeable, but apparently she said the predicted pressure pattern was closer than the initial pressure pattern; no one would have said that of the predictions he presented in this book in 1922! In his response to Charney, Richardson first congratulated him and his collaborators on their remarkable progress, and then, after describing his test, commented that 'this, although not a great success of a popular sort, is anyway an enormous scientific advance on the single, and quite wrong result,... in which the calculations of Richardson (1922:1) ended'.

Besides continuing his tireless scientific life at Kilmun, Richardson also worked hard in the house and garden, because his pension left little money for luxuries or for employing others to help. He remained a great experimental innovator all his days; he constructed an amazing heating system of pipes supported by wires in the central hall-way of his house, he used very smelly chemical tests before choosing paints for decorating the home, he had a great system for making jam (to economize on the rationed allowance of sugar) and he devised, with Dorothy, ingenious ways of consuming the vegetable harvests as they ripened, such as dishes of broad beans for breakfast, lunch and tea at a certain period in August! Dorothy continued to provide fun and hospitality in their large home to a local youth club and to many friends and relatives who visited them.

Richardson had often spoken about the possibility of returning to Cambridge; and in the summer of 1953, after seeing an advertisement for a Research Fellowship at his old College of King's, he applied, in order that he could use a 'library larger than those in Glasgow'.

However during September 1953 he felt 'very old and tired', as he wrote to Quincy Wright in the USA. The letter was characteristically argumentative and he expressed again his long held disbelief in the stability of a balance of power. He died in his sleep on 30 September 1953.

His wife Dorothy moved from Kilmun to live with her sister in England and died in a road accident in 1956.

Concluding remarks

This brief account and personal view of Richardson's life and work shows the extraordinary originality of his research, the intense care which he took over it (for example ensuring the tracing paper did not shrink when a light was brought close to it!) and the novelty of its presentation. I hope that other readers of his papers will find them as stimulating as I did in reading them for these volumes. Some of their ideas are still worth pursuing and continue to give insights into current research controversies; they also contain useful empirical data that have still not been scientifically 'digested'.

After reading the body of his work it is interesting to consider the methods of his research. It was generally driven by his need to obtain quantitative answers to scientific and practical problems; in every case the models and methods he constructed were rich enough to suggest new hypotheses which could be proved or disproved by comparing with other data. Where new techniques and advances in mathematics and science were relevant, he eagerly used them, and with extraordinary insight and originality; even as late as his 1953(:2) paper on 'The submissiveness of nations' he applied the recent research of Dame Mary Cartwright on nonlinear differential equations to analyse when a balance of armaments might or might not exist. He applied his robust quantitative approach in fields which had not previously been studied in this way, notably approximations, computational meteorology, quantitative experimental psychology, mathematics of wars, and nonsmooth mathematical functions. Another of his characteristics was to look for *analogue models*. Just as analogue methods of computing have disappeared with the advent of digital computers, so it is no longer scientifically conventional to use a physical or mechanical model of a phenomenon (such as Richardson's physical models of the mind or of population concentration) rather than an abstract mathematical model. This is probably why those aspects of his work tend nowadays to be disregarded.

Since many of his scientific investigations were driven by practical problems and his humanitarian interests one might ask to what extent his research has been applied in practice. His researches on numerical methods, meteorology and mathematics still have so many applications, and in so many fields outside those he was considering, that a complete list would be long and tedious. (A number have been given in this introduction and in the specialist reviews by Fox and Charnock, in Volume 1).

A major reason why the list is so long is because it has been possible to apply Richardson's results and methods to practical problems even without understanding the scientific and mathematical reasons behind them.

Just to list the results he expressed concisely in formulae makes the point; Richardson's method for extrapolation; the Richardson Number *Ri* for turbulent flows in the presence of buoyancy forces; the Richardson 'four-thirds' laws for the diffusion of clouds of particles in turbulent flows; his concept of eddy motions, now called the 'Richardson' cascade; his scales for pain and hearing; Richardson's model of arms races; the scale for the magnitude of wars and the formula for their frequency; and finally the index (now called a fractal dimension) as a measure of the irregularity of curves. Some of these celebrated formulae and concepts were completely new, but others synthesized, and neatly expressed, the results of different strands of research both of his own and of other scientists. In fact they continue to provide a useful basis for analysing the results of experiments and theory even now, more than 50 years later.

His numerical methods, especially on extrapolation, led to concepts and specific methods which were eventually incorporated into the array of numerical techniques that enables computers to solve the differential equations involving practical problems fast (but still not usually 'in a morning') and to an accuracy that can be defined in advance. These methods are used in all areas of science, technology and more recently, social sciences and commerce.

Similarly the results of his studies of turbulence are applied to fields far removed from the lower atmosphere and surface waters where he performed his experiments, notably to turbulence artificially generated in engines, or accidentally generated in fires and other natural disasters. The Richardson criterion and the Richardson 4/3 diffusion law continue to intrigue physicists and mathematicians; but in the meantime the formulae provide the essential basis for calculating many natural and industrial flows.

In experimental psychology Richardson helped introduce quantitative techniques that are now used in many applications wherever the sensory perception of physical conditions or influences are the crucial factors in fields such as medicine, environmental health and planning, industrial hygiene and ergonomics.

However it is much harder to evaluate his work on how hostile attitudes and actions of nations change when wars are likely, and his results on the statistics of international and civil warfare and their relation to geographical factors. As Rapoport (1968, p. 45) and Ashford (1985) have remarked, Richardson's work has profoundly influenced academic studies of conflict especially those taking a quantitative approach; but, one might add, it may also be influential as an antithesis for studies which are based on the individual 'eddies' of history. Richardson's way of publishing his models in this field in learned journals and then interpreting the results in terms of the current international situation and government policy has been continued by these researchers. As to whether

Richardson's work has influenced government policy in international affairs, all one can say with certainty is that in the largest countries (especially the USA and the USSR) the arms race, as well as the planning and prevention of wars, have been significantly influenced by various kinds of mathematical modelling, some of which resemble the methods introduced by Richardson and others between 1915 and 1935 (e.g. Bennett 1988).

One cannot help feeling that the author of the paper in 1951 'Could an arms-race end without fighting?' would be mightily pleased with the recent remarkable decisions by the former USSR and other countries to begin ending the nuclear arms race. But at the same time, the Quaker scientist who discovered the law describing the frequencies of different levels of conflicts would have noted, with sorrow, that recent history gives no hint that his inexorable formula for the frequency of smaller scale conflicts has ceased to apply.

Introduction

As an undergraduate, Lewis Richardson privately decided 'to spend the first half of my life under the strict discipline of physics, and afterwards to apply that training to researches on living things' (Gold 1954). This volume contains his papers in the second category, but the move from physics was gradual rather than abrupt. It was marked by his attendance at night classes in the psychology department at University College London, leading to a B.Sc. degree in psychology in 1929 at the age of 48, three years after he was elected FRS; and by his early retirement from the headship of Paisley Technical College in 1940 'in order further to investigate the causes of wars'.

The papers in this volume fall into two quite distinct groups. The first group is concerned with the quantifying of individual mental events and sensations, these papers all appearing between 1928 and 1937. The other, much larger, group is described by Richardson as 'studies of the causation of wars with a view to their avoidance'; these papers range in date from 1919 to 1953. Richardson's contributions in these fields will be reviewed in turn. This introduction deliberately concentrates on giving in broad outline the more important of Richardson's ideas and findings, with a minimum of technicality. The detail, and the wealth, of his methods, evidence and arguments are apparent in the papers that follow.

The quantifying of mental events and sensations

BY E. C. POULTON

Richardson appears first to have quantified mental events like imagery, mental effort and pain, which have no obvious physical correlates. He was subsequently given advice by F. C. Bartlett (later to become Professor Sir Frederic Bartlett FRS), who was the editor of the *British Journal of Psychology* in which five of Richardson's articles were published. The advice was to study sensations, which can be related directly to the physical stimuli that produce them. This was a shrewd suggestion. A physicist is well qualified to devise appropriate physical stimuli and to measure them accurately. Also the quantitative judgments can be used to produce standards for industry.

Richardson's first published article in psychology (1928:3) reported a novel method of determining the two-point threshold on the skin of the forearm. This is the smallest separation that can be felt as two points. The two points were stimulated simultaneously using compasses. Instead of measuring the threshold directly, Richardson asked his two observers (one of whom was his wife, Dorothy) to estimate the distance between the points in millimetres. The threshold was the greatest average separation that was judged to be a single point. Richardson described four different methods for deciding which point this was.

For his interval judgements of lightness and colour, Richardson (1929:4) devised two methods of eliciting responses. For the lightness or darkness of browns, he called black

0 and white 100. The 36 observers judged the lightness or darkness by selecting the appropriate number between 0 and 100. For colour, Richardson presented a horizontal line 100 mm long. The line was labelled 'white' at one end and 'scarlet' at the other. The 323 observers made a single judgement each of the degree of redness of a pink, by marking the line at the corresponding point. (This method was suggested by his nephew, the actor Sir Ralph Richardson; it is now widely used.)

At that time, Richardson (1930:3, 1932:4) found it difficult, if not impossible, to convince some of his contemporaries of the validity of measuring sensory magnitudes. Thus Myers (1931, p. 252) wrote in his *Text-Book of Experimental Psychology, Part 1*: 'we are powerless to measure sensation strengths at all'. Today it is hard to believe that such a controversy could have arisen. The controversy appears to centre on a few sentences quoted out of context from the writings of the Harvard philosopher and psychologist William James in his *Principles of Psychology* (1890, Vol. 1, pp. 545–6). In these sentences James criticizes two of Fechner's assumptions: '(1) that the just-perceptible increment is the *sensation unit*, and is in all parts of the scale the same (mathematically expressed, $\Delta s = \text{const.}$); (2) that all our sensations consist of sums of these units'. Arguing against (2), James wrote: 'To introspection, our feeling of pink is surely not a portion of our feeling of scarlet'. But this does not mean that sensory magnitudes cannot be measured. About three lines later James said 'when we take a simple sensible quality like light or sound, and say that there is now twice or thrice as much of it present as there was a moment ago, ... we mean that if we were to arrange the various possible degrees of the quality in a scale of serial increase, the *distance*, *interval*, or *difference* between the stronger and the weaker specimen before us would seem about as great as that between the weaker one and the beginning of the scale. *It is these RELATIONS, these DISTANCES, that we are measuring and not the composition of the qualities themselves*, as Fechner thinks' (italics and capitals in the original). This is exactly what Richardson (1929:4) measured in his investigations of lightness and pinkness.

For good or ill, Richardson's most influential article in experimental psychology appears to be that on the direct measurement of loudness (Richardson and Ross, 1930:2). At that time the loudness level of a sound was specified in phons. The level in phons is defined as the intensity of a 1 kHz tone in decibels re. 20 $\mu N/m^2$ (0.0002 $dynes/cm^2$) that matches the sound in loudness. Richardson wanted to devise a direct measure of loudness. He asked his observers to judge the ratios of the loudnesses of pairs of sounds, which he measured in terms of telephone voltage. After a thorough investigation on himself as observer, he enlisted eleven new observers. The observers included Myers, who had stated in his textbook that sensation strengths cannot be measured. The frequencies offered to different observers ranged from 0.55 to 1.1 kHz.

Richardson plotted the judged ratios of the pairs of sounds on the ordinate of a graph, against the ratio of the two telephone voltages on the abscissa using a log–log plot. As the threshold is approached, the function must turn downwards. This is because the threshold is a finite point or region on the logarithmic physical scale of the abscissa. But the threshold lies at zero on the logarithmic subjective scale of the ordinate, which is an infinite distance downwards. Thus the plotted functions of the ten observers other than Richardson and Myers were concave downwards. For these ten observers, Richardson measured only the slope of the roughly linear part of the function at the top. The relation is:

$$(\text{loudness}) = (\text{constant}) \times (\text{voltage})^n$$

where n ranges from 0.24 to 0.73 for different observers, with a mean of 0.44.

In America Richardson's direct method of judging ratios was used almost immediately in psychoacoustics. But it was 20 years before the method came to be used extensively by psychologists. Eventually Stevens (1975, p. 15) listed the slopes on a log–log plot of 33 different stimulus dimensions, only five of which have familiar physical units. All the slopes were determined in Stevens' laboratory at Harvard using ratio judgements.

However, for stimulus dimensions like loudness, where observers do not have familiar units, judging ratios is not a valid procedure. Suppose one group of untrained observers judges the differences in the loudness between pairs of sound intensities, and another group judges the ratios of the loudness between the same pairs. The rank order of the judged sizes of the differences is found to be identical to the rank order of the judged sizes of the ratios. (Birnbaum, 1980). This should not happen, because for example $(7-3) > (4-1)$ but $7/3 < 4/1$. It appears that both groups judge the sizes of the differences, but the group that is told to judge the ratios uses a logarithmic or ratio scale of numbers instead of a linear scale. Thus the difference judgements are consistent with matching two intervals and with bisecting an interval, whereas the ratio judgements are not. This happens because without familiar units of magnitude, the observers can only discriminate between the stimuli, using their improvised subjective units of equal discriminability. A scale of equal discriminability does not have an obvious zero corresponding to the threshold. It cannot be used to express valid ratios because you cannot judge a ratio without knowing where the zero lies. By contrast, differences can be judged using a scale of equal discriminability, because in judging differences the position of the zero is irrelevant. Thus it is the ratio judgements that are invalid. The ratio judgements can be said to have a logarithmic response bias (Poulton 1989, Chapter 6).

Richardson's main theoretical article in experimental psychology (1930:4) drew analogies between mental processes and electric sparks. It has not been followed up. Richardson suggested that the intensity of trying to think corresponds to potential difference; intensity of imagery corresponds either to electric current or to radiated energy; and cleverness corresponds to the ionization produced by X-rays or γ-rays. In some cases he described analogies to physiological processes like neural reflexes and the rate of transmission of impulses along a nerve. Richardson also designed electric circuits containing two neon lamps, which demonstrated reciprocal inhibition and other phenomena (1937:2, in Volume 1). He hoped that once physicists had worked out the physical equations, they could be used by physiologists and psychologists. This has not happened.

Models were not commonly used by experimental psychologists until 30 years later, with the advent of the computer. A typical present day model differs from Richardson's; it comprises an arrangement of labelled boxes connected by arrows. The models are used to help their authors to describe and think about the phenomena that they represent, and so to suggest new investigations. This black box approach to modelling is related more closely to the old theories of mental faculties than to Richardson's physical analogies.

The causation of wars

BY IAN SUTHERLAND

The other papers in this volume, which supplement and extend Richardson's books *Statistics of Deadly Quarrels* and *Arms and Insecurity*, are concerned with mathematical and statistical studies of the psychology of international relations and the nature and causation of wars. They divide naturally into three subgroups, reviewed here separately, which correspond to the themes of the two books and studies of other aspects of international relations.

Statistics of deadly quarrels

In 1940 Richardson started compiling a list of wars and other deadly quarrels throughout the world that ended in or after 1820, together with detailed information on the population groups that participated, their aims and the results of the conflict. He took his information from a large number of historical sources. He classified these quarrels primarily according to their magnitude, which he defined as the logarithm to the base 10 of the total number of persons killed as a result of the quarrel. This enabled him to include on a single manageable scale the Second World War (magnitude 7.4) and a murder, involving only one death (magnitude 0). Richardson progressively added to, revised and updated this list. A first version was published in the microfilm of *Statistics of Deadly Quarrels* (1950:4) and the final version (August 1953) in the book (1960:1). By that time it included 105 wars of magnitude 3.5 or more (3,163 or more deaths) that ended between 1820 and 1952, and Richardson believed this total to be complete or very nearly complete. For smaller wars (magnitude 2.5 to 3.5) he recognized that the list was incomplete (209 quarrels in the final version), and for smaller quarrels still only fragmentary data were available. The number of murders world-wide was estimated from police statistics.

Richardson made a series of ingenious statistical studies of these conflicts and their associated characteristics, and reached a large number of informative general conclusions. He first examined the frequency of quarrels according to their magnitude (1941:1, 1948:4). The complete data for the larger wars showed that the logarithm of the frequency increased linearly with decreasing magnitude. The murders were plotted on the same diagram, and the introduction of a slightly curved extension linked the two parts of the graph acceptably. Moreover, information on frequency and magnitude for certain groups of smaller quarrels was in accord with and fully supported the validity of this link. Of the total of 59 million quarrel-dead estimated from the graph to have occurred between 1820 and 1945, the heaviest losses of life occurred at the ends of the scale (36 million in the two World Wars, and 10 million in connection with murders). The total represents about 1.6 per cent of the deaths from all causes during this period.

A study of the distribution of wars in time (1944:2, 1945, 1950:4) showed that the numbers of wars beginning each year, and also the numbers ending each year, agreed closely with a Poisson distribution. This applied both to Richardson's own list and to another list, covering a much longer period (432 years), compiled on different principles by Wright (1942). A more detailed subdivision of the latter material showed irregular variations in the risk of conflict during the period. Richardson's own list did not reveal any general trend towards more or fewer fatal quarrels during the period 1820–1940, even when the doubling of the world population between these dates was taken into account. There was however a tendency for large wars to become more frequent and small wars less frequent.

The number of nations on each side of a war was examined in 1946:1,:2. Wars with one nation on each side were by far the commonest type, the frequency of encounters which involved more nations decreasing rapidly with increasing complexity. Richardson explored a progression of theories in the attempt to account for the observed frequencies. He eventually found reasonable agreement with history by supposing that each dispute interested only eight nations, and that the probability that any two of them came to war with each other about it was uniformly 0.35. This clearly failed to explain the small number of larger wars involving more than eight nations, and Richardson explicitly introduced geography, postulating in a further theory differing probabilities of war between nations according to whether they had land contact and whether they ranked as land powers or sea powers. This theory succeeded in explaining the rare occurrence of very complicated wars, but failed insofar as the observed probability that two local powers were in conflict was not constant (as postulated by the theory), but increased with the number of powers involved, as if fighting were infectious. Richardson's final conclusion was that 'a chaos, restricted by geography, and further modified by the infectiousness of fighting' offered the best description of the observed pattern of the number of nations on each side of a war.

The next group of studies appeared in the microfilm of *Statistics of Deadly Quarrels* (1950:4, not reproduced here; the chapter references below are to the book (1960:1). See the 'Notes on *Statistics of Deadly Quarrels*' below, p. 537). An examination of the characteristics of named nations (Chapter V) showed that throughout the period 1820–1945 new belligerents were continually appearing, and at a fairly constant rate. Participation in wars was widespread, with Britain (29) France (21) and Russia (18) heading the list. Germany (or Prussia) took part in 10 and Japan in 9 of these wars. Richardson did not attempt to resolve the 'obscure and controversial distinctions between aggression and defence' but pointed out that no nation could have been the aggressor in more wars than those in which they took part. The involvement of states in external wars increased with the number of their frontiers, suggesting that geographical contiguity played some part in quarrels.

Richardson found (Chapter VI) that civil wars decreased in frequency with the duration of previous common government and that wartime alliances tended to persist. He concluded that common government, and comradeship in war, both appeared to have a pacifying influence. The frequency of retaliatory wars between nations decreased with the duration of intervening peace, as if there was a slow period of forgiving and forgetting. In Chapter VII Richardson found that marked economic disparity between the opposed belligerents, indicative of a 'class war', was noted by historians in only eleven of 83 wars, whereas 33 of these wars were for territory.

Chapters VIII and IX deal respectively with the associations between languages and wars and religions and wars. It is relatively simple to enumerate pairs of belligerents according to their languages or religions. But to decide whether diversity of language, or of religion has been a cause of war, some estimate is needed of the *expected* numbers of conflicts between groups speaking particular languages, or following certain religions. This problem is much complicated by the fact that, by virtue of the occurrence of civil wars, the conceivable belligerent groups are not the same as political states. Richardson therefore developed a number of theoretical models involving pairs of population groups of a defined size, which led him to upper and lower limits for the expected numbers of pairs of opposed belligerents with the same language (or religion), on the basic assumption that sameness of language (or religion) was irrelevant. He concluded that there were fewer wars in which both sides spoke one of the Chinese languages, and more in which both sides spoke Spanish, than would be expected from the total populations speaking those languages. 'For the other chief languages the statistics neither confirm nor refute Zamenhof's belief that a common language would have a pacificatory effect.' As regards religions, there were many fewer wars than expected between those adhering to the Chinese mixture of religions (Confucian – Taoist – Buddhist); the figures 'suggest but do not prove' that Christianity incited wars between its adherents, and that the Moslem religion prevented wars between its adherents. However, there were many more wars between Christians and Moslems than would have been expected if religious differences had played no part.

In several of these studies Richardson was confronted with the problem of assessing geographical opportunities for fighting. When studying the number of nations on each sde of a war (1946:1, also Chapter X) and the frequency of external wars (Chapter V) he used the number of frontiers between a state and its neighbours for this purpose, but this was of no help in studying the associations of wars with languages and religions (Chapters VIII and IX) or the relative frequency of civil and external wars (Chapter VI), where the sizes of the populations speaking the same language, adhering to the same religion, or sharing a common government, had to be taken into account. For the studies of languages and religions (which he had made in 1942) Richardson had used what he later referred to as a 'rather crude geometry' of equal-sized population groups to reach approximate answers. His ingenious practical solution to both aspects of the problem was to develop a method of 'mapping by compact cells of equal population' (1951:2) in which the globe was covered with a theoretical network of cells, each containing the same number of persons, with at most three cells meeting in a point, with no cell completely surrounding another, with national frontiers lying in the edges of cells, and with each cell being 'compact', that is, about as broad as long. Richardson enumerated the total numbers of cells in the world, and the numbers of civil, foreign and other boundary edges to those cells, for cell populations of 10^7, 10^6, 10^5 and 10^4.

In 1952:2, Richardson returned to and developed a topic he had already touched upon, namely whether geographical contiguity between population groups provided a sufficient explanation of the frequency of occurrence of wars between them, or whether some other influence was operating. He showed first that if the population of the world was regarded as divided into equal-sized groups, each liable to quarrel with its neighbours, the actual numbers of quarrels were far fewer if the groups were small (or far more numerous if the groups were large) than would be expected from the geographical opportunities for conflict, suggesting the existence of some local pacifier. A study of the duration of fighting in wars of different magnitudes, taking the geographical contiguity of the belligerents into account, confirmed the existence of this local pacifying influence. Its nature was less easy to discover. But Richardson's paper on contiguity (1951:2) provided him with measures of the separate geographical opportunities for civil and foreign conflict, and permitted the conclusion that some influence suppressed civil fighting relative to foreign fighting. Thus the local pacifying influence appeared to be the habit of obedience to a common government; however, intermarriage, a common language, a common religion and a tendency to dislike foreigners may also have had local pacifying effects which were obscured by their correlation with common government. Intermarriage as a pacifier was discussed more fully in 1950:2.

Arms and insecurity

Richardson's formulation of mathematical descriptions of the psychology of international relations and of arms races had long preceded his statistical studies of deadly quarrels. His first publication in this field (1919:1) was written towards the end of the First World War; his last major contribution (1953:3) was published just after his death.

His initial approach was explicitly psychological, but based on communal rather than individual psychology. In *Mathematical Psychology of War* (1919:1) he examined and expressed in mathematical terms the various influences that affect the 'vigour-to-war' or 'warlike striving' of a nation, which he suggested is 'largely, though not entirely, an instinctive reaction to the stimulus of the warlike striving of the opposing side'. After obtaining a mathematical expression for the relationship between the warlike activity and the vigour-to-war of a single nation (equation 9), Richardson considered in turn the numerous separate influences, mostly instinctive, that contributed to, or detracted from, the vigour-to-war, and expressed the contribution of each in mathematical terms. A long expression (equation 33) represented the total vigour-to-war due to the ensemble of instinctive influences. This was then incorporated into a further version (generalized to include the effects of free will and higher influences) of the relationship between the warlike activity and the vigour-to-war (equation 34). A series of approximations, neglecting influences that were likely to be small, led to an equation (38) for one of two opposed nations, 'the x-side. It has been assumed that the relations for the y-side are similar in form but with different values of the constants'. This pair of differential equations, derived during the First World War to describe the psychological interactions of two nations, preparing for or waging war, represents the basis for all Richardson's later mathematical studies of the relationships between nations.

Stimulated by 'the present regrettable rearmament', Richardson resuscitated the earlier equations in 1935 in the context of two opposed nations in peace time. He added a constant term to each equation in 1935:3 and introduced his standard later notation, namely:

$$\mathrm{d}x/\mathrm{d}t = ky - \alpha x + g, \tag{1}$$
$$\mathrm{d}y/\mathrm{d}t = lx - \beta y + h. \tag{2}$$

Equation (1) expresses in mathematical terms a statement that the rate of increase of the 'defences' of the first nation, $\mathrm{d}x/\mathrm{d}t$, is proportional (with a positive 'defence coefficient'

k) to the size of the 'defences' of the second nation, y, which are interpreted by the first nation as 'threats'; at the same time the rate of increase is restrained by the 'fatigue and expense' to the first nation of maintaining its own defences, which is proportional to x with a positive coefficient α; finally the rate may be influenced by some grievance ($g > 0$) or feeling of contentment ($g < 0$) of the first nation towards the second. Equation (2) expresses the corresponding statement about the rate of increase of the 'defences' of the second nation, with different values for the coefficients.

'The international situation is thus represented by a point (x, y) in a plane. Let us think of this point as a particle moving in accordance with the equations. If the particle be tending towards plus infinity of both x and y, then war looms ahead. But if the particle be going in an opposite direction the prospect is peaceful' (1935:3). Richardson presented in this paper the main characteristics of the solution to these equations. They define two linear 'barriers', intersecting at a 'point of balance' at which $\mathrm{d}x/\mathrm{d}t = \mathrm{d}y/\mathrm{d}t = 0$. The point (x, y) is confined to the sector defined by one of these barriers, drifting towards the point of balance if $kl < \alpha\beta$ and away towards infinity and instability if $kl > \alpha\beta$. 'That is to say; the international regime is unstable if and only if the product of the two defence coefficients exceeds the product of the two fatigue-and-expense coefficients'. It may be noted that the grievances do not enter into the stability criterion.

In 1938:2,:3 Richardson introduced an 'improved description' of x as 'threats minus cooperation' (to take into account the theoretical possibility of a negative value for this variable) and suggested that it might be measured in terms of a balance between warlike expenditure and foreign trade. Using such a measure, Richardson showed in 1938:3 that the rise in the combined expenditure on armaments of France and Russia, and of Germany and Austria–Hungary, between 1909 and 1913 was in remarkably close conformity to the linear trend predicted by his equations.

The monograph 'Generalized Foreign Politics' (1939:1) presents a full exposition of these matters, together with a generalization of his theory, first to three nations (or groups of nations) and then to n nations, with the set of linear equations:

$$\mathrm{d}x_i/\mathrm{d}t = g_i + \sum_{j=1}^{n} k_{ij}\, x_j \quad (i = 1, 2, \ldots, n). \tag{3}$$

Richardson derived a criterion for stability for n nations of unequal sizes (that is, unequal in their warlike potential) but with perfect communications (that is, with their interactions unaffected by geographical distance), and with equal fatigue-and-expense coefficients. This criterion for stability may be expressed in general terms as the fatigue-and-expense coefficients being collectively greater than the defence coefficients. (Mathematically, the system is stable if the eigenvalues of the matrix $(\mathbf{k}_{ij})$ are all negative.) He estimated the coefficients of the matrix for the nine main great powers in 1935, and confirmed that this represented an unstable situation, the largest eigenvalue being positive. He also showed that the trend of the defence expenditure of these nations between 1929 and 1937 was in close conformity with the size of this eigenvalue.

The studies of arms races were taken a stage further in the microfilm of *Arms and Insecurity* (1947:2, not reproduced here; the chapter references below are to the book (1960:2). See the 'Notes on *Arms and Insecurity*' below, p. 419). A historical analysis (Chapter VI) suggested that unstable arms races first appeared between 1877 and 1900, perhaps as a consequence of the application of science to war material. The European arms race of 1909–13 was re-examined in more detail, including data for Britain and

Italy, and allowing for trade expenditures between the Triple Entente and the Triple Alliance (Chapters VII, VIII). Again there was close conformity between fact and theory.

The arms race of 1929–39 was also re-examined in great detail for ten nations, Poland being added to the previous nine (Chapters XIX, XX). Because of the currency instabilities after the First World War it was no longer possible to express the national defence expenditure of each country in terms of gold, as Richardson had done for the earlier arms race. His solution was to replace the annual defence expenditure of each country by a figure for the number of persons who could earn that total amount in a year at the current wage-rate for a semi-skilled engineer in that country, a statistic he refers to as the 'warlike worktime' (or 'warfinpersal', war finance per salary). The revised matrix of defence and fatigue-and-expense coefficients in 1935 still implied instability. It also, through another eigenvalue, implied the existence at that time of two opposed groups of nations, with France, Russia, Czechoslovakia and Britain on one side, and Japan, Italy and Germany on the other, the USA, Poland and China not being clearly on one side or the other. The inference, from data for 1935, of an eventual alliance between Britain and Russia against Germany is of particular interest in view of the surprise non-aggression pact between Russia and Germany in August 1939 and the equally unexpected attack of Germany on Russsia in June 1941.

In *Arms and Insecurity* Richardson also studied the whole period between the two world wars in considerable detail, to see whether his linear theory (for two groups of nations) held during the period of disarmament, followed by apparent stability before the second arms race (Chapter XXIII). He considered Germany on the one hand and France, Britain, Poland, Czechoslovakia and Russia on the other. From 1919 to 1933 the point corresponding to $(x+y, x-y)$ appeared to have moved along a barrier to a point of balance, where it remained from 1922 to 1931, and then began to move away along another barrier, apparently the second barrier in the same field. Between 1933 and 1934, however, the trend changed abruptly, and the point then moved along what appeared to be a barrier in a new field, away from a different point of balance. The change would correspond to an abrupt alteration in the defence coefficients and in the fatigue-and-expense coefficients to new constant values, at the time when the Nazis came to power in Germany.

Further consideration of the trend between 1919 and 1930 led Richardson to the conclusion that it could not be explained entirely in terms of his basic equations, and he turned to a nonlinear modification, first advanced in 1939:1, to incorporate the concept of 'submissiveness', namely the postulated tendency of one nation to submit to another if confronted with a disparity in armaments:

$$\mathrm{d}x/\mathrm{d}t = ky\{1-\sigma(y-x)\}-\alpha x+g, \tag{4}$$
$$\mathrm{d}y/\mathrm{d}t = ly\{1-\rho(x-y)\}-\beta y+h, \tag{5}$$

where σ and ρ are positive 'coefficients of submissiveness'. With ρ positive (for Germany) and $\sigma = 0$ these equations fit the period of disarmament, the ten-year pause near the point of balance, and the turn into the arms race from 1930 to 1933. However, alternative explanations of the turn into the arms race were offered by the trade depression or the fading of 'war-weariness', or a combination of these with submissiveness, and no discrimination between these explanations was possible (Chapter XXIV).

Richardson's main interest in later papers was the further exploration of the concept of submissiveness. He accepted that submission of one nation to another could occur after military defeat, or in the presence of a gross disparity in armaments, in conformity with his equations, but was much more concerned with the possible occurrence, and with the consequences, of submissiveness between two nations nearly equal in their military capability. A first approach to the solution of equations (4) and (5) in this situation encouraged him to think that the operation of submissiveness from the early stages of an arms race could bring the development of that arms race to an end 'without fighting', that is, with a reduction of the armaments on both sides (1951:4,:5). However, the full treatment of these two equations in 1953:2 did not confirm this interpretation. The theoretical consequences of this formulation of submissiveness were complex, without parallel in history, and did not include a lasting reduction in armaments. Richardson therefore considered an alternative formulation; in this the operation of submissiveness in a developing arms race would lead to a 'thoroughly stable' end-point with large and equal armed forces on the two sides, a theoretical outcome which had the outbreak of the First World War as evidence against it. Richardson concluded that he could find no evidence of submissive effects between two equal and 'rested' groups of nations.

In his final paper (1953:3), following a useful summary of his theory and its application to events from 1907 to 1939, Richardson analysed the disarmament from 1945 to 1948 and the first few years of the arms race that began in 1948. Disarmament was more rapid than after the First World War, and by 1950 the arms race was already showing signs of developing differently from the previous two. He also traced the similar spirals of warlike expenditure in Britain before, during and after each of the two world wars, plotting the mean annual expenditure against the annual change. Richardson's linear theories relate to the periods of armament and disarmament on the left of these spirals, not to the periods of major fighting on the right. He commented: 'For the spiral as a whole, I have not at present any formula to offer.'

Other aspects of international relations

In three papers on voting strengths in an international assembly (1918, 1926:2, 1953:1) Richardson advocated a logical system for determining the number of votes to which a state should be entitled, based on the state's international rather than its national activities. Such a system would be preferable, he claimed, to that based, both in the League of Nations and in the United Nations, on the equality of sovereign states.

In a complex paper on war-moods (1948:2) (it contains 195 numbered equations or statements) Richardson explored the changes in population moods before during and after a war. He suggested that in individuals these follow a progression of pairs of feelings, conscious and unconscious, towards the other nation, these feelings being friendliness, hostility and war-weariness in different combinations. He developed a series of mathematical equations, based on epidemic disease theory, to describe transitions in the population from one dual mood to the next, and compared their implications with historical facts before, during and after the First World War. Interestingly the equations, derived from this alternative psychological approach, for the phase leading up to a war were of the same linear form as those derived from the previous and different psychological approach to the theory of arms races, with terms corresponding to the defence coefficients, the fatigue-and-expense coefficients and the grievances. However, this parallel indicated the need to broaden the concept of the original fatigue-and-expense coefficients to include other restraining influences, and Richardson refers to them thereafter as 'restraint coefficients' (Statement (73)). The initial theoretical formulations ignored 'gregariousness', or the postulated tendency of individuals to act together. Richardson modified his equations to include the concept, and examined the evidence for the contributory effects of gregariousness both in the First World War and the Sudeten crisis of September 1938. In 1949:4 Richardson discussed the persistence of national hatred and the changeability of its objects.

Several of Richardson's papers from 1950 onwards consisted mainly of expositions of his earlier findings for general or specific audiences (1950:2,:5, 1951:3, 1952:1), to compensate in part for the fact that *Statistics of Deadly Quarrels* and *Arms and Insecurity* were available only on microfilm.

Richardson's achievement in his studies of the causation of wars

One of the factors contributing to the slow recognition and appreciation of Richardson's scientific studies of conflict during his lifetime was the limited availability of his key publications (1919:1, 1939:1, 1947:2, 1950:4). It was only with the printing of *Statistics of Deadly Quarrels* and *Arms and Insecurity* after his death that his work became more widely known and has been developed, particularly in the USA (see Nicholson, below).

This publication of Richardson's collected papers achieves three main objectives. Firstly it enables the books *Statistics of Deadly Quarrels* and *Arms and Insecurity*, printed only in 1960 but written mainly between 1940 and 1943, to be viewed in their correct chronological setting at the centre, not at the end, of his output on the causation of wars. This reveals the stimulus that was provided by contemporaneous world events to Richardson's development of the theoretical framework for his studies, to his close and continuous monitoring of relevant statistical information, historical as well as contemporary, and to his comparisons of events with the implied consequences of his theories. Obvious examples are the writing of *Mathematical Psychology of War* (1919:1) while Richardson was with the Friends' Ambulance Unit in France during the final months of the First World War; his pointing out of the theoretical implications of the failure of the Disarmament Conference (1935:2,:3); and his analysis of the course of the arms race from 1929 to 1937, based on numerical data for nine great powers, and published in June 1939, before the outbreak of the Second World War (1939:1).

Secondly, this collection, by including the monographs *Mathematical Psychology of War* and 'Generalized Foreign Politics', clarifies both the genesis and the evolution of Richardson's thoughts and analyses on arms races and disarmaments. In particular, his basic pair of linear equations is seen to derive not from a set of empirical postulates but from an attempt to assess comprehensively the effects of all the relevant psychological motivations between opposed nations. Thirdly, the collection makes accessible Richardson's later work on submissiveness and arms races, as well as his studies of war-moods and other aspects of international relations. As a result it has become practicable for the first time readily to study the whole of Richardson's output on conflict and to assess his achievement in this field.

Richardson adopted the standard scientific approach to each problem, with a theoretical formulation based on current understanding or belief about the underlying factors, and an exploration of the implications of the theory. He then investigated whether observed data accorded with, or failed to conform, to the theoretical expectations. He consistently applied his own version of Occam's razor, namely '"Formulae are not to be complicated without good evidence" or briefly "try out the easiest formulae first"'. He introduced modifications or a different theory only if the earlier theory did not adequately fit the facts. The study of the number of nations on each side of a war (1946:1) provides a particularly good example of the progressive modification of an initial formulation, and also of Richardson's recognition that a theory that fitted the facts did not necessarily reflect the true underlying processes.

In a field in which experimentation is impossible, the scientist has to analyse and draw what conclusions he can from observational data. The pattern revealed by Richardson's descriptive statistical analyses of wars between nations is basically one that he termed 'chaos', both in time and space. The timing of the outbreaks of war, and of peace, is apparently random, and new belligerent groups continually arise. The probability of war between any two nations appears to be modified by geographical contiguity, by the possession of a navy, by differences in language or religion, and common government and intermarriage both appear to be pacifying influences. However, these historical and geographical associations are not so strong that they enable one to assess with any precision how likely, or unlikely, it is that any particular nation or group of nations will become involved in a war within a stated period. In this respect these analyses are no more or less informative than the findings of most observational enquiries in any field – they reveal much of interest, but because of a multiplicity of relevant underlying factors, little of aetiological or predictive value.

Modelling is the other scientific technique which may be of value in a complex situation in which experimentation is impossible, and this is Richardson's main approach to the analysis of international relations. His basic model is explicitly derived from a consideration of the group psychology of two nations opposed in war (1919:1). The starting-point is his recognition (some may prefer to call it an assumption) of the *mutuality* of the relations between the nations – the idea that the set of attitudes and actions of one nation towards another interacts with, and helps to determine, the set of attitudes of the second nation towards the first. Whether one accepts this as obvious or not, it is the fundamental assumption from which Richardson's theories grew. He attributed (in several contexts) his realization of the importance of mutuality to Bertrand Russell's pamphlet: *War the Offspring of Fear* (1914). Ashford (1985, pp. 63–4) has however suggested that this may have been a misattribution by Richardson, and he offers an alternative source for the idea (Anon, 1914).

Richardson simplified his original very complex equations by discarding, at least provisionally, terms known or likely to be small, resulting in a basic pair of linear differential equations as an approximate description of the psychological interaction of two warring nations. From 1935, Richardson applied these equations specifically to the situation of two or more nations or groups of nations preparing for war, and 'in 1938 ... hit on a way of testing the formulae quantitatively' by using expenditure on armaments as a measure of a nation's defences. In this context his basic equations showed remarkable success both in describing the course of the arms races of 1909–13 and 1933–8, and in predicting from pre-war data the eventual alignment of the nations during the Second World War.

The same model (with different constant values for the coefficients) also 'fitted' the period of disarmament and subsequent stability from 1919 to 1930, but not the turn into the second arms race from 1930 to 1933. With the addition of terms expressing the submission of Germany to the Allies after the First World War, the theory (with constant coefficients) 'fitted' the whole period from 1919 to 1933, when an abrupt change in the coefficients to new constant values was required for the theory to fit the arms race from 1933 onwards. Richardson's further theoretical explorations of submissiveness between two equal and 'rested' nations, however, led him to two important conclusions – that there was no evidence of the occurrence of submissiveness during the arms races of 1909–13 and 1933–8, and little hope, even if submissiveness occurred, that an arms race, once started, could end with a reduction of armaments on both sides.

The ability of this relatively simple deterministic model, derived from considerations of psychological motivation and mutual interaction, to describe so closely the course of the two arms races and the military expenditure of the two sides between the wars, is very impressive. But the applicability of the model to other situations is less certain. In particular, Richardson had no opportunity to test its validity in an improving rather than deteriorating situation between two groups of nations (except in the special case of the two short periods of disarmament).

The advent of nuclear weapons is a major new factor. Richardson made no special mention of nuclear weapons (nor of their postulated deterrent effects) in his later writings, but he saw no reason to modify his basic equations, nor to expect the course of a nuclear arms race to differ materially from one involving only non-nuclear weapons (Ashford 1985, p. 224). However the third major arms race, between the USA and the USSR, having lasted many years longer than the first two, and having threatened on several occasions to end in war, had by 1990 apparently come to a halt without fighting, and was being replaced by a disarmament process. The signs of this were already apparent in the progress made in disarmament talks during the years preceding the dramatic political changes in Eastern Europe in the autumn of 1989. The contrast with the theoretical expectations from Richardson's equations, and the distinctive contribution (if any) of nuclear weapons, both deserve further examination.

In the complex field of meteorology, Richardson's aims were to gain sufficient understanding of the physical processes, and sufficient relevant information, to permit forward predictions of the weather. Through his pioneer work, and with the aid of electronic computers, these aims have now largely been achieved in relation to short-term forecasting. In the immensely more complex field of international relations, with his 'studies of the causation of wars with a view to their avoidance', Richardson's aims were not simply understanding and prediction using relevant information – there was the additional hope, distant though it may have been, of control as well. In relation to these demanding aims, Richardson has taken the first large step. His descriptive statstical analyses of the frequency, timing and observed characteristics of the wars between nations or groups of nations greatly increase our understanding of wars and their background, although they cannot help much with either prediction or control. The success of his theoretical models emphasizes the important roles of psychological motivation and mutual interaction in international relations, and advances our understanding of arms races considerably. Richardson claimed that 'the equations are merely a description of what people would do if they did not stop to think'. It was one of his unfulfilled hopes that his findings would become known to and accepted by the world's statesmen, who would then be in a position to avoid or overcome the instinctive reactions leading to arms races, and replace them by efforts of will leading to greater cooperation between nations. However, Richardson's models unfortunately provide no practical guidance to the statesman on how to resolve an arms race without fighting and how to replace it with peaceful cooperation.

Richardson was the first to subject international relations in general, and wars and arms races in particular, to mathematical and statistical analysis, and he did so with great competence and ingenuity, and immense single-handed labour. His analyses provide us with new perspectives, encouraging us to view human conflict globally rather than from a national or partisan standpoint. His scientific approaches are as thought-provoking, and as topical in their relevance to the current international situation, as they were when they were made. Throughout, he gives the reader plenty to think about, and plenty to build upon. His achievement has been to demonstrate the power of scientific method to illuminate and increase our understanding of conflict between population groups, and to point the way to further scientific effort, directed towards the improvement of international relations.

Later developments

BY MICHAEL NICHOLSON

When Richardson died in 1953, his work on peace and war was virtually unknown and almost totally so among those professionally involved in the study of international relations, whose subject matter he so imaginatively dealt with. Moyal (1949) made some further studies of the distribution of wars in time (1944:2) and Richardson incorporated his findings in *Statistics of Deadly Quarrels* (Chapter III), Horne (1951) queried Richardson's theoretical formulation of submissiveness (1951:4) and Richardson replied (1951:5). His work was occasionally referred to (Penrose 1952). In general it was neglected.

This did not last for long. In 1956, James Newman produced two excerpts from Richardson's work in *The World of Mathematics* along with an admiring commentary. Shortly after, Rapoport (1957) wrote a long and detailed article 'Lewis F. Richardson's Mathematical Theory of War' in the newly-founded *Journal of Conflict Resolution.* These, followed by the publication of Richardson's two books (1960:1,:2) and further expositions in Rapoport (1960) and Boulding (1962) alerted a constituency that was highly receptive but had previously been ignorant of Richardson's work. From its earlier obscurity it now appeared that, ironically, Richardson's work had been published at just the right time. Many scholars, primarily in the United States, were dissatisfied with the way international relations as an academic discipline had developed and were anxious to put it on to a firmer theoretical and scientific footing so that it could look more firmly in the eye better established social sciences, such as economics and psychology. In the social sciences as a whole at this time there was a move to put theory on a more rigorous and systematic basis. Social scientists were increasingly concerned with the careful testing of hypotheses, while the development of theory in mathematical form was growing apace. Richardson's model of the arms race provided one such theory of a process which is clearly important in the international system. This delighted the economists, mathematicians and physicists who, worried by the failure of academic international relations to offer explanations of, much less solutions to, the appalling problems of the age, were eager to see something more rigorously expressed than the vague verbal generalizations which characterized the discipline of the time.

A great deal of effort has gone into the development and the testing of the Richardson model, to some degree because of its historical primacy. For example, work on *n*-party arms races and arms distributions by Schrodt (1978, 1981) has made important contributions to our understanding of stability in the international system. Other work on more general systems of hostility and friendliness in the international system by Zinnes and her collaborators (1983) has been developed as one of the natural successors to the Richardson model and explicitly avowed as such. These are two important contributions but there are many others. A surprising factor which makes the Richardson model less applicable than might be thought is that the current state of the debate in an admittedly rapidly changing field seems to favour the view that the interactive element in arms races is relatively small. The evidence and analysis favours the view that arms races are much more the consequence of factors internal to the states and nations than the result of inter-state interactions and hostility. This is not always true. The Middle East since the Second World War and the pre-1914 naval arms race fit in with a Richardson model rather nicely, as does the nonnuclear element in the NATO/Warsaw Pact arms race. Surprisingly, though, these seem to be the rarities.

However, this is not the only issue. Richardson's model was the first deductive model to be rigorously stated in the discipline of international relations. It would be surprising if the first effort were to be overwhelmingly successful. Along with the work which came from the theory of games, also avidly seized upon, Richardson's work formed the basis of a significant formal modelling tradition in international relations which has proved of great importance (Nicholson, 1989). As such, it has been very influential in the development of deductive theory in international relations.

The other aspect of Richardson's work was his vast compilation of data (without any help from a grant-giving body). Along with that of Quincy Wright in his immense *Study of War* (1942), these were the original large-scale sets of data collected in international relations. In *Statistics of Deadly Quarrels* the data are primarily on wars in which the incidence of war is related *inter alia* to religions, language, contiguity of states and economic factors. In *Arms and Insecurity*, where he develops his theory of arms races, he also tries to test it with data concerning armaments, trade and other relevant factors. This approach has had many followers also. Wilkinson (1980) up-dated Richardson's work and reconsidered some of his conclusions. Useful though that study is, it is not the most important of the post-Richardson developments. There has been massive work on expanding the data about the international system, particularly on the period since 1815, the end of the Napoleonic wars. The availability of relevant data is clearly a prerequisite for the serious testing of hypotheses. If one is to make generalizations about, say, arms races and war, then there needs to be a body of reliable data against which to do so. Richardson, a more careful statistician than Wright, with a greater concern for precise definition and analysis, recognized this and performed his remarkable one-man feat of data collection. This has now been greatly extended. Cioffi-Revilla (1990) gives an account of the various major projects, of which the Correlates of War Project at the University of Michigan is perhaps the most well known and widely used. While no single individual has shown the same mixture of assiduity in the collection of data and ingenuity in its analysis as Richardson, the teams who have followed him, and been directly inspired by his work, have collectively produced impressive results, even if the conclusions to be drawn are still controversial.

Richardson's reputation as a social scientist working on war turned, just after his death, from that of a barely known eccentric into the acknowledged founding father of a discipline. Indeed he was the founder of two of the three main strands of work in the social science tradition – formal theory and statistical testing. Only about the third strand, rational choice and game theory, did he have little to say. Ironically, in Britain he is still regarded by the mainstream of the international relations profession, still dominated by historians, as being on the fringe. For explanations of this we have to delve into the sociology of knowledge. However, few issues of the major American journals which deal with international theory do not have a reference to Richardson somewhere.

References

Anon. 1914 *The International Industry of War*. London: Union of Democratic Control.

Ashford, O. M. 1985 *Prophet or Professor? The Life and Work of Lewis Fry Richardson*. Bristol and Boston: Adam Hilger.

Batchelor, G. K. 1952 Diffusion in a field of homogeneous turbulence II. The relative motion of particles. *Proc. Camb. Phil. Soc.* **48**, 345–62.

Bennett, P. (ed.) 1988 Mathematical modelling of conflict and its resolution. *Proc. IMA Conference*, Cambridge, December 1986. Oxford: Clarendon Press.

Birnbaum, M. H. 1980 Comparison of two theories of 'ratio' and 'difference' judgments. *J exper. Psychol.* (gen. Sect), **109**, 304–19.

Boulding, K. E. 1962 *Conflict and Defense: a General Theory*. New York: Harper and Row.

Campbell, L. and Garnett, W. 1882 *Life of James Clerk Maxwell*. London: Macmillan.

Chapman, S. 1965 Foreword to *Weather Prediction by Numerical Process* by L. F. Richardson. New York: Dover.

Charney, J. G., Fjörtoft, R. and von Neumann, J. 1950 Numerical integration of the barotropic vorticity equation. *Tellus* **2**, 237–54.

Cioffi-Revilla, C. 1990 *The Scientific Measurement of International Conflict: Handbook of Datasets on Crises and Wars 1945–1988*. Boulder and London: Lynne Reinner Publishers.

Drysdale, C. V., Ferguson, A., Geddes, A. E. M., Gibson, A. H., Hunt, F. R. W., Lamb, H., Mitchell, A. G. M., Taylor, G. I. and Goodwin, G. 1936 *The Mechanical Properties of Fluids*. London: Blackie.

Eversley, D. 1988 Unpublished letter to Mrs V. Bottomley, MP.

Floud, R. C. 1973 *Introduction to Quantitative Methods for Historians*. London: Methuen.

Freud, S. 1933 *Civilization, Society and Religion*. Penguin Freud Library **12**, 349–62. London: Penguin.

Gold, E. 1954 Lewis Fry Richardson. *Biographical Memoirs of Fellows of the Royal Society* **9**, 217–35.

Horne, M. R. 1951 Could an arms-race end without fighting? *Nature, Lond.* **168**, 920.

Hunt, J. C. R., Poulton, E. C. and Mumford, J. E. 1976 The effects of wind on people; new criteria based on wind tunnel experiments. *Building & Environment* **11**, 15–28.

James, W. 1890 *Principles of Psychology*. London: Macmillan, New York: Henry Holt.

Kolmogorov, A. N. 1941 The local structure of turbulence in incompressible viscous fluid for very large Reynolds numbers. *Comptes Rendus Acad. Sci. U.R.S.S.*, 301–5.

Kolmogorov, A. N. 1962 A refinement of previous hypotheses concerning the local structure of turbulence in a viscous incompressible fluid at high Reynolds number. *J. Fluid Mech.* **13**, 82–5.

Lanchester, F. W. 1916 *Aircraft in Warfare, the Dawn of the Fourth Arm*. London: Constable.

McDougall, W. 1901 On the seat of the psycho-physical processes. *Brain* 24, 577–630.

McDougall, W. 1905 *Physiological Psychology*. London: Dent.

Mandelbrot, B. B. 1982 *The Fractal Geometry of Nature*. New York: W. H. Freeman.

Moyal, J. E. 1949 The distribution of wars in time. *Jl R. Statist. Soc.* **112**, 446–9.

Myers, C. S. 1931 *A Text-Book of Experimental Psychology, Part* 1. Third edition. Cambridge: University Press.

Newman, J. R. (ed.) 1956 *The World of Mathematics*. New York: Simon and Schuster.

Nicholson, M. 1989 *Formal Theories in International Relations*. Cambridge: University Press.

Oboukhov, A. M. 1941 On the distribution of energy in the spectrum of turbulent flow. *Izv. Akad. Nauk S.S.S.R.* Ser. Geogr. i. Geofiz. **5**, 453–66.

Penrose, L. S. 1952 *On the Objective Study of Crowd Behaviour*. London: H. K. Lewis & Co.

Poulton, E. C. 1989 *Bias in Quantifying Judgments*. London: Lawrence Erlbaum Associates.

Rapoport, A. 1957 Lewis F. Richardson's mathematical theory of war. *Journal of Conflict Resolution* **1**, 249–99.
Also in *General Systems Yearbook* 1957, **2**, 55–90.

Rapoport, A. 1960 *Fights, Games and Debates*. Ann Arbor: University of Michigan Press.

Rapoport, A. 1968 Foreword to *On War* by Clausewitz. London: Penguin.

Runge, C. 1895 Ueber die numerische Auflösung von Differentialgleichungen *Math. Ann.* **46**, 167–78.

Russell, B. 1914 *War the Offspring of Fear*. London: Union of Democratic Control.

Sagan, C. 1980 *Cosmos*. New York: Random House.

Savage, L. J. 1954 *The Foundations of Statistics*. New York: Dover.

Schlichting, H. 1935 Turbulenz bei Wärmeschichtung. *Z. angew. Math. Mech.* **15**, 313–38.

Schrodt, P. 1978 Richardson's *n*-nation model and the balance of power. *American Journal of Political Science* **22**, 364–90.

Schrodt, P. 1981 *Preserving Arms Distributions in a Multi-polar World: A Mathematical Study*. Denver: Monograph Series in World Affairs. University of Denver.

Sheppard, W. F. 1899 Central-difference formulae. *Proc. Lond. math. Soc.* **31**, 449–88.

Stevens, S. S. 1975 *Psychophysics: Introduction to its Perceptual, Neural and Social Prospects*. New York: John Wiley and Sons.

Stommel, H. M. 1985 *Prophet or Professor?* by Oliver M. Ashford. *Bull. Amer. met. Soc.* **66**, 1317.

Taylor, G. I. 1959 The present position in the theory of turbulent diffusion. *Adv. Geophys.* **6**, 101–11.

van der Pol, B. 1926 On relaxation-oscillations. *Phil. Mag.* (6) **61**, 978–92.

von Neumann, J. and Morgenstern, O. 1944 *Theory of Games and Economic Behaviour*. Princeton: University Press.

Wald, A. 1950 *Statistical Decision Functions*. New York: Dover.

Wilkinson, D. 1980 *Deadly Quarrels: Lewis F. Richardson and the Statistical Study of War*. Berkeley and London: University of California Press.

Wright, Q. 1942 *A Study of War*. Chicago: University Press.

Zinnes, D. A. and Muncaster, R. G. 1983 A model of inter-national hostility dynamics and war. *Conflict Management and Peace Science* **6**, 19–37.

Note on the arrangement and presentation of the papers

The papers included in each volume are arranged as far as possible in chronological order of composition. This is particularly desirable for a full appreciation of the development of Richardson's work on wars and arms races in this volume, because of the considerable difficulty he experienced in getting this work published promptly or widely during his lifetime. The chronological sequence covers all Richardson's publications, not only the papers reproduced in these two volumes. The sequence within each year is indicated by numerical affixes; this ordering differs in many instances from that in earlier bibliographies using affixed letters.

The chronological order is normally indicated adequately by the nominal month and year of publication. The exceptions are a few reports presented at, or circulated prior to, scientific discussion meetings, which are referred instead to the date of the meeting (for example 1915, 1929:2, 1932:4, 1938:2); contributions to official meteorological reports, referred to the year in which the observations ended (1920:1); posthumously published and previously unpublished papers, referred to the date of completion (1936, 1951:2, 1951:3, 1953:1); and, of particular importance, the two posthumous books *Statistics of Deadly Quarrels* and *Arms and Insecurity*, which are referred primarily to the dates of the original microfilm editions (1950:4 and 1947:2).

An introductory page before each reprinted or newly-printed paper gives a standard abbreviated publication reference according to the World List of Scientific Periodicals, with first and last original page numbers, followed by fuller bibliographical information, and any necessary editorial notes.

Richardson's three books, *Weather Prediction by Numerical Process* (1922:1, reprinted 1965:1), *Statistics of Deadly Quarrels* (1960:1) and *Arms and Insecurity* (1960:2), are not reproduced here. It is also unnecessary to reproduce the earlier microfilm editions of *Statistics of Deadly Quarrels* (1950:4) and *Arms and Insecurity* (1947:2, 1949:2), although they are not widely available, because the published texts are so close to the original versions. Full notes on these two publications appear in the appropriate places in this volume, namely under 1950:4 for *Statistics of Deadly Quarrels* (p. 535), and 1947:2 for *Arms and Insecurity* (p. 417), and these give details of the few changes that were made in the later editions.

1913:1

THE MEASUREMENT OF MENTAL 'NATURE' AND THE STUDY OF ADOPTED CHILDREN

The Eugenics Review
London
Vol. IV, No. 4, January 1913, pp. 391–394

THE MEASUREMENT OF MENTAL "NATURE" AND THE STUDY OF ADOPTED CHILDREN.

Different estimates of the amount of intrinsic worth latent among the poor lead to such opposite political aims as those of the Labour Party on the one hand and Mr. G. P. Mudge[1] on the other. At the First International Eugenics Congress, Professor Loria told us that germinal worth and economic position were independent. A widespread belief to the contrary is, perhaps, a part cause of the general disapproval turned upon anyone who marries into a lower social class.

As it is strength of mind rather than of muscle which earns large incomes, I will confine the following statements to mental and moral qualities.

I venture to think that the whole question will be greatly clarified if we adopt a definition of Nature which is directly applicable to sociological investigations. The customary definition of Nature as "that which is contained in the fertilized ovum," or "that which is inborn," has no immediate application. No one proposes to measure the mental qualities of new-born babies as a step in eugenic research. A definition more satisfactory than "that which is inborn" would be "the growth attained under standard conditions of nurture and personal effort." We may apply this definition to biometric methods as follows:—

Let Q be a mental quality in the parent.

Let W be the parents' effort to develope this quality in himself.

Let E_1 E_2 . . . E_n be diverse environmental conditions affecting this quality in the parent.

Let q, w, e_1 . . . e_n have like meanings for the child.

Then Professor Pearson[2] speaks of the total correlation between Q and q as if it were a measure of the strength of inheritance. I submit that the proper measure is the partial correlation between Q and q for constant values of W_1 E_1 . . . E_n and w_1 e_1 . . . e_n. To calculate this partial correlation we require to know, among other total coefficients, that between the

1 "A plea for a more virile sentiment," Mendel Journal.

2 Prof. Pearson. Nature and Nurture, Dulau & Co., 1910.

environments of the parent and of the child; also that between the environment of the child and the development of the parent; and neither of these, as far as I am aware, has been published. Some, however, of the total correlations involved in the partial coefficient have been determined,[1] namely, r_{Qq}, and a large number of the type r_{qe}, between the teacher's opinion of the child and the state of its home. Belonging to the same type r_{qe} are a further set of correlations which it would be very interesting to know, namely, those between teachers' opinion on the one hand, and home traditions, conversation, interests, and discipline on the other. It is not unlikely that these coefficients will be larger than the very low values found[2] for correlations between mental capacity and such things as state of teeth, cleanliness, clothing or nutrition. For mind may be expected to feed primarily on mental food. In view of the above I submit that the size of the correlation which measures mental inheritance is still quite uncertain, although the valuable labours of Professor Pearson and his school have led us partway towards its determination. Comparisons by the correlation method such as Professor Pearson has made[3] between the strength of Nature and Nurture for mental qualities are not yet justified.

Moreover, in addition to mathematical complications, it is very difficult to find out exactly what mental influences a person has been subject to; still more difficult to compare them with any standard education. And the same applies with even greater force to the personal efforts made by the individual considered. On this account the foregoing definition of Nature, as the limit of growth under standard conditions, is likely to lead to uncertain measurements which will disturb any application whether Galtonian or, preferably, Mendelian.

A more hopeful line of advance appears to me to be that taken by Mr. Cyril Burt in his novel argument with reference to his instrumental tests.[4]

There is, however, another way. By going into training a boy may add an inch to his chest measurement, but not twelve inches. By caring for sweet peas a gardener can increase their growth, but only up to a certain point. In a graphic passage in "Hereditary Genius" Francis Galton describes the way in which competition gradually forces a young man to realize that there is a limit to his general ability, beyond which no effort, learning, or enthusiasm will carry him. Correspondingly the measure of Nature which I wish to propose is "*the growth attained when conditions of environment and personal effort have been so favourable, throughout the whole time of growth,*

[1] [2] Prof. Pearson. Nature and Nurture, Dulau & Co., 1910.

[3] loc. cit. p. 27.

[4] Eugenics Review, July, 1912, p. 181.

that a limit is reached." Such conditions may be spoken of briefly as "saturating conditions," and we may accordingly say that "*Nature is saturated growth.*" The advantage of this definition is that it does not require any measurement of the environmental and personal conditions provided they are sufficiently favourable. In this sense it is an absolute definition. When it can be applied much trouble and uncertainty are avoided.

One or two books would suffice to give an imbecile such elementary literary knowledge as he could grasp, whereas the style of a Lord Macaulay might lack some element had he been deprived of a single one of a hundred of his favourite authors; so in general the greater the innate power, the more favourable the conditions, which will be necessary if growth is to reach its limit.

A saturated growth of any one faculty or combined group of faculties is moreover often inconsistent with a saturated growth in other directions.

So much for the definition. Next as to two applications of it. Let us seek for cases of saturating conditions.

1. The sons of a man who is eminent in any line usually have the benefit of his experience, and in addition a good deal of stimulus to emulate him; so that if they show no capacity for their father's work it is a pretty clear proof that their nature differs from his. Thus difference between brothers is a surer indication of the strength of heredity than is resemblance. It would be useful to collect cases in which a lady's father and husband are both eminent in the same profession; her brothers and sons have then grown up in saturating conditions as regards this profession and we can trace the inheritance of any lack of capacity for it through her. The studies in Galton's "Hereditary Genius" are somewhat inconclusive because there is not sufficient evidence as to the amount of latent ability in the population outside the specially successful families which he considered.

2. When mental defect occurs in one member of an otherwise normal and well-brought up family, and when the defect cannot be attributed to accident or ill-health, we may be pretty certain that this defect is a true measure of the individual's nature in this particular direction. For it is supposed that we are dealing with cases in which some training has been provided with the intention of overcoming the defect, and that the encouragement by parents and the teasing by schoolfellows have stimulated the will of the defective one till this has been saturating also. On this account, I think that the researches which have been published on the inheritance of mental defect deserve a degree of confidence which cannot at present be placed in those on the inheritance of ability.

Now we may apply the foregoing considerations to compare the nature of different classes of society. For if offspring of poor parents, adopted when newly born into well-to-do and well-educated families, turn out markedly different from the birthright members of those families then the presumption is that the dullness, of whichever is the duller, is a saturated growth. If on the other hand they all turn out much alike there is no proof that growth is saturated for any of them. There remains the presumption that the conditions have been much alike for all the members of one family and we get a more uncertain but still useful comparison of native worth, as pointed out above. A thorough study of a hundred such cases of adopted children would do more to reveal the nature of the poorer classes than statistics of 100,000 poor persons brought up in poverty. The author hopes that a co-operative study of this kind can be arranged among the adherents of the Eugenics Education Society.

L. F. RICHARDSON.

1918

INTERNATIONAL VOTING POWER

War & Peace
London
Vol. 4, No. 53, February 1918, pp. 193–196

Notes:

1. Richardson refers to this paper later by various titles.
2. This topic is continued in 1926:2 and in 1953:1.

International Voting Power

AT the close of the war and afterwards there will, in all probability, be general political congresses, attended by many nations. The *object of this essay* is to discuss various proposals for the relative voting strength in such meetings.

*Historical.**—Before the war, voting of any kind had been the exception in diplomatic congresses. Unanimous agreement was the only form of assent which was considered compatible with national sovereignty and independence. A rule of this kind made it possible for a small state to block proposals agreeable to the rest of the congress. Unanimity was found to be so inconvenient that the following ways of doing without it had been put into practice :—

(*a*) Unanimous agreement between a limited number of States, as for example, in the original formation of the Postal Union.

(*b*) For certain very special and clearly defined purposes a simple majority vote has been accepted as binding on all parties concerned. For example, in some of the affairs of the Postal Union and of the Sugar Union.

(*c*) The Hague Conference of 1907 adopted a scheme for the election of judges to the proposed prize court, which virtually gave States relative voting strengths varying from twenty to unity. The scale is reproduced in Column 2 of the table given below. The proposal, not being ratified by all the States, never took effect.

Various Proposals for Fixed Scales have appeared in print, for instance :—

(*a*) The Fabian Research Committee suggests that some such scale as that referred to above in connection with the Prize Court should be agreed upon for general purposes.

(*b*) Lord Bryce, and others in their "Proposals for the Prevention of Future Wars," suggest a similar but less finely divided scale, "say, three to each of the Great Powers, and one at least to each of the rest."

(*c*) The Central Organization for a Durable Peace, in its Draft Treaty, Art. 6, proposes that "Germany, the United States of America, Austria-Hungary, France, Great Britain, Italy, Japan and Russia shall have each three votes. The other states shall have only one." This scale is intended to be used for a variety of defined purposes, as will be seen by looking at Articles 11, 22, 24, 49, 61, 89, 94, 100, of the Draft Treaty.

(*d*) M. Paul Otlet in "Les Problèmes Internationaux et la Guerre" makes a very interesting suggestion which may be mentioned here, although it lies a little to one side of our path. It is as follows :—The political wills of individuals are to-day organized twice over in two quite different ways : first, nationally in the democratic states; secondly, internationally in the growing inter-state organizations of finance, labor, science, trade, and the like. A world-parliament, such as Otlet proposes, in representing individuals, should pay attention to the ways in which their wills are already organized. To this end Otlet proposes two Chambers, the first Chamber being composed of representatives of nations, the second Chamber of representatives of international organizations. For instance, one third of the second Chamber might be composed of the representatives of international labor organizations, another third going to capitalists, and the remaining third to science, art, letters and religion.

A serious objection to all these proposals to agree upon a fixed scale of voting strength is that they provide no means for revision. For example fifty years ago Japan was not a great power. This warns us that any fixed scale of voting strength might become out of date even in fifty years. The international organization would then find itself up against a demand for a redistribution of votes. And, votes having by then become much more important than they are now, such a demand might lead to war; unless some rational way of settling the question had previously been thought out and accepted.

Published proposals for keeping the voting strength always in step with reality.

(*a*) Umano "Essai sur une Constitution internationale," 1907, says "this international assembly will be composed of representatives, sent by each of the nations in the group in number proportional to the mean strength of industry, population and wealth, which, revised from time to time on the definite results of budgets and of statistics, will be attributed to each nation and which will make each nation to appear and to weigh in the assembly in accordance with its real power."

(*b*) M. Otlet proposes a voting strength depending on the following national characteristics : population, birth-rate, longevity, literacy, production of wheat, foreign trade, area of territory, climate, mines and forests, accessibility of territory.

What are the fundamental ideas behind these various proposals to consider states as having different voting strengths, whether fixed or adjustable? There appear to be two fundamental ideas :—

(*a*) That the most highly civilized states ought to have most power. This would appear to have been M. Otlet's idea, in introducing a characteristic such as literacy into his index. But it must be admitted that there would be likely to be serious differences of opinion as to what constitutes "height" of civilization, so that it would be correspondingly difficult to reach to any internationally agreed voting strength on this basis.

(*b*) There is the second fundamental idea. For

* For details see "International Government," by L. S. Woolf (Allen & Unwin); "The Framework of a Lasting Peace," by L. S. Woolf; and "Les Problèmes Internationaux et la Guerre," by P. Otlet (Rousseau et Cie).

why, to take an instance, should China, enormous in population and large in territory, be graded so low? It is not altogether because Chinamen are considered to be less wise or less happy than Europeans. It is mainly because they stay at home and mind their own affairs. In a word, because they are not internationalized.

If we were to trust this principle and to extend it, it would lead us to conclude that, in forming an index of voting strength in an international congress, we should ignore all purely internal national characteristics, such as internal trade, area of territory, number of inhabitants, literacy; and should consider only the corresponding relations with other states—namely foreign trade, accessibility of territory, number of travellers entering and leaving and foreign post.

This does not imply that a self-contained nation is of smaller importance, but only that its affairs are more suitably dealt with in its own national parliament than in an international one.

The rest of this essay is devoted to the working out of this second fundamental idea (*b*). The other basis—that the most highly civilized states ought to have most power—has been put aside as being too much a matter of conflicting opinion.

The national characteristics to be used in calculating an index of voting strength must each satisfy three main requirements :

(*a*) As just stated they must be measures of the internationality of a State's life.

(*b*) They must be capable of being expressed numerically.

(*c*) The numerical value must be practically ascertainable to the satisfaction of all the States concerned.

Expenditure on Armaments. This has been, and now is, a principal factor of the " power " of a State in the world and, on that account, might be used in calculating the voting strength. On the other hand it may be said that when the stage of voting has arrived that of armaments will be beginning to pass away; and that an individual is not given votes in a national parliament in accordance with the number of firearms that he keeps in his house.

Even if, in the future, national armies and navies become the elements of an international police force, yet nations would almost certainly wish their voting strengths to depend on the shares of the burden of international police duty, which they severally bear. That could be arranged by considering the cost of " international police duty " as a " contribution to the international exchequer " and by using the latter in forming the index of voting strength.

On the whole it seems best to exclude armaments as a characteristic to be used for the index of voting strength. As a matter of interest, the expenditures on armaments of the different nations, prior to the war, are given in the table in Sec. 10 on p. 195, for comparison with the voting strengths. The sum of the annual costs of land and sea forces has been taken in preference to either cost separately.

Foreign trade, reckoned in money, appears to be a very suitable measure of a State's internationality. The sum of imports and exports, less imports intended for re-exportation, has been taken in the table below. The statistics required are already available.

Foreign travel. It is proposed to take the number of persons entering and leaving each State during the year. If this were done a temporary visitor would be counted twice over, once on entering and once on leaving, whilst an emigrant or immigrant would be counted once only.

Shipping is at first sight a possible measure of internationality. It was taken into account in the Hague scale of votes in connection with the Prize Court, for the reason that the Court had to deal with ships. It is a measure which would be advantageous to the British Empire. On the other hand, to arrange the scheme so as to make one's own nation come out on top, is a temptation to be avoided. Impartiality must be attempted. Now if shipping is taken as a characteristic, it must, in justice to land-locked states, be balanced by taking also railway communications to the exterior. A discussion would then arise as to the relative importance of shipping and railways. Now their effects are summed up in foreign trade and foreign travel. It seems best, therefore, to reject both shipping and railways, whilst retaining foreign trade and travel.

Foreign Post. Much numerical information is to be found in the Statistique Général published by L'Union Postale Universelle at Berne. The question here arises as to what one should take from among letters, post cards, printed matter, business circulars, samples of merchandise and parcels. Now it is a general principle that the more things that are included, provided that they are all more or less suitable, the steadier will be the average. On the other hand, samples of merchandise or parcels may be dealt with under foreign trade. Also the United States of America has special arrangements for parcels. Therefore it seems best to draw the line before these two, including all written and printed matter. Post between colonies and the home country should be excluded if they vote as a single state. The Statistique Général of the Postal Union does not show how much of colonial post goes to the home country and in consequence the data in the table below are inexact, but not to a great extent.

Area of Territory has been suggested as a suitable characteristic by Otlet. Otlet, however, makes a second suggestion to the effect that the voting power should vary inversely as the " utilizable area " of the state and directly as the product of the population into the foreign trade. For it might be argued that the fact that a state has a large area for a small population, is a sign that it ought to give up some of its territory to over-crowded states, and therefore that its voting power should be weakened as a warning. However, as area of territory is not a measure of internationality, it has been left out of consideration in what follows.

Contribution to the International Exchequer. Before the war these contributions were confined to such insignificant and scattered payments, as the joint financial support of central bureaux for posts, telegraphs, agricultural information, weights and measures and the like. But these organizations had been rapidly increasing. That votes and taxation should, to some extent, be mutually dependent is a recognized principle, to be found, for instance, in the British constitution. Some special international organizations have adopted it, for example the Institute of Agriculture. Thus contribution to the international exchequer seems a very suitable characteristic for the purpose. But the numerical data relating to the state of affairs before the war, are likely to be a poor indication of probably much greater contributions in the future. As was mentioned above under " Armaments," the cost of " international police duty " would be included here.

Capital invested abroad might also be taken as a characteristic.

The export of ideas would appear to be a suitable characteristic if it could be expressed numerically; but there is the difficulty. The number of foreign copyrights, of foreign patents, would be some indication. So also would be the number of books exported. Again the number of Nobel prizes which have gone to a nation is some indication of its productivity in the higher forms of thought.

To sum up: The following national characteristics have been noted as suitable: foreign trade, foreign travel, foreign post, contribution to international exchequer, capital invested abroad and, if a way could be found of measuring it, the export of ideas.

The Mathematical Formula.

The voting strength must be expressed as depending on the characteristics by means of a formula. Let no one say that this is an attempt to reduce politics to mathematics. The proper voting strength is, and must always remain, a matter of judgment, of opinion. But a well-chosen mathematical formula may help it to become a matter of considered and agreed opinion. In the choice of a formula we may be guided by a good old rule in applied mathematics, that of taking the simplest formula which makes sense. The vagueness of our general principles would make a mockery of any mathematical complication. Now the simplest formulæ are: the simple average, the weighted average and the product of powers. To any of which a constant may be added.

Let us consider these in turn. But first let us be clear about the numerical expression of the characteristics. It is only the *relative* voting powers of states that matter; so that we may divide all the votes by the same quantity without producing any effect on majorities and minorities. Accordingly, for the sake of neatness, let us express each national characteristic as a percentage of its total for the world. For example, before the war, the foreign trade of the United States of America was about 738 millions sterling annually. The sum of the foreign trades of all the independent states of the world came to about 6,000 millions sterling annually. So that the United States of America had about 12.3 per cent. of the total. This 12.3 is taken as the numerical value of the characteristic for the U.S.A. Let all the other characteristics be similarly reduced to percentages of their world totals. Then let p stand for foreign trade, q for travel, and r, s, t, u, etc. for the several other characteristics selected. Let V be the voting strength.

The Product. $V = p \times q \times r \times s \times t \times u.$

This vanishes if any characteristic vanishes. Thus if a state lacked one characteristic, all its others would be as nought. Such a formula would not make sense in such a case, and it should therefore be dismissed. A product of powers such as $V = p^a \times q^b \times s^c \times t^d \times u^e$ where a, b, c, d, e, are constants, is no better and should also be dismissed.

The weighted average $V = Ap + Bq + Cr + Ds + Et + Fu$ where A, B, C, D, E, F, are numbers, known as "weights," which are fixed once and for all. The object of having "weights," is in order to give larger weight to the more suitable characteristics and conversely. But if any characteristic were really very much less suitable, it had better be excluded altogether. And, if this has been done, there is a presumption that the weights do not differ so very much among themselves. And then, as there is no obvious way of finding out which weight should be small and which large, we had better simply make them all equal, thus coming back to :—

The simple average $V = \frac{1}{n}(p+q+r+s+t+u+\ )$

where n is the number of characteristics used. It would be a good plan to have a large number of characteristics, so that should one of them be less suitable or should become unsuitable in course of time, its bad effect would be swamped by the others. There is another advantage in having many characteristics. There will no doubt be a tendency for states to force up the characteristics on which their voting strength depends. Possibly, for instance, to write more foreign letters than they would otherwise have done, in order to get more votes. Now, if foreign post were the sole characteristic chosen, a nation might enormously increase its voting strength, by devoting its energies to superfluous foreign correspondence; an action which would be regarded by its neighbors as a piece of reprehensible trickery. But suppose on the contrary that the characteristics chosen numbered not one, but say ten, twenty or fifty, and that they represented broad aspects of national life. An ambitious nation could not direct its energies into fifty diverse and broad channels without a real increase of national activity. What was

STATES	FIXED SCALES OF VOTING STRENGTH Proposed by: Hague, 1907 for Prize Court		Central Organisation for a Durable Peace		Lord Bryce's Group		ANNUAL EXPENDITURE ON ARMAMENTS		CHARACTERISTICS FOR SELF-ADJUSTING SCALE for the epoch 1905 to 1913: FOREIGN TRADE		FOREIGN TRAVEL		FOREIGN POST		CAPITAL INVESTED ABROAD		CONTRIBUTION TO INTERNATIONAL EXCHEQUER		VOTING STRENGTH AS AVERAGE OF PERCENTAGE CHARACTERISTICS	STATES (repeated from Column I.)
	Actual	Per cent	Actual	Per cent	Actual		Millions of francs	Per cent	Millions Sterling	Per cent		Per cent	Millions of Missives	Per cent		Per cent		Per cent		
Column Number I	II	III	IV	V	VI	VII	VIII	IX	X	XI	XII	XIII	XIV	XV	XVI	XVII	XVIII	XIX	XX	XXI
Reference	1		6		5		12		13				14							
Austria Hungary	20	7.0	3	4.9	3	Percentage is indeterminate since total is not specified	453	3.8	238	3.9			745	16.9						Austria Hungary
British Empire	20	7.0	3	4.9	3		2,290	19.2	03[illegible]	17.2			546	12.4						British Empire
France	20	7.0	3	4.9	3		1,408	11.8	566	9.5			396	9.0						France
Germany	20	7.0	3	4.9	3		1,656	13.8	876	14.6			771	17.6						Germany
Italy	20	7.0	3	4.9	3		618	5.2	222	6.7			179	4.1						Italy
Japan	20	7.0	3	4.9	3		481	4.0	101	1.7			71	1.6						Japan
Russia	20	7.0	3	4.9	3		1,665	14.0	268	4.5			271	6.2						Russia
U.S.A.	20	7.0	3	4.9	3		1,472	12.3	738	12.3			728	16.5						U.S.A.
Spain	12	4.2	1	1.6	One vote at least each		267	2.2	80	1.3			104	2.4						Spain
Netherlands	9	3.2	1	1.6			105	0.9	503	8.4			103	2.3						Netherlands
Belgium	6	2.1	1	1.6			63	0.5	324	5.4										Belgium
Denmark	6	2.1	1	1.6			47	0.4	65	1.1			39	0.9						Denmark
Greece	6	2.1	1	1.6			31	0.3	13	0.2										Greece
Norway	6	2.1	1	1.6			27	0.2	42	0.7			28	0.6						Norway
Portugal	6	2.1	1	1.6			65	0.5	23	0.4			28	0.6						Portugal
Sweden	6	2.1	1	1.6			117	1.0	75	1.2			39	0.9						Sweden
China	6	2.1	1	1.6			210	1.8	115	1.9										China
Roumania	6	2.1	1	1.6			72	0.6	41	0.7			28	0.6						Roumania
Turkey	6	2.1	1	1.6			251	2.1	48	0.8			22	0.5						Turkey
Argentina	4	1.4	1	1.6			98	0.8	140	2.3			74	1.7						Argentina
Brazil	4	1.4	1	1.6			234	1.9	120	2.0										Brazil
Chile	4	1.4	1	1.6			91	0.7	51	0.9			16	0.4						Chile
Mexico	4	1.4	1	1.6			56	0.5	51	0.9										Mexico
Switzerland	3	1.1	1	1.6			45	0.4	126	2.1			160	3.6						Switzerland
Bulgaria	3	1.1	1	1.6			41	0.3	15	0.2			15	0.3						Bulgaria
Persia	3	1.1	1	1.6					16	0.3			8	0.2						Persia
Colombia	2	0.7	1	1.6			10	0.1	7	0.1										Colombia
Peru	2	0.7	1	1.6			28	0.2	14	0.2			9	0.2						Peru
Uruguay	2	0.7	1	1.6			16	0.1	17	0.3			26	0.6						Uruguay
Venezuela	2	0.7	1	1.6			8	0.1	7	0.1										Venezuela
Serbia	2	0.7	1	1.6			28	0.2	10	0.2										Serbia
Siam	2	0.7	1	1.6					14	0.2			Rest unknown or omitted							Siam
Cuba	1	0.4	1	1.6					51	0.8										Cuba
Say 12 others	1 each	4.5	1 each	19.7			others omitted		others omitted											
Totals	885	100.3	61	98.9			11,961	99.9	6,005	100.1			4,300	99.9						

disgraceful trickery if there was only one characteristic would have to be admitted as fair competition if there were fifty characteristics.

Should a constant be added so as to make

$$V = K + \frac{1}{n}(p+q+r+s+t+u+\ \ ) \quad ?$$

The addition K we suppose to be the same for all states. It would have the effect of a bonus to the smaller states. The theory that all states are sovereign and equal is equivalent to putting $V = K$ and so ignoring p, q, r, s, etc. altogether. This theory has to some extent been acted upon in the past. Therefore, to avoid making too sudden a break with tradition, it might be well to admit K, perhaps for a term of years limited to say fifty. The numerical value of K would require a special decision.

THE FICTITIOUS EQUALITY OF EIGHT GREAT POWERS BEFORE THE WAR.

It is seen in the annexed table, that the voting strengths of the eight great powers do not come out even approximately equal, when calculated by the characteristics proposed. Whether we consider foreign trade or foreign post, Japan had less than one-third of that of the United States of America; and the same applied to armaments. This inequality might be supposed to be a sign that we had chosen an unsuitable formula in the simple average. Now we have seen that the weighted average would appear to be a suitable formula, if there were any process for determining the weights. So let us assume that the equality of the eight great powers, at a date near 1910, represented the true judgment of competent men, and let us use this equality to determine the weights in the weighted average. To make the problem determinate, since there are seven given equalities, we can find the ratios between eight weights. That is to say, we must choose eight characteristics. This done, it would be a straightforward task of solving eight simultaneous equations. As that would take much time and paper let us consider first a specially simplified illustration, as follows. Let Japan have the same voting strength as Russia and for either of them let voting strength = (foreign trade + x × (foreign post) where x is the weight which we have to determine. Then referring to the statistics in the table, we have

Japanese voting strength = $17+16x$
Russian voting strength = $45+62x$

Whence $17+16x=45+62x$ from which it follows that $x=-0.61$. The result of applying this negative value of x to the states which are not among the equal great, would be to make some of their voting strengths negative, for instance, those of Spain, Switzerland and Uruguay. But that would be absurd. There is no reason to suppose that this absurdity would be avoided, if we considered eight states, instead of only two, as equal. Such studies give one the impression that the equality of eight states is very much of a fiction, and that it could be replaced by something more natural.

How would the division of a state into two, as Norway and Sweden divided, or the combination of several states in one, as in the South African Union, affect the joint voting strength of the peoples concerned? If divided they might of course vote diversely. But supposing that they all remained of one mind upon a particular question, would their total voting strength be affected by division or agglomeration? Well, for example, the trade between Norway and Sweden changed, on their separation, from internal trade, which we have supposed not to count in voting strength, to foreign trade, which we have counted. A similar statement applies to most of the other characteristics selected above, namely to foreign travel, foreign post, capital invested abroad, export of ideas. The remaining characteristic, namely the contribution to the international exchequer, might or might not give this effect. But on the average it appears highly probable that subdivision would increase the joint voting strength of a people, and conversely that agglomeration would diminish it. Thus, other things being equal, small states would tend to have greater voting strength per inhabitant than large states. This conclusion may be expressed in another way by saying that small states tend naturally to be more highly internationalized than large states. Now it is with a view to protecting the interests of small states that the constant K in the above formula has been suggested. But it is now seen that the inhabitants of small states would still have an automatic advantage, even if K were, in course of time, reduced to zero.

LEWIS F. RICHARDSON.

Richardson, L. F. (1919)
Mathematical Psychology of War
Oxford: W. Hunt

1919:1

MATHEMATICAL PSYCHOLOGY OF WAR

Oxford: Wm Hunt

Published privately in typescript (3 + 50 pp.) and reset for this volume

(Title page dated 25 February 1919)

Notes:

1. A few typing errors were corrected when transferring this essay to print
2. This topic is continued in 1935:2

Mathematical psychology of war

By

Lewis F. Richardson

Dedicated to my Comrades of the motor ambulance convoy known as S. S. Anglaise 13, in whose company this essay was mainly written.

25 February 1919

To be obtained from WM HUNT, 18 Broad Street, Oxford. Price 5/– nett

The cooling advice which we get from others when the fever-fit is on us is the most jarring and exasperating thing in life. Reply we cannot, so we get angry; for by a sort of self-preserving instinct which our passion has, it feels that these chill objects, if they once but gain a lodgment, will work until they have frozen the very vital spark from out of all our mood and brought our airy castles in ruin to the ground. Such is the inevitable effect of reasonable ideas over others – *if they can once get a quiet hearing*; and passion's cue accordingly is always and everywhere to prevent their still small voice from being heard at all. 'Let me not think of that! Don't speak to me of that!' This is the sudden cry of all those who in a passion perceive some sobering considerations about to check them in mid-career. 'Haec tibi erit janua leti', we feel. There is something so icy in this cold-water bath, something which seems so hostile to the movement of our life, so purely negative, in Reason when she lays her corpse-like finger on our heart and says, 'Halt! give up! leave off! go back! sit down!' that it is no wonder that to most men the steadying influence seems, for the time being, a very minister of death.

The strong-willed man, however, is the man who hears the still small voice unflinchingly, and who, when the death-bringing consideration comes, looks at its face, consents to its presence, clings to it, affirms it, and holds it fast, in spite of the host of exciting mental images which rise in revolt against it and would expel it from the mind. Sustained in this way by a resolute effort of attention, the difficult object ere long begins to call up its own congeners and associates and ends by changing the disposition of the man's consciousness altogether. And with his consciousness, his action changes, for the new object, once stably in possession of the field of his thoughts, infallibly produces its own motor effects'.

(William James, in the chapter on The Will in his *Psychology*).

Contents

I Introduction

The death-grip of opposing forces which the war developed out of the comparatively mild rivalries of peace, demands our best efforts to understand it.

The view here put foward is that the warlike striving of either side is largely, though not entirely, an instinctive reaction to the stimulus of the warlike striving of the opposing side. By an instinct is here meant an inborn tendency to perceive a certain state of affairs, and thereupon to feel in a particular way, and to act towards a corresponding end. A tendency, that is to say, which one might easily follow without considering it; and to resist which, if one judged it desirable to resist, would require an effort of will.

We may make this view of the matter more definite by putting it into symbols. Let A_E be the warlike activity of the Entente, A_G the warlike activity of the Germanic alliance. Then the instinctive tendency which has been referred to above, may be expressed by two such equations as the following, in which t is the time, κ_E and κ_G are positive constants, and d is the operator of the differential calculus so that dA_E/dt means the rate of increase of A_E with time:

$$\frac{dA_E}{dt} = \kappa_E A_G \tag{1}$$

$$\frac{dA_G}{dt} = \kappa_G A_E \tag{2}$$

In other words, a war, once set afoot, is continued and augmented by the tendency to mutual reprisals.

To see further what these equations imply, let us deduce some of their consequences. If, at any instant, both A_G and A_E were zero, then, by the equations (1) and (2), their rates of increase would be zero, so that they would both remain eternally nil. This is rather like the view of those who say that armaments tend to produce war, and that if armaments could be abolished, wars would be less likely to arise. Equations (1) and (2) are in fact an exaggeration or caricature of the view that armaments tend to produce war. The equations must be modified to avoid this exaggeration, by taking account of the desires, ambitions, and rivalries which exist and which make for war independently of the reciprocal instinctive stimulation of the kind represented by (1) and (2). The rivalries are perhaps mainly those of a powerful few, who are guided by intelligent self-interest after the manner in which the Utilitarians supposed all men to be guided, and who influence the suggestible instinctive Many.

A further logical deduction from equations (1) and (2) is that if A_E and A_G were at any instant greater than zero, then they would both increase, becoming large without limit as time went on. It is true that armies and military expenditure increased as the war proceeded, but there are limits to all things human, and so we must modify the equations to take account of fatigue.

Other improvements in the equations will also be attempted. The final results are set out in section VIII2. They are led up to by the intervening argument.

II An apology for the use of mathematics

To have to translate one's verbal statements into Mathematical formulae compels one carefully to scrutinize the ideas therein expressed. Next the possession of formulae makes it much easier to deduce the consequences. In this way absurd implications, which might have passed unnoticed in a verbal statement, are brought clearly into view and stimulate one to amend the formula. An additional advantage of a mathematical mode of expression is its brevity, which greatly diminishes the labour of memorizing the idea expressed. If the statements of an individual become the subject of a controversy, this definiteness and brevity lead to a speeding up of discussions over disputable points, so that obscurities can be cleared away, errors refuted and truth found and expressed more quickly than they could have been, had a more cumbrous method of discussion been pursued. Mathematical expressions have, however, their special tendencies to pervert thought: the definiteness may be spurious, existing in the equations but not in the phenomena to be described; and the brevity may be due to the omission of the more important things, simply because they cannot be mathematized. Against these faults we must constantly be on our guard. It will probably be impossible to avoid them entirely, and so they ought to be realized and admitted.

Mathematical expressions are in general use in various parts of Sociology, for example in Economics[1], in Anthropometry and in all cases where statistics of large masses of mankind have to be described. It therefore seems reasonable to enquire whether mathematical language can also express the behaviour of people in another case in which they also act in large groups, namely in war.

It is sometimes supposed that mathematical expressions can be used to describe only the action of objects which follow laws of a rigid, mechanical, deterministic type in all particulars. That this is not necessarily so, is shown by the fact that Prof. William James used mathematical expressions to illustrate his defence of the view that our wills are partially free. His example has been followed in this essay in this respect.

A fundamental rule of scientific method is Ockham's 'Razor' to the effect that: 'Entities are not to be postulated without necessity'. For shaving off the superabundant growth of mathematical uncertainties and difficulties I have frequently used an analogous rule: 'Formulae are not to be complicated without necessity'. For example, if observation shows nothing except that two quantities increase and decrease together, and vanish together, then one quantity has been taken as a simple multiple of the other. Quadratic, cubic or other more complicated terms have not been introduced without clear evidence to show that they were necessary. The formulae set down are therefore at best only rough approximations. Indeed on account of the difficulty of defining the fundamental quantities, there remains a general vagueness, which may scandalize some of those who have been trained in the exact sciences, but which, in the author's opinion, does not deprive the formulae of meaning, interest and suggestiveness.

In a treatise on mathematical physics it is customary to state hypotheses and then to deduce the consequences. In this essay a very different use is made of mathematical symbols. The successive formulae are not usually deduced from those which precede. Rather each formula has been mentally compared with the miscellaneous facts known to the author, and the succeeding formula is often an improvement, a higher synthesis in the Hegelian sense, and not a deduction.

1. *Vide* Marshall's *Principles of Economics* V ed. p. 101 for an appreciation of the service rendered to Economics by mathematics.

III Presuppositions

Professor McDougall, in his very interesting book on *Social Psychology*, 'propounds a theory of action which is applicable to every form of animal and human effort, from the animalcule's pursuit of food or prey to the highest forms of moral volition'.[1] While not accepting this theory as adequate, it will be convenient to attempt to state it as a starting point. If I understand it aright, its central idea is that mankind, all animals, (and plants also) are purposive. They strive towards ends, overcoming obstacles by persistence with varied effort. The animal may or may not be conscious of the aim of its striving. Consciousness of the aim ranges from a dim awareness that something or other is due to happen, to an objective clearly thought out in detail. Some of these aims are inherited and are then called instinctive. Thus observation of human beings and of the higher mammals indicates that mankind has some eight primary instincts. Each of these instinctive dipositions has a threefold structure; it is a tendency to perceive certain objects and thereupon to feel and to act in certain ways. As when, on hearing a very loud unexpected noise, we feel a momentary shudder of fear, and we start as if to take shelter. McDougall's list of primary instincts runs as follows. The name of the instinct is put first and after it that of the corresponding emotion. Flight and concealment with fear. Repulsion with disgust. Curiosity with wonder. Pugnacity with anger. Self-abasement with subjection. Self-assertion with elation. The parental instinct with tender emotion. And lastly the sex instinct.

In addition to the eight foregoing instincts, Prof. McDougall also treats, as fundamental, some other instincts of less well-defined emotional tendency, namely those of gregariousness, acquisition and construction. He also takes, as axiomatic, various other general innate tendencies. Among these there is a group of three divided according as they depend on thought, feeling, or action, namely: suggestibility, in the sense of accepting ideas on authority independently of evidence; sympathy, in the sense of experiencing an emotion because it has been expressed by another person; and thirdly of imitation of the actions of those with whom we have to do. These three tendencies are concerned in the occasional unreasoning spread of panic among a crowd, and with the tendency to uniformity in public opinion generally. Other fundamental innate properties discussed by McDougall are the impulse to rivalry and play, the formation of habits by practice and the tendency to feel pleasure and pain.

From these innate beginnings Prof. McDougall traces the gradual development of the mind of the growing child under the influence of his social environment. The instincts appear in their simplest form at the time of life at which they mature; but even then they are seldom as simple as the corresponding instincts of animals. Afterwards they become interwoven, and attached to new objects of thought, so as to form complicated emotional dispositions. For instance he considers the parental instinct to be the root also of brotherly love. An example, of the formation of a disposition, which has a special bearing on war, may be quoted at some length. It was published[2] before the oubreak of the present war. It deals with 'the case of anger roused by an insulting blow and restricted in its expression by fear. Up to the time of the incident, I had been, we may suppose, as nearly as possible indifferent to my assailant; that is to say his presence had evoked in me no well-defined feeling or attitude. But after the painful incident, I

1. *Social Psychology*. Methuen & Co., Ltd. XII edition, page 352.

2. *Psychology*. Home University Library, page 115.

cannot think of him without fear, or anger, or both, and without desiring both to avoid him and to get the better of him in some way. Suppose, now, that circumstances repeatedly bring us together, and that his behaviour on such occasions is that of a bully covertly reminding me of the past insult that I dare not avenge. My attitude of blended anger and fear is renewed on each such occasion, and being thus confirmed and rendered permanent, it becomes a full-blown sentiment of hatred ...'. This is a linking of the idea of a particular person with the instinct to destroy and to feel angry, and with the instinct to flee away and to feel fear. 'The effect of such linkage is not only that whenever the object of the sentiment is forced upon my attention, my thinking of him is coloured or suffused with these emotions, but also that I am rendered peculiarly apt to think of him. If I pass by a crowd of which he is a member, my eye singles him out and watches him furtively; if we both have occasion to attend the same board meeting, I am acutely aware of him and of all he says and does, though I may avoid glancing at him; if I overhear his name mentioned by others in conversation, I am all agog to hear what is said. And this may continue in spite of my best efforts to cast out this demon of hatred and to resume my former attitude of indifference. Again, all my thinking of my adversary is biased by my attitude; whatever I hear to his discredit I accept and retain, and I attribute his actions to the meanest motives; until by repetition of this process of selective thinking under the guidance of the specialized conative tendency, I come to think of him as a monster of iniquity'.

In many particulars the above description would apply to the way in which the newspapers of each side have written of the other side during this war.

To return to general social psychology, McDougall attempts to show how the moral and religious sentiments are formed by association of the innate mental tendencies, which have been enumerated above, with various objects of thought, and especially with the individual's idea of himself or herself, under the influence of the society in which he or she grows up. McDougall writes (page 382) 'Our "sense of duty" is, in short, at the lower moral level our sense of what is demanded of us by our fellows, and, at the higher moral level, it is our sense of what we demand of ourselves in virtue of the ideal of character that we have formed'. With a scientist's desire to economize hypotheses, McDougall is anxious to avoid introducing, as independent causes, either an inner free self acting on these instincts, or a moral sense, or a conscience, or anything in the nature of God. McDougall repeatedly uses such words as higher, better, wrong, lower, or unjustifiable, and they give a moral tone to his whole discourse; but one does not see how these terms can logically derive any meaning from his restricted hypotheses – they appear to have been imported from an outside source. His theory is therefore no argument for making these restrictions. It leaves us free to accept instead the traditional belief in an inner free self acting on the brain (as McDougall has elsewhere suggested,[1] by altering the resistance of the synapses) and to believe that this inner self may put itself into communication with the Divine through prayer,[2] and may possibly also communicate with other persons by telepathy.[3]

Other general properties of the mind, to which we shall need to refer later, are distraction, warming-up, learning-by-practice and forgetfulness. They might have been mentioned earlier, together with habit. With regard to distraction, there is the familiar

1. *Psychological Psychology*.
2. *The Meaning of Prayer* by H. E. Fosdick, Student Christian Movement.
3. *Human Personality* by Frederick Myers.

difficulty about 'doing two things at once'. Either of them tends to distract attention from the other. A comical example is the difficulty many people experience in simultaneously patting their head and rubbing their chest with a circular motion. On the other hand there are plenty of diverse actions which can be done easily and sweetly together, as for instance to sing while the hands ply some familar routine, as women often do in factories even when working hard on a 'piece-rate'. Again other actions are not merely indifferent, but might well be called 'serial' because the first paves the way for the second; thus to know much about any machine, for example about a gun, makes one eager to see it work. Now whether two particular courses of thought, feeling, and action are 'rival' or 'indifferent' or 'serial' can be found out by experience and observation. It is an observed fact that anger and tender emotion when directed towards the same object, are rival in this sense. To love a person is almost automatically to cease to hate him and vice versa. Subsequently, whichever of these two feelings has obtained a footing by displacing its rival, it is increased by 'warming-up' and by practice, and tends to become habitual; in contrast with the rejected alternative, which becomes progressively more completely obliterated by forgetfulness. These properties of the mind tend to intensify loves and hatreds. In wartime many other agencies work with them.

Several mental dispositions or tendencies may be illustrated by comparing thought to the flow of water down a bank of soft mud. Wherever water has flowed, a channel is formed, which is analogous to a habit tending to direct any future flow. Wherever the flow ceases in any channel, the mud, owing to its softness, begins to close in, obliterating the groove and thus illustrating forgetfulness. The phenomena of distraction require us to suppose that the supply of water is limited, so that the flow down any one channel can only be made at the expense of the flows down all the other channels together. The Will may be compared to a person who can block up certain channels or scoop out others; but, to make the analogy fit, we must suppose that some of the channels are out of his reach, and that even in those which he can reach, the amount of closing or enlarging, which he can do, is very limited. On this analogy the instincts are to be represented by capacious channels in a harder mud, or in a rock even, so that they do not tend to close up when disused. Across the upper ends of many of these instinctive channels the Will has constructed a barrage of mud. But should a sudden rush of water wash away this barrage, then the instinctive channel, by its depth and breadth, draws all the water to itself, draining the others. Thus it is when a man loses his temper. His struggles to regain equanimity may be compared to pouring mud into the torrent flowing down the instinctive channel. At first the mud may perhaps be washed away as fast as he can pour it in; but as the supply of water becomes exhausted, the barrage is restored and the remaining flow diverted to other channels.

McDougall's theory of the development of purposes, by the attachment of instincts to new objects, is contrasted by him with the fundamental hypothesis of the older economists, according to which man was a rational creature guided mainly by intelligent self-interest. While admitting that the older economists greatly exaggerated this side of human nature, we must also admit that it is a side which exists, and exists perhaps especially among those, such as philosophers or the heads of businesses, who are accustomed to take long views. This type of man has his instincts well under control and is quick to realize logical necessities.

IV Certain general relations

IV1 *The Vigour-to-War*

The present theory attempts to describe the changes of the Purposes or Intentions of the two opposing sides under the various influences to which they are subjected. This purpose or intention is considered as varying from Eagerness-for-War at one end of the scale, to Peace-At-Any-Price at the other end. It is thus a 'quantity', not measurable exactly, but yet having enough of the characteristics of largeness, smallness, positiveness or negativeness about it to justify one in denoting it by an algebraic symbol. The symbols chosen are V_E for the Entente, V_G for the Germanic alliance, or, where the side does not require to be specified, simply V. It is difficult to find a completely satisfactory name for V. The neat alliterative phrase the 'Will-to-War' comes to mind; but unfortunately the appearance of the word Will in this phrase gives a suggestion that Volition has something to do with it; and that is unnecessary and confusing. Because when, for example, somebody hits me violently on the nose, my tendency to personal combat with him is not a matter of volition at all, but simply of automatic instinct, which the Will has to struggle to resist. Better names for V would be 'Warlike Striving' or 'Vigour-to-War', and by the latter name V is denoted in what follows. But it will be convenient to restrict V to mean the instinctive part of the Vigour-to-War, denoting by H and F the remaining parts due to higher influences and freewill respectively.

IV2 *Conditions at the moment of victory or defeat*

From the military point of view that side is considered beaten whose vigour-to-war first sinks to zero. When the peace is 'by understanding' it might be argued that the vigours-to-war of the two sides reach zero simultaneously at the moment of signing the peace treaty. In contrast to this, one side might continue to invade territory, while the other was willing and eager to make peace. From a non-military point of view that side may well at long last appear as the victor, which has given to humanity most and best.

It is remarkable that rifles, guns, grenades, poison gas, tanks, battleships, submarines and aeroplanes have not made war more deadly than it was when men fought with bows and arrows, swords and spears, horse-chariots and rowing-boats. The limit at which a side admits defeat, as measured by percentage casualties, has remained about the same, thus showing that it depends not on mechanical contrivances, but on human capacity for endurance.

IV3 *Notation for the two opposing sides*

It will be convenient to distinguish them by subscript letters

E for the Entente
G for the Germanic Alliance

and to use for a quality such, for example, as the vigour-to-war the same symbol V, distinguishing it as V_E or V_G. In fact if any psychological feature is noticeable on one side, it seems best to look for a corresponding feature, possibly much greater or much less, existing on the other side; and to have a symbol ready for it. Thus the differences between the two sides in character and temperament will be left to be expressed by the numerical values of certain algebraic symbols. Lest any extreme nationalists should be scandalized at seeing the same symbol used to represent the ideas of the enemy, even with a different subscript, it may be well to point out that the appropriate method of expressing their dislike of the enemy, in this notation, is to assign low values to the symbols for desirable qualities and high values to the symbols for undesirable qualities, when they bear the enemy's subscript. This aims to be first a method of discussion, in order that, in the course of years, it may lead ultimately to a statement of agreed fact.

It is sometimes convenient, in general discussions, to distinguish the two sides by the suffixes x and y. It is then to be understood that x may be either of E or G provided that y is the other one. By this device the number of equations, that have to be written down, is halved.

IV4 *The individual and the government*

Before proceeding to consider the behaviour of nations, it will be well to take note of the diversity of individuals in any one nation – rich and poor, brave and cowardly, annexionists and pacifists, dutiful and selfish, pertinacious and easily fatigued, internationlists and nationalists, mystical and matter-of-fact; we can hardly lump them all together without a blurring and confusion of vital issues. On the other hand, to take separate note of all the diversities of mankind. would lead us into bewildering complexities. So, avoiding infinitesimals, let us seek a division into a few significant finite differences. We can perhaps hardly do better than to follow the customary division into (i) the rich, (ii) the comfortable, (iii) the poor. Any symbol will, when necessary, be distinguished by a subscript 1 for the rich, a subscript 2 for the comfortable, or 3 for the poor. The subscript-number thus increases as the type of individual becomes more numerous, and as his possessions become less. A symbol with no subscript-number is a joint one for all social classes combined. By the vigour-to-war of a side we must here understand the effective vigour as organized in the Government, or Group of Governments of that side. It is not to be confused with an average vigour which might be estimated by the counting of heads. The vigour-to-war is some sort of balance or resultant – perhaps obtained by biased scales – of the various vigours in the side. The vigours of individuals are the motive power, and their effectiveness depends partly on their numbers, partly on their influence with their government. So let

n be the number of individuals in any class

i be the influence of the average individual with his government

v be the vigour-to-war of the average individual.

Accordingly we may put

$$V_E = n_{1E}\, i_{1E}\, v_{1E} + n_{2E}\, i_{2E}\, v_{2E} + n_{3E}\, i_{3E}\, v_{3E} \tag{3}$$

$$V_G = n_{1G}\, i_{1G}\, v_{1G} + n_{2G}\, i_{2G}\, v_{2G} + n_{3G}\, i_{3G}\, v_{3G} \tag{4}$$

In a 'perfectly democratic' country, governed on the principle of 'one man one unit of influence', i_1 would be equal to i_2 and to i_3. In this connection 'influence' and 'votes' are not by any means equivalent. Such a 'perfectly democratic country', like the 'perfect fluid' discussed by mathematicians, does not actually exist. In all actual countries the influence i_1 of an average rich man, is greater than that i_3 of an average poor man. But i_3/i_1 will probably be larger in countries commonly called democratic than in those known as autocratic. Even in the latter i_3 is not negligible; for however autocratic a government may be in form, it must always in fact pay great attention to the contentment of the governed. For example the fact that the German Government did not deign to reply to a certain Socialist interpellation in the Reichstag is no proof that the socialists were without influence on their government. (Contrary to *Daily Mail*'s assertion, May 1918). Despite its lack of morality towards outsiders the German autocracy looked after its people.

IV5 *Warlike Activity*

A nation is stirred up if it knows, or imagines, that a neighbouring nation has a warlike intention or 'vigour-to-war' against it; but it is much more profoundly stirred up if its neighbour's intention is made evident by acts of invasion, bombardment, and the like. We may group all such acts under the general name of 'Warlike Activity' and denote them by the letter A.

IV6 *Relation of the Vigour-to-War to the Warlike Activity*

What then is the relation between the vigour-to-war V and the warlike activity A of the same side? We had an instance of their relationship in the British Isles in 1914, September, for then the national vigour-to-war was very large and yet business went on much as usual; so that A remained small in comparison with its value twelve months later. Evidently the relation between V and A depends on warming-up to new tasks, on the formation of habits by practice, and on the forgetfulness and fatigue which set a limit to the acquirement of skill. It may be said that, in so far as a thought or action is instinctive, it is not facilitated by practice, nor caused to be forgotten by disuse. The nervous connections which it involves are fairly permanently organized. Now the general war-aim 'to injure the enemy' is certainly partly instinctive; but its practical working out is by way of rifle-shooting, grenade throwing, gunnery, aviation and many other processes, which are not at all instinctive, and which can only be learnt by practice.

To such small extent as the warlike activity A is itself instinctive it may be taken as directly proportional to the vigour-to-war V, so that

$$A = \alpha V, \quad \text{where } \alpha \text{ is a positive constant} \tag{5}$$

But to the much larger extent to which warlike activity is a matter of instruction and drill, we may reasonably say that what the vigour-to-war does, is to cause the warlike activity to increase. To express this idea in symbols, bearing in mind the general principle that 'Formulae are not to be complicated without necessity', we may put

$$\frac{\mathrm{d}A}{\mathrm{d}t} = \beta V, \quad \text{where } \beta \text{ is a positive constant} \tag{6}$$

Then fatigue comes in to set a limit to A in the following manner. In the absence of any vigour-to-war, fatigue would cause a diminution of warlike activity; not indeed a sudden cessation, for men would tend to carry on by mere habit; but a diminution more rapid according as A were larger and therefore more exhausting. This may be expressed by the formula

$$\frac{\mathrm{d}A}{\mathrm{d}t} = -\gamma A, \quad \text{where } \gamma \text{ is a positive constant} \tag{7}$$

However it may be noticed that a very little warlike activity, say ε in amount, is not tiring but is rather pleasing to a nation as a whole; so that we ought to replace (7) by

$$\frac{\mathrm{d}A}{\mathrm{d}t} = -\gamma(A-\varepsilon) \tag{8}$$

Now no one of equations (5), (6) and (8) is complete in itself. To describe any actual situation they must be combined. This can be done by differentiating (5) with respect to time and then writing in the additional terms from (6) and (8), with the result that

$$\frac{dA}{dt} = \alpha\frac{dV}{dt} + \beta V - \gamma(A - \varepsilon) \tag{9}$$

Notice that this equation yields a credible deduction when applied to the conditions in the British Isles in the autumn and winter of 1914. Then V, the vigour-to-war, was very large, so that the term βV was also large. On the other hand the warlike activity A was already far too large to be pleasant, so that A was greater than ε. The changes in V were slight, so that we may neglect $\alpha dV/dt$, especially as the coefficient α is small.' Then, as A was increasing, it is clear, from the equation (9), that βV must have exceeded $\gamma(A - \varepsilon)$. Later on, in 1917 say, the warlike activity of the British Empire had risen to such a large value that $\gamma(A - \varepsilon)$ had become equal to βV and so, in accordance with equation (9), A had ceased to increase. And yet a stimulus to the vigour-to-war of the British, such as that caused by the great German offensive of March 1918, by increasing V and βV, set on foot a new increase of A, of which the visible sign was the raising of the military age-limit from 41 to 50 years.

Thus in many respects equation (9) correctly describes the relation of the vigour-to-war to the warlike activity; a relation direct as far as the term $\alpha dV/dt$ is concerned, due to practice as far as the term βV is concerned, and due to fatigue (and perhaps also to forgetfulness) in respect to the term $\gamma(A - \varepsilon)$.

Will the same equation also serve to describe the changes at the times of armistice, peace and demobilization? The signing of the armistice was the signal for a sudden drop in the popular vigour-to-war, on the Entente side, to a lower, but not negligible, level. On the same day fighting ceased, which we would represent by a sudden drop in A_E and A_G. The suddenness was due to the fact that it was done by order, carried out with a military habit of obedience. It cannot be expressed by an equation, such as (9), which slurs over the distinction between rulers and ruled, and which takes account only of a vigour-to-war common to both of them.

In the periods of peace which have followed the wars of the past, the warlike activities of nations have been represented by their armaments. If equation (9) is to apply to such a period, then as dA/dt was then small it may be said that armaments have been kept up at such a level A that $\gamma(A - \varepsilon)$ was just equal to βV, where V was the small vigour-to-war caused by lingering suspicion and distrust of their neighbours.

IV7 *Freewill*

It will now be shown how the equations may be generalized to admit of one of the current views about the freedom of the Will. As this is a subtle and controversial subject, the author is compelled to point out that there are other views. With this preface, it may be said that if Will were all-powerful, science would be impossible. But a limited Will, such as we know, leaves room for a limited science, which may state what will happen according as the person in question tries to perform the action in question, tries not to perform it, or makes no effort. But *whether* he tries, has to be left indeterminate. How can we symbolize this? In what follows we have many equations expressing the effects of different influences on the vigour-to-war. Now, these equations are mostly partial, in the sense that they suppose circumstances, other than the one immediately under consideration, to remain unchanged. In particular they suppose that the volitional efforts, of the individuals concerned, do not change. And so, after all the instinctive effects have been collected together to form the combined vigour-to-war V, we may add on a freely willed portion F, which may be either positive or negative; and then it will be $(V+F)$ and not V alone which will be connected with the warlike activity, so that equation (9), as thus modified, will read

$$\frac{\mathrm{d}A}{\mathrm{d}t} = \alpha\frac{\mathrm{d}(V+F)}{\mathrm{d}t} + \beta(V+F) - \gamma(A-\varepsilon) \tag{10}$$

and F is left indeterminate, except that its magnitude will have to lie within a limited range.

IV8 *Belief*

The remarkable contrast of belief between the two sides of the fighting line – obvious facts for us, appearing as deliberate lies to the Germans, and vice versa – is to be attributed (i) partly to different sources of information, emphasizing or repressing different aspects, (ii) but partly also to the unifying attribute of the human mind, which tends to make people believe statements which harmonize with the course of action in which they are engaged, and to disbelieve those which do not do so. (*Vide* Stout's *Manual of Psychology* page 678), (iii) the divergence of beliefs between the two combatants, when thus started, is further intensified by mass-suggestion, the gregarious tendency to say, and thence to think, the same as those with whom one has to do. This makes opinion on each side more uniform, by silencing the discordant voices.

An artist of my acquaintance, Mr Eric Robertson, tells me that if you take a parti-coloured object, such as a map drawn with red roads, blue rivers, green woods and yellow boundaries upon a sheet of white paper, and if you place it at a slight distance, so that it is possible to ignore its details, and if you ask a number of people to tint sheets of paper to match the average tint of the map, you will find that some of them match it as greenish, some as yellow, others as pink, others as of a blue tint. In a similar way, looking at Germany from a distance, the structure of her life appears blurred, and some see her as composed of perpetrators of atrocities, others as of deluded patriots, according to the temperament and associations of the one who sees. The truth exists nevertheless. 'Too often is it said that there is no absolute truth, but only opinion and private judgment; ... Of such scepticism mathematics is a perpetual reproof'.[1]

An explanation is necessary about short intervals of time. The equations in the essay are written as if an act committed by the enemy – e.g. the sinking of the *Lusitania* – aroused immediately a state of feeling in the population as a whole. Actually of course half a day or so commonly elapses before the man in the street reads what those who control the newspapers see fit to tell him; and a considerably longer time passes before the man in the ploughed field comes to believe the same information. These intervals are, nevertheless, short compared to the duration of the war, and they have been neglected.

Mr Bertrand Russell has sent me the following note about the making of opinion. It is expressed in a different notation to that used elsewhere in this essay.

1. Mr Bertrand Russell, from whom this remark is quoted (*Mysticism and Logic*, p. 71) refers of course to strict deductive mathematics, which is of quite a different type from the symbolism used in this essay as an aid to memory and induction.

Problem: To produce in two nations a mutual will to war

For the sake of mathematical simplicity, take two nations, say Tibet and Uruguay, which have never heard of each other. We will call the two x and y. Let each contain one individual, and only one, who is actuated by aggressiveness towards the other; call them I_x and I_y, Let us assume that each is able, makes a fortune, and invests it in newspapers. I_x fills his newspapers, to begin with, with unpleasant descriptions and a biassed history of y; I_y does likewise. Hence x and y feel moral reprobation towards each other: call these Mo_x and Mo_y. I_x then writes articles which, by a slight misquotation, can be made to appear as if they urged x to attack y; I_y makes the misquotation. Hence arises in y the fear of being attacked, rousing a defensive will-to-war D_y. By the same methods a defensive will-to-war D_x is produced in x. Since x does not know of the misquotation, D_y appears as an aggressive will-to-war, and thus heightens D_x, which in turn heightens D_y. And so on. In this way, in the absence of a positive will-to-peace, the two individuals I_x and I_y may suffice to drive their nations into war.

V Various influences affecting the Vigour-to-War

V1 *Conquest, casualties and the destruction of wealth*

The loss of Alsace-Lorraine in 1871 bit much more deeply into the French popular mind than did the indemnity which France paid at the same time. Territory is an enduring and visible sign; and the first generation of those who have emigrated, in order to escape foreign rule, does not forget. Conquests have thus a large 'value' in terms of vigour-to-war, for both sides. Let P_x and P_y be the increases which the vigour-to-war of the two sides would undergo if conquests could return to the pre-war frontiers. Then, as one side's joy, in this matter, is the other side's sorrow, P_x and P_y are, at any instant, of opposite signs. Whether the dis-satisfaction of the side which loses is more or less than the satisfaction of the side which gains, depends on the special circumstances of the case. Only if the conquests are very large, one may expect from the principle of the satiability of wants, which has been elaborately studied by the economists, that the winning side, being glutted with territory, will probably derive less satisfaction from the last portion annexed, than the losing side will derive dissatisfaction, so that a very large annexation will probably decrease the sum of human happiness.

Next as to casualties, let B_x and B_y stand for the rates at which casualties were occurring. These rates cannot be negative. There is almost certainly a positive correlation between B_{E} and B_{G}, as they increased together wherever hard fighting was going on.

Also let L_{E} and L_{G} be the rates at which wealth was being destroyed. There is again almost certainly a positive correlation between L_{E} and L_{G}.

Now as to the relation of the warlike activity to the foregoing quantities. Since the object of the military leaders of the x-side is to cause casualties, destruction of wealth, and loss of territory to their opponents, it follows that B_y and L_y must increase with A_x and also with the military skill of the x-side. But we must distinguish. Mere drilling and arming does not cause casualties to the enemy, so only that part of A_x should be counted, which represents actual fighting. Call this part J_x, and let activity in preparation for fighting be denoted by R; so that

$$A_x = J_x + R_x \tag{10a}$$

Then B_y the rate of casualties on the y-side increases as J_x increases. Now if the x-side wants mainly to conquer or to reconquer territory, it may do so with small casualties on either side, if its enemies do not resist, that is to say if J_y is small. This suggests that the rates of casualites on the two sides are

$$B_y = \chi'_x J_x J_y; \quad B_x = \chi'_y J_y J_x$$

where χ'_x and χ'_y are positive constants. If this be so, the casualties on the one side would be a fixed multiple of those on the other, and the correlation between them would be unity. That is no doubt an over-simplified view of the affair, but it is perhaps the best that can be managed without going into much complicated detail. We shall require the casualties expressed in terms of A thus

$$B_y = \chi'_x (A_x - R_x)(A_y - R_y) \qquad (11)$$

$$B_x = \chi'_y (A_y - R_y)(A_x - R_x) \qquad (12)$$

In armies that are not growing rapidly, the relation of training to fighting is more or less fixed, so that by a change of constants from χ' to χ we may put

$$B_y = \chi_x A_x A_y; \quad B_x = \chi_y A_y A_x \qquad (13), (14)$$

as applicable to armies that have attained their full development.

Next as to rate of destruction of wealth. This is partly the loss of the services which men in the army would have rendered had they remained as civilians, partly the cost of the munitions exploded daily, and partly also the value of the buildings and land which are spoilt in the fighting. Altogether we shall probably not be far wrong in taking the ratio of loss of wealth as given by

$$L_x = a' R_x + a'' J_x + b'' J_y \qquad (15)$$

Or, as before, in armies which have attained a steady condition, R is proportional to J; so that, by a change of constants we may put:

$$L_x = a A_x + b A_y \qquad (16)$$

V2 *War as a joyful adventure, the satisfaction of a long-suppressed instinct*

Two brilliant essays on this subject come to mind, one by R. L. Stevenson on Sir Richard Grenfell, the other by William James on a Substitute for War. We may also refer to Edward Carpenter who says of voluntary recruits (*The Healing of Nations*, pp. 144–145):

> The gay look on their faces, the blood in their cheeks, the upright carriage and quick, elate step – when compared with the hang-dog, sallow, dull creatures I knew before – all testify to the working of some magic influence... It is simply escape from the hateful conditions of present-day commercialism and its hideous wage-slavery into something like the normal life of young manhood – a life in the open under the wide sky, blood-stirring enterprise, risk if you will, co-operation and *camaraderie*. These are the inviting, beckoning things...

He might have mentioned also: the approving smiles of the fair sex. The love of adventure has, in most cases, been surfeited with excess of fighting. In this generation the appetite tends to sicken. What would have just satisified it, would perhaps have been a nice short summer campaign, with 'fighting from 2 to 4 on Tuesdays, Thursdays and Saturdays' and then England, Home and Glory.

Many who dislike campaigning and hate war are yet susceptible to the appeal of the ancient proverb 'Dulce et decorum est pro patria mori'.

With this youthful love of adventure may be grouped the opinion of older men that war 'is salutary, necessary, and is the only national tonic that can be prescribed', as Lord Roberts said.[1] Similar statements may be found in the writings of the Prussian Bernhardi. Similar, at least in its relation to the warlike activity, is the 'professional' desire to put design, training and preparation to the test. This also contributes to ε.

V3 *The defence of that which is held dear*

Let us consider some of the countries separately.

GREAT BRITAIN. The universality and strength of the defensive disposition in Great Britain may be illustrated by the following facts, if it be necessary to illustrate anything so obvious:

(i) Britain's entry into the war was advocated by *The Times* newspaper in the critical days of August 1914 on the principle of: Our Turn Next if the Germans crush France.

(ii) Tribunals have very generally appealed to this disposition, in their efforts to persuade conscientious objectors to fight, by means of such questions as: 'What would you do, if your mother were attacked by a bear?'

(iii) The German offensive of March and April 1918 caused the military age to be raised from 41 to 50.

FRANCE. To the vast majority of French people this war was obviously defensive: 'Les Boches sont chez nous' – there was no more to be said.

ITALY. The German advance into Italy in November 1917 caused certain Italian politicians to change from being against the war, to being supporters of it. The Germans imagined that they 'saw an Italy demanding a separate peace, and found that they had produced the very opposite effect'. (*Le Journal*, Paris C.M.[2] VII 131).

GERMANY. There is doubt in many English minds as to whether the same instinctive disposition exists in Germany, and so some evidence has been collected to show firstly that the Germans often write or speak about it, secondly that certain neutrals refer to this instinct in Germany. To those who are fixed in the belief that all these writings are merely a piece of organized hypocrisy, it is difficult to make any convincing answer, except to say that it would be very astonishing if our present enemies, alone among nations, should prove to be exempt from this very widespread instinct.

(i) On 31 July 1914, at 7 a.m. von Moltke, chief of the German General Staff, speaking over the telephone to a General near the Russian frontier, is reported to have said 'I must have tangible proofs if they (the Russians) are mobilizing against us in reality. Before I have them, I cannot get an order for mobilization through' (*Vossische Zeitung* 11 & 12 Sept. 1917, *Camb. Mag.* VII 10).

(ii) In Germany 'on the outbreak of war every expression of jealousy was culled from the British Press, and every German knew by heart that monstrous leading article in which the *Saturday Review* called (11 September 1897) for a

1. Lowes Dickinson *The Choice Before Us*, p. 74.

2. Note C.M. VII.131 is a contraction for *Cambridge Magazine*, Vol. VII, page 131.

war with Germany, because "the destruction of her trade would add millions to our national income"' (quoted from *A League of Nations*, by H. N. Brailsford, p. 27). The object of the German press in publishing these instances of British jealousy, was presumably to satisfy a craving for justification for a war already begun, and also to stimulate the defensive instinct of the German people. What isolated and ancient instances sufficed to produce the effect they desired!

(iii) *Kölnische Zeitung* of 9 Sept. 1917 (Quoted in *Camb. Mag.* VII, 111) says 'We entered the war, or rather we were swept into it, with the intention of defending our existence as a people and a state, and of creating preliminary conditions for a future in which we could work for the blessings of peace without interference from without... It does not alter the character of a war of defence that we invaded the enemy's country in order to spare our own'.

(iv) The rage of the German press over the Entente governments joint reply to President Wilson in 1917, January, especially about the clauses dealing with Austria, Constantinople and the German colonies.

(v) On 15 May 1917, in the Reichstag, Scheidemann, the leader of the Majority Social Democrats, declared 'If the English and French Governments were, as the Russian Government has already done, to renounce annexations, and the German Government were then to desire to continue the war for conquest aims, then, you may rely upon it, you will have a revolution in the country'. (*Camb. Mag.* VII 227).

(vi) In the *Volksstimme* (Frankfurt, Social Democrat) of 8 January 1918, Herr Quarck wrote: 'Of course, nobody amongst us does think of striking. If for no other reason, then simply because we are threatened on all sides' (*Camb. Mag.* VII, 348).

(vii) Theodore Wolff in the *Berliner Tageblatt* of 24 Sept. 1917 wrote 'Thus we were told that resistance to the lust of conquest might weaken the national spirit, and, to escape the reproach of luke-warmness, many, who might have spoken, remained dumb'. (*Camb. Mag.* VII 39).

(viii) Bolo financed with German money the robustly patriotic French *Journal*. Some entertain the view that the German Government wished him to make annexationist proposals appear in France, in order to stimulate the vigour-to-war of the German proletariat.

(ix) The German Majority Socialists, in conference at Wurzburg in October, 1917, passed by an overwhelming majority a resolution which 'enjoins the Socialist Reichstag deputies in future as in the past to make the voting of war credits contingent on whether such vote is demanded by the interests of national defence' (*Camb. Mag.* VII 136).

(x) An appeal published in *Vorwaerts* of 1 Nov. 1917 by the Executive of the German Social-Democrat Party contains this phrase: 'Ready at any moment for peace, the German people fights and suffers in order to protect the homeland but not for conquests or any other gains... (*Camb. Mag.* VII 137).

(xi) See also Bevan's *German Social Democracy during the War*, Allen and Unwin, 1918.

AUSTRIA–HUNGARY

(i) Count Andrassy wrote in Nov. or Dec. 1917, 'Our aim on entering the war was certainly not annexation, but purely self-defence...' (*Camb. Mag.* VII 267).

(ii) The following is an extract from a speech by Victor Adler at the close of the Austrian Social Democratic General Party meeting (*Arbeiter Zeitung* 25 Oct., 1917, quoted by *Camb. Mag.* VII 219). 'It is said we adopted too credulously the catchword of the "fight against Czarism" which was skilfully put forward by the Governments. That is not true. It is the truth that the danger of war with Czarism was a terrible one for us and for the Germans, which was thoroughly understood by us, and that we were determined to defend ourselves against it. That the Governments utilized it, of that there can be no doubt... but that we ought to have made less of it because we recognized the fault of our Goverment, that is a false conclusion. If a house is on fire the first thing is to extinguish the flames'.

NEUTRAL or ENTENTE comment on the defensive instinct of the Germanic alliance.

(i) *National Zeitung* (Basle, Switzerland) of 9 Oct. 1917 summarized in the *Cambridge Magazine* (VII 261) as follows: 'The two trump cards in the present political game are victory and peace. The "party of the Generals" in Germany promise "victory". The Reichstag majority put forward "peace by reconciliation", but this party necessarily remains the weakest so long as the Entente on their side refuse to abjure annexations'.

(ii) The *Vaderland* (Dutch, Liberal) referring to former French designs, now repudiated, on the Left Bank of the Rhine, writes of Ribot's speech: 'In Germany itself this confirmation of Dr Michaelis' revelations can naturally only form another reason for continuing the war. It will be all the same to the Germans whether the aforesaid German areas are to become French or autonomous. It was in any case intended to wrench them from Germany, which is hardly compatible with the principle of nationality. (*Camb. Mag.* VI 870 of 25 Aug. 1917).

(iii) *The World*, New York, Radical Democratic 31 Aug. 1917. 'An equally great mistake was the Paris Conference which planned a business boycott of Germany after the war as a measure of continuing reprisals... The immediate consequence of it was to rally all the German people to more vigorous effort, and it gave to the German autocracy the chance to reassert its claim to be the defender of the German people against the plots to destroy them (*Camb. Mag.* VI 941).

(iv) That the rapid growth of the German Navy in the early years of this century was stimulated at least in part by the defensive disposition, is a conclusion which may be drawn from W. J. Ashley's pamphlet 'The War and its Economic Aspects' (Oxford Pamphlets 1914), for instance on page 11 he quotes as credible the following statement from a German economist writing in 1900. '"In one way or another, from 24 to 26 millions of Germans", out of a population, at the time of some 55, "are dependent for their livelihood and work upon unrestricted import and export by water"'.

The foregoing long series of extracts intended to show that the German vigour-to-war was largely a determination to defend what they held dear, must not be taken to mean that it was exclusively of that nature. That they had some excuse in the later stages of the war for supposing that the Entente meant to destroy them, might be proved by numerous articles in Entente newspapers. A sample of one of the more serious and weighty of these may suffice. It is taken from a financial paper. (*The Statist* 11 May 1918, page 805.) 'We do not ourselves doubt, in spite of the unsatisfactory state of things at present, that we shall win in the end, and that, in fact, Germany and Austria will be ruined. Before these ends are attained ...'

Leaving now the discussion of the defensive disposition in particular countries, let us turn to the more general question: what is it that stimulates the defensive vigour-to-war, is it the acts of the enemy or his intentions? Observation shows that it is both. For we form an opinion of a foreigner according to the same modes of thought by which we form an opinion of a neighbour; and by British law, or by the Code Napoleon, individuals are judged by their acts mainly, but also in part by their intentions as well. The same applies in non-legal matters, for instance Mr Antony, in Galsworthy's *Strife* Act I, 'What sort of mercy do you think you'd get, if no one stood between you and the continual demands of labour? This sort of mercy. [He puts his hand up to his throat and squeezes it]'.

We have already denoted by $-P_x$ the vigour-to-war of the x-side due to the loss of territory. This $-P_x$ arises for the most part from the defensive disposition. In addition, almost everything we possess, is held more or less dear, when an outsider is seen to be destroying or threatening it; so that the vigour-to-war or warlike activity of the one side immediately stimulates the defensive impulse of the other. Thus if V_x^{i} be the partial vigour-to-war due to the defensive disposition alone, we shall not be far wrong if we take V_x^{i} to be given by the following expression

$$V_x^{\text{i}} = -P_x + \zeta_x V_y + \eta_x A_y \tag{17}$$

where ζ and η are positive constants. Of these ζ may be called the 'suspicion coefficient' because it depends largely on surmise rather than on reliable unbiased information.

V4 *Vengeance*

Anger, the primary emotion which forms part of the instinct of pugnacity, is very prominent in war. It is stimulated by the warlike acts of the enemy. It forms one of the emotions which are potential in the defensive disposition, which has just been considered; it is also an essential part of the vengeful emotion, to which we now turn. There have been many expressions of vengefulness in the press on both sides.

A vengeful man reckons up his wrongs. The mere cessation of the warlike activity of the enemy would not appease him. Past debts must first be wiped out. This reckoning up, when expressed in mathematical language, would take the form of an integration. Thus if V_x^{ii} be the vigour-to-war of the x-side, due to vengefulness alone, then V_x^{ii} depends mainly on casualties. But V_x^{ii} is certainly not proportional to B_x, the rate of casualties; V_x^{ii} is more probably proportional to $\int B_x dt$ taken from the beginning of the war. This integral has already been denoted by C_x. The corresponding accumulated total of destroyed wealth has been denoted by M_x. It appears then that we may take the vigour-to-war, due to vengefulness alone, as of the form

$$V_x^{\rm ii} = \theta_x C_x + \lambda_x M_x \tag{18}$$

where θ and λ are positive constants.

Now we have already discussed the relation of casualties and of the destruction of wealth to warlike activity, and have arrived at the expressions

$$C_x = \int B_x \mathrm{d}t = \int \chi_y'(A_y - R_y)(A_x - R_x)\,\mathrm{d}t \tag{19}$$

$$M_x = \int L_x \mathrm{d}t = \int (aA_x + bA_y)\,\mathrm{d}t. \tag{20}$$

The integrals are to be reckoned from the beginning of the war.

These can be substituted in the expression for $V_x^{\rm ii}$ the partial vigour-to-war due to vengeance. When this is done it appears as a logical deduction that a side desires vengeance on its enemies for losses partly caused by warlike activity of its own. But that paradox is in accordance with experience. For instance if you hurt your knuckles by hitting another man on a hard button, you are apt to be annoyed with him, for wearing such buttons. This tendency has led, I suspect, to an exaggerated picture in Entente newspapers of the wanton destruction caused by the Germans in their retreat through France in October 1918. I refer to wanton destruction as distinct from the ordinary military operation of blowing up places of refuge such as cellars.

V5 *Rivalry or the maintenance of national prestige*

A determination not to be beaten, but to win through and thereby show that we are the better men. This emulation may safely be assumed to be a deep-rooted instinct, which no nation is without. As an indication of its distant ancestry, we have the fact that it may be observed even in dogs. Its influence on the present war may be illustrated by the views of Professor Krückmann, who, writing in the Berlin Junker paper *Kreuz Zeitung* on 20 Sept. 1917, said (*Camb. Mag.* VII.9) '*Unless we keep Belgium no human being will believe that we have won* – even the German people themselves will not believe it. *We do not only want to win, we want everyone in the whole world unmistakably to recognize the fact*...the latter is as important as the former... *We are fighting for our future political prestige*... that prestige would be England's if we surrendered Belgium'.

Similarly on the Entente side some say that the military pride of Germany will never be humbled until she has been forced to give up Alsace-Lorraine. In contrast with these writers may be mentioned Mr W. E. Gladstone, who was apparently scarcely at all influenced by the desire to win for winning's sake; for, after the British were defeated by the Boers at Majuba Hill, he was satisfied on obtaining, from the Boers, the terms which he had previously wanted, and did not consider it necessary, in addition, to inflict on them a military defeat. (See his *Life* by Morley, Book VIII, Ch. III.)

How can the emulative impulse be expressed in symbols? It is not like the economic wants, discussed in section V6 below, because its aim is not the gain of a statable quantity of any permanent thing. To quote Prof. Krückmann again (*loc.cit.*) 'In the prolonged struggle between Germany and England that country will win which possesses one sack more of coal than the other'. It is the *difference* which he regards, and coal may be taken as a picturesque illustration of war materials in general. But the imponderables cannot be neglected here any more than elsewhere, in fact less so, for emulation is essentially a question of morale, as is shown by Foch's proverb 'Une bataille perdue, c'est une bataille q'on a cru perdre'. (A lost battle is a battle which one has believed to have lost.) So that we reach the conclusion that the Germanic desire to win for winning's sake is a desire to keep $(V_{\rm G} - V_{\rm E})$ always positive and equal say to $K_{\rm G}$, where K_G is the margin of superiority which is considered proper in Germany. Neither side however is directly aware of the vigour-to-war of its adversaries. The best it can do is to form an estimate from certain outward and visible signs such as their resistance in battle, the tone of their newspapers, private reports of visitors to their country and the like. Let us represent the erroneousness of this estimate by writing $V_G - \iota V_E$ in place of $V_G - V_E$ where ι would be unity if the estimate were perfect, then perhaps it would be correct to say that V_x^{iii}, the partial vigour-to-war due to emulation alone, is given by

$$V_x^{iii} = u(\iota V_y - V_x + K_x) \tag{21}$$

where u is a positive constant the 'emulatance'.

Note that this equation implies that V_x^{iii} would become negative if the total V_x were larger than $\iota V_y + K_x$; that corresponds to a state of self-satisified slackness.

V6 *Business advantages*

The economic influences here discussed, do not include the privations of a population as a whole, which are grouped along with casualties under Pain. In the present section the discussion is confined to those wants of a powerful minority, which, before the war, created a situation which has been brilliantly described in Brailsford's *War of Steel and Gold.*[1] As examples during the war may be mentioned a manifesto issued by the Six Associations of German Industrialists in May 1915, against which we may perhaps set the activities of the Anti-German League. Suffice it to say here, that these wants are much more thought-out than the instinctive impulse to defence, although they also may be rooted in the acquisitive, constructive and self-assertive instincts of business men. These wants also are satiable by concessions or indemnities that are clearly foreseen by those who want them; concessions which the enemy probably would not, but yet conceivably might, grant if he were undefeated. In this respect they are in contrast with anger which makes the defeat of the enemy the essential condition of its appeasement.

Most business men would regard with loathing the idea of making a war for their private gain. More subtle is the temptation to take advantage of a war already in progress. Thus Milyuhov, the Russian minister, speaking to French socialists in 1916 said: 'Yes, we are waging a war of defence... Yes... that's understood... But... as we are at war... why not profit by it to attain certain results, practical, material...?' (*Camb. Mag.* VII. 346). Another well-known temptation is to camouflage war-aims. Those who feel the wants here discussed, are naturally conscious that they are a minority of the nation, and therefore they tend not to flaunt their desires too openly in front of the majority; but rather they are constantly tempted to stimulate the vigour-to-war of the majority, by such arguments as are likely to appeal to it. During the war, these special interests of a wealthy minority have largely been transformed into national interests, owing to the limitation of profits and the rationing of produce; that however may be only temporary.

Let us put Q_x for the material concession or indemnity acquired by the x-side. Since the y-side must have lost the same, it follows that

$$Q_y = -Q_x \tag{22}$$

as Q is a material, not a mental, quantity. No such equality need exist between the utilities which the two sides attribute to Q_y and $-Q_x$. The desire to obtain Q_x no doubt

1. 'Ancient imperialism levied tribute, modern imperialism exports capital at interest'.

existed, in a few minds, before the war, when Q_x was zero. Let its pre-war value in terms of vigour-to-war be W_x, and, more generally, let its value at any time be V_x^{iv}. Then if more and more Q_x is obtained, V_x^{iv} sinks from W_x towards zero, and might even become negative if acquisitions became embarrassingly large. The relation between the amount of a material thing, such as Q_x, and the sacrifices, measured by V_x^{iv}, which people are willing to undergo in order to obtain it, has been studied in a great variety of cases by the economists, and they have found that in general the sacrifice which a person makes in order to acquire a definite increase of material, is less, the more of this material the person, who receives it, already possesses. Accordingly we may expect that $-\mathrm{d}V_x^{\text{iv}}/\mathrm{d}Q_x$ will be less as Q_x increases. A function which would have these properties is

$$V_x^{\text{iv}} = W_x - p\log\left(\frac{Q_x}{q}+1\right) \tag{23}$$

where p and q are positive constants.

There are of course innumerable other functions which also have the aforementioned properties; but the function in the last equation, being perhaps the simplest, has been adopted for illustration.

V7 *War as a source of income*

What has been said about business advantages refers to the possible results of victory. But there are other motives connected not so much with victory or defeat as with warlike activity. However much they may regret the occurrence of war, many people know that the cessation of preparation for it, would be financially inconvenient to themselves. Surely this knowledge must tend to produce some vigour, if not to war, at least to arming. As an extreme example there was a scandal in which a German armament firm was accused of starting rumours in France in order to provoke preparations in Germany. Ambassador Gerard in *My Four Years in Germany* (page 77) states that: 'many of the larger newspapers are either owned or influenced by concerns like the Krupps'. But the tendency is naturally not confined to Germans nor to capitalists. Away back in 1910 or 1911 an artisan said to me in Newcastle upon Tyne: 'Elswick will never be itself again, until we have a right good war'.

Activity in preparation for war has been denoted by R. At any moment machinery and personnel are adapted to a certain value of R, say N. If R exceeds N, those engaged in preparations are overworked. If R is less than N, they are faced with the prospect of a diminished income. So manufacturers, who can control N, tend to adapt machinery and personnel somewhat in the sense of the equation

$$\frac{\mathrm{d}N}{\mathrm{d}t} = (R-N)\times(\text{a positive constant}). \tag{23a}$$

At the same time it may be suspected that there is a tendency in certain quarters to bring pressure on the Government to adapt the scale of national preparation R, so that it shall not differ too much from N, and especially when R is less than N; a pressure that is to say which, if it acted alone, as it never does, would perhaps affect R somewhat in the sense of

$$\frac{\mathrm{d}R}{\mathrm{d}t} = \frac{N-R}{h_1+R}\times h_2 \tag{23b}$$

where h_1 and h_2 are positive constants.

Of course the Governmental control is far stronger than this pressure, as is well seen now in February 1919, $\mathrm{d}R/\mathrm{d}t$ being negative although $(N-R)$ is positive. To discuss further these internal relations would lead us too far away from the main theme, which is the external conflict of nations.

V8 *The security of rulers* in respect of the *avoidance of internal trouble*, as doctors sometimes blister the chest as a counter-irritation to pneumonia

'Tsardom', as Baron Rosen says, 'was bent upon forcing war in order to stave off revolution' (*Manchester Guardian* 27 Feb. 1918, quoted by E. D. Morel). Again, one thing to be thankful for, in the British Isles, amid the general tragedy of the first month of the War, was that it had saved us from threatening trouble in Ireland.

Again as *Avanti*, the Milan socialist paper says (22 Oct. 1917, quoted *Camb. Mag.* VII.106), 'We will only remark generally, on a purely historical view of it, that no revolution ever took place in any country as long as its granaries were full, its civil life fully organized, and its armies victorious in the field'.

However, though Tsardom may have wished to start a war to stave off revolution, it would hardly be likely to have wished long to continue a war for that reason. For war is a powerful medicine; quite a small dose would suffice to cure the internal malady in question. Schemers of this kind presumably wish to maintain the vigour-of-war V_x of their own side at a value, say f, sufficient to distract attention from internal grievances, and yet not so large as to be inconvenient. Their schemes, if they act on the nation, act then somewhat in the sense of the equation

$$\frac{\mathrm{d}(V_x+H_x+F_x)}{\mathrm{d}t} = g\{f-(V_x+H_x+F_x)\} \tag{24}$$

where g is a positive constant. For this equation represents a tendency of the total vigour-to-war $(V_x+H_x+F_x)$, when for any reason it has been increased or diminished, always to return to the value f.

Some rulers, Julius Caesar or Napoleon I for example, apparently made war partly because they enjoyed it. That part of a ruler's ambition is similar to the same enjoyment among common people in so far as both can be represented by an increase in the value of ε in equation (10).

V9 *Fear*

Fear, according to Prof. McDougall, is the emotion which belongs to the instinct of flight. It has much to do with the losing of battles, but surprisingly little, at first sight, with the direction of policy. For even if part of the armies have to retreat rapidly, the leaders express themselves as calm, confident and full of courage; and the effect on the nation as a whole has usually been to rouse it, to come more vigorously to the aid of the army. For instance take the raising of the military age in Great Britain in April 1918, consequent upon the defeat of the 5th British army a few weeks before. In fact, fear of what the enemy may do, or is suspected of being likely to do, is generally a stimulant to the vigour-of-war, in the manner set forth in Bertrand Russell's pamphlet 'War the Offspring of Fear' (Union of Democratic Control). The American Ambassador, Mr Gerard, says in *My Four Years in Germany* (1917, page 57), 'To the outsider the Germans seem a fierce and martial nation. But in reality the mass of the Germans, in consenting to the great sacrifice entailed by their enormous preparations for war, have been actuated by fear'. In the last stages of a war, however, when one of the sides is reduced to exhaustion and helplessness, then fear of what the enemy may do, makes it more ready to make peace. Thus, by the action of fear, V_y or A_y tends to increase V_x if the total casualties C_x and the total destruction of wealth M_x are small; but the same V_y or A_y tends to decrease V_x if the casualties and loss of wealth on the x-side exceed a certain amount. This may be represented by the following formula for V_x^{v}, the partial vigour-to-war due to fear:

$$V_x^{\mathrm{v}} = (\mu A_y + \nu V_y)(1 - \xi C_x - \rho M_x) \tag{25}$$

in which μ, ν, ξ, ρ are positive constants which probably have different values for the two sides.

V10 *Pain*

When schoolboys struggle, the pain they inflict on each other has at first an exciting effect, stimulating them to struggle more fiercely. Only if pain is prolonged and exhausting does it make one of them cry 'Pax'. The course of this war indicates that the same is true of the pains, suffered by populations. The pain of a wound puts a wounded man into a dread of further shocks; but the knowledge that others are suffering, or the painful anxiety for their safety, on the whole has stimulated the vigour-to-war: and this indirect pain has much more than counterbalanced the direct pain, at least until a population has become so worn and starved as for example the Russians were when they made peace. Thus the effect of pain shows a reversal much like that of fear, and a very similar equation may be used to represent it, thus, if V_x^{vi} be the partial vigour-to-war due to pain

$$V_x^{\mathrm{vi}} = (1 - \xi C_x - \rho M_x)(\sigma B_x + \tau L_x) \tag{26}$$

where σ and τ are positive constants.

In this equation it is supposed that $(\sigma B_x + \tau L_x)$ is proportional to the painful feeling, while $(1 - \xi C_x - \rho M_x)$ changes from positive to negative when exhaustion is reached. Painful feeling is here made proportional to the *rate* at which casualties are occurring, by the term σB_y, with an addition for the rate at which wealth is being destroyed. But there exists also pain depending on the memory of all the losses reckoned up. In addition both the fact that territory has been lost, or that it is being lost, are also painful to think upon. If all these sources of pain be included, the last equation changes into

$$V_x^{\mathrm{vi}} = \left(\sigma B_x + \tau L_x + \upsilon C_x + \omega M_x - \pi S_x - \phi \frac{\mathrm{d}S_x}{\mathrm{d}t}\right)(1 - \xi C_x - \rho M_x) \tag{27}$$

where S_x is the area of territory gained by the x-side, and σ, τ, υ, ω, π, ϕ are positive constants. Note that all the subscripts in the last equation are alike.

I have supposed that the vigour-to-war due to fear or pain passes gradually from positive to negative. Mr Bertrand Russell sends me the following commentary which he regards 'only as a suggestion, not as always or exactly correct'. His notation differs from mine.

Mild persecutions produce obstinacy, severe ones produce submission. Since Louis XIV, rulers have generally erred in making their persecutions too mild.

The transition from obstinacy to submission is sudden. The will-to-war produced by a pain P may be represented by

$$\frac{1}{P_0 - P}$$

where P_0 is the 'critical pain'. This will-to-war increases as P approaches P_0, tending towards infinity, and then, when P surpasses P_0, it passes suddenly into a very great will-to-peace. This transition represents the moment of defeat. As P increases beyond P_0, the will-to-peace diminishes, because the will generally, and all power of action, diminish.

A formula, which avoids infinities, and yet exhibits the oscillation which Mr Russell has observed, is in his notation

$$\frac{P_0 - P}{1 + (P_0 - P)^2}$$

V11 *Fatigue*

In the equation (10), connecting the vigour-to-war with the warlike activity of the same side, an equation which runs

$$\frac{dA_x}{dt} = \alpha_x \frac{d(V_x + F_x)}{dt} + \beta_x(V_x + F_x) - \gamma_x(A_x - \varepsilon_x)$$

the term $-\gamma(A - \varepsilon)$ was inserted to represent the effect of the fatigue due to maintaining the activity A, due that is to say to drilling, fighting, munition-making and the like. There is also fatigue due to the enemy having cut off part of necessary food supplies, a fatigue which would perhaps best be represented by making γ_x increase with A_y, or say by replacing γ_x by $\gamma_x(1 + rA_y)$, where r is a positive constant, so that the equation reads

$$\frac{dA_x}{dt} = \alpha_x \frac{d(V_x + F_x)}{dt} + \beta_x(V_x + F_x) - \gamma_x(1 + rA_y)(A_x - \varepsilon_x) \qquad (28)$$

V12 *The desire for change*

Boredom on quiet parts of the front afflicted certain individuals with an intensity such that a minor wound would have been welcomed as a relief. This kind of boredom is a vague diffused discomfort due to unused capabilities, unpractised powers, skill lying idle, or, in Graham Wallas's phrase, due to a baulked disposition.[1] It prompts a man to seek change, even to fly to ills he knows not of. In war, boredom mainly takes the form of a yearning for a return to civil life, and with a slight admixture of fatigue, might go to the tune of 'Three Blind Mice' thus: I, want-to-go, home; I, want-to-go, home; No more bloody war; no more bloody war; no more bloody war; I've stood, two years of this, blasted strife; I'm tired, of going, in fear of my, life; I want, to go home to, my children, and wife; I, want-to-go, home... and so on, ceaselessly. But the excitement of a battle dispels boredom temporarily. Boredom arises, in other words, from the dim distracting effect of various activities, which are only potential, but which we should

1. *The Great Society*, by Graham Wallas.

enjoy were they real. In times of prolonged peace, one of these beckoning activities is war. That effect has already been represented by the ε in the equation (28) connecting warlike activity with vigour-to-war.

I suppose that one reason why some newspapers have so regularly denounced all German peace overtures as 'traps', is because those who controlled these papers feared that thoughts of peace would 'mind us of departed joys' and thereby intensify that boredom which distracted from the vigour-to-war.

V13 *The prospect of military success*

A side is cheered and heartened and has its vigour-to-war strengthened, if it believes that, beyond present difficulties, there lies ultimate success. For if this be not so, why have the leaders of the nations at war so frequently assured their peoples that they were confident of final victory? Again the tendency to enlarge war-aims after a victory, may be observed in the newspapers of both sides, and comes perhaps partly from an increase in the willingness to endure, although it may partly also come from the hope of obtaining more, without increased effort. Again Brailsford remarks (*A League of Nations*, p. 9) that 'The cynical game of guessing when, and on which side a neutral would intervene, taught us that military success may avail more to win allies than community of race, political sympathies, or past services'. Again the fear of being-thought-to-be-willing-to-discuss-peace, as exhibited in the Czernin–Clemenceau controversy, may perhaps arise from the knowledge that willingness to discuss peace is, in many people, a sign that they expect defeat.

The prospect of military success arouses a deep-rooted instinctive feeling which affects people in general. Those who have put their terms high naturally have to cut them down if their side is defeated. That has happened to the Pan-Germans. Those who have put their terms low, and who did not raise them when their side was victorious, may be accounted as unaffected by the prospect of success. For instance, the varying tide of great offensives to and fro across France did not greatly alter the war-aims of President Wilson, of the Union of Democratic Control, or the German Minority Social Democrats.

The question next arises: how does a nation, and especially how do its journalists, estimate this prospect? If the enemy shows signs of weakening, his ultimate defeat appears possible, but if at the moment he is very strong, that defeat may take a long time. Simultaneously each side is conscious that its own determination is liable to wear. In other words the prospect would appear to depend on the relative courses of the two vigours-to-war, regarded as functions of the time. We should here have to deal with the *total* vigour to war, $(V+H+F)$ due to instinctive and higher influences and to freewill. That side is said to be beaten, whose total vigour-to-war first sinks to zero. Now the vigour-to-war of a side may suddenly change, owing to a sudden change in the peace terms offered by the opposing side. The prospect of success therefore depends not on the momentary vigour-to-war of the adversary, but on what his vigour-to-war would be, supposing the desired peace terms could be imposed upon him. Thus the prospect of success is a function of the peace terms and cannot be stated apart from them. There is however one condition of peace which will at least serve as an example, namely the return to the territorial boundaries which preceded the war. Now we have already denoted by P_E and P_G the increases which the vigours-to-war of the two sides would undergo, on a return to pre-war frontiers. On this definition P is positive for a side which has gained territory, negative for a side which has lost. So $(P_E + V_E + H_E + F_E)$ and $(P_G + V_G + H_G + F_G)$ are what the vigours-to-war would be, on a return to pre-war frontiers. For brevity denote these two expressions by U_E and U_G respectively. Suppose the courses in time of U_x and U_y are those represented by the thick lines on the adjoining diagram. At the instant marked t_0, which side had the better prospect? Did people in general estimate x's prospect to be better because the ordinate U_x was greater, or worse because the slope $\mathrm{d}U_x/\mathrm{d}t$ was more steeply negative? Or did they combine the ordinate with the slope in judging that x's prospects were worse, because U_x would sink to zero

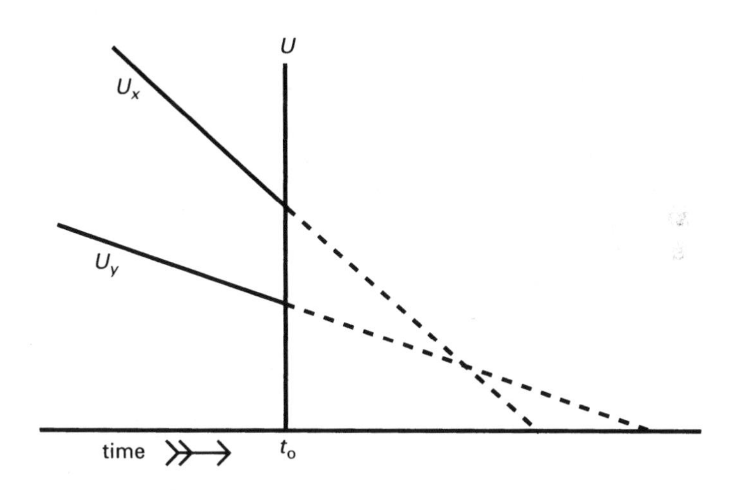

before U_y would, if both were continued along the tangents to the curves at the moment when the estimate was made. Or lastly were the courses carried forward in imagination along curved paths? In what follows it is assumed that the prospect of success is popularly estimated in some way equivalent to considering which tangent will first cut the line $U = 0$. Now, on this basis, the time, reckoned from t_0, which would elapse before U sank to zero, would be $U \div \mathrm{d}U/\mathrm{d}t$, both U and $\mathrm{d}U/\mathrm{d}t$ being taken at t_0. The difference between these times for the two sides would be one measure of the prospect. Mathematically it is an inconvenient measure, because it has $\mathrm{d}U/\mathrm{d}t$ in the denominator. So, bearing in mind that formulae are not to be complicated without necessity, I have taken instead the difference between the reciprocals of these times. Thus the 'prospect of military success' for the Entente has been taken to be

$$\frac{1}{U_G}\frac{\mathrm{d}U_G}{\mathrm{d}t} - \frac{1}{U_E}\frac{\mathrm{d}U_E}{\mathrm{d}t}$$

The prospect for the Germanic alliance was then the same quantity with its sign changed. Thus there appears to be a contribution V^{vii} to the vigour-to-war, given by

$$V_x^{\mathrm{vii}} = S_x\left\{\frac{1}{U_y}\frac{\mathrm{d}U_y}{\mathrm{d}t} - \frac{1}{U_x}\frac{\mathrm{d}U_x}{\mathrm{d}t}\right\} \tag{29}$$

where S is a positive constant which may be called the 'susceptibility to the prospect of military success'.

We have heard much of the 'swelled head' from which Germany suffered in the years just before the war. A temperate description of it is to be found in W. J. Ashley's pamphlet 'The War and its economic aspects' (Oxford 1914, pp. 4–6). 'In academic circles' he says 'the legitimate pride in German science seemed sometimes to have become almost an obsession, and to have the effect of shutting out of sight what was being done in other lands'. Swelled head, in general, would seem to be analysable into (i) an error of judgment which we can represent by making the coefficient ι in equation (21) much less than unity (ii) a resulting sense of unbounded power, which makes the prospect of success seem greater than that given by the 'level headed' estimate of equation (29).

V14 *Duty to the dead*

The greater losses a people has incurred in attempting to attain its war-aims, the more does their attainment appear to be worthy of further endurance, 'Lest' in Mr Asquith's phrase, 'so many and willing sacrifices should have been in vain'. If C_x be the total accumulated casualties, then, in the 'drag' stage of a war, because casualties tend to decrease the vigour-to-war, $\delta V_x/\delta C_x$ is negative; but it becomes less negative as C_x increases, on account of the feeling of duty to the dead.

V15 *Habituation to casualties*

Similar in form to the effect which has just been considered, although different in origin, is the mere deadening effect of habituation to continual casualties. Together they might be represented perhaps, by replacing the factor $(1-\xi C_x-\rho M_x)$ in the fear and pain equations, by $(1-\xi C_x-\rho M_x)/C_x$. This C in the denominator would imply that, at the beginning of a war, when the numerator is positive, and C is small, very small casualties would produce an enormous inflammation of popular feeling. But that is just what happens when the news of the first encounter appears on the newspapers placards.

V16 *Pity for the adversary*

In Galsworthy's wonderful study of *Strife*, which contains so many analogies to the recent war, pity for the adversary is indicated as the best motive for making peace.

In the war such pity has been felt by scattered individuals, but it has scarcely had any outward effect, being almost entirely overwhelmed by the general anger at the warlike acts of the enemy. For it appears that only the great souled, for instance Abraham Lincoln, can have simultaneously two such opposing emotions as pity and anger towards the same object. In most people the stronger of these two emotions is not added to the weaker one; it displaces it. As an extreme instance of this kind of displacement there is the behaviour of King Lear in the first scene of the play.

It would not be easy to find expressions of pity for the adversary in the daily press. There probably are some in the publications of the Fellowship of Reconciliation, if they have not been suppressed by the censor, as being likely to prejudice recruiting. On the other hand the treatment for example of wounded Germans in French aidposts, as far as the author has seen it, has been practically identical with the treatment of French wounded, no sign of anger being mixed with a routine which has its origin largely in pity.

Now, if, before the war, there had been a disastrous earthquake in Germany, there would no doubt have been practical expressions of sympathy – Lord Mayor's funds and the like – in Entente countries. Again, as soon as the German menance was removed by their commencing disarmament in Nov. 1918, President Wilson agreed to supply them with food. So, to formulate, we may say that, in the absence of anger, the pity would depend upon, and increase with, the casualties and loss of wealth which the enemy has suffered or is suffering, that is to say on C_y, M_y, $\mathrm{d}C_y/\mathrm{d}t$, $\mathrm{d}M_y/\mathrm{d}t$. So if V_x^{viii} be the, essentially negative, partial vigour-to-war due to pity, we may put

$$V_x^{\mathrm{viii}} = -\left(jC_y + kM_y + l\frac{\mathrm{d}C_y}{\mathrm{d}t} + m\frac{\mathrm{d}M_y}{\mathrm{d}t}\right) \times (\text{a function of } V_x \text{ due to anger}) \quad (30)$$

in which j, k, l, m, are positive constants.

The function of V_x due to anger must be finite when anger is zero, and zero when anger is large. The simplest function of this kind is

$$\frac{1}{1+(V_x \text{ due to anger})}$$

V17 *Racial antipathy and cohesion*

No marked racial differences occur in Great Britain, so that it is difficult for an Englishman to understand them in their Balkan intensity. Of course differences of clothing, of food, of manners, and especially of language, combine to make a man feel a stranger when he is in a foreign land. To feel a stranger is to feel depressed and out of touch with life, and is unpleasant. Thus many travellers tend to acquire a dislike of foreigners in general, coupled with a firm conviction that the ways of their own nation are better. But how can such comparatively mild feelings explain a story like following, told by the Italian Premier, Signor Orlando, addressing in Rome in 1918 the leaders of the Congress of Oppressed Nationalities.[1]

> It was night, dark and gloomy, and our own and the enemy's first lines were plunged in that silence full of mystery and menace which broods over two armies confronting each other. In the Austrian advance posts there were at one point many Czechs.
>
> *Suddenly in the darkness some one began to sing. Homer alone could have described the solemnity of the moment. It was the Czech national hymn. And then the sentinels were seen to change their positions, the soldiers in the trenches rose to their feet and stood bareheaded till the singer ended. Nothing more simple or more profound; in the night one felt the breath of epic poetry.*
>
> These men, with an enemy in front, who might, in ignorance, fire upon them, with another worse enemy behind them, who at the sight of so bold and magnificent an assertion of national feeling might well fire on them treacherously from the rear – these men feared neither open nor hidden danger, and at the voice of the Fatherland sprang to their feet.

It has been suggested that these extreme cases of antipathy involve a hereditary difference of temperament, worked upon and aggravated by bad treatment (e.g. Ireland). Whereas the milder cases depend merely on custom and tradition, and as such

1. Here quoted from *Public Opinion* of 10 May 1918.

are more easily healed. For instance no one now speaks, as they used to, of Frenchmen and Englishmen as 'natural enemies'.

Lest racial animosity be forgotten in counting up parts of the vigour-to-war we might give it as symbol V_x^{ix}.

VI Higher influences

The types of behaviour discussed in the preceding pages have been mainly instinctive reactions of a regular quasi-mechanical type. It has been found to be possible to describe these types of behaviour, at least in their broad outlines, by means of mathematical expressions. These instinctive reactions are very prominent in war time. But they do not by any means make up the whole of behaviour.

We come now to a group of influences, powerful but elusive, including intuition, reason, justice and religion. Their relations are not obviously expressible by any kind of mathematical formulae. Their discussion lies outside the narrow scope of this essay. But, as a reminder of their great importance, it will be well to have a symbol for the vigour-to-war due to them. Let this symbol be H. Then H should appear in the equation connecting warlike activity with vigour-to-war in the combination $(V_x + H_x + F_x)$ where V is the part of the vigour-to-war due to instinctive influences, H the part due to higher influences and F the part due to free will.

A few words about intuition, reason, justice and religion, in their relation to vigour-to-war, may not be out of place.

VI1 *Reason and intuition*

A distinguished reasoner, Mr Bertrand Russell, in contrasting reason with intuition, has written, 'Reason is a harmonizing, controlling force rather than a creative one. Even in the most purely logical realm it is insight that first arrives at what is new' (*Mysticism and Logic*, p. 13). In the midst of the passions of war, men find themselves reluctantly compelled to admit logical necessities, which desire cannot create, nor anger destroy.

VI2 *Justice*

The fundamental ideas of justice appear to be that human beings are, in limited and sometimes dimly realized respects, equal; and that the weak should therefore, in these respects, be protected from the strong. Of McDougall's list of primary instincts it is the protective parental instinct and also anger, which appear to be the instinctive mainsprings of justice; but these alone would not produce justice without a strong control by reason, which to continue the analogy, may be compared to the balance-wheel and escapement of the clock which indicates justice with its pointer.

As President Wilson has said (New York 27 Sept. 1918): 'Impartial justice meted out must involve no discrimination between those to whom we wish to be just and those to whom we do not wish to be just. It must be a justice that knows no favorites and knows no standards but the equal rights of the several peoples concerned'. To judge of the equality of rights, in a world where one man's food is occasionally another's poison, much sympathy and insight are needed; and for these qualities one would go to learn from the novelists of the several peoples concerned.

But some things have been very clear. Nearly all the world, including many Germans, has regarded the German attack on Belgium as a crying injustice. This attack was one of the principal causes of the enormous voluntary British enlistments in 1914.

Hear also *Vorwaerts* of 18 Sept. 1917 (summarized in *Cambridge Magazine* Vol. VII, p. 9): Whether the German Government can shut its ears to Belgium's claim for resititution is 'not merely a question of Power, and is not merely a question of political judgment, but *it is a question of Right and Conscience*'. The apostles of force may sneer at morality and conscience, but 'no policy seems to us really able which ignores these factors in international life'.

VI3 *The destruction of militarism*

The aim to establish justice is closely allied to the aim to destroy militarism if by militarism be meant a national habit of doing things because the nation is strong enough to do them, regardless of the right or wrong of the matter.[1] The aim, to destroy militarism, was also a considerable cause of British enlistments in 1914, and still operates, although the net effect of the war, in all belligerent countries, up to the time of signing of the Armistice, has apparently been to increase militarism rather than to destroy it.

That however may possibly change in a later stage. The aim to destroy militarism tends to maintain the will-to-war of a side, until the other side shows signs of clearly understanding the meaning of right and wrong; as these are understood by the first side. In this matter each side is like the audience in a theatre, sitting in a dim light and watching, over its own heads, the other side brilliantly lit up on the stage.

VI4 *Religion*

All religious beliefs, which point to a God common to the opposing sides, in so far encourage that sympathy and insight which are necessary preliminaries to justice.

Certain religious systems encourage war with the infidel, for instance Mohammedanism has done so. Other religious groups or sects discountenance war with varying degress of strictness, for instance the Buddhists, Quakers, Christadelphians and the Doukhobors do so. It is sometimes said that war will forever be an institution, because it is instinctive in man, and because instincts are unchangeable. But as a small-scale sample of what can be done in the way of the atrophy, diversion or control of these particular instincts, it is interesting to notice the remarkable strength of the restraint which Quaker influence, whether of home or of school, has commonly exerted on the warlike proclivities of individuals.

The above refers largely to religious beliefs and practices, which are taught by one generation to the next. Of another order is that direct communication with Someone spiritually greater and better than ourselves, which persons of the most various religious beliefs, or of none, have testified to experiencing.

VI5 *A sense of guilt*

There having been such general agreement throughout the world that Germany's invasion of Belgium was a crime – even Bethmann-Hollweg the German Chancellor

1. See section V13 above on the prospect of military success.

having admitted as much in a speech in August 1914 – many people on the Entente side spoke of the Entente army as a policeman and the German army as a burglar; and it was customary to say that the war ought to go on until Germany repented. But let us continue the analogy. The criminal struggles; the policeman also. They roll over and over, kicking, biting and tearing each other until both are half dead. Can one expect the burglar to repent of his crime at the moment when his face is being kicked? Such repentance is scarcely in human nature; the distractions are too great. Even so, in Germany signs of a knowledge of guilt, such for example as the book *I Accuse*, have been as faint as a whisper during the cheering of a crowd. The reason apparently is that the human mind has a limited span of attention. The instinctive disposition causes the attention to be focused on defence, and thereby excludes almost entirely any thought about previous guilt. The latter is 'inhibited by drainage' in Prof. McDougall's phrase. In 1918 October and early November, the German government was changed and the Kaiser made to abdicate. To continue the analogy it is as though the burglar cut off and cast away that one of his hands which had been guilty of the theft, while with the other hand he continued to struggle in self-defence against the policeman.

VII A list of symbols is here given for reference

A Warlike activity
B Rate of casualties
C Total casualties
E Subscript for Entente
F Freely willed vigour-to-war
G Subscript for Germanic Allies
H Higher motivated vigour-to-war
J Activity in fighting
K Margin of superiority
L Rate of destruction of wealth
M Total wealth destroyed
N A value of R
P Conquests as vigour-to-war
Q Business advantage
R Activity in preparation
S Area of territory transferred
T Tender emotion, pity
$U = P+V+H+F$
V Vigour-to-war
W *See* Business advantages
Z Anger

a, b *See* Activity and destruction
c *See* Business advantages
d Differentiator
f, g *See* Security of rulers
h *See* Section V7
i Influence coefficient
j, k, l, m *See* Pity for the adversary
n Number of peoples
p, q *See* Business advantages
r *See* Starvation and fatigue
s Susceptibility to success
t Time
u Emulatance
v Individual vigour-to-war
w *See* Tender emotion and anger
x, y Subscripts for opposing sides
z *See* Tender emotion and anger

α Instinctiveness of activity
β Practice coefficient
γ Fatigue coefficient
ε Activity which is enjoyed
ζ, η *See* Defensive dispositions
θ *See* Vengefulness
ι *See* Emulation
κ Used in Introduction
λ *See* Vengefulness
μ, ν, ξ *See* Fear
π *See* Pain
o *See* Fear
σ, τ, υ, ϕ *See* Pain
χ Military skill coefficient
ω *See* Pain

VIII Summary

VIII1 Collected equations

For ease of reference the equations have been brought together. Here they are merely collected, no attention being paid to the fact that some overlap or replace others. Later an attempt will be made to assemble them.

First rough notion

$$\frac{dA_E}{dt} = \kappa_E A_G; \quad \frac{dA_G}{dt} = \kappa_G A_E \tag{1), (2}$$

The collective vigour-to-war of a state

$$V_x = n_{1x} i_{1x} v_{1x} + n_{2x} i_{2x} v_{2x} + n_{3x} i_{3x} v_{3x} \tag{3) and (4}$$

The relation of the warlike activity to the vigour-to-war of the same side

$$\frac{dA_x}{dt} = \alpha \frac{dV_x}{dt} + \beta V_x - \gamma(A_x - \varepsilon) \tag{9}$$

Freewill

$$\text{Replace } V \text{ in equation (9) by } (V+F) \tag{10}$$

Conquests alone

$$V_x = -P_x$$

Military activity and the resulting casualties and destruction of wealth

$$B_x = \chi'_y (A_y - R_y)(A_x - R_x) \tag{12}$$

$$= \chi_y A_y A_x \text{ for established armies} \tag{14}$$

$$L_x = a' R_x + a'' J_x + b'' J_y \tag{15}$$

$$= a A_x + b A_y \text{ for established armies} \tag{16}$$

The defensive disposition alone

$$V_x^{\text{i}} = -P_x + \zeta_x V_y + \eta_x A_y \tag{17}$$

Vengefulness alone

$$V_x^{\text{ii}} = \theta_x C_x + \lambda_x M_x \tag{18}$$

Prestige alone

$$V_x^{\text{iii}} = u(\iota V_y - V_x + K_x) \tag{21}$$

Business advantages alone

$$Q_x = -Q_y \tag{22}$$

$$V_x^{\text{iv}} = W_x - p \log\left(\frac{Q_x}{q} + 1\right) \tag{23}$$

The avoidance of internal trouble

$$\frac{\mathrm{d}(V_x+H_x+F_x)}{\mathrm{d}t}=g\{f-(V_x+H_x+F_x)\} \tag{24}$$

Fear

$$V_x^{\mathrm{v}}=(\mu A_y+\nu V_y)(1-\xi C_x-\rho M_x) \tag{25}$$

Pain

$$V_x^{\mathrm{vi}}=\left(\sigma B_x+\tau L_x+\upsilon C_x+\omega M_x-\pi S_x-\phi\frac{\mathrm{d}S_x}{\mathrm{d}t}\right)(1-\xi C_x-\rho M_x) \tag{27}$$

Fatigue due to privations
Insert the term rA_y in (10) which then reads

$$\frac{\mathrm{d}A_x}{\mathrm{d}t}=\alpha_x\frac{\mathrm{d}(V_x+F_x)}{\mathrm{d}t}+\beta_x(V_x+F_x)-\gamma_x(1+rA_y)(A_x-\varepsilon_x) \tag{28}$$

The prospect of military success

$$V_x^{\mathrm{vii}}=S_x\left\{\frac{1}{U_y}\frac{\mathrm{d}U_y}{\mathrm{d}t}-\frac{1}{U_x}\frac{\mathrm{d}U_x}{\mathrm{d}t}\right\} \tag{29}$$

where $U_x=P_x+V_x+H_x+F_x$

Duty to the dead and habituation
Replace the factor $(1-\xi C_x-\rho M_x)$ in the fear and pain equations by $(1-\xi C_x-\rho M_x)/C_x$.

Pity for the adversary

$$V_x^{\mathrm{viii}}=-\left(jC_y+kM_y+l\frac{\mathrm{d}C_y}{\mathrm{d}t}+m\frac{\mathrm{d}M_y}{\mathrm{d}t}\right)\times\frac{1}{1+(V_x\text{ due to anger})} \tag{30}$$

Racial dislike V_x^{ix}
Higher influence
Replace V_x or V_x+F_x by $V_x+H_x+F_x$ wherever this has not already been done.

VIII2 *Assembling the effects*

We have so far considered the changes in the vigour-to-war which would be produced by various influences acting singly, other things remaining the same. We now have to consider what happens when all these influences act jointly. Can we simply add algebraically the partial vigours-to-war due to the separate influences? It would seem natural to make such an addition for the effects of the defensive disposition, of vengefulness, of emulation, of business ambitions, of the prospect of military success and of racial antipathy. It is not easy to say whether the effects of fear, and of pain, should also be added; when they increase the vigour-to-war they probably do so by way of anger; and as the defensive disposition and vengefulness also act by way of anger, there may be some overlapping.

The effects of a sense of duty to the dead, and of habituation to casualties have already been incorporated in the fear and pain equations. The effects of fatigue and of the enjoyment of warlike activity appear in the equation connecting the total vigour-to-war with the warlike activity. The desire to avoid internal revolution has also been expressed as affecting the total vigour-to-war. As to boredom, it appears to be impossible to express the effect of that in equations, without having symbols for activities which are rival to activity in war. There remain, of the instinctive vigours-to-war which have been mentioned above, only that due to pity for the adversary. Now the peculiarity of the latter is that it involves what the psychologists call 'tender emotion', and that to have simultaneously tender emotion and anger towards the same person or nation requires a greatness of soul which is not common. Usually one of these two emotions displaces the other. Their relation is quite different to the coexistence, which envy, for instance, can have with either affection or dislike. Let T be the tender emotion towards a certain object excited by a stimulus I. Let Z be the anger with the same object excited by a stimulus II. Then if the stimuli I and II act together, as when, for example, a special friend of ours breaks one of our special treasures, it is often observed that the rival emotions of irritation and affection come and go alternately, both mingled with pain. They seem to enhance each other, because each makes us attend to the person, who is connected in our minds with the opposite emotion. On the other hand the duration of each is shortened, by the interruption caused by the other; and in the outward manifestations to which they prompt us, they tend to neutralize one another. The foregoing refers to the mild irritations of the drawing-room; the anger produced by war is often so much more intense as to fill the mind to the entire exclusion of pity. Now a formula which represents several of these facts in a simple manner, is that which makes the stimuli I and II, acting jointly, produce a vigour-to-war equal to

$$\frac{Z}{1+wT}-\frac{T}{1+zZ} \tag{31}$$

where w and z are positive constants. For, according to this formula, the vigour-to-war is $Z-T$, that is to say: (anger) − (tender emotion), if both anger and tender emotion are small; whereas, if anger is large enough, it cuts out tender emotion altogether by the occurrence of Z in the denominator under T; and vice versa.

May we next, in place of Z in the above formula (31), insert the sum of the partial vigours-to-war which work by way of anger, namely those connected with the defensive disposition, with vengefulness and, except for an exhausted nation, with fear and pain? That would seem to be a passable expedient. Denote this sum by $\mathscr{V}$. Then

$$\begin{aligned} \mathscr{V}_x = & -P_x+\zeta_x V_y+\eta_x A_y \quad \text{for the defensive disposition} \\ & +\theta_x C_x+\lambda_x M_x \quad \text{for vengefulness} \\ & +\frac{1-\xi_x C_x-\rho_x M_x}{C_x}\left\{\begin{array}{l}\mu_x A_y+\nu_x V_y+\sigma_x B_x+\tau_x L_x+\upsilon_x C_x \\ \quad +\omega_x M_x-\pi_x S_x-\phi_x (\mathrm{d}S_x/\mathrm{d}t)\end{array}\right\} \end{aligned} \tag{32}$$

The last term, representing fear and pain jointly, is only to be included if $\xi_x C_x+\rho_x M_x$ is less than unity, that is to say only if the term as a whole is positive, as it is for a nation which is not exhausted.

Let T now stand for the pity-for-the-adversary so that according to section V16 we have

$$T_x = j_x C_y + k_x M_y + l_x \frac{\mathrm{d}C_y}{\mathrm{d}t} + m_x \frac{\mathrm{d}M_y}{\mathrm{d}t}$$

Then the joint effect of all the instinctive influences which have been discussed is

$$\frac{\mathscr{V}_x}{1+wT_x}-\frac{T_x}{1+z\mathscr{V}_x} \quad \text{for anger and tender emotion}$$

$$+\frac{1-\xi C_x-\rho M_x}{C_x}\left\{\mu A_y+\nu V_y+\sigma B_x+\tau L_x+\upsilon C_x+\omega M_x-\pi S_x-\phi\frac{\mathrm{d}S_x}{\mathrm{d}t}\right\}$$

for fear and pain, to be included only if negative

$$+u(\iota V_y-V_x+K_x) \quad \text{for emulation}$$

$$+W_x-p\log\left(\frac{Q_x}{q}+1\right) \quad \text{for business ambitions}$$

$$+V_x^{\mathrm{ix}} \quad \text{for racial antipathy}$$

$$+S_x\left\{\frac{1}{U_y}\frac{\mathrm{d}U_y}{\mathrm{d}t}-\frac{1}{U_x}\frac{\mathrm{d}U_x}{\mathrm{d}t}\right\} \quad \text{for the prospect of military success}$$

$$=V_x \tag{33}$$

where V_x is the total vigour-to-war due to the ensemble of instinctive influences. Now V_x appears on both sides of this equation, for $-uV_x$ occurs in the emulation term in the first member. Collect the terms in V_x into the second member which then becomes $(1+u)V_x$, while the emulation term is cut down to (ιV_y+K_x).

Under the heading 'The security of rulers' we arrived at an equation

$$\frac{\mathrm{d}(V_x+H_x+F_x)}{\mathrm{d}t}=g\{f-(V_x+H_x+F_x)\}$$

intended to represent the effect of schemes to distract attention from internal grievances. Such schemes, by altering the beliefs of a nation, alter the 'constants' in the long equation (33) for V_x.

Next, to the total instinctive vigour-to-war V_x, there must be added a part H_x representing the effect of reason, justice, religion and other higher influences. A further addition F_x has been made to represent the effect of freewill. This F_x is supposed to be indeterminate, except that it is confined to a limited range. Then the total $(V_x+H_x+F_x)$ is related to the warlike activity A_x by an equation, which has already been incompletely formulated in number (28) above,

$$\frac{\mathrm{d}A_x}{\mathrm{d}t}=\alpha_x\frac{\mathrm{d}(V_x+H_x+F_x)}{\mathrm{d}t}+\beta_x(V_x+H_x+F_x)-\gamma_x(1+rA_y)(A_x-\varepsilon_x) \tag{34}$$

That completes a statement of the relations for the x-side. It has been assumed that the relations for the y-side are similar in form but with different values of the constants. There are cross-connections between the two sides wherever the x and y suffixes both appear in the same equation. And especially, in addition to the equations already reviewed, there are cross-connections through the following two equations:

Rate of casualties $B_x=\chi_y'(A_y-R_y)(A_x-R_x)$ (12)

Rate of destruction of wealth

$$L_x=a'R_x+a''J_x+b''J_y. \tag{15}$$

These equations containing both the suffixes x and y are one way of expressing in part the interdependence of nations, which, in diplomatic language, would be described as sovereign and independent. International influence powerfully exists, whether we like it or not; but it has not yet been fully organized.

What of the first rough notion with which this essay began? It was that $dA_x/dt = \kappa_x A_y$. Can it be deduced, as an approximation, from the very much more elaborate equations which have been evolved from it, by a long process of inductive reasoning? Well, in equation (34) which expresses the connection between the vigour-to-war and the warlike activity, if we neglect α which depends on the small part of the activity which is instinctive, and ε which measures the small activity which is pleasurable and r which depends upon the privations due to the enemy having throttled supplies, there remains

$$\frac{dA_x}{dt} = \beta_x(V_x + H_x + F_x) - \gamma_x A_x \tag{35}$$

Next if we neglect higher influences and freewill we thereby omit $H_x + F_x$. And then if we regard the defensive vigour-to-war as so much the greatest of all the instinctive influences that we may put approximately

$$V_x = V_x^{\text{i}} = -P_x + \zeta_x V_y + \eta_x A_y \tag{36}$$

Then

$$\frac{dA_x}{dt} = \beta_x\{-P_x + \zeta_x V_y + \eta_x A_y\} - \gamma_x A_x \tag{37}$$

Finally if conquests were unimportant P_x might be neglected and in any case the 'suspicion' term $\zeta_x V_y$ is probably less than the 'direct evidence' term $\eta_x A_y$, so that, neglecting also the former, the equation comes down to

$$\frac{dA_x}{dt} = \beta_x \eta_x A_y - \gamma_x A_x \tag{38}$$

which resembles the first rough notion $dA_x/dt = \kappa_x A_y$ except that the term $-\gamma_x A_x$, representing the fatigue which sets a limit to all things human, cannot be neglected.

IX Conclusion

The previous paragraph contains equations setting out the general form of the observed relationships. In conclusion some estimates will now be made of the relative size of the various terms. These estimates are admittedly more controversial than the general statements of relations.[1] Previous to the war there existed, as causes tending to it, such purposes as: certain special business ambitions, racial jealousies, the desire for security on the part of dynastic rulers, the desire to regain Alsace-Lorraine and the love of warlike activity in itself. These desires seem insignificant in comparison with the subsequent outburst of destructive fury. But they engendered suspicion between nations, and suspicion produced armaments, and armaments in turn increased suspicion, and so on alternately, distrust and preparedness-for-war mutually increasing each other; until war broke out. From the time of the first hostile act, the

[1] Possibly numerical measures might be extracted from statistics of recruiting or of subscriptions to war-loans. But the analysis and interpretation would be beset with difficulties.

intensity of the war was increased by mutual reprisals; for all warlike measures appear almost as atrocities to those on whom they are inflicted, however justifiable and necessary they may seem to the inflicting party. In other words the largest part of the total vigour-to-war, $(V+H+F)$, was, on both sides, that due to the instinctive defensive disposition, and the second largest part, I should say, that due to vengefulness. A limit to the warlike activity was set by fatigue. This with fear, pain and boredom slowly wore down the vigour-to-war, until Germany practically confessed herself beaten. Now that fighting has ceased, the powerful distracting influence of the defensive and vengeful dispositions has abated, and pity for the adversary and what have been called the higher influences in all countries are beginning to reassert themselves.

1926:2

POWER IN THE LEAGUE OF NATIONS

The World Outlook
A Quaker survey of international life and service
London
No. 19, 7 May 1926, p. 38

The Friend
A religious and literary journal
London
New Series, Vol. **66**, No. 19, 7 May 1926, p. 384

Note: This topic is continued in 1953:1

POWER IN THE LEAGUE OF NATIONS.

WITH reference to the recent dispute about permanent seats on the Council, we may ask what is it that makes some States great and powerful relatively to others called small and weak. The old-fashioned standard answer has been (though not in these columns) that Power is Might in the threatened battle. If the world is to live peaceably we must change that idea of power. Some idea of power we must however retain, for it is no use pretending that France and Lichtenstein should be equal in influence simply because both are States. Well, then, try population as a measure of power. France is more populous than Lichtenstein. But Tibet has about the same population as Norway, and does it seem reasonable to regard them as of equal power on that ground alone? Evidently something else besides population is involved in our idea of the importance of a State. What is that elusive something? I suggest that it is Services Rendered to the rest of the world. Tibet keeps itself to itself, Tibetans do not travel and explorers find it a difficult country to enter. Norwegians travel, the rest of the world eats their canned fish, hears Ibsen's plays and Grieg's music and uses their science. We can find nothing analogous for Tibet.

Now can this vague idea of services rendered be made quantitative? Yes, there seems to be a way, open and waiting for exploration.* The statistics of imports and exports, expressed as money, are a measure of one kind of services rendered; for example, the foreign trade of Norway is about 100 times that of Tibet. The number of foreign letters passing is a measure of another kind. The amount of capital invested abroad is another. Contribution by States to the exchequers of the League of Nations and the many other international undertakings are again measures of their helpfulness.

We can even measure the intellectual services rendered by a State to the rest of the world. For honours are given to foreigners for their published work. We have then only to measure, by a suitable statistical method, which States have thus been most honoured by other States.

Now just as H.M. Board of Trade makes an "index-figure" for the cost of living by combining prices in suitable proportions, so the Secretariat could, if it would, make an index-figure for the power of each State, by combining in suitable proportions the separate measures of services rendered in goods, money and ideas. The relative importance to be attached to merchandise and ideas is of course a very controversial question that would have to be settled by the Assembly.

But the first thing needed is for several amateurs to make and publish unofficial estimates along these lines, in order that everyone may see what bearing it would have on such current questions as the Spain-Poland-Brazil-Germany competition.

LEWIS F. RICHARDSON.

* A preliminary study appeared in *War and Peace*, 1917.

Richardson, L. F. (1928)
Br. J. Psychol. **19**, 158–66

1928:3

THRESHOLDS WHEN SENSATION IS REGARDED AS QUANTITATIVE

The British Journal of Psychology – General Section
Cambridge
Vol. XIX, Part 2, October 1928, pp. 158–166

(Manuscript received 13 March 1928)

THRESHOLDS WHEN SENSATION IS REGARDED AS QUANTITATIVE

By LEWIS F. RICHARDSON.

(*Lecturer on Physics at Westminster Training College, London.*)

1. Absolute Thresholds.

1·1. *Experiments on touching the skin with two points.*

These were carried out on the inner surface of the forearm with an ebonite-pointed esthesiometer in the way described by Whipple[1], except that:

(*a*) The subject was asked to try to estimate the separation of the points in centimetres, and as an aid thereto was allowed to look at a centimetre scale with one eye. The other eye, on the side of the arm which was being investigated, was blindfolded.

(*b*) Previous practice was avoided as far as possible, and the subject was kept in ignorance of the stimuli until the end of the experiments.

(*c*) The esthesiometer was set in turn at every millimetre of separation from 0 to 100, both included, but in a random sequence. About twenty different stimuli were applied at one sitting.

The esthesiometer (= compasses) weighed 53 grams and was allowed to exert its own weight on the skin, being gently propped up almost in the vertical plane.

Fig. 1 shows sensation (y) plotted against stimulus (x). Subject Dorothy Richardson; age 41; right forearm; centre of compasses 33 cm. from finger tips; experiments at about 8 p.m.

[1] *Manual of Mental and Physical Tests*, 1914, Test 23.

One point on this diagram ($x = 0$, $y = 5$ mm.) shows that a stimulus by a single point was felt as two points. G. M. Whipple[1] remarks:

"It will be found that some subjects will, nevertheless, occasionally answer 'two' when but one point is given. This is the not uncommon *Vexirfehler*, or esthesiometric paradox, which is well recognised as a source of difficulty in esthesiometry."

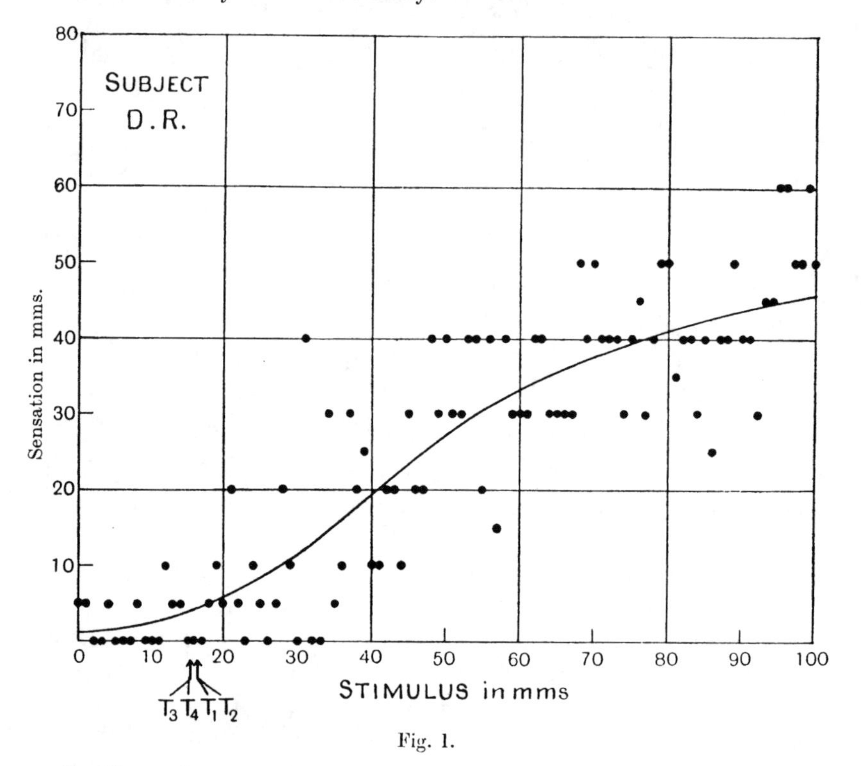

Fig. 1.

In Fig. 1, however, the point ($x = 0$, $y = 5$ mm.) is not isolated. It has neighbouring observational points, which, by keeping it company, make it seem quite ordinary. Thus one result of regarding the sensation as quantitative has been to do away with a paradox.

Fig. 1 differs from a normal Bravais correlation diagram in two important ways:

A. The frequency is cut off by both coordinate axes.

B. The mean of an array of sensations corresponding to a narrow range of stimuli is a property of the subject. Not so the mean of an array of stimuli corresponding to a narrow range of sensations; for

[1] Whipple, *Manual of Mental and Physical Tests*, 1914, I, 247.

although the stimulus-range of this array is fixed by the observer, the frequency in it might be piled up at one end or the other by the experimenter. Thus only one of the two regression curves is a property of the observer, namely the curve through the mean sensations. The statistical methods given by K. Pearson[1] on 'the general theory of skew correlation and non-linear regression' seem to be exactly appropriate; but less laborious and less thorough methods are used here.

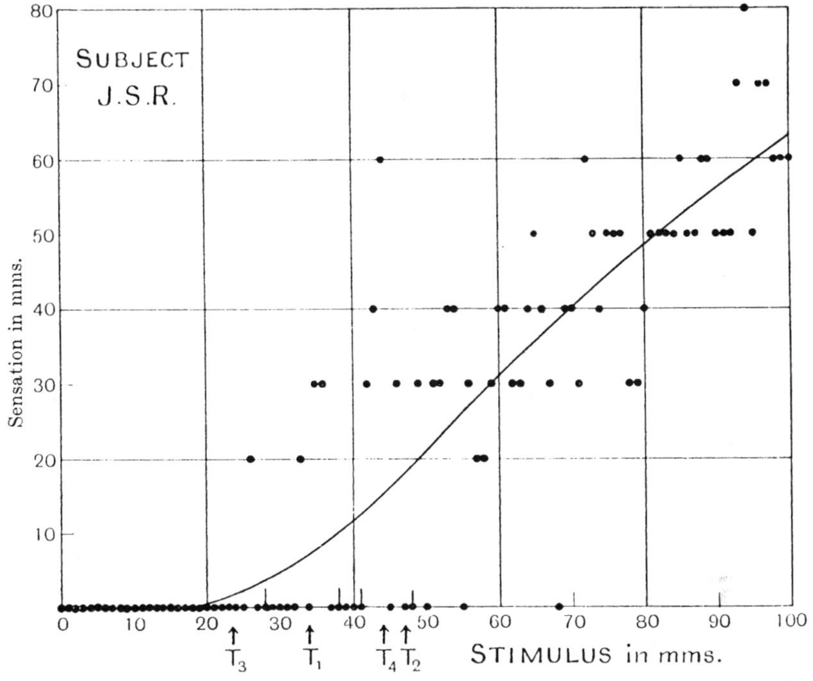

Fig. 2. The four dots on the base line at 28, 38, 41, 48 mm. have upward strokes attached to signify that the sensation was 'qualitatively unlike single touch.'

The stimuli at 7, 78, 93 mm. were applied after J. S. R. had seen the rest of the diagram. This was regrettable, but he thought it made no difference.

Fig. 2 shows the estimates of James S. Ross, age 35, Lecturer on Education at Westminster Training College; left forearm; centre of compasses 32 cm. from finger tips; time 2 p.m. just after a meal.

On comparing these subjects it is seen that both underestimate the separation. The most notable contrast is that J. S. R. gave no estimates between 0 and 20 mm., whereas D. R. gave 21 such. It may be asked

[1] K. Pearson, *Drapers' Co. Research Memoirs, Biometric Series*, II, No. XIV (Dulau and Co.), 1905.

whether this difference between the observers is not so much a difference in the sensations they experience as a different interpretation of the same sensation; but the question is hard to decide.

On the classical view, the purpose of the experiment is to determine one point called the absolute threshold on the horizontal axis of Fig. 1 or Fig. 2. The position of this mark evidently requires some convention to define it. Four different conventions T_1, T_2, T_3, T_4 will now be discussed. The corresponding points are indicated on the diagrams.

T_1 is defined to be half the greatest stimulus for zero sensation.

T_2 is defined to be half the greatest stimulus for zero sensation plus half the least stimulus for non-zero sensation. T_2 was suggested by the classical 'limiting method.'

An objection to both T_1 and T_2 is that they depend on extreme variates and that extremes are regarded by statisticians as unreliable. Now T_3 and T_4 avoid this objection.

T_3 is defined to be the mean of the stimuli for zero sensation ('method of mean error').

T_4 is defined to be a stimulus that exceeds as many non-zero sensations as zero sensations exceed it. This definition may only constrain T_4 to lie in a short range of stimulus equal to the least step used in the tests; if so, by a further definition the midpoint of the range is taken as T_4. T_4 was suggested by the, rather different, 'method of serial groups.'

But an objection to both T_3 and T_4 is that they are liable to displacement by the experimenter, for, as already mentioned, the experimenter can pile up the frequency at any part of a range of stimulus fixed by the subject. To make T_3 and T_4 reliable, some further convention, such as uniform distribution of stimuli, is necessary.

The values found are in mm.

Subject	T_1	T_2	T_3	T_4
D. R.	16·5	16·5	15·3	15·5
J. S. R.	34	47	23·5	44

Again the significance of any statistic depends on its size relative to the corresponding scatter. Any conventional absolute threshold is necessarily about as large as the range of stimuli corresponding to zero sensation, and because this connection is necessary, therefore this range cannot serve as a test of the significance of the threshold. Instead we might suitably take, as a test, the range of stimuli corresponding to a fixed sensation near to, but other than, zero. For instance, Fig. 2 shows that, for a sensation of 30 mm., J. S. R. had a range of stimuli from 35 to 79 mm. This range is as much as 0·65 of the range for zero sensation, and therefore *any threshold, which we could define, might be significant as some kind of mean of many estimates, but could not be significant for each estimate separately.* (*Anti-paradox.*) Fig. 1 emphasizes this point of view.

Any threshold, being a single point on the x-axis, is a very inadequate representation of the scatter diagram.

If we abandon the idea of thresholds and use instead that of the 'regression curve' passing through the means of arrays of sensation, we obtain the curve shown in Fig. 1; the ranges of stimuli defining the arrays being 0 to 10, 11 to 20, 21 to 40, 41 to 70, 71 to 100 mm. It is suggested that *this regression curve is a better description of the facts than any threshold.* But, like any other mean, it needs to be supplemented by statements of the scatter, such as the present diagrams or Pearson's scedastic and clitic curves (*loc. cit.*).

The corresponding regression curve for Fig. 2, using the same limits for the arrays, is also shown; but with diffidence as to its suitableness, on account of the absence of sensations between 0 and 20 mm.

For those who prefer numbers, the same facts are set out in the following table.

Arrays of Sensation.

All numbers are millimetres.

Stimulus bounds	Stimulus means	Sensation means		σ_A		σ_B	
		D. R.	J. S. R.	D. R.	J. S. R.	D. R.	J. S. R.
0 to 10	5·0	2	0	2	0	2	0
11 to 20	15·5	4	0	4	0	4	0
21 to 40	30·5	12	5	12	10	11	10
41 to 70	55·5	31	27	10	17	9	15
71 to 100	85·5	42	53	9	12	9	9

Here σ_A is the standard deviation from the mean of the array. If the observed points all lay exactly on the regression curve, $\sigma_A{}^2$ would not vanish, but would be approximately equal to $\frac{1}{12}$ (square of sensation-range of regression curve within the array). To correct for this effect of the width of the array, σ_B shown in the table is defined by

$$\sigma_B{}^2 = \sigma_A{}^2 - \tfrac{1}{12}\,(y\text{-range of regression curve})^2.$$

It seems to be impossible to escape entirely from conventions in statistics. Pearson's convention is that the regression curve is of the form

$$y = b_0 + b_1 x + b_2 x^2 + b_3 x^3 + b_4 x^4,$$

in which b_0, b_1, b_2, b_3, b_4 have to be determined. The present convention is that the arrays have the above-mentioned bounds in stimulus.

1·2. *Supporting evidence taken from the work of others.*

Al Sufi, who died in A.D. 986, centuries before there were any telescopes, wrote in his *Description of the Fixed Stars*[1]:

"Many people believe that the total number of fixed stars in the sky is some thousand and twenty-five; but this is a manifest error. The ancients observed only this number of stars, which they divided into six classes, according to magnitude. They put the most brilliant in the first magnitude; those which are slightly smaller, in the second; those which are a little smaller than them, in the third; and so on to the sixth. As to those which are below the sixth magnitude, they found that they were too numerous to be counted; that is why they left them out. It is easy to convince oneself of this. If one looks attentively at a constellation, the stars of which are well known and recorded, one finds in their intervals many other stars which are not counted at all. Let us take for example Cygnus: it is composed of seventeen internal stars, the first on the beak, the last on the foot, the most brilliant on the tail, the others on the wings, the neck and the breast; in addition below the left wing are two stars which do not in any way form part of the figure. Between these different stars, if you look carefully, you will see a multitude of others, so small and so crowded that one cannot possibly determine their number. It is the same with all the other constellations."

E. M. Smith and F. C. Bartlett experimenting "On listening to sounds of weak intensity[2]" found that the range of variation of the threshold caused by variation of the experimental procedure was two or three times greater than the least value of the threshold. See their Table IV, p. 113. Although they sometimes observed a "positive silence" (p. 145) which was "something else than sound or the cessation of sound" yet they record also subjective sensations (p. 148) occurring in the absence of stimuli.

R. W. Pickford in experiments on "The perception of almost inaudible sounds[3]" found that any sounds which are critically perceived have to be distinguished from the intrinsic hum of the cerebro-auditory apparatus. His subjects sometimes thought that they had heard the sound when no stimulus was presented.

[1] Al Sufi, *Description des Étoiles fixes*, tr. Schjellerup (St Petersburg, 1874), p. 40. The English is mine. I am indebted to Dr E. B. Knobel, F.R.A.S., for the information that this is the most explicit of the ancient works on star magnitudes.

[2] This *Journal*, x, 1919–20, 101–129 and 133–168.

[3] Pickford, *Ibid.* XVII, 1926–27, 222.

Oscillations of mental efficiency in general have been discussed by Spearman[1].

A much-investigated threshold is that for the discrimination of two points touching the skin. G. M. Whipple[2] quotes Foucault's distinctions between eight possible perceptions and four types of observer. The threshold he defines by a convention (*loc. cit.* p. 249) concerning two errors in ten trials.

G. H. Thomson[3] in a research on the same question, after rejecting experiments in which catch errors were frequent, left the few remaining catch errors out of the calculation. In a previous research[4] he instructed his subjects to answer 'one' unless two separate points could be felt.

1·3. *Contrary Views.*

Fechner's view is represented in Fig. 3. The curve is $y = k \log x$, where k is independent of x. Ordinates drawn downwards from the

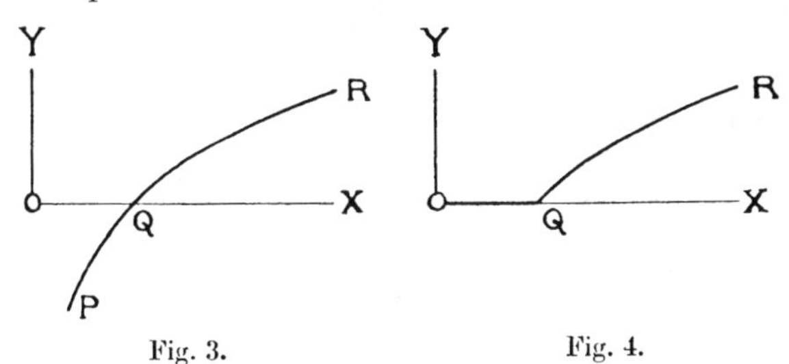

Fig. 3. Fig. 4.

line OX measure, according to Fechner[5], the depth of unconsciousness. Some who have rejected[6] the portion PQ of the curve because it is not observed by introspection, still suppose OQR in Fig. 4 to be a satisfactory[7] description. There is then an angle at Q, at which $y(x)$ is not differentiable. In mathematics differentiability is defined precisely[8]. But the scattered dots of Figs. 1 and 2 cannot prove in any precise sense the existence of the angle at Q in Fig. 4. And for mathematical purposes it is much more convenient if $y(x)$ may, without contradicting the facts, be differentiable to all orders. For these reasons some function that is 'analytic' will be adopted in the manner suggested by the regression curve in Fig. 1.

[1] Spearman, *Abilities of Man*, Ch. XIX.

[2] Whipple, *Manual of Mental and Physical Tests*, 2nd ed., I, 256.

[3] This *Journal*, VI, 1913–14, 432–448.

[4] *Ibid.* V, 1912–13, 239, l. 1.

[5] Fechner, *Psychophysik*, 1860, II, 39.

[6] McDougall, *Body and Mind*, Methuen, 1911, pp. 172, 173.

[7] Compare Howell, *Text Book of Physiology*, 1908, p. 257 and Starling, *Principles of Human Physiology* (1926), p. 393.

[8] G. H. Hardy, *Pure Mathematics*, Camb. Press, 1925, Art. 110.

2. DIFFERENCE THRESHOLDS.

The chief peculiarity may be stated thus[1]. If a, b, c denote intensities of three stimuli of the same kind such that $a < b < c$ and if α, β, γ are the sensations produced respectively by a, b, c, then by choosing $(c - b)$ and $(b - a)$ suitably small it can be arranged that α is indistinguishable from β, and β is indistinguishable from γ, yet nevertheless α can be distinguished from γ.

If in the previous sentence 'indistinguishable' meant that the sensations were evidently exactly equal, then the phenomenon would indeed be most peculiar, and not expressible in mathematics.

But if 'indistinguishable' means that the two sensations are not obviously unequal, then the phenomenon is of the statistical type familiar in errors of observation.

The work of Urban[2] shows clearly that we must accept the latter view for lifted weights, because he found that when one stimulus was a fixed standard, the probability of a decision 'equal' varied with the other stimulus in a gradual manner.

3. SHARP THRESHOLDS.

The upper and lower bounds of the frequency of audible vibrations have also some scatter when the same observer makes repeated trials; but, in contrast with the threshold for double touch, the range of inaudible frequency beyond the treble and the range of inaudible wavelength beyond the bass are both enormous in comparison with the scatter.

The photoelectric threshold in physics is sharp in comparison with the range of frequency between it and zero[3].

4. CONCLUSION.

The word threshold is commonly used to denote two quite distinct relations. (i) A beginning at a point which is definitely not zero. (ii) A beginning confused by a vagueness, uncertainty or scatter. Discrimination of double touch belongs to the latter class. The best representation of it seems to be the regression curve giving mean sensation

[1] Following G. F. Stout, *Manual of Psychology*, 1915, pp. 303–304. T. Smith, *Nature*, Feb. 25, 1928, has drawn attention to "An Optical Paradox" of this kind. N. R. Campbell and T. Smith discuss it on April 7.

[2] F. M. Urban, *Arch. f. d. ges. Psychol.* v, 15, 16.

[3] O. W. Richardson, *Roy. Soc. Proc.* A, CXVII, 1928, 720.

as a function of stimulus; together, if the observations are sufficiently numerous, with the scedastic and clitic curves of K. Pearson. Paradoxes then disappear.

The new facts in this paper are the two hundred opinions kindly given me by Mr J. S. Ross and by my wife.

(*Manuscript received* 13 *March*, 1928.)

Richardson, L. F. (1929)
J. gen. Psychol. **2**, 324–52

1929:3

IMAGERY, CONATION AND CEREBRAL CONDUCTANCE

The Pedagogical Seminary and Journal of General Psychology
Worcester, Massachusetts
Vol. II, Nos. 2 and 3, April–July 1929, pp. 324–352

(Received for publication 9 May 1928)

Offprinted from *The Pedagogical Seminary and Journal of General Psychology*, 1929, **2**, 324-352.

IMAGERY, CONATION, AND CEREBRAL CONDUCTANCE*

From the Department of Physics of Westminster Training College, London

LEWIS F. RICHARDSON

1. INTRODUCTION

The object of this research has been to make a mathematical description of the time changes of the intensity of imagery, including the ups and downs, the refusals, delays, sudden onsets, fadings, recoveries, and disappearances of the intensity of relevant imagery, which comes, or fails to come, in response to trying-to-think with various intensities.

It is a problem more in induction than in deduction.

If the description which is finally attained should accord with modern physiology, so much the better; if not, it still may be true psychology.

"Trying-to-think" is an active variety of attention. The outward signs are to be seen at examinations, and include some or all of: tense facial muscles, slight frown, eyes fixed and half closed, restless finger movements. We get the experience whenever we try to solve a puzzle of any kind. "Imagery" is used in its ordinary psychological sense.

No progress can be made in the desired direction unless it is possible to estimate numerically the intensities of imagery and of trying-to-think, purely by introspection. Fechner[1] asserted the possibility of thus estimating sensations. But he made other indefensible assumptions. W. James[2] and C. S. Myers[3] have contradicted him. G. F. Stout[4] saw no objection to sensations being measured, but added that it had not yet been done. The question has been dis-

*Received for publication by William McDougall of the Editorial Board, May 9, 1928.

[1]Fechner, G. T. Psychophysik. Leipzig: Breitkopf, 1860.

[2]James, W. Principles of psychology. New York: Holt, 1901. See Vol. I, p. 547.

[3]Myers, C. S. A text-book of experimental psychology. Cambridge: Cambridge Univ. Press; New York: Longmans, Green, 1911, 1925. See p. 252.

[4]Stout, G. F. A manual of psychology. London: Clive, Univ. Tutorial Press, 1915. See p. 307.

cussed at a Symposium.[5] For present purposes the best evidence that intensity of sensations can be numerically estimated is contained in a paper called "Loudness and Telephone Current" by the present writer and J. S. Ross.[6]

Once numbers have been assigned to introspective intensities, an algebraic treatment of those numbers becomes possible. And if we ever slip into loose but convenient phrases like "the square of the intensity of imagery" or "the logarithm of the intensity of trying," it must be understood that the correct though cumbrous form would refer instead to the square or logarithm of the number attached as a label to the intensity.

Let then the *intensity of trying to think* be represented by a number which we denote by v.

Let the *intensity of relevant imagery* be measured by a number symbolized by m.

Let us begin by studying m in relation to v and to time t.

2. Experiments

2a) *Consistency*

It would, of course, be desirable to survey mental types. But in a preliminary study like this it is more important that all the observations should fit together into one set of "brain constants"; therefore they have all been made on a single subject, namely, the writer.

It is less difficult to estimate an intensity than to estimate the uncertainty of that estimate; and in this paper there are no estimates of uncertainty. But for sensations, on the advice of Mr. F. C. Bartlett, some estimates of uncertainty will be found in the paper on "Loudness and Telephone Current."[6]

2b) *Persistence, Success, and Its After-Effects*

Experiment 1A. A chronograph[7] having three pens writing on a moving band of paper was arranged so that a standard clock marked seconds by one pen. The subject in another quiet room pressed a

[5]Myers, C. S.; Hicks, G. D.; Watt, H. J.; & Brown, W. Are the intensity differences of sensation quantitative? *Brit. J. Psychol.*, 1913-14, **6**; I, 137-154; II, 155-174; III, 175-183; IV, 184-189.

[6]To be published in this journal. See other supporting evidence in the *Brit. J. Psychol.*, 1928 or 1929, under the titles, "Thresholds When Sensation is Regarded as Quantitative" and "Quantitative Mental Estimates of Light and Colour" by L. F. Richardson, also a further work by R. S. Maxwell.

[7]Cambridge Sci. Instr. Co., List No. 88, Instrument 8854 is similar.

key with his right hand whenever he was trying to think and pressed another key with his left hand whenever he had relevant imagery. These keys operated electrically the other two pens. A dictionary or index was opened at random. Looking at the top word, the subject tried to think of its opposite. As soon as he had succeeded, however imperfectly, he looked at a different word and tried for its opposite. Several such distractors were sometimes inserted. Finally, he tried again to think of the opposite of the first word.

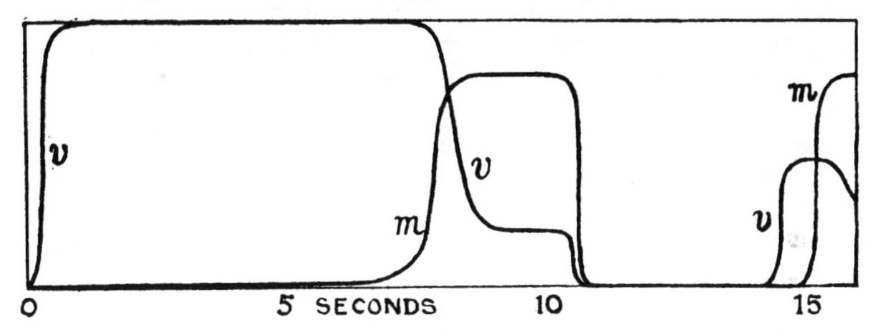

FIGURE 1
(See qualifications in *2d*)

Table 1 shows that the "opposite," previously thought of, comes up much more rapidly at the second trial, especially if the interval between the two trials is not more than 15 seconds.

TABLE 1

First stimulus word	Time of trying in secs.	Opposite thought of	Beginning of renewed trying on original stimulus seconds from start	Duration of renewed trying before idea revived seconds	Number of uninvited recurrences
censorship	2.8	liberty	7.3	0.1	1
dramatic rights	14.1	no dramatic rights	23.5	0.4	1
half profits	3.1	whole profits	11.6	0.5	0
imposition	9.1	half-holiday	17.0	0.9	0
misprints	2.0	correctness	10.4	0.6	0
shoulderhead	5.0	tailpiece	13.8	0.7	0
belpless	2.5	competent	20.1	1.6	0
counterplot	9.6	friendly treatment	42.0	3.8	0
orange	3.1	lemon (silly)	34.5	2.1	1

Figure 1 shows the intensities given by introspection for a typical experiment. We have to find equations for these curves.

2c) *The Imageless Suspense*

A referee who read an earlier version of this paper objected that "the author wrongly supposes that what is thought of must necessarily occur in the form of imagery." This objection deserves a careful answer. Now the relation of imagery to thought has been reviewed by Spearman.[8] After re-reading these chapters I reply that the subject of the present experiments ordinarily thinks with inner speech. Whether by means of it, or merely to the accompaniment of it, is not easy to say. Visual images also occur and are used, but they are dim, inaccurate, and sometimes difficult to evoke. Tactile images can be vivid, but are seldom used. Images of any sort are usually present "on business." Occasional images intruding irrelevantly are neglected and usually die of neglect. But the decision as to what is relevant seems often to precede by a fraction of a second the inner word expressing the decision; so that, during that short time, thought is either imageless, or its image is unnoticed, or it might be said that the thought is unconscious. This agrees with T. V. Moore's statement that ". . . . there is a distinct interval of time between the perception of the meaning of a word or picture and any other subsequent imagery."[9] In solving problems as in Experiment 1 this imageless suspense may lengthen out to several seconds, as in the early part of Figure 1, which has been likened in the physiological note (2*n*) to the summation period. But extensive trains of imageless thought do not occur in me. Unlike the subjects of Martin,[10] if I try to suppress all images altogether, thought dwindles and I drift towards sleep.

This describes age 46; at age 15 visual images were numerous, bright, and easy to evoke, while appropriate words were difficult to find.

Another well-known psychologist has objected that the phrase "trying-to-think" should be replaced by "trying to evoke images." The proposed alteration seems to me unsuitable. In the experi-

[8]Spearman, C. E. The nature of intelligence and the principles of cognition. New York: Macmillan, 1923. See chaps. XII and XIII.

[9]Moore, T. V. Image and meaning in memory and perception. (Psychol. Stud. Cath. Univ. Amer.) *Psychol. Monog.*, 1919, **27**, No. 119. Pp. 296.

[10]Spearman, *op. cit.*, p. 190.

ments, I tried to think in my habitual way, and as a consequence images happened.

But it must be admitted that the great variety in the use or disuse of imagery by different persons[11] does restrict the application of the present experiments.

Experiment 1B. To observe what, if anything, happened during the imageless suspense, the following routine was adopted. The eyes were suddenly turned to a printed word, often the last word of a paragraph in a newspaper, and a stop-watch was started as the word was understood. The subject immediately tried to think of the opposite. The watch was stopped as soon as the auditory verbal image of the opposite was quite clear, or as soon as the search was abandoned as desperate. Immediately after stopping the watch and before looking at it, and before recording the opposite, the subject considered the following questions:

A. Did anything other than trying-to-think precede either the clear verbal image of the opposite or the abandonment of the task? If question A was answered "yes" then:

B. Did this something include a faint auditory image of the whole opposite word or phrase?

C. Did it include the first part of the opposite, occuring detached from the rest?

D. Did it include a feeling, not expressed in words, but meaning the same as "it is about to begin" or "something is arriving"?

E. Did it include a word or words proposed as opposite but rejected as unsuitable?

If the answer to A was "no" then the answer to B, C, D, and E were taken to be "no."

The results of 50 trials are given in Table 2. The least

TABLE 2

Question	A	B	C	D	E
yes	30	3	4	0	27
doubtful	2	0	1	1	0
no	18	47	45	49	23

time was 1 second, the greatest 28 seconds, the median 5 seconds. Visual images frequently occurred but were not systematically recorded. The image which was finally accepted as the right one,

[11] *Ibid.*, p. 182.

was occasionally heard earlier in some faint or abbreviated form (3 answers "yes" to B, and 4 to C) but usually it began almost as suddenly as a spoken word.

A description, at first sight appropriate, was that the imagery "had been rising continuously since the beginning of trying, and its sudden appearance was due to it crossing a threshold." To understand this more clearly a separate research was made on "Thresholds When Sensation is Regarded as Quantitative,"[12] and the conclusion was made that this threshold is essentially a beginning which goes differently in successive trials so that we like to represent its average progress by a "regression curve" passing at each instant through the mean intensity of imagery in many trials; that it would be impossible to prove that this regression curve took off from the time axis at an angle; and that it is quite as true, and mathematically much more convenient, to suppose that the curve rises without an angle as in Figure 1.

2*d*) *Thought, Imagery, and the Cut-off*

After drawing Figure 1, I became sceptical about the long flat top of the *m*-curve between 8.5 and 10.5 seconds.

Experiment 1C. This portion of the curve was re-examined by repeating Experiment 1A without distractors and trying to maintain the image of the opposite. Times were taken by a stop watch. It was found as the result of 15 trials that the auditory image of the opposite word came to an end in an ordinary conversation time, of the order of 1 second. If the word was still thought of, the auditory image often repeated itself many times without appreciable gaps; sometimes a vowel was long drawn out; sometimes relevant visual images alternated with the auditory images. This conglomerate is perhaps what some writers call "thought." If so, the thought is like jam containing in it certain recognizable berries which are like the images. And the cut-off shown at 10.6 seconds in Figure 1 is the cut-off of the thought, for individual images faded inevitably and could not be sustained by *v*.

2*e*) *The Technique of "Brief and Feeble" in Experiments on Forgetting*

By forgetting is here meant the unconscious process which begins with ceasing to think about anything, and ends with a difficulty

[12] *Brit. J. Psychol.*, 1928, Oct.

in recalling it. The experiments are for times usually less than one minute. Now Experiment 1 shows that trying-to-think, when it succeeds, has a persistent effect in shortening the delay at the repeated trial, so if we want to test the progress of forgetting we must:

(1) Make a suitably feeble effort to try to remember;

(2) Keep up this effort for only a brief time. Here 1.5 seconds has been allowed;

(3) Provide distraction during the period extending from ceasing to think up to the test;

(4) Take note of any uninvited remembrances which may occur, and regard such observations as suspect until the contrary is proved.

An established and elaborate technique, designed to eliminate from memory experiments the effects of various disturbances, is described by C. S. Myers.[13] But it relates chiefly to intervals of minutes, hours, or days, whereas the problem has been to adapt it to an interval of 10 seconds. Accordingly, various disturbances, such as fatigue, practice, special interests, and repressed complexes, have had regretfully to be left to be partially eliminated by mere averaging. From a comparison of the dots and circles in Figure 2, it looks as if the special precautions described in (1), (2), (3), (4), above, might be important; although there is an alternative ex-

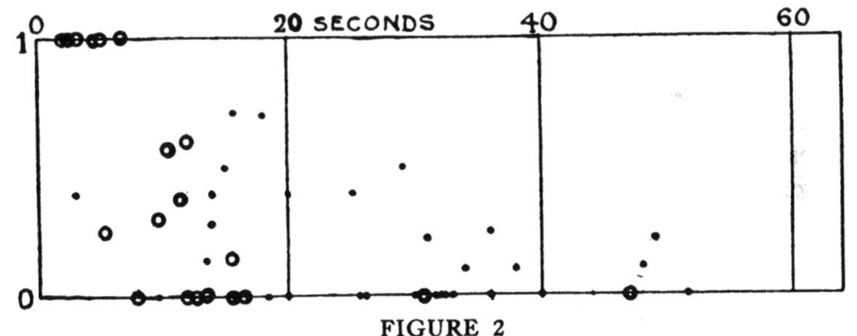

FIGURE 2
The time-scale is the same for Figures 2, 3, 4, and 5

planation (see 2*i*). Distraction has been recommended by G. M. Whipple[14] for a similar purpose.

---

[13]Myers, C. S. A text-book of experimental psychology. Cambridge: Cambridge Univ. Press; New York: Longmans, Green, 1911, 1925. See Vol. I, Chaps. 12 and 13.

[14]Whipple, G. M. Manual of mental and physical tests. Baltimore: Warwick & York, 1915, 1921. See Vol. II, p. 158.

2f) *Forgetting*

Experiment 2. An interesting book was read in the subject's ordinary way by visual perception, accompanied by auditory word images. The last line of the page, previously kept hidden, was noticed by reading it once only, but more slowly, and with close attention during 3 or 4 seconds. As the eye passed the midpoint of the last line a stop-watch was started. The reading was continued attentively down the next page until the end of the first paragraph, when the watch was stopped and the last line recalled as far as it came up quite easily within about 1½ seconds. Prolonged or intense effort to recall was purposely avoided. The diagram (Figure 2) shows the fraction of the words in the last line which were correctly remembered after various times of reading in 50 trials. A complication was that the words sometimes came up as auditory images before they were invited. Such observations are distinguished as dots from the circles which represent observations in which no uninvited images occurred; but the dot represents the voluntary remembrances subsequent to the uninvited.

Disregarding the dots, Figure 2 shows that all lines of words do not behave alike—yet alike in so far as the fractional number of words correctly remembered was unity up to 5 seconds, and zero beyond 16 seconds. The mean time of transition was about 9 seconds.

Of the intensity of the memory images, Figure 2 tells us nothing. By introspection it seemed at first that the words emphasized by silent, slow reading were either remembered in an ordinary conversational tone, or else not at all, never as a yell nor as a whisper. Repeated experiments, however, revealed images that developed rather more slowly than the rest. Once there was an unmistakable dimness, neither all nor nothing; this occurred at 10.4 seconds, about the time when Figure 2 shows that most synapses would be closing. Introspection thus gives curves like Figure 3.

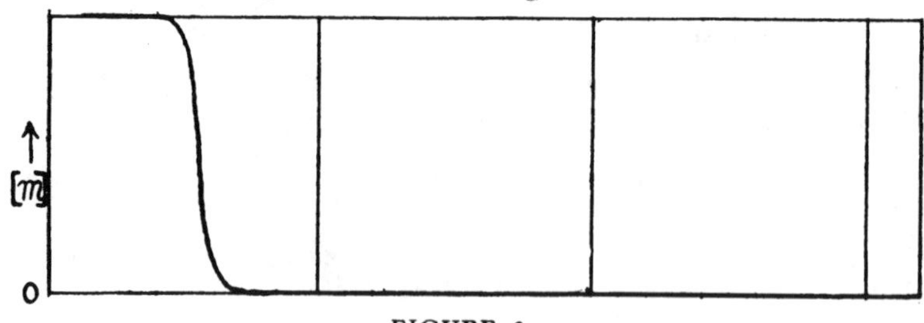

FIGURE 3

This 10-second period is possibly the same as the "specious present" of William James.

Whereas a single experiment of Type 1A gives the *whole* curve of Figure 1, a single experiment of Type 2 gives only two points on the curve of Figure 3, namely, the starting point at 0 seconds, and the variable finishing-point. The ordinate of this curve might be called "readiness for imagery," it being understood that readiness is measured by the intensity of imagery attained after not more than 1.5 seconds of feeble trying. A small square bracket as in [*m*] is used to denote readiness.

2*g*) *Fading and Recovery*

Experiment 3. This was like Experiment 2 except that the last line of the page was read 20 times. Usually the stop-watch was set going at the middle of the 20th reading. As a result of 49 trials it was found that the auditory imagery was usually most intense at the first reading. Understanding was at its maximum at the first to the fourth reading, according as the passage was obvious or subtle. Boredom was most intense at about the seventh reading; after that one became resigned, and imagery faded gradually. The fraction of the initial intensity to which it had faded at the 20th reading was estimated in 41 trials, as in Table 3.

TABLE 3

Fraction of initial intensity	0.1	0.2	¼	0.3	⅓	0.4	0.5	0.6	0.7	0.8	0.9	1.0
Number of trials	3	4	1	10	2	7	5	3	0	4	2	0

The median fraction is near 1/3 of initial intensity. After reading the last line for the 20th time, the subject continued immediately to read the next page in the ordinary way, but with close attention to insure distraction. Stopping the watch anywhere, the subject then tried briefly and feebly to recall the last line of the previous page. Figure 4 shows, for 39 trials, the fractional number of words remembered as a function of the time. The dots and circles have the same meaning as in Figure 2. The average time of forgetting is seen in Figure 4 to be about 30 seconds, as against 10 seconds in Figure 2.

The recovery from fading was studied in 14 of the aforesaid trials by estimating the intensity of the voluntarily recalled image as

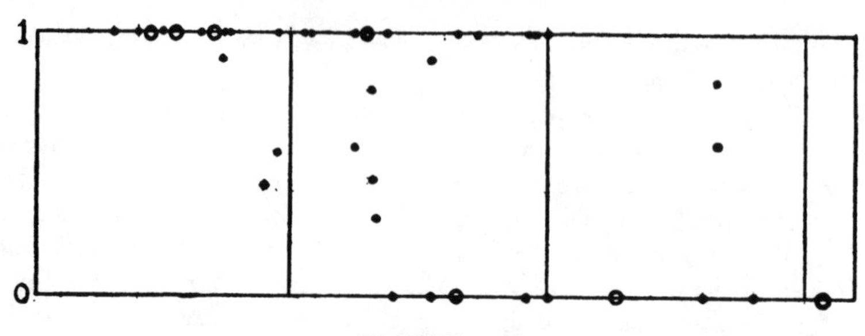

FIGURE 4

a fraction of the maximum normal intensity. In this work a stop-watch was found to be too distracting, and lines of print were used as timekeepers. They were compared with the stop-watch by reading them again immediately afterwards. These 14 trials are each represented by one straight line in Figure 5. The numbers 0, 1, 2, 3, at the ends of these lines are the number of uninvited recurrences preceding the voluntary recall. All the words were remembered except in three of these 14 trials, and even in those more than half were remembered.

The estimates of auditory intensity were for the average of the words in the line of print from which individual syllables may vary.

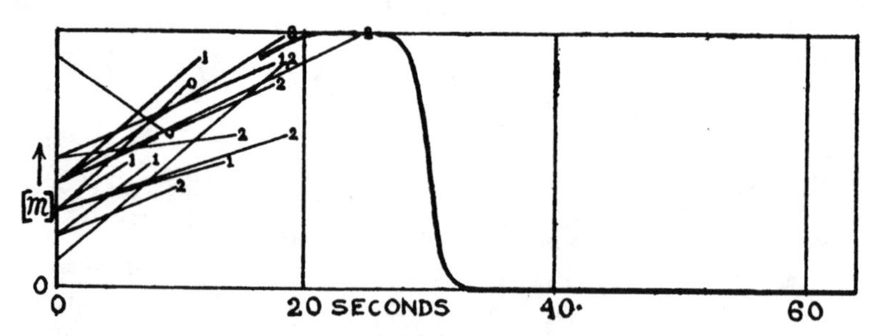

FIGURE 5

In Figure 5 we see two unconscious processes, first recovery from fading, then forgetting. Much use will be made of this diagram when formulae are to be fitted. But as several psychologists found it to be unconvincing, Experiment 3 has been repeated in 58 trials, spread over 4 months, since Figures 4 and 5 were drawn.

Twenty-two of these new trials were free from uninvited recurrences, and have been plotted in a diagram like Figure 4 (circles) but not printed. The longest time at which all words were remembered was 26.6 seconds. The shortest time at which no words were remembered was 12.7 seconds. There were five of these trials, for which some but not all words were remembered, and the mean of their times was 23 seconds. This is an estimate of the "closing time." But it may be too small, as there were too few observations at longer times. The previous estimate from Figure 4 was 30 seconds.

For the purposes of Figure 5, on the contrary, we should accept only those trials in which all the words were correctly remembered after about 1.5 seconds of feeble trying. These 28 trials have been classified according to the number of uninvited recurrences in each of them, and are shown in Table 4. Each of these classes has been further subdivided into long times and short times, putting equal numbers of trials into the two compartments, and sharing equally an odd middle trial when it was present.

TABLE 4

Number of uninvited recurrences per trial		0	1	2, 3, or 5
Number of trials		10	9	9
Plotted on figure		5A	5B	5C
Mean intensity at 20th reading as fraction of maximum		0.3_0	0.3_2	0.4_2
Shorter times	Mean slope as fraction of maximum intensity per second	0.052	0.027	0.027
	Time of mean midpoint, seconds	3.9	6.0	7.5
Longer times	Mean slope as fraction of maximum intensity per second	0.017	0.022	0.010
	Time of mean midpoint, seconds	8.7	10.4	14.5
The numbers in the last 4 rows are plotted in Figure 5D as		circles	dots	triangles

Two alternative standards were employed as aids to estimating the intensity of the faded imagery. Thus in some trials, after reading the last line of the page 19 times, the previous line was read once and assumed to be an unfaded standard; then the last line was read for the 20th time during which the stop-watch was started. These observations are marked on Figures 5A, 5B, 5C, by single arrowheads. In other trials a less disturbing standard was arranged by sitting at such a distance from a gas fire or from a clock that its sound matched the intensity of unfaded imagery. These are marked by double arrowheads. In the future a telephone emitting a steady sound may be used, as in the research on "Loudness and Telephone Current."

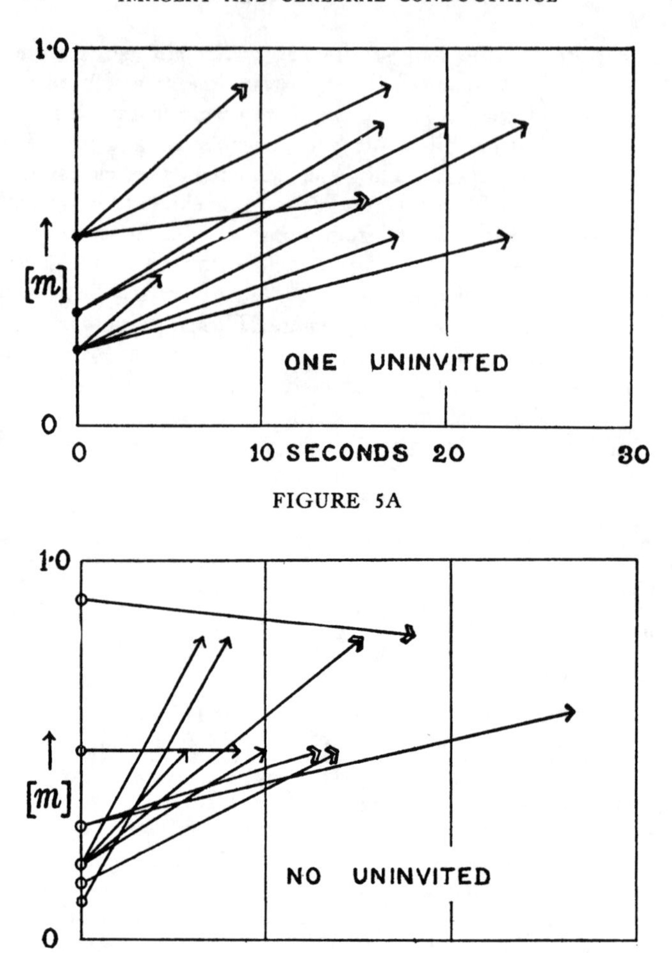

FIGURE 5A

FIGURE 5B

For simplicity in Figures 5, 5A, 5B, 5C, the lines are drawn straight, but they probably ought to be curves convex upwards, for when in Figure 5D the mean slope is plotted against the time of the mean midpoint, we see that the mean slope decreases as time increases.

Do words or syllables recover completely from fading before

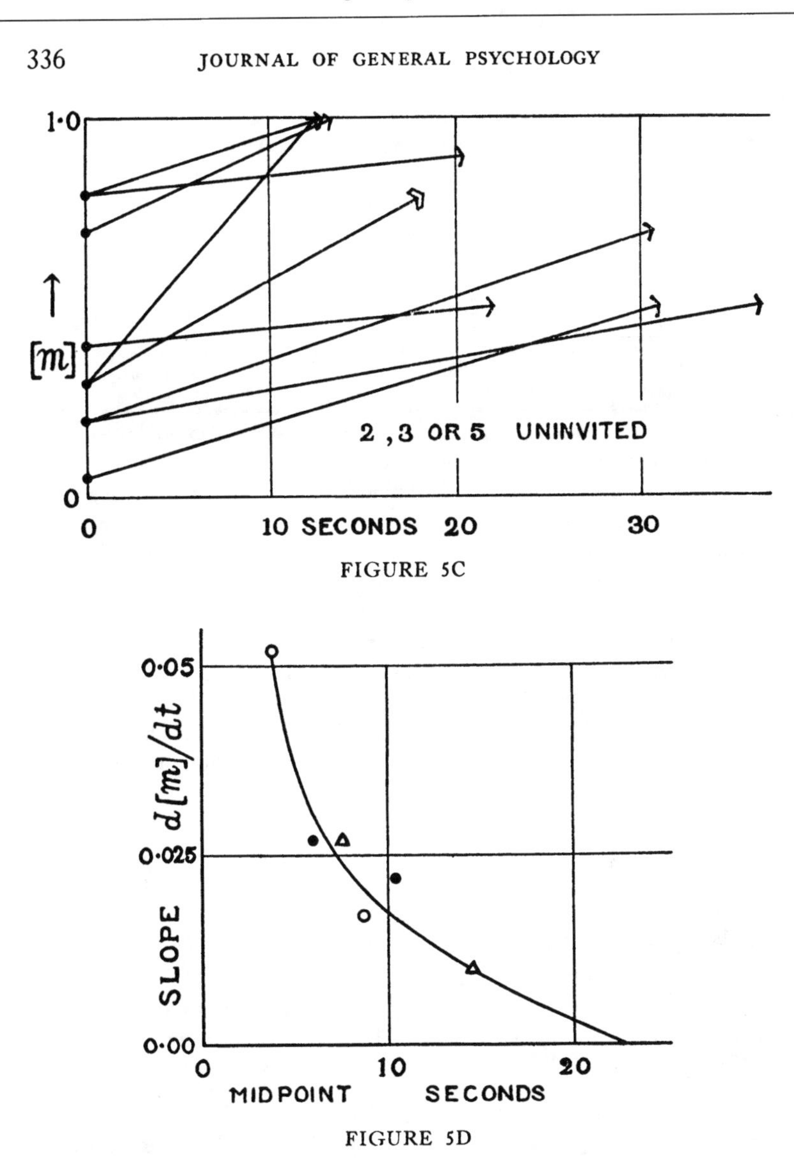

FIGURE 5C

FIGURE 5D

they begin to be forgotten? In Figure 5C are three arrows which actually meet the top line, but in Figures 5A and 5B there are none. Some arrows are horizontal. Among these 58 new trials of Experiment 3 there are 15 in which some words of the line were forgotten when the intensity of the remainder was estimated. These 15 trials have been divided into two groups according as the fractional number of words remembered was large or small. The group means are given in Table 5. From these facts it is evident that a whole line of words does not completely recover from fading before it

TABLE 5

Number of trials	Fractional number of words remembered	Intensity of remainder as fraction of maximum intensity
7½	0.7_8	0.8_3
7½	0.3_6	0.7_7

begins to be forgotten. As to individual words and syllables, the observations could be explained by incomplete recovery before forgetting. But an equally valid explanation would be that separate words all recover perfectly but that their closing times are scattered over 10 seconds, and each word is near to its maximum for a briefer time, say 1 second.

In Figure 5D a curve drawn through the circles, dots, and triangles, tends to cut the base line somewhere between 19 and 24 seconds. When it cuts, then $d\,[m]/dt = 0$, in which $[m]$ is the readiness for imagery. Then $[m]$ is stationary, actually a maximum. This time 19 to 24 seconds agrees roughly with the "closing time" of 23 to 30 seconds, estimated from the fractional number of words remembered.

2*h*) *Completer Fading*

Experiment 4. It was noticed in Experiment 3 that any slight pause in the 20 readings allowed the imagery to recover a little from fading. It seems probable that one word of the line was thus recovering while another word was being read. Therefore, to insure completer fading only one word should be repeated.

The word *jolly* was imagined auditorily 100 times in rapid succession, the eyes being closed. This experiment was far from jolly to perform.

2*i*) *Analysis of the Uninvited Remembrance*[15]

(1) Not every phrase tends thus to recur. Emphasis seems to be the cause.

[15]Similar processes have been observed in reflex arcs: Dreyer, N. B., & Sherrington, C. S. Brevity, frequency of rhythm, and amount of reflex nervous discharge as indicated by the reflex contraction. *Proc. Roy. Soc.* (*B*), 1917-19, **90**, 270.

TABLE 6

Time from start (secs.)	Repetitions from start: Intentional	Unintentional	Intensity of auditory imagery	Word recalled
0	1	0	1 (standard)	*jolly*
–	20	0	0.5	"
110	100	0	0.02	"
Now read a novel to distract				
720	100	13		"
780	101	?	0.05	"
Distracting reading continued				
1920	102	21	0.6	"
Distracting reading continued				
5100	103	38	0.6	*often* then *jolly*
Meal, conversation				
7320	104	40		*often* then *hopeful* then after a few seconds *jolly*
Reading				
8580	105	40	1.0	*jolly* after a few seconds of hard trying

(2) *Experiment 5.* To study the time relations only the last word of a page was emphasized, and the time of each uninvited recurrence was recorded by electric key and chronograph. The book was *The Story of Prague* by Count Lützow (Dent,, 102). The last word was spoken aloud, the page was turned, and the reading continued.

TABLE 7

Page	Word	Times in seconds from reading aloud to beginnings of successive occurrences					
25	does	2.4	5.6	8.7	12.3	36.3	
26	Prague	3.3	8.1	12.5	16.0	21.2	29.7
30	to	2.8	6.6	9.7	16.3	26.5	
		(The word *took* occurred in the distractive reading and became entangled with the emphasized word *to*)					
31	the	3.8	9.6	15.9			

The phenomenon differed from the familiar damped oscillation described by the linear equation

$$\frac{d^2x}{dt^2}+2\epsilon\frac{dx}{dt}+\mu^2x=0$$

(in which μ and ϵ are independent of t), because the pauses were longer than the disturbances, and because

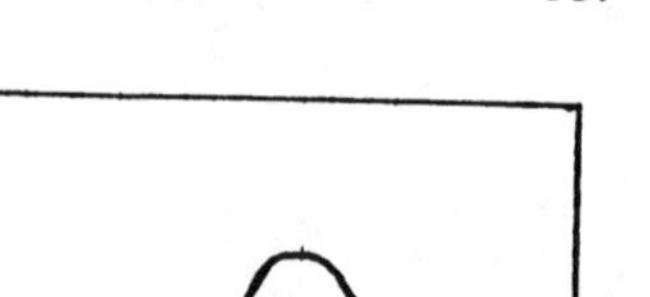

FIGURE 6

the second recurrence was often more intense than the first or third. Figure 6 is typical of the introspections.

(3) It may be suggested that the readiness for imagery oscillates. To find whether this was so, two experiments were made.

Experiment 6A. Just after an uninvited image. An uninvited auditory image occurred during distracting reading 2 seconds after the emphasis. As soon as the image had ceased I tried to recall it, and got it at once. This shows that the readiness for imagery had not decreased much. The experiment was repeated on 14 words, and the invited image, closely following the uninvited, was obtained 27 times, more times than the number of words because some words led to 3 recurrences. Once only was there confusion instead of the invited image.

Experiment 6B. Just previous to an uninvited image. This differed from 6A in that the observer tried to recall the word just before the second uninvited image was due. This experiment was much more tricky than 6A. It was tried 24 times with the results given in Table 8. Reviewing Experiments 6A and 6B, it is seen

TABLE 8

	Number of trials
Story too interesting, experiment forgotten	1
No clear uninvited images, experiment impossible	3
Confusion came instead of invited image	1
Invited image obtained	
(*a*) just before uninvited	9
(*b*) but no uninvited followed	4
(*c*) coincident with uninvited	3
(*d*) mistimed giving sequence of Experiment 6A	3
Total	24

that only twice out of 27 + 24 trials did the image fail to come when invited. This was during about 10 seconds after the emphasis. It is clear evidence that the readiness for imagery did not decrease much.

If there is a "refractory phase,"[16] it seems from these experiments to be shorter than half a second.

There was a remarkable difference in pleasantness between Experiments 6A and 6B. To have an invited image about 0.7 seconds *after* an uninvited image was as pleasant as repetition in music; but to have them in the reverse time sequence was often mildly perplexing, disconcerting, with a feeling of being bumped or knocked about inside the head.

(4) The disturbance is very conative, being, as well as the auditory image, a feeling like "it is important" or "soon I must," but vaguer and not expressed in words. In one experiment no image of the emphasized word appeared, but at 1.8, 6, 12, and 19 seconds there was a brief tension in the forehead and a desire to get on. The recurrences were more marked when reading an intellectual book *Research and the Land* by V. E. Wilkins than when reading *Labrador Days* by W. T. Grenfell or *The Constant Nymph* by Margaret Kennedy, which two are emotionally exciting. The latter were more effective distractors.

(5) The dots in Figures 2 and 4 lie to the right of the circles. Two alternative explanations (*a*) and (*b*) are possible:

(*a*) The closing-time varies greatly from causes that have nothing to do with the uninvited images, but delay in closing affords more chance for uninvited images to intervene.

(*b*) The conative urge which the subject tries unsuccessfully to prevent acts in the same way as intentional trying-to-think, in so far as both make memory persist.

Similarly, in recovery from fading, on comparing Figure 5A with Figure 5B and Figure 5C, it looks as if the uninvited recurrences delayed the recovery as in explanation (*b*). But when in Figure 5D we see the circles, dots, and triangles all near the same curve, it looks as if time were the only thing of consequence—as in explanation (*a*).

No way has been found to decide between the alternative explanations (*a*) and (*b*).

2j) *The Maximum Intensity of Imagery*

It has already been noted in Experiment 2 that words read silently in the ordinary way were usually remembered in an ordinary conversational tone, sometimes more faintly, but never louder.

[16] Lucas, K. Conduction of nervous impulse. New York: Longmans, Green, 1917. See chap. VI.

Experiment 7. This was like Experiment 2, except that, when taking note of the last line by eye, it was imagined as a yell. This was mainly auditory, but there were also muscular sensations or images near the eardrum. After distraction, the remembered image sometimes came up as a yell, sometimes in a conversational tone. Once the image was a yell without any words. This suggests that the image of a yell has a different mechanism from that of conversation. If so, we had better not mix them. The experiment was discontinued after 12 trials.

Experiment 8 was like Experiment 2 except that the last line was read aloud, and in 19 trials gave results much like Experiment 2, except that the duration of memory was slightly prolonged to about 15 seconds.

2k) *The Equilibrium or Threshold[17] Intensity of Trying-to-Think*

Is there a threshold such that if we try less intently imagery does not arise, however long we persist in trying; whereas if we try more intensely than this alleged threshold, then relevant imagery sooner or later occurs?

When thoroughly overworked, one may try for two hours and yet fail to solve a mathematical problem which comes out in five minutes after a holiday. In a dream one may even try persistently but fail to recall the name of a close acquaintance.

Experiment 9. Threshold depends on difficulty. On an afternoon when listless and drowsy, I found it impossible to persist in trying so feebly as not to succeed in recollecting the names of relations familiar since childhood. Yet a harder problem, the proof of the relation between latus rectum and the shortest radius vector of a parabola, which proof I had known for twenty-seven years, was sought after for several minutes without any apparent progress. On switching on two or three times the previous intensity of trying-to-think, the proof became obvious after a few seconds, and left me awake.

Experiment 10. Estimates of intensity of trying. It has been shown by Experiment 2 that readiness for imagery is near its maximum up to six seconds after the silent emphasis. Now stopping the distracting reading at the end of the first line of the new page, that is, after about three seconds, I planned to try so feebly as just

[17]The word threshold is unfortunately ambiguous. See: Richardson, L. F. Thresholds when sensation is regarded as quantitative. *Brit. J. Psychol.*, 1928, Oct.

not to recall the emphasized words in spite of persistent trying. It was difficult to be feeble enough. It could sometimes be managed by almost falling asleep while reading the distracting line. After 10 trials it was estimated that the threshold intensity of trying was 1/5 of the "feeble" trying of Experiment 2 and perhaps 1/15 of "examination pressure," which is about the same as the intense trying of the opposite test in Experiment 1. But this experiment, No. 10, is the most difficult and uncertain of Experiments 1 to 14.

Definition. Let the unit intensity of trying-to-think be defined to be the threshold value when readiness for imagery is at its maximum with respect to time in the course of recovery and forgetting. This unit is natural to the individual; but there is no obvious way of comparing the units of two individuals. When expressed in this unit let the intensity be noted by V.

The result of Experiment 9 is then that $V=5$ during Experiment 2, and $V=15$ in the early part of Experiment 1.

2l) *Rapid Alternations of Intensity of Attention*

Experiment 11. In contrast to the persistence for five or more seconds as in Experiment 1 and Experiment 10, I now tried to alternate from strong to feeble trying as rapidly as posible. It could be done in about two seconds. This fluctuating attention was applied to a memory which was at or near its maximum of readiness for imagery in the course of the curve of Figure 5. This was contrived by reading a book, noticing the last uncommon word of the page silently as in Experiment 2, turning to the next page, reading one line only, and then trying to remember; attention being in the sequence *strong-weak-strong* in some experiments, and *weak-strong-weak* in others. In 15 trials it was observed that the intensity of imagery usually went up and down with the intensity of trying. Simple proportionality of m to v seemed the best general description. But in a few trials m was not variable or else less variable, as if $m \propto v^{1/2}$ certainly not to v^2. Also the changes in m may have lagged a few tenths of a second after those of v.

2m) *The Behavior of Simple Vowels*

Experiment 12. Speed of fading and recovery. On trying intensely to experience the auditory image of *a* in *father*, the image faded almost to nothing in 20 seconds. This is faster than the fading of words. Recovery from fading was also faster, for it occurred in five seconds during which *oh* was imagined as a distraction.

Experiment 13. Persistent unfading images. While in bed in a quiet room I imagined a vowel sound like *oo* in *good* very faintly, and kept it going for 200 seconds without any fading. In every in- or out-breathing the sound of the rush of air in the nostrils drowned the faint image, but it emerged again with its previous intensity.

Similarly, *a* as in *father* was kept going steadily for about 90 seconds at a feeble intensity. At greater intensities it faded. At less intensities it fell asleep, leaving the rest of me awake. There seemed to be only one intensity which could be kept steady for many seconds.

These persistent images may reasonably be supposed to occur when the subject adjusts his intensity of trying to the threshold value.

Experiment 14. But we have seen in Experiment 9 that for words the threshold increases with the difficulty of the thought. Accordingly, I tried to make the vowel image easy, by previous fading; thus: *a* as in *father.* Strong trying was maintained for 20 seconds during which the image faded to nothing. Immediately I dropped the intensity of trying to a sleepy attention. For two seconds there was no image, then it began rather suddenly and persisted for five seconds. This experiment on vowels seems at first sight to contradict Experiment 9 on more complicated relations. But they may perhaps be harmonized if we suppose that fading differs from forgetting by involving in the brain either a different chemical substance, or a different microscopic structure, or both.

2*n*) *A Connection with Physiology*

Since the present paper was first drafted, a work by Sybil Cooper and D. Denny Brown has been published.[18] Their Figure 9, showing the tension of a monkey's muscle as a function of time during and after the electrical stimulation of its cortex, has several resemblances to the course in time of the present introspections. Certain features of their observations and of mine are set out below in two columns so that corresponding features come in the same row.

---

[18]Cooper, S., & Brown, D. D. Responses to stimulation of the motor area of the cerebral cortex. *Proc. Roy. Soc.* (*B*), 1927, **102**, 222-236.

S. Cooper and D. Denny Brown	Present Paper
(1) Steady series of break shocks applied to point of motor cortex. Rhythm 12.5 per second.	(1) Steady intensity of trying-to-think. Smooth not rhythmic.
(2) Tension of corresponding muscle.	(2) Intensity of relevant auditory imagery.
(3) Summation period 1.8 seconds in Figure 9.	(3) Period of trying before image comes. Shown as 7 seconds in Figure 1; but depends on the task.
(4) "If a stimulus be reapplied within 30 seconds after a previous stimulus, the summation period is markedly shortened . . . ." (p. 226)	(4) This shortening is shown on the right of Figure 1. It was conspicuous "especially if the interval between the two trials was not more than 15 seconds."
(5) Recruitment in Figure 9. Jerky with the rhythm of the stimulus.	(5) Gradual rise at about 7 seconds in Figure 1. Smooth just as v is smooth.
(6) Suddenness of relaxation after stimulus stopped. Figure 9.	(6) Suddenness of cessation of imagery after trying-to-think stopped, shown at 10.6 seconds in Figure 1.
(7) Clonus. Figure 9. Period 0.5 seconds.	(7) Uninvited remembrance. Figure 6. Period 3 to 6 seconds.

3. Mathematical Description

3*a*) *Leading Ideas*

It is not proposed here to deduce a theory from accepted principles. Instead, a set of equations will be stated as a pure assumption; their consequences will then be deduced and shown nearly to fit the experimental facts. The search for assumptions that have this property has been a long one. It would weary the reader were I to detail: what simplicities have been found inadequate; how theories that have perished have suggested experiments that survive; what hydraulic and electric models have been imagined and then eliminated; what elaborate calculations have been found to misfit the facts. But a few of the ideas, which suggested assumptions for trial, and have not appeared in Section 2 will now be mentioned.

(1) Formulae should not be complicated without good evidence.

(2) The introspected curves, Figures 1, 5, 6, are not reminiscent of the curves of that large part of physics which is described by differential equations that are linear.[19] Thus probably we must

[19]Riemann, G. F. B. (Ed. by H. Weber.) Die partiellen Differentialgleich ungen der mathematischen Physik. Braunschweig: Vieweg, 1910.

seek non-linear equations. In the related physiology, non-linearity is necessary, for Adrian[20] records that "The effects of two stimuli presented simultaneously or in rapid succession may be quite different from the sum of effects of each stimulus presented singly."

(3) We are conscious of having forgotten, but we are not conscious of the process of forgetting. Recovery from fading is likewise an unconscious process. So besides the introspectable variables v and m it is suitable to introduce non-introspectable variables.

(4) The shortest time with which we are concerned in introspection is, say, 0.05 seconds. The more rapid changes observed in experiments on nerves must somehow be smoothed out before they can enter a psychological description. Thus Adrian[21] gives reasons for believing that the number of impulses in a second in a sensory nerve is proportional to the intensity of sensation.

(5) McDougall,[22] whose *Physiological Psychology* was the original stimulus of the present research, has assembled many reasons for believing that important non-introspectable variables are the conductances of the synapses in the cerebrum. The present research began in 1918 by an attempt to translate into equations the properties of the synapse as listed on pp. 31-32 of that book.

Garnett[23] has made wide application of the synapse theory to education.

(6) Anyone who has switched on a Nernst lamp[24] and watched its dim persistent glow rising suddenly to splendor will probably admit that this process resembles the switching of attention on to a problem, the persistence in spite of little light, and at last a sudden appropriate image.

The Nernst Lamp behaves thus because its filament, like a spark gap, conducts more readily, the more rapidly energy is being dissipated in it. It would melt, if not stabilized by a series resistance. The dim foreglow is due to a separate heater.

("The Analogy between Mental Images and Sparks" is a title of a continuation of the present research.)

---

[20]Adrian, E. D. The basis of sensation. London: Christophers; New York: Norton, 1928. See p. 117.

[21]*Ibid.*, p. 114.

[22]McDougall, W. Physiological psychology. London: Dent; New York: Macmillan, 1905. See also paper in *Brain*, 1901.

[23]Garnett, J. C. M. Education and world citizenship. Cambridge: Cambridge Univ. Press, 1921.

[24]Winkelmann, A. A. Handbuch der Physik. (3rd ed.) Leipzig: Barth. See Vol. IV, p. 674.

3b) *Formal Assumptions and Deductions*

The assumptions are equations (1) and (2)

$$M = V \frac{4S}{(S+1)^2} \qquad (1)$$

$$k \frac{dS}{dt} = \frac{(SV^n - 1)}{(SaV^n + 1)} \quad \frac{S}{(S+1)^2} \qquad (2)$$

in which t is time in seconds, V is intensity of trying-to-think, M is intensity of relevant imagery, both V and M being measured in units peculiar to the person, as will be seen; k, n, a are "brain constants" independent of V, M, S, t. The non-introspectable variable is S, and it is very like the conductance of the synapse in McDougall's theory. But if we are not concerned with physiology then nothing need be assumed about S, beyond what is explicit in equations (1) and (2). We could eliminate S by solving (1) in the form

$$S = \frac{2V}{M} - 1 \pm 2\sqrt{\frac{V^2}{M^2} - \frac{V}{M}} \qquad (3)$$

and substituting in (2). But the result is cumbrous. So the presence of S in the equations is not an extravagance, for it tends to economy of thought.

Let us now examine the consequences of these assumptions in a variety of special circumstances.

Case 1. $V = 0$, *that is to say, we are not attending at all to the words in question,* though we may be thinking about other things.

Equation (1) implies that M vanishes when V does. This is generally observed, but the uninvited remembrance of Section 2*i* is an exception; equation (1) does not describe it.

Again when $V = 0$ equation (2) becomes

$$k \frac{dS}{dt} = -\frac{S}{(S+1)^2} \qquad (5)$$

The integral of (5) is

$$\log S + 2S + \tfrac{1}{2}S^2 = -\frac{t}{k} + \text{(arbitrary constant)} \qquad (6)$$

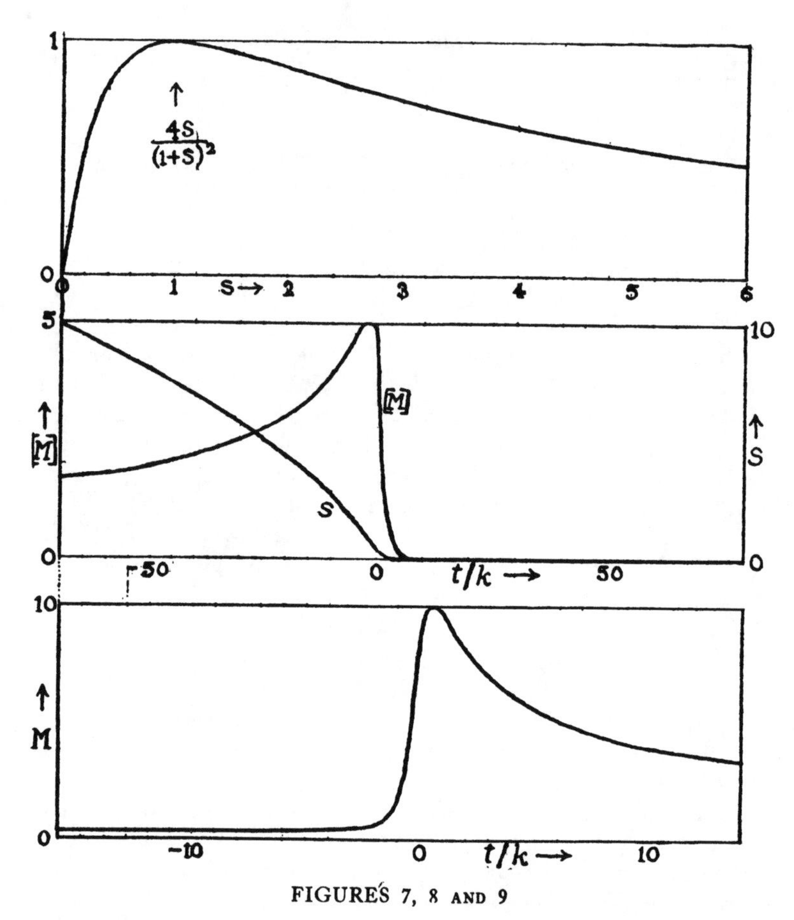

FIGURES 7, 8 AND 9

The course of S in time is shown in Figure 8. If it be supposed that at various stages of this course the observer tests progress by application of $V = 5$ for a time so brief (see the technique of 2*e*) as not seriously to introduce the terms in V in equation (2) then the observed M will be found by eliminating S between (6) and (1). This readiness for imagery is thus found by substituting $V = 5$ in (1) and so is

$$[M] = 5 \frac{4S}{(S+1)^2} \tag{7}$$

The course of $[M]$ when $V = 0$ is also shown in Figure 8, which should be compared with Figure 5. In both are seen the slow recovery from fading, the maximum, the sudden fall (forgetting), and the lingering faint remembrance represented by the asymptotic approach to zero.

Equation (6) shows that when V is constantly zero and $t \rightarrow \infty$ then $S \rightarrow 0$. That is to say $S = 0$ corresponds to complete forgetting, small S corresponds to moderate forgetting.

The early part of the $[M]$ curve in Figure 8 is concave where Figure 5D shows that it should be convex. In view of the uncertainty of the introspections, this defect is not disastrous. The time scale of Figure 5 agrees best on the whole with that of Figure 8, when k is about 0.5.

The calculated maximum is perhaps no more peaky than the introspected would be if syllables were observed individually (see $2g$).

The completer fading of Experiment 4 is beyond the left edge of Figure 8. It was estimated in that experiment that potential intensity of imagery recovered from 0.02 times to 1.0 times its maximum in about 7000 seconds. The equations (1) and (5) would make the corresponding change occur $t/k = 20{,}000$. Hence $k = 7/20$ second.

Case 2. V is independent of time and not zero.

The integral of (2) is then

$$t/k + \text{const.} = \tag{8}$$

$$S += \tfrac{1}{2}aS^2 + \frac{(1+2aV^n+a)}{V^n}S +$$

$$\left(2+aV^n+V^n+\frac{1+2aV^n+a}{V^n}\right)\frac{1}{V^n}\log_e\left(S-\frac{1}{V^n}\right)$$

$$\left.\begin{array}{l}\text{and in particular for } a = 0.2,\ n = 2,\ \text{const.} = 0\\ \text{together with intense trying represented by } V = 10\end{array}\right\} \tag{9}$$

we have

$$\frac{t}{k} = \frac{S^2}{10} + 0.412S + 1.22412 \log_e (S—0.01) —\log_e S \tag{10}$$

and M is found by eliminating S between (10) and (1). This was done numerically. As a function of time M is shown in Figure 9, which should be compared with Figure 1. In both we see the long, almost imageless suspense, the sudden rise of an image followed by slower fading. In Figure 9 the initial M is not so small as it should be. To make the time scale of Figure 9 agree with that of Figure 1, it is necessary to make $k = 0.5$ or thereabouts. This is of the same order as $k = 0.5$ or $7/20$ deduced from the recovery from fading.

Comparing (10) with (6) it is seen that the time taken to revive an image under intense trying $V = 10$ is longer, according as the previous time, during which the image had not occurred because of inattention ($V = 0$), is longer. That agrees on the average with everyday experience, although there are many exceptions.

Again when S is initially the same S_0, (6) and (1) show that the image may be developed either by intense effort in a short time, or by less effort spread over a longer time,[25] provided $S_0 V^n > 1$.

Case 3. $V = S^{-1/n}$ (11)

Then from (2) it is seen that $dS/dt = 0$ (12)

Hence by (11) and (1) it follows that $dM/dt = 0$ (13)

This corresponds to the threshold or equilibrium intensity of V described in 2*k*. According to (11) the threshold V is larger as S is smaller, that is to say, as the idea is more difficult. This agrees with the introspections on words, but not with Experiment 14 on simple vowels.

Case 4. S is in the neighborhood of unity. (14)

A graph of $4S/(S+1)^2$ as a function of S is Figure 7. It shows a maximum at $S = 1$, so that if S lies anywhere between 0.6 and 1.6 we have approximately

$$M = V \qquad (15)$$

And although by (2), S will change, yet because of the maximum, (15) may remain nearly true for a short time. This is in rough agreement with the results of Experiment 11, except that it does not show a lag of a few tenths of a second.

Case 5. The units of V and M.

Referring back to the definition of unit V in Section 2*k* we see that to verify it from equations (1) and (2) we must first chose S to be

[25] Cf. Spearman, C. E. The abilities of man. New York: Macmillan, 1927. See pp. 334-335.

such that readiness for imagery is at its maximum when S varies. Hence from (1), $S = 1$; next we must insert this value of S into the factor, the vanishing of which in (2) fixes the threshold, that is to say, into $SV^n—1 = 0$. It follows that $V = 1$, as required.

This done, the equations themselves fix the unit of M. For when $V = 1$ and $S = 1$ it follows from (1) that $M = 1$. That is to say: The intensity of imagery is said to be unity when readiness for imagery is at its maximum in the course of fading, recovery, and forgetting, and when intensity of trying is adjusted to the very feeble threshold value. An image of unit intensity is accordingly quite faint.

4. Conclusion

Equations (1) and (2) may be said to score about 13 major agreements with introspection on words, as against 4 minor misfits. For simple vowels Experiments 12 and 13 could be described by the same equations with altered constants, but Experiment 14 misfits.

So far, equations (1) and (2) have been proved to describe only the auditory images of the author. But in view of the great range of facts coordinated by McDougall's synapse theory, with which these equations are allied, there is hope that with appropriate values of the brain constants k, a, n, the equations may extend to many sorts of images and of people. In particular, they could probably be used in describing sleep and hypnotism.

The experiments and equations are restricted to times less than 10^4 seconds and indeed usually to less than 30 seconds, so that their relation to reminiscence and obliviscence over several days has not been studied.

The equations describe a mechanism which we are free to manage.

Technical College
Paisley, Scotland

L'IMAGE, LA FACULTÉ DE VOLITION ET DE DÉSIR, ET LA CONDUCTANCE CÉRÉBRALE

(Résumé)

On montre que plusieurs des changements observés dans l'intensité de l'image auditive des mots sont décrits par les équations suivantes:

$$M = V \frac{4S}{(S+1)^2} \qquad (1)$$

$$k \frac{ds}{dt} = \frac{(SV^n-1)}{(aSV^n+1)} \cdot \frac{S}{(S+1)^2} \qquad (2)$$

où t est le temps en secondes, V est l'intensité de l'effort de penser, M est l'intensité de l'image appropriée, et k, n, a, sont des "constants cérébraux" indépendants de t. Il n'est pas nécessaire de rien supposer sur S que ce qui est explicite dans les équations, car S est éliminé avant que les équations soient comparées à l'introspection. Mais S ressemble beaucoup à la synapse de la théorie de McDougall.

Pour voir la signification des équations, considérons quelques exemples, donnant les phénomènes psychologiques entre crochets. Supposons que $V = 0$ (manque d'attention) pendant que t s'accroît. Ainsi (2) montre que S décroît toujours vers zéro (l'oubli); et (1) montre que $M = 0$ (l'oubli est un processus inconscient).

Quand $S \ll 1$ (presque oublié), si V devient tout d'un coup égal à 10 et y reste (essai difficile persistant), les équations veulent dire que M reste petit pendant quelque temps (incertitude sans image) et ensuite tout d'un coup s'accroît au maximum (problème résolu), et après, M décroît toujours (devenant plus faible).

Les équations décrivent un méchanisme qu'on est libre de contrôler.

RICHARDSON

EINBILDUNG, ANTRIEB, UND GEHIRMLEITUNGSKRAFT

(Referat)

Es wird gezeigt, dass viele der beobachteten Veränderungen in der Intensität der hörbaren Wortdarstellung in folgenden Gleichungen beschrieben werden:

$$M = V \frac{4S}{(S+1)^2} \qquad (1)$$

$$k \frac{ds}{dt} = \frac{(SV^n - 1)}{(aSV^n + 1)} \cdot \frac{S}{(S+1)^2} \qquad (2)$$

wobei t di Zeit in Sekunden darstellt. V ist die Intensität des Denkversuchs, M ist die Intensität der bezüglichen Einbildung, K, n, a, "Gehirnkonstante" die unabhängig von t sind. In betreff des S braucht man nichts weiter anzunehmen als was in den Gleichungen erscheint, da S ausgeschieden wird ehe die Gleichungen mit Introspektion verglichen werden. Doch S ist der Leitungskraft der Synapse in McDougalls Theorie sehr ähnlich.

Um die Bedeutung der Gleichungen zu verstehen, betrachten wir einige Beispiele, mit Beziehung auf die psychologischen Erscheinungen in Klammern, Sagen wir $V = 0$ (Unaufmerksamkeit) während t zunimmt. Dann zeigt (2), dass S stetig bis zum Nullpunkt abnimmt (Vergessen); und (1) zeigt dass $M = 0$ (Vergessen ist ein unbewuster Vorgang).

Wenn nun, falls S 1 (fast vergessen), V plötzlich gleich wird und bleibt (anstrengendes dauerndes Versuchen), so deuten die Gleichungen an, dass M längere Zeit ganz klein ist (Vorstellungslose Spannung) und dass M dann plötzlich bis zum Maximum emporsteigt (gelöstes Problem), und dass darauf M stetig abnimmt (verschwindet).

Die Gleichungen beschreiben einen Mechanismus den wir willkürlich beherrschen.

RICHARDSON

ОБРАЗЫ, ПОБУЖДЕНИЯ и ЦЕРЕБРАЛЬНАЯ ПРОВОДИМОСТЬ.

(Реферат).

Многие из наблюдавшихся изменений в интенсивности слуховых словесных образов могут быть описаны следующими уравнениями:

$$M = V\frac{4S}{(S+1)^2} \quad \ldots \ldots \ldots \ldots \ldots \quad (1)$$

$$k\frac{dS}{dt} = \frac{(SV^n-1)}{(SV^n+1)} \cdot \frac{S}{(S+1)^2} \quad \ldots \ldots \ldots \quad (2)$$

в которых t означает время в секундах, V—интенсивность интеллектуального усилия, M—интенсивность образа, а k, n—мозговые константы, независимые от t. О величине S можно сказать лишь то, что выявлено в формуле, так как S элиминируется до проверки уравнения самонаблюдением. Однако, S очень близка к проводимости синапсов в теории Мак-Дауголла.

Для того, чтобы разобрать формулу, мы приведем несколько примеров ее приложения, указывая в скобках соответствующие психологические явления. Если $V=0$ (отсутствие внимания) при растущем t, то (2) указывает, что S постепенно доходит до нуля (забывание), а (1) показывает, что $M=0$ (забывание—бессознательный процесс).

Когда $S << 1$ (близко к забытому), V внезапно делается равным 10 и остается таким (перманентные настойчивые пробы), уравнение указывает на незначительность M для некоторого промежутка времени (состояние, лишенное образов) и что M внезапно доходит до максимума (решение задачи) и уже потом уменьшается (угасание).

Уравнение описывает здесь некоторый произвольный механизм.

Ричардсон (Richardson).

Richardson, L. F. (1929)
Br. J. Psychol. **20**, 27–37

1929:4

QUANTITATIVE MENTAL ESTIMATES OF LIGHT AND COLOUR

The British Journal of Psychology – General Section
Cambridge
Vol. XX, Part 1, July 1929, pp. 27–37
(Manuscript received 16 August 1928)

QUANTITATIVE MENTAL ESTIMATES OF LIGHT AND COLOUR

BY LEWIS F. RICHARDSON.

1. INTRODUCTION.

PSYCHOLOGY will never be an exact science unless psychic intensities can be measured. Some authorities say that such measurement is impossible*. Many estimates of the intensity of auditory imagery have been made by the present author†, but some psychologists were incredulous. A research was therefore pursued, in collaboration with J. S. Rose, M.A., on Loudness and Telephone current‡, and is found to show that the estimates were significant, the mean uncertainty of an individual estimate being the ratio 1 : 1·6. In the present paper the enquiry is extended to visual sensations, the question being: How do people agree or differ in their estimates of visual sensations? A further paper by R. S. Maxwell, M.A., deals with the functional relation between stimulus and red sensation. Guesses at mid-tints have been studied statistically by Karl Pearson§ some thirty years ago.

* W. James, *Principles*, 1901, I, 546; also C. S. Myers, *Textbook of Experimental Psychology*, 1925, I, 252. But contrast Brown and Thomson, *Mental Measurement*, Ch. I.

† L. F. Richardson, "Conation, Imagery and Cerebral Conductance," *Journal of General Psychology* (Clark University Press, U.S.A.), April 1929.

‡ L. F. Richardson and J. S. Ross, "Loudness and Telephone Current," accepted by aforesaid journal.

§ Karl Pearson, "Contributions to the Mathematical Theory of Evolution," *Phil. Trans.* A, CLXXXVI, 392.

2. Numbering browns.

1927, Oct. 11. On a rectangle of brown paper, 54 × 39 cm., were mounted four paper squares each 14 × 14 cm. and severally white, buff, brown and black. The whole was set on a drawing-board and illuminated by gas-filled tungsten lamps, shining nearly normally on it. It was arranged to be viewed from a distance of about 140 cm., and rather from one side.

Thirty-six men, studying for the Intermediate B.Sc., at Westminster Training College, were given, during an examination in practical physics in which each student did in turn the same 36 tasks of four minutes each, the following written instruction:

"Stand in the chalk ring and look at the drawing-board. On a scale of lightness and darkness, call the black square 0, the white 100. What numbers do you then give to (*a*) the background, (*b*) the brown square, (*c*) the paler square?" The time allowed to each man was four minutes. They were kept in ignorance of each other's opinions till all had been recorded. As far as I could ascertain, no one of them had had any previous training or practice in the assignment of numbers to tones. No one complained that the task was impossible; but one man put down sizes by mistake.

The paper was then covered to protect it from dirt and light.

1927, Oct. 15. The same men being seated in a class-room, I said to them that about half the class had been of one opinion as to the order of the tones and about half of the other; and that I should like to have their considered judgment. The board was mounted in view of all and the room illuminated by gas-filled tungsten lamps, daylight being excluded. On the blackboard I wrote *o* black, *a* background, *b* brown square, *c* paler square. I also told them to look at the contrasts.

At this second attempt they all gave the same sequence except one man, and when questioned he immediately conformed.

The results are set out below.

		a		*b*		*c*
Oct. 11.	Means for the 21 students who maintained the same sequence on Oct. 15 ...	34·5		52·9		86·8
Oct. 15.	For all 36 students:					
	Means	29·3		50·4		83·8
	Least	15		30		60
	Greatest	55		70		96
	Standard deviation from mean	12		12		11
			a, b		*b, c*	
	Correlation coefficients (product moment)		0·7		0·6	
	Probable error of correlation		0·06		0·07	

3. LOCATING PINK ON A MENTAL SCALE OF REDNESS.

(1) *Preparations and methods.*

This enquiry was provoked by the remark of William James: "To introspection, our feeling of pink is surely not a portion of our feeling of scarlet*."

The *white* used was that of a pure white cardboard sold for mounting pictures. The *scarlet* was Messrs Winsor and Newton's 'scarlet poster card.' It nearly matches the scarlet variety of mercuric iodide.

The pink was provided in different ways in the apparatuses known as 'Triangles' and 'Wheel.'

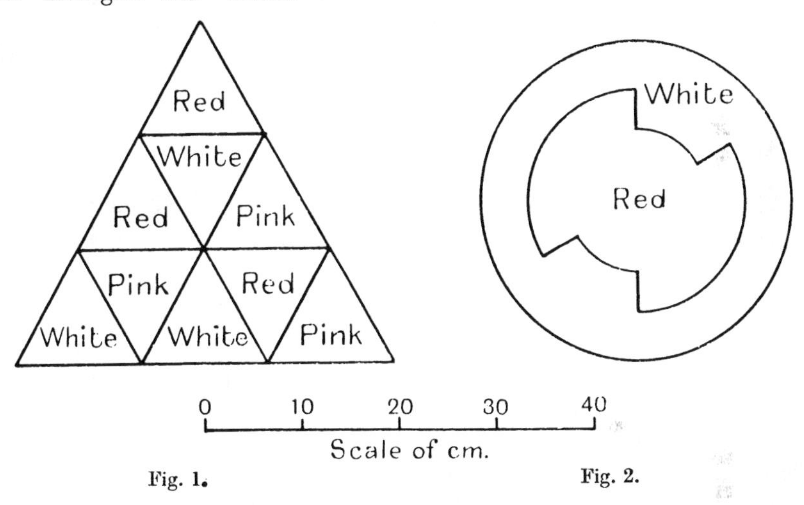

Fig. 1. Fig. 2.

Triangles. Pale red drawing paper, sold by the Old Water Colour Society, was cut into equilateral triangles 13 cm. in the side, and was mounted with equal scarlet triangles on the white card in the pattern shown in Fig. 1. This pattern has three advantages: (i) the white, pink and scarlet occupy equal areas; (ii) each is in contact with the other two; (iii) such contacts are triplicated so as to help the observer to eliminate any slight variations of illumination across the pattern.

In a separate experiment the pink paper was compared with mixtures of the white and the scarlet on a colour wheel. It matched for darkness when the angle of red was 210, 217, 210, 245†, 245† by electric light, 205 by daylight, the rest of the 360° being white. When thus matched

* W. James, *Principles of Psychology*, 1901, I, 546.

† 245 by a poster artist, Mr Clifford Gabriel; other estimates by author. We agreed about saturation and yellowness.

for darkness the pink paper matched also for saturation but was decidedly too yellow.

(Note added 1928, Dec. 21.) The reflectivities of the white, pink and red corners of the set of triangles shown in Fig. 1 have now been measured on a photometer for six ranges of wave-length. As a standard white, a slab of plaster of Paris was prepared. It was cast on plate-glass from powder sold by the British Drug Houses, Ltd.; the dark spots were cut out and refilled; and the polish, visible at nearly grazing incidence, was removed by fine emery cloth. The plaster was dried before use. This white surface, *A* in Fig. 3, was set up at one end of a photometer-bench and was illuminated by light diverging from *C*, a hole 0·2 cm. diameter, on which light from a white hot tungsten ball was concentrated by lenses. The cone of diverging light was so narrow that scarcely any of it fell

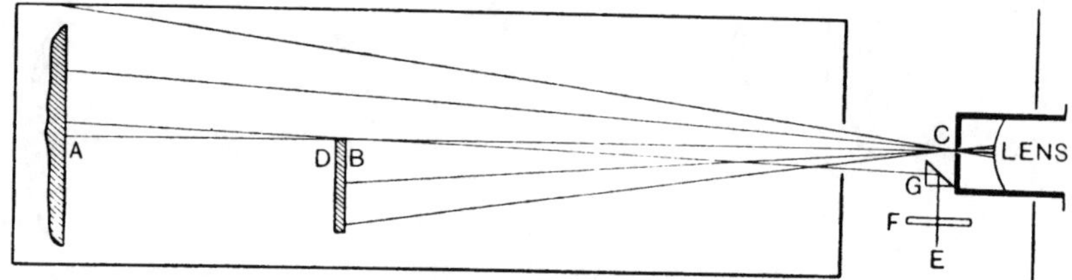

Fig. 3. Diagram of Photometer.

on the black materials which enclosed the apparatus. The triangle for test was placed at *B* on a stiff mount. The back *D* of the mount was covered with black paper; so were also the five triangles not in use. The triangle and the plaster were viewed by an eye *E*, through a colour filter *F*, from near the hole *C*. To get the viewpoint near enough to *C* it was necessary to use the reflecting prism *G*, when *B* was near *C*. The distance *AB* was varied until *B* matched *A*. The reflectivity of *B* relative to that of *A* was then assumed to be $(BC/AC)^2$ in which *BC* and *AC* are distances. The range of wave-lengths transmitted by the filter was measured separately by a spectrometer.

The results were as follows:

Colour	Range of wave-lengths in microns	Reflectivities if plaster has unit reflectivity: White mounting board x	Pink paper O.W.S. y	Scarlet poster card W. & N. z	$\frac{210z+150x}{360}$
Blue	0·42 to 0·48	0·81	0·48	0·11	0·40
Green	0·47 to 0·54	0·83	0·46	0·08	0·39
Green	0·51 to 0·58	0·84	0·50	0·09	0·40
Orange	0·59 to 0·60	0·86	0·59	0·32	0·55
Red	0·61 to 0·67	0·87	0·79	0·72	0·78
Red	$0{\cdot}65_5$ to $0{\cdot}69_5$	0·91	0·77	0·76	0·82

The uncertainty of the measured reflectivity is about $\pm$ 0·02. In connection with the colour-wheel estimates, the function $(210z + 150x)/360$ is given in the last column; it should be compared with y. The excess of orange and green in y is obvious.

Wheel. The pink is here made by rapid alternations of the scarlet and the white on a disc driven by an electric motor at a speed sufficient to prevent any flicker. The white and scarlet were nominally the same as those used for the triangles. Fig. 2 shows the disc. In the middle zone, the angle occupied by scarlet was $240° \pm 0{\cdot}5°$; that is $\frac{2}{3}$ of the whole. The advantages will be discussed later.

Procedure and instructions. Those about to observe the wheel were carefully kept in ignorance of the angles of red and white, and were allowed to see the wheel when it was rotating only.

Whereas for the browns, the observers were asked to assign numerals, for the pinks they were asked to divide a segment of a line 100 mm. long. This plan was suggested to me by Mr Ralph D. Richardson to make judgment easier, and I believe it does so. The typewritten instructions were as follows:

Comparing colour intervals.

Please look at the white, pink and scarlet (triangles). In degree of redness is the pink nearer to the white or to the scarlet?

Below is a line marked 'white' at one end and 'scarlet' at the other. Please put a mark on this line to show where you feel that the pink comes on this scale of redness.

Sign here　　　　　　　　　　　　　　Date

Daylight } cross out one.
Electric }

If the observer showed difficulty in understanding the typed instructions, a verbal explanation was sometimes added to the following effect: "If the pink looks almost the same as the white, put a mark here (pointing to end of line labelled white). If the pink looks almost the same as the red, put a mark here (pointing to end of line labelled scarlet). If the pink is equally like the red and the white, put a mark in the middle of the line. But it is your own opinion that I want. As far as I know, there is no 'right answer.' Everyone's opinion is equally good."

1929:4 *Br. J. Psychol.* **20**, 27–37

If the observer put the mark very far away from the rest of the group I sometimes discussed the matter further and was surprised to find that these extreme estimates were generally not due to any misunderstanding of instructions. One paper was self-contradictory and was rejected.

Conversion to numbers. Throughout the scale is

White = 0 Scarlet = 100

The 'estimate' for the pale red is defined to be the percentage of the length of the line from the end marked white up to the mark. The lengths were measured by me, using a scale engraved on glass. Throughout the arithmetic, numbers are rounded off to the nearest unit percentage, or if exactly midway, then to neighbouring integer which is even.

(2) *Tabulated estimates.*

Table I. *Triangles.*

	Men									Women			
Group reference ...	1	2	3	4	5	6	7	8	9	10	11	12	13
Daylight or electric ...	*d*	*d*	*e*	*d*	*e*	*d*	*d, e*	*e*	*d, e*	*e*	*d*	*e*	*d*
No. who tried and failed	0	0	0	1	0	0	0	2	0	1	1	0	0
No. who made estimates	33	15	25	8	9	8	4	9	11	25	10	6	6
Estimates: Greatest ...	46	40	76	59	67	33	21	51	39	60	76	52	35
Estimates: Median ...	36	33	37	34	34	26	18	29	32	36	26	34	28
Estimates: Mean ...	36	30	40	34	37	26	17	30	30	37	36	36	26
Estimates: Least ...	20	12	25	19	13	18	12	8	17	14	20	25	14
Standard deviation of individual from group-mean	6	7	11	11	17	5	3	10	6	11	20	9	7

Group index:

Men
1. Class for Intermediate B.Sc.
2. Class for Diploma in Education
3. Class for drawing (Board of Education)
4. Teaching Staff.

(1–4: Westminster Training College, London.)

5. Quakers. Social gathering. London, N.W. Ages 20 to 70.
6. Art students. Roy. Coll. Art. Teachers' Course, Post Diploma.
7. Other teachers and inspector of drawing and painting.
8. Fellows of the Royal Society.
9. Unclassified; mostly university graduates.

Women
10. Quakers, the same as 5, but women.
11. Art students, the same as 6, but women.
12. Unclassified, mostly relatives of group 9.

13. Class in Advanced Psychology Univ. Coll. London. Mixed sexes.

Supplementary Index for Wheel:

Men 14. Class for Intermediate B.A., Westminster Training Coll., London.
Women 15. Wesleyan Missionary Society's Garden Party, London.
Men 16. Unclassified men.

Table II. *Wheel.*

	Men								Women	
Group reference	1	14	2	4	6	7	8	16	11	15
No. who tried and failed	0	0	0	0	0	0	1	0	1	0
No. who made estimates	29	16	18	6	8	2	12	20	10	36
Estimates: Greatest	58	62	83	63	74	59	78	65	89	86
Estimates: Median	40	40	40	48	52	.	35	42	66	44
Estimates: Mean	39	41	46	46	54	.	44	43	59	47
Estimates: Least	26	29	29	24	37	37	24	27	24	15
Standard deviation of individual from mean of group	7	9	15	14	11	.	15	10	20	16

Note. The same individual is counted once only for the triangles and once for the wheel, even if, logically, he belongs in several groups.

For those who are interested in frequency-distributions the following summary is collected in ranges of 5 units.

Table III. *Frequencies.*

Centre of range of estimate	Triangles excluding group 13: Men	Triangles excluding group 13: Women	Wheel: Men	Wheel: Women
5	0	0	0	0
10	3	0	0	0
15	2	2	0	1
20	11	4	0	0
25	13	5	4	2
30	24	4	17	3
35	34	10	19	9
40	21	7	29	4
45	6	1	11	4
50	2	5	5	6
55	1	0	8	1
60	2	1	9	1
65	2	0	4	5
70	0	0	2	5
75	1	2	1	3
80	0	0	1	0
85	0	0	1	1
90	0	0	0	1
95	0	0	0	0
Total persons	122	41	111	46
Mean estimate	34	36	43	50
Standard deviation	10	13	12	17

(3) *Causes of difficulty and of variation.*

Observers have mentioned many causes. Although I have not thoroughly explored any variations, the following notes may be useful to future investigators.

(1) *Obliquity of view.* Care was taken to choose papers equally matt, so that the pink seemed to me in about the same relation to white and scarlet from whatever angle the card was viewed.

(2) *The change from daylight to electric,* might be expected to affect

the triangles. The means are: daylight 36, 30, 34, 26, 36, mean 32·5; electric 40, 37, 30, 37, 36, mean 36, and the difference is scarcely significant.

With the colour wheel any such effect is probably less.

(3) *Confusion between darkness and saturation.* Several subjects were perplexed about this for the triangles. Some observers stated that they estimated by darkness, but I have not rejected their estimates on that account. With the wheel the estimate would probably be much the same whether judged by darkness or saturation.

(4) *Pink not of same hue as scarlet.* The *Concise Oxford Dictionary* describes pink as a "pale red slightly inclining to purple," and scarlet as a "brilliant red colour inclining to orange." Thinking that these contrary inclinations might perhaps have been all or part of the difficulty that James felt, I tried, in choosing the pale triangles, to make them of the same hue as the red. But evidently I overshot the mark, for many subjects objected, stating that the 'pink' was apricot, buff, brownish, flesh coloured, terracotta, too yellow, too opaque, too grey. Indeed two observers in Group 8, who said that the task was impossible, gave this difference of hue as their reason.

The colour wheel was made to avoid this difficulty and at the same time to give much the same result whether the observer judged by saturation, as the instructions indicate, or by darkness. Yet even when the pink was made by mixing the white with the scarlet on a colour wheel an Art Student remarked "pink is slightly purple—probably slightly off the true scale" and six others in the same group made similar if less explicit comment. One other subject said that the pink was too blue.

These experiences suggest that people vary a good deal in their sensitivity to light waves of lengths between 0·6 and 0·7 micron.*

(5) *Half closing the eyes.* Two observers stated that this moved the pink nearer to the white.

(6) *Distance.* Two skilled observers remarked that the estimate increased with distance, roughly thus:

	1 m.	2 m.	6 m.
Triangles, electric light ...	15	—	21
Wheel	25	36	49

But on sorting the estimates of Group 1 on triangles any effect of distance was lost in the general scatter. The distances used were between 1 and 5 metres.

* As is well known. See Sir J. Herbert Parsons, *An Introduction to the Study of Colour Vision* (Cambridge University Press), 2nd ed. p. 193.

(7) *Bias by pigments.* One observer, who gave the lowest estimate in Group 5, remarked that he was a dyer and was influenced by a knowledge of the strength of dye required to make the paper. Accordingly I asked the Art students to put from them all thought of quantities of pigment and to judge the colours simply mentally. But, having tried, about half of them protested that they could not thus put aside their experience.

(8) *Practice.* Groups 6 and 7 have of course had most practice, but also most opportunity for bias by pigments. No one remarked on having tried this sort of task before; and I provided no practice beyond the repetitions here recorded. That is to say, the group that made most repetitions was No. 1, the Intermediate Science class, and their successive standard deviations were:

	1927 Oct. 11 Browns	1927 Oct. 15 Browns	1928 June 11 Pink wheel	1928 July 12 Pink triangles
		12	—	—
	Confused	12	$6{\cdot}_6$	$6{\cdot}_2$
		11	—	—

Much practice would be a desirable preliminary to future researches.

(9) *Quick changes.* One observer of the triangles marked his first impression as 8, changing during about three minutes to 26.

Another estimated the triangles at 29 by first impression and added that "reasoning would tend to bring the mark half-way—if not nearer to scarlet."

Another said of the wheel that the pink "appeared to become paler the longer the colours were viewed."

Another agreed with this last statement.

(10) *Abnormalities of colour vision.* That eyes may differ is shown by an educationalist, aged 77, who viewed the triangles by electric light and reported that by right eye he saw scarlet and gamboge; estimate 27. By left eye, which is partly blind to red, he saw brick red and lemon yellow; estimate 13. Other estimates by men who said that their vision was peculiar were:

Abnormality	Triangles daylight	Triangles electric	Wheel
Unknown type ...	—	26	—
Unknown type ...	38	—	31
Blind for red ...	—	34	—
Weak for red ...	—	—	24
Weak for green ...	—	—	30

(11) *Sex.* The means for men and women are shown separately at the foot of Table III. On the wheel women see the pink redder than do the

men and the difference is perhaps significant, for the means are $43 \pm 0{\cdot}8$, $50 \pm 1{\cdot}7$ where the numbers following the $\pm$ sign are 'probable errors.'

(12) *Age.* The observers mentioned in Tables I, II, III were all over 17. A few children have also looked at the triangles and at two other wheels. Below 12 years, estimates were sometimes very high, as if erratic.

(13) *Persistence of individual differences.* This was sought for by plotting correlation diagrams in which the co-ordinates of a point are the estimates of the Triangles and of the Wheel made by the same person. This was done for Groups 1, 2, 6, 11. But there was no significant correlation. We might expect to find persistent individual differences if the observers had much more practice. For the browns there was the significant correlation stated in § 2.

(14) *Reasoning.* Mr Dudley Heath made the interesting suggestion that those art students who decline to make a decision in this and similar problems are often those who are also argumentative; and that they could place the mark on the line if they would trust to their first impressions, instead of to reasoning.

4. Reply to an objection.

The quotation from James continues "...nor does the light of an electric arc seem to contain that of a tallow candle in itself." Even if we could abstract brightness as a quality of the light, it would be difficult to compare the brightnesses, because the arc is dazzling.

Similarly it is difficult to compare the loudness of a pistol-crack with that of a cannon-roar[1] because the roar is deafening.

Here is another specimen. It is very difficult to say how many times less painful it is to be pricked by a pin than to have a tooth extracted without anaesthetic; because the dental pain is overwhelming.

But these are no arguments against the estimation of less violent sensations.

James in this passage was perhaps more concerned to emphasize the words *portion* and *contain*. That is another question. There is no evidence that any of the present observers counted up little bits of pinkness, and no objection to James's remarks about addition.

5. Acknowledgements.

My best thanks are due to the many observers for their careful opinions; and also especially to those who helped to organize the work,

[1] C. S. Myers, *Textbook of Experimental Psychology* (1925), Part I, 252.

namely Dr H. H. Dale, Sec. R.S., Prof. Flügel and Mr Raper of University College London, Mr Wellington and Mr Dudley Heath of the Royal College of Art, and to Principal Workman, Mr Patten and Mr J. S. Ross of Westminster Training College.

The arithmetic was performed on a machine lent by the Government Grant Committee of the Royal Society.

6. SUMMARY.

(1) Of 164 persons who were asked to estimate quantitatively the redness of pink paper, only five said that the task was impossible. With a pink made on a colour wheel only two out of 159 found the task impossible.

(2) The pink, made by mixing $\frac{2}{3}$ of scarlet with $\frac{1}{3}$ of white on a colour-wheel, was judged mentally to be at the following positions on a scale in which 0 is white, 100 is scarlet. Mean estimate by men 43, by women 50. There is thus a remarkable disagreement between angles and mental estimates. Mr R. S. Maxwell, M.A., is exploring this further.

(3) The standard deviation of an individual estimate from the mean for the same sex is about 13 of these units. Many possible causes of this variability have been discussed.

(*Manuscript received* 16 *August*, 1928.)

Richardson, L. F. and Ross, J. S. (1930)
J. gen. Psychol. **3**, 288–306

1930:2

LOUDNESS AND TELEPHONE CURRENT

The Journal of General Psychology
Worcester, Massachusetts
Vol. III, No. 2, April 1930, pp. 288–306

(Received for publication 5 October 1928)

Offprinted from *The Journal of General Psychology,* 1930, 3, 288-306.

LOUDNESS AND TELEPHONE CURRENT*

From the Departments of Physics and Education of Westminster Training College, London

L. F. RICHARDSON AND J. S. ROSS

INTRODUCTION

William James described Psychology as "the hope of a science." There would be more hope if psychic intensities could be measured.

The custom nowadays is to measure accurately stimuli and movements, but to leave psychic intensities quite unmeasured. This seems to us ill-balanced.

A recent important book by Dr. Harvey Fletcher (4*a*) describes a long series of thorough investigations into hearing made by the Bell Telephone Laboratories. Dr. Fletcher regularly uses the phrase "sensation-level" to denote a scale, the units of which are occasionally (Figure 71) called "sensation units" but more often "decibels." The decibel is defined (p. 69) so that when two acoustic waves have powers of J and J_0 microwatts, the former is to exceed the latter by $10 \log_{10} J/J_0$ decibels. From our point of view, the decibel is a special measure of stimulus, and not a unit of sensation; so that what Dr. Fletcher calls "sensation level" is not a scale of sensation, except perhaps at its zero point, which is defined to be the threshold. Fletcher (p. 226) further defines the "loudness" of any pure tone to be the number of decibels above the threshold, of a comparison tone of 1000 vibrations per second, which sounds equally loud. This is a mixed procedure, equality being judged psychologically, decibels measured physically. As such, Fletcher's scale of "loudness" is entirely different from the purely introspective scale used by our observers.[1]

A former research by one of us (8) depended chiefly on numerical mental estimates of the loudness of auditory imagery. Two eminent psychologists who read that work in draft found it difficult to believe that those numbers could have any genuine significance. In-

*Received for publication by Carl Murchison of the Editorial Board, October 5, 1928.

[1]Note added in January, 1930.

deed C. S. Myers (7, p. 252) had stated that sensations could not be measured. We have set out to explore this question by asking people to listen to telephones supplied in sequence with various measured electric currents or voltages.

The experimenter, in one room, regulated the current, signalled its onset, and measured its strength. The observer, in another quiet room, wore the headphones and wrote down his estimates of loudness. A pleasant loudness was first chosen as standard and called 1.00, and the corresponding standard current was applied before and after each unknown. The signals, made by a small electric lamp, were one flash for "standard begins"; two flashes for "unknown begins"; three flashes for "rest begins." The observer was kept in ignorance of the relative strengths of successive stimuli until the end of the last of his estimates as here published.

There is reason to believe (3, Figure 20) that, when the frequency is kept fixed, the current in a telephone is proportional to the voltage between its terminals and to the amplitude of the aerial wave produced; so the sound-energy per volume of air, being proportional to the square of the amplitude, is proportional to the square of the current and to the square of the voltage. The telephones chosen, Messrs. S. G. Brown's F type, had no adjustment; the sensitivity was fixed.

Current Generator

This is shown in Figure 1. It is of a well-known type (11, Figure 83). The particular valves, coils, and voltages were altered several times during the course of the experiments, but we always used 5 to 17 kilograms of copper in the tuning-coil τ_1, τ_2 and avoided iron except for a few woodscrews, the object being to avoid harmonics. Again the coils υ and σ were never closely wrapped round τ_1, τ_2, but only placed end to end. Accordingly the coupling was loose, and the frequency nearly independent both of the high tension and of the power output.

The condensor ι (Telegraph Condensor Co. 14, 1797, 177*) allowed the capacitance to be adjusted by steps of about 0.02 microfarad from 3.5 microfarads downwards.

Frequency Measurements

In the earlier experiments (March, April, May, and June) the telephone ω was clamped just above the water-tube resonator, and

TABLE 1

PARTICULARS OF GENERATOR

Ref. to Figure 1	Inductance henries Self	Mutual	Resistance for D. C. ohms	Ref. to Figure 1	Triodes
March					
υ	0.10		42.		Benjamin short-
				ρ	path redspot.
π_1, π_2	0.30		21.		200-volt light-
				ϕ	ing mains.
σ	12.		2.3 x 10^3		
June and July					
			21.		
υ	0.30	0.02			
π_1, π_2	0.18		5.	ρ	Mullard P. M., 254.
		0.09			
σ	12.		2.3 x 10^3	ϕ	Batteries for this only.

the quarter wave-length observed, allowing for the end-correction. The corresponding frequency was found from Blaikley's (1, p. 549) value for the speeds of sound in dry air in tubes.

As the air was not dry, these values were checked in July by simultaneous measurements with a stroboscope ψ illuminated by a flickering neon lamp χ. The high tension battery ϕ was made just

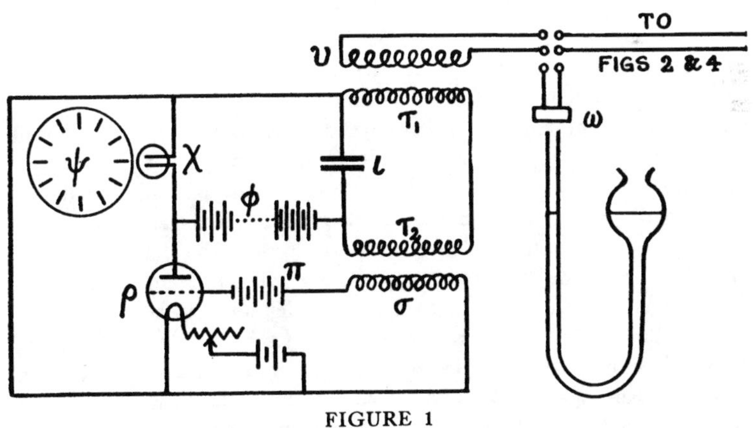

FIGURE 1

GENERATOR AND FREQUENCY METERS

insufficient to flash the lamp when the filament of the triode was cold. When this filament was heated, the oscillatory voltage of the coils, τ_1, τ_2, caused the neon lamp to flicker. The stroboscope had 100 and 360 holes and was turned by a gramophone motor (H. M. V.). The frequencies obtained in this way were self-consistent to about 1 in 500 and showed that the water-resonator values required no revision. Nevertheless in July we measured all frequencies by the stroboscope.

As a further check, the square of the reciprocal of the frequency was plotted against the capacitance, which had been measured by Mr. R. S. Maxwell, using a ballistic galvanometer, and by Messrs. J. R. Slater and C. V. Stephens, using a Carey Foster null method. The points were found to lie on a straight line, as theory indicates.

Harmonics. With loud sounds in the telephone the octave was heard faintly. Its presence was confirmed by the resonance tube; but the stroboscope did not show it. Very likely the octave is absent in the tuning coils, τ_1, τ_2, but is produced in the telephone as theory leads one to expect (6, p. 845). No other harmonic was noticed.

Test of Validity of Estimates in the Research on Conation, Imagery, and Cerebral Conductance

The Scale of Loudness. The scale of loudness was simply that which seemed intuitively natural to the observer (L. F. Richardson). As far as he could remember, it was the same as the scale which he had used during the preceding 10 months in estimating the loudness of inner imagery, as described in the paper aforesaid. The scale must have involved experience of measuring and numbering many other sorts of physical quantities. But as far as he is aware, it did not depend on any experience of physically measured sound intensities. For this particular experience happened to be scanty and amounted only to (*a*) a 2-hour experiment on noise made by rubber balls falling from measured heights (5 years ago) and (*b*) about 10 observations in November, 1926, on the telephone currents that produced sounds just audible, pleasant, and too loud. The observer took care never to listen to telephone currents changing in ratios known to him, during the period July 1, 1927, to June 15, 1928. The estimates of the loudness of inner imagery had been posted to America on May 2, 1928, so as to be quite free from sophistication by any knowledge of telephone currents.

Regulating and Measuring the Telephone Current. The arrangements are shown in Figure 2. When the observer was listening, the terminals at η were joined 2 to 3 and 1 to 4, leaving 5 and 6 disconnected.

a) The alternating electromotive force was applied to the telephone circuit by means of a variable mutual inductance ξ_1, ξ_2. The latter was of the Campbell type made by Messrs. R. W. Paul and believed to be subdivided to an accuracy of 1 in 1000 or better. The resistances and inductances of all parts of the circuit were measured. The well-known theory of the transformer (10, p. 274) then showed how the current depended on the mutual inductance. There are altogether three mutual inductances. But each of them is small relative to the geometric mean of the self-inductances of the same pair of coils. That is to say the couplings were all loose. Accordingly the telephone current would have been simply proportional to the variable mutual inductance were it not for the fact that the resistance and self-inductance of the telephone circuit varied automatically as more turns of wire were switched into ξ_2. A rather long calculation showed that

$$\text{(current) varies as } M\ (1—bM)$$

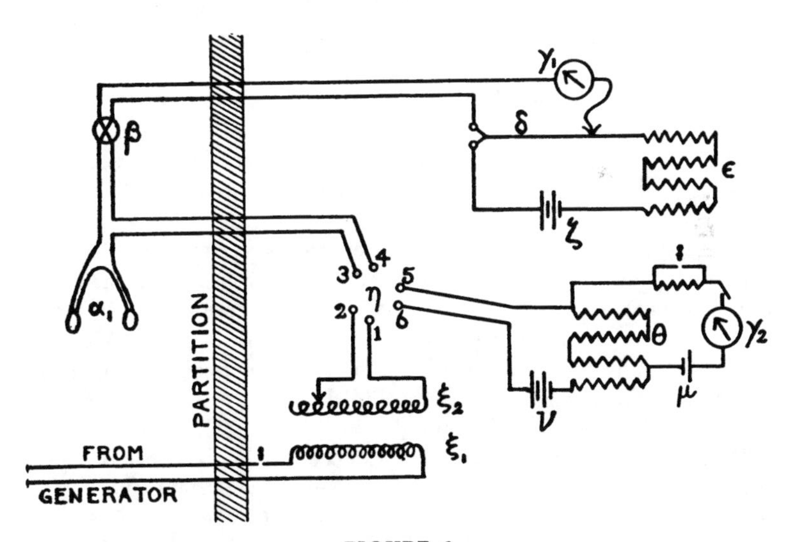

FIGURE 2
Current Regulation and Measurement for 6-Ohm Headphone

where M is the mutual inductance in henries and b is a constant estimated to be roughly 15 henry^{-1} in May and June.

b) *Absolute measurements.* About half of the telephone currents at each sitting were also measured absolutely by the vacuo-junction β in Figure 2. This was the most sensitive type sold by the Cambridge Instrument Co. and the headphones (Messrs. S. G. Brown's F type, 6 ohms) were selected of low impedance so as to require a current large enough to be thus measured. After each sitting the connections at η were changed by disconnecting 1 and 2 and joining 3 to 6 and 4 to 5. The voltage of the accumulator ν was measured in terms of the standard Weston cell μ and the accumulator ν was then used with the resistance box θ, having N.P.L. corrections, to supply known direct currents to the vacuo-junction taking account of the resistance of the heater and headphones. The electromotive force from the vacuo-junction was measured in March simply by a galvanometer, the scale of which was calibrated, as just described, at 21 points. But in May and June we preferred to balance the E.M.F. on the manganin slide wire δ which formed part of the potentiometer γ_1, δ, ϵ, ζ. The potentiometer was standardized after each sitting by the ν, θ, μ, γ_2 circuit. The "virtual" alternating current was taken to be that direct current which on the mean of the two senses of flow produced the same E.M.F. in the vacuo-junction as did the alternating current.

c) *Comparison of the two methods of measuring current.* The mutual inductance was very easily adjusted and read. It had a range of 1 to 10000 microhenries, which was ample. It was read on every occasion when the listener made an estimate, and is accurate.

The vacuum junction was wearisomely slow in arriving at a definite state; and occasionally it appeared afterwards that sufficient time had not been allowed. The galvanometer-zero drifted a little.

The fainter sounds corresponded to galvanometer-deflections that were too small to be measured. The vacuo-junction was used at every mental estimate of loudness in March and at about half of those at each sitting in May and June.

What we want mainly is the ratio between a pair of telephone currents occurring within about 15 seconds of one another and, as the most accurate way of obtaining this ratio, we fitted to the collection of absolute measurements by the vacuo-junction the formula

$$j = km(1 - bm)$$

in which j is the "virtual" current in milliamperes. m is the mutual inductance in millihenries, and k and b are independent of m. Then if subscript 1 denotes the standard, the required ratio is

$$\frac{j}{j_1} = \frac{m(1 - bm)}{m_1(1 - bm_1)}$$

Or expanding by the logarithmic series

$$\log_{10}\frac{j}{j_1} = \log_{10}\frac{m}{m_1} - 0.434\left\{b(m - m_1) + \frac{b^2}{2}(m^2 - m_1^2)\ \ldots\right\}$$

The observed values of b were

March 26—$b=0.023$
March 27—$b=0.037$
June 7—$b=0.012$
June 15—$b=0.018$ } We took b as 0.015 for all May and June.

In this way we consider j/j_1 is given with an accuracy of 2% or better whenever $j/j_1 < 1$ and an uncertainty not more than 10% at the worst. This is less than the uncertainty of the mental estimates.

Routine of Tests. The experimenter applied the standard currents before and after each test current. The tests were in a random sequence quite unknown to the observer. The observer made his estimate in the first 5 seconds after the unknown began, occasionally revising it on rehearing the standard. Telephones, if removed, were placed in the same position by feeling the lobes of the ears. On March 26 the observer could hear clicks when the inductance dials were turned. On all subsequent days, to prevent this clue, a switch was opened during the adjustment of the inductance.

Results. The best summary is Figure 3, but further particulars are necessary. For in the research on "Conation, Imagery and Cerebral Conductance" (8) the unit of loudness was taken to be the loudness of L. F. Richardson's ordinary inner speech, and faded imagery was recorded chiefly in the range 1.0 to 0.1 of that unit, excluding estimates of "zero." To fit this range on to the scale of telephone current he made comparisons in the course of the routine tests. These results are shown in the last three columns of Table 2.

From the mean values at the foot of each column it is seen that unfaded inner speech corresponded to $0.16/0.57=0.28$ of our standard current. The corresponding vertical line on Figure 3 is located on the abscissa at $\log_{10} 0.28 = \bar{1}.45$. This vertical corresponds

TABLE 2

Sitting		Number of estimates	Frequency per second	Virtual milliameters in S. G. Brown's F type 6 ohm headphones				
				Standard	Just inaudible	Phone relative to inner-speech: Rather fainter	About the same	Rather louder
March	26	23	357	0.59		0.10	0.08	0.20, 0.16, 0.24
	27	20	352	0.62		0.03	0.11	
May	31	12	354	0.77	0.010	0.06	0.32	
June	2	12	355	0.49		0.04		0.49
	7	12	355	0.52	0.003		0.23	
	9	12	356	0.51	0.006		0.09	
	14	4	350	0.52			0.13	0.51
	15	12	355	0.52		0.11		
		Means	354	0.57	0.006	0.07	0.16	0.32

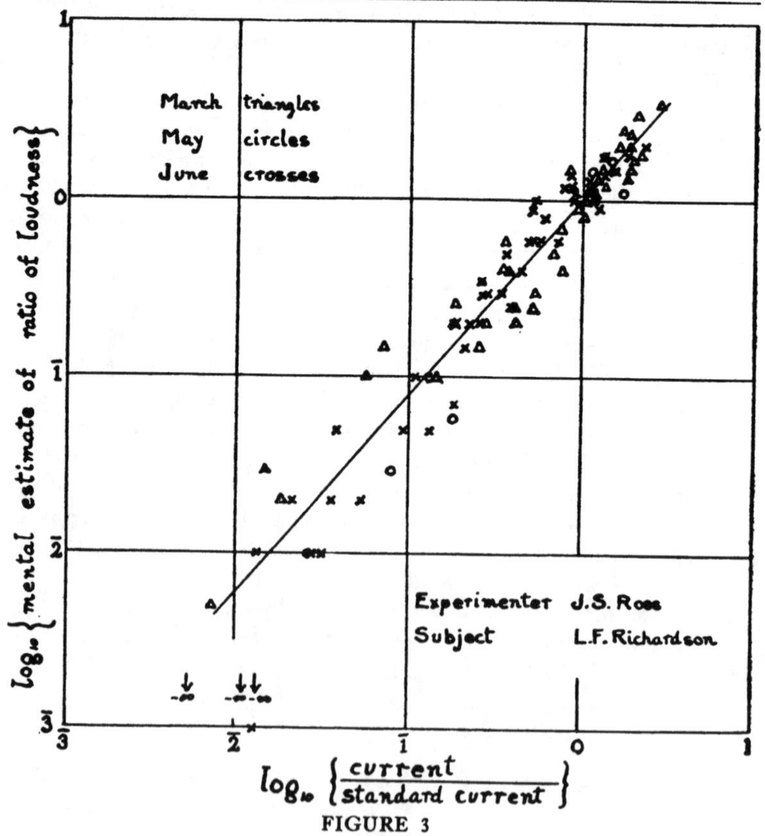

FIGURE 3

1930:2 *J. gen. Psychol.* **3**, 288–306

to a mean log_{10} (ratio of loudness) $= \bar{1}.40$. Therefore the range chiefly used in the research on imagery lies on the ordinates of Figure 3 between $\bar{1}.40$ and $\bar{2}.40$. This is a part of the diagram where the regression is linear, where the slope of the central line corresponds to the relation

$$(Loudness) = (Constant) \times (Current)^{1.1}$$

and where the mean deviation of log_{10} (current-ratio) from the central line is found to be 0.19, corresponding to an uncertainty of the current in the ratio 1.6. This uncertainty is sufficiently small to justify the *mean* values of the work on imagery.

Difficulty in estimating loudness seems to increase (*a*) with absolute faintness, but also (*b*) with greater separation from the standard. As in the research on imagery, the standard was nearer, the uncertainty may be a little less than that just calculated.

Apart from zero, only two estimates in that research fell below the range just considered; they were 0.05 and 0.02 of the loudness of normal inner speech, and on Figure 3, the corresponding points are located on the ordinate at $\bar{2}.10$ and $\bar{3}.70$. It is difficult to know whether the regression is still linear in this region, for what is just inaudible depends on the quietness of the room.

Quick Tests of Several Persons

Regulating and Measuring Voltage. Taking a hint from the

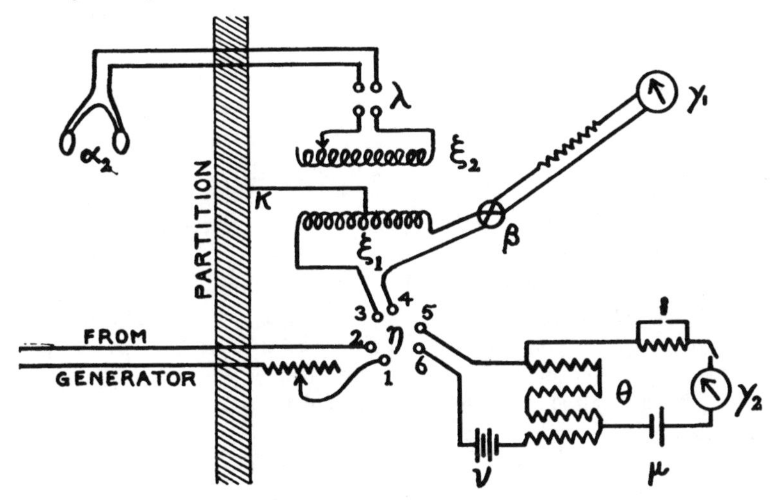

FIGURE 4
Voltage Regulation and Measurement for 4000-Ohm Headphone

sound division of the National Physical Laboratory, we rearranged the apparatus as shown in Figure 4. The torpid vacuo-junction β was placed in the primary circuit ξ_1 of the inductometer, and was there used to see that the primary current was kept constant, and to measure that constant. Telephones α_2 were chosen having so large an impedance that in comparison we could neglect the variation of the impedance of the secondary of the inductometer, a variation which caused the *b*-complication mentioned on page 294. The standardizing circuit ν, μ, θ, γ_2 is shown in both Figures 2 and 4, as it was used with both arrangements.

The telephones and the inductometer were 5 meters from each other and the nearest was 10 meters from the generator. A large search coil was used to make sure that stray fields were negligible.

With the telephone of large impedance a sound was heard even when the mutual inductance of ξ_1, ξ_2 was zero. The sound persisted when one telephone terminal was disconnected, but ceased when both were disconnected. The telephone being connected again, the sound was abolished when a wire from the midpoint of the primary of the inductometer was held in the mouth of the person who wore the headphones. As some people might dislike a wire in the mouth, the midpoint of the primary ξ_1 was connected instead by κ to a waterpipe which was almost as good. This stray current behaves as if it were due to two capacitances in series (*a*) between the coils of the inductometer and (*b*) between the telephones and the head. It sometimes made a sound as loud as that due to 30 microhenries of mutual inductance.

The effective voltage applied to the headphones is calculated as (*effective current in primary*) 2π (*frequency*) (*mutual inductance*) thus ignoring any harmonics. They are known to be very weak.

The "voltage-ratio," defined to mean (test voltage) / (standard voltage), is thus simply equal to (test mutual inductance) / (standard mutual inductance).

Routine. The routine for the experimenter was as follows. Double pole switch λ (in Figure 4) opened; ξ_2 set at standard, galvanometer γ_1 read; one tap on signalling key; λ closed; waited 15 seconds; λ opened; set ξ_2 to a different value; two taps on signalling key; λ closed; waited 15 seconds during which the mutual inductance was recorded; then the cycle began again.

To the observer seated in the adjoining room this routine appeared as: silence, one flash; standard sound during 15 seconds, 10 to 15 sec-

onds' silence, two flashes, a different sound the loudness of which must be estimated, 10 seconds' silence, one flash; standard sound again, and a second estimate of the previous unknown.
Circles show estimates made at the beginning of the test sound, crosses show those made at the beginning of the subsequent standard. Where cross and circle coincide the effect is like a black dot.

In each diagram the louder part to the right of the center is approximately straight. The slope of a well-fitting straight line, and the mean deviation from the line, are set out in Table 3.

Specimen observation. Observer Professor J. C. Flügel, the well-known psychologist. Various pitches over a range of 3 octaves were offered, and he chose one. Frequency 1100 per sec. Eleven preliminary tests were disregarded at the observer's request. Millivolts across telephone were 2.5 when inductance was 100 microhenries.

Serial number	Mental estimates: When unknown began	Mental estimates: When standard began again	Observer's notes	Mutual inductance microhenries (standard = 100)
12	20.0	10.0		6000
13	0.2	0.4		6
14	4.0	2.5		600
15	?	0.5		60
16	20.0	15.0		10,000
17	0.05	0.10		2
18	2.5	2		100
19	7.0	7.		1000
20	0.2	0.4		10
	interval for luncheon			
21	1.4	1.6		150
22	5.0	4.0	interval seemed long	1500
23	0.25	?		1.5
24	?	0.25	surprised	3
25	8.0	10.0		3000
26	3.0	2.0		300
27	0.0	0.5		30
28	15.0	15.0		10,000
29	0.05	0.2		10
30	1.0	1.0	surprised	101
31	0.005	0.01		1
32	0.1	0.2		10
33	3.0	3.0		1000
	a few minutes rest			
34	0.1	0.2		3
35	0.02	0.05		1.5

Results. In Figure 5 the coordinates have the following meaning:
$X = log_{10}$ (test voltage) / (standard voltage).
$Y = log_{10}$ (ratio of loudness to standard)

Explanation of Table 3. n is the slope of a straight line that fits well the diagram of the type of Figure 5 for the louder sounds, namely, for half the diagram. For this region loudness=(constant) $\times$ (voltage)^n.

Δ is the mean deviation of log_{10} (loudness ratio) from the aforesaid line for the same part of the diagram. In forming this mean,

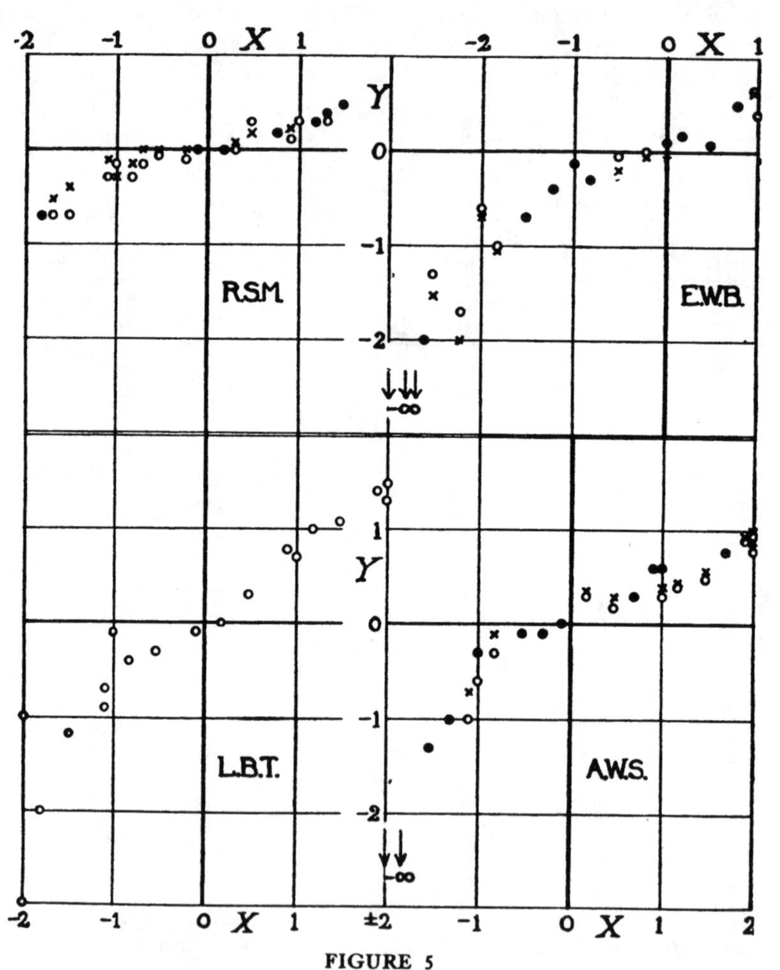

FIGURE 5
LEGENDS X AND Y ARE DEFINED ON PAGES 299-300

1930:2 *J. gen. Psychol.* **3**, 288–306

the first and second estimates of the same test sound were counted as two if they differed, as one if they coincided.

Y_0 is the intercept made by the same straight line on the Y axis.

TABLE 3

SUMMARY OF OBSERVATIONS USING S. G. BROWN'S F. 4000-OHM HEADPHONES

Date 1928	Observer	Age	Sex	Effective electromotive force						
				Just-unpleasant-ly loud (volts)	Standard (mili-volts)	Just inaudible (millivolts)	n	Δ	Y_0	Frequency per sec.
July										
10	R. S. Maxwell		M	0.08	2.5	0.03	0.24	0.06	0.04	765
13	E. O. Martin		M	?	2.4	?	0.40	?	?	1108
17	E. O. Martin		M	?	2.4	< 0.04	0.65	0.16	0.18	1108
13	A. W. Spurr		M	0.2$_0$	2.5	0.04	0.38	0.10	0.10	550
13	L. B. Trant		M	0.2$_0$	2.4	< 0.03	0.73	0.07	0.00	1108
13	J. S. Ross	35	M	0.2$_4$	2.4	< 0.03	0.55	0.10	–0.18	1108
16	E. W. Barrett		M	> 0.2$_1$	21.3	0.04	0.40	0.10	0.12	765
17	J. C. Flügel	44	M	?	2.5	< 0.03	0.53	0.11	0.14	1108
20	R. Aspinwall		M	0.2$_4$	2.4	?	0.26	0.06	0.00	550
27	P. C. Austin		M	0.1$_2$	2.1	< 0.02	0.34	0.07	0.00	1100
Aug.										
17	C. S. Myers	55	M	?	1.9	< 0.02	0.37	0.06	0.05	767
Mean							0.44	0.09	0.04	

Fechner takes "stimulus" to be proportional to energy, that is, to j^2, where j is the telephone current.

If, with Fechner, we suppose that

$$l = k \log_{10}(j^2/a)$$

l being loudness and k and a being constants and l_0, j_0 the standard values then

$$l_0 = k \log_{10}(j_0^2/a)$$

whence it follows that

$$l/l_0 - 1 = \frac{2k}{l_0} \log_{10}(j/j_0)$$

On plotting l/l_0 against $\log_{10}(j/j_0)$, the last formula indicates a straight line, but actually we find a curved distribution for the subjects Richardson, Maxwell, Trant, Barrett; and presumably for all, for the above Fechnerian formulae imply that

$$log_{10}(l/l_0 = log_{10} \left\{ 1 + \frac{2k}{l_0} log_{10}(j/j_0) \right\}$$

that is, in the coordinates of Figure 5,

$$Y = log_{10} \left\{ 1 + \frac{2k}{l_0} X \right\}$$

This is a curved locus, but actually we find for all subjects that the $(X_1 Y)$ graph is practically straight when X is not too small.

Replies to Objections

W. James (5) quotes Stumpf: "One sensation cannot be a multiple of another. If it could, we ought to be able to subtract the one from the other, and to feel the remainder by itself. Every sensation presents itself as an indivisible unit" (p. 547). This true remark of Stumpf's lies dangerously near to falsity, as may be seen from the following parody: "One mountain cannot be twice as high as another. If it could, we ought to be able to subtract the one from the other and to climb up the remainder by itself. Every mountain presents itself as an indivisible lump." In the parody we have inserted a confusion between a concrete thing, mountain, and its measurable quality, height. There is risk of a similar confusion with sensations. Sensations cannot be subtracted, but the numbers representing their intensities can.

As tending to prove that sensations cannot be measured, it has been asserted by C. S. Myers (7) that: "The cannon roar is louder than the crack of a pistol, but we cannot express the former experience in terms of the latter" (p. 252).

Although our opinion is not based on special experiments with cannons and pistols but only on memories of the war, we are inclined to agree with Myers. But our reasons are special, namely, (*a*) that the roar of a cannon is temporarily deafening to anyone standing near it; (*b*) that the roar differs from the crack not only in loudness but also in duration, in mean pitch and in intense unpleasantness; and that these other differences distract attention from loudness. There is nothing in these special difficulties which can be regarded as an objection to the estimation of loudness in favorable circumstances.

The Logic of Measurement

Dr. N. R. Campbell in a recent analysis of the logic of measurement (2) divides magnitudes into classes A and B:

A-magnitudes are those which are measured by physical processes resembling the addition of equal numbers; for example by the placing of meter-rods end to end, or by the putting of equal weights into the same pan of a balance, or by the connecting of equal electric resistances in series.

B-magnitudes are those which are derived from A-Magnitudes by calculation.

We have to consider whether the mental estimates fit into Campbell's classification. One observer, L. B. Trant, who is the college organist, reported that he "thought of how many organ-pipes would produce a sound like that." C. S. Myers, after estimating four unknowns, said, "How I do it, I don't know"; but after further introspection he recorded, at the 12th unknown, that he thought of "the strength of the stimulus to produce such a difference"; at the 18th, that he "imagined two equal sounds"; at the 20th, that he "imagined five sounds together." No other observer reported anything like addition. Most failed to describe the way in which they arrived at their numbers. J. S. Ross "hazarded guesses." L. F. Richardson "could not detect (*a*) any mental addition of imagined loudnesses, (*b*) nor any calculation." For him certainly, and perhaps for most of the others, the mental estimates are neither A-magnitudes nor B-magnitudes in Campbell's sense.

Fechner (4, p. 60) assigned numbers to sensations by counting increments in some sense equal. The scale of sensations, thus numbered, would be an A-magnitude, if it conformed to the other laws of measurement as set out by Campbell.

Summary

1) Of 11 persons tested, none has refused to assign numbers to loudnesses, though several found it difficult or complained that they were "only guessing."

2) Sounds that were unpleasantly loud were produced by telephone currents that were about 10,000 times stronger than those which produced sounds almost inaudible. The corresponding sensations usually vary in a less ratio, but for some subjects by as much as 3000. Because these ratios are so great, it is convenient to plot logarithms of current j and of loudness l.

3) The points on the (*log j, log l*) diagram for any one person, although scattered, lie conspicuously near to a smooth curve.

4) The shape of this curve has the following features common to all the persons tested:

a) For moderate and loud sounds it is nearly straight, implying that (loudness) = (constant) × (current)^n.

b) For fainter sounds the slope *d log l/d log j* becomes greater except for two observers (C. S. M. and L. F. R.).

c) A portion of one branch of a hyperbola, having one asymptote inclined with a slope nearly equal to *n*, and the other parallel to the axis of *log l*, would fit many observations well.

d) For inaudible sounds $log\ l \rightarrow -\infty$, while *log j* may vary in a large range. There is thus an infinite gap between groups of observed points. We have not investigated the threshold in detail but it is probably "confused by a vagueness, uncertainty, or scatter" (9).

5) Persons differ considerably in the index *n*. It has ranged from 0.24 (for R.S.M. in Table 3) to 1.1 (for L.F.R. in Figure 3), with a mean of about 0.5. That is to say, a telephone current 10 times stronger than the standard is, on the average, judged to produce a sound three times as loud, but the extreme estimates are 1.7 times and 13 times as loud as the standard. Two explanations are possible. Suppose two fixed telephone currents, the same for all persons. Then we may have either (*a*) sensations the same for all persons, divergent estimates wrong, or (*b*) sensations really different for different persons, estimates showing this fact.

We do not know which of these two explanations, (*a*) or (*b*), is the more correct.

6) The scatter of the estimates of loudness is measured by a mean deviation of 0.09 in $log_{10}l$ from the best fitting curve, for all sounds that are not too faint.

7) Two observers imagined stimuli in superposition. But for one observer at least, and perhaps for most, the loudness is neither an A-magnitude nor a B-magnitude in the sense that these terms are used by N. R. Campbell.

8) Sensation, as Fechner numbered it, is perhaps an A-magnitude. There is no reason to expect Fechner's scale to agree with the present intuitive scale. And if, in conformity with Fechner (4, pp. 20-40; 175-182), we measure the stimulus by the density of sound-energy per volume, we have not found any subject for which

(intuitively estimated sensation) = *k log* (stimulus/constant) where *k* is a constant. Whether Fechner's law is true for the scale defined by him is a separate question.

9) The decibel scale of "loudness" as defined by Dr. Harvey Fletcher (12) is not a scale of sensation.

10) Persons insensitive to frequency-differences ("tone deaf") can estimate loudness with no more scatter than those who have good musical ears (see Figure 5 in which R.S.M. and L.B.T. are sensitive to frequency differences, while E.W.B. and A.W.S. are insensitive).

11) The observations in the section on Test of Validity of Estimates in the research on "Conation, Imagery, and Cerebral Conductance" (8) tend to justify the averages of the previous estimates of L. F. Richardson on auditory imagery by showing that the mean uncertainty of an individual estimate is in the ratio 1.6.

12) We are indebted to Mr. F. C. Bartlett for the suggestion that it was desirable to make experiments on sensations. Mr. Albert Campbell freely gave up his expert advice on alternating currents. We wish to thank the observers for their care in making estimates.

REFERENCES

1. BLAIKLEY, —. In Barton Textbook of Sound. London: Macmillan, 1922. Pp. 687.
2. CAMPBELL, N. R. Measurement and calculation. London: Longmans, Green, 1928.
3. COHEN, B. S., ALDRIDGE, A. J., & WEST, W *J. Instit. Elec. Eng.*, 1926.
4. FECHNER, G. T. Elemente der Psychophysik. Vol. 1. Leipzig, 1860.
4*a*. FLETCHER, H. W. Speech and hearing. London: Macmillan, 1929. Pp. xv+331.
5. JAMES, W. Principles of psychology. Vol. 1. New York: Holt, 1901.
6. JEWETT, F. B. Article telephony. In Dictionary of Applied Physics. Vol. 2. London: Macmillan.
7. MYERS, C. S. Text book of experimental psychology. Vol. 1. London: Cambridge Univ. Press 1925.
8. RICHARDSON, L. F. Imagery, conation, and cerebral conductance. *J. Gen. Psychol.*, 1928, **2**, 324-352.
9. ————. Threshold when sensation is regarded as quantitative. *Brit. J. Psychol.*, 1928, **19**, 158-166.
10. RUSSELL, A. Alternating currents. (2nd ed.) Vol. 2. London: Cambridge Univ. Press.
11. TURNER, L. B. Outline of wireless. Cambridge: Cambridge Univ. Press, 1921.

Technical College
Paisley, Scotland

Westminster Training College
London

L'INTENSITÉ DES SONS ET LE COURANT TÉLÉPHONIQUE

(Résumé)

On a fait cette étude dans le but de constater (1) si l'on peut faire des estimations numériques de l'intensité, et, cela étant possible, (2) le rapport fonctionnel entre l'intensité moyenne estimée et le courant téléphonique, (3) la constance des estimations successives d'un individu, et (4) la différence entre les individus dans leurs estimations.

L'expérimentateur, dans une chambre, a réglé un courant produisant une note musicale d'une fréquence constante et d'une intensité variable, a signalé son approche au moyen de lumières, et a mesuré sa force; pendant que l'observateur, dans une autre chambre, a écouté un appareil téléphonique, et on lui a demandé d'ecrire, si possible, des estimations numériques de l'intensité des sons du test comme multiples de l'intensité d'un son étalon, sonné avant et après chaque test. Le courant téléphonique a subi des variations entre 10.000 et 1.

Les réponses aux questions citées en haut ont été celles-ci:

(1) Tous les onze observateurs ont réussi à donner des numéros aux intensités.

(2) Pour les sons qui n'ont pas été trop faibles,

(intensité)=constant x (ou courant ou voltage)^n

Pour les sons faibles, l'intensité a été généralement moindre que celle trouvée par extrapolation de cette formule appliquée aux sons de grande intensité.

(3) La constance des estimations individuelles, mesurée par "l'écart moyen" Δ $\log_{10}$ (intensité) de la courbe individuelle moyenne, mise en moyenne Δ=0,09 pour la partie de plus grande intensité.

(4) n a varié de 0,24 à 1,1 avec une moyenne de 0,5.

Pour l'observateur (L. F. R.) de l'article sur "Conation, Imagerie et Conductance Cérébrale," et pour les sons faibles notés là, n = 1,1: et l'écart moyen de $\log_{10}$ (courant) de la ligne moyenne a été de 0,19 pour une intensité fixe.

RICHARDSON ET ROSS

TONSTÄRKE UND DER TELEPHONSTROM

(Referat)

Der Zweck dieser Untersuchung war zu entdecken (1) ob man die Tonstärke in Zahlen abschätzen kann, und (wenn dies mölich ist) (2) das funktionelle Verhältnis der mittlern also abgeschätzten Tonstärke zu dem Telephonstrom, (3) wie konsequent eine Vp. in solchen auf einander folgenden Abschätzungen wäre, und (4) in wie fern die Vpn. variieren oder übereinstimmen.

Der VI., in einem Zimmer, regulierte einen Strom der einen musikalischen Ton von konstanter Häufigkeit doch variieren der Instensität hervorbrachte, signalisierte dessen Ansatz durch das Licht und bemass dessen Stärke; während die Vp., in einem andern Zimmer, am Telephon horchte, und wenn möglich die Tonstärke in Zahlen aufzeichnen sollte als Multiplum der Tonstärke des Masstabs, der vor und nach jedem Test angegeben wurde. Der Telephonstrom variierte von 10,000 bis 1.

Die Antworten auf diese Fragen waren wie folgt:

(1) Allen 11 Vpn. gelang es die verschiedenen Tonstärken in Zahlen auszdrücken.

(2) Für Töne die nicht zu schwach waren (Tonstärke) = Konstant x (entweder Strom odor *voltage*)^n. Für schwache Töne war die Tonstärke gewöhnlich geringer als die welche durch die Extrapolisierung nach dieser Formel den lauten Tönen angepasst waren.

(3) Persönliche Konsequenz, bemessen nach der "mittlern Abweichung" von Δ in $\log_{10}$ (Tonstärke) von der persönlichen mittlern Kurve, im Durchschnitt $\Delta=0.09$ für den lautern Teil.

(4) n erstreckte sich von 0.24 bis 1.1 mit einem Durchschnitt von 0.5. Für den Beobachter (L. F. R.) der Abhandlung über "Antrieb, Einbildung und Gehirnleitung" und dür die schwachen Töne die dort erwähnt werden, gibt $n=1.1$; und die mittlere Abweichung des $\log_{10}$ (Strom) von der mittlern Linie was 0.19 für eine bestimmte Tonstärke.

RICHARDSON UND ROSS

Richardson, L. F. and Maxwell, R. S. (1930)
Br. J. Psychol. **20**, 365–7

1930:3

THE QUANTITATIVE MENTAL ESTIMATION OF HUE, BRIGHTNESS OR SATURATION

A reply to Mr T. Smith's criticisms

The British Journal of Psychology – General Section
Cambridge
Vol. XX, Part 4, April 1930, pp. 365–367

(Manuscript received 30 December 1929)

Note: This was a reply by Richardson and Maxwell to: Smith, T. (1930). The quantitative estimation of the sensation of colour. *Br. J. Psychol.* **20**, 362–4. Smith had commented on earlier papers of Richardson (1929:4) and Maxwell (*Br. J. Psychol.* **20**, 181–9)

[From THE BRITISH JOURNAL OF PSYCHOLOGY (GENERAL SECTION),
Vol. XX. Part 4, April, 1930.]

THE QUANTITATIVE MENTAL ESTIMATION OF HUE, BRIGHTNESS OR SATURATION.

A REPLY TO MR T. SMITH'S CRITICISMS.

By L. F. RICHARDSON and R. S. MAXWELL.

There seems to be a fairly general agreement that the classification of 'formless visual sensations' (by which we mean uniform expanses of colour, white, grey or black) requires three independent qualities. Parsons[1] names them (1) hue = tone = Farbenton, (2) brightness = luminosity = Helligkeit, (3) saturation = purity = Sättigung. Hereinafter we shall name them hue, sensation-brightness and saturation. The unqualified word 'brightness' is confusing; for it is used by Mr Smith to denote a stimulus in some sentences, a sensation in others. We also have formerly not been sufficiently careful in this matter. The word 'tone' is ambiguous, for artists use it to mean degree of brightness or darkness, and so we shall avoid it. This classification is not quite universal, for Myers[2] uses four qualities: hue, intensity, saturation and brightness. However, we shall here follow the majority in supposing that three independent qualities are enough.

Pink, as popularly understood, differs from scarlet in being (1) a brighter sensation, (2) less saturated. It may or may not differ slightly, (3) in hue.

This statement can be made clearer by reference to water-pigments spread on white paper. On looking at a mixture of Chinese white with Messrs Winsor and Newton's alizarin scarlet, we experience a sensation that is brighter and less saturated than, and almost of the same hue as, that produced by the alizarin scarlet alone in a thick coat. The former sensation would ordinarily be called pink.

It is possible to produce a sensation which differs only in saturation from that due to a thick coat of alizarin scarlet laid on white paper. The appropriate stimulus can be arranged by adjusting a mixture of the same scarlet with Indian ink and burnt umber. But the result is not pink in

[1] Sir John Herbert Parsons, *An Introduction to the Study of Colour Vision*, p. 32, Camb. Univ. Press, 1924.

[2] C. S. Myers, *A Text Book of Experimental Psychology*, I, 71, Camb. Univ. Press, 1925.

any ordinary sense. It would have to be called a pinkish brown. The cover of this *Journal* is an example of pinkish brown; the cover of Parsons' *Colour-Vision* is another, but less bright and more saturated.

"Our feeling of pink is not a portion of our feeling of scarlet" thus refers to sensations which differ in saturation, as well as in brightness. In testing the truth of this dictum, both sensation-brightness and saturation must be varied; they must be varied together, not independently; and the observer must be asked to make a judgment on the effect of sensation-brightness and saturation jointly. We asked the observers to judge on a scale of 'redness'; for in framing instructions popular words had to be used.

It may be said that such a procedure is a muddle; for it would be more scientific to try quantitative mental estimates of sensations when hue, sensation-brightness and saturation were varied separately one at a time. We agree, and hope that such experiments will be made. But they would not test James' dictum. That is why we deferred them.

A possible defence of James not mentioned by Mr Smith, would be to point out that we have replaced James' doublet 'pink and scarlet' by a triplet 'white, pink and scarlet.' To use a geometrical phraseology we have dealt with sensation-intervals; James may possibly have intended sensation-points.

It is generally agreed that formless visual sensations are best classified by arranging their representative points in a volume. For example, C. S. Myers[1] gave a diagram in the shape of two cones, base to base, with white at one vertex, black at the other vertex, the greys along the axis. So that sensation-brightness increases steadily along the axis, saturation increases with distance from the axis and hue depends on angular position round about the axis.

We now[2] add the further, and controversial, idea that the distances of the representative points can be so chosen that when point B lies on the straight line AC, the ratio of lengths $AB : BC$ becomes equal to the ratio of the corresponding sensation-intervals as judged by an average observer.

From the way in which the pink stimulus was made on our colour wheels, it was evident that as the energy in the light-waves increased, owing to increasing angle of the white sector, so also the distribution of that energy would spread into wave-lengths shorter than 0·6 micron.

[1] C. S. Myers, *An Introduction to Experimental Psychology*, p. 18, Camb. Univ. Press, 1911.

[2] Apparently in agreement with Peddie, *Nature*, p. 791, Nov. 23rd, 1929.

Sensationally it appeared to our eyes that the pink was intermediate between the white and the scarlet, both in saturation and in sensation-brightness. We purposely arranged it so. Thus we think that the white, pink and scarlet as exhibited by our colour wheels lay in an axial plane of Myers' cones, and in a straight line in that plane.

Let the representative points be A for white, B for pink, C for scarlet. Let their normal projections on to the axis of the cones be A_1, B_1, C_1, and their projections on to the basal plane be A_2, B_2, C_2. Then because ABC is straight, therefore $AB : BC = A_1B_1 : B_1C_1 = A_2B_2 = B_2C_2$. If the observer judged by sensation-brightness alone (as Mr Smith alleges), he would estimate $A_1B_1 : B_1C_1$. If he judged by saturation alone, as the instruction about 'redness' may perhaps have suggested to some observers, he would estimate $A_2B_2 : B_2C_2$. If he judged his sensations without troubling to analyse them into saturation and brightness, he would estimate $AB : BC$. But all these ratios are equal. This may explain how Mr Smith's formula, derived from considerations about sensation-brightness, comes to be such an excellent fit to observations involving saturation.

Was the influence of the experimenter in our observations misdirected or excessive? We think that it was neither. In Richardson's research some eminent observers were quite willing to try the experiment; and the mean of their estimates (see Group 8) for the wheel was 44 per cent., in close agreement with the mean of 43 for all the men, and not far from the mean of 40 for all the schoolboys in Maxwell's research.

Mr Smith remarks, "their fundamental question was put in such a way as to ensure an answer apparently unfavourable to William James." The word 'ensure' is an exaggeration, for observers were at liberty to refuse to put any mark on the line, and the numbers of those who did so refuse are published. Mr Smith might truly have said 'suggest' instead of 'ensure.' But as the task of estimation requires some effort of attention, it is necessary to persuade the observer to try.

In spite of Mr Smith's criticisms, we continue to think that, by attacking James' dictum about pink and scarlet, we have cleared the way towards a more scientific study of quantitativeness in sensations of various kinds. The rare mistakes of a genius like James are apt to enslave the rest of us.

[1] Richardson, "Quantitative Mental Estimates of Light and Colour," this *Journal*, xx, 1, 27–37.

(*Manuscript received* 30 *December*, 1929.)

1930:4

THE ANALOGY BETWEEN MENTAL IMAGES AND SPARKS

The Psychological Review
Lancaster, Pennsylvania
Vol. 37, No. 3, May 1930, pp. 214–227

(Manuscript received 5 November 1929)

Richardson added a note to this paper in:
The Psychological Review
Lancaster, Pennsylvania
Vol. 37, No. 4, July 1930, p. 364

(Manuscript received 22 May 1930)

This note is appended to the main paper (below, p. 206)

Note: Richardson's own studies on the behaviour of neon lamps are fully described in 1937:2 (in Volume 1)

THE ANALOGY BETWEEN MENTAL IMAGES AND SPARKS

BY LEWIS F. RICHARDSON

The Technical College, Paisley, Scotland

INTRODUCTION

W. McDougall [1] in a paper 'On the seat of the psychophysical processes' argued in 1901 that the passage of the synapse is a discontinuous process comparable with sparking between the knobs of a Wimshurst machine, and that each spark generates a pulse of sensation.

That such a process may be of great social importance is suggested by the following quotation from Graham Wallas:

". . . the formation of the original hypothesis, the inventive moment on which successful action depends, must take place in an individual brain. If we wish to estimate the real possibility of using the ever-growing mass of recorded fact for the guidance of organised social action, we must think, not of the long rows of tables and microscopes in a scientific laboratory, nor of the numbered stacks of books and maps in the British Museum or the Library of Congress, but of a minister or responsible official when he has put back his books on their shelves, has said goodbye to his last expert adviser, and sits with shut eyes at his desk, hoping that if he can maintain long enough the effort of straining expectancy some new idea will come into his mind." [2]

That description suggests the electric strain in a spark-gap before the spark passes.

The long suspense ending in the sudden decision is not restricted to important people, nor even to humans. For Köhler [3] frequently observed that a hungry chimpanzee, which

[1] W. McDougall, *Brain*, 1901, 24, p. 577.

[2] Graham Wallas, The great society, Macmillan & Co., 1914, p. 17.

[3] W. Köhler, The mentality of apes, Kegan Paul, 1925, pp. 60, 111, 112, 180, 184, 200.

had tried and failed to reach the only visible food, sat down and gazed quietly about during several minutes, and then suddenly obtained the food by a different movement in a few seconds.

The author's attention was drawn again to the analogy by the following known experiment:

A commercial neon-lamp was connected across a high tension battery and the voltage was reduced by steps of about 2 volts. The lamp came thus to a voltage at which it did not light immediately but lit after the voltage had been applied for some time. In successive trials at the same voltage the suspense varied irregularly between 5 and 69 seconds. This fickle behaviour, so unusual in physical things, reminds one of the fickle behaviour of one's mind when the same problem has to be solved on the blackboard once a year for many years in succession.

Numerous researches on the neon-lamp have been published, and the object of this study is to see whether they contain anything helpful for psychology.

At the outset I wish to disclaim four ideas:

I. It is not suggested that the mind is entirely a mechanism, but rather that the mind is connected with a mechanism which we try to manage.

II. It is not suggested that a spark-gap becomes conscious when a spark passes.

III. It is not suggested that there are ionized gases in the brain.

IV. It is not suggested that the electric current in a wire is at all like the impulse in a nerve axon. The analogy to be developed has to do with synapses and not with axons.

On the contrary the working ideas of this enquiry are:

(*a*) That the time-changes of imagery in the mind are so like the time-changes of electric current in a neon-lamp when exposed to a potential difference near to its sparking potential, that if we can find an equation, which describes the behaviour of the lamp with respect to time, we should be able to find psychological names for the symbols, names which will make the equation a description of mental events. There is a presumption that the symbol called potential-difference in the theory of the lamp must be renamed the intensity of trying to think. And that the symbol called energy-radiated may be renamed imagery. But let us hold these first ideas for revision as we go on.

(*b*) That although an attempt to fit equations of this kind directly to introspective data has already been made [4] with some success, yet we are more likely to reach the truth if we let the present analogy develop in its own way, delaying comparisons with the published equations until the present enquiry has matured.

(*c*) That there are ionized liquids in the brain.

Statement of the Analogy

The analogy is divided into parts numbered by Roman numerals. For each part, the physical property of the spark is lettered *A*, while the corresponding aspect of the mind is lettered *C*, and physiological analogues, when noticed, are lettered *B*. Any facts that seem contrary to the analogy are lettered Z, and a reason is stated for neglecting them.

I. *A*. For a given spark-gap there is a critical voltage such that lower voltages do not cause a steady spark to pass however long they may be applied.

I. *C*. For a given problem there is a critical intensity of trying-to-think below which an answer will never be attained. For the sort of problems that we are accustomed to solve in say one minute, the critical intensity may escape our notice because it is a state of drowsy inattention below our ordinary wide-awake level.[5]

I. *Z*. The solution of problems during inattention, has been entertainingly described by Poincaré,[6] and has been experienced by many people. Dr. Maxwell Garnett has suggested to me that this inattentive discovery is essentially connected with the functioning of two or more synapses in series. Such a series-effect, being complicated, is left over for future investigation.

[4] L. F. Richardson, Imagery, conation and cerebral conductance, *J. Gen. Psychol.*, 1929, **2**, 324.

[5] See reference 4, pp. 341, 342.

[6] H. Poincaré, Science and method, Nelson & Sons, pp. 46–63.

II. *A.* For voltages slightly above the critical voltage, there is a delay or retardation (Funkenverzögerung) between the application of the voltage and the occurrence of the spark.[7]

II. *C.* The answer to a problem often occurs to us after a few seconds of vain attempt to think.

III. *A.* The electric current during the delay is usually at first very small and varying very slowly. Then it increases rapidly to the spark.

III. *C.* It is a common experience that, at the beginning of an attempt to solve a problem, no result is attained, and none seems to be arriving. A few seconds later the answer arrives rather suddenly.

IV. *A.* When a given spark-gap is subjected to a fixed voltage until a spark occurs after delay δ_1 seconds, then to no voltage during T seconds, then to the original voltage again during δ_2 seconds, after which time a second spark occurs, it is observed [8] that:

δ_2 is on the average equal to δ_1 if T is long enough, but δ_2 is on the average less than δ_1 if T is short enough.

IV. *C.* If we attempt to solve the same problem twice in succession with, between the two attempts, an interval T during which we think about other things, it is observed that the delay δ_2 at the second attempt is less than the delay δ_1, at the first attempt unless the interval T has been so long as to allow us completely to forget the answer.[9]

V. *A.* For a given initial condition of the electrodes and of the gas, the mean delay varies approximately inversely as the excess of the actual voltage above the critical voltage.[10]

V. *B.* The latent period for a reflex is shorter, the stronger the stimulus, within a middle range.[10a]

V. *C.* In emergencies and examinations when people want quick answers some try harder to think.

[7] K. Zuber, *Ann. der Physik*, 1925, **76**, p. 231; also W. Braunbek, *Zsch. f. Physik*, 1926, **36**, p. 582.

[8] W. Braunbek, *Zsch. f. Physik*, 1926, **36**, p. 583.

[9] L. F. Richardson, *op. cit.*, p. 326.

[10] W. Braunbek, *Zsch. f. Physik*, 1926, **36**, p. 604.

[10a] C. S. Sherrington, Integrative action of the nervous system, Constable, 1915, pp. 19–21.

VI. *A.* When all the conditions are macroscopically alike, the spark-delay varies at successive applications of the same voltage. Zuber and Braunbek have published statistics of observed times.[11]

VI. *C.* Mental measurement always involves a scatter which requires statistical treatment. As evidence it is sufficient to note that a well-known text book with the title 'The essentials of mental measurement' [12] consists chiefly of statistical theory and practice.

VII. *A.* A neon-lamp exposed to a steady voltage, just below its sparking potential, is occasionally observed to light up feebly and go out again of its own accord (Vorentladungen). Braunbek's statistical theory explains this.[13]

VII. *C.* Introspection reveals, during the suspense of trying to solve a problem, occasional images so fragmentary and fleeting that they can hardly be recognised.[14]

VIII. *A.* A neon-lamp, placed in parallel with a capacitance K and the combination of these two being arranged in series with a wire-resistance R and a battery B having a voltage considerably exceeding the sparking potential, is known to produce oscillations.[15] The explanation is that the lamp flashes almost as soon as the capacitance has filled up to the critical level. The lamp then drains the reservoir and goes out. No self inductance is involved. The current, though periodic, is far from sinusoidal. The pauses between flashes can be longer than the flashes themselves. Frequencies between 0.003 sec^{-1} and 95000 sec^{-1} have been obtained in this way.[16]

VIII. *C.* Automatic repetition is prominent in mental phenomena. For instance:

(i) Repeated notes and trills in music.

(ii) McDougall[17] showed that a single momentary stimulus to the retina generates a train of pulses of visual sensation of

[11] See footnote (7).

[12] By W. Brown and G. H. Thomson (Camb. Univ. Press).

[13] *Zsch. f. Physik*, 1926, **39**, p. 22.

[14] L. F. Richardson, *op. cit.*, pp. 327–329.

[15] S. O. Pearson and H. St. G. Anson, *Phys. Soc. Proc.*, 1922, **34**, p. 204. J. Taylor and W. Clarkson, *Phys. Soc. Proc.*, 1924, **36**, p. 269.

[16] U. A. Oschwald and A. G. Tarrant, *Proc. Phys. Soc.*, 1924, **36**, p. 265.

[17] W. McDougall, *Brit. J. Psychol.*, 1904–1905, **1**.

diminishing intensity, the train being long in proportion to the intensity of the stimulus.

(iii) When, in reading a novel, the last word of a page was strongly emphasized, that word recurred uninvited during the reading of the next page. The word sometimes recurred at fairly regular intervals of about six seconds. But the phenomenon differed from a damped harmonic oscillation because the pauses were longer than the disturbances and because the second recurrence was often more intense than the first or third.[18]

VIII. *B.* In the sensory nerves electric impulses have been found following one another at a rate of the order of 50 per second. Although periodic, they are far from sinusoidal, and therefore quite unlike the oscillation of a tuned circuit containing inductance and capacity. This is clearly shown in some curves computed by Adrian [19] from his own photographs. Again the pauses are often longer than the disturbances.

To complete the analogy we should have to seek in the brain for some things having differential equations with respect to time like the equations of a neon-lamp, a wire-resistance and a capacitance, but there is no need to look for anything having time-equations like those of an inductance. The synapse is strongly suggested as the part to compare with the neon-lamp. To make a rash suggestion: the cell body of a neurone looks like a capacitance.

IX. *A.* The spark-delay varies inversely as the ionization in the spark-gap due to external causes, such as X-rays, when the voltage and other circumstances are maintained constant.[20]

IX. *C.* Delay in arriving at an answer is less as the individual is cleverer, other circumstances being the same. Cleverness is here used in the sense defined by Garnett.[21]

[18] L. F. Richardson, Imagery, conation, and cerebral conductance, *J. Gen. Psychol.*, 1929, **2**, pp. 337 to 339.

[19] E. D. Adrian, The basis of sensation, Christophers, 1928, p. 52, Fig. 6D.

[20] W. Braunbek, *Zsch. f. Physik*, 1926, **36**, p. 604.

[21] J. C. M. Garnett, *Brit. J. Psychol.*, 1919, **9**, also *Proc. Roy. Soc. A.*, 1919, **96**, pp. 91 to 111.

Interim Summary

(1) The preceding comparisons suggest that we should take the symbol called voltage in the equations of the spark-gap, and rename it 'intensity of trying to think.'

(2) Also that we should apply the name 'intensity of relevant imagery' to one or other of several symbols:

(*a*) Either to the symbol for electric current;

(*b*) Or to the symbol for any physical quantity, such as the time-rate of energy, which is proportional to the current when the voltage is fixed.

(That word 'relevant' is connected with philosophic problems which are here ignored.)

(3) Also that we should apply the name 'cleverness' to the symbol for the intensity of ionization due to external causes such as X-rays, γ-rays or cosmic rays.

(4) So far I have taken known facts about the neon-lamp and have found their mental analogues. Encouraged by success, I next tried to reverse the process, taking known mental phenomena and using them to predict the results of experiments on neon-lamps. Accordingly X. *C* precedes X. *A*.

X. *C*. One of the most remarkable of mental phenomena is the extinction of a thought by its successor, a process which occurs in everyday thinking, reading or conversation. Spearman includes it in universal mental competition.[22] McDougall [23] has a very interesting and comprehensive theory of the same phenomena described as 'inhibition by drainage.' Garnett [24] has applied McDougall's theory to education, calling it one of the five laws of thought.

X. *B*. An apparently similar phenomenon is well known to physiologists as reciprocal inhibition [25] and the drainage theory is one among many put forward to explain it.

A critical survey of facts and theories of inhibition was

[22] C. Spearman, The abilities of man, Macmillan, 1926, Ch. VIII.

[23] W. McDougall, *Brain*, 1903, 26, p. 153.

[24] J. C. M. Garnett, Education and world citizenship, Camb. Press, 1921, p. 70.

[25] Sherrington, Integrative action of the nervous system, p. 201. K. Lucas, The conduction of the nervous impulse, Longmans Green, 1917, p. 94.

given by Bayliss.[26] He believed in the existence of directly inhibitory fibres or end-organs. Dr. R. D. Gillespie tells me that a corresponding direct inhibition, not caused by distraction, finds its place in psychopathology. L. B. Turner [26a] has given an electrical model showing such simple inhibition. But none of these facts is opposed to the view that there exists also another kind of inhibition, namely by distraction, as in X. *C*.

X. *A*. From the mental phenomena in X. *C* it seemed reasonable to predict:

(1) That it would be possible to connect two neon-lamps to the same battery so that the lighting of one lamp should extinguish the other; the change-over depending on the properties of the lamps, and not *merely* on the movement of a metal 2-way switch.

(2) That resistance, obeying Ohm's law, would be required in the circuit.

(3) Possibly also capacitance.

(4) But not inductance.

(5) That under some circumstances, analogous to reverie, the second lamp would start of itself.

(6) That under other circumstances, analogous to hard purposive thinking, the second lamp would require an extra voltage, analogous to the Will, to start it.

The connections were not indicated by the analogy.

Of these predictions Nos. (1), (2), (4), and (6) were fulfilled by the apparatus shown in Fig. (1). In it *P* and *Q* were two neon-lamps.

They were prepared, from commercial Osglim lamps marked 200–220 V 5 W, by cutting open the brass caps, removing the coils of wire and making soldered connections. In each lamp one electrode was a disk, the other a spiral. For probable impurities such as hydrogen in the neon, see J. W. Ryde.[27]

The wire-coils *R* and *S* were each of 1.50×10^4 ohms, and

[26] Sir W. M. Bayliss, Principles of general physiology, Longmans Green, 1918, Ch. XIII.

[26a] L. B. Turner, Outline of wireless, Camb. Univ. Press, 1921, p. 92.

[27] J. W. Ryde, *Proc. Phys. Soc.*, 1924, 36, pp. 278–279.

U, also of wire, had 9.74×10^4 ohms. There was a main battery, of voltage *V*, which could be applied steadily by putting in the wander-plug *K*. An extra battery of voltage *W* could be applied temporarily by a tapping contact either at the point *x* or at the point *y*. The experiments were made in a room lit by tungsten lamps and free from daylight or radioactive matter. The behaviour of the apparatus depended greatly on the voltage *V*.

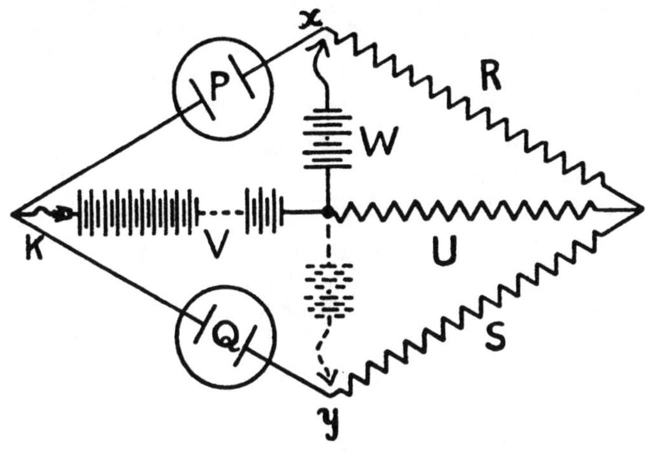

FIG. 1. Electrical Model illustrating a Mind having a Will but capable of only Two Ideas. See Analogies X., XI., XII., XIII.

V = 154 Volts. The lamps having been rested overnight, insertion of the plug *K* caused lamp *Q* to light while *P* remained extinct. Contact at *x*, applying the extra *W* of 12 volts for about a second, caused both lamps to flash, *P* more brightly than *Q*; and when contact at *x* was withdrawn, *P* remained alight but *Q* extinct, although the main voltage *V* was applied steadily all the time. Thus the lighting of *P* had inhibited *Q*. Again a temporary contact for about a second at *y* left *Q* alight but *P* extinct. And so on to and fro many times in succession.

It may be objected that something like a switch is used to apply the extra voltage *W* first at *x*, then at *y*. The reply is that *W* merely starts the inhibition, which is maintained by the lamps and resistances. Similarly we often have to make an effort to begin thinking about, say, a difficult theory;

but once we have well begun, the theory goes without effort, and rival ideas vanish.

McDougall [28] realized that to explain reciprocal inhibition something was required beyond mere drainage through resistances obeying Ohm's law; and he described a hydraulic model involving distended india-rubber tubes and valves controlled by springs. Now the stretched rubber is, in its equation of motion, like an electrical capacitance; and the inertia of the water is like self inductance. So McDougall's model involves two things omitted in the experiment of Fig. 1. The properties of the neon-lamp which are effective in causing reciprocal inhibition would seem to be

(1) The existence of a sparking potential. McDougall's valves had a corresponding pressure at which they opened.

(2) The fact that the sparking potential is lowered by the immediately previous passage of current and rises again while no current passes. McDougall (*loc. cit.*, p. 133, footnote) attributed exactly the opposite properties to the valves or to the synapses. He wrote ". . . each valve should tend to return to the closed position the more strongly, the longer it has been open." But indeed an Osglim lamp seems to show this opposite behaviour after longer times such as 20 minutes. See a paper by Shaxby and Evans [29] and discussion thereon.

Garnett [30] has emphasized the view that Will acts by reinforcing excitement; this is very like the function of the battery W in Fig. 1.

Besides reciprocal inhibition, the apparatus of Fig. 1, showed several other analogies for which further Roman numerals are assigned.

XI. *A*. At $V = 154$ volts only one lamp lit when the plug K was inserted. But which lamp it was, depended on their recent history. If both had been rested for several minutes then Q lit. But if the lamp P had been set going by a brief touch at x from W and left glowing for, say, a minute, then

[28] W. McDougall, Physiological psychology, J. M. Dent, 1911, p. 132.

[29] Shaxby and Evans, *Proc. Phys. Soc.*, 1922, 34, pp. 254, 278.

[30] J. C. M. Garnett, Education and world citizenship, Camb. Press, pp. 99, 101.

taking out plug K and inserting K again after a second, caused P to relight instead of Q.

XI. *C.* The glowing of Q is like an instinctive thought that predominates, unless P has been made easier by recently thinking P.

XII. *A.* On setting the main voltage to the lower value $V = 143$ volts, the insertion of K did not light either lamp. W was 12 volts as before. Temporary contact for one second at y caused Q to light but not P. Temporary contact for one second at x left both lamps extinct.

XII. *C.* This is like a drowsy person who by a momentary effort W can just think the easy thought Q. But if he turns his attention to the more difficult thought P, he falls asleep altogether.

XIII. *A.* When $V = 154$ volts, $W = 12$ volts, it was occasionally observed that a temporary contact at x or y left both lamps lit. When this happened a more prolonged contact at x or y was found to leave the lamp on the same side alight and opposite lamp extinct.

XIII. *C.* This may be compared with a mental conflict and its resolution by an effort of Will. I am indebted to Mr. J. S. Ross for this comparison.

XIV. *B.* Adrian [30a] has found that, in sensory nerves, impulses follow one another at a rate comparable with 100 per second.

XIV. *A.* To imitate this, however imperfectly, a rotating commutator Z, driven by a motor M as in Fig. 2, was arranged to interrupt the current about 130 times per second. The insulation-resistance between M and Z exceeded 10^{10} ohms. The lamps P and Q and the resistances R, S, U were the same in figure 2 as in figure 1. But the extra battery W was not used. The voltage of V was set at 179 volts by the aid of a voltmeter E. On making contact at K on fourteen occasions, the lamp P lit on 8 occasions alone, and Q lit alone on the other 6 occasions. When Q lit no further change was observed. But—and this is the interesting point—when P lit and the apparatus was left untouched, then after a

[30a] E. D. Adrian, The basis of sensation, Christophers, 1928.

few seconds *P* went out and *Q* simultaneously lighted up. This was observed on all but one of the 8 occasions on which *P* lit first. The delays, with *P* glowing, before the automatic change from *P* to *Q*, were distributed thus in seconds 1, 15, 7, 2, 3, 7.

XIV. *C.* This resembles the automatic change of imagery that occurs in *reverie.* Thus the experiment of XIV. *A* fulfils the prediction made in X. *A* (5) except that I have not obtained the automatic change with a steady voltage, but only when the interrupter was running.

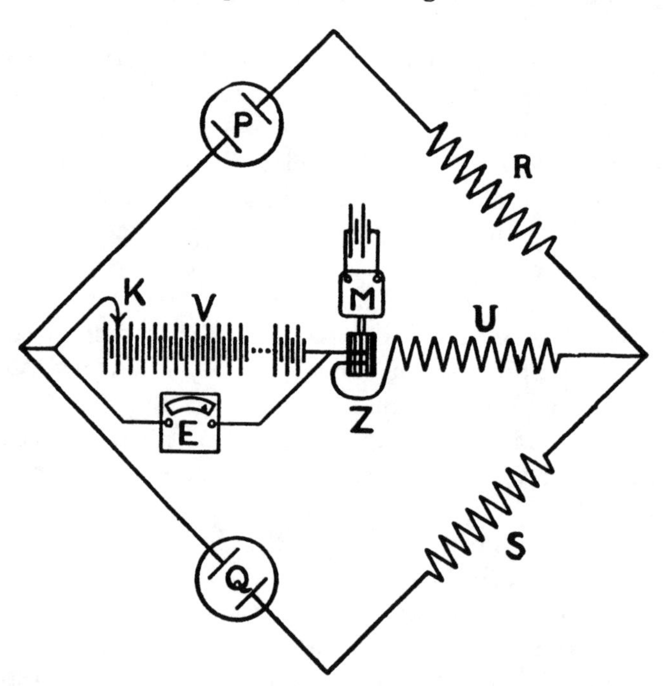

FIG. 2. Intermittent Current and Reverie. See Analogies XIV., XV.

XV. *A.* In an experiment similar to XIV. *A* it was found that the times during which *P* remained glowing before the automatic interchange from *P* to *Q* were greatly prolonged by placing about 0.0002 grm. of radium near *P* while *Q* was protected by lead screens.

XV. *C.* Ideas for which a person is clever tend to occupy his mind.

XVI. *B.* A characteristic property of the synapse is one-way conduction.[31]

XVI. *A.* Unsymmetrical spark-gaps, which have a higher sparking voltage in one direction than in the other, are well-known.[32] Experiments are in prospect with neon-lamps of this type.

XVI. *C.* Mental one-way conduction, of a complicated kind, occurs when a person, who knows the alphabet perfectly forwards, finds it difficult to repeat it backwards. But the details of this analogy are obscure.

Summary and Abstract

This is a development from an analogy between sensations and sparks published by McDougall in 'Brain' of 1901. The strain of expectation, which we feel when we wait for the solution of a problem, is here compared with the electric intensity in a spark-gap before the spark passes. The time of imageless suspense is compared with the spark-delay, noting that each of them is fickle. The least intensity with which we must attempt any problem if we are ever to solve it, is compared with the minimum sparking potential. Cleverness, in J. C. M. Garnett's sense, is compared with the ionization of the spark-gap by X-rays or γ-rays. Having found nine analogies, of which the foregoing are a sample, the author then predicted, from McDougall's theory of inhibition by drainage, that it would be possible to connect two spark-gaps so that the passage of one spark should inhibit the other; the change-over either (*a*) requiring an extra stimulus, as in purposive thinking; or (*b*) occurring automatically, as in reverie. Both predictions have been verified by using neon lamps as spark-gaps. Altogether sixteen, mutually consistent analogies have been collected. These analogies show that when physicists shall have agreed among themselves about the theory of sparking, their formulæ will be of great interest to both physiologists and psy-

[31] C. S. Sherrington, Integrative action of the nervous system, Constable, 1915, p. 14.

[32] Siegbahn, The spectroscopy of X-rays, Oxford Press, 1925, pp. 75, 76. Also Townsend, Electricity in gases, Oxford Press, 1915, pp. 378–391.

chologists. For the physical equations will provide a description of the analogous psychological events on giving the following new names to the symbols: Intensity of trying-to-think for potential difference, intensity of imagery for either electric current or energy radiated, cleverness for ionization due to X- or γ-rays.

[MS. received November 5, 1929]

NOTE ON 'THE ANALOGY BETWEEN MENTAL IMAGES AND SPARKS'

The analogies suggested in my article which appeared in the May issue of this REVIEW (pp. 214–227) may be amplified by the following addition to number V.

V.Z. Others, notably Spearman and Coué, deprecate trying harder. But the Coué-Baudouin 'law of reversed effort' may be compared with the tendency of a high potential to cause sparks in wrong places, where the insulation is weak. For the 'reversed' effect is a thought or action from which the patient wishes to escape; that is to say, one which he has recently experienced; and this recency has left that synapse with enough residual conductance for the 'reversed' discharge to begin. (C. Spearman, The abilities of man, New York, Macmillan, 1927, p. 334; C. Baudouin, Suggestion and autosuggestion, London, Allen & Unwin, 1922, pp. 121–127.)

LEWIS F. RICHARDSON

THE TECHNICAL COLLEGE
PAISLEY, SCOTLAND

[MS. received May 22, 1930]

1932:2

DETERMINISM

Nature
London
Vol. 129, No. 3252, 27 February 1932, p. 316
(Letter dated 13 February)

316 *NATURE* [FEBRUARY 27, 1932

Determinism

Is it really necessary to appeal to anything so *recherché* as Heisenberg's Principle of Indeterminacy in order to justify anything so familiar as personal freedom of choice? This question arises on reading Sir Arthur Eddington's interesting address in NATURE of Feb. 13. Consider any one of the laws of physics commonly verified in the laboratory, say $T = 2\pi\sqrt{l/g}$ for a simple pendulum. If, while one student is observing the pendulum, another student were to knock it about, the observations might misfit the formula. And so in general: the accepted laws of physics are verified only if no person interferes with the apparatus. We cannot interfere with the moon, because it is so massive and so far away: and that is part of the reason why the motion of the moon is almost deterministic; the 'almost' referring to the extremely small Heisenbergian indeterminacy. But there is no great mass or great distance to prevent John Doe interfering with his own brain in the act of making his decision to buy a house from Richard Roe.

LEWIS F. RICHARDSON.

The Technical College, Paisley,

Feb. 13.

Richardson, L. F. (1932)
In: *Report of a joint discussion on vision*, 112–14; 116
London: The Physical Society

1932:4

THE MEASURABILITY OF SENSATIONS OF HUE, BRIGHTNESS OR SATURATION

In: *Report of a Joint Discussion on Vision held on June 3rd 1932 at the Imperial College of Science by the Physical and Optical Societies*
London, The Physical Society, pp. 112–114; 116

112

THE MEASURABILITY OF SENSATIONS OF HUE, BRIGHTNESS OR SATURATION

BY LEWIS F. RICHARDSON, B.Sc. (Psychol.), D.Sc. (Physics)

ABSTRACT. It has been shown to be possible to make direct intuitive mental estimates of the intensities of various visual sensations. This opens up the prospect of a Psychophysics which shall be quantitative as to both sensation and stimulus.

S, E
J, R

THERE are at least three ways of measuring sensations of hue, brightness or saturation as distinct from stimuli S: (E) by counting small equal-appearing intervals; (J) by counting just-perceptible intervals; (R) by directly estimating the *ratio* of unequal intervals both much larger than the least perceptible. This R method is not yet as well known as it deserves to be.

Before there were photometers, Al Sufi[1], about the year 964 A.D., estimated the brightness of more than 1000 stars. His description of his method seems to indicate that he judged by equal-appearing intervals E. C. S. Pierce[2] compared Sufi's estimates and many others with the Harvard photometric measurements and found general agreement with Fechner's law, $E = \text{constant} \times \log S$.

But Fechner[3] had made some queer assumptions, such as the existence of negative sensations "below the threshold," which provoked people by reaction to deny that formless visual sensations were quantitative at all. This seems to me an excessive reaction. Yet several otherwise excellent men of science have gone to that extreme[4]. Stumpf[5] well expressed a commonly felt difficulty when he wrote: "One sensation cannot be a multiple of another. If it could, we ought to be able to subtract the one from the other, and to feel the remainder by itself. Every sensation presents itself as an indivisible unit." Stumpf's remark seems convincing until one compares it with the following parody, "One mountain cannot be twice as high as another. If it could, we ought to be able to subtract the one from the other and to climb up the remainder by itself. Every mountain presents itself as an indivisible lump[6]." In the phraseology of N. R. Campbell[7], all that Stumpf asserted was that sensation is not an "A-magnitude" like mass or electric resistance, measurable by the juxtaposition of unchanging objects. That, being admitted, leaves open the question whether sensation is, like viscosity or magnetic intensity, what Campbell calls a "B-magnitude," or whether, as I suggest, sensation belongs to some more primitive type of magnitude which has hitherto escaped classification.

Feeling sure from my own experience that the ratio of large unequal intervals could be estimated (method R) I asked 323 people to estimate the redness of certain pinks by putting a mark on a segment of a line labelled red at one end and white at the other. Three hundred and sixteen of them succeeded in making an estimate[8]. The standard deviation of an individual's single estimate from the mean for the same sex was about 13 per cent of the range. The women estimated the pink as redder than did the men.

R. S. Maxwell[9] pursued the enquiry much further by mixing on a colour-wheel the same white and red in different proportions and asking 35 schoolboys of average age 15·8 years to make mental estimates of the redness of the sensation (method R) by putting a mark on a segment of a line. A clear functional relation appeared, which was well described by the formula

$$(S - 156)(R + 56) = -8736,$$

where S is the angle of red as a percentage of 360°, and R is the mean mental estimate as per cent from white as zero towards red as 100.

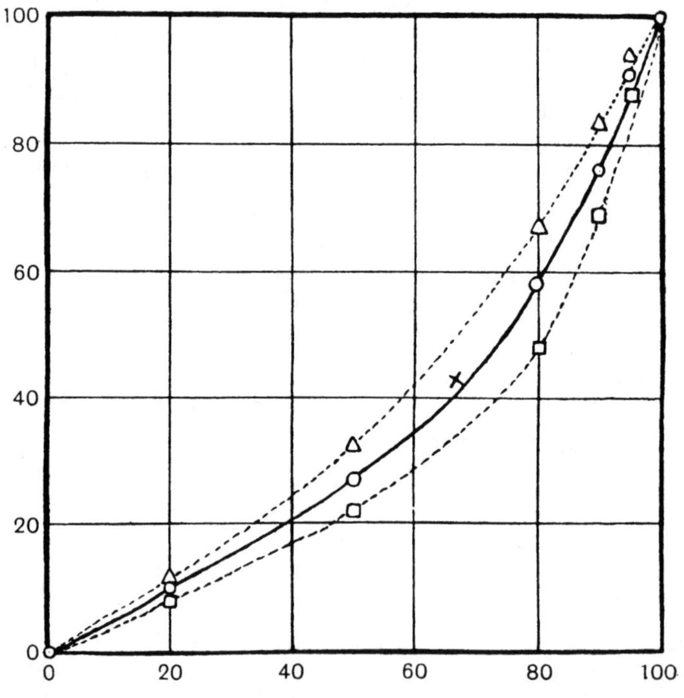

Reproduced from figure 2 of a paper by R. S. Maxwell, *Brit. J. Psychol.* **20**, 186 (1929), by the courtesy of the Cambridge University Press.

There is still much to be found out about the relations between R and E, as to whether they agree, and if not which is better. From an investigation by T. Smith[10], taken in conjunction with its discussion by Richardson and Maxwell it looks as if the R method for colours might agree with Fechner's logarithmic law for moderate ranges. (For loudness the R method definitely disagreed[6].) The simplicity and intuitive directness of the R method commends it for psychological purposes.

It has been shown that sensations are vaguely quantitative. Accordingly the measure of a stimulus, however precise, is for psychological purposes no substitute for the intuitive estimate of a sensation, vague though it be. The reliability of instruments, when we wish to measure sensations, is like the reliability of a paranoiac, who, when he is asked a particular question, invariably gives the same wrong answer. For example, a spectrograph records a spectrum extending continuously past the wave-length at which sensation ceases. Regarded as a psychologist the spectrograph is evidently suffering from a delusion.

Nevertheless when paints or dyed cloth are to be sold, with the risk of legal disputes about conformity to specification, it is essential that the standards should be acceptable to all. And as individuals differ in colour-vision and in aesthetic opinions, the standardizing institutions are led to measure stimuli, not sensations; to do physics as being easier than psychology, although the ultimate purpose of colour is usually aesthetic. In the words of K. S. Gibson, "Spectrophotometry... is the fundamental basis of all work of this kind(11)."

A most important principle in the philosophy of measurement is Einstein's that "All our space-time verifications invariably amount to a determination of space-time coincidences" and "As all our physical experience can be ultimately reduced to such coincidences, there is no immediate reason for preferring certain systems of co-ordinates to others, that is to say, we arrive at the requirement of general co-variance(12)."

On considering the R method there appears to be nothing in it like coincidences, at least for some estimators. We must, I think, conclude that Einstein's coincidence principle, while of the first importance for physics, does not necessarily apply to psychology.

REFERENCES

(1) Al Sufi, translated by Schjellerup as *Description des Magnitudes des Étoiles* (1874).

(2) C. S. Pierce. *Annals of the Astronomical Observatory, Harvard*, **9**, 7 (1878).

(3) G. T. Fechner. *Psychophysik*. Leipzig 1860.

(4) W. James, *Principles of Psychology*, **1**, 545–9 (1901); C. S. Myers, *Text-Book of Experimental Psychology*, **1**, 252 (1925); J. Guild, *Nature*, **129**, 454 (1932).

(5) Stumpf, quoted by James, *loc. cit.* p. 547.

(6) L. F. Richardson and J. S. Ross. *Journ. Gen. Psychol.* **3**, 301 (1930).

(7) N. R. Campbell. *Measurement and Calculation*. (Longmans, Green, 1928.)

(8) L. F. Richardson. *Brit. Journ. Psychol.* **20**, 27–37 (1929).

(9) R. S. Maxwell. *Brit. Journ. Psychol.* **20**, 181–9 (1929).

(10) T. Smith, also L. F. Richardson and R. S. Maxwell. *Brit. Journ. Psychol.* **20**, 362–7 (1930).

(11) K. S. Gibson. *Dict. App. Phys.* **4**, 746. Compare J. Guild, *Nature*, **129**, 454 (1932).

(12) A. Einstein. *Ann. d. Phys.* **49** (1916), translated by W. Perrett and G. B. Jefferey as *The Principle of Relativity* by A. Einstein and others, pp. 117–18 (Methuen).

AUTHOR'S reply. While I agree with Mr Guild that the magnitudes of sensations are a kind of magnitude different from those of stimuli, yet it seems to me quite a mistake to attempt as he does to forbid the study of the functional relation between the magnitudes of sensations and those of stimuli. He does so for the reason that they are philosophically non-comparable phenomena. Nevertheless Mr R. S. Maxwell has found the functional relationship in one instance and many others will no doubt be discovered later.

Richardson, L. F. (1933)
Br. J. Psychol. **23**, 401–3

1933:4

A QUANTITATIVE VIEW OF PAIN

The British Journal of Psychology – General Section
Cambridge
Vol. XXIII, Part 4, April 1933, pp. 401–403

(Manuscript received 18 October 1932)

[From THE BRITISH JOURNAL OF PSYCHOLOGY (GENERAL SECTION), Vol. XXIII. Part 4, April, 1933.]

A QUANTITATIVE VIEW OF PAIN.

By LEWIS F. RICHARDSON.

The observations here recorded were made on the pain caused by cellulitis of the right thumb, which hurt me with varying intensities for seven weeks. The word 'pain' is used, following Wohlgemuth (1), for the sensation localized in the hand and distinct from the accompanying feeling of 'unpleasure.' Of course pain is really five-dimensional, the intensity being a function of one co-ordinate of time and three of space; but that view is too complicated for a beginning; and so, ignoring throbbings, let us specialize on pains that are constant in time, or nearly so. Further to simplify the scheme, it is possible to ignore the spatial distinction between a toothache and stomach-ache, and so to establish a one-dimensional sequence of pain-intensity, by studying the effects of pain in distracting from thought, and of thought in distracting from pain. Thus, in sequence of increasing intensity there were:

(A) a pain so slight that it was perceived only when carefully attended to,

(B) a pain which was obvious when attended to, but which ceased entirely when other matters were thought of,

(C) a severe pain which became more bearable when one engaged in lively conversation, or read an exciting story, or concentrated on a mathematical problem, or decided a difficult administrative question,

(D) a still more intense pain, which was little, if at all, diminished by the sort of distractions which were effective for (C). On some occasions the intensity of the pain (D) seemed to be somewhat increased by attempted distraction,

(E) a pain so intense that the rôles played in the process of distraction by pain and thought were interchanged; for the pain arrested the movement of thought, leaving a stationary mental state consisting of just oneself and pain. But this I experienced for only a few seconds at a time, during surgical operations. The phenomenon is vividly described by Stout (2).

There is another way of defining a 'critical intensity' of pain, here denoted by α, namely as that pain which just prevents one from going to sleep at the time and under the circumstances in which one does ordinarily fall asleep when free from pain. My observations showed that the intensity α was more than (C) and less than (D): sleep and distraction both became unattainable at about the same intensity of pain.

This observation accords with the theory that the physiology of falling asleep is similar to that of distraction. Important in that connection is Pavlov's work on sleep in dogs (3) and McDougall's discussion thereof (4).

As part of a series of researches into the quantitativeness of sensations, I attempted to assign numerical values to the typical intensities described above. The unit was defined so that α equalled 100. The pain (A) was produced by biting the uninjured left thumb. Then in the course of many days, it was estimated that (A) = 3, (B) = 20, (C) = 70 decreasing on distraction to 30, (D) = 200, (E) = 500. For other sensations it has been found that estimates by different people vary widely, and the same is to be expected for pain (5, 6).

To those who are accustomed to regard the process of measurement as being necessarily either (i) the marking and counting of units, or (ii) calculation using such counts (7), I should explain that there was no counting here nor any calculation. Instead ratios were estimated in a manner so direct as to elude analysis, and were expressed as numbers by means of past experience with centimetre scales. This intuitive directness, and the fact that so many people can estimate ratios of colour, loudness or other sensations, suggests that there may be in most mental experience a primaeval vague quantitativeness, out of which the refined conceptions of modern physical measurement have been gradually evolved.

There is a second critical intensity here denoted by β, namely the intensity of pain which spoilt one's control of voluntary muscles. As far as I can tell, β seemed to be about the same as the intensity (E) which arrested the movement of thought; for it was observed (i) that a pain of intensity about 70 to 150 continued for several days with but little ill effect on voluntary muscles, but (ii) a pain of intensity 300, which approached the intensity 500 at which pain prohibited thought, had after six hours' continuance begun to split the personality into a querulous child making uncontrolled movements *versus* an ashamed but helpless spectator, who had an awful sense of further disruption impending. But further disruption did not occur, for at this stage the physician gave me morphia.

The unpleasure accompanying the pain appeared, when intense, to consist of, or to be indistinguishably mingled with, visceral sensations. Wohlgemuth regarded them as distinct, but I did not separate them. Although the intensity of unpleasure rose and fell more or less with that of the pain, yet the changes of unpleasure sometimes lagged in time behind the changes of pain. For example, on having gauze drainers pulled out from surgical cuts, the pain ran up suddenly to intensity (E), thought stopped for a second or two, meanwhile intense unpleasure came on as a feeling of being all mixed up inside the torso. Then in the course of the next hour or so the pain decreased rapidly to (D) or (C), while the unpleasure decreased much more slowly; so that there was a stage at which the unpleasure was more conspicuous than the pain. But when on the contrary the pain had been nearly steady all day at intensities (C) or (D), then the unpleasure was less conspicuous than the pain; and it was the pain not the unpleasure that kept me awake. The feeling 'all mixed up inside' is reported in connection with unpleasure and fear by one of Conklin and Dimmick's observers (8).

Food had an astonishingly large effect on the pain. To the well-fed patient the presence of food in the stomach was of much less intellectual interest than the story which he was reading, yet breakfast regularly decreased the pain from 70 to 30, a decrease as great as could be achieved by close attention to the story.

The introspections recorded in this paper were made at times when the observer was free from the influence of any drug such as morphia, chloroform or nembutol,

In conclusion I formulate some questions for other observers, having already given my own answers above.

(i) Is $C < \alpha < D$?

(ii) Is $\beta = E$?

(iii) What is the ratio α/A?

(iv) What is the ratio β/α?

REFERENCES.

(1) WOHLGEMUTH, A. (1919). "Pleasure-unpleasure." *Brit. J. Psych. Monograph Supplements*, II, 217, 218.

(2) STOUT, G. F. (1915). *Manual of Psychology*, p. 314. London: University Tutorial Press, Ltd.

(3) PAVLOV, I. P. (1927). *Conditioned Reflexes.* Oxford University Press.

(4) MCDOUGALL, W. (1929). *Journ. General Psych.* II, 231.

(5) RICHARDSON, L. F. and ROSS, J. S. (1930). "Loudness and telephone-current." *Journ. General Psych.* III, 288.

(6) RICHARDSON, L. F. (1929). "Quantitative mental estimates of light and colour." This *Journal*, XX, 27.

(7) CAMPBELL, N. R. (1928). *Measurement and Calculation.* London: Longmans, Green.

(8) CONKLIN, V. and DIMMICK, F. L. (1925). "An experimental study of fear." *Amer. J. Psych.* XXXVI, 96.

(*Manuscript received* 18 *October*, 1932.)

Richardson, L. F. (1933)
Proc. Phys. Soc. Lond. **45**, 585–8

1933:5

THE MEASUREMENT OF VISUAL SENSATIONS

Reply by Dr L. F. Richardson

The Proceedings of the Physical Society
London
Vol. 45, Part 4, No. 249, 1 July 1933, pp. 585–588

Note: This was a reply by Richardson in the discussion on: Campbell, N. R. (1933). The measurement of visual sensations: a criticism of Dr L. F. Richardson's proposed method of measuring sensations by 'mental estimates'. *Proc. Phys. Soc. Lond.* **45**, 565–71

Reply by Dr L. F. RICHARDSON:

1. Physicists are accustomed to estimate mentally, or if you prefer to guess, tenths of a scale division; also to guess tenths of a second when using a clock beating whole seconds. But let no one suppose that I advocate an extension of guessing in practical physics; for I really admire the growth of instrumental precision. The present controversy is not about physics but is about logic, psychology and metaphysics.

2. If anyone desires to read the theory of order and quantity set out with clarity and charm let me recommend *The Fundamental Concepts of Algebra and Geometry* by John Wesley Young†. Of course we must all agree with Dr Campbell that "a process of assigning numerals is not regarded as measurement unless it fulfils conditions more stringent than the mere representation of order." But mental estimation gives more than mere order. This is well known in connexion with estimating tenths of a scale-division. Again, if you saw an ordinary street of houses all alike numbered consecutively 1, 2, 3, 14, 25, 638, 639, would you not feel that although the order was correct the numbering was silly? Would you not feel this immediately without needing to measure distances with a tape? If so you become immediately aware of more than mere order. Dr Campbell admits this himself in regard to yellowness in his § 10 and indeed he proposes an ingenious difference-method for assigning numerals.

The serious question is not whether sensory events are quantitative, but how accurately they are quantitative. That can be found out by psychophysical experiment, but never by mere argument. Here follows a suggestion for a crucial experiment. Let four stimuli A, B, C, D, steadily increasing from A to D in some quality, be adjusted so that the successive intervals of sensation appear equal when the stimuli are considered in adjacent pairs, namely AB with BC and then BC with CD. How nearly does the interval of sensation from B to D then appear to be twice the interval of sensation from A to B? The accuracy of this result is known to be good

† Macmillan, New York (1911).

when the sensation is that of lengths drawn on paper, but for loudness, colour, warmth, taste and pain there is I believe as yet very little definite information about accuracy in this sense, although there is already a good deal of information about accuracy of repetition of estimates of various sensations.

3. I disown the remarks of the fictitious Dr Richardson introduced by Dr Campbell in his § 3, and give instead my own conviction that the sensation of warmth is quite distinct from thermodynamic temperature. Similarly for colour. What should we think of a fictitious person who could not distinguish qualitatively between a wave-length and the sensation of colour which it produces?

4. Important in Dr Campbell's argument is his doctrine about the sameness of things. He uses the phrase "the same thing" at least ten times, and in a very peculiar and confusing sense. Thus in § 5 he states that "two methods measure the same thing, if the order of the numerals assigned by one to the members of a group is always the same as the order of the numerals assigned by the other to those same members." Because, as x increases, x^3, $\tan^{-1} x$, $\exp x$ all steadily increase, therefore according to Dr Campbell, x, x^3, $\tan^{-1} x$, $\exp x$ all measure "the same thing." Neither in mathematics nor in common speech is it customary to say that x and x^3 measure the same thing. This doctrine of Dr Campbell's is metaphysical and reminds one of Kant's doctrine of the alleged things-in-themselves behind phenomena. To prevent confusion we ought to have a technical term, and I propose in future to call Dr Campbell's "same thing" a "same N.R.C. thing."

I am now able to state that whereas sensation and stimulus are entirely distinct, yet under specially constant circumstances they are approximately the same N.R.C. thing.

5. A fundamental distinction between sensation and stimulus is that each person has his own immediate experience of sensation, but only an inferred and usually co-operative knowledge of stimuli. Hamlet's epigram on the certainty of immediate experience* may be crudely brought up to date as follows:

Doubt how electrons behave;
Doubt if anything really is there;
Doubt whether sound is a wave;
But never doubt you hear,
Unless of course you really *are* not quite certain.

This comic anticlimax supports the thesis, for "not-being-quite-certain" is a personal mental state, very different from a stimulus.

6. The plainest evidence that sensation is not, in ordinary parlance, the same thing as the stimulus external to the human body is afforded by a study of the illusions, Gestalt-phenomena†, eidetic images† and hallucinations. For example a sensation of brightness may be produced either by a train of waves entering the pupil or

* Act 2, Scene 2.

† E. R. Jaensch, *Eidetic Imagery* (Kegan Paul, London, 1930); W. Köhler, *Gestalt Psychology* (Bell & Sons, London, 1930).

by digital pressure on the outside of the eyeball. Now if the sensation is the same as each of these different stimuli, it would follow that the stimuli are the same, which is absurd. It may be replied that each kind of sensation is in one-to-one correspondence with a special state of affairs in the brain; but even so, a state of affairs in the brain is not what we mean ordinarily by the "stimulus".

7. With regard to the minor variations in the relation between stimulus and sensation, that is to say setting aside hallucinations and illusions of the more extreme kind and considering only such minor affairs as simultaneous and successive contrast, Dr Campbell gives two strangely different arguments. *Firstly:* when these minor variations are negligible, he regards this in his § 5 as evidence that a mental estimate of a sensation and the measure of a stimulus are the measures of the same N.R.C. thing. I agree, but find the statement rather dull. *Secondly:* when these minor variations are not negligible, Dr Campbell in his § 8 considers that the variations indicate that the mental estimate is not an estimate of a sensation. In other words he accuses the observers of not knowing their own minds. Most people know their own minds more or less, though not perfectly. Dr Campbell condemns the whole procedure because some of the details are unclear. I think it will be much wiser to go patiently on with the experimental elucidation of these variations.

In the last paragraph of his § 8 Dr Campbell attends to the chief difficulty of psychology, namely the privacy of the individual mind. It is indeed a formidable difficulty. Personally I refrain from assuming either of the criteria which he mentions and rejects. It is not convenient to have to remember the individual estimates of hundreds of observers; so I prefer instead to state their mean and their standard deviation. This is a harmless, customary and useful procedure and there is no justification for Dr Campbell's violent condemnatory words.

8. With regard to the functional relation between stimulus and the average of the estimates of sensation, the credit of obtaining the interesting curve, which Dr Campbell attributes to me, is really due to the Rev. R. S. Maxwell, B.Sc.

In discussing this curve Dr Campbell asserts in his § 7 that "Physicists do not regard a law as a true numerical law, unless its form is simple and/or explicable by a theory; if it does not fulfil that condition, it is a mere empirical law." This statement of Dr Campbell's contains, I think, more than one misleading antithesis. For the opposite to "true" is "false," not "empirical." Also the opposite to "simple" is "complicated," not "empirical." For example, Stefan's radiation law was in the year 1880 empirical and simple and true for black bodies, before its thermodynamic theory was published by Boltzmann in 1884. Again, the law by which the tides at any particular port are predicted is fairly true, decidedly complicated and partly empirical. Disagreeing as I do with Dr Campbell's fundamentals on laws, it is useless for me to argue about his deductions on this subject.

9. A restriction of the meaning of the word "measurement" so that it should apply only to what Dr Campbell has named A-magnitudes and B-magnitudes is recommended by several speakers. Such a conventional restriction might suitably be left to the decision of the Committee appointed by Sections A and J of the British

Association in 1932. But I must point out that Dr Campbell formulated his valuable classification of types of magnitude before he had sufficiently considered the existence and properties of mental estimates. Dr Houstoun in this Discussion has mentioned excellent reasons for not thus restricting the meaning of "measurement." Might we not suitably say that mental estimates are "C-magnitudes," and that all magnitudes are measured?

10. Taking each person's word "measurement" in his own sense, I am happy to be in general agreement with the remarks of Mr R. J. Bartlett, Dr W. D. Wright, Capt. Hume, the Rev. R. S. Maxwell and Dr Houstoun and am much interested in the facts brought into comparison by Dr Beatty.

11. As to bias, I agree with Mr Guild that experience of painting, dyeing or of colorimetry would be likely to distort mental estimates of colour. But ordinary adolescents should be much less biased.

12. Mr T. Smith has much to say about the difficulties experienced by himself and some colleagues and mentions by contrast the lesser scatter in the estimates of Mr Maxwell's schoolboys. But the existence of great learning and intelligence in some persons, who find difficulty in making estimates, does not, I think, discredit the estimates of those who make them easily. For among nearly 400 persons, whom I have asked to make estimates, there was no obvious connexion between refusal on the one part and learning and intelligence on the other. Also it is known that various sensory accomplishments are distributed among the population in a manner having no important correlation with general intelligence. According to G. M. Whipple's *Manual of Mental and Physical Tests** this is so for the discrimination of lifted weights and of double touch on the skin. Some very intelligent people are colour-blind, as John Dalton was. An honours B.Sc. is no guarantee that its possessor can sing in tune. Sometimes logical propensities are actually a disqualification. Thus Sir Francis Galton found that imagery was weaker among those accustomed to much abstract thought than among women and children†. Again Mr Dudley Heath of the Royal College of Art remarked, in connexion with estimates of pink, that those art-students who decline to make a decision in this and similar problems are those who are also argumentative; and that they could place the mark on the line if they would trust to their first impressions, instead of to reasoning‡.

* (Warwick & York, Baltimore). † See W. James, *Principles of Psychology*, **2**, 51–57.
‡ *Brit. J. Psych.*, General Section, **20**, 36 (1929).

Richardson, L. F. (1935)
Nature, Lond. **135**, 830–1

1935:2

MATHEMATICAL PSYCHOLOGY OF WAR

Nature
London
Vol. 135, No. 3420, 18 May 1935, pp. 830–831
(Letter dated 11 April)

Note: This topic is continued in 1935:3

(*Reprinted from* NATURE, *Vol.* 135, *page* 830, *May* 18, 1935.)

Mathematical Psychology of War

As NATURE has encouraged scientific workers to think about public affairs, I beg space to remark that equations, describing the onset of the War, and published under the above title* in 1919, have again a topical interest, in connexion with the present regrettable rearmament. In revised form :

$$\frac{dx_1}{dt} = k_{12}.x_2 - \gamma_1 x_1 + \Delta_1 ;$$

$$\frac{dx_2}{dt} = k_{21}.x_1 - \gamma_2 x_2 + \Delta_2 .$$

The suffixes 1 and 2 refer to the opposing nations, or groups of nations. The symbol x denotes the variable preparedness for war ; t is the time ; k is a 'defence-coefficient' and is positive and more or less constant ; γ is a 'fatigue and expense' coefficient and is also positive and moderately constant. Lastly, Δ represents those dissatisfactions-with-treaties, which tend to provoke a breach of the peace.

If Δ_1, Δ_2, x_1, x_2 could all have been made zero simultaneously, the equations show that x_1 and x_2 would have remained zero. That ideal condition would have been permanent peace by disarmament-and-satisfaction. The equations further imply that mutual disarmament without satisfaction is not permanent, for if x_1 and x_2 instantaneously vanish, $dx_1/dt = \Delta_1$ and $dx_2/dt = \Delta_2$.

Unilateral disarmament corresponds to putting $x_2 = 0$ at a certain instant. We have at that time :

$$\frac{dx_1}{dt} = - \gamma_1 x_1 + \Delta_1 ;$$

$$\frac{dx_2}{dt} = k_{21} . x_1 + \Delta_2 .$$

The second of these equations implies that x_2 will not remain zero ; later, when x_2 has grown, the term $k_{12}.x_2$ will cause x_1 to grow also. So unilateral disarmament is not permanent, as Germany has shown us.

A race in armaments, such as was in progress in 1912, occurs when the defence-terms predominate in the second members of the equations. We have then approximately

$$\frac{dx_1}{dt} = k_{12}.x_2, \qquad \frac{dx_2}{dt} = k_{21}.x_1,$$

and both x_1 and x_2 tend towards infinity.

* Obtainable from Geneva Research Center, 2 Place Chateaubriand, Geneva, price 5s. post paid. Few copies remain.

I submit that the equations do describe, at least crudely, the way in which things have been done in the past. As to the future, while indicating the desirability of disarmament-and-satisfaction, they suggest that such a condition might easily become unstable, and that there is a need for controlling terms of a quite novel type. More strength to the statesmen who are trying to provide such !

LEWIS F. RICHARDSON.

38 Main Road,
Castlehead, Paisley.
April 11.

Printed in Great Britain by FISHFR, KNIGHT & Co., LTD., St. Albans

1935:3

MATHEMATICAL PSYCHOLOGY OF WAR

Nature
London
Vol. 136, No. 3452, 28 December 1935, p. 1025
(Letter dated 4 December)

Note: This topic is continued in 1938:2 and 1939:1

DECEMBER 28, 1935 NATURE 1025

Letters to the Editor

The Editor does not hold himself responsible for opinions expressed by his correspondents. He cannot undertake to return, or to correspond with the writers of, rejected manuscripts intended for this or any other part of NATURE. *No notice is taken of anonymous communications.*

NOTES ON POINTS IN SOME OF THIS WEEK'S LETTERS APPEAR ON P. 1030.

CORRESPONDENTS ARE INVITED TO ATTACH SIMILAR SUMMARIES TO THEIR COMMUNICATIONS.

Mathematical Psychology of War

THIS is relevant to the Naval Conference. A letter in NATURE of May 18 was intended to show, by reference to historical facts, that a certain pair of equations did represent, at least crudely, the behaviour of nations prior and subsequent to the Great War. In order to proceed to deductions, it is convenient now to change the notation. The hypothesis aforesaid is that

$$\frac{dx}{dt} = ky - \alpha x + g \quad . \quad . \quad . \quad . \quad . \ (1)$$

$$\frac{dy}{dt} = lx - \beta y + h \quad . \quad . \quad . \quad . \quad . \ (2)$$

in which x denotes the variable preparedness for war of one group of nations, y that of the opposing group ; t is the time ; k and l are 'defence coefficients' ; α and β are 'fatigue-and-expense coefficients' and g and h are measures of dissatisfaction with the results of treaties. Each of $k, l, \alpha, \beta, g, h$ is regarded as temporarily constant. Of these, k, l, α, β are each positive ; but g and h may have either sign, and will be positive for the 'have-nots', negative for the 'haves'. The international situation is thus represented by a point (x, y) in a plane. Let us think of this point as a particle moving in accordance with the equations. If the particle be tending towards plus infinity of both x and y, then war looms ahead. But if the particle be going in an opposite direction the prospect is peaceful.

If dx/dt be zero, the equation (1) represents a straight line ; similarly for dy/dt and equation (2). The intersection (x_0, y_0) of these two straight lines is the point of balance of power. It is given by

$$x_0 = (kh + \beta g)/(\alpha\beta - kl),\ y_0 = (lg + \alpha h)/(\alpha\beta - kl) \quad (3), (4)$$

But, unfortunately for those who live under a policy of balance of power, the point of balance has several awkward possibilities. First, by a rare chance, $\alpha\beta$ might equal kl so that the lines would be parallel and no point of balance would exist. Secondly, it is quite probable that the point of balance may be situated in one of the quadrants where x or y or both are negative ; so that balance could be attained only by negative preparedness for war, that is, by positive preparedness for co-operation. Thirdly, although the particle is undoubtedly in equilibrium when it coincides with the point of balance of power, yet that equilibrium may be unstable.

To investigate stability the equations have been solved by a standard method via the auxiliary equation

$$m^2 + (\alpha + \beta)\,m + \alpha\beta - kl = 0 \ ; \quad . \quad . \quad . \ (5)$$

and it is found that the particle, wherever it begins, will ultimately arrive at the point of balance (x_0, y_0) if

$$kl < \alpha\beta, \quad . \quad . \quad . \quad . \quad . \ (6)$$

whereas the particle will ultimately go off towards infinity if

$$kl > \alpha\beta. \quad . \quad . \quad . \quad . \quad . \ (7)$$

That is to say : the international régime is unstable if, and only if, the product of the two defence coefficients exceeds the product of the two fatigue-and-expense coefficients.

In the unstable régime, the particle will tend towards one or other of two opposite infinities according to where it begins. We may classify the starting points by making a transformation to polar coordinates (r, θ) centred at the point of balance, so that

$$x - x_0 = r\cos\theta, \quad y - y_0 = r\sin\theta \quad (8), (9)$$

For then

$$\frac{d\theta}{dt} = l\cos^2\theta + (\alpha - \beta)\cos\theta\sin\theta - k\sin^2\theta, \quad . \quad . \ (10)$$

whence it follows that there are always two real lines given by

$$2k\tan\theta = \alpha - \beta \pm \sqrt{(\alpha - \beta)^2 + 4lk} \quad . \quad . \ (11)$$

on which $d\theta/dt = 0$. The particle cannot cross these lines. They form barriers dividing the plane into four angular portions. In particular, if $\alpha = \beta$ and $l = k$ the barriers are at $\theta = \pm\pi/4 \pm \pi$. When the balance of power is unstable $(kl > \alpha\beta)$ the barrier at $\theta = -45°, +135°$, or in general at

$$2k\tan\theta = \alpha - \beta - \sqrt{(\alpha - \beta)^2 + 4lk}$$

is of profound significance, for it divides the international plane into two portions, one of a drift towards war, the other of a drift towards co-operation.

The 'constants' are liable to secular change. The progress of engineering and chemistry has notably decreased α and β, so that if l and k were unchanged, the international régime would now be much more unstable than it was a century ago. The defence coefficients k and l are influenced by the suspicion or goodwill distributed by print or broadcast.

Over all, it must be remembered that the equations describe the rivalries of two groups having populations of the same order of magnitude. In the new tentative practice of 'collective security', one population is many times greater than the other, and a quite different course of events may be expected.

LEWIS F. RICHARDSON.

38 Main Road,
Castlehead, Paisley.

Dec. 4.

Richardson, L. F. (1937)
Br. J. Psychol. **28**, 212–15

1937:1

HINTS FROM PHYSICS AND METEOROLOGY AS TO MENTAL PERIODICITIES

The British Journal of Psychology – General Section
Cambridge
Vol. XXVIII, Part 2, October 1937, pp. 212–215

(Manuscript received 10 May 1937)

Note: Richardson made a critical assessment of Dr S. J. F. Philpott's researches on mental periodicities in 1952:4

[From THE BRITISH JOURNAL OF PSYCHOLOGY (GENERAL SECTION), Vol. XXVIII. Part 2, October, 1937.]

HINTS FROM PHYSICS AND METEOROLOGY AS TO MENTAL PERIODICITIES

By LEWIS F. RICHARDSON

Fresh interest in mental periodicity has been aroused by the researches of Flugel (1928, 1934) and Philpott (1932).

There is a formal resemblance between psychology and meteorology. In both sciences an enormous collection of observational data has been open to public inspection for a long time. In both sciences the attempt to find order among the miscellany of the observations has proceeded slowly with many failures. In both sciences there are too many circumstances varying simultaneously for the comfort of the mathematician who would like to make a neat, simple theory. Both sciences use correlation coefficients freely. But, judging by the extent to which observations are explained by theories, it appears that meteorology is ahead of psychology, perhaps by the work of a generation.

With this preface, a warning issued by the editor of the *Quarterly Journal of the Royal Meteorological Society* (Sir Gilbert Walker, F.R.S.) may be of interest to psychologists. He (Walker, 1936) "would draw their attention to another duty laid upon him by Dr Jeffreys in his recent book on *Earthquakes and Mountains*. On p. 54, after drawing attention to the two criteria for reality in periods and lamenting the frequency of the errors to which their neglect gives rise, he remarks: 'It is to be wished that editors of journals would make it an absolute rule not to publish papers on periodicities if these criteria are not applied; if the results are not significant they are worthless, and if they are significant opinion is prejudiced against them in advance.' If, as is probably true, ninety-five per cent of the periods announced are non-existent, it will save costs in printing as well as valuable time spent in reading, if such a rule is enforced."

That is the stern rule at which meteorologists have arrived, after half a century's experience of periodicities. On enquiry from Sir Gilbert Walker for particulars of the "two criteria for reality" he kindly sent me the references under his name at the end of this paper.

Turning now to physics, we notice that periodogram analysis (Schuster, 1906) on which the aforesaid dread "criteria of reality" are based was worked out in those decades when the vibrating systems regarded as typical were those like the pendulum, the spring-and-mass or the inductance and capacitance. When properly stated periodogram analysis is, of course, a set of theorems in pure mathematics, quite independent of any alleged mechanism. Nevertheless it was generally supposed that the periods, when found, were to be explained by a mechanism of the spring-and-inertia type. The general characteristic of such mechanisms is that the period is well defined, not much affected by damping, nor much affected by outside small disturbances. Accordingly one of the best methods for finding the period was to count the number of periods in a long interval of time.

In 1926 B. van der Pol set before us a new type of oscillation which he named relaxational (van der Pol, 1926). It does not depend on inertia, but on capacitance and resistance. Its wave form is characterized by regions where the curvature is much greater than in any part of a sine curve of the same wavelength and same range of oscillation. Its period is subject to outside control to a much greater degree than is that of the spring-and-inertia vibration. The solutions of van der Pol's differential equation,

$$y'' - \epsilon\,(1 - y^2)\,y' + y = 0 \qquad \text{......(1),}$$

are of course strictly periodic. But just there the mathematics is a misleading idealization of the facts. Moreover the class of relaxation oscillations is much wider than those represented by that equation. To apprehend the essential distinction between pendular and relaxational oscillations we must attend to the physics. In the rising phase of a relaxation oscillation a capacitance of some sort is continuously filling. At a certain level of fullness the resistance breaks down and a sudden outflow occurs. The process then repeats itself. In each period there is an unstable phase when the capacitance is filled nearly to the breaking level; so that a small external disturbance will cause a breakdown prior to its normal time. In the swing of a pendulum there is no unstable phase, nor does one occur in the electrical oscillation of a condenser and self-inductance. B. van der Pol (1930) has drawn attention to a number of biological oscillations that are relaxational, notably the heart beat. It seems likely that many mental oscillations involve an unstable phase and for that reason their periodicity is variable. The widespread popularity of syncopated music is a sure indication that there is something in the human nervous system that oscillates with rhythm that is not regular.

But, to return to physical illustrations, a neon lamp arranged with a condenser, resistance and battery to flash after the manner of Pearson & Anson (1922) usually flashes with a period that is not steady.

The reason is interesting. If the time-interval since the previous flash has been sufficiently long, the electric current has entirely ceased. The current u will not depart from zero until the voltage ϕ across the lamp exceeds a critical value ϕ_e. The voltage meanwhile is rising in a continuous predictable sweep. But du/dt is proportional to $u(\phi-\phi_e)$. So that, even when $\phi>\phi_e$, the rate of growth of current is indefinitely slow, unless some accidental disturbance, such as incident radiation, gives the current a start. If suitable radiation is abundant, the period is fairly steady; if radiation is very scarce, the period becomes extremely erratic. [Braunbek (1926), Valle (1936), L. F. Richardson (1937)]. *Query*: Do cosmic rays start the release of stored brain-energy and so occasion sudden thoughts?

Again the Abraham-Bloch multivibrator will, when left alone, run with a period steady to 30 parts in 100,000; but it will also follow with an accuracy of 1 part in 100,000 a period differing from its natural period by ± 4000 parts in 100,000 supplied to it by a valve-controlled tuning fork (Dye, 1924). This is in contrast with the behaviour of pendulums. To control a pendulum, the period of the controlling impulse would have to agree much more closely with the free period of the pendulum; or else proportionately more energy would have to be supplied.

We may describe this contrast by saying that, as regards the interval between pulses, relaxation oscillations are docile, whereas spring-and-inertia oscillations are obstinate.

When observational data are to be analysed in the search for relaxation oscillations the established methods of periodogram analysis are unsuitable, for they depend on steadiness persisting over many periods: on action at a distance in time. Instead, following Faraday, we need to study action in a neighbourhood of time. Udny Yule (1927) has made an important step in this direction, although he thought in terms of pendulums. Yule (1927) contrasted two types of disturbance of simple harmonic motion. In the first type a pendulum swings regularly but the observations have random errors. The observed curve is spiky but the underlying period revealed by periodogram analysis is steady. In the second type of disturbance it is imagined for illustration that the observations are made with precision but that boys throw peas at the pendulum from opposite sides. The observed curve is smooth, but the phase and amplitude wander. For the reduction of the latter kind of observations Yule has devised a statistical method which is an approximation to the fitting of the differential equation

$$\frac{d^2y}{dt^2}+2k\frac{dy}{dt}+\omega^2y=0 \qquad \ldots\ldots(2)$$

to each successive small part of the curve.

Yule's data were given at equal intervals of t, the values of y being, say, $y_0, y_1, y_2 \dots y_n$. What he did in practice was to use the method of partial correlation to determine b_1 and b_2 in the regression equation

$$y_n = b_1 y_{n-1} - b_2 y_{n-2} \qquad \text{......(3).}$$

In this way k and ω can be determined.

If instead of (1) the phenomenon is described by van der Pol's equation,

$$\frac{d^2v}{dt^2} + \alpha\,(1 - v^2)\,\frac{dv}{dt} + \omega^2 v = 0 \qquad \text{......(4),}$$

a procedure rather similar to Yule's is indicated, but the regression will not be linear.

If this were done it would of course still be necessary to submit the derived statistics to the test of criteria of reality appropriate to the new method. But by computing in a neighbourhood, instead of over a long time, we should avoid the risk of having a real oscillation rejected as spurious merely because it had an unstable phase.

REFERENCES

BRAUNBEK, W. (1926). *Zschr. f. Physik.* XXXIX, 8.
DYE, D. W. (1924). *Philos. Trans.* A, CCXXIV, 259–301.
FLUGEL, J. C. (1928). *Brit. J. Psychol. Monog. Suppl.* No. 13.
—— (1934). *Indian J. Psychol.* pp. 1–48.
PEARSON, S. O. & ANSON, H. St G. (1922). *Proc. phys. Soc.* XXXIV, 174–6, 204–14.
PHILPOTT, S. J. F. (1932). *Brit. J. Psychol. Monog. Suppl.*, No. 17.
RICHARDSON, L. F. (1937). *Proc. Roy. Soc.* A.
SCHUSTER, A. (1906). *Philos. Trans.* A, CCVI, 69.
VALLE, G. (1936). *Sci. Abst.* A, CLXVII, Jan.
VAN DER POL, B. (1926). *Phil. Mag.* LI, 978–92.
—— (1930). *l'Onde Électrique*, IX, 293–312.
WALKER, Sir GILBERT T. (1936). *Quart. J. Roy. Met. Soc.* LXII, 1–2.
—— (1925). *Quart. J. Roy. Met. Soc.* LI, 337–46.
—— (1928). *Quart. J. Roy. Met. Soc.* LIV, 26 (correcting the previous paper).
—— (1927). *Mem. Roy. Met. Soc.* I, 119–26.
—— (1930). *Mem. Roy. Met. Soc.* III, 97–101.
YULE, G. UDNY (1927). *Philos. Trans.* A, CCXXVI, 267–98.

(*Manuscript received* 10 May 1937)

Richardson, L. F. (1938)
Rep. Br. Ass. Advmt Sci., 488–9

1938:2

GENERALISED FOREIGN POLITICS

British Association for the Advancement of Science
Report of the annual meeting, 1938
Cambridge, August 17–24
Transactions of Section J – Psychology pp. 488–489

Notes:

1. This is an abstract of Richardson's presentation to the meeting on 22 August 1938. The full report, which was 'expected to appear as a Monograph Supplement to British Journal of Psychology' (same report p. 1543), is 1939:1
2. At this meeting Richardson showed the diagram of the arms race of 1909–13, subsequently published in 1938:3

Dr. L. F. Richardson, F.R.S.—*Generalised foreign politics* (11.30).

Love and hate are alike in this: that the chief stimulus to either is any sign of the same feeling in the opposite person, or nation. The simplest mathematical expression of this mutual instinctive stimulation is

$$dx/dt = ky, \qquad dy/dt = kx \tag{1}$$

where k is a positive constant, t is time and x and y, when positive, are the intensities of hate in the two persons and, when negative, the intensities of love. The solution of (1) is

$$x = Ae^{kt} + Be^{-kt}, \qquad y = Ae^{kt} - Be^{-kt} \tag{2}$$

where A and B are constants; so that, as t becomes large, x and y both increase with the same sign. The point $x = 0, y = 0$ is a point of balance; but the balance is unstable. The instinctive drift may lead in one or other of two opposite directions.

The expression can be made more lifelike by the introduction of positive fatigue-and-expense coefficients α and of positive or negative grievances g so that

$$dx/dt = k_{12}y - \alpha_1 x + g_1, \qquad dy/dt = k_{21}x - \alpha_2 y + g_2 \tag{3}$$

The terms in α have a stabilising effect; so that instability only occurs if

$$k_{12} \,.\, k_{21} > \alpha_1 \, \alpha_2 \tag{4}$$

There are still two opposite kinds of drift. For nations, the positive infinity may be called 'war'; and we have to find a name for the negative infinity, which is obviously not tranquil exclusiveness, but may be called 'united organisation,' or 'close co-operation.'

Foreign trade is the commonest form of co-operation between nations. Alfred Marshall in his discussion of foreign trade, when near to zero, gave an argument closely related to equations (3). Accordingly, on looking for something objectively measureable which might serve to represent x and y, the author has formed two statistics of 'threats minus co-operation.' One is

$$\psi = (\text{population}) \log_{10} \left\{ \frac{(\text{warlike expenditure})}{(\text{foreign trade}) \,.\, (\text{constant})} \right\}$$

the constant being adjusted to make ψ zero for an average nation in the year 1926.

So far it appears that assumptions of type (3) are a credible approximation to actuality. Accordingly they have been generalised for n nations thus

$$\frac{dx_i}{dt} = g_i + \sum_{j=1}^{j=n} k_{ij}\, x_j \qquad (i = 1, 2, 3 \ldots n)$$

and the criterion of stability has been deduced. In many cases there is an instinctive barrier separating two regions of opposite drift; the barrier goes through a point of balance, the position of which depends on the grievances g.

The relation between g and objective facts is very peculiar. A halting apology may be received as though it were a fresh insult.

The practical conclusion is that the traditional policy of the balance of power is now futile, because the balance is unstable. To bring the point (x, y) into the region where the instinctive drift goes towards more co-operation, two actions can be taken: (i) the barrier can be heaved to the positive side by abolishing grievances g, and (ii) the point (x, y) can be heaved to the negative side by decreasing threats. Neither of these actions is instinctive; both would require national efforts of will.

1938:3

THE ARMS RACE OF 1909–13

Nature
London
Vol. 142, No. 3600, 29 October 1938, pp. 792–793
(Letter dated 18 September)

Note: The 'much fuller discussion' referred to is 1939:1

The Arms Race of 1909–13

THE interaction of fear with cost and with grievances was represented in a letter in NATURE (Dec. 25, 1935) by the equations

$$dx/dt = ky - \alpha x + g \ , \quad dy/dt = lx - \beta y + h, \qquad (1), (2)$$

in which t is time and α, β, k, l are positive constants, and g and h are positive or negative constant grievances. In that publication, x was described as the variable preparedness for war of the first group of nations, y that of the second. But on further consideration, it appears that the description of x given in 1935 needs to be modified by taking into account the co-operation between the groups of nations which goes on as trade, travel and correspondence concurrently with their mutual threatenings. The improved description of x is in general terms

x = threats minus co-operation.

In the hope of reaching quantitative measures, let us try

$$x = U - U_0 \ , \quad y = V - V_0, \qquad (3), (4)$$

in which U is the annual defence budget for the first group of nations, V that for the second group, and U_0, V_0 are measures of co-operation. U, V, U_0, V_0 are all here expressed in million pounds sterling. As a tentative assumption, let U_0, V_0 be regarded as constants during the arms-race (5).

In 1909 France was allied with Russia, Germany with Austria-Hungary. These two pairs of nations were very roughly equal, so that we may simplify (1) and (2) by putting $k = l$, $\alpha = \beta$. Then, by addition, it follows that

$$\frac{d(x+y)}{dt} = (k-\alpha)(x+y) + g+h \, ;$$

and by substitution of (3), (4), (5)

$$\frac{d(U+V)}{dt} = (k-\alpha)(U+V) + \{g+h - (k-\alpha)(U_0+V_0)\}.$$

This implies that when $d(U+V)/dt$ is plotted against $U+V$, we should expect a straight line of slope $k-\alpha$. The statistics for Austria-Hungary have been taken from the "Statesman's Year Book", those for the other countries from a pamphlet by Per Jacobsson[1]. The accompanying diagram shows that the four points lie close to a straight line of slope $k-\alpha = 0{\cdot}73$ year^{-1}. Furthermore, by a short extrapolation, the line cuts the axis of zero $d(U+V)/dt$ at $U+V = 194$. This 194 million sterling is the amount of defence expenditure, by the four nations concerned, that would just have been mutually forgiven in view of the amount of goodwill then existing.

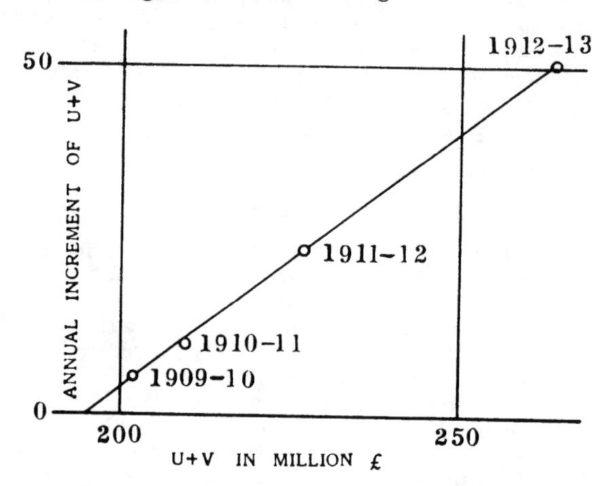

It is, to say the least, a remarkable coincidence that the trade between these opposing pairs of nations was on the average 206 millions sterling, close to 194.

A much fuller discussion is due to appear under the title "Generalised Foreign Politics". In particular, the assumption that U_0, V_0 were constant cannot be expected to remain valid for long periods of time. Also the budgets are variously stated.

Lewis F. Richardson.

38 Main Road,
Castlehead,
Paisley.
Sept. 18.

[1] Jacobsson, "Armaments Expenditure of the World," *The Economist*, London.

1939:1

GENERALIZED FOREIGN POLITICS

A study in group psychology

The British Journal of Psychology Monograph Supplements XXIII
Cambridge: University Press
[June] 1939, pp. vi + 91

(Preface dated 12 December 1938, with a postscript dated 19 March 1939)

Notes:

1. During the Second World War this monograph was considerably enlarged and extended to become the book *Arms and Insecurity* (1947:2, 1949:2, 1960:2); most of the monograph was incorporated unchanged into the book. See Note 3 of the 'Notes on *Arms and Insecurity*' (below p. 419)
2. The particular topic of 'submissiveness' (below p. 282) is developed in 1944:1 and in *Arms and Insecurity* (Chapters IV, XXIV); and then in 1951:3, 4, 5 and 1953:2
3. In the index at the end of this monograph the pagination is that of the original

THE BRITISH JOURNAL OF PSYCHOLOGY

MONOGRAPH SUPPLEMENTS

General Editor: JAMES DREVER, M.A., B.Sc., D.Phil.

XXIII

GENERALIZED FOREIGN POLITICS

by

LEWIS F. RICHARDSON
B.Sc. (Psychol.), D.Sc. (Physics), F.R.S.

CAMBRIDGE
UNIVERSITY PRESS
LONDON: BENTLEY HOUSE
CHICAGO
The University of Chicago Press
(Agents for the United States)
BOMBAY CALCUTTA MADRAS
Macmillan
TOKYO
Maruzen Company Ltd

GENERALIZED FOREIGN POLITICS

A study in group psychology

by

LEWIS F. RICHARDSON

B.Sc. (Psychol.), D.Sc. (Physics), F.R.S.

CAMBRIDGE

AT THE UNIVERSITY PRESS

1939

PRINTED IN GREAT BRITAIN

PREFACE

For the convenience of readers of various experience, the gist is explained at the beginning without mathematics, and again at the end; the interaction of two nations is analysed by elementary graphical methods; and the more difficult mathematics is collected in § 2·11 and §§ 6–13, where it can be skipped.

The author has tried to make a piece of objective science, avoiding, with what success others will judge, the bias due to his English ancestry, Quaker upbringing and long residence in Great Britain; to all of which influences he is affectionately attached. He has often been in doubt as to how to vote in Parliamentary Elections.

The seed from which the present monograph grew was perhaps the Hon. Bertrand Russell's remarkable pamphlet *War, the Offspring of Fear*, published by the Union of Democratic Control soon after the outbreak of the Great War. During the latter half of that hateful struggle the author was a member of a motor ambulance convoy (S.S.A. 13) attached to the 16th French infantry division. Thus for two years the author was 'not paid to think', and so had abundant opportunities for meditation. One result was an essay on the *Mathematical Psychology of War*. This was published in typescript on 25 February 1919 (Wm Hunt, 18, Broad Street, Oxford), and was almost entirely inductive: that is to say mathematics were used to describe phenomena, but hardly any consequences were deduced.

For the next sixteen years the author neglected the theory, as international affairs were comparatively tranquil. Then in 1935, after the failure of the Disarmament Conference, the author felt it to be time to reconsider and to republish the theory. In the version of 1919 an attempt had been made to tell the whole truth; and in consequence the equations were complicated and intractable. In 1935 the theory of war was omitted and the statements concerning peace were selected and simplified so as to make deductions convenient. The first of these results were published, thanks to the editor of *Nature*, on 18 May and 25 December 1935. During the years 1937–8 the typescript grew. It was sent successively to three editors, and after each rejection was clarified and extended.

The present Part II was made in response to a criticism by Prof. F. C. Bartlett, F.R.S., who suggested that a run of empirical data was needed. In that connexion the author wishes to acknowledge helpful advice and information received from Mr Alex Wilson, Miss G. A. Gegg and other members of the headquarters staff of the League of Nations Union.

An abstract was read to the British Association in August 1938. The present publication may correct some misleading newspaper reports of that occasion.

The author is grateful to the British Psychological Society, via Prof. Drever, for the honour of a place among their monographs, and for a grant of £50 towards the cost of publication. Mr Babington Smith, a psychologist and mathematician, has kindly consented to read the proofs. It is fortunate that the mathematics will be set in type by the Cambridge University Press.

L.F.R.

12 *December* 1938

Postscript, 19 March 1939. Important statistics have come to hand while the sheets were at the Press. The *Armaments Year Book* 1938 has been used in the revision of Figs. 15–19. Subsequently Balogh's work on German expenditure was brought to the author's notice by Mr Douglas Jay and was introduced on pp. 47, 78, 79 together with a device for smoothing gold roubles.

L.F.R.

CONTENTS

PART I. INTRODUCTION; TWO NATIONS

PART II. STATISTICAL DISCUSSION OF SEVEN GREAT POWERS FOR 1922–38

CONTENTS

PART I. INTRODUCTION; TWO NATIONS

§1. GENERAL OBJECTIONS

§ 1·1. Freewill.

Chairman. 'I now call on Dr Richardson to explain his science of foreign politics.'

Critic. 'Sir! I beg to move the previous question: that we do not waste our time on such an absurdity. How can anyone possibly make scientific statements about foreign politics? These are questions of right, of loyalty, of power, of the dignity of free choice. They touch a little on law, but are far beyond the reach of any science.'

Author. 'I admit that the discussion of free choice is better left to the dramatists. But nowadays science does usefully treat many phenomena that are only in part deterministic, witness the many social applications of statistics and the astounding progress of theories as to the probable position of an electron.'

Critic. 'The electron? That surely is irrelevant. On glancing at your summary I see mathematical equations with the symbol t in them. Does t stand for time?'

Author. 'You have guessed rightly.'

Critic. 'Can you predict the date at which the next war will break out?'

Author. 'No, of course not. The equations are merely a description of what people would do if they did not stop to think. Why are so many nations reluctantly but steadily increasing their armaments as if they were mechanically compelled to do so? Because, I say, they follow their traditions which are fixtures and their instincts which are mechanical; and because they have not yet made a sufficiently strenuous intellectual and moral effort to control the situation. The process described by the ensuing equations is not to be thought of as inevitable. It is what *would occur if instinct and tradition were allowed to act uncontrolled*. In this respect the equations have some analogy to a dream. For a dream often warns an individual of the antisocial acts that his instincts would lead him to commit, if he were not wakeful.'

Chairman. 'I think we might go on.'

2 TWO NATIONS

§ 1·2. Lack of detail.

This theory is about general tendencies common to all nations; about how they resent defiance, how they suspect defence to be concealed aggression, how they respond to imports by sending out exports; about how expenditure on armaments is restrained by the difficulty of paying for them; and, lastly, about grievances and their queer irrational ways, so that a halting apology may be received as though it were an added insult.

A rule of the theoretical game is that a nation is to be represented by a single variable, its outward attitude of threatening or co-operation. So the great statesmen, who collect, emphasize and direct the national will, need not be mentioned by name. This is politics without personalities. It must seem, to most newspaper readers, to be a miserably meagre way of treating the subject.

A like objection was probably made to the early geometers in ancient Greece, thus:

Poet. 'I am really sorry, my dear geometer, to see that you spend your time in thinking about abstractions. Your straight lines, how flat and dull they are, your triangles, how angular and ugly, your circles how empty and lifeless! The world is not filled with geometrical forms, but with mountains, waves, grass, grapes, children, and with many other beauteous living things.'

Geometer. 'But, alas, I have no power to think accurately about mountains, waves, grass, grapes and children. My own children, I confess, puzzle me extremely; I never know what they will do next. But about straight lines, triangles and circles, I, and many others who have learnt the science, can reason clearly and definitely. This power grows among the Greeks. It has much developed since the time of our grandfathers. There is a hope that some day we shall have a geometry of waves, perhaps even of grapes and of children.'

Whether a science should be made in detail or in broad outline is sometimes a question not of truth but of convenience. A well-known illustration is afforded by the various theories of gases, all of them true but appropriate to different purposes. It is correct to regard the wind as composed of about 10^{19} molecules per c.c. moving in all directions with speeds averaging about 10^4 cm. sec.$^{-1}$ For other purposes it is more convenient and sufficiently correct to regard the wind as composed of eddies in a continuous fluid endowed with viscosity. The process of smoothing away the detail may be carried to a further stage by ignoring the forms and motions of the eddies while taking statistical account of their average effect by means of an eddy viscosity, in the manner familiar to meteorologists (Brunt, 1934).

A similar freedom to choose either detail or else generalities may be expected in the social sciences. Foreign affairs as they appear day by day in the newspaper: the text of the despatch, the facial expression of the ambassador as he comes away from the important interview, the movements of warships, these may be likened to the eddying view of a wind. Whereas the theory here presented may be likened to an account of the general circulation of the earth's atmosphere. Just as in meteorology it has been found necessary to endow the air with an eddy viscosity to compensate for the motions which we cannot study in detail, so in the social sciences it is to be expected that there will be a need to regard the social group as endowed with properties, analogous to viscosity, arising from the lack of agreement in the purposes of the individuals which compose the group. Although no symbol called 'social viscosity' occurs explicitly in the following equations, yet they are of viscous type in so far as the 'velocity' not the 'acceleration' is proportional to the 'forces'. The effects of social viscosity must be regarded as included in the other constants.

§ 2. LINEAR THEORY OF TWO NATIONS

§ 2·1. Introduction.

Replies to objections of a general character having now been given, we are ready for the gist of the problem.

The various motives which lead a nation in time of peace to increase or decrease its preparations for war may be classified according to the manner of their dependence on its own existing preparations and on those of other nations. For simplicity the nations are here regarded as forming two groups. There are motives such as revenge or dissatisfaction with the results of treaties; these motives are independent of existing armaments. Then there is the very strong motive of fear which moves each group to increase its armaments because of the existence of those of the opposing group. Also there is rivalry which, more than fear, attends to the difference between the armaments of the two groups rather than to the magnitude of those of the other group. Lastly, there is always a tendency for each group to reduce its armaments in order to economise expenditure and effort.

What the result of all these motives may be is not at all evident when they are stated in words. It is here that mathematics can give powerful aid. But before the mathematician can get to work on the problem, the data must be stated precisely; and precision may seem inappropriate to sociology. Would it not be more politic to remain content with a modest vagueness? No! For unless our statements are in terms so precise that it would be possible in them to make a definite mistake, it will be impossible in them to make a definite advance in science.

Permit me to discuss a generalized public speech, fictitious but typical of the year 1937. The Defence Minister of Jedesland, when introducing his estimates, said:

> The intentions of our country are entirely pacific. We have given ample evidence of this by the treaties which we have recently concluded with our neighbours. Yet, when we consider the state of unrest in the world at large and the menaces by which we are surrounded, we should be failing in our duty as a Government if we did not take adequate steps to increase the defences of our beloved land.

We have now to translate that into mathematics. At first sight there might seem to be no way of doing it, and on second thoughts perhaps too many ways. But we can shave the problem clean with Ockham's razor. This principle in its usual form runs 'entities are not to be multiplied without necessity'. Mathematical physics has progressed by trying out first the simplest formulae that described the broad features of the early experiments and by leaving complicated formulae to wait for more accurate experiments. By 'entities' let us understand terms and coefficients in formulae, and let us restate Ockham's rule as 'formulae are not to be complicated without good evidence' or briefly 'try out the easiest formulae first'.

Now the simplest representation of what that generalized defence minister said is this

$$dx/dt = ky, \tag{1}$$

where t is time, x represents his own defences, y represents the menaces by which he is surrounded and k is a positive constant, which will be named a 'defence coefficient'. Let us for simplicity assume that what he euphemistically called 'surroundings' is in fact a single nation. Its defence minister asserts similarly that

$$dy/dt = kx. \tag{2}$$

The solution of this pair of differential equations is

$$x = Ae^{kt} + Be^{-kt}, \tag{3}$$

$$y = Ae^{kt} - Be^{-kt}, \tag{4}$$

where A and B are arbitrary constants. Because k is positive, therefore $e^{kt} \to \infty$ as $t \to \infty$. The system described by these equations is unstable. It must be understood that the equations are valid only for small or moderate disturbances. But can the international system really be inevitably unstable? Surely not. We have left out the cost of armaments which has a restraining effect. So let the equations be improved into

$$dx/dt = ky - \alpha x, \tag{5}$$

$$dy/dt = lx - \beta y, \tag{6}$$

where α and β are positive constants representing the fatigue and expense of keeping up defences, and k and l are positive defence coefficients, which we now regard as possibly unequal.

Let us compare this statement with some opinions of statesmen. Sir Edward Grey, who was British Foreign Secretary when the Great War broke out, afterwards wrote (Grey, 1925, vol. I, p. 92):

> The increase of armaments that is intended in each nation to produce consciousness of strength, and a sense of security, does not produce these effects. On the contrary, it produces a consciousness of the strength of other nations and a sense of fear. The enormous growth of armaments in Europe, the sense of insecurity and fear caused by them—it was these that made war inevitable.... This is the real and final account of the origin of the Great War.

Sir Edward Grey's statement is symbolized by the terms in k and l. Compare also Thucydides' account of the cause of the Peloponnesian war: 'The real though unavowed cause I believe to have been the growth of Athenian power, which terrified the Lacedaemonians and forced them into war;...' (Jowett, 1881, p. 16). When this opinion of Sir Edward Grey's was quoted by Mr Noel Baker, M.P., in the House of Commons on 20 July 1936, Mr L. S. Amery, M.P., said in reply:

> With all respect to the memory of an eminent statesman, I believe that statement to be entirely mistaken. The armaments were only the symptoms of the conflict of ambitions and ideals, of those nationalist forces, which created the war. The War was brought about because Serbia, Italy, Rumania, passionately desired the incorporation in their States of territories which at that time belonged to the Austrian Empire and which the Austrian Government were not prepared to abandon without a struggle. France was prepared if the opportunity ever came to make an effort to recover Alsace-Lorraine. It was in those facts, in those insoluble conflicts of ambitions and not in the armaments themselves that the cause of the War lay.

Mr Amery's objections should, I think, be met, not by leaving out Sir Edward Grey's terms, but by inserting additional terms, namely,

g and h, to represent grievances, provisionally regarded as constants, so that the equations become

$$dx/dt = ky - \alpha x + g, \tag{7}$$

$$dy/dt = lx - \beta y + h. \tag{8}$$

These equations were published in 1935 (Richardson, 1935*a, b*), as a simplification of an earlier theory (Richardson, 1919).

The preparations for war of the two groups may be regarded as the rectangular co-ordinates of a particle in an 'international plane', so that every point in this plane represents one conceivable instantaneous international situation. The differential equations are then the equations of motion of the particle.

A hint will now be taken from Plato's advice concerning astronomy (*Republic*, VII, § 530). We shall pursue foreign politics with the help of problems, just as we pursue geometry.* Statistics will follow in § 2·10.

If g, h, x, y are all made zero simultaneously, the equations show that x and y remain zero. That ideal condition is *permanent peace by disarmament and satisfaction*. It has existed since 1817 on the frontier between U.S.A. and Canada, also since 1905 on the frontier between Norway and Sweden. (For the treaties see *League of Nations Armaments Year-book*, 1937, pp. 918, 923).

The equations further imply that *mutual disarmament without satisfaction* is not permanent, for if x and y instantaneously vanish, $dx/dt = g$ and $dy/dt = h$.

Unilateral disarmament corresponds to putting $y = 0$ at a certain instant. We have at that time

$$dx/dt = -\alpha x + g, \quad dy/dt = lx + h.$$

The second of these equations implies that y will not remain zero if the grievance h is positive; later, when y has grown, the term ky will cause x to grow also. So, according to the equations, unilateral disarmament is not permanent. This accords with the historical fact that Germany, whose army was reduced by the Treaty of Versailles in 1919 to 100,000 men, a level far below that of several of her neighbours, insisted on rearming during the years 1933–6.

A *race in armaments*, such as was in progress in 1912, occurs when the defence terms predominate in the second members of the equations. If those were the only terms we should have

$$dx/dt = ky, \quad dy/dt = lx,$$

and both x and y would tend to the same infinity, which, if positive, we may interpret as war. But for large x and y linearity may fail.

* But tentatively.

§ 2·2. The opposite to war.

The generalizing spirit of mathematics is very suggestive. It draws our attention to the possibility of 'negative preparedness for war' and invites us to assign a name to it, and to enquire whether the above general statements still hold true when the signs of the variables are changed. As a preliminary let us consider, not nations, but only two people, and let us compare quarrelling with falling in love. If hatred may be regarded as negative love, these two activities are opposite. Yet there are important resemblances between them. The chief stimulus to falling deeply in love is any sign of love from the other person; just as the chief stimulus to becoming more annoyed is any insult or injury from the other person. When quarrelling is represented by $dx/dt = ky$, and $dy/dt = lx$, these same two equations with k and l still positive represent falling in love when x and y are both negative.

Now returning to the study of nations, we notice that the classical antithesis of 'war or peace' is not appropriate here. For war is an intense activity, whereas peace, in the sense of a mere tranquil inattention to the doings of foreigners, resembles zero rather than a negative quantity. Negative preparedness for war must mean that the group directs towards foreigners an activity designed to please rather than to annoy them. Thus a suitable name for negative preparedness for war seems to be 'co-operation'.*

Just as armaments provoke counter armaments, so assistance evokes reciprocal assistance; for example, imports and exports tend to equality. Also there is a tendency to reduce co-operation on account of the fatigue and expense which it involves. Thus it appears that the general statements remain broadly true when the preparedness changes from positive to negative. The most extensive form of co-operation is foreign trade. There exist also more than 500 international associations, having little or no connexion with one another, but listed by the League of Nations (1936). The extreme form of co-operation, corresponding to the infinity opposite to war, would appear to be a world-state, as imagined for instance by H. G. Wells in his book *The Shape of Things to Come* (1933).

Quarrelling is here regarded as a positive activity, making friends as a negative activity. That happened because the equations were first written during the Great War. It might be better now to reverse the convention of signs. But the change has not been made in this monograph.

* See F. C. Bartlett's plan of important research (Carmichael, 1939).

8 TWO NATIONS

§ 2·3. Grievances.

The effect of a change in g has not been easy to observe during the last 19 years; for although some of the smaller grievances that existed in 1919 have since been removed, yet other greater grievances have remained unremedied. In private life a halting apology may be received as though it were an added insult; and it would seem that the partial removal of a great national grievance may sometimes stimulate effort for its total abolition. Such a psychological effect cannot be properly described by linear equations. But we can go some way towards its description by saying that g is not a continuous variable; but is one that changes only in large steps. (Compare Seton-Watson, 1938, p. 98.)

One of the best examples of the settlement of a dispute by the League of Nations occurred in October 1925 when the Greeks invaded Bulgaria. The invasion was stopped by the League Council. The Council's Commission under Sir Horace Rumbold fixed the blame and Greece paid £45,000 damages. The dispute was settled towards the close of 1925 (Drummond, 1930; Garnett, 1936). Let us see whether this caused any diminution of Bulgarian armaments. Here are the annual budget expenditures on defence by Bulgaria. The financial year then ran from 1 April to 31 March:

Year	1924–5	1925–6	1926–7	1927–8	1928–9	1929–30	1930–1
In millions of leva	1325	1222	1201	1032	1039	1078	1033
In millions of old gold U.S.A. dollars	9·65	8·90	8·66	7·44	7·48	7·77	7·43

It is seen that, after a delay of about $1\frac{1}{4}$ years, a distinct diminution occurred and persisted.

It will be interesting to see whether the attempt at appeasement in the Mediterranean that was made in March 1938 will result in a reduction of Italian and British expenditure on defence.

§ 2·4. Units and order of magnitude of the constants.

In accordance with the well-known theory of physical dimensions (Porter, 1929) it is obvious that, whatever be the units of x and y, those of α and β are reciprocals of a time. Physicists would call α^{-1} and β^{-1} '*relaxation times*'. For if y and g were permanently zero,* equation (7) would simplify to $dx/dt = -\alpha x$, from which it can be deduced that $\log_e x = -\alpha t + \text{const}$. That is to say x would decrease in the ratio 2·718 whenever αt increased by unity. The question therefore is: How long time would be required for a State's armaments to be reduced in the ratio 2·718 if that State had no grievances and if no

* Together with $l=0$, $h=0$.

other State had any armaments? These circumstances are hypothetical. The answer must allow for the time required for the public to become convinced of the existence of the supposed circumstances, also for the opposition of those who would be done out of paid jobs; these effects are examples of the 'social viscosity' mentioned above. Would a reasonable guess for Great Britain be the lifetime of a Parliament, say $\alpha^{-1} = 5$ years, $\alpha = 0{\cdot}2\ (\text{year})^{-1}$? It may be objected that in 1919 Great Britain effected demobilization more rapidly. The answer to this objection is that, in many other parts of science, linear equations are valid only for moderate disturbances from equilibrium, and that we can hardly expect them to apply to so violent a disturbance as the War.

It seems likely that the fatigue and expense coefficient may be of the same 'order of magnitude' ($\div$ or $\times\,3$) for all nations.

We now turn to the defence coefficients k and l. Provided that x is measured in the same units as y, it is obvious that k and l are each the reciprocal of a time. To estimate this time, let us take a hypothetical case in which $g = 0$ and $y = y_1$, a constant, so that

$$dx/dt = ky_1 - \alpha x.$$

At the instant when $x = 0$,

$$\frac{1}{k} = \frac{y_1}{dx/dt},$$

so that $1/k$ is the ratio of the distance y_1, which x has to go, to its speed dx/dt. Accordingly we may call $1/k$ the 'apparent catching up time in an arms race from zero with no grievances'. The word 'apparent' is inserted because it must not be implied that any catching up will necessarily occur. The notion is: 'Beginning at the speed we are going now, we will be equal to them in $1/k$ years if they stand still and we do not slow down.' To estimate $1/k$ we may take the German re-armament during 1933–6. It did start nearly from zero. They did catch up in about 3 years, in spite of the slowing effect of α. So $k = 0{\cdot}3\ \text{year}^{-1}$; very roughly, for they had a strong grievance.

But suppose that the x-nation were much more populous and industrially organized than the y-nation. The apparent time required for the x-nation to catch up with y would be much shorter than the apparent time for y to catch up with x. That is to say: *other things being equal the defence coefficient is proportional to the 'size' of the nation*; size being measured, not by area of land, but as some increasing function of population and industry.

The defence coefficient of the x-nation is also, for positive x and y, increased by any suspicions and decreased by any neutrality, confidence or pacifism that the x-nation may have.

§ 2·5. Equilibrium.

If dx/dt be zero, the equation (7) represents a straight line; similarly for dy/dt and equation (8). The intersection (x_0, y_0) of these two straight lines is the *point of balance of power*. It is given by

$$x_0 = (kh + \beta g)/(\alpha\beta - kl), \quad (9)$$

$$y_0 = (lg + \alpha h)/(\alpha\beta - kl). \quad (10)$$

But, unfortunately for those who live under a policy of balance of power, the point of balance has several awkward possibilities. First, by a rare chance, $\alpha\beta$ might equal kl so that the lines would be parallel and no point of balance would exist. Secondly, it is quite probable that the point of balance may be situated in one of the quadrants where x or y or both are negative; so that balance could be attained only by negative preparedness for war, that is, by trade or some other positive form of co-operation. Thirdly, although the particle is undoubtedly in equilibrium when it coincides with the point of balance of power, yet that equilibrium may be unstable.

§ 2·6. Stability.

Considerable insight into the meaning of the equations can be gained with the aid of elementary mathematics. Let us shade with lines parallel to the x-axis those portions of the international plane for which dx/dt is positive. Similarly, let us shade with lines parallel to the y-axis wherever dy/dt is positive (see Figs. 1–8). The equilibrium lines are not actually drawn on these diagrams, but they are visible nevertheless as the boundaries of the shaded areas. A region shaded both ways is one where the drift is towards hostilities. A region shaded neither way is one where the drift is towards more co-operation. Very much depends on whether these drifts lead towards or away from the point of balance. If towards we may expect the balance to be stable; if away, unstable. Now dx/dt is positive on that side of the line

$$ky - \alpha x + g = 0 \quad (11)$$

on which y is greater. Similarly dy/dt is positive on that side of the line

$$lx - \beta y + h = 0 \quad (12)$$

on which x is greater. If the slope dy/dx of the line (12) exceeds that of the line (11), there will be a region in which both dx/dt and dy/dt are

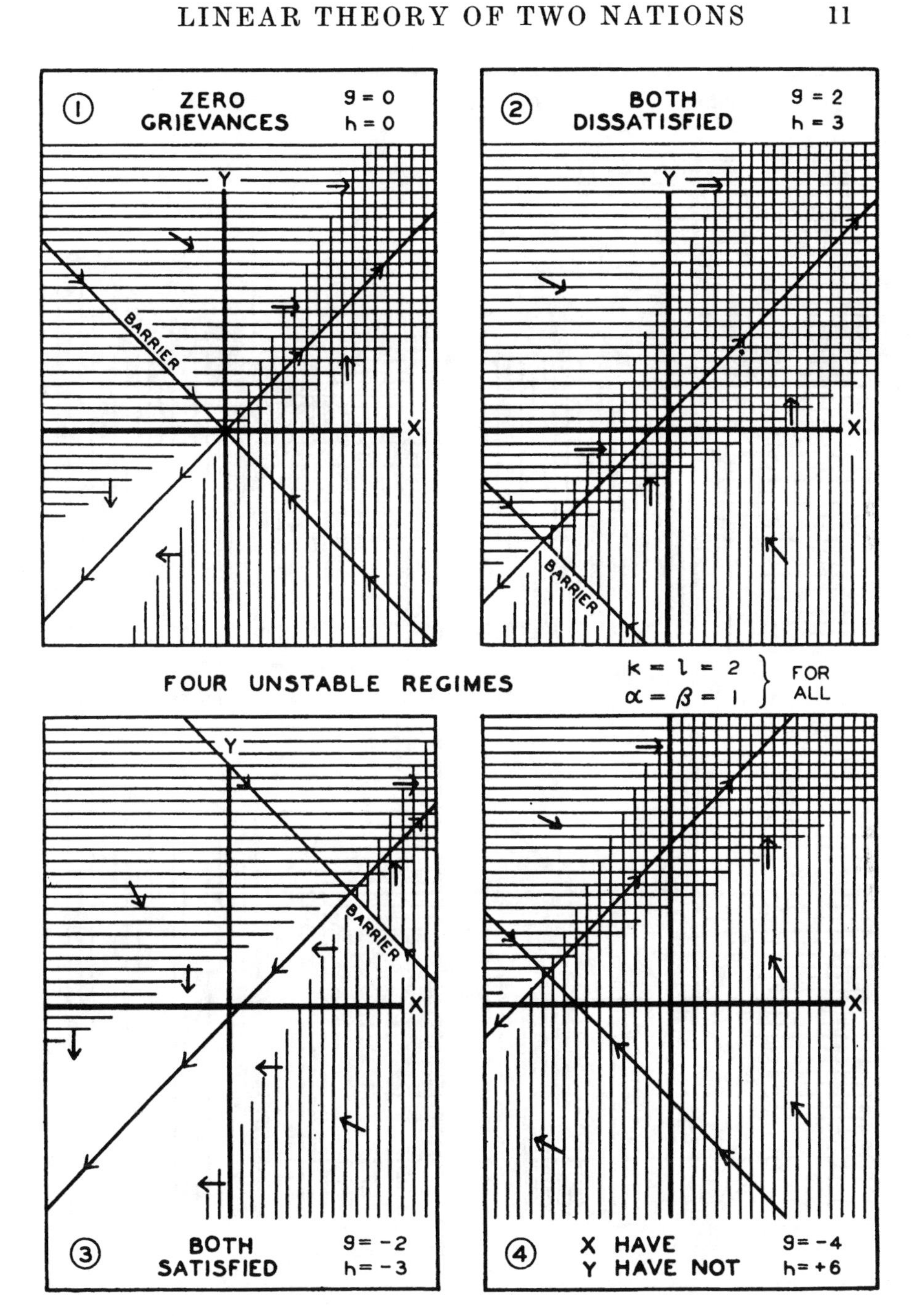

FOUR UNSTABLE REGIMES

12 TWO NATIONS

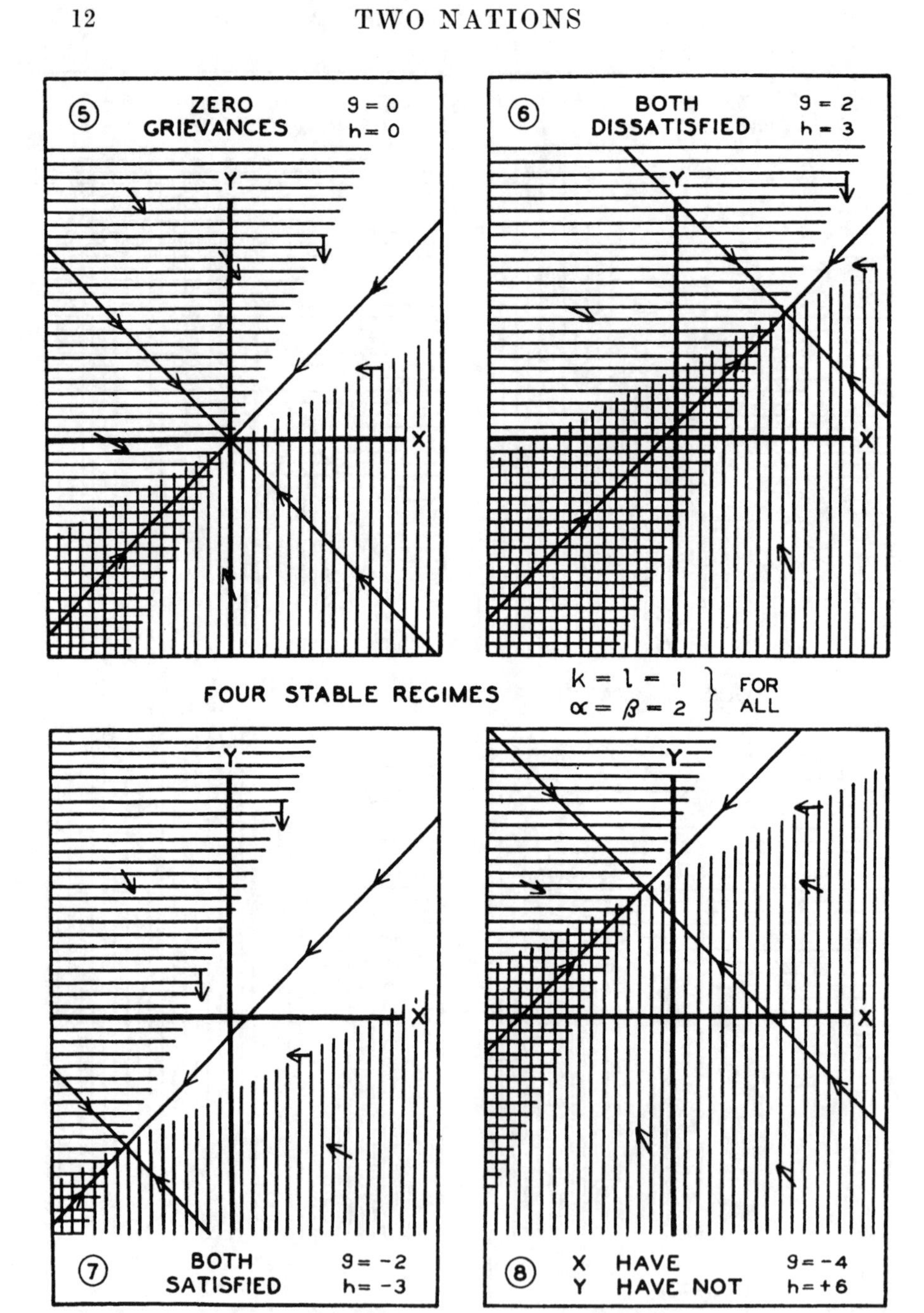

positive lying on that side of the point of balance (x_0, y_0) on which $x > x_0$ and $y > y_0$; and accordingly the regime will be unstable. But for (11),

$$dy/dx = \alpha/k,$$

and for (12),

$$dy/dx = l/\beta.$$

Instability therefore occurs when $l/\beta > \alpha/k$, that is, when

$$lk > \alpha\beta. \tag{13}$$

By a similar argument it is easily shown that the regime will be stable if

$$lk < \alpha\beta. \tag{14}$$

§ 2·7. Description of the diagrams Nos. 1–8.

In Figs. 1–4 are shown four unstable regimes with different values of the grievances g and h. The arrows indicate the direction of motion but not its speed. It is particularly important to notice that the denominators in equations (9) and (10) are negative in the unstable regime so that when both grievances g and h are positive the point of balance lies in the quadrant where x and y are negative. Four cases of stability are shown in Figs. 5–8. Here the denominators of (9), (10) are positive, so that when both grievances are positive, x_0 and y_0 are positive. When g and h have opposite signs one must refer to equations (9) and (10) to get any idea about the position of the point of balance.

In Figs. 1–8 the unit of x and y is four spaces between rulings. The diagrams have been drawn with k, l, α, β equal to either 1 or 2 for simplicity; whereas it has previously been argued that the values of these constants are of the order of $0{\cdot}3\ \text{year}^{-1}$. To bring the diagrams into agreement with this mean value, all that we need to do is to choose 5 years for our unit of time when we are using these diagrams. For $0{\cdot}3\ \text{year}^{-1} = 1{\cdot}5\ (5\ \text{year})^{-1}$.

§ 2·8. The barriers. (Not to be confused with trade-barriers.)

In the unstable regime, the particle will tend towards one or other of two opposite infinities according to where it begins. We may classify the starting points by making a transformation to polar co-ordinates (r, θ) centred at the point of balance, so that

$$x - x_0 = r\cos\theta, \tag{15}$$

$$y - y_0 = r\sin\theta. \tag{16}$$

For then it may be deduced from (7) and (8) that

$$d\theta/dt = l\cos^2\theta + (\alpha-\beta)\cos\theta\,\sin\theta - k\sin^2\theta, \tag{17}$$

whence it follows that there are always two real lines given by

$$2k \tan \theta = \alpha - \beta \pm \sqrt{\{(\alpha-\beta)^2 + 4lk\}}, \tag{18}$$

on which $d\theta/dt = 0$. The particle cannot cross these lines. They form barriers dividing the plane into four angular portions. In particular, if $\alpha = \beta$ and $l = k$, the barriers are at $\theta = \pm\frac{1}{4}\pi \pm \pi$. When the balance of power is unstable ($kl > \alpha\beta$) the barrier at $\theta = -45°, +135°$, or in general at

$$2k \tan \theta = \alpha - \beta - \sqrt{\{(\alpha-\beta)^2 + 4lk\}} \tag{19}$$

is of profound significance, for it divides the international plane into two portions, one of a drift towards war, the other of a drift towards closer co-operation.

The fact that the international situation is in some ways more easily discussed in the variables r, θ than in the variables x, y is an illustration of the extent to which the doings of nominally sovereign and independent nations are really controlled by their neighbours.

§ 2·9. International trade regarded as opposite to war.

International co-operation in all its varieties, including trade, was introduced into this wartime theory as an afterthought. But the possibility of more co-operation is the most practically hopeful feature of the situation.

International trade was discussed by Marshall (1923) with the aid of a diagram very similar to ours. For simplicity it is imagined that there are two nations trading only with one another. By ignoring credit, the problem is reduced to the barter of bales of goods. Each bale is supposed to have required the same amount of labour and capital for its production. By an x-bale is meant a bale produced by the x-nation. The co-ordinates (x, y) of a point represent the number of x-bales and y-bales respectively that are interchanged in a standard time, say a year, between the x-nation and the y-nation. Marshall draws two curves, one showing the terms on which the x-nation would be just willing to trade, the other showing the terms on which the y-nation would be just willing. Where the curves intersect, both are willing; and the intersection is called by Marshall a point of equilibrium.

Although he does not write a pair of simultaneous differential equations corresponding to our (5), (6), he does give an economic argument about direction of motion on either side of the curves, and thus he deduces the stability of the point of equilibrium by a geometrical argument somewhat similar to that used above.

It will suffice here to consider the diagram which Marshall calls normal, ignoring various exceptional possibilities. In order to express the fact that international trade between nations is on the whole an activity to their mutual advantage, and is thus emotionally opposite to preparations for war between them, the direction of Marshall's co-ordinates must be reversed before comparison with ours. Accordingly that part of Fig. 9 where x and y are both negative has been obtained by taking Marshall's normal Fig. 10, reversing its co-ordinates,

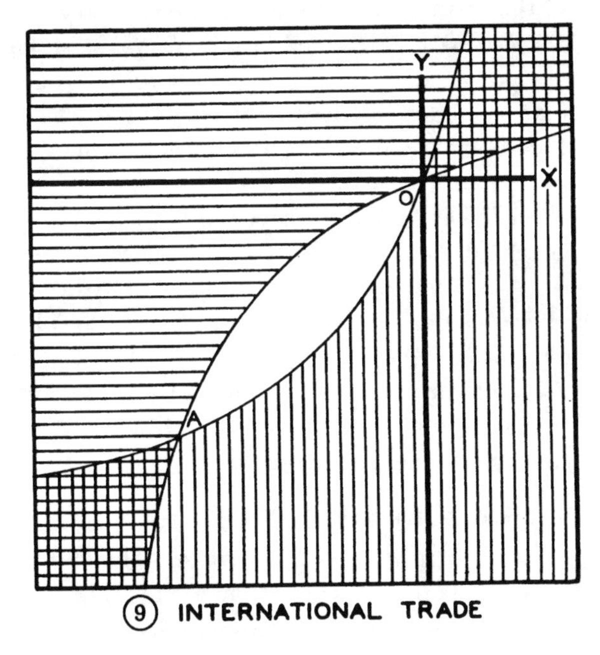

(9) INTERNATIONAL TRADE

and adding shading on our convention to represent Marshall's statements about direction of motion. The shading shows that the point A of balance of trade is stable. Now comes the remarkable feature. If we make the assumption that the trade curves may be produced, without any sudden change of direction, through the point of zero trade, we see that zero trade is an equilibrium unstable on both sides. On one side the tendency is towards more trade until A is reached; on the other side the tendency is towards an increase of negative trade, which may be interpreted as mutual recriminations, threats and injuries. In fact the neighbourhood of the origin of Fig. 9 is very like that of Fig. 1.

Usually both threats and co-operation are going on simultaneously; so that, as we have restricted ourselves to a single variable x for one group, that variable must represent the net result, the excess of threats over co-operation; and similarly for y.

Good reasons were given in § 2·2 for believing that the coefficients k and l remain of the same sign, while the variable x and y change from positive to negative, that is from threats to co-operation. It is also almost obvious that α, β, g, h retain their signs when x and y reverse. But the assumption, that k, l, α, β retain the same numerical values while x and y change sign, has had no better basis than the desire for simplicity, and the absence of evidence to the contrary. Let us now attempt a direct estimate of k and l in the co-operative region. The reciprocals k^{-1} and l^{-1} are times, the 'catching-up times' for co-operation. If nation A does a friendly service to nation B, how long time elapses before some equivalent service is returned? For trade between nations, 3 months is called a 'short-term credit' and 20 years a 'long-term credit'. The boundary between long and short is put by the Bank of England at 1 year; by the Commercial and Financial Chronicle of U.S.A. at 5 years (*League of Nations Statistical Year-book*, 1936, pp. 306, 312). The mean is 3 years, and its reciprocal 0·3 year^{-1}. So, without going into details, we reach the interesting conclusion that k and l for trade are of the same order of magnitude as k and l for war preparations of a large nation like Germany. But for a small nation k for war preparations would be smaller, as shown in § 2·4.

§ 2·10. The European arms race of 1909–14.

Amplifying a letter in *Nature* of 29 October 1938 a statistical test of the foregoing general ideas will now be made for Europe during the period 1909–14. France was allied with Russia, Germany with Austria-Hungary. Neither Italy nor Britain was in a definite alliance with either party (see G. P. Gooch, 1913). The two opposing alliances were roughly equal in size; so let us assume that $k=l$, also for simplicity put $\alpha=\beta$. The equations of motion are then

$$dx/dt = -\alpha x + ky + g,$$

$$dy/dt = kx - \alpha y + h.$$

By addition we obtain

$$d(x+y)/dt = (k-\alpha)(x+y) + g + h,$$

which involves a single dependent variable $(x+y)$.

We have seen in § 2·9 that the objective measure of x or y cannot be simply armaments, but must be more like

$$x = (\text{threats}) - (\text{co-operation}).$$

Let U, V be the annual defence budgets of the two alliances, both expressed in the same monetary unit. Let us try out the assumption that

$$x = U - U_0, \quad y = V - V_0,$$

where U_0, V_0 represent the co-operations, which merely for simplicity are tentatively assumed to have remained constant. It follows that

$$\frac{d(U+V)}{dt} = (k-\alpha)\left\{U + V - \left[U_0 + V_0 - \frac{g+h}{k-\alpha}\right]\right\}.$$

Table I. *The arms race of* 1909–14

Defence budgets expressed in millions of £ sterling

	1909	1910	1911	1912	1913
France	48·6	50·9	57·1	63·2	74·7
Russia	66·7	68·5	70·7	81·8	92·0
Germany	63·1	62·0	62·5	68·2	95·4
Austria-Hungary	20·8	23·4	24·6	25·5	26·9
Total $= U + V$	199·2	204·8	214·9	238·7	289·0

Time rate $= \Delta(U+V)/\Delta t$	5·6	10·1	23·8	50·3
$(U+V)$ at same date	202·0	209·8	226·8	263·8

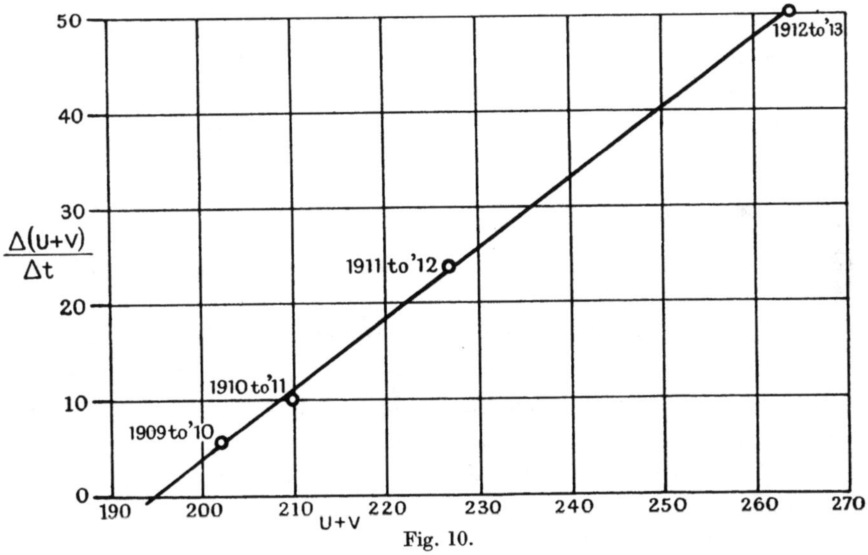

Fig. 10.

The term in [ ] is a constant. So if $d(U+V)/dt$ is plotted against the contemporaneous $(U+V)$ we should expect to get a straight line. The statistics are set out in Table I. Those for France, Germany and Russia are taken from a digest by Per Jacobsson (1929?); those for Austria-Hungary from the *Statesman's Year-books*. The annual increment of $(U+V)$ is plotted against the mean $(U+V)$ for the two years used in forming the increment. The four points lie close to a straight line. The mere regularity of these phenomena shows that foreign politics had then a rather machine-like quality, intermediate between the predictability of the moon and the freedom of an unmarried young man.

The slope of the line gives $k-\alpha = 0{\cdot}73$ year^{-1}. From vague data it was deduced in § 2·4 that α was probably of the order of 0·2 year^{-1} for all nations and that for Germany $k = 0{\cdot}3$ year^{-1}, but that k was proportional to the size of the nation. The alliances which we are now studying had each about twice the size of Germany; so that it appears reasonable to interpret $k-\alpha = 0{\cdot}73$ year^{-1} as implying $k=0{\cdot}9$, $\alpha=0{\cdot}2$, for either of these alliances.

If we extrapolate the observations along the straight line to the point where $d(U+V)/dt$ vanishes, we find there $U+V =$ £194 million sterling. That is to say

$$U_0+V_0-\frac{g+h}{k-\alpha} = 194.$$

As love covereth a multitude of sins, so the goodwill between the opposing alliances would just have covered £194 million of defence expenditure, on the part of the four nations concerned. Their actual expenditure in 1909 was £199 millions; and so began an arms race which led to the Great War.

Let us next look among the statistics of foreign trade to see if there is any that may reasonably be regarded as representing a goodwill of £194 million sterling. For this purpose we want the trade of each alliance with the other alliance, excluding other parts of the world. The statistics supplied by Germany have been extracted from the official *Statistisches Jahrbuch* for 1915; the other statistics from the *Statesman's Year-books*. The value of the goods transferred is published both by the country of origin and the country of destination. These two publications seldom agree. Following the practice of a recent report of the League of Nations (1936 Ser. *L.o.N.P.* 1937, II, A 21) 'as exporters are frequently unable to state where the goods exported are actually consumed, especially in the case of staple products, import statistics have been used'.

Table II shows that the sum of the imports and exports of one alliance from and to the other alliance had annual values of £172 to £230 millions, which are close to the goodwill of £194 millions.

It also appears that the intertrade was mainly increasing from 1909 to 1913. Ought we therefore to assume that U_0 and V_0 were increasing? A similar question is whether the great trade slump of 1931 diminished the international goodwill. The discussion will be postponed to § 4.

Although it was a theory that led the author to notice that

$$\Delta(U+V)/\Delta t = 0\cdot73\,(U+V-194),$$

yet this equation describes a historical fact, true independently of theories.

Table II. *Trade between pairs of nations*

Published as imports by the country of destination, and now expressed in million £ sterling

Motion of merchandise ...	1909	1910	1911	1912	1913
France to Germany	23·7	24·9	25·6	27·0	28·5
Germany to France	26·2	32·6	38·8	36·0	42·4
France to Austria-Hungary	4·0	4·7	4·7	5·0	4·7
Austria-Hungary to France	[3·3]	[3·5]	3·5	4·2	4·1
Russia to Germany	66·8	67·9	80·0	74·8	69·7
Germany to Russia	37·5	46·6	50·4	54·9	68·0
Russia to Austria-Hungary	7·5	7·0	8·7	9·5	8·4
Austria-Hungary to Russia	2·8	3·6	4·0	3·4	3·7
Totals	171·8	190·8	215·7	214·8	229·5
Increments		+19·0	+24·9	−0·9	+14·7

Note: Numbers in [ ] are extrapolated.

§ 2·11. Mathematical appendix.

General solution. Mathematicians will require to see the foregoing geometrical argument confirmed by standard methods.

Given

$$dx/dt = -\alpha x + ky + g, \tag{1}$$

$$dy/dt = lx - \beta y + h, \tag{2}$$

we first remove h and g by shifting the origin to

$$x_0 = \frac{kh+\beta g}{\alpha\beta - kl}, \tag{3}$$

$$y_0 = \frac{lg+\alpha h}{\alpha\beta - kl}. \tag{4}$$

It is convenient to write

$$X = x - x_0, \tag{5}$$

$$Y = y - y_0. \tag{6}$$

Then the trial substitutions $X = C_1 e^{\lambda t}$, $Y = C_2 e^{\lambda t}$, where C_1, C_2, λ are constants, show that if λ_1, λ_2 are the two roots of

$$\lambda^2 + (\alpha+\beta)\lambda + \alpha\beta - kl = 0, \tag{7}$$

the general solution of (1), (2) is

$$x = x_0 + Ae^{\lambda_1 t} + Be^{\lambda_2 t}, \tag{8}$$

$$y = y_0 + \frac{\lambda_1+\alpha}{k} Ae^{\lambda_1 t} + \frac{\lambda_2+\alpha}{k} Be^{\lambda_2 t}, \tag{9}$$

where A, B are arbitrary constants specifying the position at one given time.

Because kl is positive, the roots λ_1, λ_2 are real. (10)

Because α and β are positive $\lambda_1 + \lambda_2 < 0$. (11)

Also $\lambda_1 \lambda_2 = \alpha\beta - kl$ and may be of either sign. If $\alpha\beta > kl$ both λ_1 and λ_2 are negative; so that as $t \to \infty$, $x \to x_0$ and $y \to y_0$. This has been called stability. (12)

If $\alpha\beta < kl$ the roots are of opposite signs and as $t \to \infty$, $x \to \infty$ and $y \to \infty$. This has been called instability. (13)

Normal co-ordinates and barriers. The phrase 'normal co-ordinate' will be used to denote a linear function of the original variables x and y with coefficients so chosen that only a single exponential function of time is involved. We have already, in the simple case of § 2·10, seen how the transformation to $x+y$ allowed us easily to compare the theory with fact. It remains to treat the general case. Let equations (1) and (2) be multiplied respectively by constants p, q and let the products be added (Forsyth, 1929, *Treatise*, p. 348, Ex. 4). Then

$$\frac{d}{dt}(pX + qY) = X(-\alpha p + ql) + Y(kp - \beta q). \tag{14}$$

Next let p, q be chosen so that the coefficients of X and Y in the first member are proportional to those in the second. This will occur if there is a common constant, ν say, such that

$$-\alpha p + ql = \nu p, \quad kp - \beta q = \nu q. \tag{15}$$

These two equations are consistent if

$$\begin{vmatrix} (-\alpha - \nu), & l \\ k, & (-\beta - \nu) \end{vmatrix} = 0, \tag{16}$$

whence $$(\alpha + \nu)(\beta + \nu) - kl = 0,$$

so $\nu = \lambda$ as given by (7). To each λ corresponds a pair p, q. Let them be connected by a common suffix. Then

$$\frac{d}{dt}(p_1 X + q_1 Y) = \lambda_1 (p_1 X + q_1 Y), \tag{17}$$

so that after integration the normal co-ordinate $p_1 X + q_1 Y$ is expressed by a single exponential, thus

$$p_1 X + q_1 Y = A_1 e^{\lambda_1 t}, \tag{18}$$

where A_1 is an arbitrary constant. We have seen that the roots are real so that $A_1 e^{\lambda_1 t}$ cannot change sign. Therefore

$$p_1 X + q_1 Y = 0 \tag{19}$$

is a *barrier* which the particle, if guided only by the instinctive drift, cannot cross. The other normal co-ordinate is given by replacing suffix 1 by suffix 2. We have already noticed the barrier in § 2·8. The quadratic (7) in λ can be shown to agree with the quadratic which determined the barrier in § 2·8, namely,

$$0 = l\cos^2\theta + (\alpha-\beta)\cos\theta\sin\theta - k\sin^2\theta.$$

For if ν be eliminated between the two equations (15) we have

$$0 = lq^2 + (\beta-\alpha)pq - kp^2, \tag{20}$$

so that the quadratics agree if $p/q = -\tan\theta$. (21)

The orbits. The tracks along which the representative particles move in the international plane are found by eliminating t between equations (8) and (9). It is then found that

$$\frac{[X\sqrt{\{l(\lambda_1+\alpha)\}} - Y\sqrt{\{k(\lambda_1+\beta)\}}]^{\lambda_1}}{[X\sqrt{\{l(\lambda_2+\alpha)\}} - Y\sqrt{\{k(\lambda_2+\beta)\}}]^{\lambda_2}} = \text{constant}. \tag{22}$$

The constant is fixed by the starting point. This expression has been confirmed by elimination of dt between (1) and (2) which gives

$$\frac{dY}{dX} = \frac{lX-\beta Y}{kY-\alpha X}, \tag{23}$$

whence it follows that

$$(Y-XV_1)^{V_1k-\alpha} = (\text{constant})\,(Y-XV_2)^{V_2k-\alpha}, \tag{24}$$

where V_1, V_2 are the roots of

$$0 = l + (\alpha-\beta)V - kV^2. \tag{25}$$

After a number of steps we regain (22).

§ 2·12. Theorem on rivalry.

Suppose that what moves a government to arm is not the absolute magnitude of other nations' armaments but the difference between its own and theirs. This effect was stated in some detail in an earlier work (Richardson, 1919, p. 25) under the heading 'Rivalry'. Its simplest expression is

$$dx/dt = k_1(y-x) - \alpha_1 x + g, \tag{26}$$

$$dy/dt = l_1(x-y) - \beta_1 y + h, \tag{27}$$

where k_1, l_1 are positive constants called 'emulatances'; α_1, β_1 are positive, and g, h have the same meaning as in § 2·1. It is seen that the equations are of the same form as (7) and (8) of § 2·1 except that α is replaced by α_1+k_1, and β by β_1+l_1. The condition of stability now is

$$k_1 l_1 \gtreqless (\alpha_1+k_1)(\beta_1+l_1),$$

that is

$$0 \gtreqless \alpha_1\beta_1 + \alpha_1 l_1 + \beta_1 k_1. \tag{28}$$

Only the lowest sign is true; and this corresponds to stability. Conclusion: *two nations affected only by rivalry, cost and fixed grievances, in the stated manner, form a system that is necessarily stable.*

As in fact arms races do occur, we have here a proof that rivalry as formulated is not an adequate description of the interaction between nations. But no doubt it has some effect, and therefore α and β as used in § 2·4 are of the form $\alpha_1 + k_1$, $\beta_1 + l_1$, where α_1, β_1 are *pure* fatigue and expense coefficients.

§ 2·13. Physical analogy.

Although linear differential equations occur in many parts of physics, the author has not met the present pair in any publication, at least not with the actual signs of the defence coefficients.

A non-linear pair which is rather similar, namely,

$$dn/dt = a + bnn' - cn, \quad dn'/dt = a' + b'nn' - c'n' \qquad (29), (30)$$

appears in a paper on 'The ignition of explosive gases' (Mole, 1936), a title not inappropriate to the international situation.

§ 3. SUBMISSIVENESS

The author has previously (1919) attempted to set out the truth in much fuller detail; but the resulting equations were so entangled as to be difficult to manipulate or to remember. So, going to the opposite extreme, he next sought the greatest possible simplification that would retain the gist of the matter, and thus arrived at the pair of linear equations given above. Now a moderate increase of detail and a moderate increase of complexity will be allowed.

It seems particularly desirable to include some mention of submissiveness. Nations would not keep up armaments unless they believed them to be effective. The prevailing notion seems to be that armaments are for keeping other people in order. Each nation seems to think, 'If we threaten our neighbours, they will not make trouble.' There must be a considerable body of experience behind this widespread belief. The policy of threats, however unkind it may have been, must have been effective. In the experience of many people it worked in the nursery and in the school playground. It has worked in military experience, in various parts of the world, all through history. The strange fact is that it does not work now in Europe. If Germany threatens, France does not back down. Quite the contrary. And vice versa. Why is there this contradiction between the theory, by which the maintenance of armaments is defended, and the actual result?

Let us attempt a generalization of the equations to include submissiveness:

```
y-axis
| C        B     B     B
| C      B          B
| C        B
|    B                 A
|        A     A       A
|______________________ x-axis
```

In the region $A \ldots A$ of the international plane we continue to regard dx/dt as negative, because the threat from y is small, so that armament does not appear to be worth the cost of upkeep. In the region $B \ldots B$ the threat from y is provoking, and so dx/dt is positive. This also was represented by the linear equations. The innovation now made is to introduce the region $C \ldots C$ in which dx/dt is negative because the threat from y is overwhelming and resistance appears to be hopeless.

It is evident that after this proposed modification the equations will no longer be linear; and that in consequence they may be difficult to solve. It is also evident that the required change of sign must be expressible in a great many different ways, which, though not mathematically equivalent, may yet be sufficiently alike for the purpose in hand.

There are several criteria for the choice of a formula:

(i) The linear equations seem to be correct near the origin. So that the new equations should tend to the form already given as x and y tend to zero. Certainly there should be no discontinuity at the origin. Terms in y/x or x/y cannot be admitted.

(ii) No amount of threats will change the cost of armaments from positive to negative. The reversal of sign should therefore be confined to the defence coefficients, which in the linear equations were regarded as positive constants.

(iii) The overwhelmingness of the threat depends not on y absolutely, but on y relative to x.

(iv) Our search is not simply for truth, but for truth in an easily understandable expression.

(v) The three quadrants in which x and y are not both positive are of great practical interest, but will require a separate discussion.

We may satisfy the above criteria by writing

$$dx/dt = ky\{1-\sigma(y-x)\} - \alpha x + g, \tag{31}$$

$$dy/dt = lx\{1-\rho(x-y)\} - \beta y + h, \tag{32}$$

in which $t, x, y, k, l, \alpha, \beta, g, h$, have the same meaning as for the linear equations but σ and ρ are new constants representing 'submissiveness'. We assume that σ, ρ are both positive.

The effects of σ and ρ can be shown by quite simple mathematics if we distinguish the regions of the international plane in which dx/dt, dy/dt are positive by shading them parallel to the corresponding component velocity, as we did for the linear equations.

The loci $dx/dt = 0$, $dy/dt = 0$ are the conic sections

$$0 = dx/dt = \sigma kxy - \sigma ky^2 - \alpha x + ky + g = \phi, \text{ say,} \tag{33}$$

$$0 = dy/dt = -\rho lx^2 + \rho lxy + lx - \beta y + h = \psi, \text{ say.} \tag{34}$$

Because two conics can cut one another in at most four points, there can be at most four points of equilibrium. It suffices to discuss the first conic, since the second can be obtained from it by merely interchanging x with y and changing the constants. The condition that $\phi = 0$ should degenerate into a pair of straight lines is that its three-rowed determinant should vanish, and this works out to

$$0 = \tfrac{1}{4}\sigma k(\alpha^2 - \alpha k - \sigma kg). \tag{35}$$

Unless this special condition chances to be satisfied the conic is a hyperbola, because the two-rowed determinant of its quadratic term is $-\tfrac{1}{4}\sigma^2 k^2$ which is negative.

The centre of the hyperbola is at

$$\bar{x} = \frac{2\alpha - k}{\sigma k}, \tag{36}$$

$$\bar{y} = \frac{\alpha}{\sigma k} \tag{37}$$

which do not involve the grievance g.

The asymptotes of this hyperbola are parallel to the lines $y = 0$, $x = y$; and are thus in the same two directions for all values of the constants.

When transferred to its centre by writing $x = \bar{x} + x'$, $y = \bar{y} + y'$ the hyperbola is

$$\frac{\sigma^2 k^2}{\alpha^2 - \alpha k - \sigma kg}(x'y' - y'^2) = 1. \tag{38}$$

The length R of the real semiaxis of the hyperbola is equal to whichever of the two following expressions is real,

$$R = 2{\cdot}20\frac{\sqrt{(\alpha^2 - k\alpha - \sigma kg)}}{\sigma k}, \quad 0{\cdot}910\frac{\sqrt{(-\alpha^2 + k\alpha + \sigma kg)}}{\sigma k}. \tag{39}$$

As $\sigma \to 0$, $R \to \infty$. If g were zero, one branch of the hyperbola would pass through the origin for all values of σ.

SUBMISSIVENESS 25

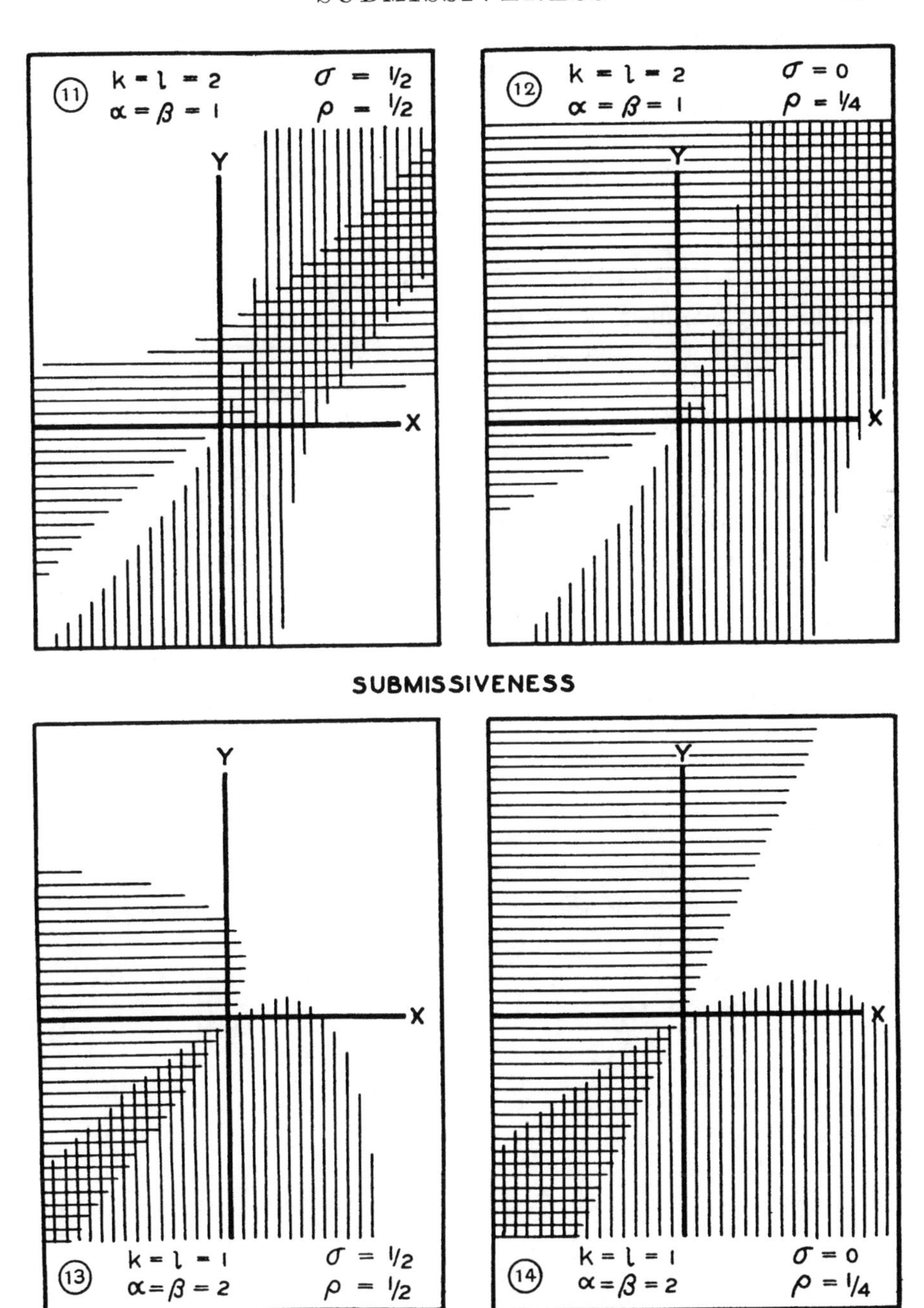

Note: the unit of x and y is four spaces between rulings.

TWO NATIONS

Some consequences of assumptions (31), (32) concerning submissiveness are shown in Figs. 11–14. For simplicity the grievances g and h are zero in all these four diagrams.

Near the origin the field of Fig. 11 is the same as that of Fig. 1, but at a distance from the origin the field is much affected by the equal submissiveness of the two groups. There are three points of balance. That at the origin is unstable. The other two are 'whirlpool' points around which the particle circulates. This is suggested by the shading, and has been verified by the method outlined below. The two whirlpools circulate opposite ways.

Fig. 13 depicts the same two submissivenesses as Fig. 11 but with a stable centre, which is locally identical with that of Fig. 5. The shaded areas in Fig. 13 are bounded by two hyperbolae, the other branches of which lie beyond the top right-hand corner of the diagram. It is perhaps unnecessary to attribute any psychological meaning to these outlying areas, which would be shaded if the diagram were large enough; for they are very inaccessible.

In Figs. 12 and 14 the X group is pertinacious, the Y group submissive. The point of balance in Fig. 12 is unstable, resembling that of Fig. 1; whereas in Fig. 14 the point of balance is stable like that of Fig. 5. The other branch of the hyperbola in Fig. 14 lies above and to the right of the diagram.

Equations of the type of (31), (32) have been discussed by Forsyth (1900). A convenient approximation (Kostitzin, 1937) in the neighbourhood of any point (x_1, y_1) may be obtained by writing $x = x_1 + x'$, $y = y_1 + y'$ and by neglecting $x'x'$, $x'y'$, $y'y'$. We thus obtain equations linear in the local co-ordinates x', y'. We can then adapt the general linear solution of § 2·10 to the equations in x', y' by making appropriate changes in the constants.

PART II. STATISTICAL DISCUSSION OF SEVEN GREAT POWERS FOR 1922–38

§ 4·1. Guiding principles and available statistics.

The variables x and y of Part I were originally apprehended as an expression of facts, including public speeches, made common knowledge by the newspapers. For the arms race of 1909–14 an astonishingly good agreement between statistics and theory was obtained in § 2·10 by the assumption that

$$x = (\text{annual defence budget}) - (\text{constant}).$$

That agreement may have been peculiar to 1909–13. In view of the apparent complexity of foreign politics, the statistical measure of x certainly deserves further discussion.

Before we can attempt to measure anything we must have a preliminary ideal of what we wish to measure. But it would be most unwise to give that ideal a rigid definition at the outset, because the experience gained during the process of measurement should be allowed to react upon and refine the ideal.

So here let us carry over from Part I the preliminary ideal that x when positive is to be some sort of measure of a nation's warlike preparations and that, when negative, x is to be some sort of measure of the amount of that nation's co-operation with the rest of the world.

The League of Nations published in 1923 a *Statistical Enquiry into National Armaments* (Ref. A 20, 1923, IX), and has continued the enquiry annually since 1924–5 in the form of an *Armaments Year-book* containing, for about sixty nations, particulars of military systems in peace time. Here we find numbers of effective personnel, sizes of guns, ships by categories; in fact far too much detail for present purposes. But there is also:

(i) A summary of budget expenditure on national defence with notes of extra-budgetal loans. Pensions are mostly excluded. The money is expressed in each nation's currency.

(ii) The index of wholesale prices and sometimes also of retail.

In the *Statistical Year-book* and the *Monthly Bulletin of Statistics* of the League of Nations are given:

(iii) Imports and exports for each nation in its own currency.

(iv) Rates of monetary exchange between nations.

In the *Year-book of Labour Statistics* (1937, I.L.O. Geneva) there is also

(v) Information about wages in most countries from 1927 to 1936.

The problem is to find some function of these five kinds of statistics which shall correspond to our psychologically apprehended x.

The stability of peace depends not only on what nations do, but on what they are suspected to be about to do. A contemporary example is the mutual suspicion between Germany and Czechoslovakia during 1938, 19–27 May. We can separate the objective part from the subjective part by regarding the effect of the y-nation on the x-nation as $k_{12}y$, where y represents the objective, statistically measurable, doings of the y-nation, while the defence coefficient k_{12} involves any mood and misunderstandings that the x-nation may have when it regards y.

We have the following six guiding principles:

(i) To attempt at first to be objective and emotionally detached. At a later stage it may be necessary to correct by the coefficient k for an excess of reasonableness in the statistic.

(ii) National currencies must either be eliminated or expressed in a common unit.

(iii) The statistic should be proportional to the size of its nation, other things being the same.

(iv) As co-operation goes on simultaneously with threatenings, they partly compensate one another.

(v) Various effects that are not analysed in detail can be eliminated by considering deviations from a mean.

(vi) Allowance may be made for accumulation and obsolescence.

§ 4·2. Standards of economic value.

The ethical or political value, positive or negative, of armaments is regarded as a separate question to which we shall return in the sequel. The problem of economic value may be subdivided into:

(i) The comparison of different places at the same time. For this the exchange rates on London, Paris and New York are available.

(ii) The comparison of different times at the same place. For this purpose there are available (1) gold, (2) indices of wholesale prices, except since 1930 for Russia, (3) indices of wages.

When times two thousand years apart are to be compared, gold is a nonsensical standard, and so in various degrees are some present articles of wholesale commerce such as potatoes, iron, coal or tobacco, which were formerly scarce or unknown. A general discussion of values by Professors Allyn Young and Wilbur M. Urban (*Encycl. Brit.* 1929, Articles 'Value', 'Value, Theory of') ends without recommending any definite standard. The present author adheres to the old view that the fundamental economic standard over centuries of time is human effort, say an hour's work of a human being of specified type, preferably of average intelligence. R. G. Hawtrey defends a standard of this kind (but 'unskilled') in the *Art of Central Banking*, p. 319, Longmans, 1932.

In the statistical enquiry published by the League of Nations in 1923 (Ref. A 20, 1923, IX) the changes with time of each national currency were eliminated by dividing the successive defence budgets by the corresponding index numbers of wholesale prices; and the results were stated with an accuracy of 1 %.

Ten years later a Technical Committee of the League of Nations reported that:

> The impossibility of comparing armaments of the different countries on the basis of their armament expenditure arises primarily from the fact that there is no satisfactory common measure between these expenditures. Even in periods when the national currencies are stable, the internal prices of the different countries may vary as a result of circumstances, and no conversion of the successive figures of expenditures into any currency selected as a standard will indicate the variations in internal prices and in the cost of armaments themselves. When, as at the present time, the currencies themselves are subject to fluctuations while internal prices remain relatively stable, the expression in a selected currency at a given date of the figures of expenditure of the various states produces absurd results if any attempt is made to compare on this basis either the successive expenditures of one state or the separate expenditures of various countries.

This is a quotation from a report published on 8 April 1933 (*L.o.N. Armaments Year-book*, 1936, p. 1030) after a decade during which budget expenditures on defence had on the whole been steady. The Committee gives weighty reasons why a comparison cannot be made with an accuracy of 1 or even 20 %. But since they reported the actual changes have been so much greater that the Committee's warning need not now deter enquiry.

In recent publications of the League of Nations many comparisons are made by way of 'old U.S.A. gold dollars', that is to say by exchange rates on New York assuming that the dollar had a fixed value from 1920 to 1932, and that from February 1934 to 1938 it had 20·67/35·00 of its old value. The gap during 1933 is smoothed over by exchange rates on Paris, which at that time was on the gold standard.

The old U.S.A. gold dollar is equivalent to 1·5046 g. of fine gold.

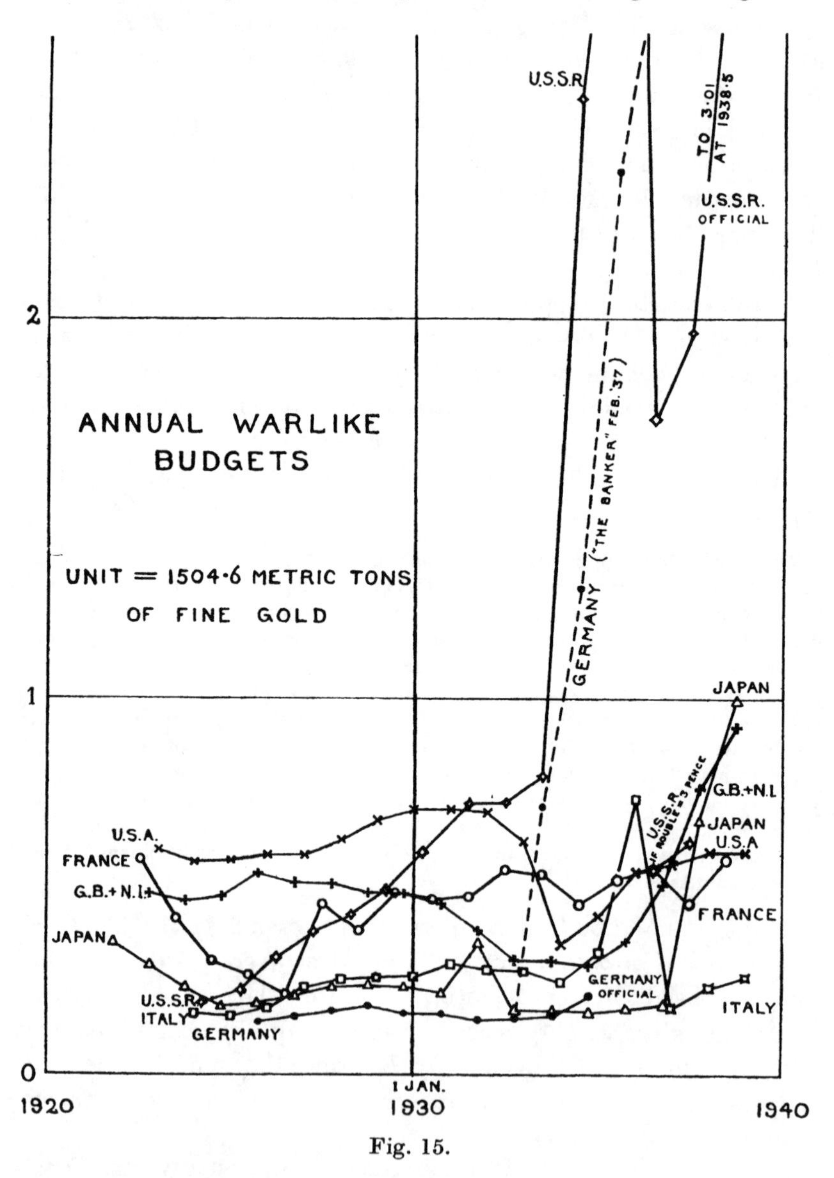

Fig. 15.

For the period 1923–38, gold seems to be the best of the readily available standards, and it has been used in the present enquiry. Although it would be more scientific to use, instead of gold, a standard based on human effort; and although such a standard could probably be concocted from the published statistics of wages, yet the method of computation would require prolonged consideration.

§ 4·3. 'Defence' budgets.

These are shown in Fig. 15 as annual expenditure expressed in the old gold U.S.A. dollar (= 1·5046 g. of fine gold). The data have been collected mostly from the summary tables of the *League of Nations Armaments Year-book* (hereafter briefly referred to as *L.N.A.Y.B.*); but certain important qualifications must be mentioned.

The Italian budget as given in *L.N.A.Y.B.** excludes the cost of the Abyssinian war. But, for present purposes, to present the cost of an ideal peace establishment in time of war would be absurd. For the attention or alarm awakened in the world by actual fighting between nations *A* and *B* surely equals or exceeds that caused by the expenditure by *A* and *B* of the same amount of money on mere preparations. So either estimates, albeit rough, of the cost of wars have been included; or else wartime budgets have been rejected as misleading.

Pensions arising from the war of 1914–18 are everywhere excluded from the *L.N.A.Y.B.*

There are puzzling questions concerning ordinary pensions, colonial expenditure, civil aviation, capital loans and receipts appropriated in aid. For these the *L.N.A.Y.B.* give particulars, but their summaries only tend gradually towards a consistent plan. It has sometimes been necessary to correct the earlier summaries to the plan of the later. The data appear in various stages from first estimate to audited account. The latest available has been used, but occasionally only voted estimates have been published (e.g. France, 1924–6). All the annual *L.N.A.Y.B.* have been collated, including that published in Jan. 1939.

The apparent accuracy of statement varies with the country. The *L.N.A.Y.B.* give the French expenditure to 1 part in 100,000. But, in view of the aforesaid puzzling questions, the real accuracy is probably more like 1 % at best. This is quite sufficient for psychological purposes. Indeed, the Russian changes have been so great that an uncertainty of 30 % would be unimportant in them.

A plausible objection to the present use of lump-sum defence budgets is that different nations spend their money in such different ways; some have no navy, some have conscript and others have voluntary armies. If we were to attempt a detailed analysis of budgets, we should have to decide which of the various forms of defence expenditure were the more and which the less efficacious. By using only total defence

* Except the 1938 *L.N.A.Y.B.*

budgets we allow those awkward questions to be decided for us by the experts of the nation concerned; for they spend their money in the way that seems to them most efficacious.

A more serious objection is that the cash value of unpaid compulsory service ought to be estimated and added to the budget. The author agrees that it ought; and only the difficulty has deterred him from the attempt. But the amount that should thus be added may not be so great as might at first appear. For a conscript has to be fed, uniformed, armed, and in peace time housed, instructed, and kept in a good temper at the expense of the State. How much of his equivalent wages would be left after he had paid for rent, food, clothing, wine, tobacco and the cinema?

Besides these general qualifications there are others that affect particular nations.

Great Britain and Northern Ireland. The Middle East is included but not other parts of the British Empire. For 1929–30 their several defence budgets amounted to the following percentages of that for Great Britain and Northern Ireland: *India* 38, *Canada* 4, *Australia* 4, *South Africa* 1, *New Zealand* 1, *Irish Free State* 1. The *Middle East* decreased from 5 % in 1924–5 to 2 % in 1933–4.

The earlier issues of *L.N.A.Y.B.* up to 1929–30 included ordinary British pensions. Later issues exclude them. For consistency, pensions have been subtracted from the earlier summaries. Contrariwise statistics published in answer to Parliamentary questions on 21 July 1936 exclude the Middle East and include ordinary pensions and so disagree with those used here.

If instead the defence expenditure of the whole British Empire had been taken, it would have been necessary in § 4·4 to take the trade of the Empire with the rest of the world.

France. Pensions are excluded. There is a mystery about 1927, for which year the *L.N.A.Y.B.* dated 1930–31 published a closed account of 11,574 million francs, while the next later issue of *L.N.A.Y.B.* gave instead only an estimate of 7728. The closed account has been used here. Similar remarks apply to 1928 and 1929. Various extra-budgetal armaments loans, as detailed in notes in the *L.N.A.Y.B.*, have been added to the summaries. The amounts expended are usually stated. But since 1937 there remains a balance of $19{,}992 \times 10^6$ fr. which, in the circumstances, has been assumed to be all spent in 1938.

The French financial year has twice been changed between 1929 and 1933. These irregularities have been compensated in the preparation of the present data.

Germany. Up to 1932–3 the numbers have been taken from *L.N.A.Y.B.* with the exclusion of the cost of 'execution of the Peace Treaty alias war-charges, disarmament and dismantling of fortresses'. But for the next three years they are indirect estimates published in *The Banker* for February 1937 (London Financial News, Ltd.).*

Japan. The *L.N.A.Y.B.* state that their summaries refer to the general account only and that no details regarding the special accounts are available. There is no explicit statement in *L.N.A.Y.B.* as to whether the cost of wars with China is included or not. But other evidence is given in the annual *Surveys of International Affairs* (London Roy. Inst. Internat. Affairs). The Japanese attacked Manchuria on 18 September 1931. Between March and November 1932 special 'Manchurian incident bonds' were issued to the total of 323,701,000 yen (*Survey*, 1932, p. 429). This sum is about half the concurrent annual expenditure on 'defence' as given in *L.N.A.Y.B.* The Japanese expenditure for 'Manchurian affairs' is given thus (*Surveys*, 1933, p. 471, and 1934, p. 644):

1932–3	288 million yen
1933–4	185 ,,
1934–5	160 ,,
1935–6	171 ,,

It is very difficult to make out whether these sums are included in the *L.N.A.Y.B.* defence budgets or not. I have assumed that from 1932 to 1937 the *L.N.A.Y.B.* includes the cost of fighting in China but I have added the aforesaid 324 million yen to the *L.N.A.Y.B.* summary for 1931–2. Again for 1937–8 it appears proper to add 2560 million yen to the estimate 1409 in *L.N.A.Y.B.* This 2560 is the cost of the recent China 'incident' as given in the budget of 17 February 1938 according to the *Glasgow Herald*.

Italy. The earlier summaries in the *L.N.A.Y.B.* have been corrected to the plan of the later, by the removal of pensions.

The Abyssinian war lasted formally from 2 October 1935 to 1 May 1936. In May 1937 the Italian finance minister stated that the cost of conquering and taking over Abyssinia was 12,111 million lira, as covered by the budgets of 1934–5 and 1935–6 (Keesing's *Contemporary Archives*, 20–21 May 1938). The part of this sum falling in the budget prior to 30 June 1935 is given as 975 million lira (H. V. Hodson in *Survey of International Affairs*, 1935, vol. II, p. 437, London Roy. Inst. Internat. Affairs). Therefore 975 and (12,111 − 975) have been added to the *L.N.A.Y.B.* budgets for 1934–5 and 1935–6 respectively, for they

* For revised values see p. 46.

are stated to exclude 'the extra-ordinary expenditure for the war in Ethiopia', except in the 1938 *L.N.A.Y.B.*

U.S.A. Pensions are excluded. Air service is included since 1925–6. The estimate of $289 millions for 1938–9 and $934 millions for 1937–8 are taken from Keesing's *Contemporary Archives* for 5 January 1938.

U.S.S.R. The Russian expenditure since 1931 appears larger than that of any other country. This comparison is obtained by taking one rouble to equal 0·515 old gold dollar from 1925 to 1935 and 0·117 old gold dollar in 1936 and 1937, these being official rates of exchange.*

The financial year changed in 1930. For three months October to December 1930 a special budget is mentioned in *L.N.A.Y.B.* but the sum is not stated. The gap has therefore been filled in by interpolation.

§ 4·4. Statistics of foreign trade.

These, expressed in old gold U.S.A. dollars, are shown in Fig. 16. The numbers have been extracted from the *International Statistical Year-books*, from their successors the *Statistical Year-books*, and from the *Monthly Bulletin of Statistics*, all of the League of Nations. The trade statistics are much plainer than the defence budgets, so that only a few particulars need explanation. The statistics are for special trade, which means that imports for re-export are excluded. They are for merchandise only, excluding bullion and specie, at least in the recent year-books. The data in earlier year-books have where necessary been corrected by subtraction of bullion and specie before being used in the present research.

Is the world slump of 1930–2 to be regarded as a refusal by the nations to co-operate with one another? Or was it, like the weather, uncontrollable? Von Haberler in his study of *Prosperity and Depression* (1936, Geneva, League of Nations), allows that Central Banks do have some influence, but (p. 261) denies that they have an effective control. In the preface A. Loveday states that 'our knowledge of the causes of depressions has not yet reached the stage at which measures can be designed to avert them'. That is a question about objective fact. But the international situation was affected not only by what the slump really was; but by what it was variously believed to be. We cannot escape from this subjective problem. For the slump certainly did give rise to fierce discontent with home governments, resulting in the ejection from office of Hoover in U.S.A., of the Labour Government in Britain, of Brüning in Germany; and some of the discontent spread out against foreign governments. These accusations against rulers

* For smoothed rates see pp. 78, 79.

may have been unfair, nevertheless the effect of the slump on international relations was in the same direction, though probably not so intense as if the slump had been a deliberate refusal of co-operation. This objective-subjective problem will crop up again in connexion with the statistics ψ and χ.

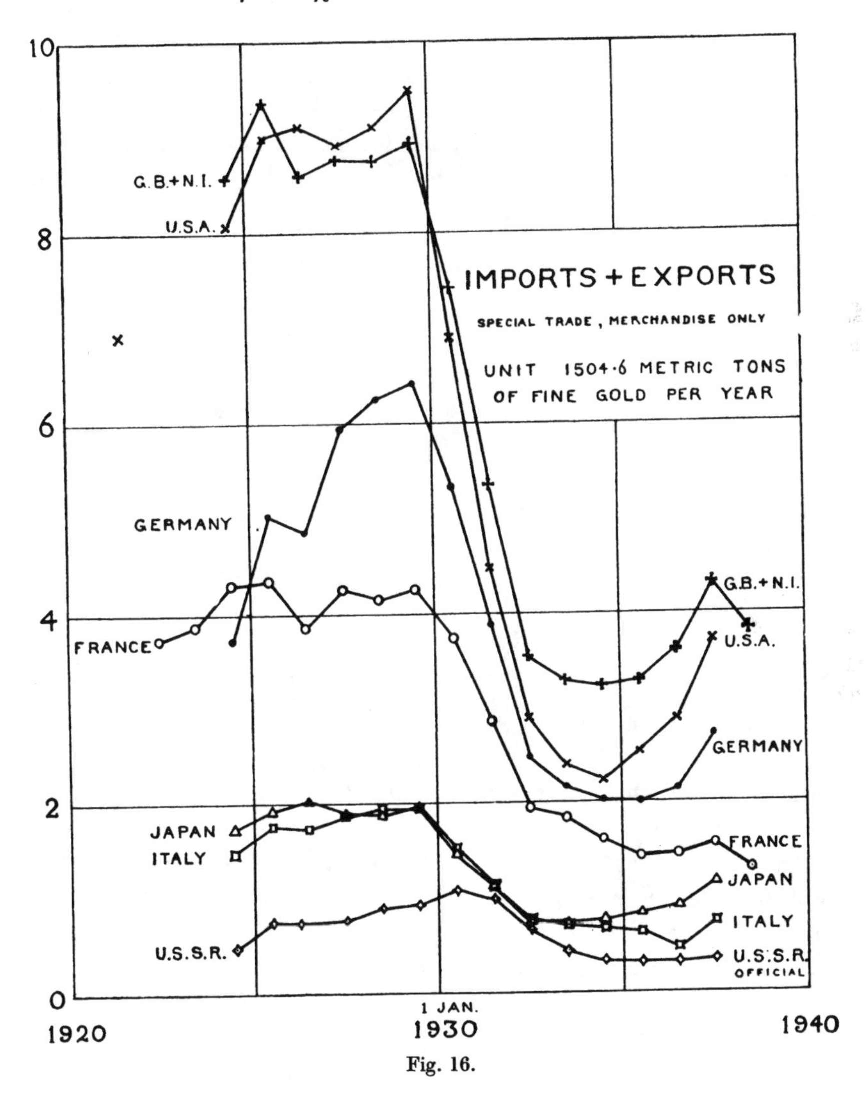

Fig. 16.

§ 4·5. Custom as a standard.

When we have to examine statistics of very complicated phenomena it is usually wise to refer them to some condition that can be called average or customary or steady or equilibrium; because by so doing we eliminate many unanalysed effects. We do not thereby assert that custom is sacred and should be permanent; it may be more honoured in the breach. We merely use custom as a reference mark from which to measure changes. No true world equilibrium has existed in international affairs since 1919; for Germany could not be considered as in equilibrium during the first decade, because she was abnormally disarmed by the Treaty of Versailles. But between the years 1920 and 1931 budgetary expenditure on defence remained fairly steady, except in Russia where general industrial production was increasing much more rapidly than elsewhere. The index numbers of industrial production (*L. of N. Statistical Year-book*, 1935–6, pp. 173–4) run as follows:

Year	Russia	The rest of the world
1926	55·7	92·4
1935	288·2	92·8

The only factual standard available seems to be the world-mean. Let u denote any nation's annual expenditure on 'defence' and m denote the sum of its imports and exports, both u and m being in the national currency. When converted to old gold dollars let these become U and M respectively. Let Σ sum for all the nations of the world. Two estimates of ΣU are given in *L.N.A.Y.B.* 1935, Annex II. For consistency with the rest of the present work, that based on mean annual rates on New York is preferred here (Table III).

Table III. *World-totals. Unit* 10^9 *old gold dollars*

Year ...	1925	1926	1927	1928	1929	1930	1931	1932	1933	1934	1935
Defence ΣU	3·5	3·6	3·9	3·9	4·1	4·2	4·0	3·5	3·3	3·6	—
Trade ΣM	64·5	61·9	65·1	67·4	68·6	55·6	39·7	26·9	24·2	23·3	23·8

§ 4·6. First method of compensation of threats by co-operation. Statistic ψ.

This problem, so far as the author is aware, is one on which no published theory is available. So various statistics have been tried and some rejected.

Statistic ψ is defined for each nation by

$$\psi = n \log_{10}\left(\frac{wu}{m}\right),$$

where n is the population in millions, u is the annual 'defence' expenditure, m is the annual sum of imports and exports expressed in the same unit as u. The constant w is adjusted so as to make $\psi = 0$

when world totals for 1926 are inserted in the formula in place of national values. That is to say

$$w = \left(\frac{\Sigma M}{\Sigma U}\right)_{1926} = \frac{61 \cdot 9}{3 \cdot 6} = 17 \cdot 2.$$

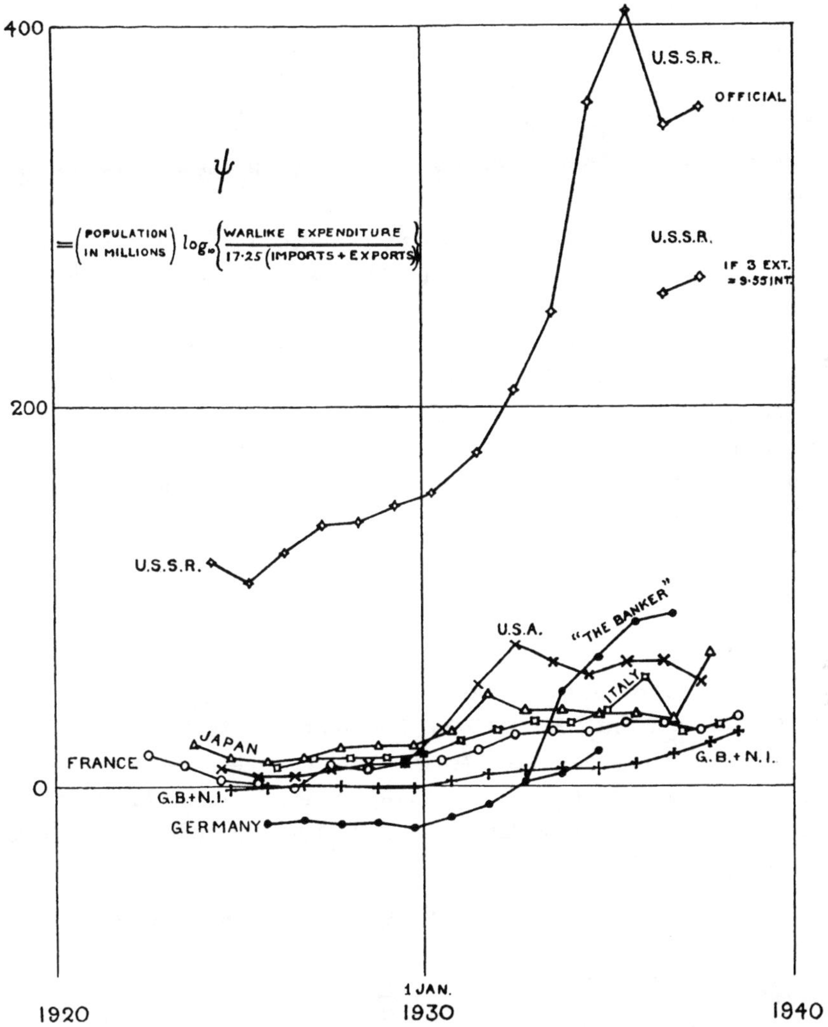

Fig. 17.

The year 1926 was chosen as being moderately calm both politically and industrially. Graphs of ψ are shown in Fig. 17.

The merits of ψ are that:

(i) ψ brings together three of the most important characteristics of a nation.

(ii) ψ is positive for a nation whose warlike preparations preponderate over its foreign trade; ψ is negative for a nation whose foreign trade preponderates over its warlike preparations. The logarithm was introduced in order to make both signs possible.

(iii) ψ is zero for an average nation in the standard year 1926.

(iv) National currency is eliminated and yet the gold standard need not be introduced, except in finding w. Even that could be avoided by a modified world-constant w' defined by $w' = \text{mean}\ (m/u)$.

(v) ψ is proportional to the size of the nation, in the sense of population.

Because the line $\psi = 0$ represents the world mean, we might have expected this sample of the nations to be equally distributed on the positive and negative sides of $\psi = 0$. Actually the curves are nearly all on the positive side. But it must be borne in mind that these seven nations, together with Austria-Hungary, made up the eight 'great powers' of the years 1900–14 and are still so regarded. Whence it appears that a 'great power' is not simply a nation with a large population, but one that is also more warlike or less co-operative than the world mean.

Judging only by Fig. 17 it appears that the most co-operative of these seven nations during the years 1925–32 was Germany; the reason being that she had a large trade, but an army and navy restricted by treaty.

Of these seven States, the one that on the average has kept its ψ closest to the world mean is Great Britain and Northern Ireland.

Critics may object that the British Empire should have been treated as a whole. This is done, at least roughly, in the following table for the year 1929–30. The trade figure is taken from the official *Statistical Abstract for the British Empire*:

		Great Britain and Northern Ireland	Whole British Empire
Million £	u	99·3	148
	m	1762	2339
$\log_{10}\left(\frac{17{\cdot}2u}{m}\right)$		$-0{\cdot}013_5$	$+0{\cdot}036_8$
Population, millions		45·8	457
ψ		$-0{\cdot}6$	$+16{\cdot}9$

Thus for the whole British Empire ψ had a value near to those of Italy, Japan, France and U.S.A. in 1929–30.

The startling phenomenon shown in Fig. 17 is the high value of ψ for Russia. But it may be argued that Russian ψ when calculated from the official rates appears larger than it really was, because the rouble used for foreign trade was in effect larger than the nominally identical rouble used inside Russia. A. Z. Arnold (1937, p. 447) states that in 1924 the internal price level was alleged to be too high for the external exchange rate; and that from 1931 to 1935 foreign visitors to Russia paid in their own currencies at the official rate of exchange in special Torgsin shops, open to foreigners only; and that the rouble prices in Torgsin shops were lower (at times tenfold) than the rouble prices in shops open to Russians. For the years 1936–7 L. H. Callendar (1938) has asserted that the purchasing power of the rouble for food and necessities of life is equivalent to $3d.$ in England, whereas the official rates used in calculating ψ are equivalent to 1 rouble $= 9{\cdot}55d.$ for the same years. Therefore ψ is also shown on the assumption that 3 external roubles $= 9{\cdot}55$ internal roubles.

On the contrary, M. Ruheman (1938), after residence in Russia, maintains that when a Russian professor's whole domestic economy is considered, including his wife's earnings, and various things that are free to Russians, the official exchange rates are about right.

For Germany ψ as shown in Fig. 17 may also be too large for a similar reason, but to a less extent. In recent years the price of books in marks to foreigners has been 25 % less than to Germans, and there has been a similar discount on the exchange rate for tourists.

Peculiarities of ψ that may be regarded as defects are:

(*a*) Population is not a good measure of a nation's international importance. This becomes evident when we compare two nations having about the same population, but one of them shut-in, like Tibet, the other sea-going like Norway.

(*b*) ψ makes no allowance for accumulation and obsolescence. When we seek to remedy this defect we require a permanent standard of economic value.

(*c*) ψ is much affected by the fluctuations of world trade.

Let us now seek another statistic, designed to avoid the peculiarities (*a*), (*b*) and (*c*).

§ 4·7. Accumulation and obsolescence.

What comforts allies and alarms rivals is not exactly the annual expenditure U on 'defence' but the accumulated size of a war establishment, which may be expressed in money as its capitalized value P. So at least it may be argued. The relation between U, P and t may be assumed in peace time to be of the form familiar in connexion with the upkeep and depreciation of a manufactory; that is to say

$$\frac{dP}{dt} = U - cP, \tag{1}$$

where c is a constant expressing the obsolescence of training and of equipment; and cP is the annual cost of mere upkeep. By the 1930 Naval Treaty a battleship becomes out of date after 26 years and a destroyer after 13 years. Service in the army or navy for 21 years is called long. Aeroplanes become out of date in a much shorter time. The value of c ranges from 0·04 year^{-1} to perhaps 0·2 year^{-1} for different ingredients. Strictly these different ingredients should be accounted separately. But in the present work they will be lumped together and c will be taken as 0·1. Given U and c, the problem is to find P. After 1918 there was a general demobilization during which, for most nations, P decreased. It is generally believed that P is everywhere rising again in 1939. Therefore at some time between 1918 and 1939 it must have happened that P was a minimum, and then $dP/dt = 0$, for we regard the variation of P as continuous. At that time it follows that

$$P = \frac{U}{c} = P_0, \text{ say.} \tag{2}$$

Thus we can determine P_0 if we can recognize the year of minimum P. It is not to be confused with minimum U. Some evidence about the dates of minimum P for the several nations is given in the sections headed 'defence' of the article on those nations in the *Encyclopaedia Britannica*, 14th ed. The Disposal Board in London had sold three-quarters of the surplus British war stores by the late summer of 1920. After the signature of the Washington Naval Treaty on 6 February 1922 by U.S.A., the British Empire, France, Italy and Japan, these powers reduced their navies. The personnel of the British navy had a minimum of 99,000 at the end of 1924 (Cmd. 2718 of 1926). The tonnage of the French navy continued to decrease until 1924. In 1925 the Locarno treaties left a sense of appeasement in Europe. In January 1925 Japan agreed with Russia to withdraw Japanese troops from northern Sakhalin. In January 1928 France arranged to reduce compulsory military serivce from 18 months to 12 months beginning on 1 November 1930. In August 1928 the Kellogg Pact was signed in Paris. The world total of warships by categories is given since 1928 in the *L.N.A.Y.B.* 1937, p. 1060. Least values for years 1928–36 occurred in the following years: battleships and battle-cruisers 1932, coast-defence ships and monitors 1936, cruisers 1931, aircraft carriers 1928, flotilla leaders, destroyers and torpedo boats 1935, submarines 1928. Altogether it appears that there was a long period extending from about 1924 to 1931, during which P was nearly stationary for the nations U.S.A., Great Britain, France, Italy and Japan; so that for them we shall not be far wrong in assuming $dP/dt = 0$ in 1927 or 1928. But for Russia the minimum seems to have occurred as early as 1923 and for Germany about 1925.

Given c, P_0 and U, the differential equation

$$\frac{dP}{dt} = U - cP$$

has next to be solved for P by a numerical process that will give P in each successive year. To do this let us make the harmless assumption that the rate of expenditure U remains constant during each budgetary year, but changes from one year to another. It may then be deduced strictly that

$$P_{t+1} = e^{-c}P_t + \left(\frac{1-e^{-c}}{c}\right) U_{t+\frac{1}{2}},$$

where the suffixes denote time in years. In particular for $c = 0{\cdot}1$ this formula becomes

$$P_{t+1} = 0{\cdot}9048P_t + 0{\cdot}952U_{t+\frac{1}{2}} \quad \text{for working forwards,}$$

and $$P_t = 1{\cdot}1052P_{t+1} - 1{\cdot}052U_{t+\frac{1}{2}} \quad \text{for working backwards.}$$

These are the formulae that have been used. Any error ϵ in the initial P_0 will cause an error in P that decreases as time goes on, falling to $0{\cdot}35\epsilon$ after 10 years.

A puzzling question which arises in connexion with Japanese and Italian wars is how to capitalize their cost F. In calculating ψ it seemed proper to include F in U. But can F be said to accumulate? Is it not wasted? I have assmed that

$$dP/dt = (U - F) - cP.$$

It is usually $(U - F)$ that is given in the *Armaments Year-books*.

In Fig. 18 graphs of P are shown. On comparison with the graphs of U in Fig. 15 it is seen that capitalization has effected a smoothing, without much alteration in the order of the nations. On the average P is about ten times U. This is a direct consequence of the choice of $c = 0{\cdot}1$ as the rate of depreciation.

42 SEVEN GREAT POWERS

The same capitalizing calculation, with $c = 0{\cdot}1$, has been applied to ΣU the world-total of defence budgets, as given in § 4·5. Here are the results expressed in 10^9 old gold U.S.A. dollars:

Date	1924·5	25·5	26·5	27·5	28·5	29·5	30·5	31·5	32·5	33·5	34·5
ΣP	39·8	39·3	39·0	39·0	39·0	39·2	39·5	39·5	39·1	38·5	38·2

For the world the minimum of ΣP was assumed to have occurred in 1927. It is seen that ΣP was almost constant from 1925 to 1934.

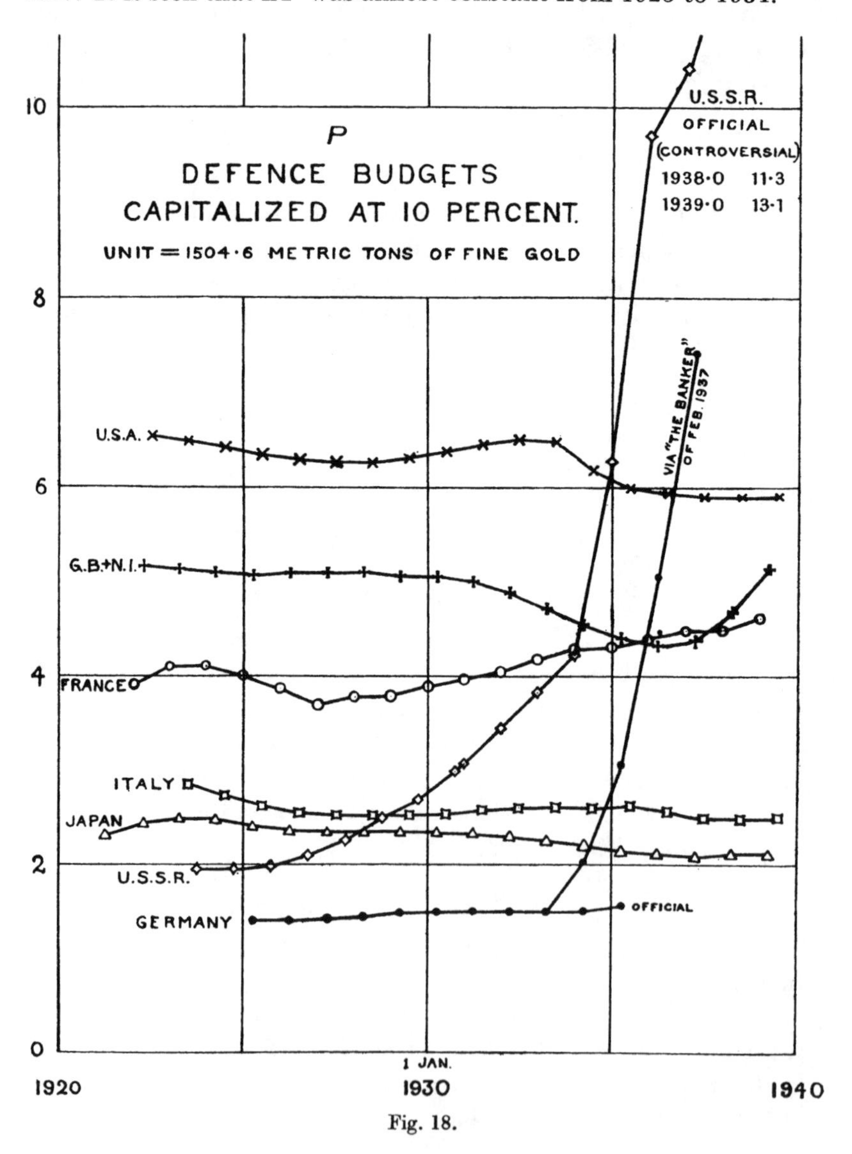

Fig. 18.

§ 4·8. Second method of compensation. Statistic χ.

To avoid the difficulties (*a*), (*b*), (*c*) mentioned at the end of § 4·6 let us proceed as follows. We take the mean $\bar{M}$ of imports plus exports for each nation for the ten years 1925–34.

	G.B. + N.I.	France	Germany	Italy	Japan	U.S.A.	U.S.S.R.
$\bar{M} =$	$6{\cdot}73_0$	$3{\cdot}28_3$	$4{\cdot}43_9$	$1{\cdot}41_0$	$1{\cdot}45_5$	$6{\cdot}45_2$	$0{\cdot}76_1$

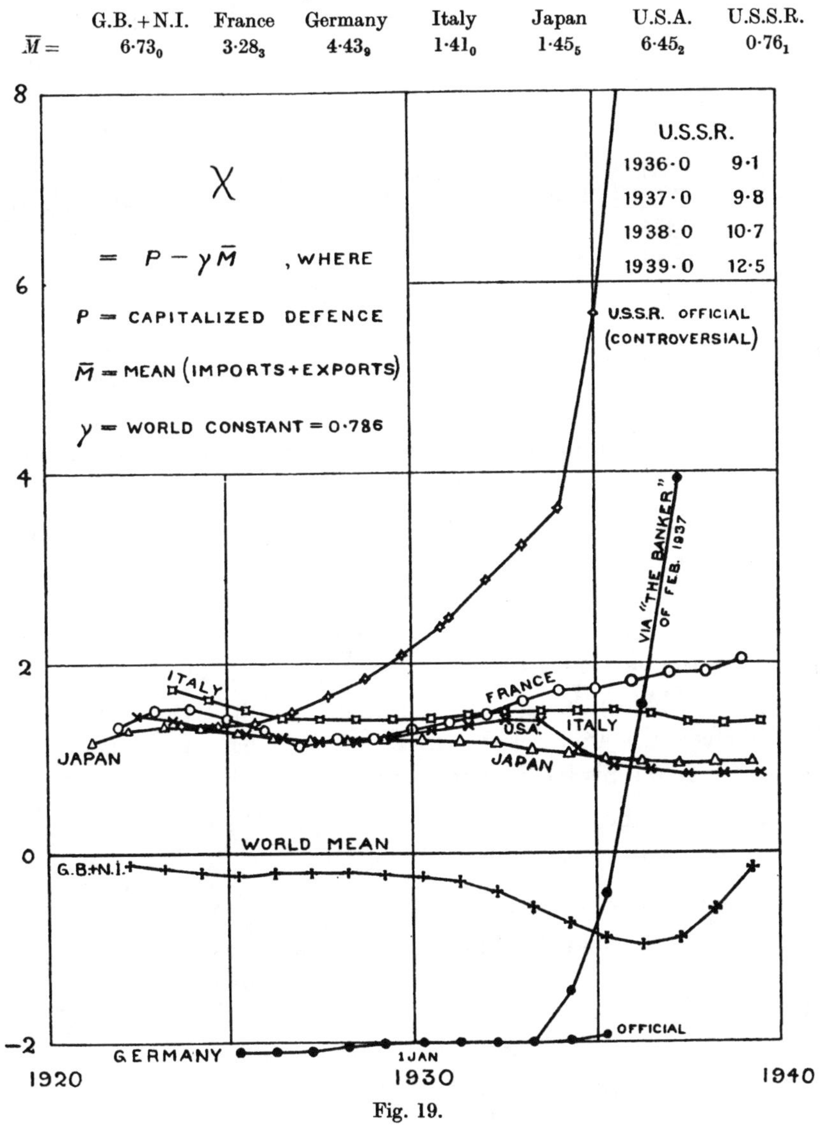

Fig. 19.

The unit is 10^9 old gold dollars. We then define χ thus:

$$\chi = (P - \gamma \bar{M}),$$

where P is the capitalized value of armaments calculated in the manner described in § 4·7 and expressed in the same unit as $\overline{M}$, and γ is a world-constant defined thus:

$$\gamma = \frac{\overline{\Sigma P}}{\overline{\Sigma M}},$$

Σ denoting a world-total, $\overline{\Sigma M}$ being the average for the same ten years 1925–34, and $\overline{\Sigma P}$ being the average for the eleven years $1924\frac{1}{2}$–$34\frac{1}{2}$ which have the same centre as the ten-year period.

It is found that $\overline{\Sigma M} = 49{\cdot}7_3$ and $\overline{\Sigma P} = 39{\cdot}1_0$, both in 10^9 old gold dollars. Hence $\gamma = 0{\cdot}786_2$.

Fig. 19 shows the course of χ in time. The separate graphs for χ are of course of the same shape as those for P, but shifted individually to various levels by the term $\gamma\overline{M}$.

The statistic χ has the following advantages:

(i) χ is positive for a nation whose warlike preparations predominate over its foreign trade, and negative for a nation whose foreign trade predominates over its warlike preparations.

(ii) The preparations aforesaid are capitalized values, allowing for accumulation and obsolescence.

(iii) The doubly averaged value of χ taken both over the world, and over the ten years 1925–34, is zero.

(iv) But χ has purposely not been defined so as to make its world-mean constantly equal to zero; for to do so would be to preclude by definition the very time-changes which we wish to investigate.

(v) Other things being the same, the separate terms γM and P which make up χ are measures of the international importance of the nation, more impressive than its population.

(vi) χ is not affected by the world slump of 1931.

(vii) National currency is eliminated.

The defects of χ are that:

(*a*) χ involves the gold standard, which is not as constant in time as a standard ought to be, and may be misleading when applied to Russia.

(*b*) The rate of depreciation, $c = 0{\cdot}1$, is a rough estimate of a mean of values ranging from 0·04 to 0·2 or more.

§ 4·9. Comparison of the statistics ψ and χ.

Although the definition of χ looks decidedly different from that of ψ, yet there are broad general resemblances between their graphs in

Figs. 17 and 19. In both the line for Germany begins in the negative region and rises across the lines of five other nations. In both, Great Britain and Northern Ireland is on the average the nearest to the zero line. In both, U.S.S.R. appears in the upper part of the diagram. These resemblances are welcome, because ψ and χ were both defined to be measures of 'threats minus co-operation'. Of the two, the definition of χ appears to be the more objective and equitable. The astonishing increase of χ for U.S.S.R. was followed by an increase of χ for Germany and that by an increase of χ for Great Britain and Northern Ireland. But there is nothing alarming about the χ curves for France, Italy, Japan, and U.S.A., except that their ordinates are positive. In this respect χ is in contrast with British newspapers and broadcasts, which in 1937–8 gave a much gloomier view of the international situation. The explanation of this contrast seems to be that χ is derived from P and that in forming P the cost F of actual fighting by Japan and Italy was not accumulated. That is to say F was regarded as having gone to waste; and so P expresses power to fight, but gives no hint of recent aggressions nor of intention to fight. The newspapers and broadcasts, on the contrary, hint at the intentions of nations.

§ 4·10. Sensitive indicators.

Besides ψ and χ which, when available, certainly give a good deal of information, there are some other features of the monetary statistics which may be regarded as sensitive indicators of a nation's anxiety or aggressiveness in contrast with its contentment; or of the changes in these emotional states. Unfortunately they fail to distinguish between anxiety and aggressiveness.

(1) *Whether the actual defence expenditure is more or less than the estimate.* In the years 1922 and 1927, when it was commonly said in England that 'there's not going to be another war for a long time', the actual was less than the estimate. Thus in millions sterling including both ordinary pensions and Middle Eastern defence:

Year ...	1922–3	1923–4	1924–5	1925–6	1926–7
Estimate	150·4	131·6	123·0	127·5	122·8
Closed account	127·0	119·0	121·2	127·0	121·8

Whereas in 1935–7, a time of growing anxiety, the actual exceeded the estimate, thus excluding pensions but including the Middle East:

Year ...	1935–6	1936–7
Estimate	109·1	160·7
Closed account	122·3	172·3

(2) *Rapid increases of defence expenditure.* The most notable are those of Russia in 1933–5, and of Germany about a year later.

(3) *Refusal of information.* Germany withheld information as to defence expenditure after 1935.

(4) *Camouflage of facts.* The German official figures for 1933–5 are much less than unofficial estimates. Thus in millions of Reichsmarks:

Year	1933–4	1934–5
Official German	673	944
From *The Banker* of February 1937	3000	5500

The French expenditure on defence has, in recent years, been scattered over a number of different accounts and Ministries, as is shown by the notes in the *L.N.A.Y.B.*

Postscript on German expenditure. Balogh (1938, p. 477) estimates 'about Rm. 8·5 and Rm. 13 milliards for total extraordinary Government expenditure in the two financial years 1936–37 and 1937–38 respectively'. Also (on p. 478) 'The aggregate extraordinary expenditure since the beginning of 1933 (including civil work creation) thus seems to be somewhere between Rm. 31 and Rm. 42 milliards, according to which hypothesis we accept for estimating total borrowing'.

Summary in milliard Rm.

Year ...	1933–34	34–35	35–36	36–37	37–38
The Banker	3·0*	5·5	10·0	12·6	—
Balogh	[Total for 3 years 9·5 to 20·5]			8·5	13·0

* Including strategic roads.

PART III. THEORY OF THE INSTABILITY OF PEACE BETWEEN SEVERAL NATIONS

§ 5. INTRODUCTION

MENTOR. 'Do the monetary statistics of § 4 uphold the generalizations from politicians speeches in § 2?'

AUTHOR. 'Yes, certainly in some ways. Russia has for long proclaimed that she is afraid of the capitalist powers, and Germany has repeatedly expressed her fear of communism. Look at the way in which their curves for ψ or χ go rushing up together! Russia seems to have begun it.'

MENTOR. 'But what about the other nations?'

AUTHOR. 'The theory of § 2 relates to a pair of nations. Before your question can be answered, we must extend the theory to n nations. I propose to assume that

$$\frac{dx_i}{dt} = g_i + \sum_{j=1}^{j=n} \kappa_{ij} x_j, \quad (i, j = 1, 2, 3, \ldots, n)$$

where x_i is a measure of "threats minus co-operation" for the ith nation, and g_i and κ_{ij} are constants.'

MENTOR. 'That seems to beg many questions.'

AUTHOR. 'We are constrained by the logic of classification to believe that dx_i/dt can depend only on (i) x_i, (ii) x_j, where $j \neq i$, and (iii) other things. My assumptions are (a) that the dependence on the x is a proportionality, and (b) that the "other things" may be represented by a constant g_i.'

MENTOR. 'In Volterra's theory of competing species, on which I heard an interesting lecture by Dr I. M. H. Etherington, the first term is $d \log x_i/dt$ instead of your dx_i/dt. Volterra's formulation is also better for war than yours, because the rate at which a nation can increase its armaments depends on how much armament industry it has got going.'

AUTHOR. 'I beg to differ. For the rate at which a nation can increase its warlike preparations depends also largely on its general industry and population. These I take into account by making the defence coefficient κ_{ij} proportional to the "size", in a suitable sense, of the nation indicated by the first suffix. For the analysis of the stability of peace it is proper to regard the size as constant. Volterra's relations between predator and prey are quite unlike those here studied, because in the former the populations may change in large ratios, by the predator eating the prey.'

MENTOR. 'But surely there is much evidence that the equations which describe biological phenomena are not linear; for example, relaxation oscillations and the all-or-nothing response. The latter occurs also in politics. Two or three days before he fled from Abyssinia in 1936, Haile Selassie made a speech confident of victory. This suddenness of defeat, this switch from all-response to a nothing-response, is not like the phenomena which in physics are represented by linear differential equations.'

AUTHOR. 'That is quite true. But please remember that I am not making a theory of victory and defeat but only a theory of the stability or instability of peace. For the description of that gentler phenomenon linear equations may be expected to yield a useful approximation.'

MENTOR. 'But why not aim at accuracy instead of at a mere approximation?'

AUTHOR. 'Because, if we had accurate non-linear equations, it is likely that the only possible formulae of solution would be attained by local linear approximations; as was done in § 3 for submissiveness. So linear theory is a necessary preliminary. Solutions could indeed also be obtained by numerical processes; but for n variables those would be very laborious.'

MENTOR. 'Your x_i is a scalar. But the outward attitude of a nation, is it not really a vector, having $(n-1)$ components $x_{i1}, x_{i2}, \ldots, x_{in}$ directed severally to the other nations? Then instead of κ_{ij} we should need a symbol with four suffixes, like those tensors in Relativity.'

AUTHOR. 'You may be right. I rely merely on Ockham's razor in the form, "Try out the easiest first". I regard x_i as an objective fact, κ_{ij} as involving intentions and suspicions directed from one nation to another.'

MENTOR. 'Many people will distrust your use of mathematics because they will say that it gives a false appearance of precision to a subject essentially nebulous. You make, as it were, a firm bold outline of a cloud.'

AUTHOR. 'Yes, we must be on our guard against mistaking precision in the deductive method for precision in the data and results. But we are accustomed to that fictitious accuracy in business. Thus a man may buy a house for £964. 13*s*. 7*d*. and expect his banker to keep the record of the transaction to the precise penny, although he knows that he would have been willing to pay £50 more. We relish our bacon although the machine that sliced it was much more accurate than it need have been.'

MENTOR. 'Is your mathematical treatment original?'

AUTHOR. 'Except possibly for the process of §13·2 the pure mathematics required is nearly all to be found scattered among the textbooks. I had to collect and arrange it. The application to politics is, I believe, original.'

§6. THREE NATIONS (OR THREE GROUPS OF NATIONS)

Before discussing n nations we should discuss three. For three nations can be illustrated geometrically in a way that a larger number cannot. Moreover, it was an observed fact during 1914–18 that the nations of the world tended to agglomerate into three groups, one of them being the Neutrals. The same occurred in 1938. See p. 80.

The international situation is represented by the position of a particle specified by rectangular co-ordinates x, y, z. By an obvious generalization of the equations for two nations let us assume that the equations of motion of the particle are, for instinctive behaviour,

$$\left.\begin{aligned} dx/dt &= -\alpha_1 x + \kappa_{12} y + \kappa_{13} z + g_1, \\ dy/dt &= \kappa_{21} x - \alpha_2 y + \kappa_{23} z + g_2, \\ dz/dt &= \kappa_{31} x + \kappa_{32} y - \alpha_3 z + g_3, \end{aligned}\right\} \tag{1}$$

where x represents, when positive, the net outward attitude of the first nation in the sense of warlike preparations minus co-operation; and y represents a similar quantity for the second nation and z for the third. The three α are positive 'fatigue and expense coefficients'. The six κ are 'defence coefficients' and are positive. The g represent the effect of grievances concerning the treaty situation.

For symmetry it is sometimes convenient to re-letter α thus,

$$-\alpha_1 = \kappa_{11}, \quad -\alpha_2 = \kappa_{22}, \quad -\alpha_3 = \kappa_{33}, \tag{2}$$

because many consequences depend upon the matrix

$$\begin{bmatrix} -\alpha_1, & \kappa_{12}, & \kappa_{13} \\ \kappa_{21}, & -\alpha_2, & \kappa_{23} \\ \kappa_{31}, & \kappa_{32}, & -\alpha_3 \end{bmatrix} \text{ which we can now write briefly as } [\kappa_{ij}] \tag{3}$$

or on the determinant of this matrix, which will be denoted by $|\kappa_{ij}|$. (4)

§6·1. The point of balance.

If in (1)

$$dx/dt = 0, \quad dy/dt = 0, \quad dz/dt = 0 \tag{5}$$

simultaneously, we are left with a set of three linear algebraic equations in x, y, z. Ordinarily, that is to say unless $|\kappa_{ij}| = 0$, these equations determine a point say (x_0, y_0, z_0) appropriately called the equilibrium point or point of balance. It satisfies

$$\left.\begin{aligned} 0 &= -\alpha_1 x_0 + \kappa_{12} y_0 + \kappa_{13} z_0 + g_1, \\ 0 &= \kappa_{21} x_0 - \alpha_2 y_0 + \kappa_{23} z_0 + g_2, \\ 0 &= \kappa_{31} x_0 + \kappa_{32} y_0 - \alpha_3 z_0 + g_3. \end{aligned}\right\} \tag{6}$$

Let each of equations (6) be subtracted from the corresponding equation of (1) and let

$$x - x_0 = X, \quad y - y_0 = Y, \quad z - z_0 = Z, \tag{7}$$

so that capital letters are co-ordinates reckoned from the point of balance. It follows that

$$\left.\begin{aligned} dX/dt &= -\alpha_1 X + \kappa_{12} Y + \kappa_{13} Z, \\ dY/dt &= \kappa_{21} X - \alpha_2 Y + \kappa_{23} Z, \\ dZ/dt &= \kappa_{31} X + \kappa_{32} Y - \alpha_3 Z. \end{aligned}\right\} \tag{8}$$

The equations (8) are convenient because they do not contain the grievances g, which are regarded as constants between crises.

We want to know whether the particle will move to the point of balance and stay there; or whether the particle will move off to infinity, and if so in which direction. The problem is complicated, because, as we have seen in two dimensions, it is quite possible for the particle to describe an orbit partly approaching and partly receding from the point of balance. Instead of tracing orbits in detail we shall find it economical of thought to mark out certain special lines and planes.

§6·2. General solution in normal co-ordinates. Barrier planes.

We here extend the method of normal co-ordinates which worked so pleasantly in §2·10, §2·11.

Let equations (8) be multiplied respectively by undetermined constants l, m, n and let the products be added so that

$$\left.\begin{aligned} d(lX + mY + nZ)/dt = {} & (-\alpha_1 l + \kappa_{21} m + \kappa_{31} n)\, X, \\ & + (\kappa_{12} l - \alpha_2 m + \kappa_{32} n)\, Y, \\ & + (\kappa_{13} l + \kappa_{23} m - \alpha_3 n)\, Z. \end{aligned}\right\} \tag{9}$$

Next let us try to choose l, m, n so that the two linear functions of X, Y, Z that appear in (9) may be constant multiples of one another. This will happen if the same quantity λ occurs in each of the three following equations:

$$\left.\begin{aligned} \lambda l &= -\alpha_1 l + \kappa_{21} m + \kappa_{31} n, \\ \lambda m &= \kappa_{12} l - \alpha_2 m + \kappa_{32} n, \\ \lambda n &= \kappa_{13} l + \kappa_{23} m - \alpha_3 n. \end{aligned}\right\} \qquad (10)$$

The matrix of the second members is the transpose of $[\kappa_{ij}]$. The condition that the three equations of (10) may be consistent for non-zero l, m, n may be written transposed as

$$\begin{vmatrix} (-\alpha_1-\lambda), & \kappa_{12}, & \kappa_{13} \\ \kappa_{21}, & (-\alpha_2-\lambda), & \kappa_{23} \\ \kappa_{31}, & \kappa_{32}, & (-\alpha_3-\lambda) \end{vmatrix} = 0. \qquad (11)$$

This is a cubic in λ having roots λ_1, λ_2, λ_3, say, of which one must be real and the other two must be either both real or both complex. Corresponding to each root there is a set of ratios $(l:m:n)$, real or complex, determined by (10) such that (9) takes the simple form

$$d(lX+mY+nZ)/dt = \lambda(lX+mY+nZ), \qquad (12)$$

the integral of which is

$$lX+mY+nZ = Ae^{\lambda t}, \qquad (13)$$

where A is an arbitrary constant. So that if the roots λ_1, λ_2, λ_3 are distinct and if we use the same suffixes to distinguish the corresponding l, m, n, A, we have three distinct integrals each containing only one exponential function of time, thus

$$l_1X+m_1Y+n_1Z = A_1e^{\lambda_1 t}, \qquad (14)$$

$$l_2X+m_2Y+n_2Z = A_2e^{\lambda_2 t}, \qquad (15)$$

$$l_3X+m_3Y+n_3Z = A_3e^{\lambda_3 t}. \qquad (16)$$

The first members of (14), (15), (16) are the '*normal co-ordinates*'. We must now distinguish different kinds of roots.

(i) If there is a *positive root*, say λ_1, the corresponding normal co-ordinate $l_1X+m_1Y+n_1Z$ goes off to plus or minus infinity as $t\to\infty$. To which infinity the normal co-ordinate goes, depends on the sign of A_1, for $A_1e^{\lambda_1 t}$ cannot change sign. That is to say $l_1X+m_1Y+n_1Z=0$ is, in this case, a *barrier plane*, which is of great political importance because it separates two regions where the instinctive drift leads to opposite infinities. Like other instinctive barriers in private life, it could presumably be crossed by an effort.

(ii) If there is a *negative root*, the corresponding normal co-ordinate tends to zero as time goes on. There is again a barrier plane, but as the situation is quiescent, the barrier is of less political interest than that corresponding to a positive λ.

(iii) If λ_1 is a complex root, let one accent denote the real part and two accents denote the coefficient of $i = \sqrt{-1}$, in any quantity; so that $\lambda_1 = \lambda' + i\lambda''$. There must be a conjugate root $\lambda' - i\lambda'' = \lambda_2$ say. Also let

$$l_1 = l' + il'', \quad m_1 = m' + im'', \quad n_1 = n' + in''.$$

It follows from (10) that l_2, m_2, n_2 are conjugate to l_1, m_1, n_1; that is to say

$$l_2 = l' - il'', \quad m_2 = m' - im'', \quad n_2 = n' - in''.$$

So if we add equations (14) and (15) and put $A_1 = A_2 = B$ we obtain

$$l'X + m'Y + n'Z = Be^{\lambda' t} \cos \lambda'' t. \tag{17}$$

Similarly by subtraction of (15) from (14), where C is a constant,

$$l''X + m''Y + n''Z = Ce^{\lambda' t} \sin \lambda'' t. \tag{18}$$

The equation (16) must then correspond to a real root and requires no transformation.

(iv) *Equal roots* often occur in the sequel accompanied by other forms of mathematical degeneracy. Their discussion is best taken for the special cases that arise, as generalities are difficult.

Summary for all kinds of unequal roots. If a normal co-ordinate were to be tending to infinity, the phenomena which it briefly describes would be certain to attract much public attention. Therefore in the subsequent pages there will be much discussion about the greatest of the real parts of the roots. A real root is here regarded as the real part of a complex root having zero imaginary part. If this greatest real part is positive, at least one normal co-ordinate is tending towards infinity, and the system will be called unstable. On the contrary if the greatest real part is negative, no normal co-ordinate can tend to infinity and the system will be called stable, for it will settle at the point of balance of power. What happens when roots are equal is left for discussion in special cases.

The general solution for the nations separately. It will usually be possible to solve three normal equations for X, Y, Z separately; and if so X, Y, Z will usually each involve all three of the functions of time which appear in the second members of the normal equations. When the theory is being compared with historical facts it is important that the functions of time should be as simple as possible; and for that purpose normal co-ordinates are much more convenient than X, Y, Z separately. See p. 78.

Finally we must not forget that X, Y, Z are measured from the point of balance x_0, y_0, z_0 as given by (6) above.

Special Case I

Three nations having the same coefficients

$$dX/dt = -\alpha X + \kappa Y + \kappa Z,$$
$$dY/dt = \kappa X - \alpha Y + \kappa Z,$$
$$dZ/dt = \kappa X + \kappa Y - \alpha Z.$$

By addition,

$$d(X+Y+Z)/dt = (2\kappa - \alpha)(X+Y+Z).$$

By subtraction,

$$d(X-Y)/dt = (-\kappa - \alpha)(X-Y).$$

So normal co-ordinates are

$$X+Y+Z = Ae^{(2\kappa-\alpha)t},$$
$$X-Y = Be^{(-\kappa-\alpha)t},$$
$$Y-Z = Ce^{(-\kappa-\alpha)t},$$

where A, B, C are arbitrary constants. We might equally well have chosen $(X-Z)$ as a normal co-ordinate. As time goes on $(X-Y)$ and $(Y-Z)$ tend to zero; while $(X+Y+Z)$ tends to zero if $2\kappa < \alpha$, but to infinity if $2\kappa > \alpha$. Thus $2\kappa = \alpha$ is the criterion of stability.

In this case the determinant has a pair of equal roots

$$\lambda_2 = \lambda_3 = -\kappa - \alpha;$$

so that we might have expected from the general theory in the textbooks that the solution would contain terms in $te^{\lambda_2 t}$. But the substitution of X, Y, $Z = (A, B, C)\, te^{\lambda_2 t}$ into the differential equations shows that such terms cannot occur in this case.

§ 6·3. Radii along which the particle can move.

We enquire whether there can be a real quantity λ such that

$$dX/dt = \lambda X, \quad dY/dt = \lambda Y, \quad dZ/dt = \lambda Z, \tag{19}$$

for, if so, the particle is moving in a straight line through the point of balance. Substitution of (19) in (8) gives

$$\left.\begin{aligned} (-\alpha_1 - \lambda) X + \kappa_{12} Y + \kappa_{13} Z &= 0, \\ \kappa_{21} X + (-\alpha_2 - \lambda) Y + \kappa_{23} Z &= 0, \\ \kappa_{31} X + \kappa_{32} Y + (-\alpha_3 - \lambda) Z &= 0. \end{aligned}\right\} \tag{20}$$

The condition of consistency of these three equations is (11) as before. To each real root of (11) there corresponds a radius along which the particle can move. If the root is negative the particle moves towards the point of balance, if positive away. The direction of the radius is given by the common intersection of the three planes (20). When λ is positive this direction will be called the 'axis of instability'.

§ 6·4. Bounds to the roots λ.

The equation (11) having the roots λ_1, λ_2, λ_3, when expanded in powers of λ is

$$\lambda^3+\lambda^2(\alpha_1+\alpha_2+\alpha_3)+\lambda\{(\alpha_2\alpha_3-\kappa_{23}\kappa_{32})+(\alpha_3\alpha_1-\kappa_{31}\kappa_{13})+(\alpha_1\alpha_2-\kappa_{12}\kappa_{21})\}$$
$$-|\kappa_{ij}|\equiv f(\lambda),\ \text{say}=0. \tag{21}$$

Then

$$\lambda_1+\lambda_2+\lambda_3=-(\alpha_1+\alpha_2+\alpha_3)<0, \tag{22}$$

so that *the real parts of the roots cannot all be positive.* (23)

As $\lambda\to\infty$, $f(\lambda)\to\infty$ also $f(0)=-|\kappa_{ij}|$. Therefore if $|\kappa_{ij}|>0$ there is at least one positive real root and the particle will go towards infinity. (24)

Similarly as $\lambda\to-\infty$, $f(\lambda)\to-\infty$. So if $|\kappa_{ij}|<0$ then $f(0)>0$ and there is at least one negative real root; a fact which however does not tell us anything about the ultimate destination of the particle. (25)

The coefficient of λ in (21) depends on the stability of the separate pairs of nations, as was proved in §§ 2·6 and 2·11. This leads us to interesting special cases.

Special Case II

If each possible pair of nations would alone be unstable, then by § 2·11

$$\kappa_{23}\kappa_{32}>\alpha_2\alpha_3,\quad \kappa_{31}\kappa_{13}>\alpha_3\alpha_1,\quad \kappa_{12}\kappa_{21}>\alpha_1\alpha_2.$$

Now $|\kappa_{ij}|$ when expanded in the elements of its first row is

$$|\kappa_{ij}|=\alpha_1(\kappa_{23}\kappa_{32}-\alpha_2\alpha_3)+\kappa_{12}(\alpha_3\kappa_{21}+\kappa_{23}\kappa_{31})+\kappa_{13}(\kappa_{21}\kappa_{32}+\alpha_2\kappa_{31})$$

>0 in this case. So there is at least one positve root. Equation (21) takes the form

$$0=\lambda^3+P_2\lambda^2-P_1\lambda-P_0,$$

where P_2, P_1, P_0 are all positive. By Descartes' rule of signs, there is not more than one positive real root. Thus we conclude that: *If each of the three pairs of nations be separately unstable then the triplet is necessarily unstable and there is a single radius along which the particle will move to one or other of two opposite infinities.* (26)

Special Case III

If each possible pair of nations would alone be stable then

$$\alpha_2\alpha_3 > \kappa_{23}\kappa_{32}, \quad \alpha_3\alpha_1 > \kappa_{31}\kappa_{13}, \quad \alpha_1\alpha_2 > \kappa_{12}\kappa_{21}.$$

But this does not fix the sign of $|\kappa_{ij}|$. So Q_2, Q_1, Q_0 being positive, equation (21) takes the form

$$0 = \lambda + Q_2\lambda^2 + Q_1\lambda \pm Q_0.$$

It follows that: *If each of the three pairs of nations be separately stable there remains the possibility that the triplet of nations may be unstable.* (27)

This is essentially an abstract theorem. We cannot observe nations in detached pairs. So even a hint from other forms of life is acceptable; as many instincts are common to men and animals. Those who observe dogs tell me that dogs which will play peaceably in all possible pairs may yet quarrel when three are present together.

§ 6·5. Variation of the length R of the radius vector from the point of balance.

A study of dR/dt leads to an *a fortiori* theorem concerning stability, which is especially convenient when generalized for n nations. For obviously the particle cannot go off to infinity if dR/dt is everywhere negative.

We have $$R^2 = X^2 + Y^2 + Z^2. \qquad (28)$$

Let L, M, N be direction-cosines at the point of balance so that

$$LR = X, \quad MR = Y, \quad NR = Z. \qquad (29)$$

We multiply the three equations (8) respectively by X, Y, Z, add the products and divide their sum by R^2. Thus we obtain

$$d\log R/dt = -\alpha_1 L^2 - \alpha_2 M^2 - \alpha_3 N^2 + (\kappa_{23}+\kappa_{32})MN + (\kappa_{31}+\kappa_{13})NL + (\kappa_{12}+\kappa_{21})LM = \phi(L,M,N), \text{ say.} \qquad (30)$$

The quadratic form in the second member may be, and for convenience is, regarded as having the symmetrical matrix $[\bar{\kappa}_{ij}]$, where

$$\bar{\kappa}_{ij} = \tfrac{1}{2}(\kappa_{ij}+\kappa_{ji}). \qquad (31)$$

We will call $[\bar{\kappa}_{ij}]$ the 'symmetrized matrix'.

On the axis of any one of the co-ordinates X, Y, Z, two out of L, M, N are zero and so $d\log R/dt$ as given by (30) is negative. (32)

The question is whether $d\log R/dt$ is negative for all L, M, N. This can be answered by transforming $\phi(L,M,N)$ to its principal axes by a rotation of rectangular co-ordinates such that simultaneously

$$L^2+M^2+N^2 = L_1^2+M_1^2+N_1^2 \qquad (33)$$

and $$\phi(L, M, N) = \mu_1 L_1^2 + \mu_2 M_1^2 + \mu_3 N_1^2. \tag{34}$$

It is proved in text-books on solid geometry that μ_1, μ_2, μ_3 are the roots, necessarily real, of the 'discriminating cubic' in μ, that is of

$$\begin{vmatrix} (-\alpha_1 - \mu), & \bar{\kappa}_{12}, & \bar{\kappa}_{13} \\ \bar{\kappa}_{12}, & (-\alpha_2 - \mu), & \bar{\kappa}_{23} \\ \bar{\kappa}_{13}, & \bar{\kappa}_{23}, & (-\alpha_3 - \mu) \end{vmatrix} = 0. \tag{35}$$

Other names for μ_1, μ_2, μ_3 are the 'latent roots' or 'characteristic roots' of the symmetrized matrix $[\bar{\kappa}_{ij}]$.

Thus while the necessary and sufficient condition of stability is that the real parts of the roots $\lambda_1, \lambda_2, \lambda_3$ *of* (21) *must all be negative, a sufficient condition is that the real roots* μ_1, μ_2, μ_3 *of* (35) *should all be negative.*

If μ_1, μ_2, μ_3 are not all of the same sign there is a quadric cone $\phi(L, M, N) = 0$ on which $d \log R/dt$ is zero.

Multiplication of (30) by R^2 gives in view of (34)

$$\tfrac{1}{2} dR^2/dt = \mu_1 X_1^2 + \mu_2 Y_1^2 + \mu_3 Z_1^2,$$

where $$X_1 = L_1 R, \quad Y_1 = M_1 R, \quad Z_1 = N_1 R.$$

Whatever be the signs of μ_1, μ_2, μ_3 there are quadric surfaces on which dR^2/dt is constant, and X_1, Y_1, Z_1 are measured along the principal axes. For the discussion of equal roots and other mathematically 'degenerate' cases, reference may be made to W. H. Macaulay, 1930, *Solid Geometry*, Cambridge University Press.

Special Case IV

The given matrix is symmetrical $\kappa_{ij} = \kappa_{ji}$. Then

$$\lambda_1, \lambda_2, \lambda_3 = \mu_1, \mu_2, \mu_3.$$

If these roots are distinct there are just three radii along which the particle can move. Also l, m, n as determined by (10) are proportional to X, Y, Z as determined by (20). That is to say the planes which the particle cannot cross are normal to the radii along which it can move, and the latter coincide with the principal axes of the quadric surface on which dR^2/dt is constant.

Special Case V

We imagine two pugnacious nations between which the defence coefficient $\kappa = 3$ and one neutral nation towards or from which $\kappa = 1$, the unit of time being 10 years. The fatigue and expense coefficient is taken to be $\alpha = 2$ the same for all the nations. Thus equation (11) is

$$0 = \begin{vmatrix} (2+\lambda), & -3, & -1 \\ -3, & (2+\lambda), & -1 \\ -1, & -1, & (2+\lambda) \end{vmatrix}.$$

The roots are found to be $\lambda = -5, \quad -2{\cdot}5616\ldots, \quad 1{\cdot}5616\ldots.$
As there is a positive root the triplet of nations is unstable. The matrix is symmetrical. Along two of the principal axes of the quadric cone, on which $d \log R/dt$ is zero, the particle moves towards the point of balance; on the third principal axis the particle moves away from the point of balance in either direction. The direction of this 'axis of instability' is found from (10) by substitution of $\lambda = 1{\cdot}5616$ and is such that $X : Y : Z = 1 : 1 : 0{\cdot}5616$. The most interesting feature of this axis is that along it X, Y, Z have all the same sign, but that the effect on the neutral nation is less than the effects on either of the pugnacious nations. This is shown more in detail by working out the normal co-ordinates which are found to be

$$X - Y = Ae^{-5t} \rightarrow 0,$$
$$X + Y - 3{\cdot}5616Z = Be^{-2{\cdot}5616t} \rightarrow 0,$$
$$X + Y + 0{\cdot}5616Z = Ce^{1{\cdot}5616t} \rightarrow \pm\infty.$$

Thus by suitable choice of different κ we may expect to be able to explain the different rates of increase of the statistics ψ or χ for the various nations discussed in Part II.

Special Case VI

One pacifist nation and two pugnacious nations. There is a small but convinced minority which says that the best way to meet aggression is by non-resistance. Let us try out this proposal in mathematics. Suppose that the z-nation has, when x and y are positive, zero defence coefficients directed towards the x- and y-nations, in spite of the fact that they have positive defence coefficients directed towards it. Pacifists are like other people in returning favour for favour; so it is only for positive x and y that κ_{31} and κ_{32} vanish. For simplicity let the problem be restricted to positive x and y. Let the matrix be

$$[\kappa_{ij}] = \begin{bmatrix} -\alpha, & \kappa, & \kappa \\ \kappa, & -\alpha, & \kappa \\ 0, & 0, & -\alpha \end{bmatrix}$$

which is unsymmetrical. The latent roots, as given by (11) are, in order of magnitude

$$\lambda_1 = -\alpha - \kappa, \quad \lambda_2 = -\alpha, \quad \lambda_3 = -\alpha + \kappa.$$

They are all real, and two at least are negative. After a long time the terms in $\exp(\lambda_3 t)$ predominate over those in λ_1, λ_2. A radius along which the particle can move when the terms in λ_1 and λ_2 have faded out is, according to (10),

$$X = Y, \quad Z = 0.$$

The system is unstable if $\kappa > \alpha$ and then X and Y depart indefinitely from zero. The mathematics, however, gives no hint as to whether the actions of the pugnacious nations are directed towards each other or towards the pacifist nation. We feel the need for the double-suffixed x_{ij} which the mentor advocated. It is introduced in the next section.

§ 7. DIRECTED INTENTIONS

To penetrate further into the last problem, let us now regard grievances as directed; g_{ij} being the grievance felt by the ith nation against the jth nation independently of armaments. Thus the grievances now form a matrix $[g_{ij}]$. As the theory is not intended to apply to civil strife, we may assume that no nation harbours a grievance against itself, and therefore that $g_{ii} = 0$.

The warlike preparations of each nation are now also supposed to be directed. Accordingly x_{ij} is the preparation of the ith nation against the jth nation. Although x_{ii} might be allocated to the maintenance of internal order, it seems doubtful if its relations will fit into the scheme appropriate to x_{ij} in general; and so it will be assumed that $x_{ii} = 0$ for foreign politics.

It will be convenient to retain the symbol x_i for the total warlike preparation of the ith nation so that

$$x_i = \sum_{j=1}^{j=n} x_{ij}. \tag{1}$$

Other nations are supposed not to have definite information about the intentions of the ith nation; they only guess and suspect; and therefore they are disturbed by the total x_i rather than by its component parts. Thus the new sort of defence coefficients require only two suffixes, not as the mentor expected four. But as they are a new sort, they will be denoted by q not κ.

It will further be assumed for simplicity that the fatigue and expense coefficient α_i is independent of direction, and thus requires only a single suffix.

Accordingly the assumptions are that, for $i, j = 1, 2, 3, \ldots, n$,

$$dx_{ij}/dt = -\alpha_i x_{ij} + q_{ij} x_j + g_{ij}, \tag{2}$$

in which

$$x_{ii} = 0, \quad g_{ii} = 0 \quad \text{and} \quad q_{ii} = 0. \tag{3}$$

This is where q behaves differently from κ; for $\kappa_{ii} = -\alpha_i$. For brevity let

$$g_i = \sum_{j=1}^{j=n} g_{ij}. \tag{4}$$

Then on summing (2) for all j we obtain in view of (1), (3), (4)

$$\frac{dx_i}{dt} = -\alpha_i x_i + \sum_{j=1}^{j=n} q_{ij} x_j + g_i, \tag{5}$$

which is the assumption on which all the previous mathematics except that in § 3 has been based. Thus the present view of directed intentions gives more detail, without in any way contradicting what has already been set out, except for submissiveness.

Special Case VI reconsidered

Let it be assumed that the pacifist nation, numbered 3, is itself contented; but has only recently been converted to pacifism, and has by its past conduct left grievances against it in nations 1 and 2, so that

$$[g_{ij}] = \begin{bmatrix} 0, & 0, & g \\ 0, & 0, & g \\ 0, & 0, & 0 \end{bmatrix}. \tag{6}$$

Whereas

$$[q_{ij}] = \begin{bmatrix} 0, & \kappa, & \kappa \\ \kappa, & 0, & \kappa \\ 0, & 0, & 0 \end{bmatrix} \tag{7}$$

as before; and

$$\alpha_1 = \alpha_2 = \alpha_3 = \alpha \tag{8}$$

as before. The equations of motion (2) when written out in full are

$$dx_{12}/dt = -\alpha x_{12} + \kappa(x_{21} + x_{23}), \tag{9}$$

$$dx_{13}/dt = -\alpha x_{13} + \kappa(x_{31} + x_{32}) + g, \tag{10}$$

$$dx_{21}/dt = -\alpha x_{21} + \kappa(x_{12} + x_{13}), \tag{11}$$

$$dx_{23}/dt = -\alpha x_{23} + \kappa(x_{31} + x_{32}) + g, \tag{12}$$

$$dx_{31}/dt = -\alpha x_{31}, \tag{13}$$

$$dx_{32}/dt = -\alpha x_{32}. \tag{14}$$

From (13) and (14) it follows that

$$(x_{31} + x_{32}) = Ae^{-\alpha t}, \tag{15}$$

where A is an arbitrary constant. When (15) is substituted in (10) and (12) they become both of the form

$$dy/dt = -\alpha y + A\kappa e^{-\alpha t} + g, \tag{16}$$

where y is either x_{13} or x_{23}.

The solution of (16) is

$$y = \frac{g}{\alpha} + (B + A\kappa t)\, e^{-\alpha t}, \tag{17}$$

where B is an arbitrary constant. Thus as time goes on the aggressive preparations of the pugnacious nations towards the pacifist nation settle down to a constant $x_{13} = x_{23} = g/\alpha$ which depends on old standing grievances.

Meanwhile the two pugnacious nations interact with each other; for the substitution of (17) into (9) and (11) gives

$$dx_{12}/dt = -\alpha x_{12} + \kappa x_{21} + \kappa \left\{\frac{g}{\alpha} + (B + A\kappa t)\, e^{-\alpha t}\right\}, \tag{18}$$

$$dx_{21}/dt = -\alpha x_{21} + \kappa x_{12} + \kappa \left\{\frac{g}{\alpha} + (B + A\kappa t)\, e^{-\alpha t}\right\}. \tag{19}$$

After sufficient time the terms in A and B become negligible and then the system (18) and (19) is unstable if $\kappa > \alpha$. But supposing $\kappa > \alpha$ we cannot tell without further data and analysis whether the two pugnacious nations fight each other or form a close alliance, for that depends on the initial values of x_{12}, x_{21} which are not given; and we have restricted the problem to x_1, x_2 positive at the outset. In (18) and (19) there is an effective grievance $\kappa g/\alpha$ of the first two nations against one another. It would vanish if they had no grievance g against the third nation. Is this the phenomenon, well known for individual persons, of the unreasoned spreading of moods?

§ 8. GENERAL THEORY OF n NATIONS

It has been shown in § 7 for n nations that the detailed analysis of directed intentions is consistent with a more summary analysis which starts from the assumption

$$\frac{dx_i}{dt} = g_i + \sum_{j=1}^{j=n} \kappa_{ij} x_j, \quad (i = 1, 2, 3, \ldots, n) \tag{1}$$

where g_i is a dissatisfaction with treaties. If $i \neq j$ then κ_{ij} is a defence coefficient and is positive. If $i = j$ then κ_{ij} is a fatigue and expense coefficient and is negative. We shall sometimes write $-\kappa_{ii} = \alpha_i$ so that $\alpha_i > 0$. (2)

We first remove the g_i. Let the equations

$$0 = g_i + \sum_{j=i}^{j=n} \kappa_{ij} x_j \tag{3}$$

be solved, if possible, for the x. The solution exists provided the determinant

$$|\kappa_{ij}| \neq 0. \tag{4}$$

Let the values of x which satisfy the equilibrium condition (3) be

$$x_{0i}, \quad (i = 1, 2, \ldots, n). \tag{5}$$

By subtracting each equation of the set (3) from the corresponding equation of the set (1) and by writing

$$X_i = x_i - x_{0i} \tag{6}$$

we obtain

$$\frac{dX_i}{dt} = \sum_{j=1}^{j=n} \kappa_{ij} X_j, \quad (i = 1, 2, \ldots, n) \tag{7}$$

which do not contain the g, which have been assumed constant.

§ 8·1. Normal co-ordinates and barriers.

We extend the methods of §§ 2·10, 2·11 and 6·2.

Let (7) be multiplied by a set of undetermined constants l_i and let the products be summed for $i = 1, 2, \ldots, n$. Then

$$\frac{d}{dt} \sum_i l_i X_i = \sum_i l_i \sum_j \kappa_{ij} X_j = \sum_j X_j \sum_i l_i \kappa_{ij}. \tag{8}$$

Next, if possible, choose the l_i so that the linear functions of $X_1, X_2, \ldots, X_n$ in the two members are simply proportional to one another. The coefficient of X_j in the first member is l_j and in the second member $\sum_i l_i \kappa_{ij}$. The proportionality will hold if λ is the same in each of the n equations

$$\lambda l_j = \sum_{i=1}^{i=n} l_i \kappa_{ij}, \quad (j = 1, 2, \ldots, n). \tag{9}$$

The matrix of the coefficients of the l_i is

$$\begin{bmatrix} (-\alpha_1 - \lambda), & \kappa_{21}, & \kappa_{31}, & \ldots, & \kappa_{n1} \\ \kappa_{12}, & (-\alpha_2 - \lambda), & \kappa_{32}, & \ldots, & \kappa_{n2} \\ \cdots & \cdots & \cdots & \cdots & \cdots \\ \kappa_{1n}, & \kappa_{2n}, & \kappa_{3n}, & \ldots, & (-\alpha_n - \lambda) \end{bmatrix} \tag{10}$$

The condition that the n equations (9) may be consistent for non-zero l is that the determinant D of the matrix (10) should vanish. (11)

$D = 0$, regarded as an equation for λ, has n roots $\lambda_1, \lambda_2, \ldots, \lambda_n$ in the complex domain. They are known as the latent roots of the matrix

$[\kappa_{ij}]$. For each distinct root λ_r the equations (9) determine a corresponding set of ratios $l_1 : l_2 : l_3 : \ldots : l_n$ which for distinctness we may write

$$l_{r1} : l_{r2} : l_{r3} : \ldots : l_{rn}$$

and (8) becomes

$$\frac{d}{dt} \sum_i l_{ri} X_i = \lambda_r \sum_i l_{ri} X_i, \tag{12}$$

the integral of which is

$$\sum_i l_{ri} X_i = A_r e^{\lambda_r t}, \quad (r = 1, 2, \ldots, n) \tag{13}$$

where A_r is an arbitrary constant. The first member of (13) is called a normal co-ordinate.

If λ_r is real, $A_r e^{\lambda_r t}$ cannot change sign and so

$$\sum_i l_{ri} X_i = 0 \tag{14}$$

is a *barrier* separating opposite instinctive drifts, when λ_r is real. The treatment of complex roots proceeds as in § 6·2.

The n equations (13) can usually be solved for each of $X_1, X_2, \ldots, X_n$ separately. Finally $x_i = X_i + x_{0i}$, where x_{0i} is the co-ordinate of the point of balance.

Equal roots are best left for discussion in special cases.

§ 8·2. Bounds to the roots.

Because the exact calculation of the roots might in a given problem for large n be extremely laborious, certain theorems about bounds are welcome. The expansion of D in powers of λ begins and ends thus

$$0 = \lambda^n + \lambda^{n-1}(\alpha_1 + \alpha_2 + \ldots + \alpha_n) + \ldots + (-1)^n \mid \kappa_{ij} \mid .$$

Therefore $\quad \lambda_1 + \lambda_2 + \ldots + \lambda_n = -(\alpha_1 + \alpha_2 + \ldots + \alpha_n) < 0.$

Next there is a theorem (MacDuffee, 1933, *The Theory of Matrices*, Theorem 18·5. Berlin: Julius Springer) that the real parts of $\lambda_1, \lambda_2, \ldots, \lambda_n$ lie between the greatest and least of $\mu_1, \mu_2, \ldots, \mu_n$ where the μ are the latent roots of the symmetrized matrix $[\frac{1}{2}(\kappa_{ij} + \kappa_{ji})]$ and the μ are necessarily real (MacDuffee, 1933, Theorem 18·31). Thus if it can be shown that the matrix $[\frac{1}{2}(\kappa_{ij} + \kappa_{ji})]$ has no positive latent root it will follow *a fortiori* that the system is stable.

§ 8·3. Idealized Models.

That is enough about the general theory. Let us now turn to particular applications. Strange to say it is to the advantage of realism that mathematicians customarily replace the actual world by various idealized models. For they choose models that can be analysed with ease; and thus they are free to think about the resemblances or misfits between the model and the actual world. If, with a solemn feeling of the importance of things as they really are, we were to admit the irregularities of the actual world into the statement of our problems, we should in consequence have to attend to enormous elaborations of mathematics in the process of solution, whereby our attention would for a long time be distracted away from the actual world.

Special Case VII

n nations such that each of the defence coefficients is equal to κ and each of the fatigue and expense coefficients is equal to α. The equations of motion are

$$dX_1/dt = -\alpha X_1 + \kappa X_2 + \ldots + \kappa X_n,$$
$$dX_2/dt = \kappa X_1 - \alpha X_2 + \ldots + \kappa X_n,$$
$$\ldots\ldots\ldots\ldots\ldots\ldots\ldots\ldots\ldots\ldots\ldots\ldots$$
$$dX_n/dt = \kappa X_1 + \kappa X_2 + \ldots - \alpha X_n.$$

By addition

$$d(X_1 + X_2 + \ldots + X_n)/dt = \{(n-1)\kappa - \alpha\}\{X_1 + X_2 + \ldots + X_n\}.$$

Therefore

$$X_1 + X_2 + \ldots + X_n = Ae^{\{(n-1)\kappa-\alpha\}t},$$

where A is an arbitrary constant. Thus $(X_1 + X_2 + \ldots + X_n)$ is a normal co-ordinate which will tend to that infinity which has the same sign as A if $(n-1)\kappa > \alpha$; but will tend to zero if $(n-1)\kappa < \alpha$.

Instead by subtraction of any two of the equations of motion

$$d(X_i - X_j)/dt = -(\alpha + \kappa)(X_i - X_j),$$

whence

$$X_i - X_j = B_{ij}e^{-(\alpha+\kappa)t},$$

where B_{ij} is an arbitrary constant. Thus the several nations all tend to have equal X_i after a long time. Their x_i may however remain unequal for $X_i = x_i - x_{0i}$ in which x_{0i} is the co-ordinate of the point of balance.

The sum $(X_1 + X_2 + \ldots + X_n)$ and the $n-1$ differences $X_i - X_j$ suffice to express every individual X in terms of the two exponentials and the arbitrary constants. It can be shown that the determinant is

$$D = \{-\lambda - \alpha + (n-1)\kappa\}\{\lambda + \alpha + \kappa\}^{n-1}.$$

The presence of $n-1$ equal roots $\lambda_2 = -\alpha - \kappa$ might lead one to expect in X terms of the type

$$(C_0 + C_1 t + C_2 t^2 + \ldots + C_{n-2}t^{n-2})\,e^{\lambda_2 t}$$

(Goursat, 1925, vol. II, p. 506) and it is remarkable that in this case the constants $C_1, C_2, \ldots, C_{n-2}$ all vanish. (Compare Thomson and Tait, *Nat. Phil.* 1879, Art 343*m*.)

§9. THE EFFECT OF IMPROVED COMMUNICATIONS BETWEEN n EQUAL NATIONS

Special Case VIII

Let us consider a world of n equal nations. Among them there is no pacifist nation which, being hated, hates not. All the sentiments of love and hate are mutual, so that $\kappa_{ij} = \kappa_{ji}$. The nations are regarded as arranged in a ring, say around the equator. The relation between any nation and its neighbours is the same all around the equator, and the same to the east as to the west. We propose to compare various worlds, all of this general type, but differing in the range of communications, so that in one world, resembling remote antiquity, only immediate neighbours interact; in another world, resembling the present, all nations interact, but with an intensity that diminishes as their separation increases; in a third world, resembling a future towards which improvements in aviation seem to be leading, all nations interact equally, regardless of distance.

It economizes mathematics, however, to solve the general problem first, and to put in the interesting specializations later. So let the typical row of the determinant be

$$\ldots, \kappa_3, \kappa_2, \kappa_1, c, \kappa_1, \kappa_2, \kappa_3, \ldots,$$

where the κ are as yet unspecialized and where for brevity

$$c = -\alpha - \lambda. \tag{1}$$

All the c lie on the principal diagonal. The determinant is best imagined as wrapped round a cylinder so that the last column comes immediately to the left of the first column. The row j may be obtained from row 1 by rotating it $(j-1)$ places about the axis of the cylinder. Such determinants are called 'cyclic'. If n is odd, the highest suffix to κ is $\frac{1}{2}(n-1)$ and $\kappa_{\frac{1}{2}(n-1)}$ occurs twice. If n is even, the highest suffix is $\frac{1}{2}n$ and $\kappa_{\frac{1}{2}n}$ occurs once only. For definiteness consider the case of even n.

Let $$\omega^n = 1, \tag{2}$$

so that $$\omega = \cos\theta + \sqrt{(-1)}.\sin\theta,$$

where $$\theta = 2s\pi/n \quad \text{and} \quad s = 0, 1, 2, \ldots, (n-1). \tag{3}$$

After mutiplying the rows respectively by $1, \omega, \omega^2, \ldots, \omega^{n-1}$ (Scott and Mathews, 1904, p. 102) and adding the products we obtain in the elements of the summed row a common factor

$$c+\kappa_1(\omega+\omega^{-1})+\kappa_2(\omega^2+\omega^{-2})+\ldots+\kappa_{\frac{1}{2}n-1}(\omega^{\frac{1}{2}n-1}+\omega^{1-\frac{1}{2}n})+\kappa_{\frac{1}{2}n}\omega^{\frac{1}{2}n}$$
$$= -\lambda-\alpha+2\{\kappa_1\cos\theta+\kappa_2\cos 2\theta+\ldots+\kappa_{\frac{1}{2}n-1}\cos(\tfrac{1}{2}n-1)\theta\}+\kappa_{\frac{1}{2}n}\cos s\pi. \quad (4)$$

The determinant vanishes for any value of λ which makes this factor vanish, and there are n such values of λ corresponding to the n values of θ. These roots λ are all real, as we knew already from the fact that $\kappa_{ij}=\kappa_{ji}$. We are chiefly interested in the greatest of the n roots, λ_1 say; for the system is stable or unstable according as λ_1 is negative or positive. Since all the κ are positive, λ will have its greatest value when each of the cosines is unity; and this occurs when $\theta=0$, $s=0$. Therefore

$$\lambda_1 = -\alpha+2(\kappa_1+\kappa_2+\ldots+\kappa_{\frac{1}{2}n-1})+\kappa_{\frac{1}{2}n}$$

when n is even. Similarly when n is odd it may be proved that

$$\lambda_1 = -\alpha+2(\kappa_1+\kappa_2+\ldots+\kappa_{\frac{1}{2}n-1}).$$

Let $\sum_j q_{ij}$ denote the sum of all the defence coefficients belonging to any one nation; that is the sum of all the elements in a row, except c. Then whether n be odd or even

$$\lambda_1 = -\alpha+\sum_j q_{ij}. \quad (5)$$

This is the general solution. Now for subcases:

Subcase A. 'Remote Antiquity.' Each of the n nations interacts only with its two immediate neighbours. The typical row is

$$\ldots, 0, 0, 0, \kappa_1, c, \kappa_1, 0, 0, 0, \ldots$$

all the other elements being zero. So

$$\lambda_1 = -\alpha+2\kappa_1.$$

Subcase B. 'The present.' All nations interact but with an intensity that decreases in geometric progression, with the increase of their separation measured around the equator. The typical row is

$$\ldots, r^3\kappa_1, r^2\kappa_1, r\kappa_1, \kappa_1, c, \kappa_1, r\kappa_1, r^2\kappa_1, r^3\kappa_1, \ldots,$$

where $0<r<1$. It is easily proved that

$$\lambda_1 = -\alpha+2\kappa_1\frac{1-r^{\frac{1}{2}(n-1)}}{1-r} \quad \text{for odd } n,$$

$$\lambda_1 = -\alpha+2\kappa_1\frac{1-\frac{1}{2}(r^{\frac{1}{2}n-1}+r^{\frac{1}{2}n})}{1-r} \quad \text{for even } n.$$

Subcase C. 'The future.' Communications are perfect, so that the typical row is

$$\ldots, \kappa_1, \kappa_1, \kappa_1, c, \kappa_1, \kappa_1, \kappa_1, \ldots$$

all the other elements being κ_1, whence

$$\lambda_1 = -\alpha + (n-1)\kappa_1.$$

This subcase is the same as Special Case VII.

Comparison of the subcases. We must consider whether α, κ_1, $\sum_j q_{ij}$, which have been regarded as constants, can really be the same before and after an extension of the range of communications.

To simplify, let this problem be restricted to the growth of aviation in the present century. There is no obvious reason why α, an internal property of the nation, should be much affected by long-range aviation. So tentatively, let us regard α as unchanging.

In all the subcases κ_1 denotes the defence coefficient between *closest* neighbours. *If κ_1 remains the same while aviation increases, the formulae show that λ_1 would increase; so that increase of aviation would tend towards more instability or less stability or a change from stability to instability.*

But is it not more likely that aerial contact with distant nations somewhat distracts attention from closest neighbours, and thereby decreases κ_1? An extreme form of this distraction hypothesis would be to suppose that $\sum q_{ij}$ remains the same while aviation increases. If so, growth of aviation would have no effect on international stability.

The truth may lie between these extremes.

§ 10. RIVALRY WHEN COMMUNICATIONS ARE PERFECT AND NATIONS EQUAL

Special Case IX

In accordance with § 2·12 suppose that the equations of motion are

$$\frac{dX_i}{dt} = -\alpha' X_i + \kappa' \sum_{j=1}^{n} (X_j - X_i), \quad \begin{cases} i = 1, 2, 3, \ldots, n \\ j \neq i \end{cases}$$

where α', κ' are positive constants. The criterion of stability is found from Special Case VII by the substitutions

$$\kappa = \kappa', \quad \alpha = \alpha' + (n-1)\kappa'$$

and therefore is

$$\kappa' \gtreqless \frac{\alpha' + (n-1)\kappa'}{n-1}.$$

Only the lowest sign is true; and therefore, *when the interaction depends only on rivalry as formulated, the system is necessarily stable.*

§ 11. THE EFFECT OF VARIOUS SIZES OF NATIONS IN A WORLD OF PERFECT COMMUNICATIONS

It has already been remarked on pp. 9, 18 that the defence coefficient is proportional to the 'size' of a nation in the sense of its industry and population. Owing to mathematical difficulties it is not possible to discuss various sized nations unless the communications are perfect. The relation of size may be introduced by the following problem.

Special Case X

§ 11·1. The formation of an alliance.

Suppose that the last two of the n equal nations mentioned in Special Case VII form an alliance with one another so that, on account of their mutual confidence $\kappa_{n,n-1}$ and $\kappa_{n-1,n}$ become zero for positive x_{n-1}, x_n to which the problem is restricted. For example in March 1938, after the formation of the Berlin-Rome axis, Italy was not perturbed by the presence of German troops at the Brenner Pass.

When the alliance is in operation the equations of motion are

$$dx_1/dt - g_1 = -\alpha x_1 + \kappa x_2 + \kappa x_3 + \ldots + \kappa x_{n-1} + \kappa x_n,$$

....

$$dx_{n-1}/dt - g_{n-1} = \kappa x_1 + \kappa x_2 + \kappa x_3 + \ldots - \alpha x_{n-1} + 0,$$

$$dx_n/dt - g_n = \kappa x_1 + \kappa x_2 + \kappa x_3 + \ldots + 0 - \alpha x_n.$$

Let a single variable for the allied nations be defined by

$$x_{n-1} + x_n = x'_{n-1}.$$

Then by adding the last two equations and re-arranging the others we obtain a set of $(n-1)$ equations between $(n-1)$ dependent variables, thus

$$dx_1/dt - g_1 = -\alpha x_1 + \kappa x_2 + \ldots + \kappa x_{n-2} + \kappa x'_{n-1},$$

....

$$dx_{n-2}/dt - g_{n-2} = \kappa x_1 + \kappa x_2 + \ldots - \alpha x_{n-2} + \kappa x'_{n-1},$$

$$dx'_{n-1}/dt - (g_{n-1} + g_n) = 2\kappa x_1 + 2\kappa x_2 + \ldots + 2\kappa x_{n-2} - \alpha x'_{n-1}.$$

The presence of 2κ in the last row conforms with the previous statement that the defence coefficient is proportional to the size; for as two equal nations have allied, it seems natural to regard the size as doubled.

Let us next enquire whether the formation of the alliance has made the international system more stable or less stable. To answer this question we need to factorize the $(n-1)$ rowed determinant

$$D = \begin{vmatrix} c & \kappa & \kappa & \dots & \kappa & \kappa \\ \kappa & c & \kappa & \dots & \kappa & \kappa \\ \dots & \dots & \dots & \dots & \dots & \dots \\ \kappa & \kappa & \kappa & \dots & c & \kappa \\ 2\kappa & 2\kappa & 2\kappa & \dots & 2\kappa & c \end{vmatrix},$$

where
$$c = -\alpha - \lambda.$$

This can be done by subtraction of the second column from the first, then of 3rd from 2nd, then of 4th from 3rd and so on to the last. Afterwards expansion in elements of the first column yields a recurrence formula for a determinant D_n of this type having n rows, thus

$$\frac{D_n}{(c-\kappa)^{n-2}} = \frac{D_{n-1}}{(c-\kappa)^{n-3}} - \kappa(2\kappa - c),$$

whence it follows that

$$D_{n-1} = (c-\kappa)^{n-3}\{c^2 + (n-3)\,\kappa c - (2n-4)\,\kappa^2\}.$$

The $(n-1)$ roots of $D_{n-1} = 0$ accordingly consist of $(n-3)$ roots each

$$\lambda = -(\alpha + \kappa) < 0$$

and a pair of real roots

$$\lambda = -\alpha + \tfrac{1}{2}\kappa\{(n-3) \pm \sqrt{(n^2 + 2n - 7)}\}.$$

Stability during the alliance depends on the greatest root λ_1' say, which is found by taking the upper sign before the square root.

Prior to the alliance there were n equal nations and the greatest root, λ_1 say, has been given in Special Case VIII, Subcase C, thus

$$\lambda_1 = -\alpha + \kappa(n-1).$$

It follows that

$$\frac{2(\lambda_1' - \lambda_1)}{\kappa} = \sqrt{(n^2 + 2n - 7)} - (n+1)$$

$$= \sqrt{(n^2 + 2n - 7)} - \sqrt{(n^2 + 2n + 1)} < 0.$$

Therefore $\lambda_1' < \lambda_1$. That is to say *the formation of this alliance has an effect in the sense of stabilization of the international system, provided κ remains unchanged.*

Special Case XI

§ 11·2. The n nations have various sizes; the communications are perfect; and all the fatigue and expense coefficients are equal.

As before let

$$c = -\lambda - \alpha, \tag{1}$$

then

$$D = \begin{vmatrix} c & \kappa_1 & \kappa_1 & \cdots & \kappa_1 \\ \kappa_2 & c & \kappa_2 & \cdots & \kappa_2 \\ \kappa_3 & \kappa_3 & c & \cdots & \kappa_3 \\ \cdots & \cdots & \cdots & \cdots & \cdots \\ \kappa_n & \kappa_n & \kappa_n & \cdots & c \end{vmatrix}. \tag{2}$$

By subtracting the first column from each of the others and then expanding D as a linear function of the elements of the first column (Burnside and Panton, 1904, 2, p. 57; also Scott and Mathews, 1904, p. 48) we can show that

$$D = [(c-\kappa_1)(c-\kappa_2)\dots(c-\kappa_n)]\left\{1+\frac{\kappa_1}{c-\kappa_1}+\frac{\kappa_2}{c-\kappa_2}+\dots+\frac{\kappa_n}{c-\kappa_n}\right\}. \tag{3}$$

When cleared of fractions, D is a polynomial in c of degree n. There is a remarkable partitioning of the roots of $D(c)=0$ between the square and curly brackets, depending on the presence of equal roots.

Subcase A. All the κ are distinct. Without further loss of generality they may be arranged thus

$$\kappa_1 > \kappa_2 > \kappa_3 > \dots > \kappa_n. \tag{4}$$

We consider changes of sign of $D(c)$. If κ_s be any one of $\kappa_1, \kappa_2, \dots, \kappa_n$ then

$$D(\kappa_s) = (\kappa_s-\kappa_1)\dots(\kappa_s-\kappa_{s-1})\,\kappa_s(\kappa_s-\kappa_{s+1})\dots(\kappa_s-\kappa_n). \tag{5}$$

But

$$D(0) = (-1)^n\kappa_1\kappa_2\dots\kappa_n(1-n), \tag{6}$$

and

$$D(\infty) \text{ is positive,} \tag{7}$$

and

$$D(-\infty) \text{ has the sign of } (-1)^n. \tag{8}$$

Therefore as $D(c)$ is continuous, and $n>1$, the equation $D(c)=0$ has one negative root c_1, and $(n-1)$ positive roots, one lying between each pair of consecutive $\kappa_1, \kappa_2, \dots, \kappa_n$. (9)

The factor in the square bracket in (3) does not vanish at any of the roots of $D(c)=0$ and therefore all n roots are given by the vanishing of the factor in the curly bracket. This factor, when cleared of fractions, is of degree n in c. Insertion of (1) gives

$$1 = \frac{\kappa_1}{\lambda+\alpha+\kappa_1}+\frac{\kappa_2}{\lambda+\alpha+\kappa_2}+\dots+\frac{\kappa_n}{\lambda+\alpha+\kappa_n} = \phi(\lambda) \text{ say.} \tag{10}$$

The system is unstable if there is a positive λ that satisfies $\phi(\lambda)=1$, stable if only negative λ satisfy $\phi(\lambda)=1$. Now for $\lambda > -\alpha$, $\phi(\lambda)$ is a positive monotone decreasing function of λ; and $\phi(\infty)=0$. Therefore the criterion of stability of n nations all unequal depends on $\phi(0)$, thus

$$\frac{\kappa_1}{\alpha+\kappa_1}+\frac{\kappa_2}{\alpha+\kappa_2}+\dots+\frac{\kappa_n}{\alpha+\kappa_n} \underset{\text{stable}}{\overset{\text{unstable}}{\gtreqless}} 1. \tag{11}$$

Subcase B. Some of the κ are equal. Let there be $r+1$ equal κ, namely

$$\kappa_s = \kappa_{s+1} = \kappa_{s+2} = \dots = \kappa_{s+r}. \tag{12}$$

By (9) there is a root $c=\kappa_s$ repeated r times. Also (3) takes the form

$$D = [(c-\kappa_1)(c-\kappa_2)\dots(c-\kappa_{s-1})(c-\kappa_s)^r(c-\kappa_{s+r+1})\dots(c-\kappa_n)]$$

$$\times\left\{1+\frac{\kappa_1}{c-\kappa_1}+\dots+\frac{\kappa_{s-1}}{c-\kappa_{s-1}}+r\frac{\kappa_s}{c-\kappa_s}+\frac{\kappa_{s+r+1}}{c-\kappa_{s+r+1}}+\dots+\frac{\kappa_n}{c-\kappa_n}\right\}. \tag{13}$$

The square bracket vanishes for the repeated root; and the curly bracket, when just cleared of fractions, is of degree $(n-r)$, just sufficient to determine the unequal roots.

In view of (1), the only root that can give rise to instability is $\lambda>0$, $c<0$. This cannot be a repeated root; for all the other roots c are positive.

So the greatest of the roots λ is the greatest of the $(n-r)$ roots of (10) when the restriction (12) has been inserted. If there is more than one set of equal κ similar arguments will apply. *Whether there be equal κ or not, the criterion of stability is the inequality* (11).

Special Case XII

§ 11·3. The effect of inequality on stability.

Would a world composed of equal nations be more stable than a world composed of the same number of unequal nations? This question has no answer unless we are given the relations between κ and α in one world and those in the other. So far κ has been regarded as proportional to the size of a nation; and this assumption worked out consistently in the problem of alliance. In our present state of ignorance, the most reasonable tentative assumption is that

$$\kappa_1+\kappa_2+\kappa_3+\dots+\kappa_n = \text{a world constant}, \tag{14}$$

during changes of size of nations by secession of parts, which immediately ally with other nations. Let it also be assumed that the α are unchanged. (15)

It will suffice to consider

$$\lambda + \alpha > 0, \quad \lambda = \lambda_1 \tag{16}$$

as that is the only case in which instability can occur. In the problem so defined, λ depends on $\kappa_1, \kappa_2, \ldots, \kappa_n$ but not on the process whereby the κ have arrived at those values. We are therefore at liberty to choose a process that is mathematically convenient, disregarding the question whether it is politically possible. It is convenient to equalize pairs. Initially let $\kappa_r \neq \kappa_s$ and finally let both become κ', which, to satisfy (14), must be such that

$$\kappa_r + \kappa_s = 2\kappa'. \tag{17}$$

Initially let λ satisfy (10) that is $\phi(\lambda) = 1$. When λ is fixed

$$\frac{d^2}{d\kappa^2}\left(\frac{\kappa}{\lambda + \alpha + \kappa}\right) = -\frac{2(\lambda + \alpha)}{(\lambda + \alpha + \kappa)^3} < 0 \tag{18}$$

and therefore $\kappa/(\lambda + \alpha + \kappa)$ is a 'concave' function of κ (Hardy, Littlewood and Pólya, 1934, pp. 70–7). Accordingly

$$\frac{\kappa_r}{\lambda + \alpha + \kappa_r} + \frac{\kappa_s}{\lambda + \alpha + \kappa_s} < 2\frac{\kappa'}{\lambda + \alpha + \kappa'}. \tag{19}$$

So the equalization of these two nations would make $\phi(\lambda) > 1$ if λ were unchanged. To redress the balance, λ must be increased. By repeated equalization of pairs, all the nations may be brought to the same size, and at each step in the process λ is increased. Infinitely many steps may be necessary.

The final approach to perfect equality is best investigated by another method. Let us seek a relation between $\kappa_1, \kappa_2, \ldots, \kappa_n$ which will keep λ fixed for arbitrary infinitesimal variations of the κ subject to the restriction (14) which gives

$$0 = \sum_{i=1}^{n} d\kappa_i. \tag{20}$$

The differentiation of (10) gives

$$0 = \sum_{i=1}^{n} \frac{(\lambda + \alpha)\, d\kappa_i}{(\lambda + \alpha + \kappa_i)^2}. \tag{21}$$

Let (20) be multiplied by an undetermined Lagrangian multiplier M and subtracted from (21). All but one of $d\kappa_1, d\kappa_2, \ldots, d\kappa_n$ are arbitrary. Let the coefficient of that one be made to vanish by suitable choice of M. It follows that

$$\frac{\lambda + \alpha}{(\lambda + \alpha + \kappa_i)^2} - M = 0 \quad \text{for every } i = 1, 2, 3, \ldots, n$$

and therefore $\kappa_1 = \kappa_2 = \kappa_3 \ldots = \kappa_n$ is the condition that λ should be stationary.

Conclusion. On the assumptions that:

(i) there are in the system any fixed number of nations,

(ii) all the defence coefficients of any one nation are the same,

(iii) the fatigue and expense coefficients are fixed and all equal,

(iv) the world-total of defence coefficients is fixed,

it follows that the greatest root λ_1 is both maximum and stationary when the nations are equal; that is to say equality makes for greatest instability or least stability.

As a check on this conclusion, let us return to a world composed of only two nations. The criterion as proved in § 2·6 is $\kappa_{12}\kappa_{21} \gtreqless \alpha_1\alpha_2$ where the top sign corresponds to instability. Now let $\kappa_{12}+\kappa_{21} = 2b$, a constant, so that we may write $\kappa_{12} = b+\epsilon$, $\kappa_{21} = b-\epsilon > 0$. Then $\kappa_{12}\kappa_{21} = b^2-\epsilon^2$ implying that when the inequality ϵ increases, the system moves towards stability.

§ 11·4. The greatest stability.

Suppose that, in search of as much stability as possible, we imagine the interchange of territory and population between nations to be so regulated as to make λ_1 decrease as much as possible. Consider any pair of non-zero κ_r and κ_s restricted by $\kappa_r+\kappa_s =$ a fixed total, and suppose for definiteness that $\kappa_r > \kappa_s$. Because $\kappa/(\lambda+\alpha+\kappa)$ is a concave function of κ we can prove that, if $\epsilon > 0$,

$$\frac{\kappa_r}{\lambda+\alpha+\kappa_r}+\frac{\kappa_s}{\lambda+\alpha+\kappa_s} > \frac{\kappa_r+\epsilon}{\lambda+\alpha+\kappa_r+\epsilon}+\frac{\kappa_s-\epsilon}{\lambda+\alpha+\kappa_s-\epsilon}.$$

This can be seen from a graph:

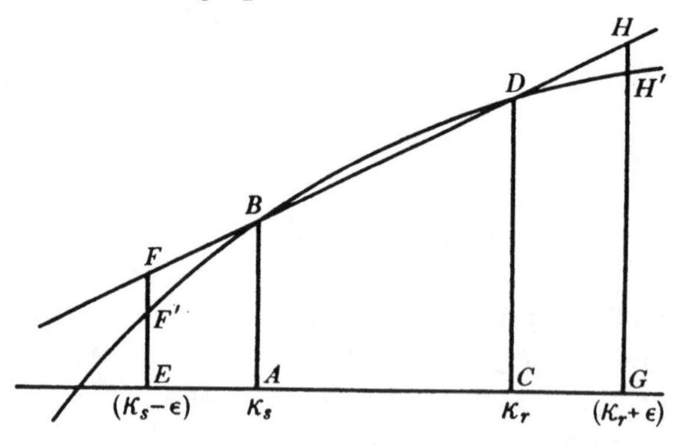

for $$AB+CD = EF+GH > EF'+GH'.$$

So when the κ are made more unequal by transferring ϵ from one to the other, $\phi(\lambda)$ as defined by (10) is decreased; and afterwards to redress the balance, λ must be decreased.

Let this type of transfer be repeated between pairs selected in any way. After each step, λ will be decreased. The decrease in λ will not come to an end while it is still possible to make any two nations more unequal in size. So the end of the process will be reached when, and not until, all but one of the nations have been reduced to zero size, and there is a single world-nation having

$$\kappa = \kappa_1 + \kappa_2 + \ldots + \kappa_n.$$

If we wish still to describe the final state as a problem in n nations we can say that there remain $(n-1)$ token nations of zero size. The corresponding determinant is

$$D = \begin{vmatrix} c & \kappa & \kappa & \ldots & \kappa \\ 0 & c & 0 & \ldots & 0 \\ 0 & 0 & c & \ldots & 0 \\ \ldots & \ldots & \ldots & \ldots & \ldots \\ 0 & 0 & 0 & \ldots & c \end{vmatrix} = c^n = (-\alpha - \lambda)^n.$$

So the n roots are each $\lambda = -\alpha$ and that world is necessarily stable. But we must not forget a hypothesis, that runs through the whole of this generalized foreign politics, namely that a nation is any set of people who are held together by a common loyalty, so that they act together in foreign politics. If the above arguments are to be valid, the interchanges of territory and population must at each step be made with the consent of those who are transferred.

§ 12. TENTATIVE HYPOTHESIS CONCERNING CHANGES DURING THE TWENTIETH CENTURY

Many people wish for stability; but many also dislike the idea of the single world-nation, which was indicated in § 11 as giving the greatest stability. So it may come to pass that the world will for most of the time be content with just enough stability. There are infinitely many sets of sizes of nations which make a world of the type considered in § 11 just stable. For, given any just-stable world, another could be formed from it by making one pair of nations more equal and another pair less equal. This proposition fails in an exceptional case when a world of equal nations happens to be just stable.

During an era when aviation is increasing we have seen in § 9 that a just stable world will in time probably become unstable. When instability occurs it will lead either to a war or to a peaceful rearrangement; after which the world will again be just stable. Thus, while not forgetting the assumptions that have been made on the way, we are led to the tentative hypothesis that: *on the average over the peaceful part of a long period such as a century, during which aviation is increasing, the world may be expected to be just stable*; that is to say the time-average of the greatest root λ_1 will be zero.

This hypothesis opens up the possibility of the determination of the defence coefficients q_{ij} in terms of the fatigue and expense coefficients α_i by insertion of the actual sizes of nations in a formula for $\lambda_1 = 0$. The obstacle in the way of this procedure is that formula (11) on p. 70 relates to an idealized world in which communications are perfect; and that the mathematical problem of imperfect communications between n unequal nations has not yet been solved. The author wishes to draw the attention of mathematicians to this problem.

§13. GREAT POWERS DURING THE PERIOD 1929–37

Leaving the idealized worlds that have algebraic elements in their matrix $[\kappa_{ij}]$ let us now try to represent the actual world by a matrix having numerical elements.

§ 13·1. Statement of the problem.

We first fill in the fatigue and expense coefficients, making each equal to 0·2 year^{-1}, that being the value suggested in § 2·4. A check on this estimate will be obtained at a later stage by comparison of the theoretical and observed rates of drift towards larger armaments.

We next attend to the sizes of the nations, as expressed by population, steel production and electric energy generated, each being reckoned as a percentage of the total for the world.

	Population at 1934 Dec. 31 %	Steel production in 1935 %	Electric power in 1934 %	Mean %
Czechoslovakia	0·7	1	1	1
China	21·6	0?	0?	7
France	2·0	6	4	4
Germany	3·2	17	9	9
G.B. and N.I.	2·3	10	6	6
Italy	2·1	2	3	2
Japan	3·3	4	4	4
U.S.A.	6·1	35	32	24
U.S.S.R.	8·2	13	6	9

In § 11, where communications were assumed to be perfect, it did not matter whether we took the size of the nation i to be proportional to q_{ij} or to $\sum_j q_{ij}$. But, in the actual world, communications are imperfect and we must reconsider the relation of size to q_{ij}. The effect on the i-nation of an increase of armaments of the j-nation must surely be more, if the i-nation is defensive against the j-nation only, than if the i-nation is defensive against several others. Accordingly it seems proper to take the size s_i of the i-nation as setting an upper bound to $\sum_j q_{ij}$, the sum along its row. Say $\sum_j q_{ij} \leqslant ws_i$, where w is a world constant.

For example, the U.S.A. is the nation of greatest size; but, on account of its policy of non-interference, we had better take its $\sum_j q_{ij}$ to be much less than the size of U.S.A. would permit. Similar remarks apply to China.

We have next to estimate individual q_{ij}. Where the i and j nations are geographically inaccessible to each other we may put $q_{ij}=0$. Again if they are in alliance $q_{ij}=0$. But if suspicion and enmity obviously exists between them, q_{ij} will be large. We bring from § 2·10 the presumption that, for two nations as much at enmity as France and Germany were in 1909 to 1913, we should put $q_{ij}=0{\cdot}45$. Various milder degrees of suspicion or neutrality will correspond to $0 < q_{ij} < 0{\cdot}45$. The individual q_{ij}, when assigned, have to add up along their rows to totals having the aforesaid relations to the national size. These considerations control the elements of the matrix $[q_{ij}]$ but do not fix them definitely. There is, however, a great advantage in having a definite statement to discuss, even if it is partly wrong; for the wrongness may be revealed by the discussion. To fix the individual q_{ij} is an act partaking not only of science, but also of art, or perhaps, alas, of caricature. The author makes the assumptions shown in the following table, with apologies to all concerned. To avoid decimals the elements κ_{ij} have each been multiplied by ten; a procedure equivalent to taking ten years as the unit of time.

Assumed matrix κ_{ij} *for the year* 1935

Unit of time = 10 years

Suffix	Czechoslovakia	China	France	Germany	G.B. and N.I.	Italy	Japan	U.S.A.	U.S.S.R.	$\sum_j q_{ij}$
1. Czechoslovakia	−2	0	0	1	0	0	0	0	0	1
2. China	0	−2	0	0	0	0	2	0	1	3
3. France	0	0	−2	2	0	2	0	0	0	4
4. Germany	2	0	2	−2	1	0	0	0	4	9
5. G.B. and N.I.	0	0	0	2	−2	3	1	0	0	6
6. Italy	0	0	1	0	2	−2	0	0	1	4
7. Japan	0	2	0	0	0	0	−2	2	2	6
8. U.S.A.	0	0	0	1	1	1	2	−2	1	6
9. U.S.S.R.	0	1	0	4	1	1	2	0	−2	9
$\sum_i q_{ij}$	2	3	3	10	5	7	7	2	9	

§ 13·2. Solution of the problem.

Those who have access to a Mallock machine might prefer to determine all the latent roots of the matrix. But our practical concern is mainly with the ultimate destination, or oscillatory state, to which the instinctive drift would carry the system, and that depends, as shown in § 8·1, on the greatest real part to be found among the latent roots.

There is an easy method of treating the matrix, which answers the question of stability and, if the system is unstable, gives the ratios $X_1 : X_2 : X_3 : \ldots : X_n$ after a long time. The first notion of the method is to solve the differential equations by numerical steps starting from any initial values of the X. But that would be too tedious. The process may be abbreviated by taking a little care to choose initial X which are not likely to be far from the axis of instability. A common-sense device is to take the initial X proportional to the $\sum_j q_{ij}$ of the respective nations. Another abbreviation becomes possible because we do not require the course of X_i in time, but only the ultimate state. There is therefore no need to make the time-steps small, nor to centre them. It was found in practice that convergence was good if $\Delta t\, dX_i/dt$ was of the order of $\frac{1}{2}X_i$ for every i. If Δt is chosen too small the process is too slow. If Δt is chosen too large, spurious oscillations may be generated. But there is a wide practical range for Δt and it is unnecessary to enquire closely about its best value. In passing from one approximation to the next, we substitute the approximate $X_1, X_2, \ldots, X_n$ into the differential equation and observe the resulting misfits. The arithmetical operations may be compared with such handicrafts as polishing a concave mirror or focusing a spectrometer by Schuster's method. An abstract theory of the convergence of such processes is not necessary; because the interim result is tested at each stage. Arithmetical approximations of this type were used by Richardson (1910); and recently by Southwell and his school (1937) under the name of 'relaxation processes'. For convergence see Richardson (1910), Temple (1939).

We are only concerned with the ratios $X_1 : X_2 : \ldots : X_n$, so that a constant multiplier does not matter. If $X_i^{(r)}$ is the rth approximation, the next step is

$$X_i^{(r+1)} = cX_i^{(r)} + dX_i^{(r)}/dt,$$

where c is the same for all i and is chosen so that $cX_i^{(r)}$ is of the order of twice the term to which it is added. This process gives the ultimate ratios $X_1^{(\infty)} : X_2^{(\infty)} : \ldots : X_n^{(\infty)}$ if they exist. For these ultimate ratios

$$\lambda_1 = \frac{1}{X_i^{(\infty)}} \frac{dX_i^{(\infty)}}{dt} = \frac{\sum_i dX_i^{(\infty)}/dt}{\sum_i X_i^{(\infty)}} \text{ and so } \lambda_1 \text{ is found.}$$

For comparison with statistics we must find the normal co-ordinate corresponding to λ_1. It is convenient as a preliminary to explain the relations in the geometrical language applicable to three nations. The ultimate ratios $X_1^{(\infty)} : X_2^{(\infty)} : X_3^{(\infty)}$ define an axis of instability. The corresponding normal co-ordinate may be thought of as a plane moving always parallel to the barrier plane. The plane is normal to the axis if $\kappa_{ij} = \kappa_{ji}$. If we imagine the moving plane as dusted with the moving particles, it must be noted that although the particles continue to lie in the plane they may move about in it. Any arbitrary plane

$$P = a_1 X_1 + a_2 X_2 + a_3 X_3$$

may usually be expressed as a linear function of the normal co-ordinates N_1, N_2, N_3 say

$$P = b_1 N_1 + b_2 N_2 + b_3 N_3.$$

As time goes on N_1 predominates over N_2 and N_3. So by taking an arbitrary plane and tracing its future we ultimately arrive at a plane parallel to the most unstable normal co-ordinate if one exists. The above remarks about abbreviation apply here also. To apply these ideas to n nations let the rth approximation to the barrier be $P_r = 0$, where $P_r = \sum_i l_i^{(r)} X_i$. We can easily compute the coefficients of the X_i in dP_r/dt from the matrix because

$$\frac{dP_r}{dt} = \sum_i l_i^{(r)} \frac{dX_i}{dt} = \sum_i l_i^{(r)} \sum_j \kappa_{ij} X_j,$$

and having obtained these coefficients we form the next approximation thus

$$P_{r+1} = cP_r + dP_r/dt,$$

where c is a number chosen as explained above. This process is very like that used in finding the ultimate ratios, but with the rows and columns of the matrix interchanged. We need a suitable initial P_0. It will do to take $l_i^{(0)} = \sum_i q_{ij}$ which is the sum up the columns instead of along the rows. As the matrix is approximately symmetrical another suitable initial P_0 would be found by taking $l_i^{(0)} = X_i^{(\infty)}$ which in three dimensions corresponds to taking the initial plane normal to the axis of instability.

Finally
$$\lambda_1 = \frac{1}{P_\infty} \frac{dP_\infty}{dt}.$$

After a few pages of arithmetic, it was found that the system defined by the assumed matrix is unstable, that the dangerous root is $\lambda_1 = 0{\cdot}43$ year^{-1}, and that the corresponding normal co-ordinate is approximately

$$6X_1 + 7X_2 + 8X_3 + 20X_4 + 11X_5 + 12X_6 + 12X_7 + 4X_8 + 20X_9 = Ae^{0{\cdot}43t},$$

where the coefficients are percentages of their total and the suffixes are those at the left-hand margin of the matrix.* The coefficient in $4X_8$ for U.S.A. is small because U.S.A. is rather detached from the politics of other nations.

The ultimate ratios are approximately

$$X_1 : X_2 : X_3 : X_4 : X_5 : X_6 : X_7 : X_8 : X_9$$
$$= 3 : 9 : 8 : 18 : 12 : 8 : 12 : 13 : 19.$$

§ 13·3. Comparison of the solution with fact.

We are now ready to apply to the statistics for the years 1929–37 a generalization of the method which was successfully applied in § 2·10 to the arms race of 1909–13. That is to say, let us first assume that

$$X_i = U_i - (\text{an unknown constant})_i, \quad (i = 1, 2, \ldots, 9)$$

where U_i is the annual defence budget expressed in a unit that is the same for all nations. Old gold U.S.A. dollars will be used here. Next from the U_i we form a weighted mean with weights as near as may be to those of the dangerous normal co-ordinate. Let us take the weights deduced from the assumed matrix and form with them the mean

$$s = \tfrac{1}{100}\{6U_1 + 7U_2 + 8U_3 + 20U_4 + 11U_5 + 12U_6 + 12U_7 + 4U_8 + 20U_9\}.$$

Next we need to plot $\Delta s/\Delta t$ against the contemporaneous s in order to determine λ. But we encounter a difficulty that did not occur in 1909–13. Throughout that earlier period the rates of monetary exchange between the four nations concerned were steady. But sterling was devalued in 1931, the yen in 1932, the U.S.A. dollar in 1933, the koruna in 1934, and in the largest ratio, the rouble in 1936, lastly the franc in 1936 and 1937. Conversion to a gold standard, while keeping the mean level correct, may introduce peculiar oscillations, which would ruin a method depending on annual increments Δs. For example, compare the steady increase of the Russian defence budgets in roubles, with their up and down swing in gold. Official estimates are marked ϵ. Compare also the steady increase of rouble wages.

	1931	1932	1933	1934	1935	1936	1937
10^6 roubles	1,404	1,412	1,547	$5{,}000_\epsilon$	$8{,}200_\epsilon$	14,816	$17{,}481_\epsilon$
10^6 old gold \$	723	727	796	2,575	4,223	1,739	1,968
Daily wages Rbl.	4·11	4·88	5·18	5·94	7·55	9·23	10·15

A similar though less effect is seen in the U.S.A. defence budgets thus:

	1931–2	1932–3	1933–4	1934–5	1935–6
10^6 \$	699	642	544	712	913
10^6 old gold \$	699	621	347	421	539

* More accurately $6{\cdot}_5X_1 + 6{\cdot}_9X_2 + 8{\cdot}_3X_3 + 20{\cdot}_6X_4 + 10{\cdot}_7X_5 + 11{\cdot}_6X_6 + 11{\cdot}_5X_7 + 3{\cdot}_7X_8 + 20{\cdot}_2X_9$.

A smoothed Russian gold budget is therefore (§ 4·2) defined to be

$$\frac{\text{(budget in roubles)}}{\text{(daily wages in roubles)}} \times \text{(constant)},$$

where the constant ($=2\cdot062$) is chosen to make the total for the seven years 1931–37 equal to that found from the official gold values of the rouble.

The defence budgets are shown in the adjoining table together with their weighted mean s. They have been obtained in the manner described in Part II.

Defence budgets

Expressed in a common unit, namely 1·5046 metric tons of fine gold; and, where necessary, interpolated linearly to the calendar year

Calendar year	Czecho-slovakia	China	France	Germany	G.B. and N.I.	Italy	Japan	U.S.A.	U.S.S.R.	Weighted mean s
1929	54	95	481	168	484	259	234	690	518	326
1930	51	83	468	161	461	277	222	704	616	340
1931	53	69	475	149	404	288	208	702	704†	349
1932	49	68	545	149	327	280	325	660	597†	335
1933	47	71	531	571*	304	264	170	492	616†	393
1934	46	73	454	1147*	295	268	166	374	1736†	720
1935	67	71	518	2122*	339	542	172	483	2240†	1065
1936	90	58	549	2482*	470	466	184	552	3310†	1364
1937	100	$\geqslant 62$	459	2942*	699	198	552	579	3551†	1536
1938	135	—	571	—	894	249	934	597	—	—

* *Germany*: From 1933–34 to 1935–36 these values are derived from *The Banker* of February 1937, for 1937–38 from Balogh; and for 1936–37 from the mean of the two. See p. 46.

† *Russia*: These numbers have been smoothed as described above.

The differences from year to year are irregular, as we might have expected from the many devaluations of currency. But by using 4-year intervals we may make the significant deductions that:

from 1929 to 1933, $\Delta s/\Delta t = 17$ for mean $s = 349$,
from 1933 to 1937, $\Delta s/\Delta t = 286$ for mean $s = 1016$.

The slope of the line joining these two points on an $(s, \Delta s/\Delta t)$ diagram is

$$\frac{286-17}{1016-349} = 0\cdot40 \text{ year}^{-1} = \lambda.$$

This is not far from $\lambda_1 = 0\cdot43$ year^{-1} deduced from the assumed matrix. The two λ could be made to agree either by increasing the fatigue and expense coefficients from 0·2 to 0·23 year^{-1}, or instead by some slight decrease in the defence coefficients.

It makes one gloomy to think that the world is, or so recently was, unstable. But it is cheerful now to understand better the causes of drift. May diagnosis be followed by cure!

80 THE INSTABILITY OF PEACE

§14. LESSONS FROM THE EUROPEAN CRISIS OF SEPTEMBER 1938

(i) *Formation of three groups.* On 26 September it looked as if the nations had arranged themselves in three groups, for this particular dispute, one group including Germany, Italy, Japan, Hungary, Poland and Insurgent-Spain, another group including Czechoslovakia, France and Morocco, Great Britain and the British Dominions, Indian princes, Russia, Government-Spain, Jugoslavia, Turkey and China, the third group being the neutrals and including U.S.A., Switzerland, Scandinavia and many other states. So that on 26 September the simpler theory of three groups in §6 might have sufficed instead of the more elaborate treatment for n separate nations in §8.

(ii) *Simultaneity of threats and co-operation.* Both threats and co-operation between the opposing sides went on simultaneously. Threats included the German large-scale manoeuvres from 13 August onwards, the French calling up of reservists on 6 September, the Czecho-slovakian mobilization on 27 September, British naval mobilization in the night of 27–28 September. Co-operation included negotiations between the Czechs, Sudetens and Lord Runciman during August, Mr Chamberlain's very vigorous conciliation in September, Signor Mussolini's intervention and the resulting four power meeting in Munich, on 29 September. It is very difficult to say whether threats or co-operation predominated; unless one begs the question by saying that because the settlement was peaceful, therefore co-operation must have predominated.

(iii) *The distinction between threats and warnings.* Mr Chamberlain in reporting to the London Parliament his conversation with Herr Hitler at Berchtesgaden on 15 September, mentioned that 'At one point he complained of British threats against him, to which I replied that he must distinguish between a threat and a warning, and that he might have just cause of complaint if I allowed him to think that in no circumstances would this country go to war with Germany, when, in fact, there were conditions in which such a contingency might arise' (Parliamentary Debates, Commons 28 September, p. 14).

Pending an official explanation, may we suppose that a warning is a threat restricted to operate only in specially defined circumstances? If you point a pistol at a man: that is a threat. But if, in pointing the pistol, you say, 'I will shoot you if, and only if, you press button A': that is a warning.

(iv) *The immediate effects of severe threats.* On 26–28 September Great Britain realized, by the issue of gas masks, by the digging of trenches in the parks, by plans for the evacuation of London, that war would not only be the affair of men at a front, but would come home to town-dwellers, by aerial bombardment. The same sort of general awareness of danger seems to have been experienced in Paris, and of course earlier in Czechoslovakia. Herr Hitler also knew that the British navy was mobilized. A consequence of these imminent threats, was a readiness to negotiate on 28 September; witness the general relief, when it was announced in the House of Commons on 28 September that Germany, Italy, France and Britain would negotiate at Munich.

(v) *The after-effects of severe threats.* Those who hope to frighten other nations into peaceable submission should have expected the general alarm of 26–28 September to be followed by a decrease in armaments. There was indeed a sudden demobilization of the British navy on 2 October in consequence of the peaceful settlement. But according to the debate in the London Parliament there was to be a maintenance or speeding up of British rearmament, partly because Britain had a new commitment to defend Czechoslovakia, but also so as to be stronger next time: i.e. an after-effect of fear. The long-term changes are summarized in the table on page 82. They make it evident that the main thesis of the present work, namely that k and l are positive in

$$dx/dt = ky - \alpha x + g, \quad dy/dt = lx - \beta y + h,$$

is upheld by the observed after-effects of recent threatenings. Consider especially Germany, which has got rid of a grievance, and yet is going to extend fortifications; and the United States of America, which was not directly threatened, but has responded by deciding to increase its Atlantic fleet from 14 warships to approximately 60 (*Glasgow Herald*, 19 October).

(vi) *The selection of facts represented by our differential equations.* The positive defence coefficients in this monograph represent the after-effect, not the immediate effects. That is proper: for in the course of the years it is the after-effects that count most. Yet the fact that the immediate and the after-effects are opposite to one another is enough to explain the widespread misunderstanding of the defence problem.

(vii) *Before and after the crisis.* Changes from July 1938 to mid-

82 THE INSTABILITY OF PEACE

November 1938 omitting short-term changes during the crisis of 14–30 September.

State	Change in the Treaty situation	Change in armaments
G.B. and N.I.	New commitment re Czechoslovakia	Increase (Parl. Debates, 3–6 Oct.)
France	No formal change, but an ally (Czechoslovakia) weakened	Plenary financial and economic powers to Government, 5 Oct. 12 Oct. extra £13 million for armaments. (Athlone broadcast)
Italy	No change except prospective guarantee of Czechoslovakia	
Germany	Sudeten grievance almost ended	9 Oct. Hitler announced an extension of fortifications in the west of Germany, and mentioned by name British statesmen who had expressed the wish to threaten Germany (*Glasgow Herald*, 10 Oct.)
U.S.A.	None	5 Nov. annual defence estimates increased by $300 million (*Glasgow Herald*, 7 Nov.)

PART IV. DISCUSSION ON APPLICATIONS

CRITIC. 'So you have got your fine equations finished have you?'

AUTHOR. 'Science is never finished. But I have at least got an approximation which is mathematically tractable and politically interesting.'

CRITIC. 'And what do you conceive to be the use of it?'

AUTHOR. 'It shows how threats or co-operation interact with grievances and cost.'

CRITIC. 'But surely practical statesmen understand that already; although they express the connexions in plain language, whereas you prefer an obscure symbolism.'

AUTHOR. 'Judging by their public speeches, some understand it and some do not. Many of them in effect assume that what I have called the "defence-coefficients" are negative, whereas the mathematics show that the existing arms race could not have developed unless the defence-coefficients were positive.'

CRITIC. 'Well, that is a point certainly.'

AUTHOR. 'Moreover the plain language of a public speech is far too clumsy a tool for the investigation of the stability of n nations.'

CRITIC. 'I still don't like the fatalistic look of your mathematics. The worst disservice that anybody can do to the world is to spread the notion that the drift towards war is fated and uncontrollable.'

AUTHOR. 'With that I agree entirely. But before a situation can be controlled, it must be understood. If you steer a boat on the theory that it ought to go towards the side to which you move the tiller, the boat will seem uncontrollable. "If we threaten", says the militarist, "they will become docile." Actually they become angry and threaten reprisals. He has put the tiller to the wrong side. Or, to express it mathematically, he has mistaken the sign of the defence-coefficient.'

CRITIC. 'But how can experienced statesmen possibly be mistaken about the sign of an important political effect?'

AUTHOR. 'They are not altogether mistaken. They attend to the immediate effects of fear and they ignore the after-effects, which are of the opposite sign, and in the long run more important.'

CRITIC. 'Fear? The British nation is not influenced by the contemptible emotion of fear.'

AUTHOR. 'In view of the revelation by psychoanalysis of repressed tendencies in individuals, I permit myself to doubt whether a boast of national fearlessness could be upheld for any nation. However, in order to avoid controversy on what is not essential, I had better try to express the reactions in behaviouristic language which makes no reference to any emotion, base or noble. Accordingly let me say that some statesmen go wrong because they attend to the immediate effects of severe threats and they ignore the delayed effects, which are of opposite sign, and in the long run more important.'

CRITIC. 'Yes, but you, admittedly, are only discussing the stability of peace; whereas the gallant colonel is thinking of winning a war, by applying such overwhelming violence that other people's defence-coefficient will change from positive to negative.'

AUTHOR. 'Which in fact amounts to this: by preparing to win a war, he would cause a war whereby most people would lose.'

CRITIC. 'A serious objection to your description is that it does not mention intelligent aggression planned by a leader as moves in a game of chess.'

AUTHOR. 'That is admitted; for a description of predictable tendencies cannot include anything so unpredictable as moves in a game of chess. Yet the acts of a leader are in part controlled by the great instinctive and traditional tendencies which are formulated in my description. It is somewhat as if the chessmen were connected by horizontal springs to heavy weights beyond the chessboard.'

CRITIC. 'But a silent dictator has a great advantage in playing chess against half a dozen democracies who discuss future moves in loud voices.'

AUTHOR. 'That silence and those discussions are slow in having effect on the instinctive and traditional tendencies. But I wonder whether the general statistical effect of unified command could somehow be represented by attributing more social viscosity to democracies than to dictatorships?'

CRITIC. 'Another thing that irks me is that you mix up friendly feelings with foreign trade and that you would, I suppose, not pause to distinguish malice aforethought from murder. That surely is a confused sort of thinking.'

AUTHOR. 'I should prefer to call it "lumped thinking" in the sense in which radio engineers talk of the lumped potentials. (Turner, 1931.) The economists also use an index of general price level which is of somewhat the same nature. But you can have the intentions represented separately from the actions, if you wish. For in 1919 I treated them separately. It is better so. But in that way we should have twice as many equations.'

CRITIC. 'No thank you! There are quite enough! I should like to know whether your theory throws any light on the problem of so-called collective security?'

AUTHOR. 'For simplicity § 2·10 dealt with two nearly equal groups each held together by its own loyalty. When one group is enormously more populous than the other, it was to be expected *a priori* that the minority would feel submissive. When y is very much greater than x the factor $\{1-\sigma(y-x)\}$ is negative, although the submissiveness σ is small. The experiment of 50 nations applying sanctions to Italy showed that σ was nearer to zero than had been anticipated; and confirmed my belief that the linear equations were a good approximation. Great Powers have submitted to conquest, but seldom to threats or sanctions.'

CRITIC. 'What answer can there be to force but force? Tell me that!'

AUTHOR. 'If there really were no answer to force but force, then although it is hundreds of years since the time of Robert the Bruce, the relations between Scotland and England might still be describable in partisan language thus: A noble army of Scots fully equipped with the latest means of defence might still be encamped on the north side of the Cheviots facing a dastardly horde of English cruelly armed with their wicked weapons of destruction. But all that hate has evaporated; and not by conquest.'

CRITIC. 'How then do you think that peace ought to be maintained?'

AUTHOR. 'National self-righteousness is not enough. Si vis pacem para justitiam. In the words of R. B. Gregg (1936), "Peace is a by-product of the persistent application of social truth, justice, and strong intelligent love....The price of peace is the price of justice."'

CRITIC. 'Is not international justice rather difficult to define?'

AUTHOR. 'The practical definition, to quote Sir Norman Angell (1938), is by third-party judgment; of course after consideration of facts and precedents.'

CRITIC. 'But, concerning the Spanish civil war, the rest of the world took sides about equally; so that justice could not be ascertained.'

AUTHOR. 'It must, alas, be admitted! Perhaps the idea of justice belongs to a degree of organization which inter-state affairs have not yet attained. Let us speak therefore, more realistically, of the removal of grievances. Let us try to make the grievances g_1, g_2, g_3, ..., g_n if possible all zero or negative; remembering that some of them are not continuous variables, but can be changed, if at all, only in big jumps. It is also important to decrease threats. If contentment comes, security will follow.'

CRITIC. 'But, on your own showing, the abolition of grievances does not make the international regime stable. For instability, you say, occurs for two nations when and only when the product of the two defence coefficients exceeds the product of the two fatigue and expense coefficients. There is no mention of grievances in that statement.'

AUTHOR. 'Quite so! And the same is true for n nations: the condition of stability makes no mention of g_1, g_2, g_3, ..., g_n. The traditional policy of the balance of power is now useless because the balance will not stay put. I wonder whether it is the ever-growing technique of chemists and engineers that has made the balance unstable by cheapening the mass-production of weapons and thereby decreasing the fatigue and expense coefficients $\alpha_1, \alpha_2, ..., \alpha_n$? Something might perhaps be done by propaganda to diminish the defence coefficients κ_{ij} and thereby to give ballast to foreign politics.* For the near future however it seems that we shall have an international system that is unstable.'

CRITIC. 'So is there no hope?'

AUTHOR. 'Oh, but there is! For there are two opposite infinities towards one or other of which an unstable regime tends to drift. Which way the drift is going depends on how the particle that represents the present situation lies in relation to a certain barrier that passes through the point of balance. This barrier is not a fixture; for its position depends upon the grievances. When, by the efforts of diplomatists, the grievances are diminished or made negative the barrier is thereby moved. For example, when the change is from Fig. 4 to Fig. 3 on page 11 most of the dangerous double-shaded region is moved out of the picture. And, if the armaments are not too threatening, the barrier can thus be moved across the particle that represents the international situation. This done, the instinctive drift goes the other way, as in the middle part of Fig. 3, towards a united organization instead of towards war.'

CRITIC. 'If the main thing is to get rid of grievances, why do you also say "decrease threats"?'

* The International Advertising Convention, held in Glasgow on 27 and 28 June 1938.

AUTHOR. 'Because the distance that the barrier must be moved in order to be put across to the favourable side of the particle, depends of course on where the particle is, that is to say on the amount of co-operation or threatening in progress. One must distinguish here between three types of armament as Jonathan Griffin (1936) has emphasized: bombing aeroplanes that threaten foreigners sleeping peaceably in their homes, anti-aircraft guns that threaten only invaders, and air-raid shelters that in fact threaten no one although they may alarm those whom they are designed to protect. In a roundabout way the bombing aeroplanes are a danger to the nation that owns them.'

CRITIC. 'But every reasonable person knows that British bombers will never be used except against an aggressor.'

AUTHOR. 'Arms races are not made by reason and knowledge, but by instinct and suspicion.'

CRITIC. 'The word grievance suggests that nations strive only in gloom and depression to remedy intolerable wrong; whereas in fact they sometimes strive in pride and joy to extend an empire or enhance prestige.'

AUTHOR. 'Yes; a more thorough psychological analysis should be made of the g_i, which in the formulae are simply temporary constants, positive or negative. The work of Durbin and Bowlby (1939) relates to both κ and g.'

CRITIC. 'So when grievances are negative and there are no threats, I suppose that everybody will do just as they please and the roast pigs will run about crying, "eat me".'

AUTHOR. 'It pleases you to jest; but of course we both know that in a united world organization there will always be the need for compromise, discipline, law and police.'

88

REFERENCES

AMERY, L. S. (1936). *Parliamentary Official Report*, Commons, 20 July.

ANGELL, SIR NORMAN (1938). *The Great Illusion—Now*, p. 234. Penguin Books, Ltd.

ARNOLD, A. Z. (1937). *Banks, Credit and Money in Soviet Russia.* New York: Columbia Univ. Press.

BAKER, NOEL (1936). *Parliamentary Official Report*, Commons, 20 July, p. 130.

BALOGH, T. (1938). *Economic Journal*, **48**, 461–97.

Banker, The (Feb. 1937). London: Financial News Ltd.

BRUNT, D. (1934). *Physical and Dynamical Meteorology*, Chs. xii, xiii. Camb. Univ. Press.

BURNSIDE, W. S. and PANTON, A. W. (1904). *Theory of Equations.* London: Longmans.

CALLENDAR, L. H. (1938). Letter in *Nature* of 12 Feb.

CARMICHAEL, D. M. (1939). *Brit. J. Psychol.* **29**, 206–31.

DRUMMOND, SIR ERIC (now Earl of Perth) (1930). *Ten Years of World Co-operation.* Geneva: League of Nations.

DURBIN, E. F. M. and BOWLBY, J. (1939). *Personal Aggressiveness and War.* London: Kegan Paul.

FORSYTH, A. R. (1900). *Theory of Differential Equations*, vol. **3**. Camb. Univ. Press.

—— (1929). *Treatise on Differential Equations.* London: Macmillan.

GARNETT, J. C. MAXWELL (1936). *Organizing Peace*, pp. 110, 111. London: League of Nations Union.

GOOCH, G. P. (1913). *History of our Time.* London: Thornton Butterworth.

GOURSAT, E. (1925). *Cours d'Analyse*, **2**, 505–12. Paris: Gauthier-Villars.

GREGG, R. B. (1936). *The Power of Non-violence*, p. 248. London: G. Routledge and Sons, Ltd.

GREY, VISCOUNT (1925). *Twenty-five Years*, **1**, 92. London: Hodder and Stoughton.

GRIFFIN, JONATHAN (1936). *Alternative to Rearmament.* London: Macmillan.

VON HABERLER (1936). *Prosperity and Depression.* Geneva: League of Nations.

HARDY, G. H., LITTLEWOOD, J. E. and PÓLYA, G. (1934). *Inequalities.* Camb. Univ. Press.

INTERNATIONAL LABOUR OFFICE (1938). *Year-book of Labour Statistics.* Geneva.

JACOBSSON, PER (1929?). 'Armaments Expenditures of the World.' *The Economist.* London.

JOWETT, B. (1881). *Thucydides.* Oxford: Clarendon Press.

KEESING (1933–9). *Contemporary Archives.*

KOSTITZIN, V. A. (1937). *Biologie Mathématique*, pp. 31–5. Paris: Armand Colin.

LEAGUE OF NATIONS (1923). *Statistical Inquiry into National Armaments*, A 20, 1923, IX.

—— (1924–5)...(1938). *Armaments Year Book.*

—— (1925)...(1928). *International Statistical Year-book.*

—— (1929)...(1936). *Statistical Year-book.*

—— *Monthly Bulletin of Statistics.*

—— (1936). *Repertoire des Organizations Internationales.*

REFERENCES

MACAULAY, W. H. (1930). *Solid Geometry*. Camb. Univ. Press.

MACDUFFEE, C. C. (1933). *The Theory of Matrices*. Berlin: Springer.

MARSHALL, A. (1923). *Money, Credit and Commerce*, Appendix J. London: Macmillan.

MOLE, G. (1936). *Proc. Phys. Soc.* **48**, 857–64.

PLATO (1895). *Republic*, Book VII, Sec. 530. Trans. by Davies and Vaughan. London: Macmillan.

PORTER, A. W. (1929). *Encyclopaedia Britannica*, 14th ed. **22**, 853–60.

RICHARDSON, L. F. (1910). *Philos. Trans.* A, **210**, 318, § 3·2.

—— (1919). *Mathematical Psychology of War*. Oxford: Wm Hunt. (Out of print.)

—— (1935*a*). *Nature, Lond.*, **135**, 830.

—— (1935*b*). *Nature, Lond.*, **136**, 1025.

—— (1938). *Nature, Lond.*, **142**, 792.

ROYAL INSTITUTE OF INTERNATIONAL AFFAIRS. Annual *Surveys of International Affairs*. London.

RUHEMAN, M. (1938). Letter in *Nature* of 30 April.

SCOTT, R. S. and MATHEWS, G. B. (1904). *Theory of Determinants*. Camb. Univ. Press.

SETON-WATSON, R. W. (1938). *Britain and the Dictators*. Camb. Univ. Press.

SOUTHWELL, R. V. and others (1937–8). *Proc. Roy. Soc.* A, **159**, 315; **161**, 155; **164**, 447; **168**, 317.

Statistical Abstract for the British Empire, for each of the ten years 1927–1936. London: H.M. Stationery Office.

TEMPLE, G. (1939). *Proc. Roy. Soc.* A, **169**, 476–500.

THOMSON, SIR W. and TAIT, P. G. (1879). *Treatise on Natural Philosophy*. Camb. Univ. Press.

TURNER, L. B. (1931). *Wireless*, p. 197. Camb. Univ. Press.

VOLTERRA, V. (1931). *Théorie Mathématique de la Lutte pour la Vie*. Paris: Gauthier-Villars.

WELLS, H. G. (1933). *The Shape of Things to Come*. (The book, not the film.)

SUBJECT INDEX

CAMBRIDGE: PRINTED BY WALTER LEWIS, M.A., AT THE UNIVERSITY PRESS

Richardson, L. F. (1941)
Nature, Lond. **148**, 598

1941:1

FREQUENCY OF OCCURRENCE OF WARS AND OTHER FATAL QUARRELS

Nature
London
Vol. 148, No. 3759, 15 November 1941, p. 598
(Letter dated 15 October)

Notes:

1. This letter was later described by Richardson as 'a new start independent of' his previous studies of wars. It marks the beginning of the analyses that were gathered together in the book *Statistics of Deadly Quarrels* (1950:4, 1960:1). See the 'Notes on *Statistics of Deadly Quarrels*' (below p. 537)
2. The 'full account ... nearly ready for publication' did not appear until 1948:4

598 NATURE NOVEMBER 15, 1941, VOL. 148

Frequency of Occurrence of Wars and other Fatal Quarrels

IN order to investigate the causes of wars by counting occurrences, let the magnitude of any war be defined to be the logarithm, to the base ten, of the number of persons who died because of that quarrel. This definition has the advantage that it applies, not only to what are ordinarily called wars, but also to all kinds of fatal quarrels, including insurrections, frontier incidents, riots and murders.

The numbers of wars of various magnitudes, which ended from 1820 to 1929 A.D. inclusive, have been counted, after laborious search in works on history. The number of murders is an estimate from the statistics of crime. Between the wars and the murders, in the range between magnitudes 2·5 and 0·5, there were certainly many fatal quarrels; but statistics of them are scanty, presumably because such incidents are mostly too small to be history and too large to be crime. The results are:

Ends of range of magnitude	$7\pm\frac{1}{2}$	$6\pm\frac{1}{2}$	$5\pm\frac{1}{2}$	$4\pm\frac{1}{2}$	· · · ·	$0\pm\frac{1}{2}$
Observed number of fatal quarrels	1	3	16	62	· · · ·	10^7

It is seen that the numbers of wars in successive equal ranges of magnitude are nearly in agreement with the geometrical progression 1, 4, 16, 64; but that when this progression is continued, it gives 16,384 for the number of murders instead of the observed 10^7.

These remarkable relations call for explanation. A full account of this and cognate matters is nearly ready for publication elsewhere.

LEWIS F. RICHARDSON.

38 Main Road,
Castlehead,
Paisley.

Oct. 15.

Richardson, L. F. (1941)
Nature, Lond. **148**, 784

1941:2

MATHEMATICAL THEORY OF POPULATION MOVEMENT

Nature
London
Vol. 148, No. 3765, 17 December 1941, p. 784
(Letter dated 3 December)

Notes:

1. Despite the final statement in this letter, Richardson published nothing further on this topic, and no manuscript is known. Ashford (1985, p. 175) suggested that two manuscript articles on gregariousness, left incomplete by Richardson, may represent part of the 'fuller account'
2. There is a brief reference to this letter in *Statistics of Deadly Quarrels* (p. 157)
3. The frequency distribution of towns according to their population size is discussed in 1948:4 (below pp. 512–3)

784 NATURE DECEMBER 27, 1941, VOL. 148

Mathematical Theory of Population Movement

AMONG the obvious motives of mankind are the tendencies to seek company and to seek living-space. If we were to regard these tendencies as being in simple opposition to one another, we should expect the population to be able to remain uniformly spread over any uniform piece of land; and the familiar contrast between town and country would then appear, to the theoretical mind, as a mystery requiring explanation. We may, however, seek a hint as to why people concentrate into towns from Sir James Jeans's theory of why matter concentrates into stars[1]. For his theory is also concerned with two opposing tendencies: to draw together by mutual gravitation and to spread out by pressure.

Let ρ denote the density of the astronomical matter, supposed initially uniform, let $s = \delta\rho/\rho$ be its concentration at any time t and place, let p be its pressure and γ the constant of gravitation. Then Jeans showed that deviations from uniformity occur in accordance with the equation

$$\frac{d^2s}{dt^2} = 4\,\pi\gamma\rho s + \nabla^2\left(s\frac{dp}{d\rho}\right). \quad . \quad . \quad . \quad (1)$$

The essence of Jeans's theory is that the opposition between gravitation and pressure is not simple: the former is represented by a term in s, the latter by a term in $\nabla^2 s$. These considerations led me to inquire whether the existence of towns could be explained by

$$\frac{ds}{dt} = \gamma\rho s + \left(\frac{\partial^2}{\partial x^2} + \frac{\partial^2}{\partial y^2}\right)\left(s\frac{dp}{d\rho}\right). \quad . \quad . \quad (2)$$

in which ρ, supposed initially uniform, is the number of persons per square kilometre, $s = \delta\rho/\rho$ as before, x and y are horizontal co-ordinates on a flat portion of the earth, γ is a constant expressing gregarious attraction; and p is called pressure of population. The social equation (2) has been made of viscous type by replacement of the astronomical d^2s/dt^2 by ds/dt.

Whereas Jeans began the astronomical theory with γ known and p clearly understood, and thence deduced the spacing of the stars, we have to begin the social theory at the other end, and work backwards to find out more clearly what γ and p mean. Equation (2), in which $dp/d\rho$ is an unknown constant, explains why the population does not remain uniformly spread. For the amplitude of a standing wave of s either grows or diminishes, according as the wave-length is greater or less than a critical length. It can be deduced that $dp/\gamma d\rho$ is of the order of magnitude of the ratio of the number of persons in a country to the number of towns in it. Further, $dp/d\rho$ is seen to play the part of a diffusivity in equation (2). From the observed time of dispersal of concentrations having diameters much less than the distance between towns, it can be estimated that $dp/d\rho$ is of the order of 10^6 cm.^{2} sec^{-1}. Whence it follows that γ is of the order of 10 or 10^2 cm.^{2} sec.$^{-1}$ person^{-1} for normal people.

A fuller account is ready for publication as part of a book.

LEWIS F. RICHARDSON.

38 Main Road,
Castlehead,
Paisley.
Dec. 3.

[1] Jeans, Sir James, "Astronomy and Cosmogony" (Camb. Univ. Press, 1929)

Richardson, L. F. (1944)
Nature, Lond. **154**, 240

1944:1

STABILITY AFTER THE WAR

Nature
London
Vol. 145, No. 3903, 19 August 1944, p. 240
(Letter dated 1 July)

Notes: The topic of submissiveness, following the theoretical presentation in 1939:1, is continued in *Arms and Insecurity* (Chapters IV, XXIV) and then in 1951:3, 4, 5 and 1953:2

Stability after the War

THE programme of the United Nations, as broadcast at the end of 1943, may at its briefest be stated thus: the Germans and Japanese are to be defeated and disarmed, and thereafter watched for so long a time as may be necessary by an armed force controlled by the United Nations, who are firmly resolved to remain united.

Historical facts, taken alone, do not predict the future; they do so only if they are combined with some hypothesis. A general form of hypothesis for the quantitative interaction of two entities is a pair of simultaneous differential equations, having time t as independent variable. Almost the simplest of such pairs is

$$dx/dt = g - \alpha x + ky, \qquad dy/dt = h + lx - \beta y, \qquad (1)$$

where $g, h, \alpha, \beta, k, l$, are constants. The most relevant of the quantitative historical facts are the numbers of persons engaged on war-preparations in the opposing groups. Let these be x and y. It is then found that the equations (1) are capable of describing the European x and y for the years 1908–13 and again for 1933–38, during the greater part of the arms-races[1]. Moreover, the constants have psychologically intelligible names, thus: g, h, grievances and ambitions; k, l, defence coefficients; α, β, fatigue-and-expense coefficients. This analysis emphasizes, what is also obvious to common sense, namely, that if the several nations, now united, were to attempt to regain their former so-called freedom, sovereignty and independence, then after the present War had faded out of mind, disastrous arms-races would be likely to develop between them.

The other essential part of their programme is the submission of the defeated. Equations (1) are too simple to describe defeat or submission. For this purpose it is necessary to introduce at least the quadratic term in the constant ρ, which has been called a 'submissiveness', thus[2]

$$\left.\begin{aligned} dx/dt &= g - \alpha x + ky, \\ dy/dt &= h + lx\{1 - \rho(x - y)\} - \beta y. \end{aligned}\right\} \qquad (2)$$

It is very ominous that the turn in the year 1930 from the long pause into the arms-race can be described by equations (2). I am not saying that no other motives were operative. On the contrary, it seems almost certain that the great trade depression and the fading of war-weariness were involved. But I do say that there is no warrant either in those facts or in that theory for the belief that Europe will be permanently stabilized by submissiveness in the presence of grievances, ambitions, defensiveness, and the dislike of fatigue and expense. Balance of power, according to the theory, may be of various types, some stable, some unstable. The need for some other and more binding motive is clearly indicated.

LEWIS F. RICHARDSON.

Hillside House, Kilmun,
Argyll. July 1.

[1] *Nature*, **135**, 830; **136**, 1025; **142**, 792; and much yet unpublished.
[2] "Generalized Foreign Politics", 23 (Camb. Univ. Press, 1939).

Richardson, L. F. (1944)
Jl R. statist. Soc. **107**, 242–50

1944:2

THE DISTRIBUTION OF WARS IN TIME

Journal of the Royal Statistical Society
London
Vol. 107 Parts III–IV, 1944, pp. 242–250

Notes:

1. Richardson sometimes refers to this paper subsequently according to the year of actual publication, which was 1945
2. Moyal, J. E. (1949) made some further analyses of these data in: The distribution of wars in time, *Jl R. statist. Soc.* **112**, 446–9
3. Richardson's paper reappeared, considerably revised and extended, and incorporating Moyal's findings, in *Statistics of Deadly Quarrels* (Chapter III)

THE DISTRIBUTION OF WARS IN TIME

By LEWIS F. RICHARDSON

A LIST of " Wars of modern civilization " from 1480 to 1940 A.D. has been published by Quincy Wright (1942, Appendix XX). He states that his list:

> " is intended to include all hostilities involving members of the family of nations, whether international, civil, colonial, or imperial, which were recognized as states of war in the legal sense or which involved over 50,000 troops. Some other incidents are included in which hostilities of considerable but lesser magnitude, not recognized at the time as legal states of war, led to important legal results such as the creation or extinction of states, territorial transfers, or changes of government."

As a preliminary to any statistical treatment of Wright's list, a decision had to be made concerning four of the largest wars, namely the Thirty Years War, the French Revolution, the Napoleonic Wars, and the First World War, because Wright lists them both as wholes and as parts. In order to make the wars in the collection less unequal in size, I included all the parts and omitted the wholes, except two wholes which are shown as beginning before any of their named parts. The First World War is thus reckoned as five wars.

Each calendar year can be characterized by the number 0, 1, 2, 3, 4 of wars which began in it, as shown in Appendix I. The number of years of each character has been counted (by three observers, M. W., E. D. R. and L. F. R.) with the following result:

TABLE I

Years from 1500 *to* 1931 A.D.

Number of outbreaks in the year	0	1	2	3	4	>4	Total
Number of such years	223	142	48	15	4	0	432
Poisson law	216·2	149·7	51·8	12·0	2·1	0·3	432·1

By the Poisson law is here meant the statement that there were

$$Ne^{-\lambda}\lambda^x/x! \quad . \quad . \quad . \quad . \quad . \quad . \quad . \quad . \quad . \quad (1)$$

years each containing exactly x outbreaks of war. The law was fitted to the observations by first equating N to the observed total number 432 years; and then determining λ by R. A. Fisher's Principle of Maximum Likelihood, according to which λ equals the mean number of outbreaks per year; that is

$$\lambda = 299/432 \quad . \quad . \quad . \quad . \quad . \quad . \quad . \quad . \quad (2)$$

It is seen at a glance that there is a considerable resemblance between the historical facts and the Poisson law. This resemblance suggests that we may perhaps find out something about the causes of wars by thinking of their outbreaks in connection with other phenomena known to be described at least approximately by the Poisson law. Such include: factory accidents, deaths by kick from a horse, and the emission of alpha particles from radioactive substances. In the standard deduction of the Poisson law from the sequence of terms of $(p + q)^k$ where $p + q = 1$, by making pk constant while $k \to \infty$, there is no mention of time. We are at liberty to suppose that the occasions on each of which the probability of the event is p, are distributed in time as may best suit the phenomena under consideration. For the alpha particles it is suitable to suppose that the occasions are uniformly spread in time t, so that the probability of a particle escaping during the differential dt is λdt. For the wars that is not quite suitable, because there is a seasonal effect: wars in the north temperate zone have ordinarily begun in spring or summer (Wright, 1943, p. 224). Again for the wars we need not insist on there being infinitely many occasions. The limit as the index $k \to \infty$ is preferred in theory because it leads to a simple formula, and not because of any better agreement with observation than, say, $k = 1000$. But maybe $k = 20$ or 100 might be large enough to explain the observations on wars. The hypothesis so far is that *in any year there were the same large number* k *of occasions on which a war might have broken out, and the same small probability* p *of its doing so on each occasion, so that*

$$pk = \lambda \qquad (3)$$

This statistical and impersonal view of the causation of wars is in marked contrast with the popular belief that a war can usually be blamed on one or two named persons. But there are similar contrasts in other social affairs; the statistics of marriage, for example, are in contrast with any biographer's account of the incidents that led two named persons to marry each other.

The main result has now been stated. It is evidently of considerable interest, and may well move us to make a critical examination of details. Let us turn to such inter-related questions as: whether the deviations from the Poisson law are significant; whether Quincy Wright's definition of a war is the most suitable for statistical purposes; and whether λ fluctuates over periods such as a generation.

For the purpose of applying Karl Pearson's χ^2 test to the deviations between theory and observation as shown in Table I, the years containing 3 or more outbreaks have been grouped together, so as to avoid small frequencies, thus making four groups altogether. A slight approximation was admitted by not redetermining λ. Then $\chi^2 = 2{\cdot}4$. There are two constraints, one for N and one for λ. So Fisher's number of degrees of freedom is $4 - 2 = 2$. Accordingly $P(\chi^2) = 0{\cdot}3$; indicating that the deviations can reasonably be dismissed as chance.

Which of the alternative theoretical foundations of the χ^2 test are historians likely to accept? Will their detailed knowledge of fact permit them to imagine a hypothetical infinite collection of wars, from which the actual wars are a random sample? Or will they be able to experience degrees of belief that behave according to reasonable quantitative rules? Personally, I have difficulties both ways, and hope for some new foundation, not yet revealed. In the meantime methods like χ^2 may almost be regarded as justified, apart from any fundamental theory, by their credible application to a great variety of phenomena.

Founding on the doctrine of reasonable belief, Jeffreys (1939, pp. 256 to 260) has provided a method for testing whether a distribution deviates significantly from the Poisson law in the direction of the negative binomial. This test can deal with small numbers; an advantage over χ^2. Jeffreys' formula (p. 258 (20)), when applied to Table I, yields $k = 2{\cdot}77$; and this, according to Jeffreys

(p. 357), indicates that we need not further consider the negative binomial. It was otherwise with factory accidents. (Greenwood and Yule, 1920, quoted by Kendall, 1943, pp. 124–126.)

In order to search for other types of deviation from the Poisson law it is desirable to group whole n ımbers of years together so as to form successive "cells" of equal duration. Let the duration of a cell be τ years. Let there be n consecutive cells.

Then the whole run of years $= n\tau = N$, say . . . . (4)

Let $f(x)$ be the observed number of cells which each contained exactly x outbreaks . . . . (5)

For example Wright's data can be arranged thus:

Table II.

n = 144 *cells of* $\tau = 3$ *years each.*

x	0	1	2	3	4	5	6	7	>7	Total
$f(x)$	18	39	37	29	12	5	2	2	0	144
Poisson	18·1	37·5	38·9	26·9	14·0	5·ᴗ	2·0	0·6	0·2	144·0

The row headed "Poisson" contains the numbers

$$\frac{N}{\tau} e^{-\tau\lambda}(\tau\lambda)^x/x\,! \quad . \quad . \quad . \quad . \quad . \quad . \quad . \quad . \qquad (6)$$

a formula which can be deduced from (3).

The Method of Maximum Likelihood gives $\tau\lambda = 299/n = 299\tau/432$ for every τ, so that λ *automatically* comes to the same value 299/432 whatever cells we take. But the agreement of $f(x)$ with the Poisson law is not automatic, and so is a genuine test. In Table II the agreement again appears good to inspection.

The run of 432 years was chosen because $432 = 3^3 \times 2^4$, and so can be divided into cells of a great variety of durations. For example:

Table III

n = 8 *cells of* $\tau = 54$ *years each.*

Here $f(x)$ vanishes for all except the following x.

x	16	31	32	35	42	44	45	54	Total
$f(x)$	1	1	1	1	1	1	1	1	8

In Table III $f(x)$ is no longer a useful conception. It would be more convenient to regard the sample as specified by the eight numbers in the row headed x.

This type of specification can be made general. For any τ let the sample consist of the n numbers

$$(x_1, x_2, x_3 \ldots x_n) \quad . \quad . \quad . \quad . \quad . \quad . \quad . \quad . \qquad (7)$$

For $\tau = 1$ many of the 432 numbers coincide; for example, 223 of them are zero; but that causes no unclarity.

A concise test for rejecting a Poisson law.

The χ^2 test cannot be applied to compare the observations in Table III with their Poisson expectations, because all the numbers are small.

In a Poisson population the variance is equal to the mean. This equality is intimately related to rarity or improbability. For the Poisson law is the limit of the term-sequence of the expansion of the binomial $N(p + \chi)^k$ where $p + q = 1$ as $p \to 0$ and $pk \to \lambda$. But for the binomial the mean is kp and the variance kpq. So the equality of the mean and variance of a binomial law, for any k, entails that $q = 1$, $p = 0$; that is zero probability.

These considerations suggest that we should test our observed sample $(x_1, x_2, \ldots, x_n)$ by forming the *statistic g defined by*

$$m = \frac{1}{n}\sum_{i=1}^{i=n} x_i, \qquad g = \frac{1}{m(n-1)}\sum_{i=1}^{i=n}(x_i - m)^2. \quad . \quad . \quad (8)\ (9)$$

Then g has the following pleasant properties. All or most of them are known, but a collection may be helpful.

(i) The expectation of g in samples from a Poisson population is unity, for any n . . . . (10)

(ii) Provided that the mean m is large (say $m > 5$) the distribution of $(n - 1)g$ in samples from a Poisson population is the same as that of χ^2 for $n - 1$ degrees of freedom. More explicitly let $L_{>g}$, in which the inequality sign appears in the suffix, denote the probability that in samples of n from a Poisson population, g would be greater than its observed value . . (11)
Then

$$L_{>g} = P(\chi^2, n - 1) \quad . \quad . \quad . \quad . \quad . \quad . \quad . \quad . \quad (12)$$

where $\chi^2 = (n - 1)g$ and P is the probability for $n - 1$ degrees of freedom; given, say, in Fisher's table.

(iii) The χ^2 mentioned in (ii) is appropriate to testing the *hypothesis that, apart from chance deviations, the* n *numbers* $x_1, x_2, \ldots, x_n$ *would all be equal* . . . . (13)

It is thus quite distinct from the χ^2 applied above to Table I. Both Fisher (1936, p. 60) and Jeffreys (1939, p. 259) mention this special form, namely, $\chi^2 = \frac{1}{m}\sum_i (x_i - m)^2$ in connection with the Poisson law. The simple hypothesis (13) is certainly consistent with the Poisson law, but can hardly be said to be equivalent to it; for the type of chance deviations is not specified in (13). Yet there is another hidden connection. For the Poisson law, or something approximately equivalent to it, is used in some proofs of $P(\chi^2)$ (*e.g.*, Kendall, 1943, pp. 290–291). The forms and usages of χ^2 are so multifarious that in order to avoid confusion it is desirable to have here a distinctive symbol such as g defined in (9).

(iv) Just where the property (ii) is lacking because the mean m is not large, the gap is made good because in the present data n is then large, and we can prove that for samples from a Poisson population

$$\text{var } g = \frac{2}{n}, \text{ for large } n \quad . \quad . \quad . \quad . \quad . \quad . \quad . \quad (14)$$

by the methods in Kendall's (1943) " Advanced Theory of Statistics," Ch. 9.

(v) Jeffreys' test for a deviation from the Poisson law towards the negative binomial is a function of n and g, as he indicates (1939, pp. 258–9).

(vi) There are various simple relations between g and a coefficient Q^2 introduced by Lexis, but variously interpreted by his successors (Keynes, 1921, Ch. XXXII). Lexis included a factor to represent $(1 - p)$ in the binomial variance. This factor is purposely omitted from g. Irwin (1932, p. 506–8) explains the relation of Q^2 to one of the χ^2. Fisher (1936, p. 83) puts Q where the others put Q^2.

(vii) If the observed g is incredible for a sample of n from a Poisson population then the Poisson law must be rejected. But conversely, if g is credible, all that we can say is that *so far* there is no objection to a Poisson law; a more searching test might still reject it.

Resumed discussion of the distribution of wars among cells of various durations.

In Table III the distribution among cells of 54 years gives $g = 3{\cdot}56$, $L_{>g} < 0{\cdot}001$. We must therefore reject both the Poisson law, and the simple law (13) of uniformity apart from chance. Some historical explanation of the departures from uniformity, or at least of their extremes, is therefore required. The 54 year cells that contained most and least outbreaks, extended from 1824 to 1877 A.D. for most, and from 1716 to 1769 A.D. for least.

For the same data arranged as in Table I we have $g = 1{\cdot}093$ for $n = 432$. But by (10) and (14) we expect, from a Poisson population, $g = 1 \pm \sqrt{\frac{2}{432}} = 1 \pm 0{\cdot}068$, the number after the ambiguous sign being the standard deviation. The hypothesis (13) of uniformity-apart-from-chance, and the Poisson law are so far both quite credible for cells of $\tau = 1$ year. We have indeed already verified the Poisson law more thoroughly by a different χ^2 applied to Table I.

How is it that deviations from uniformity-apart-from-chance show so clearly in 54 year cells, but hardly at all in 1 year cells? For much the same reason that the fat and the lean become inconspicuous when the meat is minced.

To push this type of analysis to its extreme in the other direction let us enquire whether wars on the average became more frequent.

TABLE IV

$n = 2$ *cells of* $\tau = 216$ *years each.*

Dates	1500 to 1715 A.D.	1716 to 1931 A.D.
x	$x_1 = 143$	$x_2 = 156$

Here $g = 0{\cdot}565$ for one degree of freedom, and so by (12) and Yule's table (Kendall, 1943, p. 444), $L_{>g} = 0{\cdot}45$.

That is to say *Wright's list indicates an absence of any steady drift towards more or fewer wars* . . . . (15)

Cells of some other durations have also been examined by the same tests. The results are collected in the following *summary*.

TABLE V

Distribution of 299 *outbreaks of war among the* 432 *years extending from* 1500 A.D. *to* 1931 A.D. *according to Quincy Wright's list.*

Number of cells, n	Duration of each cell, τ years	g	$L_{>g}$	Was distribution uniform apart from chance ?	Did Poisson law hold ?
2	216	0·565	0·45	yes	evidence lacking
4	108	3·29	0·021	unlikely	unlikely
8	54	3·56	$<$0·001	no	no
16	27	2·249	$0{\cdot}00_3$	no	no
48	9	1·538	0·010 *	no	no
144	3	1·058	0·28	yes	yes
432	1	1·093	0·09	credible	yes

* By Wilson and Hilferty's extension of $P(\chi^2)$ tables.

The search for periods differs from the present type of analysis in various respects. To begin with, the former is applied to the sums of the columns, but the latter to the sums of the rows, of a Buys–Ballot table (Kendall, 1945, Art. 3). Nevertheless, a brief allusion to periods may be in place here.

Quincy Wright (1942, Ch. IX) has discussed fluctuations in the intensity of war. He remarks (p. 230) that: " A fifty-year war period has often been noticed." But he does not mention any test of its statistical significance. The need for caution concerning this alleged period has been emphasized by Kendall (1945, Art. 57).

Another collection of wars, defined more objectively

What one person might call a war, another might call a mutiny, or even an incident. The great obstacle to any scientific study of quarrels is contradictory evidence from the opposing sides. Wright's definition of a war involves the notions of its " legality " and " political importance " which are always matters of opinion, and can be permanently controversial. By contrast, the number of persons who died because of the quarrel, although often deliberately mis-stated, is in almost all cases ascertainable by ten years after the end of the fighting within a factor of three times more or less, which is small in comparison with the whole range. The war dead were taken to include all those, on both sides, whether armed personnel or civilians, who were killed fighting, or drowned by enemy action, or who died of wounds or from poison gas, or from starvation in a siege, or from other malicious acts of their enemies. Moreover, deaths from disease or exposure of armed personnel during a campaign were included; but not civilian deaths by epidemic disease in places far distant from the geographical location of the fighting. Before I saw Wright's list I had almost completed a card-index of fatal quarrels classified according as the number of war-dead were of the order of 10^7 or 10^6 or 10^5 or 10^4. The last group, which is the most numerous, concerns us now. Its precise range is intended to be from $10^{3\cdot5}$ to $10^{4\cdot5}$ in war-dead. In popular language it consists of small wars, revolts, insurrections and incidents, in any part of the world, regardless of their legality or their political importance. They were collected from sources too numerous to be specified here, but a list of the dates is set out in Appendix II. Wright's list, which begins in 1480 A.D., has the great merit of being longer than mine, which begins only with 1820 A.D. Where the two lists overlap, there are many divergencies between them, both of inclusion and exclusion. Yet Poisson's law shows in both. The 110 calendar years from 1820 to 1929 contained on my list 63 beginnings and, as it chanced, also 63 endings of fatal quarrels for which the war-dead were more than $10^{3.5}$ and fewer than $10^{4.5}$. When these are sorted, as in Table I, into cells of one year the result is:

TABLE VI

x	0	1	2	3	4	>4	sum
$f(x)$ beginnings	63	35	9	2	1	0	110
$f(x)$ ends...	62	34	13	1	0	0	110
Poisson	62·0	35·5	10·2	1·9	0·3	0·0	109·9

The agreement with Poisson's law of improbable events draws our attention to the existence of a persistent background of probability. If the beginnings of wars had been the only facts involved, we might have called it a background of pugnacity. But, as the ends of wars have the same distribution, the background appears to be composed of a restless desire for change.

REFERENCES

Fisher, R. A. (1936), *Statistical Methods for Research Workers.* Oliver and Boyd, Edinburgh.
Irwin, J. O. (1932), *J. Roy. Stat. Soc.*, **95**, 506–8.
Jeffreys, H. (1939), *Theory of Probability.* Clarendon Press, Oxford.
Keynes, J. M. (1921), *A Treatise on Probability.* Macmillan, London.
Kendall, M. G. (1943), *The Advanced Theory of Statistics*, Vol. I. Griffin, London.
Kendall, M. G. (1945), " On the Analysis of Oscillatory Time Series." *J. Roy. Stat. Soc.*
Wright, Q. (1942), *A Study of War.* University of Chicago Press.

APPENDICES

The following raw material has been provided at the request of Dr. L. Isserlis so as to enable any reader to make his own analyses.

APPENDIX I. *Extract from Prof. Quincy Wright's list* (1942) *of Wars of Modern Civilization.*

The columns are in pairs. The first column of each pair contains the date A.D.; the second column contains the number of wars which began in that calendar year. Much further information such as names of wars, names of belligerents, names of treaties of peace, dates of ending, number of battles, and type of war, can be seen in Wright's book.

A.D.		A.D.		A.D.		A.D.		A.D.		A.D.		A.D.	
1482	1	1490	0	98	0	06	0	14	0	22	1	1530	0
83	0	91	0	99	0	07	0	15	1	23	0	31	2
84	0	92	1	1500	0	08	1	16	1	24	1	32	2
85	0	93	0	01	0	09	1	17	0	25	0	33	1
86	0	94	0	02	0	1510	2	18	0	26	1	34	1
87	0	95	1	03	0	11	2	19	0	27	0	35	0
88	0	96	0	04	0	12	1	1520	1	28	0	36	1
89	0	97	0	05	0	13	0	21	2	29	0	37	1

APPENDIX I.—*Continued.*

A.D.		A.D.		A.D.		A.D.		A.D.		A.D.		A.D.	
1538	0	96	0	54	2	12	0	1770	2	28	1	86	0
39	0	97	0	55	0	13	0	71	0	29	0	87	0
1540	0	98	2	56	0	14	0	72	0	1830	3	88	0
41	1	99	0	57	2	15	2	73	1	31	3	89	1
42	2	1600	4	58	0	16	1	74	0	32	0	1890	0
43	0	01	0	59	0	17	1	75	2	33	1	91	0
44	1	02	0	1660	0	18	1	76	0	34	0	92	0
45	0	03	1	61	0	19	1	77	1	35	2	93	0
46	1	04	1	62	0	1720	0	78	1	36	1	94	1
47	0	05	1	63	0	21	0	79	0	37	0	95	1
48	0	06	0	64	0	22	0	1780	0	38	1	96	3
49	1	07	0	65	2	23	0	81	0	39	3	97	2
1550	0	08	0	66	2	24	0	82	0	1840	1	98	1
51	1	09	1	67	1	25	0	83	1	41	0	99	1
52	2	1610	1	68	1	26	1	84	0	42	0	1900	1
53	0	11	1	69	0	27	0	85	0	43	0	01	0
54	2	12	0	1670	1	28	0	86	0	44	1	02	1
55	0	13	0	71	0	29	0	87	2	45	0	03	0
56	0	14	0	72	2	1730	0	88	1	46	1	04	1
57	0	15	3	73	1	31	0	89	1	47	0	05	0
58	0	16	0	74	0	32	0	1790	0	48	3	06	1
59	2	17	1	75	1	33	1	91	0	49	1	07	0
1560	0	18	4	76	1	34	0	92	3	1850	1	08	0
61	1	19	0	77	1	35	1	93	1	51	0	09	0
62	1	1620	1	78	1	36	0	94	0	52	1	1910	1
63	0	21	2	79	0	37	0	95	1	53	2	11	1
64	0	22	0	1680	1	38	0	96	0	54	0	12	2
65	1	23	0	81	0	39	1	97	0	55	1	13	0
66	1	24	0	82	2	1740	1	98	2	56	1	14	4
67	2	25	2	83	2	41	0	99	1	57	1	15	1
68	0	26	1	84	0	42	0	1800	0	58	0	16	2
69	2	27	1	85	1	43	0	01	1	59	3	17	1
1570	0	28	0	86	1	44	0	02	2	1860	1	18	0
71	0	29	0	87	0	45	1	03	1	61	1	19	2
72	3	1630	3	88	1	46	0	04	1	62	1	1920	1
73	0	31	0	89	1	47	1	05	1	63	4	21	2
74	0	32	1	1690	0	48	0	06	2	64	1	22	0
75	2	33	0	91	0	49	0	07	2	65	2	23	0
76	1	34	1	92	0	1750	0	08	1	66	3	24	0
77	0	35	1	93	0	51	0	09	1	67	0	25	0
78	0	36	0	94	0	52	0	1810	1	68	1	26	0
79	1	37	1	95	1	53	0	11	0	69	0	27	0
1580	2	38	1	96	0	54	1	12	2	1870	1	28	1
81	0	39	1	97	0	55	0	13	1	71	0	29	0
82	0	1640	2	98	0	56	0	14	0	72	1	1930	0
83	1	41	2	99	0	57	0	15	3	73	1	31	1
84	0	42	0	1700	1	58	0	16	0	74	0	32	0
85	2	43	1	01	1	59	0	17	1	75	0	33	0
86	0	44	1	02	0	1760	0	18	0	76	0	34	0
87	0	45	0	03	1	61	0	19	0	77	1	35	1
88	0	46	2	04	0	62	0	1820	0	78	0	36	1
89	0	47	0	05	1	63	1	21	2	79	3	37	1
1590	1	48	3	06	0	64	0	22	0	1880	1	38	0
91	0	49	1	07	0	65	0	23	2	81	1	39	2
92	0	1650	1	08	0	66	0	24	1	82	2		
93	0	51	0	09	0	67	0	25	1	83	0		
94	1	52	1	1710	1	68	3	26	2	84	1		
95	0	53	0	11	0	69	0	27	0	85	1		

APPENDIX II. *Dates of beginning and ending of fatal quarrels which caused more than* $10^{3.5}$ *but fewer than* $10^{4.5}$ *deaths.*

The columns are arranged in triplets, in each of which the first column contains the date A.D., the second column shows the number of fatal quarrels of the specified magnitude which began, and the third column the number which ended, in that calendar year.

The following list does not, and ought not to, agree with Wright's list; for its definition is different. Many further particulars await publication.

A.D.			A.D.			A.D.			A.D.		
1820	0	0	1850	0	0	1880	0	1	1910	0	1
21	1	0	51	0	0	81	1	1	11	0	0
22	1	0	52	1	1	82	1	0	12	0	0
23	1	0	53	0	1	83	0	1	13	1	1
24	0	1	54	0	0	84	0	0	14	0	0
25	0	0	55	0	0	85	1	2	15	0	0
26	1	3	56	1	0	86	0	0	16	0	0
27	1	2	57	1	0	87	0	0	17	0	0
28	0	0	58	0	1	88	0	0	18	2	1
29	2	0	59	2	1	89	0	0	19	0	0
1830	1	0	1860	0	2	1890	0	0	1920	2	2
31	1	1	61	1	0	91	1	1	21	2	1
32	0	0	62	1	0	92	1	0	22	0	1
33	1	1	63	0	0	93	0	2	23	0	0
34	0	1	64	1	1	94	4	0	24	0	0
35	1	0	65	0	1	95	0	2	25	0	0
36	0	0	66	0	0	96	1	1	26	1	0
37	0	0	67	0	0	97	0	0	27	0	0
38	2	1	68	0	0	98	1	0	28	0	1
39	3	0	69	0	0	99	1	1	29	0	0
1840	0	2	1870	0	0	1900	1	2	1930	0	0
41	1	2	71	1	1	01	0	0	31	0	0
42	0	2	72	1	0	02	0	2	32	0	0
43	1	1	73	0	1	03	0	0	33	0	0
44	0	0	74	0	0	04	0	0	34	0	1
45	1	0	75	2	0	05	0	0	35	1	0
46	1	1	76	0	2	06	1	1	36	1	1
47	0	1	77	1	2	07	0	0	37	0	1
48	1	1	78	3	1	08	0	0			
49	0	1	79	2	1	09	2	1			

1945

DISTRIBUTION OF WARS IN TIME

Nature
London
Vol. 155, No. 3942, 19 May 1945, p. 610
(Letter dated 19 March)

Note: The publication referred to is 1944:2 (above)

Reprinted from NATURE, *Vol.* 155, *page* 610, *May* 19, 1945)

Distribution of Wars in Time

A STATISTICAL regularity in the dates of wars has been brought to notice by the following numerical process. A list was prepared of wars in the world as a whole. Each calendar year was thereby characterized by the number, $x = 0, 1, 2, 3, 4, \ldots,$ of wars which began in it. Next, the number, y, of years which had each such character was counted. A similar procedure was applied to the beginnings of peace. Here are some results :

	x =	0	1	2	3	4	>4
War,	y =	63	35	9	2	1	0
Peace,	y =	62	34	13	1	0	0
$110\, e^{-\mu}\mu^x/x!$	=	62·0	35·5	10·2	1·9	0·3	0·0

where $\mu = 63/110$. The formula is the Poisson law of improbable events. Other phenomena, known to be described by the Poisson law, include the distribution in time of the alpha particles emitted from radioactive substances[1], or of deaths by kick from a horse[2].

This impersonal account of the beginnings of war and of peace contrasts with the personal details in the newspapers and history books. In somewhat the same manner the statistics of marriage contrast with a love-story in a biography. The Poisson law is statistical in the sense that it does not predict the date of any future peace or war.

The particular set of wars summarized in the above table are fatal quarrels, which caused from $10^{3\cdot5}$ to $10^{4\cdot5}$ deaths, and which ended from A.D. 1820 to 1929 inclusive[3]. But the Poisson law, with other constants, also describes the beginnings of wars from A.D. 1500 to 1931, as set out in Prof. Quincy Wright's list[4].

A more critical account of these regularities has been accepted for publication by the Royal Statistical Society.

LEWIS F. RICHARDSON.

Hillside, Kilmun, Argyll.

March 19.

[1] Rutherford, Chadwick and Ellis, "Radiations from Radioactive Substances" (Camb. Univ. Press, 1930), 172.

[2] Bortkewitsch, quoted by Pearson, K., "Tables for Statisticians", (Cambridge University Press, 1914), lxxvii.

[3] *Nature*, **148**, 598 (1941).

[4] Wright, Q., "A Study of War" (University of Chicago Press, 1942), appendix xx.

PRINTED IN GREAT BRITAIN BY FISHER, KNIGHT AND CO., LTD., ST. ALBANS

Richardson, L. F. (1946)
Jl R. statist. Soc. **109**, 130–56

1946:1

THE NUMBER OF NATIONS ON EACH SIDE OF A WAR

Journal of the Royal Statistical Society
London
Vol. 109 Part II, 1946, p. 130–156

Notes:
1. Richardson sometimes refers to this paper subsequently according to the year of actual publication, which was 1947
2. This paper reappeared unaltered in *Statistics of Deadly Quarrels* (Chapter X)

MISCELLANEA

CONTENTS

THE NUMBER OF NATIONS ON EACH SIDE OF A WAR

By LEWIS F. RICHARDSON

Summary

THE historical facts to which this investigation relates are contained in two lists of wars. Each list is comprehensive, world-wide, and covers more than a century. The first step was to notice how many nations fought on the two sides of each war, and to classify the wars accordingly as 1 versus 1, or 2 versus 1, or in general as r versus s. The number of wars of each of these types was then counted, and denoted by $n(r, s)$. It is a bivariate distribution of a peculiar shape, unlike the ordinary correlation table. Attention was thus diverted from the peculiarities of named nations at particular times to some habits of belligerents in general. The observed frequency, when allotted half to the point (r, s) and half to the point (s, r), is found to vary as $(rs)^{-2\cdot5}$. That is to say, wars became rapidly rarer as the number, rs, of pairs of opposed belligerents increased. Various explanations, numbered I, II, . . . , XIII, have been attempted. One is that international relations might be likened to the molecular chaos in a gas, in so far as ternary encounters are rarer than binary encounters, for reasons which can be expressed by probabilities, after the manner of Bernoulli. This idea is developed in Theory VIII, but it fails to fit the historical $n(r, s)$ when 60 nations are regarded as all involved. Fewer nations would somehow fit better. Accordingly, in Theory X disputes are regarded as localized; and it is found that a tolerable agreement with the historical $n(r, s)$ is obtained by supposing that each dispute interested only eight nations, and that the probability that any two of the eight came to war with each other about it was 0·35. Such a theory, however, fails to explain the rare wars in which $r + s > 8$. To remedy this defect, the restriction of the number of interested nations is expressed in Theory XI by classifying them into local powers and world-powers. Any two local neighbours are assumed to be at war with one another during a total fraction z of the time. It is found that z increased with the number, rs, of pairs of opposed belligerents in the following manner:—

$rs =$	1	2	3	4
$z =$	0·008	0·020	0·046	0·119?

This increase of z can be explained by the infectiousness of fighting.

Any guidance to the practical statesman which this enquiry can offer is thus the indication of a chaos, restricted by geography, and further modified by the infectiousness of fighting. There are also some suggestions in Theory XIII as to a possible future effect of aviation.

PART I. HISTORICAL FACTS AND SMOOTHING

Introduction

The very practical questions whether alliances make for world-peace, and whether the great victor Powers of the Second World War are peculiarly well qualified by their past history for maintaining world-peace, can be answered only by comparison with the average behaviour of States in general during centuries. It is this general background of human nature which is here examined, not the peculiarities of named States at particular times. The discussion involves collections of facts lettered $A, B, \ldots, H, J$, and theories numbered I, II, . . . , XII, XIII.

Wars can be classified by the number of groups of people on the two sides: for example, 1 nation versus 3 nations, 2 states versus 2 other states, 1 government versus 1 set of insurgents; or, in general, r belligerent groups versus s belligerent groups. The decision as to what groups should be counted in r or s was made in accordance with the ordinary common-sense practice of historians—that is to say, token declarations of war were ignored, so also was any support that was only diplomatic or financial or by provision of arms; but, on the other hand, groups who did a notable amount of fighting were counted, even if they joined late or left early.

The type, r versus s, of each war having been thus recognized, the next step was to count how many wars there were of each of the types 1 versus 1, 2 versus 1, 3 versus 2, and so on.

The author has abstracted, from numerous works on history, a card-index and list of fatal quarrels that ended since A.D. 1820 in all parts of the world. This collection is subdivided according to the number of war-dead on all sides jointly. A fatal quarrel that caused 10^{μ} deaths is said to be " of magnitude μ." For most purposes fatal quarrels of magnitude less than 3·5 have been omitted, because, in popular language, they would seldom be called wars. This collection is fundamental for the present purpose, and is so far not fully published; but extracts from it appear in the present paper as Collections A, D, F, D, J.

There have been a very few wars which cannot be regarded as conflicts between two sides. For example, there was a three-sided fight in Galicia in 1846, but it was only of magnitude 3. A larger exception to the proverb, " Your enemy's enemy is your friend," was the Mexican revolution of 1910–20, which was of magnitude 5. F. A. Kirkpatrick (1938, p. 349) describes it as chaotic. It has been omitted from the present study of two-sided wars.

The number of wars of each type

Obviously the counted numbers should be arranged in a table of rows and columns in the (r, s) plane. That process has been applied to Richardson's list with the following result.

Collection A. Wars of all magnitudes greater than 3·5 *which ended from* 1820 *to* 1939 *A.D. inclusive*

s	$r=1$	2	3	4	5	6
6						0
5					0	0
4				0	0	0
3			0	1	0	0
2		3	2	1	1	0
1	42	24	5	5	2	1

There were also beyond bounds of this table:
2 wars of 7 versus 1
1 war of 9 versus 1
1 war of 15 versus 5
thus making a total of 91 wars.

For many wars there is no doubt as to how the individual persons should be grouped; so that the assignment of r and s is quite simple and obvious. For example, in 1904–05 there was a war between Russia and Japan, plainly of the type 1 versus 1. On the contrary, for some other wars there is room for considerable doubt: for example, how many German groups took part in the war of 1870–71 against France? It is therefore particularly desirable to have independent opinions. Prof. Quincy Wright has provided one by publishing (1942, Appendix XX) the number of participants on each side of each war in his impressively long list. By comparison of particular wars it is found that Wright's word " participant " usually means a belligerent State selected from a restricted set of sovereign States. But he lists 34 wars as each involving only a single participant. He explains his principle thus (p. 637):

> " . . . there was the difficulty of deciding what entities were to be regarded as participants, and in this respect actual independence before or after the war rather than legal status under international law was the criterion used."

For example, Wright assigns only one participant to the American Civil War of 1861–66, because the Southern Government did not exist either before or after the war. That may be satisfactory from a special legal point of view—Prof. Wright is a professor of international law—but it is of no use for common-sense purposes, nor for those of psychology. Some amendment had therefore to be made. It would not be sufficient merely to alter Wright's wars from type 0 versus 1 to type 1 versus 1; for his principle of " actual independence before or after the war " may have had other effects. He explains his practice further thus:

" Unrecognized princes like the Aztec and Inca rulers, semi-dependencies like the Barbary states, feudal principalities like the German and Italian states nominally subject to the Empire, and successful insurgents like the United States in 1775 have been listed as independent participants. Unsuccessful revolutionists, rebels, or insurgents which lacked even *de facto* status, except during the war itself, have not been so listed, and many of the small feudal principalities of the Holy Roman Empire have been ignored on the ground that, while enjoying some legal status though short of full independence under international law, their actual political importance was very slight."

It thus becomes necessary to correct the excessive legalism whereby Wright ignored unsuccessful rebels; and yet it is desirable to maintain a due respect for Wright's opinion on other matters. A fairly satisfactory device for attaining both these ends is to omit from Wright's list all those wars which he marked *C* for civil.* They are distributed thus when expressed in the notation *r* versus *s*.

Collection B. Civil wars in Wright's list

Type	0 versus 1	0 versus 2	1 versus 1	1 versus 2	1 versus 3	1 versus 4	1 versus 6	1 versus 18	Total
Number of wars	34	1	27	10	3	1	1	1	78

Wright does not actually use the absurd symbols 0 versus 1 and 0 versus 2, but he implies them. After the rejection of these 78 civil wars, there remain in Wright's list 200 other wars which are distributed as follows. They include four large wars which Wright states both as wholes and in parts. Here the parts are omitted and the wholes are included.

Collection C. Wars not marked civil in Wright's list from A.D. 1480 *to* 1941 *inclusive*

s \ *r*	1	2	3	4	5	6
6						0
5					0	0
4				0	0	0
3			1	1	1	0
2		4	6	0	1	1
1	117	28	12	12	3	2

There were also, beyond the bounds of this table:
1 war of 7 versus 1
1 war of 8 versus 1
1 war of 11 versus 1
1 war of 16 versus 1
1 war of 20 versus 1
1 war of 7 versus 3
1 war of 8 versus 5
1 war of 20 versus 5
1 war of 33 versus 5
1 war of 35 versus 7
1 war of 9 versus 8
thus making a total of 200 wars.

The quadrantal arrangement

In the foregoing tables the observations have been spread over the octant for which $r \geqslant s$, leaving a blank space where $r < s$. When $r = s$, there is only one place in the table for the number of wars, and it is accordingly shown there; but when $r \neq s$ there are two places, such for example as (3, 1) and (1, 3); and there is no clear reason for choosing one or the other. If we could, always and infallibly, name the first aggressor, that decision would enable us to distinguish 1 versus 2 from 2 versus 1; but in fact the question "Who began it?" usually leads to controversy. Accordingly, the number of wars might be concentrated in either place, or distributed between the two places in any manner. In the following two tables the number of wars is distributed equally, half of them in each place, when $r \neq s$. The only justification that can be given for this choice at the present quite empirical stage of enquiry is that equal distribution makes the observed facts look neater, smoother, more consistent, and more intelligible. Another justification will appear when we come to theories. The two types of table will be called octantal and quadrantal; and these adjectives will be applied also to the formulæ descriptive of the two tables. It is, of course, essential to avoid confusions between the octantal and quadrantal arrangements.

To facilitate comparisons between the two sets of statistics the frequencies in each have been divided by their respective totals.

* For the suggestion of this amendment I am indebted to another lawyer, Sheriff A. M. Hamilton, K.C.

Collection D, obtained from Collection A

Outliers beyond the bounds of the table, 0·0440. Total 1·0004.

s \ r	1	2	3	4	5	6
6	0·0055	0	0	0	0	0
5	0·0110	0·0055	0	0	0	0
4	0·0275	0·0055	0·0055	0	0	0
3	0·0275	0·0110	0	0·0055	0	0
2	0·1319	0·0330	0·0110	0·0055	0·0055	0
1	0·4615	0·1319	0·0275	0·0275	0·0110	0·0055

Collection E, obtained from Collection C

Outliers beyond the bounds of the table, 0·0550. Total 1·0000.

s \ r	1	2	3	4	5	6
6	0·0050	0·0025	0	0	0	0
5	0·0075	0·0025	0·0025	0	0	0
4	0·0300	0	0·0025	0	0	0
3	0·0300	0·0150	0·0050	0·0025	0·0025	0
2	0·0700	0·0200	0·0150	0	0·0025	0·0025
1	0·5850	0·0700	0·0300	0·0300	0·0075	0·0050

Separation of essential from accidental features

Some features of the historical facts are presumably accidental. We shall avoid useless meticulosities by referring back to the counted numbers in the previous tables, and by remembering that any counted number n of wars is likely to have a standard uncertainty $\sqrt{n}$, because the distribution of wars in time is Poissonian (Richardson, 1945).

There are strong resemblances between the two sets of statistics. In both the type 1 versus 1 is the most frequent. In both the cell-frequency in general decreases with increase of r and s. In both the cell-frequency decreases more rapidly along the diagonal where $r = s$ than it does along the lines $r = 1$ or $s = 1$. In both the quadrantal type 3 versus 2 is rarer than the quadrantal type 4 versus 1, although each involves 5 belligerents. The chief disagreements between the two sets of statistics are now seen to be that in Wright's set the wars of type 1 versus 1 are commoner, and those of types 2 versus 1 and 1 versus 2 are rarer, than in the other set.

It is desirable to reject accidental features and to retain only the meaningful essence by fitting empirical formulæ to the distributions, or to their mean.

Let $n(r, s)$ be the counted number of wars in the r, s cell of the quadrant, the table having been made symmetrical in the manner described above . . . . (1)

Let $\phi(r, s)$ be the smoothed function which we seek as a substitute for $n(r, s)$ . . . (2)

Let N be the total number of wars, so that

$$N = \sum_{r=1}^{\infty} \sum_{s=1}^{\infty} n(r, s) \qquad (3)$$

The infinite upper termini here merely indicate that we must include all the field as far as it is occupied; for beyond the most complicated wars $n(r, s)$ is everywhere zero.

The first step towards the choice of a formula is to notice the shape of the isopleths of n—that is to say, the loci on which $n(r, s)$ is constant. They are hard to discern on the octantal tables, but fairly obvious on the quadrantal. So let us confine attention to the latter. On the diagonal lines that slope upwards to the left, so that along them $r + s$ is constant, n is occasionally constant, but usually is less in the middle than at the extremes. Along the rectangular hyperbolæ on which rs is constant, n is on the average moderately constant. The conclusion so far is that a function of rs would fit better than a function of $r + s$, but that a function of some variable intermediate between rs and $r + s$ would be best. That is to say: *the number,* rs, *of pairs of opposed belligerent groups is somewhat more relevant than the total number,* r + s, *of belligerent groups.* It is also desirable to notice, as forms that must be definitely rejected because their isopleths are hollow on the wrong side, $\phi = f(r^2 + s^2)$ and $\phi = f(r!s!)$. The former, and its elliptic generalizations, are associated with correlation tables. The latter often occurs in probability-theories, and tends to prevent them from fitting these facts.

The largest counted numbers have the least fractional uncertainty. The largest is N, the next is $n(1, 1)$. With these ϕ should especially agree. When we are selecting a formula, it is convenient to take $\phi(1, 1)$ as the standard, because $\phi(1, 1)$ is obvious; whereas the total

$$\sum_{r=1}^{r+s=m} \sum_{s=1} \phi(r, s) = \Phi, \text{ say} \quad . \quad . \quad . \quad . \quad . \quad . \quad . \quad . \quad . \quad (4)$$

may be troublesome to calculate or may be ill-defined. For the upper termini of the sums are connected by $r + s = m$, where m is the number of possible belligerents in the world. The difficulties of calculation have in many cases been overcome, but there remains the uncertainty arising from our ignorance of m. Certainly the number of " sovereign and independent" States was of the order of 60 in the year 1930; but some belligerents were insurgents; so that m was an ill-defined number of the order of 100. We shall be fortunate if the form of $\phi(r, s)$ is such that no appreciable increase in the sum (4) occurs when m is increased from 60 to ∞. Here follows a selected list of expressions, regarded as candidates for the position of empirical formula, together with their chief qualifications for that appointment.

Select list of candidate-formulæ

Region of the r, s quadrant	$rs = 1$	The whole field	$rs = 2$	$rs =$ 3 to 6	$rs =$ 7 to ∞	$r + s =$ 61 to ∞
Observed Ratios { Collection A	1·000	$2{\cdot}1_{67}$	$0{\cdot}5_{71}$	$0{\cdot}42_9$	$0{\cdot}16_7$	0·000
Observed Ratios { Collection C	1·000	$1{\cdot}70_9$	$0{\cdot}23_9$	$0{\cdot}33_3$	$0{\cdot}13_7$	0·000
$(rs)^{-2\cdot5}$	1·000	1·800	0·354	0;303	0·143	<0·009
$\frac{1}{2}(r + s)(rs)^{-3}$	1·000	1·977	0·375	0·361	0·241	0·034
$2r^{-2}s^{-2}(r + s)^{-1}$	1·000	1·688	0·333	0·257	0·097	<0·004
4^{1-rs}	1·000	1·684	0·500	0·184	0·001	0·000
$(rs)^{-2\cdot4}$	1·000	$1{\cdot}91_3$	0·379	0·347	$0{\cdot}18_8$	—

The calculation of the double sums to infinity, for columns 3 *and* 7 *of the above table.* The easiest of these sums is

$$\sum_{r=1}^{\infty} \sum_{s=1}^{\infty} (rs)^{-\kappa} = \sum_{r=1}^{\infty} r^{-\kappa} \sum_{s=1}^{\infty} s^{-\kappa} = [\zeta(\kappa)]^2 \quad . \quad . \quad . \quad . \quad . \quad . \quad . \quad (5)$$

where $\zeta(\ )$ is the Zeta Function of Riemann, given in the tables of Jahnke and Ende.

It was also necessary to approximate to sums by way of integrals. The following modification of the Euler–Maclaurin approximation was found to be especially appropriate for estimating remainders after the larger terms had been summed by arithmetic. Let the r, s plane be divided into square cells by lines at $r = \frac{1}{2} + j$, $s = \frac{1}{2} + j$, where j is 1, 2, 3, 4, ... Let $\phi(r, s)$ be regarded as the ordinate at (r, s) of a surface over which r and s vary continuously. In any cell the volume under the surface ϕ numerically exceeds the ordinate at the centre of the cell, because these surfaces are all concave. Therefore

$$\sum_{r=r_1}^{\infty} \sum_{s=s_1}^{\infty} \phi(r, s) < \int_{r_1-\frac{1}{2}}^{\infty} \int_{s_1-\frac{1}{2}}^{\infty} \phi(r, s) dr ds \quad . \quad . \quad . \quad . \quad . \quad . \quad (6)$$

Most books neglect to mention this shift of a half in the termini; but it brings the above inequality much nearer to equality. In any cell the volume numerically exceeds the ordinate at the centre by $\frac{1}{24}(\partial^2\phi/\partial r^2 + \partial^2\phi/\partial s^2)$ + terms in fourth derivatives; and this correction transforms by Green's theorem, but it was not required. One of the integrals used for this purpose was

$$\int_{\alpha}^{\infty} \int_{\beta}^{\infty} \frac{2 dr ds}{r^2 s^2 (r + s)} = \frac{2}{3}\left\{\frac{\alpha + \beta}{\alpha^2\beta^2} - \frac{1}{\beta^3} \log\left(\frac{\alpha + \beta}{\alpha}\right) - \frac{1}{\alpha^3} \log\left(\frac{\alpha + \beta}{\beta}\right)\right\} \quad . \quad . \quad (7)$$

The fringe beyond the sloping line $r + s = 60$ is, for some ϕ, troublesome to calculate. However, all that we need to know about it is whether it is negligible. So I have been content with rough bounds. Of the upper bounds the easiest to obtain is the sum beyond the square whose four sides are $r = 1, s = 1, r = 30, s = 30$. A closer upper bound is the sum beyond the hyperbola $rs = 59$. For those ϕ which are functions of the single variable rs, an approximation to the sum beyond $rs = 59$ is

$$\int_{rs = 59\cdot 5}^{\infty} \phi(r, s)\bar{\tau}(rs)d(rs) \quad . \quad . \quad . \quad . \quad . \quad . \quad . \quad . \quad . \quad . \quad (8)$$

where $\tau(rs)$ is the number of divisors of rs, including 1 and rs; and $\bar{\tau}$ is a smoothed value of the wobbly function * τ. I took

$$\bar{\tau}(rs) = 1\cdot 1544 + \log_e (rs) - \frac{1}{24r^2s^2} \quad . \quad . \quad . \quad . \quad . \quad . \quad . \quad . \quad (9)$$

Selection and final test of an empirical formula

It would, of course, be possible to fit two formulæ, one to Collection A, another to Collection C. But a more useful question is whether a single formula can be found to specify a hypothetical large collection, such that both Collections A and C can be regarded as random samples from it. The preceding table suggested that $(rs)^{-2\cdot 5}$ might be a suitable proportionality. It was therefore selected for a χ^2 test. For this purpose the double sum to infinity must first be made to agree with N, the observed total number of wars, by altering the formula to

$$\phi(r, s) = \tfrac{5}{8} N(rs)^{-2\cdot 5} \quad . \quad . \quad . \quad . \quad . \quad . \quad . \quad . \quad . \quad . \quad (10)$$

Each set was then given its appropriate N—namely, 91 for Collection A, 200 for Collection C. The χ^2 test was applied to the four cells for which rs is severally 1, 2, 3 to 6, 7 to ∞. The index, $-2\cdot 5$, had been adjusted, but had certainly not been chosen to suit each set separately. It seemed proper therefore to regard N as the only adjustable constant, and to take the degrees of freedom as three. The results are:

Collection	χ^2	Probability of a greater χ^2
A	4·0	0·26
C	4·4	0·22

Thus the formula (10) is satisfactory, in so far as it credibly describes both sets of observations, and is smoother, and easier to remember, than they are. Much closer agreement could be obtained if the index, instead of being fixed at 2·5, were suited to the sets separately; but such an elaboration might not be an improvement. For, where samples can differ, a perfect fit is a misleading ideal.

The total duration of wars of these types

For testing Theory VIII which follows in Part II it was necessary to make a special collection of historical facts in the form of total durations. These facts, being of wider interest than the theory which led to their collection, are here stated separately. They are derived from two lists of wars, Wright's and my own, based on different definitions and by rather different processes. The impressive fact is that these two contrasted systems lead on the average to similar results.

In Richardson's list the intention has been to record the duration of hostilities. Wright, in some cases, has placed an extraordinary emphasis on the final legal conclusion of the quarrel. Thus he makes the Napoleonic wars end with the Final Act of the Congress of Vienna, a few days before the battle of Waterloo. Also he makes the First Carlist war continue until an amnesty dated September 2nd, 1847, seven years after the end of hostilities as they are stated by Hume or by the Cambridge Modern History.

No great accuracy can be expected of statements of the duration of hostilities. Although for most fatal quarrels the duration is fairly obvious; yet for others judgment is made difficult by such features as the following;

(i) Provocative incidents prior to the main outbreak. For example, a war between the Burmese and British is here taken to have begun in September 1823, although the British declaration of war did not follow until March 1824. (E **4**, 432.)

* For advice about τ I am indebted to Prof. G. H. Hardy, F.R.S.

(ii) The outbreak of hostilities without any declaration of war, as in Armenia 1892–94.

(iii) Armistices. For example, the First World War was taken to have ended in November 1918, but, because the blockade of Germany continued with fatal consequences, June 1919 might be a more reasonable date.

(iv) Pauses, not agreed between the belligerents as armistices. The usual pause in winter has been included as part of the hostilities. On the contrary the 12-year pause from 1864 to 1876 in the war between the Chinese and the Mohammedans of Sinkiang has been omitted from the duration, because the Chinese army was then far away.

(v) Guerilla warfare continuing after the main defeat, as in South Africa from 1900 to 1902, or in Madagascar 1897.

(vi) The cessation of hostilities without any declaration of peace. This occurred between the Chinese and the Mohammedans in Sinkiang in 1864.

As examples of wars that had extremely opposite characteristics one may pick out Bolivia versus Paraguay 1930–36 as having a duration of hostilities so indefinite that it might be stated as 1·4 or as 5 years; whereas in the American Civil War of 1861–65 the hostilities lasted 4·0 years, definite to a month. Formulæ will here be fitted to the total duration of many wars; so that the uncertainty of the particulars will to a considerable extent be averaged out.

These total durations have one advantage over the numbers of wars—namely, that they enable us to give a clearer description of those wars in which some belligerents joined after the beginning, or quitted before the end. In such a case the war was divided into non-overlapping intervals which were allocated to different types. On the contrary, it can be argued that the beginnings of wars are more relevant than their durations.

As a preliminary to an analysis by the numbers of belligerent groups, r versus s, some overall summaries are of interest. According to Wright (1942, p. 651), the average duration of the wars that occurred from A.D. 1480 to A.D. 1941 was 4·3 years each. Outstanding deviations from the mean are recorded in the names of the Thirty Years and Seven Weeks Wars. The fraction of a political State's time which was spent at war has been tabulated by Woods and Baltzly (Wright, 1942, p. 653). Over the interval A.D. 1450 to A.D. 1900 this fraction averaged 0·115 for Denmark, 0·175 for Sweden, 0·235 for France, 0·250 for Great Britain, 0·275 for Austria, 0·300 for Russia, 0·305 for Turkey, and 0·330 for Spain; the mean for those 8 States being 0·248. The facts for Prussia are there given only over the shorter interval A.D. 1600 to A.D. 1900; and for it the Prussian fraction was 0·17. This should be compared with Lord Vansittart's thesis that Germany has always been the chief aggressor.

After these preliminaries let us turn to the analysis by types r versus s, from the two aforesaid lists. Richardson's list makes possible a cross-classification by magnitudes, with the following result.

Collection F. Fatal quarrels that ended from A.D. 1820 *to* A.D. 1939 *inclusive*

Magnitude	Octantal types							
	1 versus 1	2 versus 1	3 versus 1	2 versus 2	4 versus 1	3 versus 2	More complicated	All types
	Total durations in years							
7·5 to 6·5	0	0	0	0	0	0	4·3	4·3
6·5 to 5·5	28·8	0	5	0	0	0	0	33·8
5·5 to 4·5	71·0	10·2	0·8	0	2·4	0·6	1·9	86·9
4·5 to 3·5	69·2	40·6	6·6	14·5	1·3	6·6	28·4	167·2
7·5 to 3·5	169·0	50·8	12·4	14·5	3·7	7·2	34·6	292·2

The irregularity of the distribution in the central part of the above table shows that the classification has been rather too detailed. So the further discussion has been confined to the totals shown in the bottom row. An objection to the use of these totals is that they go as far as a lower bound at magnitude 3·5, which is an arbitrary choice. It is quite conceivable that if (by an enormous labour that no one has yet performed) the statistics were extended to magnitude 2·5, it would be found that the total distribution among types r versus s was notably different. The best that can be done is to *remember the arbitrariness of the lower bound.* It corresponds, very roughly, to choosing fatal quarrels that are large enough to be called wars in the popular sense of the word.

Prof. Wright's collection does not permit an analysis by magnitudes. The separate wars into which he subdivided the Thirty Years War, the French Revolution, the Napoleonic Wars and World War I are here fundamentally rearranged as non-overlapping intervals of type r versus s. The wars which he marks civil are again omitted. A few obvious slips are amended. To make the two sets of statistics comparable, each has been divided in the following table by the intervals of years to which it relates.

	Octantal types						
	1 versus 1	2 versus 1	3 versus 1	2 versus 2	4 versus 1	3 versus 2	$r+s \geqslant 6$
	Total duration divided by interval						
Collection G from Collection F. Magnitudes 7·5 to 3·5 ...	$1{\cdot}40_8$	$0{\cdot}42_3$	$0{\cdot}10_3$	$0{\cdot}12_1$	$0{\cdot}03_1$	$0{\cdot}06_0$	$0{\cdot}28_8$
Collection H, from Wright, omitting civil wars	1·107	0·229	0·133	0·034	0·116	0·049	0·244

It is seen that there is a moderate degree of concord between the two Collections G and H. The larger duration of type 2 versus 2 in Collection G arises mainly from the struggle against Rosas of Argentina 1835–52. Wright lists this war as of type 2 versus 1 from 1839 to1852. The statistics of duration agree with those of numbers of wars in Collections A and C at least in so far as in both of them the type 1 versus 1 predominates, and the type 2 versus 1 comes next.

A gloomy feeling may be engendered by the fact that the total duration of wars of the type 1 versus 1 exceeded the historical interval in the ratio $1{\cdot}_3$. It is therefore desirable to point out * that the total duration of such wars might conceivably have been much larger than it actually was. For we can imagine that the 60 States in the world were arranged in 30 wars of type 1 versus 1, wars which went on all the time. If so, the ratio would be, not $1{\cdot}_3$, but 30.

PART II. THE SIMPLER ATTEMPTED EXPLANATIONS—NAMELY, THEORIES I TO X

General preface to particular theories

An accurate and detailed record of historical events is the necessary preliminary to any forecast of the future. Yet such a record is not, in its raw state, the best guide, for two reasons:—

(i) Every set of historical events is to be regarded as a sample, and subject to the variation of samples. This doctrine has already been admitted in so far as the observed distribution $n(r, s)$ has been replaced by a smooth empirical formula $\phi(r, s)$.

(ii) Circumstances change. Unless the events can be analysed back to recognizable causes, we cannot allow for changed circumstances, such for example as the death of leading persons, or the decrease in a population, or the increase of aviation, or the innovation of atomic energy.

Let us therefore now attend to explanations which are at least intelligible. They have various commendations: some have been popular, some are plausible at first glance, and some would simplify the mathematics. But one of the tests to be applied to them is whether they give a *quantitative* agreement with the observed $n(r, s)$ distribution, and this test is found to be remarkably critical.

The contrast between an empirical formula and a theory resembles the contrast in a factory between an old reliable foreman and an imaginative graduate fresh from the university. The graduate might at first make frightful blunders, but in the course of a decade he is likely to improve the technique more than the foreman ever could.

* I am indebted to my wife for doing so.

A general characteristic of the facts is that wars involving more belligerent groups were rarer than wars involving fewer belligerent groups. We may seek a hint leading towards explanation by noticing other phenomena which have the same characteristic. Amidst the unorganized activities in the playground of a Scottish elementary school, fights of one versus one were observed to be more frequent than other fights.*

In a gas, the collision of two molecules occurs much more frequently than the collision of three (Hinshelwood, 1940, p. 142). What have these diverse fields—the world, the playground, and the gas—in common? The answer seems to be: lack of organization. Those who supported the League of Nations often complained about the international chaos. The fights in the playground began in the absence of the teachers. In his standard treatise on *The Dynamical Theory of Gases* Jeans (1916) frequently mentioned the molecular chaos. One of our interpretive ideas should therefore be chaos, with its characteristic property that complicated events are rarer than simpler events. In statistical theory chaos is analysed somewhat in the following manner: the complicated events are regarded as built up from simpler elements, some of which are statistically independent, so that the probability of their happening together is the product of their probabilities of happening separately. The more such elements, the less resultant probability. This is certainly a leading idea; but it leaves us to seek historical elements that are statistically independent.

As a model of theoretical procedure let us bear in mind Bernoulli's classic expression for the binomial distribution, namely

$$\left\{\frac{k!}{x!(k-x)!}\right\}[p^x(1-p)^{k-x}]$$

for although it is not appropriate in detail to wars, it gives nevertheless the following useful hints: we should expect a factor, analogous to p^x, expressing the probability that x elements are in the war, and another factor, analogous to $(1-p)^{k-x}$, expressing the probability that many other elements keep out of the war; we may also expect a factor, analogous to that in { }, expressing the number of mutually exclusive wars for each of which the probability is the factor in [ ].

As the application of these general ideas of Bernoulli to the encounters between nations is impeded by the complications and the passions of politics, anyone who is conversant with physics might well prefer to begin by considering their application to the encounters between molecules in a gas. I worked that as a preparatory exercise (Richardson, 1946). There x is the number 2, 3, 4, 5, . . . , of molecules in the bunch, and $k - x$ is the enormous number of molecules that might be in, but are not. Yet at low densities p is so near to zero that $(1-p)^{k-x}$ is almost unity, and may be omitted, as in the classical theory of gases. An almost similar situation for nations occurs in the following Theory XI. On the contrary, for dense gases $(1-p)^{k-x}$ is an important factor. The corresponding factor for nations is important in Theory X.

The rejection of certain large probabilities

It often happens that the total duration of all the wars of a particular specification can be explained by two different values of p, one emphasizing factors like p^x, the other those like $(1-p)^{k-x}$. That is to say, the equation for p has more than one root. Here the overall summary of Woods and Baltzy is a guide. They found that eight large Powers were each at war on the average for ¼ of the time. It seems likely therefore that the average fraction of the time spent at war between any two *named* Powers was much less than ¼, and more likely to be of the order of 0·01. For this reason various much larger alternatives should be rejected.

Equations to be solved

The Bernoullian logic of probability, when brought into connection with historical fact, leads at simplest to many equations each of the type

$$p^b(1-p)^c = a \quad . \quad . \quad . \quad . \quad . \quad . \quad . \quad . \quad . \quad . \quad (1)$$

in which a, b, c, p and $1-p$ are each positive . . . . . . (2)

The unknown is p. As equation (1) is not a common feature in the textbooks, a few remarks on its solution may be in place here. The function

$$p^b(1-p)^c = f(p), \quad \text{say}, \quad . \quad . \quad . \quad . \quad . \quad . \quad . \quad . \quad . \quad (3)$$

has a maximum at $p = b/(b+c) = p_m$, say; . . . . (4)

* Personal information from a teacher, Mrs. C. E. Martin.

and the value at the maximum is

$$f(p_m) = b^b c^c/(b + c)^{b+c} \quad . \quad . \quad . \quad . \quad . \quad . \quad . \quad (5)$$

If $a > f(p_m)$, equation (1) has no real roots, and the proposed theory receives a knock-out blow . . . . (6)

If $a < f(p_m)$, equation (1) has two real roots p_1, p_2, which for definiteness we may dispose thus

$$0 < p_1 < p_m < p_2 < 1 \quad . \quad . \quad . \quad . \quad . \quad . \quad . \quad . \quad (7)$$

Two methods are available for finding these roots.

By nomogram

Let $$\log_{10} p = \xi, \quad \log_{10}(1 - p) = \eta \quad . \quad . \quad . \quad . \quad . \quad . \quad . \quad . \quad (8)$$

where ξ and η are Cartesian co-ordinates. The identity

$$p + (1 - p) = 1 \quad \text{is then equivalent to} \quad 10^\xi + 10^\eta = 1 \quad . \quad . \quad . \quad . \quad (9)$$

This represents a fixed curve, independent of a, b, c. This curve is plotted once for all. The equation (1) is equivalent to

$$b\xi + c\eta = \log_{10} a, \quad . \quad . \quad . \quad . \quad . \quad . \quad . \quad . \quad (10)$$

which represents the straight line joining the point $\xi = 0$, $\eta = c^{-1}\log_{10} a$ to the point $\xi = b^{-1}\log_{10} a$, $\eta = 0$. Where this straight line (10) cuts the curve (9), there the equation (1) is satisfied. Let the abscissæ of the points of intersection be ξ_1 and ξ_2, the former being not the greater. Then $p_1 = 10^{\xi_1}$, $p_2 = 10^{\xi_2}$ Unequal scales for ξ and η were taken, so as to suit the prevalent values of b and c in the set of equations.

Successive approximations. We can nip each root between numbers approaching it from opposite sides. For it can be proved that any sequence $l_0, l_1, \ldots, l_n, \ldots$, which obeys the recurrence formula

$$l_{n+1}^b(1 - l_n)^c = a \quad . \quad . \quad . \quad . \quad . \quad . \quad . \quad . \quad . \quad (11)$$

converges to the lesser root p_1 from below provided that

$$0 \leqslant l_0 < p_1 \quad . \quad . \quad . \quad . \quad . \quad . \quad . \quad . \quad . \quad . \quad (12)$$

and converges to the same root from above provided that

$$p_1 < l_0 < p_m \quad . \quad . \quad . \quad . \quad . \quad . \quad . \quad . \quad . \quad . \quad (13)$$

Also any sequence $u_0, u_1, \ldots, u_n, \ldots$, which obeys the recurrence formula

$$u_n^b(1 - u_{n+1})^c = a \quad . \quad . \quad . \quad . \quad . \quad . \quad . \quad . \quad . \quad (14)$$

converges to the greater root p_2 from below or above according as

$$p_m < u_0 < p_2 \quad \text{or} \quad p_2 < u_0 \leqslant 1 \quad . \quad . \quad . \quad . \quad . \quad . \quad (15)\ (16)$$

A table of subtraction-logarithms shortens the computation.

The proofs in the four cases (12), (13), (15), (16) are much alike. For example, take the case (13). In that range of p—namely, $p_1 < p < p_m$—the function $p^b(1 - p)^c$ increases with p; so that we have from (1), (4), (11),

$$p_1^b(1 - p_1)^c = a, \quad l_0^b(1 - l_0)^c > a, \quad l_1^b(1 - l_0)^c = a \quad . \quad . \quad (17)\ (18)\ (19)$$

From (17) and (19) $$(l_1/p_1)^b\{(1 - l_0)/(1 - p_1)\}^c = 1$$

but by (13), $$(1 - l_0)/(1 - p_1) < 1.$$

Therefore $$p_1 < l_1 \quad . \quad . \quad . \quad . \quad . \quad . \quad . \quad . \quad . \quad . \quad (20)$$

Again from (18) and (19) $$l_1 < l_0 \quad . \quad . \quad . \quad . \quad . \quad . \quad . \quad . \quad . \quad . \quad (21)$$

Similarly at the next stage $$p_1 < l_2 < l_1 < l_0.$$

By induction the sequence (l_n) is monotonic decreasing, is bounded below by p_1, and has no greater lower bound . . . . (22)

Therefore p_1 is its limit.

Example: to find the greater root of $p(1 - p)^{86} = 0{\cdot}0007$. The nomogram gave $p_2 = 0{\cdot}048$ roughly. A sequence controlled by (14) began

$$u_0 = 0{\cdot}048, \quad u_1 = 0{\cdot}047973, \quad u_2 = 0{\cdot}047966.$$

Another sequence, also controlled by (14), began

$$u_0' = 0{\cdot}04790, \quad u_1' = 0{\cdot}04795, \quad u_2' = 0{\cdot}047961.$$

Therefore $$0{\cdot}047961 < p_2 < 0{\cdot}047966.$$

Symmetry. Confusions might easily occur between the octanta and quadrantal arrangements, and would ruin any theory. To prevent them, all quantities peculiar to the octantal arrangement have a star affixed. This new convention does not change the meanings of n and ϕ, for they were defined over the quadrant, but we now have an octantal n^* connected with n thus:—

$$n^*(r, s) = n(r, s) + n(s, r) \quad \text{if} \quad r \neq s$$
$$n^*(r, r) = n(r, r)$$

and similarly for ϕ^* and ϕ. It is n^* which is taken directly from history.

Various general remarks.

In the collections of historical facts the number of belligerent groups has been counted without any attention to their heroism or wickedness, and with attention to their populations only in so far as groups of insignificant population were ignored altogether. It would be of no avail to look in the statistics for distinctions of ethics or of power which had been sieved out by the method; and therefore it will be best to refer to particular states incognito as A, B, C, D . . . instead of by their usual names. All that we can hope thus to make is an analysis of the average motives of humanity in general, but the probability attached to "any particular nation," such as A or B or C or D, is usually quite different from the probability attached to "some unspecified nation," because the latter designation leaves open more choice.

The foregoing general principles can be applied in company with various special assumptions. Strongly held opinions as to the causes of wars may be heard or read on all sides. Yet I had difficulty in finding any hypotheses which would give a *quantitative* agreement with the historical facts in Collections A, C, E, F, G, H. It seemed best therefore to begin with some theories which can easily be refuted, then to introduce improvements, and so gradually to approach realism.

Theory I. The decisions of so-called "free, sovereign and independent" States are statistically independent.

States have been in the habit of claiming more freedom, sovereignty, and independence than they ever in fact possessed. For a belligerent can hardly be at war by itself alone; and States which have been threatened have mostly felt compelled to defend themselves. The assumption "let p be the probability that any state is at war" is self-contradictory; for if one State is at war, another must be.

Theory II. The wars in the world were mainly due to aggression by Germany, Italy and Japan

This theory should perhaps be mentioned because it was widely believed in Britain during the years 1936–44. It will not, however, describe the facts over the longer interval 1820–1939, as may be seen from any reasonably full list of the names of belligerents—for example, Wright's list.

Theory III. The facts can be explained by the well-known tendency to join the winning side

That tendency can indeed explain the fact that, in the octantal plan, $n^*(3, 1) > n^*(2, 2)$. For if a war of 2 versus 1 had begun, a fourth belligerent would on this theory tend to join the larger group making 3 versus 1, not 2 versus 2. Similarly the fact that $n^*(4, 1) > n^*(3, 2)$ can be explained, but this theory fails utterly to explain why wars of 1 versus 1 are the commonest type. Any general tendency to join in fights must evidently be restrained by other motives. Humanity in general is not like the Irishman who is said to have asked: "Is this a private fight, or may anybody join?"

Theory IV. The sense of justice is universal

It can be argued that wars tend to be of the type many-versus-one because justice is a widely venerated ideal, that shines above national prejudices, and so leads many nations to combine against the transgressor, who is thus left isolated. To complete the theory we have also to suppose that transgression breaks out sporadically in different places at different times, as Ranyard West has emphasized (1942). Some at least of the kaleidoscopic patterns of nations-at-war can be so explained; for example, the coalition against Napoleon in 1815. Then in 1854 the Turks, French, British and Piedmontese combined, with the approval of Austria and Prussia, against the Russians, not from any permanent affinity with one another, but because they were all indignant at the Czar's unreasonable aggressiveness. In the next war between the Russians and the Turks, 1877–78, it was the turn of the Turks to be isolated, and to be opposed by many; this time by Russians, Rumanians, Serbs, Montenegrins, Bosnians and Herzegovinians, who were all indignant about the extortions by Turkish officials and the atrocities committed by Turkish soldiers.

On this view we should not lament the fickleness of alliances, saying that it is sad evidence of the untrustworthiness of friends; instead we should rejoice over the changing scene, because it reveals the control by the higher ideal of justice. This theory undoubtedly contains some truth, but it is nevertheless an exaggeration. For if the sense of justice had been strictly universal, there would not have been any transgressors; and if only rare lapses from justice had occurred, we may reasonably suppose that nearly all wars would have been of the police-action type: society versus a criminal, many-versus-one. The United Nations Organization designed at San Francisco in 1945 is suited to a future world of that character. In the past, however, most wars were of the type one-versus-one, which suggests ordinary quarrelling, rather than police-action. We must therefore admit that the sense of justice abroad in the world was flickering, confused, local, and fallible; so that its quantitative results must be represented by probabilities.

Theories V, VI, VII, VIII

We now come to a group of theories all of which ignore geographical restrictions. They are therefore more appropriate to a future age of aviation than to the past age of land and sea forces, but, because they are simpler mathematically than those theories which express geographic restrictions, it is convenient to discuss them first.

Theory V. Mutually exclusive wars are equally probable

The number of combinations that can be formed from m nations or other possibly belligerent groups, r at a time, is $m!/\{r!(m-r)!\}$. That having been done, there remain $m-r$ possibly belligerent groups, and from them the number of combinations s at a time is $(m-r)!/\{s!(m-r-s)!\}$. The number of arrangements in wars jointly of the types r versus s and s versus r is the product of these two expressions—namely, $m!/\{r!s!(m-r-s)!\}$ provided $r \neq s$, but if $r = s$ we have, in the above argument, counted each arrangement twice. Let A, with suitable attached symbols, denote the number of distinct arrangements. Then in the octantal plan

$$A^*(r, s) = m!/\{r!s!(m-r-s)!\} \text{ if } r \neq s,$$

but

$$A^*(r, r) = \tfrac{1}{2}m!/\{r!r!(m-2r)!\}.$$

It would be troublesome thus to have different formulæ according as $r = s$ or $r \neq s$. The quadrantal plan relieves us of this difficulty because

$$A(r, s) = A(s, r) = \tfrac{1}{2}A^*(r, s) \text{ if } r \neq s$$

and so

$$A(r, s) = \tfrac{1}{2}\frac{m!}{r!s!(m-r-s)!} \text{ for both } r = s \text{ and } r \neq s \quad . \quad . \quad (1)$$

Example, $m = 4$. States named A, B, C, D. The possible wars of the type 2 versus 2 are A and B versus C and D, A and C versus B and D, A and D versus B and C; three distinct wars and no more, agreeing with $A^*(2, 2) = 4!/(2!2!0!2)$. Again, the possible wars of the types 3 versus 1 and 1 versus 3 are jointly A and B and C versus D, B and C and D versus A, C and D and A versus B, D and A and B versus C; four distinct wars and no more. This is the octantal method of counting and it agrees with $A^*(3, 1) = 4!/(3!1!0!)$.

If r and s are given, $A(r, s)$ is the number of *mutually exclusive* wars. The probability of the whole set of them is the *sum* of their separate probabilities. Thus the expression (1) for $A(r, s)$ is likely to occur in many theories. It is, however, an embarrassment to the theorist, because its isopleths have the wrong curvature. To see this, let us attend to the sloping straight lines on which $r + s$ is constant. On such a line $A(r, s)$ varies as $(r!s!)^{-1}$. The cases which occur most frequently are:—

$r+s$	Successive values of $1/(r!s!)$ along the line $r+s$ = constant								
4			1/6		1/4		1/6		
5		1/24		1/12		1/12		1/24	
6	1/120		1/48		1/36		1/48		1/120

These fractions are greatest in the middle and least at the extremes, and so are distributed in a manner opposite to the historical facts.

Moreover, $A(s, s)$ is greatest near $s = m/3$. That is to say, if $A(s, s)$ were the only consideration, and if $m = 60$, then wars of 20 versus 20 would be the mode; whereas in fact such a war has never occurred in the interval studied.

It is thus abundantly evident that there must be a restraining factor. Guided by the Bernoullian model, we may expect it to be the probability of a war of the type r versus s between *particular* States. This factor will need to overcome the tendency of $A(r, s)$ to give isopleths hollow on the wrong side.

Theory VI. The behaviour of any pair of States is statistically independent of the behaviour of every other pair of States

We have not here the crude absurdity of Theory I; for there can be war inside the pair of States A and B without necessarily involving any other pair. The historical statistics in Collections D and E, moreover, emphasize the importance of the number, rs, of pairs of opposed belligerents, because $n(r, s)$ is almost a function of the single variable rs. The present theory asserts that the probabilities, that the different pairs are pairs of enemies, combine by multiplication. This theory accordingly explains the fact that $n(r, s)$ decreases as rs increases. Whether the theory will give the observed functional form of the decrease is a further question, which will be examined.

In particular, the tendency to join the winning side, which was the basis of Theory III, can equally well be expressed as a reluctance to form pairs of opposed belligerents. For example, if a war of 2 versus 1 is going on, a fourth belligerent would make one more such pair by converting the war to 3 versus 1, but would make two more such pairs by converting the war to 2 versus 2.

Although Theory VI has these merits, there is abundant historical evidence against it in the form of relations between different pairs. The most obvious of these is an indirect result of *alliance* between States. For example, in 1914, when Britain went to war with Germany, the ancient British alliance with Portugal began to lead Portugal into the war in which otherwise she had no great interest (E **18**, 283). For our present purpose the point is that war inside the pair Britain–Germany caused war inside the pair Portugal–Germany. Another relation between pairs was that of *distraction*. For example, in 1857, conflict inside the British–Indian pair delayed hostilities inside the British–Chinese pair (E. **5**, 536) for some British troops on their way to China were diverted to India. Again for example, Bismarck, after organizing the defeat of France in 1871, encouraged her colonial enterprises in Africa and Indo-China in order to distract her from revenge (E **3**, 665). A third relation between pairs has arisen when a State having armed forces ready for one purpose has taken the opportunity to use them for another. This may be called *opportunism*. For example, in 1858, a French fleet, ready for war with China, bombarded a port in Annam (Priestley, 1938, p. 114). That is to say war inside the French–Chinese pair facilitated fighting inside the French–Annamese pair.

We must therefore amend Theory VI.

Theory VII. Because of the fickleness of alliances and the flitting of distraction and opportunities, different pairs of States behaved in a manner statically independent of one another when the facts are collected from an interval of at least a century

Quincy Wright (1942, p. 774) remarks that:

> " Alliances and confederations intended to be permanent have seldom proved reliable unless carried to the point of federation, transferring much of legal sovereignty and the conduct of external affairs to the central organs. Such a development has seldom been possible unless geographic and cultural factors have conspired to unite the group. Alliances purely for defence have broken up if the State against which they are directed ceases to be menacing."

The objections already stated against Theory VI are for the most part not valid against Theory VII. The latter therefore is worthy to be compared with fact in more detail. This, however, requires additional hypotheses, which will now be stated.

Theory VIII. A simple chaos uniform among sixty nations

This is a special case of Theory VII and is based on precise assumptions. The probability of an event is regarded as the sum of the fractions of a long historical time-interval during which the event occurred. There are assumed to be m possible belligerents in the world. Some are States, some insurgents. For brevity they are all called nations . . . . (1)

Every schoolboy knew that his own nation was remarkably different from, and better than, the rest. Yet as our present study is of mankind in general let us, in Theory VIII, try-out the simplest assumption—namely, that the probability of war between any two nations is:—

the same at all times, ... (2)
the same for all pairs of nations, ... (3)
independent of any other wars that are going on; . . . . (4)
and let us denote this constant probability by p . . . . (5)

These assumptions specify a hypothetical world which is simpler and more chaotic than the actual world. So they should do; for the art of beginning theories is to leave out of consideration all but the most essential features.

Consider a war between r nations named $A, B \ldots$ on one side, versus s nations named $Z, Y, \ldots$ on the other. This war involves rs pairs of opposed nations. The probability of war inside each pair is p by (2), (3), (5); and by (4) the pairs are statistically independent. The probability of war simultaneously inside rs pairs is therefore p^{rs} . . . . (6)

It is essential to the specification that no other nations should be in this war, but the specification ought not to exclude quite disconnected, though simultaneous, fightings. Thus the specification does not require all the remaining $\frac{1}{2}m(m-1) - rs$ pairs to have peace inside them. The notion of disconnection may, in practical politics, involve the causes of the quarrels, which have no representative in the present abstract theory. Nevertheless we can represent the usual kind of isolation of the conflict by assuming that: The nations $A, B, \ldots$ are not at war with any other nations except the nations $Z, Y, \ldots$; and vice versa . . . . (7)

That is to say, there are $(m - r - s)$ nations not at war with the s, nor with the r nations. Altogether they form $(r + s)(m - r - s)$ pairs which have peace inside them. The probability of peace inside any pair is $(1 - p)$. So the probability of peace between the belligerents and the neutrals is $(1 - p)^{(r+s)(m-r-s)}$ . . . . (8)

in the specified war, whatever wars the neutrals may have between themselves. The probability of the war of the r specified nations versus the s specified nations is, from (6) and (8) equal to

$$p^{rs}(1 - p)^{(r+s)(m-r-s)} \quad \ldots\ldots\ldots\ldots \quad (9)$$

They could at most be at war during the whole time-interval. So expression (9), being the ratio of their time at war to the whole time-interval, lies between 0 and 1 in the manner characteristic of a probability.

The number of mutually exclusive sets of specified nations of the type r versus s, was shown in connection with Theory V to be, in the quadrantal plan,

$$A(r, s) = \tfrac{1}{2}m!/\{r!s!(m - r - s)!\} \quad \ldots\ldots\ldots \quad (10)$$

Each of these sets is at war during the fraction (9) of the time. So the total duration of the wars of this type between unspecified nations, when divided by the historical interval, is expected to be

$$\tfrac{1}{2}\frac{m!}{r!s!(m - r - s)!}p^{rs}(1 - p)^{(r+s)(m-r-s)} = Q(r, s), \quad \text{say} \quad \ldots \quad (11)$$

This $Q(r, s)$ can exceed unity, and so is not a probability.

The number m of nations in the world averaged about 60 . . . . (12)

For the types of war which occurred most frequently in history we obtain from (11), on inserting $m = 60$,

$$Q(1, 1) = 1770p(1 - p)^{116} \quad \ldots\ldots\ldots \quad (13)$$

$$Q(2, 1) = 51330p^2(1 - p)^{171} \quad \ldots\ldots\ldots \quad (14)$$

The first members of these equations are given by the historical facts in Collections G and H, according to their respective definitions of "war." For the type 2 versus 1 it is necessary to divide the octantal observed numbers by two, so as to make them comparable with the theory, which is quadrantal. The equations are then soluble for p. The two roots, p_1 and p_2, were found by the processes already described, and are set out in the following table.

Type of war	Collection of facts	p_1	p_2
1 versus 1	G	0·0009	0·031
	H	0·0007	0·034
2 versus 1 / 1 versus 2	G	0·0025	0·032
	H	0·0017	0·037

. . . . . . (15)

Theory VIII will be valid if the various historical data can all be explained by the same number for p, within the uncertainty of sampling. Only one of p_1 and p_2 need be considered; but we do not know which. So far, p_2 looks the more consistent, but before we can come to any conclusion, we need to consider wars of other types. The most reliable of the historical facts is the total duration of wars of all types. This total, however, is not convenient for comparison with theory, because the theoretical $\sum_{r=1}^{r+s=60}\sum_{s=1} Q(r, s)$ is troublesome to compute, even when p is given numerically.

On the contrary, the two simplest of the theoretical relations are

$$Q(2, 2)/Q(3, 1) = \tfrac{3}{2}p, \quad Q(3, 2)/Q(4, 1) = 2p^2 \quad . \quad . \quad . \quad (16)\ (17)$$

Their great advantage is that they do not involve m. They depend on observations which are rare and correspondingly erratic. Let us therefore take the longer historical interval—namely, Wright's, as represented in Collection H. After converting the data to the quadrantal plan, we obtain:—

$$Q(2, 2)/Q(3, 1) = 2 \times 0{\cdot}03/0{\cdot}13 \quad \text{whence, by (16),} \quad p = 0{\cdot}31$$
$$Q(3, 2)/Q(4, 1) = 0{\cdot}05/0{\cdot}12 \quad \text{whence, by (17),} \quad p = 0{\cdot}46.$$

The discrepancy between 0·31 and 0·46 is not interesting, for it can be attributed to the random occurrence of wars of these rare types, but we must take seriously the mean result that p is of the order of 0·38 . . . . (18)

This is quite inconsistent with either p_1 or p_2 as found above in (15) by putting $m = 60$ together with the observed values of $Q(1, 1)$ and $Q(2, 1)$. Moreover, $p = 0{\cdot}38$ is quite incredible; for it asserts that in a very long historical interval each of the m nations was at war with each of the other $(m - 1)$ nations for 0·38 of the time. The world was, in fact, certainly not so war-ridden, if $m = 60$.

Thus Theory VIII has crashed. No one will be surprised that a theory, which began by assuming all nations to be statistically alike, should have come to a bad end; but the precise manner of its failure could hardly have been foreseen.

Theory IX, being the Theory VIII modified by the assumption that only a few of the nations were involved in wars

We might argue thus: That incredibly large value $p = 0{\cdot}38$ may have come from an aberrant sample of history. Without violence to the facts we could replace it by $p = 0{\cdot}2$. That, moreover, would be less incredible if only a few bellicose nations were concerned. Also by reducing m from 60 to say 10 or 6 we could raise the p obtained from the wars of types 1 versus 1 and 2 versus 2 up towards 0·2. As thus amended, the theory would be much more self-consistent.

We meet, however, difficulties fatal for Theory IX when we try to *name* the bellicose minority and the non-belligerent majority. In fact it is impossible to name them, because belligerency was too widespread. Suppose, for example, that we choose as notable belligerents the ten nations: Britain, France, Russia, Turkey, China, Piedmont and its successor Italy, Prussia and its successor Germany, Spain, U.S.A., Japan.

Of the 91 wars that ended from A.D. 1820 to A.D. 1939 according to Richardson's list, 13 wars did not involve any one of those 10 notable belligerents. Also the total number of belligerents' names in those 91 wars, including names of insurgents, but not counting any name twice, was 108 —far in excess of ten.

Thus Theory IX fails, but it leaves us with the hint that although many nations were involved, yet the probability p was related to only a few of them at a time. That suggested the next theory.

Theory X. Localized quarrels

Most wars have been localized. Litvinov's doctrine that "Peace is indivisible" was announced to the League of Nations. It is commonly believed in 1946 that, in view of the new United Nations Organization, future threats of war will usually concern the whole world; but that was certainly not so prior to 1918.

Let us therefore try the theory that disputes occurred at random over the world; and that each dispute concerned only l possible belligerents, where l is much less than 60, and may be more like the average observed number, 5 or 6, of geographical neighbours.

In Theory VIII the elementary probability denoted by p was the fraction of a very long historical time during which the two specified nations were at war with one another. In Theory X we have again an elementary probability, but it has a different meaning. It is now the probability that the pair of nations went to war with one another about a particular dispute, and it will be denoted by θ. The conclusions of Theory X will therefore be comparable, not with the total durations of wars, but with the numbers of wars. With those modifications the argument of Theory VIII is so far applicable that we obtain a formula like that for $Q(r, s)$, but with l replacing m and θ replacing p, thus

$$\frac{1}{2}\frac{l!}{r!\,s!\,(l-r-s)!}\theta^{rs}(1-\theta)^{(r+s)(l-r-s)} = U(r, s), \quad \text{say} \quad . \quad . \quad . \quad . \quad (1)$$

Here $U(r, s)$ is not a total duration, but is proportional, for variations of r and s, to the probability of a war about a particular dispute. To obtain a number equal, instead of merely proportional, to the probability we should have to divide $U(r, s)$ by $\sum_{r=1}^{r+s=l}\sum_{s=1} U(r, s)$. But, as the double sum is troublesome to calculate, it is preferable to use proportionalities.

Let us now, if possible, choose l and θ so as to make $U(r, s)$ agree with the historical facts. As in Theory VIII, the two simplest of the theoretical results obtainable from (1) are that

$$U(2, 2)/U(3, 1) = \tfrac{3}{2}\theta, \quad U(3, 2)/U(4, 1) = 2\theta^2 \quad . \quad . \quad . \quad . \quad . \quad (2)\ (3)$$

for they do not involve l. On referring to the octantal collections A and C of numbers of wars and making the corrections to the quadrantal plan, we obtain the following:—

	From Collection A	From Collection C	
U(2, 2)/U(3, 1) =	6/5	8/12	(4)
therefore, by (2), θ =	0·8	0·44	
U(3, 2)/U(4, 1) =	2/5	6/12	
therefore, by (3), θ =	0·45	0·50	

The mean of these four values of θ is 0·55. The objection, previously noted, to large values of the probability p, does not hold against large values of the probability θ, for the unspecified probability of a dispute intervenes. But as the counted numbers of these wars are small, there is much uncertainty due to sampling. As a deviation from the mean to $\theta = 0{\cdot}80$ occurs in the sample, we cannot reject the possibility of an equal but opposite deviation to $\theta = 0{\cdot}30$.

An upper bound to θ is provided by the observed fact that wars of the type many-versus-one have been rare in comparison with those of the type one versus one. It follows from (1) that

$$\frac{U(l-1, 1)}{U(1, 1)} = \frac{1}{l-1}\left\{\frac{\theta}{(1-\theta)^2}\right\}^{l-2} \quad . \quad . \quad . \quad . \quad . \quad . \quad . \quad (5)$$

To agree with history, the first member must be near to zero when l is 7 or more. This can occur if $\theta < (1-\theta)^2$, but not if $\theta > (1-\theta)^2$. The equality $\theta = (1-\theta)^2$, which separates these two cases, is satisfied by $\theta = 0{\cdot}382$, or by $\theta = 2{\cdot}618$, of which the former only can be a probability. Accordingly θ *must be less than* 0·382 . . . . (6)

The conditions (4) and (6) pull θ opposite ways, so as to restrict it to a range between about 0·30 and 0·38. The value $\theta = 0{\cdot}35$ will be taken as typical of this range.

The simplest method for estimating l is to deduce from (1) that

$$2U(2, 1)/U(1, 1) = (l-2)\theta(1-\theta)^{l-5} \quad . \quad . \quad . \quad . \quad . \quad . \quad . \quad (7)$$

The first member was taken from the counted numbers of wars and the equation was then solved for l to the nearest integer, rejecting irrelevant roots between $l = 2$ and $l = 3$.

	A			C		
Collection of wars		A			C	
2U(2, 1)/U(1, 1)		24/42			28/117	
Assumed θ	0·3	0·35	0·4	0·3	0·35	0·4
Therefore l	8	8	8	12	11	10 . (8)

Remembering that l is the number of nations, or other possibly belligerent groups, which are concerned in each quarrel, we see at once that a theory which restricts l to be 12 or less cannot explain world-wars. But world-wars have been very exceptional. Thus $l = 11$ would exclude only 8 wars out of the 200 in Collection C; and $l = 8$ would exclude only 2 wars out of the 91 in Collection A. It is worth while, therefore, to make a further comparison between Theory X and the distribution of ordinary wars.

The most weighty of the present set of historical facts remains to be considered. It is the ratio of the total number of wars to the number of wars of type 1 versus 1. The theoretical ratio was obtained from formula (1) by laborious arithmetic after inserting for θ and l numbers that had been indicated as approximately suitable by the foregoing simpler methods. For comparison with these results of Theory X it seemed proper to truncate the Collections A and C by omitting from them the small minority of wars for which the theory offers no explanation—namely, wars involving more than l belligerents. The comparison is to be seen in the following table. The numbers in the block common to the last three columns and the lower four rows are ratios of total numbers of wars to the number of wars of type 1 versus 1.

l		8	9	10
$\frac{1}{U(1,1)} \sum_{r=1}^{r+s=l} \sum_{s=1} U(r, s)$	$\theta = 0{\cdot}35$	2·2910	1·8695	1·5635
	$\theta = 1/3$	2·2004	1·8346	1·5600
Truncated collections of facts A		$2{\cdot}1_2$	$2{\cdot}1_2$	$2{\cdot}1_4$
C		1·62	1·63	1·64

The resemblances between theory and fact in the above table are sufficient to make worth while further comparisons by way of the χ^2 test due to K. Pearson and R. A. Fisher. For this purpose the data were grouped in cells according to the number, rs, of pairs of opposed beliiegerents, thus:—

rs	1	2	3, 4, 5	>5	Total
Collection A, truncated	42	24	15	8	89
Theory X with $l = 8$, $\theta = 0{\cdot}35$ ...	38·85	22·40	18·80	8·95	89·00

It follows that $\chi^2 = 1{\cdot}24$. Three parameters have been adjusted—namely, the total and, less accurately, θ and l. So, as there are four cells, only one degree of freedom remains. Yule's table (Kendall, 1943, p. 445) gives the probability that χ^2 would exceed 1·24 to be 0·27; which indicates that *Theory X passes the χ^2 test for agreement with the truncated Collection A.* There may be some readers who mistrust the hypotheses on which Theory X is founded; but even they must admit that it has provided respectable formulæ, which, however, they will have to call empirical.

On the contrary, a satisfactory fit between Theory X and Collection C appears to be impossible for any values of θ and l. Here is the best of my attempts:—

rs	1	2	3, 4, 5	>5	Total
Collection C truncated	117	28	31	15	191
Theory X with $\theta = 0{\cdot}35$, $l = 10$ ...	122·16	39·69	15·91	13·24	191·00

From this table χ^2 comes to 18·2; and the probability that a larger value should occur by chance in the historical sample is less than 0·0001. I cannot offer any precise explanation of why Theory X should fit Collection A but not Collection C. The empirical formula $\phi(r, s) = \frac{5}{8}N(rs)^{-2\cdot5}$ fitted both of them. But of course I prefer the definition of war on which Collection A is based.

Summary on Theory X. With the exception of a small minority of very complicated wars, Theory X succeeds in explaining, not particular wars nor the behaviour of named States, but the average behaviour of the world over an interval of 120 years or more. Its hypotheses are of the simplest. It is supposed that disputes occur at random over the globe; that each dispute interests eight nations (or other belligerent groups) and no more than eight; that the probability of war about a dispute between every pair formed from those eight nations is 0·35, the same for every pair, the same for every dispute, and the same in every year. It is remarkable that such simple hypotheses lead to a quantitative agreement between the theory and the observed facts. But it is well to remind ourselves that a theorem does not prove its converse. Here, in particular, the observed facts do not prove that the aforesaid probability was always the same.

PART III. THE INFLUENCE OF GEOGRAPHY AND OF INFECTIOUSNESS

Multitude of contacts. Sea-powers counted separately

In order to amend Theory X so as to include world-wars, let us attend to the fact that some States had much more numerous contacts than others, and therefore more opportunities either for friendship or for quarrelling. We need both a collection of historical facts, appropriately classified, and an abstract theory of the probabilities. They will be stated in turn.

Collection J of facts

Aviation is now tending to put every nation into contact with every other, but for the interval A.D. 1820 to A.D. 1929, the opportunities for quarrelling were greatest for the sea-powers. Contact across land-frontiers had a similar effect, but not to the same extent. For whereas considerable power at sea put its possessor into contact with about forty sovereign States, the greatest number of land frontiers possessed by any other State was eleven for Brazil. It is thus proper to attend to the world-wide sea-Powers as a class apart; but coastal defence should here be classed along with the defence of land frontiers.

No sharp distinction between ubiquitous and local Powers existed in fact, for Portugal, Spain, Holland and Belgium had distant colonies. Yet, for the purpose of comparing fact with theory, a meaningful dichotomy can be made by decision; provided that the same classification is made in the facts, as in the theory. After consulting the comparative tables published by Q. Wright (1942, pp. 670, 671), together with articles on some national navies in the *Encyclopædia Britannica*, I took the many-contact States to be as follows: for the whole interval A.D. 1820 to A.D. 1929, Britain, France, Russia; from 1870, U.S.A.; from 1880, Italy; from 1900, Japan; from 1895 to 1918, Germany; from 1910 to 1918, Austria-Hungary. The number of such powers was thus three from 1820 to 1869, then rose gradually to a maximum of eight from 1910 to 1918, and afterwards settled at six for the rest of our interval, which ends with the year 1929. The mean number was 4·6.

The notation " r versus s " now becomes inadequate, for both r and s must be partitioned. Let the new notation be " r_1, r_2 versus s_1, s_2 ". Let the connection with the former notation be that

$$r_1 + r_2 = r, \qquad s_1 + s_2 = s.$$

Let r_1 and s_1 be the numbers of the many-contact powers defined by the above list; so that r_2 and s_2 are the numbers of any other belligerents.

The classification of wars has thus become four-dimensional. Distinctions of symmetry remain essential, for otherwise factors of 2 or $\frac{1}{2}$ will enter by mistake. The two-dimensional ideas " quadrantal " and " octantal " require to be generalized; yet we lack familiar words applicable to four dimensions. The distinctions can however be made plain in the special notation. Thus, for example, some historical facts are set beside the symbol (0, 1 versus 0, 1) because they belong to that subtype alone, but other facts are set beside the two symbols (1, 0 versus 0, 1) and (0, 1 versus 1, 0) because they belong to those subtypes, not severally, but jointly.

The following table was abstracted from Richardson's unpublished list of the wars that ended during the 110 years from A.D. 1820 to A.D. 1929, for all magnitudes greater than 3·5. The interval is here cut off ten years earlier than the final date for Collection A in order to avoid the larger effects of air-power. The statements in the last two columns are only loosely connected with one another. For the total durations refer strictly to one subtype, and so may include portions of more complicated wars; whereas in the third column each war is counted as a whole after being classified by the subtype of its most complicated phase.

In comparison with Collections A, C, G and H the present more detailed classification has many more empty compartments. They occur even among the subtypes of 2 versus 1, 3 versus 1, 2 versus 2. Any probability-theory of occurrences must take account also of non-occurrences. We

Collection J

Type r versus s	Subtypes r_1, r_2 versus s_1, s_2	Number of wars attaining that subtype	Total duration of subtype in years
1 versus 1	0, 1 versus 0, 1	22	110·4
	1, 0 versus 0, 1 } 0, 1 versus 1, 0 }	13	37·7
	1, 0 versus 1, 0	1	1·5
2 versus 1 and 1 versus 2	1, 1 versus 0, 1 } 0, 1 versus 1, 1 }	12	17·3
	1, 0 versus 0, 2 } 0, 2 versus 1, 0 }	5	15·8
	0, 2 versus 0, 1 } 0, 1 versus 0, 2 }	5	$14{\cdot}5_5$
	2, 0 versus 0, 1 } 0, 1 versus 2, 0 }	1	2·5
	1, 1 versus 1, 0 } 1, 0 versus 1, 1 }	1	$0{\cdot}2_5$
	2, 0 versus 1, 0 } 1, 0 versus 2, 0 }	0	0·0
3 versus 1 and 1 versus 3	0, 3 versus 0, 1 } 0, 1 versus 0, 3 }	1	8·5
	0, 1 versus 1, 2 } 1, 2 versus 0, 1 }	3	2·8
	2, 1 versus 1, 0 } 1, 0 versus 2, 1 }	0 *	$0{\cdot}7_5$
	2, 1 versus 0, 1 } 0, 1 versus 2, 1 }	1	0·4
	The other subtypes	0	0·0
2 versus 2	0, 2 versus 0, 2	2	14·2
	1, 1 versus 1, 1	1	0·3
	The other subtypes	0	0·0

* There was no war of sub-types 2, 1 versus 1, 0 and 1, 0 versus 2, 1 at its most complicated phase; and yet 0·75 year of the Crimean war of 1853–56 belonged to those subtypes.

could not, for example, discuss the probability of the First World War unless we distributed its occurrence over some of the many empty surrounding compartments, by a process of smoothing in four dimensions; and the choice of a suitable process would be a question of considerable difficulty. As an alternative, the following discussion is restricted to the parts of the distribution which show a tolerable regularity. There is evidently a considerable correlation between the total durations and the numbers of wars in the above table.

Inspection of the above Collection J shows at once that wars involving *only* the long range sea-Powers have been remarkably rare. There is only one in the table—namely, the Russo-Japanese war of 1904–1905. That may be because such powers were few in number, or because they were cautious of attacking anyone of like strength, or because they were peace-loving. We cannot say which, unless we have a probability-theory. One will now be offered.

Theory XI. Chaos restricted by a formalized geography

If we were to attempt to take account of all the features of the political map, we should get no further. Nevertheless a search on atlases brought to notice a few facts which suggest the basis for a theory:—

There is not any point where four independent States meet one another, nor any point where three such States meet the sea. This is a remarkable fact . . . . (1)

The number of different governments with which a State was in contact by land has ranged from 0 for Japan to 11 for Brazil, with a mean 5·4 when taken over 33 persistent States . . (2)

There is some, rather loose, connection between political geography and Euler's theorem that $f - e + v = 2$, where f, e, v are the numbers of faces, edges and vertices of a polyhedron. To make the polyhedron resemble the political map, each vertex must be shared between three faces, so that $3v = 2e$. It follows that the average number of edges surrounding a face is $6 - 12/f$. If $f = 60$ this comes to 5·8 edges, not far from the mean of 5·4 mentioned in (2). But the political map differs from the polyhedra as to curvature, lack of closure, single frontiers, encirclement and detached portions. Euler's theorem moreover represents a delicate compensation which is easily upset. It is therefore preferable to lay aside Euler's theorem and to base Theory XI on robust assumptions among which the fact (1), being simple, will be included as it stands; but the facts (2), being complicated, will be replaced by the following assumption (3).

Let our formalized world consist of:—

γ land-locked countries each touching five different States.

β local coastal countries each touching the sea, also touching five different States by land.

α world-wide sea-Powers which are in contact with so many other Powers by sea that their contacts by land alone can be ignored in comparison. . . . . (3)

At the stage of numerical computation the following values will be assumed.

$$\alpha = 5, \quad \beta = 44, \quad \gamma = 11 \quad . \quad . \quad . \quad . \quad . \quad . \quad . \quad . \quad (4)$$

because they are a fair representation of the average during the interval A.D. 1820 to A.D. 1929, insurgents being included, for they are so in Collection J.

As in Theory VIII, let the probability p of war between the two members of any pair of States be defined to be the fraction, of any very long historical interval, during which such war occurred. We now have to consider three kinds of States, which can form six kinds of pairs, to each of which we must assign a probability. Fortunately, however, it is possible to make theoretical progress with only four distinct probabilities, one of them being zero. They are defined in the following table:—

Members of the pair	Symbol for the probability p
(i) Two world wide sea-Powers	x
(ii) A world wide sea-Power and a local coastal Power	y
(iii) Two local coastal Powers (iv) A local coastal Power and a land-locked Power (v) Two land-locked Powers	} z if in contact, or if a local belligerent touched both; otherwise zero.
(vi) A world wide sea-Power and a land-locked Power	Zero, unless a local coastal belligerent put them in contact, when p becomes y.

These definitions will be referred to as . . . . (5)

It is assumed here for simplicity, as in Theory VIII, that the probabilities defined in the above table are constant in time and independent of all circumstances other than those specified in the definitions (5). . . . . (6)

In the definitions (5) of the probabilities it has been assumed that two States which have no contact can yet be at war with one another, provided that either is allied to an intermediary belligerent whose land extends from one to the other so that their armies can march across it. This principle may be briefly called "*conduction by land.*" . . . . (7)

It has only a slight effect on the probabilities x, y, z deduced from the wars of type 1 versus 1, but a large effect for more complicated wars.

There is a further question about geography: Are widely separated local Powers ever regarded as allies? In the First World War the formal list of "Allied and Associated Powers" (Q. Wright, 1942, Table 42) contains the names of Peru and Rumania. But if the fightings of Britain with Argentine and with Afghanistan had happened to coincide in time, as they nearly did in 1847 and 1848, they would presumably have been called two separate wars of type 1 versus 1, not a single war of type 2 versus 1. The answer thus seems to be that: *widely separated local belligerents are regarded as allies only if two or more world-powers are fighting each other and one at least of them is allied to both those local belligerents.* . . . . (8)

In this matter the mathematics must conform to the mental habits of historians.

The existence of States which could not possibly be at war with one another makes vast changes in the distribution of mutually exclusive wars. According to Theory V there could be war inside any pair, and the number of mutually exclusive wars of the type r versus s was therefore

$$A(r, s) = \tfrac{1}{2}\frac{m!}{r!\,s!\,(m-r-s)!}.$$

In particular, for $m = 60$ States in the world,

$$A(30, 30) = \tfrac{1}{2}60!/(30!)^2 = 5\cdot 9 \times 10^{16}.$$

It is conceivable that, in the future, aviation may make these $5\cdot 9 \times 10^{16}$ varieties of all-in war each possible. In Theory VIII the total duration of all such wars was made negligibly small by a factor p^{900}. In history no such war has ever occurred. On the present Theory XI not a single one of them was possible, because of zero values of p. So, in one way or another, all-in wars are satisfactorily dismissed. The great advantage of Theory XI is that it, unlike Theory X, admits the occurrence of wars such as those of 1914–18 and 1939–45, which were world-wide, although not all-in.

Let us first consider the *number of mutually exclusive possible wars,* alias "varieties," belonging to each of the subtypes of 1 versus 1.

Subtype 1, 0 *versus* 1, 0, that is a war between two world-wide sea-Powers. There are $\tfrac{1}{2}\alpha(\alpha - 1)$ varieties.

Subtypes 1, 0 *versus* 0, 1 *and* 0, 1 *versus* 1, 0 *jointly*, that is any war between a world-wide sea-Power and a local coastal Power. Of these there are $\alpha\beta$ varieties.

Subtype 0, 1 *versus* 0, 1, a purely local war. The first belligerent can be selected in $\beta + \gamma$ ways. This having been done, the second belligerent can, according to the assumption (3), be selected in five ways, not in $\beta + \gamma - 1$ ways. This manner of selection takes every pair twice. Thus there are $\tfrac{5}{2}(\beta + \gamma)$ varieties. We have further assumed that $\beta + \gamma = 55$, so making $137\tfrac{1}{2}$ varieties, which can be rounded off to 138, as the half is absurd. The reason for the half is that the assumption (3), that each State has five neighbours, though a reasonable approximation to the actual geography, is not exact; and that moreover there are some necessary logical relations, connected with Euler's theorem and sea-coasts, which would be too complicated to introduce into this theory.

Each of these mutually exclusive varieties involves one pair of belligerents with probability x or y or z. With reference to any war, let us use the phrase "outward probability" to denote the probability that the belligerents remain at peace with all the neutrals. These outward probabilities were found to need close attention, thus:

Subtype 1, 0 *versus* 1, 0. The two belligerents form $2(\alpha - 2)$ pairs in peace with the remaining $\alpha - 2$ long range sea-Powers. The probability of peace is $1 - x$ in each such pair, and so is $(1 - x)^{2\alpha - 4}$ for the $2\alpha - 4$ pairs jointly. Moreover, according to principle (8), any or all of the β local coastal Powers could join in this war; they form 2β pairs with the belligerents; and the probability that there is peace inside all these pairs is $(1 - y)^{2\beta}$. The outward probability is thus $(1 - y)^{2\alpha - 4}(1 - y)^{2\beta}$.

Subtypes 1, 0 *versus* 0, 1 *and* 1, 1 *versus* 1, 0. The long-range sea belligerent forms $\alpha - 1$ pairs with other Powers of its own kind, thus giving a factor $(1 - x)^{\alpha - 1}$. The local Powers which might, but have not, joined the war are the five neighbours of the local coastal Power, and no others. It is assumed that so long as they are not in the war, the possibility of land conduction of more distant powers does not arise . . . . (9)

There is accordingly a factor $(1 - z)^5$. The index of $(1 - y)$ is the sum of two parts. One part is the $\alpha - 1$ pairs formed by the local belligerent and the $\alpha - 1$ long-range neutrals. The other part is the five pairs formed between the long-range belligerent and the five neighbours of the local belligerent. If land-conduction were rejected this part of the index would be less than five because probably some of those neighbours have no sea-coast. On collecting factors we obtain $(1 - x)^{\alpha - 1}(1 - y)^{\alpha + 4}(1 - z)^5$ for the outward probability.

Subtype 0, 1 *versus* 0, 1. The countries of the two local belligerents may be imagined as irregular pentagons having a side in common. A remarkable fact of political geography is that there is practically never a point where more than three States meet. So the eight peaceful frontiers of the belligerents touch six, not eight, neighbouring States. According to the principle (7) of conduction by land, any one of those six neighbours might be, but is not, at war with either belligerent, thus giving a factor $(1 - z)^{12}$. The number of pairs formed between the two local belligerents and the long-range neutrals is 2α; but for a few of these pairs the probability of war is zero. To admit the possibility that a distant Power might attack either of the belligerents by sea, it suffices to suppose that *one* of the belligerents has a sea coast; for the other could then be attacked by land-conduction. With accuracy sufficient for a minor correction we may take the probability that neither belligerent has a sea-coast to be $\gamma^2/(\beta + \gamma)^2$. The index of $1 - y$ is accordingly $2\alpha\{1 - \gamma^2/(\beta + \gamma)^2\}$. The outward probability for this subtype is therefore

$$(1 - y)^{2\alpha\{1 - \gamma^2/(\beta + \gamma)^2\}}(1 - z)^{12}.$$

All these results are collected in the following table.

Subtypes of 1 versus 1 (10)

Subtype	Mutually exclusive varieties	Probability of each variety	Expectation * of subtype when $\alpha = 5, \quad \beta = 44, \quad \gamma = 11$
1, 0 versus 1, 0	$\frac{\alpha(\alpha - 1)}{2}$	$x(1 - x)^{2\alpha - 4}(1 - y)^{2\beta}$	$10x(1 - x)^6(1 - y)^{88}$
1, 0 versus 0, 1 0, 1 versus 1, 0	$\alpha\beta$	$y(1 - x)^{\alpha - 1}(1 - y)^{\alpha + 4}(1 - z)^5$	$220y(1 - x)^4(1 - y)^9(1 - z)^5$
0, 1 versus 0, 1	$\frac{5}{2}(\beta + \gamma)$	$z(1 - y)^{\frac{2\alpha\beta(\beta + 2\gamma)}{(\beta+\gamma)^2}}(1 - z)^{12}$	$\frac{275}{2}z(1 - y)^{9\cdot 6}(1 - z)^{12}$

* By "expectation" is here meant the total theoretical duration divided by the historical interval, when the interval is very long. This expectation can exceed unity, and so is not a probability.

The theoretical expectations in the last column of the above table were equated to the observed fractions of 110 years taken from Collection J for the respective subtypes. One of these fractions is greater than unity. Thus were obtained the following set of three equations, simultaneous in x, y, z.

$$x(1 - x)^6(1 - y)^{88} = 0\cdot 00136 \quad \ldots \quad (11)$$

$$y(1 - x)^4(1 - y)^9(1 - z)^5 = 0\cdot 00156 \quad \ldots \quad (12)$$

$$z(1 - y)^{4\cdot 8}(1 - z)^{12} = 0\cdot 00730 \quad \ldots \quad (13)$$

One root of these equations is

$$x = 0\cdot 00159, \quad y = 0\cdot 00166, \quad z = 0\cdot 00819 \quad \ldots \quad (14)$$

It was found by two sequences of approximations which approached it from opposite sides. The method resembled that already explained for the single equation $p^b(1 - p)^c = a$.

Considerable changes in the outward probabilities make only slight changes in the position of this root. For example, if the hypothesis (7) of land-conduction be rejected, the brackets in (13) become $(1 - y)^8(1 - z)^9$, and instead of (14) the root becomes $x = 0\cdot 00159$, $y = 0\cdot 00165$, $z = 0\cdot 00795$. This change is within the uncertainty of the historical facts. We shall meet a quite different relation for more complicated wars.

There remains the question whether the present set of three equations has any other root of historical significance. The search for such a root was conducted graphically, by regarding x, y, z as rectangular co-ordinates, so that each equation represents a surface, and any point common to all three surfaces represents a root. Because x, y, z are probabilities, we are not concerned with anything outside the cube bounded by their values 0 and 1. The first equation represents a cylinder which lies in the range $0 < y < 0\cdot 041_5$. The third equation represents another cylinder having its generators at right angles to those of the first. These cylinders cut one another in two separate arcs, which lie in the following ranges: $0\cdot 0081 < z < 0\cdot 014$ and $0\cdot 21 < z < 0\cdot 26$. Any z in the latter range is incredible, for it would assert that every local Power was fighting each of its five neighbours for more than 0·21 of the historical interval. In the former range of z one credible root has already been found. The search for another was pursued by plotting the section, by the plane $z = 0\cdot 014$, of the surface represented by the middle equation. It thus appeared that there may be a root near $z = 0\cdot 4_8$; but if so it is of no interest.

The only root of historical significance is that stated in (14).

A searching test of Theory XI will be to notice whether the same numerical values of the probabilities x, y, z, explain the occurrences of wars of various types. We shall come to that later.

In the meantime, and for the type 1 versus 1, z is nearly five times x. That is to say, *the world wide sea-Powers made much less pugnacious use of their opportunities for contact than did the local Powers.* The statistics here gathered do not afford any explanation; but on other grounds alternative explanations may be tentatively suggested, thus: (i) Local contact existed all the time, whereas distant contact by sea could be made or not, as desired by the sea-Power. The more persistent contact, the more opportunity for quarrels. (ii) A nation's capacity for fighting is bounded. Nations having very many contacts are likely to have been satiated with fighting; and therefore to have been, to that extent, peace-loving.

Leaving aside surmises, such as (i) and (ii), let us return to more definite investigations. The statistical analysis which was applied above to wars of the type 1 versus 1, has been extended to the types 2 versus 1 and 1 versus 2, jointly. As no new geographical principles are involved, it may suffice to state the results in tabular form. The broad second column shows the theoretical formula for the expectation. A few of the indices of $1 - y$ and of $1 - z$ are averages over slightly different cases. The third column contains the historically observed fraction of 110 years taken from Collection J. The fourth column headed "numerical expectation," will be explained presently.

Subtypes of (2 versus 1) and (1 versus 2) (16)

Subtype	Expectation of subtype	Historical fact	Numerical expectation
2, 0 versus 1, 0 1, 0 versus 2, 0	$\frac{\alpha!}{2!(\alpha-3)!}x^2(1-x)^{3\alpha-9}(1-y)^{3\beta}$	0·0000	0·0000
2, 0 versus 0, 1 0, 1 versus 2, 0	$\frac{\alpha(\alpha-1)\beta}{2}y^2(1-x)^{2\alpha-4}(1-y)^{\alpha+8}(1-z)^5$	0·0227	0·0734
1, 1 versus 1, 0 1, 0 versus 1, 1	$\alpha(\alpha-1)\beta xy(1-x)^{2\alpha-4}(1-y)^{\alpha+2\beta-4+\frac{10\gamma}{\beta+\gamma}}(1-z)^5$	0·0023	0·0012
1, 1 versus 0, 1 0, 1 versus 1, 1	$5\alpha\beta yz(1-x)^{\alpha-1}(1-y)^{2\alpha+4}(1-z)^{12}$	0·1573	0·2093
0, 2 versus 1, 0 1, 0 versus 0, 2	$\frac{5\alpha\beta}{2}y^2(1-x)^{\alpha-1}(1-y)^{2\alpha+\frac{13}{3}}(1-z)^{\frac{38}{3}}$	0·1436	0·0770
0, 2 versus 0, 1 0, 1 versus 0, 2	$10(\beta+\gamma)z^2(1-y)^{3\alpha-3\alpha\gamma^3/(\beta+\gamma)^3}(1-z)^{21}$	0·1323	0·1190
	Totals	0·4582	0·4799

If we equate each expression in the second column to the historical number in the third column and put $\alpha = 5$, $\beta = 44$, $\gamma = 11$ as before, we obtain six equations simultaneous in x, y, z; and, of course, somewhat inconsistent.

As a preliminary let us consider the first equation by itself. The previously noted objection to large probabilities is here illustrated in an extreme form. For we could explain why the subtypes (2, 0 versus 1, 0) and (1, 0 versus 2, 0) never occurred in the history, by inserting $x = 1$ in the first row of the table. But $x = 1$ would also entail the utterly false assertion that each of the five sea-Powers was fighting each of the others all the time. Instead, therefore, we must suppose that x is zero or near to it.

The last equation, when taken alone, provides an under-estimate of z. For $0 < y < 1$, and therefore $z^2(1 - z)^{21} > 10^{-4} \times 2{\cdot}405$; the solution of which is $z > 0{\cdot}019$.

The column headed "numerical expectation" was obtained by inserting

$$x = 0{\cdot}0004,\ y = 0{\cdot}0150,\ z = 0{\cdot}0200 \quad . \quad . \quad . \quad . \quad . \quad . \quad . \quad . \quad (17)$$

into the formulæ for the theoretical expectation. These numbers are a compromise, intended to fit all six of the simultaneous equations tolerably. They were obtained by trial and adjustment after the manner of the successive approximations explained on page 139, except that at each stage inconsistent results had to be averaged. The product moment correlation coefficient between the last two columns is 0·78. The chief obstacle to making the correlation nearer to unity is the disagreement between the two equations in y^2. In the history, two local Powers were fighting one sea-Power for about six times longer than one local Power was fighting two sea-Powers. In the theory the expectations of these subtypes are roughly equal. It is too soon to say whether this misfit is of political interest, or is merely an extreme effect of sampling. The distribution of wars in time has been shown to be described by the Poisson law (Richardson, 1935), a law which also describes the emission of alpha particles from radioactive substances (Rutherford, Chadwick and Ellis, 1930). That is to say, the historical facts are to be regarded as random samples, when wars are counted as wholes. The present detailed classification is likely to have increased the uncertainty of sampling. In this manner the misfits between the last two columns of the above table may be tentatively excused. Whoever says that these results are rough, should compare them with our previous blank ignorance.

The excess of z over x, which we noted for type 1 versus 1, is increased for the types 2 versus 1 and 1 versus 2. The political meaning of this difference in probability has already been discussed.

Both the probabilities y and z are greater for wars involving two pairs of opposed belligerents than for wars involving only one pair. This new phenomenon makes it desirable to study wars involving more than two pairs, in order to notice whether in them the probabilities are further increased.

Although the geographical principles are the same as before, one crucial piece of argument is worth stating. It relates to the subtypes (0, 3 versus 0 1) and (0, 1 versus 0, 3), jointly. The isolated belligerent can be chosen in $\beta + \gamma$ ways. The mutually exclusive varieties can then be classified according to the number of direct contacts which the isolated belligerent A has with its three enemies B, C, D. The statements which follow were obtained by sketching pentagons, with attention to the fact (1) that not more than three States meet at any point.

	Varieties $(\beta + \gamma)x$	Neutral neighbours
A touches *B*, *C*, and *D*. There are two subclasses:		
One of *B*, *C*, *D* touches *A*, but not the other two ...	5	7
The remaining subclass	5	6
A touches two only of *B*, *C*, and *D*:		
These two touch one another	{ 10	7
	{ 5	6
These two do not touch one another	20	8
A touches one only of *B*, *C*, and *D*:		
This one touches the other two	5	7
The remaining subclass	20	8
Total	70	

(18)

The mean number of neutral neighbours, weighted by the number of varieties, comes to 52/7. Any one of the four belligerents might have been at war with 52/7 neighbours on the average. Therefore the outward probability contains a factor $(1 - z)^{208/7}$, or practically $(1 - z)^{30}$. When a similar argument was applied to the subtypes (1, 2 versus 1, 0) and (0, 3 versus 1, 0), some slight ambiguities appeared. It seems that the geographical assumptions (1), (3), (5), (7), (8) do not quite fix the expectation for these subtypes, although they fix it satisfactorily for all the others so far considered. The ambiguity is not important; no easy device for avoiding it appeared; and so it was passed over by a factor $\beta + \varepsilon\gamma$ where ε is undetermined except that it lies in the range $0 < \varepsilon < 1$. The results are set out concisely in the table on the following page.

As before, let us assume that $\alpha = 5$, $\beta = 44$, $\gamma = 11$, and let us try to equate the theoretical expectations in the second column to the historical facts in the third column. The latter are the fractions of 110 years during which the subtypes occurred, according to Collection J. Although the eight equations so obtained are strictly inconsistent, it is not difficult to obtain a useful compromise solution, starting from the presumption that x, y, z, have values resembling those found from the wars of 2 versus 1 and 1 versus 2. Then the last equation almost fixes z; and there are two pairs of equations which almost fix y/z consistently. The column which is headed "numerical expectation" in the above table contains the results of inserting

$$x = 0{\cdot}001,\ y = 0{\cdot}009,\ z = 0{\cdot}046 \quad . \quad . \quad . \quad . \quad . \quad . \quad . \quad . \quad (20)$$

into the formulæ for the expectations. The product moment correlation coefficient between history and expectation over the eight subtypes of octantal 3 versus 1 then exceeds 0·99, which is satisfactory.

The successive increases of z from 0·008 to 0·020, and thence to 0·046, have so far accompanied increases of both $r + s$ and rs. There is accordingly a special interest in extending the investigation to the subtypes of 2 versus 2; for thereby $r + s$ will remain at 4, while rs will increase from 3 to 4. The most frequent of these subtypes in Collection J is the purely local 0, 2 versus 0, 2. The number of its mutually exclusive varieties can be deduced from the corresponding number for the subtypes (0, 3 versus 0, 1) and (0, 1 versus 0, 3), jointly, a number which we have seen to be $70(\beta + \gamma)$.

For consider any bunch of four countries touching one another, so that the conduction of troops by land would be possible. A solitary belligerent, opposed to the remaining three, can be selected from them in four ways; but a pair of allies, opposed to the remaining two, can be selected in $4 \times 3/2! =$ six ways. Therefore the number of mutually exclusive varieties of wars of the

Subtypes of (3 versus 1) and (1 versus 3) (19)

Subtype	Expectation of subtype	Historical fact	Numerical expectation
3, 0 versus 1, 0 1, 0 versus 3, 0	$\frac{a!}{3!(a-4)!}x^3(1-x)^{4a-16}(1-y)^{4\beta}$	0·0000	0·0000
3, 0 versus 0, 1 0, 1 versus 3, 0	$\frac{a!\beta}{3!(a-3)!}y^3(1-x)^{3a-9}(1-y)^{a+12}(1-z)^5$	0·0000	0·0002
2, 1 versus 1, 0 1, 0 versus 2, 1	$\frac{a(a-1)(a-2)\beta}{2}x^2y(1-x)^{3a-9}(1-y)^{a+3\beta-6}(1-z)^{\beta-1}$	0·0068	0·0000
2, 1 versus 0, 1 0, 1 versus 2, 1	$\frac{5a(a-1)\beta}{2}y^2z(1-x)^{2a-4}(1-y)^{2a+9}(1-z)^{13}$	0·0036	0·0037
1, 2 versus 1, 0 1, 0 versus 1, 2	$\frac{5a(a-1)\beta}{2}xy^2(1-x)^{2a-4}(1-y)^{2a+2\beta-2}(1-z)^{2\beta+2}$	0·0000	0·0000
1, 2 versus 0, 1 0, 1 versus 1, 2	$20a(\beta+\epsilon\gamma)yz^2(1-x)^{a-1}(1-y)^{3a+4}(1-z)^{21}$	0·0255	{ 0·0261 to 0·0327
0, 3 versus 1, 0 1, 0 versus 0, 3	$5a(\beta+\epsilon\gamma)y^3(1-x)^{a-1}(1-y)^{3a+\frac{11}{3}}(1-z)^{20}$	0·0000	0·0003
0, 3 versus 0, 1 0, 1 versus 0, 3	$70(\beta+\gamma)z^3(1-y)^{4a}(1-z)^{29\cdot71}$	0·0773	0·0772
	where $0 < \epsilon < 1$		

subtype 0, 2 versus 0, 2 is $\frac{6}{4} \times 70(\beta + \gamma) = 105(\beta + \gamma) = 5775$. The expectation of subtype 0, 2 versus 0, 2 is accordingly $5775z^4(1-y)^{20}(1-z)^{29\cdot71}$. In a previous case we obtained a lower bound to z by omitting the factor $1 - y$ and equating the rest of the expectation to the observed fraction of 110 years, which here is 14·2/110. But, just in this particular, Collection J may be an aberrant sample, because 11 of those 14·2 years come from a single war (Rosas) which Q. Wright classifies differently. The same contrast occurs between Collections G and H. Accordingly, to make sure of obtaining an underestimate of z, let us omit the eleven years and put

$$5775z^4(1-z)^{29\cdot71} = 3\cdot2/110 \quad . \quad . \quad . \quad (21)$$

This equation just misses having any real root in the range $0 < z < 1$. But if the observed duration had been 2·96 years instead of 3·2 years, as it well might, there would have been two coincident roots at $z = 0\cdot119$. We cannot escape from the conclusion that z is here larger than it was for wars of lesser rs.

Summary on Theory XI. In a general way Theory XI explains the rare occurrence of very complicated wars such as that of 1914–18; but in such a case detailed calculations would be too troublesome. The application has been made to wars involving not more than four nations. The probability, x, of fighting between two world-wide sea-Powers is found to be of the order of $x = 0\cdot001$; but is not sharply determined, because such wars have been rare. The probability, y, of fighting between a world-wide sea-Power and a local Power is found to have values $y = 0\cdot002$, 0·015, 0·009, according to circumstances. These look like sampling variations around a mean of $y = 0\cdot009$, which is distinctly greater than x. The probability, z, of fighting between two local

Powers is more closely determined than either x or y, because there have been many local wars. It is plain that z considerably exceeds x. That is to say, the world-wide sea-Powers have made much less pugnacious use of their opportunities for contact than have the local Powers. Theory XI assumes that z is the same in all circumstances, but when Theory XI is combined with geographical and historical facts, a different result emerges, thus:—

Number, rs, of pairs of opposed belligerents	1	2	3	4
Probability, z, of war between neighbours	0·008	0·020	0·046	0·119
Compare the empirical formula $z = 0{\cdot}008\,(2{\cdot}549)^{rs-1}$ ...	0·008	0·020	0·048	0·119

Thus Theory XI fails, but in a manner so regular as immediately to suggest suitable improvements

Theory XII. Chaos restricted by geography and modified by infectiousness

Fighting is infectious. When it is known that a war has begun, neutrals begin to ask: How will this affect us? Are our interests endangered? What is the right and wrong of the dispute? Ought we to join in? On the contrary, in Theories VIII and XI it was assumed for simplicity that the probability of war between any two nations was not affected by the peacefulness or belligerency of other nations. I do not apologize for this assumption. A condition for discovery in applied mathematics is that the assumptions must at first be simple. But now the comparison between Theory XI and fact has shown definitely that the assumption is wrong, and has indicated the numerical amount of the required amendment. Theory XII is simply an acceptance of that amendment; by assuming that the probability, z, of war between a pair of local powers is to increase with the number of pairs of opposed belligerents in the manner indicated by the last table. It is unnecessary to deduce the consequences, for that, in effect, has already been done; and they have been found to agree with historical fact.

Looking back at Theory III we notice that some rather similar gregarious tendencies were there mentioned and rejected as an explanation. Certainly infectiousness alone will not explain the facts; chaos alone is nearer the mark; it is only as a modifier of a geographical restricted chaos that infectiousness is admitted into Theory XII.

Theory XIII. A uniform chaos modified only by infectiousness

Although Theory XII has attained agreement with history, we may for logical satisfaction wish to see whether we could do as well by simpler hypotheses, retaining chaos and infectiousness, but ignoring geography, as aviation may make it ignorable in the future.

Let us suppose that the trouble begins with the existence of a world-wide controversy, for example as to the relative merits of communism and individual enterprise. Let our elementary probability be denoted by θ, not p, because here, as in Theory X, it is the probability of war about a dispute and not a fraction of the time. In contrast with Theory X, let us now suppose that θ increases with the number κ of pairs of opposed belligerents who are already fighting about the dispute, so that we write θ_κ. Also in contrast with Theory X, let us consider a whole world of m nations, instead of a local group of l. Let the expectation of a war of the type r versus s be denoted by $V(r, s)$ in the quadrantal plan. As in Theory X, we are concerned only with the ratios of V for different r, s. By modifying Theory VIII it is easily seen that

$$V(1, 1) = \frac{m(m-1)}{2}\theta_0(1-\theta_1)^{2(m-2)} \quad . \quad . \quad . \quad . \quad . \quad . \quad . \quad . \qquad (1)$$

which differs from Theory VIII by involving θ_1 and θ_0. If another nation joins the war it does so with probability θ_1 and then, for the neutrals, the probability rises to θ_2. Thus the expectation of a war of the quadrantal type 2 versus 1 is

$$\tfrac{1}{2}\frac{m(m-1)(m-2)}{2}\theta_0\theta_1(1-\theta_2)^{3(m-3)} = V(2, 1) \quad . \quad . \quad . \quad . \quad . \qquad (2)$$

A fourth nation can join the war in either of two ways. It may join the pair of allies making a war of 3 versus 1 with expectation, quadrantally

$$\tfrac{1}{2}\frac{m!}{3!\,1!\,(m-4)!}\theta_0\theta_1\theta_2(1-\theta_3)^{4(m-4)} = V(3, 1) \quad . \quad . \quad . \quad . \quad . \qquad (3)$$

Or the fourth nation may join the solitary belligerent, thus forming *two* new pairs of opposed belligerents; so that the expectation of a war of 2 versus 2 is

$$\tfrac{1}{2}\frac{m!}{2!\,2!\,(m-4)!}\theta_0\theta_1\theta_2{}^2(1-\theta_4)^{4(m-4)} = V(2, 2) \quad . \quad . \quad . \quad . \quad . \qquad (4)$$

Similarly

$$V(4, 1) = \tfrac{1}{2}\frac{m!}{4!\,1!\,(m-5)!}\theta_0\theta_1\theta_2\theta_3(1-\theta_4)^{5(m-5)} \quad . \quad . \quad . \quad . \quad . \quad . \quad (5)$$

From (4) and (5) we obtain

$$\frac{V(4, 1)}{V(4, 1)} = \frac{1}{6(m-5)}\frac{\theta_3}{\theta_2}(1-\theta_4)^{m-9} \quad . \quad . \quad . \quad . \quad . \quad . \quad . \quad . \quad . \quad . \quad (6)$$

whence, if $m = 60$,

$$\theta_3/\theta_2 > 330V(4, 1)/V(2, 2) \quad . \quad . \quad . \quad . \quad . \quad . \quad . \quad . \quad . \quad . \quad (7)$$

but the observed values of $V(4, 1)/V(2, 2)$ are 5/6 from Collection A, 6/4 from Collection C. Whence

$$\theta_3/\theta_2 > 275 \quad \text{or} \quad > 495 \quad . \quad . \quad . \quad . \quad . \quad . \quad . \quad . \quad . \quad . \quad (8)$$

These ratios are incredibly great. *Thus the introduction of infectiousness alone will not serve. We are also constrained to reduce the number of nations by attention to geography.* Theory XIII is an anticlimax. We are directed back to Theory XII.

Acknowledgments

Mr. Oliver M. Ashford, B.Sc., started all this enquiry by sorting the author's card-index into a form which has now become Collection J.

The author wishes to thank the Government Grant Committee of the Royal Society for the loan of a calculating machine.

References

E refers to the *Encyclopædia Britannica*, 14th edition.
Hinshelwood, C. N. 1940. *The Kinetics of Chemical Change.* Clarendon Press, Oxford.
Jeans, J. H. 1916. *The Dynamical Theory of Gases.* Cambridge University Press.
Kendall, M. G. 1943. *The Advanced Theory of Statistics.* Griffin, London.
Kirkpatrick, F. A. 1938. *Latin America.* Cambridge University Press.
Priestley, D. I. 1938. *France Overseas.* Appleton–Century Co., New York.
Richardson, L. F. 1941. Letter to *Nature* of November 15th.
——. 1945. *J. Roy. Stat. Soc.*, **107**, 242–250.
——. 1946. *Proc. Roy. Soc.*, A, **186**, 422–31.
Rutherford, Chadwick and Ellis. 1930. *Radiations from Radioactive Substances.* Cambridge University Press.
West, R. 1942. *Conscience and Society.* Methuen, London.
Wright, Q. 1942. *A Study of War.* Chicago University Press.

1946:2

CHAOS, INTERNATIONAL AND INTER-MOLECULAR

Nature
London
Vol. 158, No. 4004, 27 July 1946, p. 135
(Letter dated 23 June)

Note: The publications referred to are 1946:3 (in Volume 1) and 1946:1 (above) respectively

Reprinted from NATURE, *Vol.* 158, *page* 135, *July* 27, 1946.)

Chaos, International and Inter-molecular

STATISTICS of wars have been collected from the whole world for the 120 years beginning with A.D. 1820. Attention was directed to the number of nations, or other large belligerent groups, on each side of any war. Accordingly, wars were classified as 1 group versus 1 group, or as 2 versus 1, or as 2 versus 2, and in general as r versus s. The number of wars of each of these types was counted. The result was a fairly regular statistical distribution, having a peculiar shape. Among a total of 91 wars there were 42 of the type '1 versus 1', 24 of the type '2 versus 1', and not more than five wars of any one more complicated type. The simplest type of encounter was the most frequent.

In a gas at N.T.P. encounters of two molecules are much more frequent than encounters of three, as is well known from chemical experiments. This resemblance between a gas and the political world suggested a theory for each of them. The frequency of an encounter, of specified type, can be regarded, after the manner of Bernoulli, as the product of the following three factors. (i) The number of mutually exclusive encounters of that type. (ii) The probability that the opponents encounter one another. In this factor the probabilities for the separate pairs of opponents combine by multiplication. That is the chief reason why, in the chaos, complicated encounters are rarer than simple encounters. (iii) The probability that all the other nations, or molecules, keep out of the encounter. Strange to say, this third factor escaped the attention of the authors of the classical theory of gases. Consequently at high densities a proportion of the encounters which Guldberg, Waage and their modern successors have regarded as binary, are now shown to be ternary. How this affects the chemistry depends on whether, for molecules, 'two can be company but three none'.

Although three factors of the aforesaid sort are likely to appear in the theory of any chaos, yet their particular forms depend on circumstances; so that many varieties of chaos are conceivable. In the political world there were restrictions depending on geography and on sea-power. When they had been formulated, another effect became conspicuous, namely, the infectiousness of local fighting.

The justifications of the foregoing brief statements have been accepted for publication, those concerning gases in the *Proceedings of the Royal Society*, and those concerning the political world in the *Journal of the Royal Statistical Society*.

LEWIS F. RICHARDSON

Hillside House,
Kilmun,
Argyll.
June 23.

PRINTED IN GREAT BRITAIN BY FISHER, KNIGHT AND CO., LTD., ST. ALBANS

1947:1

SOCIAL SCIENCE IN THE GAP BETWEEN THE ROYAL SOCIETY AND THE BRITISH ACADEMY

Nature
London
Vol. 159, No. 4034, 22 February 1947, p. 269
(Letter dated 31 December [1946])

No. 4034 February 22, 1947 NATURE 269

Social Science in the Gap between the Royal Society and the British Academy

To which of the two above-named august institutions should research workers in social science look up for recognition, guidance, grants-in-aid, expert criticism, and publication? The recent Clapham Report[1], on the Provision for Social and Economic Research, was drawn up by a committee of nine whose qualifications included three F.B.A., three Litt.D., two Ll.D., one D.Sc., but no F.R.S. This literary preponderance suggests that the appropriate body is the British Academy. But a surer guide is a comparison of publications. Those of the British Academy, though full of literary grace and social interest, scarcely contain any quantitative analysis. I doubt if a correlation coefficient could be found anywhere among them. An article in celebration of Leibniz[2] does not contain any mathematical symbols. On the other hand, the publications of the Royal Society, though full of quantitative analysis, scarcely contain any direct mention of human social relations. To which body should a quantitative analysis of a social problem, therefore, be offered for publication? Apparently to neither. There is thus an awkward gap at the highest level of authority. No such gap was intended when the British Academy was formed[3] in 1902. One of its presidents, Sir Frederic Kenyon, in his presidential address on July 4, 1918, said of academies of learning: "Their normal formation is in two classes, one comprising what we now commonly call humanistic studies, i.e., history, philosophy and philology (in the widest sense of the latter term), the other mathematics and natural science. . . . In this country the Royal Society corresponds to one of these classes. The British Academy was founded to play the other part." Since 1902 social science, like other sciences before it, has become more statistical, more quantitative, more mathematical. To give only one example: there are now many published researches on the intelligence of school-children, and on its correlations with their parentage and other social conditions. The results are important for planning, and their discussion involves matrix algebra[4]. So a gap, unsuspected in 1902, has now opened wide. Fortunately the gap is bridged by a number of specialized societies and journals publishing severally statistics, psychology and economics. In the history of science some of the papers printed in specialized journals have turned out to be quite as important as those printed by the Royal Society; for example, Karl Pearson's χ^2 test appeared in the *Philosophical Magazine* for 1900. Nevertheless, a gap remains, because the specialized societies lack the breadth and the general authority of the senior bodies.

I have here emphasized only one of the aspects of a problem which has been explored in general by a Committee[5] on Scientific Research on Human Institutions appointed by the British Association. Reference should also be made to the review[6] of the Clapham Report by Prof. P. Sargant Florence.

LEWIS F. RICHARDSON

Hillside House,
Kilmun, Argyll.
Dec. 31.

[1] H.M. Stationery Office. Cmd. 6868.
[2] *Proc. Brit. Acad.*, **23**, 193.
[3] "Encyclo. Brit.", fourteenth edition, **1**, 85 (1929).
[4] Thomson, Godfrey H., Presidential Address to the British Psychological Society, 1946 (University of London Press).
[5] *The Advancement of Science*, **2**, 345 (1943).
[6] *Nature*, **158**, 368 (1940).

1947:2

ARMS AND INSECURITY

alias the fickleness of fear

Microfilm of reprint, typescript and manuscript, privately produced

(Title-page dated December 1946)
(Photographed 1 January [1947])

[2nd edition]

1949:2

ARMS AND INSECURITY

alias the fickleness of fear

Microfilm of reprint, typescript and manuscript, privately produced

(Title-page dated December 1946 revised May 1949)
(Photographed 17 May 1949)

[3rd edition]

1960:2

ARMS AND INSECURITY

A mathematical study of the causes and origins of war

Edited by Nicolas Rashevsky and Ernesto Trucco
Pittsburgh: Boxwood Press Chicago: Quadrangle Books
London: Stevens and Sons Limited

Notes on *Arms and Insecurity*

1. *Arms and Insecurity* is Richardson's main publication on the theory of arms races. He described it as 'a revision and great enlargement' of *Generalized Foreign Politics* (1939:1) and 'Stability after the War' (1944:1). Later developments are reported in 1951:3, 4, 5, 1953:2, 3.

2. In a preface dated 2 April 1949 to the second (microfilm) edition Richardson wrote:

   > In 1940 I retired, with my wife's consent, from a paid occupation in order further to investigate the causes of wars. By 1943 I had, as I thought, a book on the subject ready in typescript; but book-publishers were not willing. In spite of British professions of national peace-lovingness, I suspect that it would be much easier to find a market for a technical book about aeroplanes, or radar, or atomic energy, than for a book of equal technicality about the preservation of peace.

   The book referred to in this preface was to have been entitled *The Instability of Peace*. Because of the difficulty over publication, Richardson was advised in 1943 to offer sections of the book to learned societies, which he did (see the 'Notes on *Statistics of Deadly Quarrels*', below p. 537). The sections on arms races were however too extensive for a scientific journal, and these were published by Richardson privately on microfilm, as *Arms and Insecurity* (1947:2), followed by a slightly revised microfilm (1949:2). *Arms and Insecurity* appeared as a book only after Richardson's death (1960:2).

3. The differences between the three editions are very few and are listed below. Thus, except for Chapter XXII (added in 1949:2), virtually the entire main text of the 1960 book (Chapters I–XXIV) was first published in 1947, having apparently been completed about four years earlier, during the Second World War. Most of Chapters II, IV.4, X.1, XI, XIV–XVII and XXI.1, had originally appeared in 1939:1.

4. The following changes were made in the later editions. All the page, paragraph and chapter references (here and elsewhere) are to the third edition, the book (1960:2).

   Changes in 1949:2:

   Pp. 67–79: Section 6 (originally a brief note at the end of Chapter XVII) was expanded and Section 7 was added.

   Pp. 89–90: The paragraph 'As one of the crucial ... would not be justified' is a considerable expansion of the original wording to include the tests of significance.

   P. 164: The paragraph 'For the purpose ... shown to drift simply' was added.

   P. 193: The last 10 lines, 'Its tentative character ... XVII, XVIII' are an expansion of the original wording.

   P. 196: The paragraph 'From the imperfectly ... from Type XV' was altered.

   Pp. 232–6: Chapter XXII was added.

Further changes in 1960:2:

P. 8: The paragraph 'The product-moment correlation ... is also not verified' is an expansion of the original wording to include the tests of significance.

P. 16: The two paragraphs 'In *Generalized Foreign Politics* ... effects of these "imponderables"' replaced a short piece of dialogue comment (not reproduced here).

P. 27–8: The two paragraphs 'Aggressive intentions. A major ... (k or l)' were added.

5. In addition to the main text, the book *Arms and Insecurity* (1960:2) contains three prefatory items and a bibliography, which also appear in the book *Statistics of Deadly Quarrels* (1960:1) (see the 'Notes on *Statistics of Deadly Quarrels*', p. 537).

1947:2, 1949:2			1960:2	
Section	Abstract of contents	Page	Chapter	Page
	Detailed contents	171		
2	Armaments and security	177	I	1
3/1	Freewill	195	II	12
3/2	Arms-races between two nations	195	III	38
3/3	Submission and retaliation	236	IV	52
3/4	Miscellaneous notes on arms-races	250	V	61
3/5	Arms-races from 1820 to 1908	262	VI	70
3/6	Critical re-examination of 1908–14	272	VII	77
3/7	International trade 1908–14	308	VIII	98
3/8	Plan of enquiry for 1921–40	332	IX	111
3/9	Seven great powers 1921–1939	336	X	114
3/10	Warlike expenditure & foreign trade	347	XI	121
3/11	Standards of economic value	358	XII	128
3/12	Ten nations compared	369	XIII	136
3/13	Extending the hypotheses	383	XIV	145
3/14	Theory of three nations	389	XV	149
3/15	Directed intentions	402	XVI	160
3/16	Summary theory for *n*-nations	405	XVII	163
3/17	Successive approximations	433	[not in 1960:2]	
3/18	Demobilization	445	XVIII	184
3/19	1929 to '39 re-examined	458	XIX	192
3/20	Foreign trade, 1919 to '39	499	XX	213
3/21	Dialogue on practical politics	521	XXI	226
5/3	War fever [not in 1947:2]	530	XXII	232
6/7	Fading of war-weariness	531	XXIII	237
7/4	Submission of the defeated	553	XXIV	253
	List of references	565		
	Subject index	578		

6. The table above shows the cross-references between the 'Reader's Guide to the Microfilm', a printed card accompanying the 1949 microfilm, and the chapters of the book.
7. §3/17: 'Successive approximation to latent vectors and latent roots of a numerical matrix', is in both microfilms (with a different introductory paragraph), but was omitted from 1960:2. It has not been reproduced here because its technical content was superseded by 1950:1, 2 (both in Volume 1). This section is listed separately as 1947:3 and 1949:3 among the papers by Richardson not included in either volume.

1948:2

WAR-MOODS: I

Psychometrika
Colorado Springs, Colorado
Vol. 13, No. 3, September 1948, pp. 147–174

WAR-MOODS: II

Psychometrika
Colorado Springs, Colorado
Vol. 13, No. 4, December 1948, pp. 197–232

PSYCHOMETRIKA—VOL. 13, NO. 3
SEPTEMBER, 1948

WAR-MOODS: I

LEWIS F. RICHARDSON
HILLSIDE HOUSE, KILMUN, ARGYLL, BRITAIN

A person can be openly friendly, or hostile, or war-weary. Depth-psychology has shown that the underlying mood is also important. In the present theory the independent variable is time, and the dependent variables are the numbers of persons in different war-moods in two opposing nations. The rate of conversion of persons from one mood to another is taken to be proportional both to the number of susceptible persons and to the number of influencing persons, as in Kermack and McKendrick's theory of epidemics of cholera (15). This formulation leads to a set of differential equations nearly but not quite like those whereby Volterra (41) described the interaction between predator and prey among fish. Here the constants are fitted to the history of the First World War. The gregarious motive to follow a fashion of either peace or war is formulated as an amendment.

General Introduction to Parts I and II

This is a study of the moods of populations before, during, and after a war. The author's underlying purpose has been to aid in the maintenance of peace. For it is undeniable that the moods revealed during a war have much relevance to the instability of peace. This study unites an abstract of historical fact with the psycho-analytic doctrine that the unconscious is important, and then tests the resulting hypotheses quantitatively by comparison with events. Only broad general features of a war are considered, namely, the numbers of people concerned, the emotional drives of the majority, and those of a few important minorities. Almost all of the well-known dramatic incidents are ignored. Instead the war is described as a continuous conative process: a smoothed war, like a gale without the gusts. Even with this simplification, the treatment inevitably becomes elaborate.

Although the account is really all one continuous argument, yet to make it more readable, it has been divided into parts IA, IB, IIA, IIB, IIC, as follows:

Part IA *War-moods, conscious and unconscious*

This relates mainly to Germany, Britain, and the war of 1914-'18. Historical facts are collected and are brought into relation with the well-established psychological fact that the adult human mind functions in at least two levels. Without this simple broad concept of the sub- or unconscious, some quantitative aspects of the history would be incomprehensible. On the contrary, no use is here made of the detailed Freudian theory of the mental life of infants.

To Prof. T. H. Pear I am indebted for some revisions of Part IA.

Part IB *War-moods and the theory of epidemics*

The observation that fighting is infectious led to a hint being taken from the mathematical theory of epidemics, as published by Kermack and McKendrick. This theory relates to the inter-actions between fractions of the populations.

The reasoning of Part IB is inductive. Psychological processes such as infection by war-fever, or infection by defeatism, are fitted with mathematical descriptions containing adjustable constants. At each stage in the inductive reasoning there are, in the background of thought, countless varieties of formulas which would not contradict the historical evidence. From this plethora, in accordance with Occam's "razor," the simplest formula is chosen. Various infective processes are first studied separately in Part IB.

In Part IIA the same processes are regarded as possibly simultaneous. This leads to an organized system of quantitative hypotheses expressing the rates of change of six fractions of the population. These hypotheses are consistent with one another, they describe the facts mentioned in Part I, they are as simple as these facts permit them to be, and jointly they form a system capable of yielding predictions. Although the complete system of equations is in general unwieldy, it is nevertheless solved approximately in the successive phases from peace through war to peace again. Verifications are thus obtained by comparison with historical facts. The orders of magnitude of the constants are estimated. Rigid determinism is of course not to be expected.

Part IIB *Gregariousness and war-moods*

Critical allusion is made to W. Trotter's *Instincts of the Herd in Peace and War* (37). The motive which Trotter described is formulated, at first in isolation, and then in competition with a motive for change. It seems likely that this theory, here designed with reference to war and peace, would also illuminate the course of fashions in hats or in spelling; but such applications are not investigated. Instead the theory is applied to describe the crisis in Britain in September 1938 when Chamberlain's hope was "Peace for our time."

Part IIC publishes *Dr. E. B. Ludlam's chemical analogy to personal initiative*, by his permission.

PART IA

THE FRACTION OF THE POPULATION THAT WAS IN FAVOR OF THE WAR OF 1914-'18

Introduction

The fraction of the population that is in favor of any existing war is obviously important in a democratic country and cannot be ignored even by a so-called autocrat. Yet among the enormous mass of recorded information it is hard to find enough good quantitative measures. Let N be the total number of concerned persons in any population. We may omit young children as not concerned. Let ζ^* be the number of persons who are willing to prosecute the war either by actual fighting or by supporting it in any other way. Let us seek to estimate the ratio $\zeta^*/N = \zeta$ for two nations at different times. In so doing let us ignore, temporarily, the important psychoanalytic finding that an individual person is often in a state of internal conflict. He is here to be classified by that part of his personality which has control of his voluntary muscles. Later we shall distinguish two parts η and ρ of $\zeta = \eta + \rho$. It is necessary to pass in review a great many particular events, the narrative of which has already often been published, in order to extract, if possible, a few psychological principles that are likely to be applicable to the future.

It will suffice, for a beginning, to consider only two sample nations, Britain and Germany. The reader is requested to compare Fig. 1 with the following narratives.

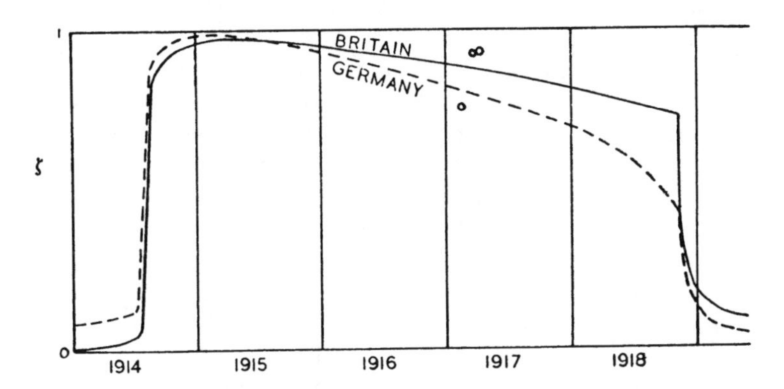

Fig. 1.—The ordinate ζ represents the fraction of the population that was overtly in favor of the war of 1914-18. This diagram is based on historical facts, not on mathematical theory.

Circles mark the British by-elections.

Statistical surveys of public opinion are not available for the First World War. As I have occasionally to rely on personal impressions, I should perhaps explain that by then I was already intensely interested in the group-psychology of peace and of war. I spent the years 1916 (Sept.) to 1919 (Feb.) in a British motor-ambulance convoy (S.S.A.13) attached to a French infantry division (16ieme), a situation which allowed me to observe the moods of the French soldiers and to compare them with official propaganda. In 1919 I published a book on the *Mathematical Psychology of War* (28).

In Britain

Looking at the situation as it appeared in the early months of 1914 one can observe that there had been in July 1911 an angry incident between Britain and Germany concerning Agadir; an arms-race had been in progress since 1909 and yearly was growing more evident; the Northcliffe Press was repeatedly calling attention to German ambition and rivalry with Britain in trade; yet it was also commonly said that nobody in Britain wanted war with Germany. Those who remember that time will perhaps agree that "nobody" may be interpreted as implying that ζ was certainly less than 0.1 and probably less than 0.01. In June 1914 the British newspapers were occupied with Ireland rather than Germany. In July 1914 they switched their attention to the Continent of Europe. After the Austrian ultimatum to Serbia on July 23, 1914, the "Times," "Daily Telegraph," and "Daily Mail" were in favor of war, while the "Manchester Guardian," "Daily News," and "Nation" argued for neutrality. See Hirst (**11**, p. 6). There had been a bitter parliamentary feud between the Liberals and Irish Nationalists on the one hand and the Unionists and Ulstermen on the other, accompanied by threats of civil war in Ireland. On July 24, 1914 Asquith reported to the House of Commons that the Conference on Ireland, summoned by the King, had failed to agree. Yet on August 3, 1914 the Unionists and Irish Nationalists offered their support to the Liberal Government in a war against Germany. F. E. Smith, who had taken a conspicuous part in the formation of the illegal Ulster army, accepted office under the government (**24**, Commons 1914, Aug. 31). Evidently ζ rose rapidly in the course of about ten days from a value near zero to a value near unity, as shown in Fig. 1. The German invasion of Belgium had a great effect in Britain. To what value ζ rose is difficult to ascertain. The official report of the debates in the House of Commons from July 20 to July 31 shows lists of members voting on criminal justice, finance, tramways, housing, Anglo-Persian oil; on whether they should continue to discuss disturbances in Dublin; on milk and dairies, agriculture, aliens, and inebriates; but for August and September, 1914 no voting lists appeared. This is a sign that there was general agreement. Yet on Aug. 3, 1914 there were at least nine speeches in the Commons asking the government to try to keep out of war. Hirst (**11**, p. 10) states that about 40 M.P.'s voted against the war. As there were about 670 members we may, by correcting "voted against" into "opposed," take Hirst's statements to imply that $\zeta = 0.94$ for the House of Commons. Out of a Cabinet of 21 members, 2 resigned in protest against the war; and $19/21 = 0.90$.

It must be remembered that Parliament was probably aroused more quickly than the population in general. On the other hand, Parliament would not have supported the declaration of war unless it had a general belief that the population was in favor, that is to say, that $\zeta > 0.5$. During August, 1914 I should say, from memory, that ζ continued to rise; and perhaps it did so for several months. As quantitative data for 1914 are scarce, we may take a hint from those elsewhere stated for 1939. The date of the maximum of ζ cannot be fixed. It may have occurred in 1914 or 1915. But the nation was never perfectly unanimous; for among 5.7×10^6 men called for military service it is stated by Graham (**10**, p. 348) that there were 16,100 genuine conscientious objectors. Whence we may suppose that $\zeta \leqslant 0.9972$. In Britain (and France) the years 1915, 1916, and 1917 were commonly described as a period of deadlock in the war. Food became scarce, prices rose, there were enormous land battles, one's friends were killed; but there was a general determination to persist. It was accompanied by an increasing weariness of war, which at the time was suppressed in the individual, but which had its effect after the armistice.

On Jan. 9, 1917 the Union of Democratic Control held a public meeting at Walthamstow in favor of a negotiated peace. The speakers were Mrs. Phillip Snowden and J. Ramsay Macdonald (afterwards Prime Minister). The meeting had proceeded for about half an hour when a band of soldiers broke it up (38, Feb. 1917, also **36**, Jan. 10, 1917, p. 7).

In the same year, the state of public opinion was tested quantitatively by three parliamentary candidates who proposed "peace-by-negotiation" at by-elections. I am indebted for the data to the Rt. Hon. F. W. Pethick-Lawrence M.P. (now Lord Pethick-Lawrence), who was one of them.

The London "Times" (**36**, Feb. 14, 1917 p. 8) reported that "Mr. Taylor was absent from the contest, being now in custody of the military authorities."

DATE & CONSTITUENCY	CANDIDATES	PROGRAMS	VOTES	$\therefore \zeta$
Feb. 13, 1917 Lancashire, Rossendale	Sir. J. H. Maden A. Taylor	Liberal Peace by negotiation	6019 1804	0.769
Mar. 20, 1917 Stockton on Tees	J. B. Watson E. Backhouse	Liberal Peace by negotiation	7641 596	0.928
Apr. 3, 1917 Aberdeen. South	Sir J. Fleming J. R. Watson P. Lawrence	Liberal Independent Peace by negotiation	3283 1507 333	0.935

It is significant that an opposite kind of anti-government candidate fared about equally badly. At Tewkesbury on May 16, 1916 Mr. W. Boosey stood as an independent candidate with a program of more effective prosecution of the war and was supported by Admiral Rose, Admiral Lord Beresford, and Mr. Leo Maxse. Mr. Boosey scored 1438 votes against 7127 to his Unionist opponent Mr. W. F. Hicks-Beach (36, May 16, 1916, p. 5). This remarkable defeat provoked the author to make a theory of gregariousness, which appears in Part IIB. A similar poll occurred in the next war on May 8, 1941 at King's Norton.

On November 29, 1917, Lord Lansdowne, a former Foreign Secretary, published in the "Daily Telegraph" an appeal for a negotiated peace. His proposal aroused considerable interest but strong opposition.

By the beginning of 1918 some decrease in the enthusiasm for the war had become evident. On Jan. 23, 1918 a conference of the Labour Party, which then was in a minority, adopted at Nottingham a resolution calling for an International Labour Peace Conference in Switzerland. The London "Times" next morning reported that "the peace question dominated the conference throughout the day." Nevertheless, for the population as a whole, ζ must surely at that date have been much above 0.5.

The armistice of November 11, 1918 was hailed with general rejoicing in Britain and France, because it implied victory and repose. There must surely have been a rapid fall in British ζ at that time.

Yet ζ certainly did not fall at once to zero; for there were some people who thought that the war ought to continue. In the "Hang the Kaiser" election campaign of December 1918 hatred was much in evidence; and in fact the war did continue, in the modified form of a blockade, which did much harm to German children, until June 1919. By the year 1922 a general tendency to forgive and forget was evident in London. But as late as 1938 I knew one lady who said she would never again speak to a German.

On piecing together these facts it is possible to draw a rough diagram of the course of ζ in time (Fig. 1).

In Germany

The value of ζ prior to the war might be expected to appear as a feature in the controversy on War Guilt. In the three articles under that title in the Encyclopedia Britannica (7), written severally by a Frenchman, a German, and a Briton, the only reference to ζ is Herr H. Lutz's statment on p. 354 that "the great mass of the peoples had no desire for war."

Mendelssohn Bartholdy (21, p. 24) remarks that a part of the governing classes in the years before 1914 had kept its faith in the Divine Government of the World and prayed for a war as a test for their bodies, as a probation for their moral courage, as a cauterizing of the sores in their body politic, as a promise of healthy frugal poverty after the period of enervating prosperity which their country had enjoyed for too many years.

This sentiment was known also in Britain: Field Marshall Lord Roberts had prescribed war as a national tonic. But in Germany an openly aggressive book had been published in 1912 by Bernhardi (1). The author has interpreted the preceding facts by drawing the prewar curve of ζ for Germany slightly above the curve for Britain.

Mendelssohn Bartholdy remarks (21, p. 26) that, during the first months of the war, the Germans experienced a unity of state and nation such as is rarely felt in time of peace. They realized that life consisted in sharing a common fate, unreservedly; and that for those who tried to preserve a hoard of their own, whether of material wealth or of skill and knowledge, the death of Ananias and Sapphira would be the appropriate punishment.

By the end of the first year and during the second year there began to be grievances about the inequality of sufferings and rewards and about the censorship which separated soldiers from their relatives at home (21, pp. 26-29).

It had become evident by the autumn of 1916 that the Supreme Command demanded more supplies of men and materials than the country could produce and still maintain its past standard of economic life and prevailing conditions of labor (21, p. 77). In 1916 the Prussian Ministry of War had come to believe that disaffection was spreading (21, p. 80). In November 1916 they made an agreement with the trade-union leaders (21, p. 81) for a form of industrial conscription known as the Hindenburg Program. This was passed by the Reichstag on Dec. 2, 1916 by 235 votes against 19 (21, p. 82).

On July 19, 1917 the Reichstag expressed by 214 votes to 116 its willingness for a peace involving no conquests and the freedom of the seas (5, of May 4, 1918, p. 224). In interpreting such votes one must remember that the Reichstag had some influence on, but no effective control over, the Chancellor or the Army.

On August 8, 1918 a German Crown Council at Spa concluded that "We can no longer hope to break the war-will of our enemies by military operations" and that "the object of our strategy must be to paralyze the enemy's war-will gradually by a strategic defensive" (7, 23, p. 773).

On Sept. 29, 1918, the German supreme command appealed to their government for an armistice; and on Oct. 3 an appeal was sent to President Wilson. On Oct. 29 the German High Sea Fleet mutinied.

Yet the armistice of Nov. 11, 1918, certainly was not in accordance with the wishes of all Germans. Hitler in particular was furiously indignant (12, Vol. I, Chap. VII).

Prospective peace-terms were almost irrelevant until the nations were tired of war

It may be objected that ζ is a function of the possible terms of peace and so is left meaningless unless those terms are specified. Such an objection attributes too much rationality to the belligerents. For in Britain and France during 1915 to 1917 it was generally felt to be almost indecent to speak of peace-negotiations. Any peace feelers put out by the enemy were denounced in those newspapers that had the large circulations as lies, as dangerous traps, or as signs of the enemy's weakness that promised an early victory. To obtain a satisfying full account of what was going on in the way of tentative negotiations or peace-aims, it was necessary to supplement the great newspapers by looking in special publications that had only small circulations, notably in the "Cambridge Magazine" and in "Commonsense."

For example on Dec. 12, 1916 Bethmann-Hollweg, then German Chancellor, proposed that the belligerents should meet in order to try to come to terms. There had been among the Socialists in Germany a favorable reception of President Wilson's idea of a League of Nations. Looking back, we may ask whether it would not have been better for all concerned if the war had ended then. Actually there followed first Lloyd George's policy of the knock-out blow, and secondly the German response in the form of unrestricted submarine warfare (**18**, Ch. 39).

It appears from the facts that fighting inhibits negotiation. To attempt both at the same time feels as unnatural as to swing the arms in phase with the legs in marching, instead of in the usual antiphase. A nation at war can hardly think about negotiation. The question whether negotiation might be in its ultimate interest is not allowed to arise. The nation continues to fight because all its energy is going that way and the valves towards negotiation are automatically shut by some general property of the central nervous system. Vice versa, the process of negotiation inhibits fighting. Diplomats in conference do not draw revolvers when they disagree.

But by the year 1918, when much weariness had developed, President Wilson's peace proposals were taken seriously.

DUAL MOODS AND THEIR TRANSFORMATIONS*

We have now to classify the population; and much depends on the choice of suitable categories. The findings of the psychoanalysts lead us to expect that the dispositions to be friendly, to fight, and to submit coexist simultaneously inside the ordinary "individual" (a word the meaning of which has altered in the same way as that of "atom" in physics); and that what happens when an individual changes from one category to another is like a change of government in the nation, insofar as a different party to the individual's internal controversy takes over control of his voluntary muscles. It is convenient to represent the mood of an individual by two words in a bracket, thus

$$\begin{Bmatrix}\text{friendly}\\ \text{hostile}\end{Bmatrix} \quad \text{or thus} \quad \{\text{friendly, hostile}\}$$

in which the upper or left-hand word names the overt mood and the lower or right-hand word names the hidden mood. A special type of bracket is used to distinguish a dual mood from a parenthesis. Of course this notation does not fully represent the complicated possibilities of human nature. For it is well known that there are vari-

* A very brief account of this was read to the British Psychological Society at Nottingham on April 20, 1941.

ous degrees of concealment. A man may keep his hostility out of his polite conversation and know of its existence; or his hostility may be disguised even in his dreams. However, the notation of dual moods certainly permits a description less ambiguous than that given by the standard phrase in the British King's speech at the opening of Parliament in peacetime: "My relations with foreign powers continue to be friendly."

It will be best to explore the possibility of a theory of dual moods before attempting anything more thorough. For applied mathematics, insofar as it is an art, is the art of leaving out everything that is not essential to a restricted purpose.

An arms-race involves conscious calculations about numbers of men, ships, and aeroplanes; but the change of mood that occurs during an arms-race can hardly be overt. For in 1913 it was said that nobody in Britain wanted war with Germany (18, pp. 32-33). Yet the fact that such a statement was felt to be necessary shows that some change of mood was occurring. A number of examples of speeches by statesmen could be quoted to show that during the years 1935-'38 they made many protestations of their peaceable intentions. The present hypothesis is that the change of mood during an arms-race is:

$$\begin{Bmatrix}\text{friendly}\\\text{friendly}\end{Bmatrix} \rightarrow \begin{Bmatrix}\text{friendly}\\\text{hostile}\end{Bmatrix}$$

Although the European arms-race of 1909-'14 passed over continuously into the outbreak of war, as far as the process was indicated by the spending of money on armaments, yet there is reason to believe that the psychological mechanism of the arms-race was distinct from that of the outbreak. Their time relations were of a different order of magnitude. It has been shown (30, 32) that the arms-race of 1909-'14 was characterized by an instability coefficient λ of the order of 0.5 year^{-1}, or by its reciprocal two years. Although this constant was derived from defence budgets, it does not involve any monetary unit and so may be compared with times derived from changes of opinion. There is evidence (18, p. 33) that the majority of Britishers changed their opinions about war with Germany during a week in 1914 between July 24 and August 4. There is thus a contrast, between two years and a week, which suggests that the psychological mechanisms were distinct.

We may begin to explain the outbreak of war by supposing that it is the change

$$\begin{Bmatrix}\text{friendly}\\\text{hostile}\end{Bmatrix} \rightarrow \begin{Bmatrix}\text{hostile}\\\text{friendly}\end{Bmatrix}.$$

This explanation accounts for the suddenness of the outbreak by stating that the hostile mood is already excited and has only to take command of the voluntary muscles. The "susceptibles" in Kermack and McKendrick's theory of epidemics **(16)** corresponds, in the present theory of war-fever, to those who are in the mood {friendly, hostile}.

The persistence of friendly feelings after the war has begun, as shown in the second bracket, is in accordance with the fact that naval men habitually take trouble and face danger to rescue defeated enemies from drowning. Armies also usually treat their prisoners in accordance with an international convention. Further evidence of the existence of a suppressed affection for the enemy is afforded by certain war neuroses. Spillane **(35**, pp. 5, 6**)** mentions "guilt over killing" as one of four or five main sources of mental stress in the front-line soldier. Wright **(42**, p. 156**)** remarks: "It must not be thought that battle dreams are always of being killed, of being buried; they are not always associated with terror of being killed, and some of the most vivid and most obstinate in their repetition are the dreams of killing, the dream of the enemy who has been bayoneted, the picture of the man who has been shot at close quarters." Calvert **(4**, p. 101**)** names three or four public executioners who committed suicide in remorse for their official duties.

Let us pass on to consider the middles and ends of wars. The author was led to the theory of dual moods by his failure, in the course of more than a hundred pages of mathematical manuscript, to explain the long steady persistent phase of a war, followed by its comparatively sudden end, as a transition of individuals from a simple mood of fighting to a mood of defeat. The difficulty was that various trial hypotheses, built after the model of Kermack and McKendrick's theory of epidemics (see **15**), either did not make the middle phase sufficiently persistent, or else did not make the final collapse sufficiently sudden. In other words the deduced curve of war-fever as a function of time was too like a Gaussian normal curve of error: it lacked the observed nearly horizontal top shown in Fig. 1. The author then recalled, from personal memories, that the change of mood which occurred in Britain and France during 1917 and 1918 consisted only slightly in an increase in the number of people who openly admitted a willingness to compromise with the enemy; but that meanwhile there was in existence a yearning for the end of the war, which increased in intensity, but was suppressed in almost all individuals by their patriotic sentiment. It will be shown that the steady persistence in the middle phase and the sudden collapse at the end are both easily explained if we regard the mood of fighting as dual, so that during the middle phase of persistence and attrition the following change is occurring:

$$\begin{Bmatrix}\text{hostile}\\ \text{friendly}\end{Bmatrix} \rightarrow \begin{Bmatrix}\text{hostile}\\ \text{war-weary}\end{Bmatrix}.$$

It is the accumulation of war-weariness in the subconscious that, on this theory, makes defeat sudden; for the end of a war corresponds to the change

$$\begin{Bmatrix}\text{hostile}\\ \text{war-weary}\end{Bmatrix} \rightarrow \begin{Bmatrix}\text{war-weary}\\ \text{hostile}\end{Bmatrix}.$$

The whole course from a friendly peace through war to a resentful peace may now be summarized thus:

$$\begin{Bmatrix}\text{friendly}\\ \text{friendly}\end{Bmatrix} \xrightarrow{\text{Arms-race}} \begin{Bmatrix}\text{friendly}\\ \text{hostile}\end{Bmatrix} \xrightarrow{\text{Outbreak}} \begin{Bmatrix}\text{hostile}\\ \text{friendly}\end{Bmatrix} \xrightarrow{\text{Attrition}} \begin{Bmatrix}\text{hostile}\\ \text{war-weary}\end{Bmatrix} \xrightarrow{\text{Armistice}} \begin{Bmatrix}\text{war-weary}\\ \text{hostile}\end{Bmatrix}.$$

It will be observed that in the two rapid changes, namely, those at the beginning and end of fighting, the personality "turns upside down," the subconscious mood becoming overt. On the other hand, in the two slow changes, namely, those during the arms-race and during the middle phase of a war, the process is confined to the subconscious. There are precedents for making these contrasts. Thus William James (**13**, pp. 237-243) explained the sudden type of religious conversion by an invasion from a subliminal mental region; and the slowness of unconscious changes was mentioned by Plato (**26**, Secs. 64, 65), a fact pointed out to me by Prof. J. C. Flugel.

The transformation {friendly, hostile} → {hostile, friendly} is hardly directly reversible

Populations have returned from the second to the first-named dual mood, but only by a roundabout process, via war-weariness and a long peace.

The evidence for this assertion is that influential public men have tried to stop wars before intense war-weariness had developed, and they have signally failed. Examples follow:

(1) Concerning the opposition of Bright and Cobden to the Crimean War in 1854, Justin McCarthy (**20**, **2**, 290) says:—"The eloquence that had coerced the intellect and reasoning power of Peel into a complete surrender to the doctrines of Free Trade, the eloquence that had aroused the populations of all the cities of England and had conquered the House of Commons, was destined now to call aloud to solitude."

(2) During the Boer War, there was in Britain a considerable "Stop-the-War" movement, but nothing came of it, at that time. Mr. David Lloyd George, afterwards Prime Minister, was an active opponent of the war. He had to escape from a wrathful public meeting in Birmingham in 1900 by disguising himself as a policeman.

(3) During the First World War, Mr. Ramsay Macdonald, the Labor Leader, and afterwards Prime Minister, wished to go to Stockholm to discuss peace terms with other socialists in the summer of 1917; but the seamen refused to take him (18, p. 1125).

(4) The Pope's proposals for peace in August 1917 ". . . were disposed of at once in a few paragraphs by the *jusqu'auboutistes* of all the belligerent countries as being pro-German or Pro-English, as the case might be, and Mr. Wilson's reply gave them their *coup de grace* (3, Sept. 22, 1917).

(5) Lenin and Stalin, who subsequently reorganized Russia, tried during the first three years of the First World War to turn aside the Russian proletariat from fighting Germans, and to direct them instead towards dispossessing the Russian bourgeoisie. They did not succeed until much war-weariness had developed in Russia (34, p. 167).

(6) A month after the beginning of the Second World War, the Germans having conquered Poland, Mr. David Lloyd George, who had led Britain to victory in 1916-'19, spoke in the House of Commons on October 3, 1939 to the effect that if the then expected peace proposals from Hitler were to be received by way of a neutral Power, and if the then neutral Great Powers Italy, Russia, and U.S.A. would cooperate in framing a settlement, the matter should not be hastily rejected. This speech of Mr. Lloyd George's aroused little sympathy and much indignation in Britain (9 of October 4, 1939).

By what process does friendliness return?

It is common knowledge that the resentment left by a war slowly passes away. By counting wars it has been shown that the number of revenges is roughly halved by each decade of delay. There must be a vast amount of scattered information as to the stages in the emotional process; but I do not know of any concise summary. Tentatively I suggest that the ordinary process during a long peace is

$$\begin{Bmatrix}\text{war-weary}\\\text{hostile}\end{Bmatrix}\ \overset{(5\text{ years})}{\rightarrow}\ \begin{Bmatrix}\text{friendly}\\\text{hostile}\end{Bmatrix}\ \overset{(30\text{ years})}{\rightarrow}\ \begin{Bmatrix}\text{friendly}\\\text{friendly}\end{Bmatrix}.$$

The latter process in the subconscious is so slow that important political changes have to wait for the uprising of a generation that did not experience the war. Much depends on how history is taught to children.

It is also obvious that the changes of mood occur at very different speeds in different individuals. Of five Boer generals who fought against the British in the war of 1899 to 1902, Botha and Smuts began, as soon as peace was declared, to organize conciliation between the Boers and British (**7**, **20**, 846) and continued steadfastly in that purpose; whereas de Wet and Beyers took the opportunity of 1914 to revolt (**7**, **7**, 296 and **7**, **3**, 490); and Herzog made bitter speeches until 1918 and continued to be distrustful of the British until the end of his life in 1942 (**7**, **11**, 527 and **36** of Nov. 23, 1942).

Analogies with disease

The common phrase "war-fever" suggests:

(i) infection, of course mental, by sights and sounds;

(ii) a rise of excitement, analogous to rise of temperature in fever;

(iii) that some people may be immune to these warlike excitements;

(iv) that others may acquire immunity, alias war-weariness, as the result of a long bout of fighting;

(v) that war-weariness may fade away after the end of the war.

But are these suggestions misleading, or are they a concise summary of reliable observation? I have certainly found facts to illustrate all five of them; but this may not be the place to state them in detail. Briefly it may be said that the rush of volunteers to enlist when war is expected, does resemble an infective process. Recently I found a quite independent piece of evidence. Numerous attempts to explain the peculiar statistical distribution of the number of nations on the two sides of a war all failed, until the infectiousness of fighting was admitted among the hypotheses (See publication 31).

PART IB

MATHEMATICAL HYPOTHESES

Hints from a theory of epidemics of disease

Kermack and McKendrick have published (**15**, **16**) a thorough mathematical theory of epidemics such as those of bubonic plague.

Let us look at this theory with a view to adapting it to politics. They give (**15**, p. 713) the following simple approximations. The total number N of persons in the population is made up of:

x, the number who have not had the disease,
y, the number who are diseased,
z, the number who have either died or recovered,
so that $N = x + y + z$.

For simplicity N is regarded as a constant, during the brief course of a severe epidemic, by neglect of births, migrations, and deaths due to other causes. It is assumed that those who have recovered are perfectly immune to the disease and that there is no other kind of immunity.

Their most interesting hypothesis is that the chance of infection is proportional to the product of the densities of two intermingling populations, one of diseased, the other of susceptible persons, so that

$$dx/dt = -Axy,$$

where t is the time and A is a positive constant.

The number removed in unit time by death or recovery is taken to be proportional to the number who are diseased, so that

$$dz/dt = By,$$

where B is a positive constant.

Starting from these simple ideas, Kermack and McKendrick eliminated x and y, and then integrated the equation for z under the restriction that $Az/B << 1$. In this way, they were able to obtain an interesting agreement with the statistics of an epidemic of plague which occurred in Bombay in 1905-'06. A remarkable consequence of their assumptions is that there is a threshold density of population, $N_0 = B/A$, below which no epidemic can occur, because a diseased person will die or recover before he has, on the average, infected another person.

The further refinement of these ideas by Kermack and McKendrick (**16**) consists partly in allowing A and B to vary during the course of an illness and partly in the removal of other restrictions. They are thus led to integral equations of Fredholm's type. But, as thoroughness makes for incomprehensibility, we shall do well to leave these generalizations aside for the present.

First notion of a mathematical theory of war-fever

Now let us try, quite tentatively and skeptically, to alter the theory of Kermack and McKendrick so as to make it applicable to war-fever. Let there be two opposing nations and let the population of each be divided into the same two fractions x and y in different moods, so that

$$x + y = 1 . \tag{1}$$

Let the changes of y follow the law of infection in its simplest form, namely,

$$dy/dt = C\, xy , \tag{2}$$

where x belongs to one nation, y to the other, and C is a constant. The solution of this pair of equations is

$$ct + \text{constant} = \log_e \left(\frac{y}{1-y} \right) , \tag{3}$$

and is graphed in Fig. 2, where time increases horizontally to the right. The curve begins on the left with a very small minority, the increase of which is barely perceptible. The curve then passes continuously through a phase of rapid change of mood among the great bulk of the population. The curve continues on the right into a phase of persistence in almost perfect unanimity. These are the features which are required for representing the outbreak of a war. The same

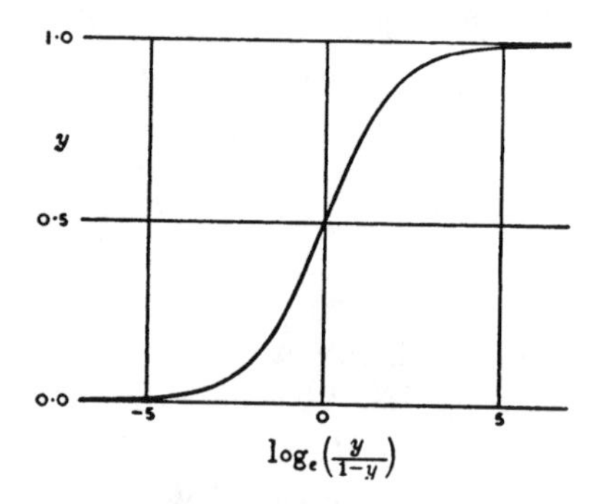

War Fever.

FIG. 2. —Simple approximation to the outbreak of a symmetrical war. The abscissa is proportional to the time. The ordinate is the fraction, of either population, overtly in favor of war.

curve, perhaps altered in scale, would represent the end of a war if the ordinate were the fraction of the population that wished to make peace. Many later developments may also be regarded as variations on this simple mathematical theme. But, as the intention is to make a theory capable of representing all the phases of a war, many questions will have to be settled before we can return to describe more thoroughly the outbreak and the end. See Cases II, IV, V, VI, VII.

How should time appear in the hypotheses?

Time, although appearing in the foregoing theme explicitly in the integral (3), appears only in the form d/dt in the hypotheses (1) and (2). The following theory is built on the same plan. It may be called *humanistic* because the only interactions which are explicitly represented are those between human beings, or *bipersonal* because two types of person interact. We know of course that summer campaigns and winter quarters depend on interactions with the sun; bombings and invasions on those with the moon; and that the weather has a potent effect. But all those variations are here regarded as detail to be smoothed out.

But there are other types of interaction, real or fancied, which cannot be regarded as detail. There are people who say that wars come to an end in God's good time; and there are a few that call a war a work of the Devil; Napoleon talked about his Destiny; some British newspapers now publish predictions connected with the planets. No such effect appears explicitly in the following theory; but faith or superstition might be taken into account as modifying the constants.

Even in a humanistic theory, the time required for the forgetting of past benefits or injuries might suitably appear in the hypotheses. The mathematics of memories which are neither very short nor very long would involve functionals, as shown by Volterra (**40**), and these would seriously complicate the treatment; whereas it is advantageous to subdue the mathematics in order to have attention to spare for the social psychology. It is proverbial that the memory of the public is short. And even when memories do persist, there is a tendency to set them aside, if they belong to bygone circumstances. For example, in November 1942 the Americans and the British were collaborating with Admiral Darlan, whom a month earlier they had reason to distrust, before they invaded North Africa. It is this customary opportunism, this concentration on the affairs of the day, that is represented in the following hypotheses, where time appears only in the operator d/dt. To describe a phenomenon is not to praise it.

Notation

Let us simplify the problem as much as possible by supposing that the struggle is between only two nations, or two alliances, whose symbols are distinguished by suffixes 1 and 2. Let N_1 and N_2 be their populations at the beginning of the war. We need not attend to births, to migrations, nor to deaths, except those caused by war.

It is now necessary to define symbols for the numbers of people in each of the dual moods. Let us regard the total population N_1 of the first nation as a constant consisting at any instant of

$$\left.\begin{array}{l} \beta_1^* \text{ persons in the mood } \{\text{friendly, friendly}\} \\ \xi_1^* \text{ persons in the mood } \{\text{friendly, aggressive}\} \\ \eta_1^* \text{ persons in the mood } \{\text{aggressive, friendly}\} \\ \theta_1^* \text{ persons who have died because of the war} \\ \rho_1^* \text{ persons in the mood } \{\text{aggressive, war-weary}\} \\ \omega_1^* \text{ persons in the mood } \{\text{war-weary, aggressive}\}\,. \end{array}\right\} \quad (4)$$

Then

$$\beta_1^* + \xi_1^* + \eta_1^* + \theta_1^* + \rho_1^* + \omega_1^* = N_1\,. \quad (5)$$

The purpose of the asterisks will be explained presently.

If a referendum were held on the question whether a war should be engaged in or continued, we may provisionally suppose that $\eta_1^* + \rho_1^*$ people would vote in favor of the war and that $\beta_1^* + \xi_1^* + \omega_1^*$ would vote together, although for different motives, against the war.

There has been a prolonged controversy as to whether mental events are quantitative. A committee of the British Association (2) reported on the question in 1938 and again in 1939. It was generally agreed that mental events may be more intense or less intense. But one party, including the author, was impressed by many experiments in which the ratios of intensities of sensations were mentally estimated. The other party was not impressed. On the present hypotheses, the state of mind of an individual belongs to one or other of five qualitative classes. To the author this "all or nothing" classification appears as an imperfection, to be tolerated for the sake of simplicity. But the present treatment may be approved by the opposite party to the controversy at least insofar as any reference to quantitative mental events is here avoided.

Pure numbers

When we have to frame hypotheses it is easiest, and therefore safest, to think of the behavior of a typical individual and of how many others there are like him. But when, in a trial working, the hypotheses were thus formulated with numbers of persons as variables, it was found that the deductions were encumbered in many places by the numbers N_1, N_2. So the author received a strong impression that the theory ought to be re-written in terms of fractions of the population. The asterisk notation allows us easily to do it in either way. For let $\beta_1^* = \beta_1 N_1$, $\xi_1^* = \xi_1 N_1$ and so for all the other variables such as

$$\omega_2^* = \omega_2 N_2\,. \quad (6)$$

Removal of an asterisk from a variable is equivalent to division by the corresponding total population. Accordingly, (5) may now be expressed more neatly as

$$\beta_1 + \xi_1 + \eta_1 + \theta_1 + \rho_1 + \omega_1 = 1\,. \tag{7}$$

This transformation from numbers of people to pure numbers is an extension to sociology of the Theory of Dimensions in physics. It is well known that valuable clues as to the behavior of material objects have been obtained from the principle that it must be possible to express the observed facts as relations between pure numbers made up by combining separate measurements in such a way that the units of measurement cancel. For our purposes a person is one of the units of measurement. Hence we may expect theoretical advantages to accrue from the cancellation of the unit of measurement in η_1^*/N_1 or the like.

Threats and self-restraint during an arms-race

We next have to consider the conversion of persons from thorough friendliness to subconscious hostility, so that ξ grows at the expense of β. One cause of this conversion may have been discontent with the treaty-situation. But when the existence of an arms-race has become obvious, these original discontents have faded into the background of attention and the foreground is occupied by the threat from the other nation, as expressed in its army, navy, and airforce, together with any indications of aggressive thoughts. So far this statement is in accordance with the linear theory of arms-races, published elsewhere by Richardson (30, 32). As armaments do not appear explicity in the present purely psychological theory, we must represent them by the number ξ^* of persons in the mood, of which armaments in peacetime are the outward and visible sign. An important effect in an arms-race is accordingly of the type

$$d\xi_1^*/dt = K\beta_1^*\xi_2^* = -\,d\beta_1^*/dt\,, \tag{8}$$

where K is a positive coefficient to be adjusted later to fit with observation; and K is assumed, in the meantime, to be a constant. There will be another pair of equations obtainable by interchanging the suffixes and by readjusting K. The product $K\beta_1^*\xi_2^*$ may be interpreted thus: there are β_1^* persons liable to change under an influence of strength ξ_2^*, and so $K\beta_1^*\xi_2^*$ is the probability of one of these persons changing in unit time. Any misinformation about the number of subconsciously hostile persons ξ_2^* in the other nation may be regarded as modifying K.

Similar but greater effects are produced by outspoken hostility. For example, the German General Bernhardi (1) published in 1912 a book "Deutschland und der nächste Krieg" in which he asserted that "Even English attempts at a rapprochement must not blind us to the real situation. We may at most use them to delay the necessary inevitable war, until we may fairly imagine that we have some prospect of success." His utterances were mentioned in the British newspapers and caused a considerable stir (7, article "Bernhardi," also 8). To represent this effect of open hostility in the expression for $d\beta_1^*/dt$, we must replace ξ_2^* by $\xi_2^* + w_{12}\eta_2^*$, where w_{12} is a weight far exceeding unity. The estimation of w_{12} is difficult, because its cofactor $\beta_1\eta_2$ is almost zero, except in one brief phase of a war.

It is arguable whether or not ρ_2^* should be included along with η_2^* in the expression above. On the one hand, it may be said that the persons counted in ρ_2^* differ from those counted in η_2^* only by a concealed distinction, so that to another nation they would all seem alike. But on the other hand it may be replied that such concealments are usually imperfect. If Bernhardi had been secretly bored with militarism, would he have troubled to write his aggressive book? Whichever way the decision goes, it makes scarcely any practical difference to the theory; for the question is whether to include or omit the product $\beta_1^*\rho_2^*$, the first factor of which is negligible in war and the second factor in the peace just before a war.

The expression so far proposed, namely,

$$d\beta_1^*/dt = -K\beta_1^*(\xi_2^* + w_{12}\eta_2^*), \tag{9}$$

implies that β_1^*, if not already zero, would always diminish if ξ_2^* or η_2^* were greater than zero. Actually nations are not so ready to take offence. They do in fact ignore small threats, for they are loath to enter upon mutual recriminations, knowing that war is expensive, disagreeable, dangerous, and, as some people think, immoral. For example, prior to 1930 the British ignored the Nazis. Therefore, to make the formula agree with historical facts, we must insert in it some expression of restraint. One proposal would be to let A' be the number of persons in the second nation who would be ignored by the first nation even if they did utter threats against it. We should then have

$$d\beta_1^*/dt = K'\beta_1^*(A' - \xi_2^* - w_{12}\eta_2^*). \tag{10}$$

But this improvement lets us in for another difficulty. For suppose that ξ_2^* and η_2^* chanced both to be zero. Then β_1^* would increase; and there is nothing in the formula to prevent β_1^* from increasing beyond N_1, which would be absurd. So we require to insert in the formula some sort of "buffer" to make $d\beta_1^*/dt$ vanish when $\beta_1^* = N_1$. This may be arranged by replacement of A' by $A'(N_1 - \beta_1^*)$. But by equation (5)

$$N_1 - \beta_1^* = \xi_1^* + \eta_1^* + \theta_1^* + \rho_1^* + \omega_1^*,$$

and in the early phase of threats prior to a war, θ_1^*, ρ_1^*, ω_1^* are all zero, so that $N_1 - \beta_1^* = \xi_1^* + \eta_1^*$. So, after rearranging the constants, we arrive at the hypothesis

$$d\beta_1^*/dt = \beta_1^*\{A_1'(\xi_1^* + \eta_1^*)\} - K_{12}'(\xi_2^* + w_{12}\eta_2^*)\}. \qquad (11)$$

When this is expressed in terms of fractions of the populations, it becomes

$$d(\beta_1 N_1)/dt = \beta_1 N_1\{A_1'N_1(\xi_1 + \eta_1) - K_{12}'N_2(\xi_2 + w_{12}\eta_2)\}, \qquad (12)$$

and, on putting $A_1'N_1 = A_1$, $K_{12}'N_2 = K_{12}$, we obtain finally on division by N_1

$$d\beta_1/dt = \beta_1\{A_1(\xi_1 + \eta_1) - K_{12}(\xi_2 + w_{12}\eta_2)\}, \qquad (13)$$

in which A_1 and K_{12} are positive and of the dimensions of reciprocals of a time. There must be equal and opposite terms in $d\xi_1/dt$ or elsewhere to keep N_1 constant.

We have been led to the term $A_1(\xi_1 + \eta_1)$ in equation (13) by considerations about patience and buffering. The meaning of

$$A_1(\xi_1 + \eta_1)$$

may be illustrated, in two different ways, by the words of individuals:
(i) When it is pointed out to a citizen that there are people in another nation who would like to make war on his nation, he may reply "Oh yes, I know there are a few; but we have fire-eaters in our own country, and we realize that they are of no importance." That is to say, he compares η_2^*, not with an absolute constant, but with η_1^*. If such a comparison were perfectly impartial, and were the only motive operating, then would A_1 equal $K_{12}w_{12}$. But it is well known that such comparisons are far from impartial, because a foreign jingo looms larger than a home-grown jingo; and consequently the "weight" $K_{12}w_{12}$ attached to the foreigner is greater than the "weight" A_1 attached to the compatriot. Thus $K_{12}w_{12}/A_1$ is a magnification of the sort referred to in the parable of the mote and the beam (**19**, Ch. 7, v. 3). The value of $K_{12}w_{12}/A_1$ is difficult to estimate. From reading the newspapers, the author ventured the preliminary guess that

$$K_{12}w_{12}/A_1 \geqslant 3\,. \qquad (14)$$

An estimate of K_{12}/A_1 will be made later: see equation (80).
(ii) In Britain in 1909, when the Conservatives asked for more bat-

tleships, saying "We want eight and we won't wait," the Liberals made reply about extravagance and jingoism. The relation between the government and the opposition in Britain was, in peacetime, often one of contrasuggestibility. What either proposed the other opposed. The term $\beta_1 A_1 (\xi_1 + \eta_1)$ in equation (13) expresses a kind of contrasuggestibility. For according to that term, the existence of η_1^* jingos would cause β_1^*, the number of internationalists, to increase.

Scarce types of human character

In this theory of European populations we have unfortunately to neglect some scarce minorities. Thus equation (13) does not describe the behavior of the Christian Pacifists; for they persist in a mood of overt friendliness when subjected to threats. But in the "Peace Ballot" (14) in Britain in 1936, they amounted to only 1 in 700 of those who returned voting papers.

The outbreak of a war

The chief effect in this phase is the rapid conversion of many persons from the mood {friendly, hostile} to the mood {hostile, friendly}, because they believe in the existence of an important number of overtly hostile foreigners. This effect is represented by

$$d\eta_1^*/dt = C_{12}'\xi_1^*(\eta_2^* + \rho_2^*) = -d\xi_1^*/dt\,, \tag{15}$$

where C_{12}' is a positive constant.

On transformation to fractions of the population, this becomes

$$d(\eta_1 N_1)/dt = C_{12}'\xi_1 N_1 N_2 (\eta_2 + \rho_2) = -d(\xi_1 N_1)/dt\,.$$

Now let $C_{12}'N_2 = C_{12}$, and we have

$$d\eta_1/dt = C_{12}\xi_1(\eta_2 + \rho_2) = -d\xi_1/dt\,, \tag{16}$$

where C_{12} is positive and the reciprocal of a time.

People, in this phase, commonly feel that the time for forbearance is past. Accordingly, in (16) there is no term to represent any restraint. This is a contrast with (13).

Hypotheses concerning casualties

The theory must take account of the war-dead in two different connections, first, as causing a small percentage change in the total population, and second, by their very important psychological effects on those who are left alive.

As to the first question, the following list is a guide to the order of magnitude of the effects. It is abstracted from Richardson (32).

The number set after the name of each country is 100 $(\theta_1^*)_T/N_1$, which was obtained by finding its deaths in the armed forces during the First World War and expressing them as a percentage of the country's total population: Serbia $8._1$, Turks of Turkey 5.7, France 3.2_7, Rumania 3.2, Austro-Hungary 3.0_1, Germany 2.9, Bulgaria 2.0, Great Britain & Ireland 1.60, Montenegro $1._5$, Italy 1.4_8, Slavs of Russia 1.2, Belgium 0.51, Greece 0.2, U.S.A. 0.1_1, Portugal 0.09, Japan 0.002.

When we consider, in connection with those statistics, the equation

$$\beta_1 + \xi_1 + \eta_1 + \rho_1 + \omega_1 = 1 - \theta_1$$

and observe that each of the numbers on the left varies across a range of at least 0.5, it becomes evident that we might almost neglect θ_1 and that we may be content to represent θ_1 by quite a rough approximation.

Other things being equal, we may expect that the rate of casualties will, like the rate of encounters of molecules in the kinetic theory of gases, be proportional to the product of the numbers of individuals of the two categories engaged. Volterra **(41)** made an assumption of that type for the encounters between competing species. Kermack and McKendrick **(15)** made the same assumption for encounters between diseased and susceptible persons.

If the whole of each nation is regarded as engaged in the conflict, we may formulate the rate of casualties as

$$d\theta_1^*/dt = E_{12}' N_1 N_2 , \tag{17}$$

where E_{12}' is adjustable to fit the empirical facts.

If, on the contrary, only those who are overtly aggressive are regarded as engaged in the conflict, we ought instead to put

$$d\theta_1^*/dt = E_{12}'' (\eta_1^* + \rho_1^*) (\eta_2^* + \rho_2^*), \tag{18}$$

so that $E_{12}'' > E_{12}$.

If we had to treat the war of two centuries ago, in which only small professional armies were engaged, we should have to introduce separate variables equal to the numbers of fighting men.

But for modern wars and psychological purposes the best simple approximation seems to be (18), because it automatically confines the casualties to the period when aggressive impulses are overt. The theory is intended to represent only the general course of a war with no attention to battles and pauses between them.

The casualties are assumed to be distributed so that in $d\eta_1^*/dt$ there is a term $- E_{12}''\eta_1^*(\eta_2^* + \rho_2^*)$ and in $d\rho_1^*/dt$ there is a term $- E_{12}\rho_1^*(\eta_2^* + \rho_2^*)$. This seems proper because the marks which distinguish the persons counted in $\eta_1{}^*$ from those in $\rho_1{}^*$ are hidden in the subconscious.

Guided by the Theory of Dimensions, let us divide (18) by N_1 and put $E_{12}{}''N_2 = E_{12}$, so that

$$d\theta_1/dt = E_{12}(\eta_1 + \rho_1)(\eta_2 + \rho_2), \tag{19}$$

where E_{12}, like A_1, K_{12} and C_{12} is the reciprocal of a time.

MENTOR: "Now you have committed the childish fallacy that Freud called the Omnipotence of Thought. Before hostilities begin, $\eta_1{}^*$ and $\eta_2{}^*$ are not zero. So you have assumed that jingoes kill each other by merely hating."

AUTHOR: "Yes, I know; but I make that error only in a place where it scarcely matters, because when η_1 and η_2 are both near zero, their product is negligible. You can, if you like, improve the hypothesis by supposing that E_{12} is to be zero if either of $(\eta_1 + \rho_1)$ or $(\eta_2 + \rho_2)$ is less than one half."

Order of magnitude of E_{12}. Physicists have to do with numbers scattered in an enormous range. When they wish to avoid unnecessary accuracy, they state a number to the nearest whole power of ten and call that its order of magnitude. Two positive numbers are said to be of the same order of magnitude when their ratio lies between $10^{\frac{1}{2}}$ and $10^{-\frac{1}{2}}$. This convention will be adopted here. A chief use of it is to enable us to decide quickly which aspects of a phenomenon are most important, and which negligible.

The approximate unanimity of a nation in war is well-known; so that, for a rough estimate, we may put $(\eta_1 + \rho_1)\ (\eta_2 + \rho_2) = 1$ and may regard $d\theta_1/dt$ as a constant during T. From (19) we then obtain

$$E_{12} = (\theta_1)_T/T = (\theta_1{}^*)_T/TN_1, \tag{20}$$

where $(\theta_1{}^*)_T$ is the total number of war-dead. More attention to $\eta + \rho$ will be given later, in Case III.

For the nations engaged in the First World War, there is a list of $100(\theta_1{}^*)_T/N_1$ above. To estimate E_{12} those numbers have to be divided by 100 and by the duration, T, which was 4.3 years for Britain, France, Germany, Austro-Hungary, 3.5 years for Italy, 3.2 years for Russia, and 1.2 years for Rumania. It then appears that:

$$E_{12} \text{ was of the order of } 10^{-2}\ \text{year}^{-1} \tag{21}$$

for most of the nations that suffered severely.

For contrast with the First World War, the following table contains particulars of a very short war, and of two wars between very unequal populations. Uncertain decimal digits are set below the line, in the manner customary in physics.

Dates	T	Nation with suffix 1	N_1	$(\theta_1{}^*)_T$	Ref. for θ	$\therefore E_{12}$
	years		millions			10^{-2} year^{-1}
1866	7/52	Prussions v.	28	10874	6, pp. 49-51	0.29
		Austrians, Bavarians, Saxons & Hanoverians		28765		
1899 to 1902	2.4	Boers v.	0.25	4×10^3	7, 21, p. 66	0.7
		British	45.	21.9×10^3	6, p. 57*	0.020
1939 to 1940	0.28_3	Finns v.	3.6_8	19576	14, p. 4089	1.88_0
		Russians	171.	48745	9 of 1940 Mar. 30	0.100_7

* Including deaths from disease.

The last two wars allow us to compare E_{21}/E_{12} with N_1/N_2; a comparison which will give us a hint as to the possible behavior of the other double-suffixed constants. For this purpose, let suffix 1 refer to the larger population.

	E_{21}/E_{12}	N_1/N_2	$(N_1/N_2)^{0.7}$
British v. Boers	35	180	37.9
Russians v. Finns	18.7	46.5	14.7

So for these very unsymmetrical wars:

$$E_{21}/E_{12} \textit{ was of the order of } (N_1/N_2)^{0.7}. \tag{22}$$

Hypothesis on the growth of suppressed war-weariness

The change in the subconscious mind from friendliness to war-weariness is counted as a transfer of persons from η^* to ρ^*. It is assumed that the change is caused by wounds to oneself, by casualties among one's acquaintances, by the growing scarcity of food and clothing and by all the other inconveniences and deprivations due to war.

It is further assumed that all these, taken together, act so that the number of persons who during dt become subconsciously wearied is proportional both to the number, η_1^*, of those who are susceptible to this change and to the number, $(\eta_2^* + \rho_2^*)$, of those who are trying to weary them. Accordingly, $d\eta_1^*/dt$ will contain a term

$$-B_{12}'\eta_1^*(\eta_2^* + \rho_2^*),$$

and $d\rho_1^*/dt$ will contain an equal but opposite term, where B_{12}' is positive, and for simplicity is regarded as constant.

These numbers of persons are next expressed as fractions of the population. On putting $B_{12}'N_2 = B_{12}$, we obtain in $d\eta_1/dt$ and $d\rho_1/dt$ the terms, respectively,

$$\mp B_{12}\eta_1(\eta_2 + \rho_2), \tag{23}$$

where B_{12} is the reciprocal of a time, and is positive.

The infectiousness of defeatism

There is much common knowledge that defeat is a matter of opinion and so can be communicated from person to person inside the same country. In 1917 and 1918, the French newspapers frequently expressed anxiety about the possible spread of defeatism. In the summer of 1940, the British law-courts began to fine or imprison persons who had made statements likely to cause alarm or despondency. A proverb attributed to Marshall Foch states that "Une bataille perdue, c'est une bataille qu'on a cru perdre."

We should summarize these facts by inserting in the expression for $d\omega_1/dt$ a term proportional to $\rho_1\omega_1$ to represent an infection of defeatism passing from the ω_1^* persons who are in the mood {war-weary, aggressive} to the ρ_1^* persons, belonging to the same nation, who are in the mood {aggressive, war-weary}.

So far we have

$$d\omega_1/dt = D_1\rho_1\omega_1, \tag{24}$$

where D_1 is a positive constant, the reciprocal of a time.

But there is another kind of infection, coming instead from the enemy. For some of those who have become subconsciously war-weary may be willing to confess to that feeling if they perceive that the enemy is admitting it also. Accordingly, we should complete the expression for $d\omega_1/dt$ to make it read

$$d\omega_1/dt = \rho_1(D_1\omega_1 + F_{12}\omega_2), \tag{25}$$

where F_{12} is the reciprocal of a time and is positive.

Final note to Part I

The discussion so far has been mainly inductive, leading from historical facts to mathematical descriptions. In Part IIA, which will follow in a later issue of *Psychometrika,* the argument will be mainly deductive, leading on from those descriptions to predictions, the credibility of which will be discussed.

REFERENCES

1. Bernhardi, F. von. Germany and the next war. Trans. by Powles, A. H. London: Edward Arnold, 1914.
2. British Association for the Advancement of Science. *Report of the Annual Meeting,* 1938, 277-334.
3. Buxton, Mrs. C. R. A weekly survey of the foreign press. *Cambridge Magazine,* Cambridge, England, 1915-1919.
4. Calvert, E. R. Capital punishment in the twentieth century. London: Putnams, 1930.
5. *Commonsense,* a London periodical current during the first world war.
6. Dumas, S. and Vedel-Petersen, K. O. Losses of life caused by war. Oxford: Clarendon Press, 1923.
7. Encyclopaedia Britannica. New York: Encyclopaedia Britannica, Inc., 1929, edition XIV.
8. German, a. J'Accuse, trans. by Gray, A. New York and London: Hodder and Stoughton, 1915.
9. *Glasgow Herald.* Daily Newspaper, Glasgow, Britain.
10. Graham, J. W. Conscription and conscience. London: Allen and Unwin, 1922.
11. Hirst, F. W. The consequences of the war to Great Britain. Oxford: Univ. Press, 1934.
12. Hitler, A. Mein Kampf, translated by Murphy, J. London: Hurst and Blackett, 1939.
13. James, W. The varieties of religious experience. London: Longmans Green, 1902.
14. *Keesing's Contemporary Archives,* weekly periodical, Bristol, England.
15. Kermack, W. O. and McKendrick, A. G. Mathematical theory of epidemics. *Proc. Roy. Soc. Lond. A,* 1927, 115, 700-722.
16. Kermack, W. O. and McKendrick, A. G. The solution of sets of simultaneous integral equations related to the equation of Volterra. *Proc. Lond. Math. Soc.,* 1936, **41**, 462-482.
17. Kruger, P. Memoirs. London: Fisher Unwin, 1902.
18. Lloyd George, D. War Memoirs. London: Odhams, 1938.
19. Matthew, Saint. Gospel, revised version. Oxford: Univ. Press, 1891.
20. McCarthy, J. A history of our own times. London: Chatto and Windus, 1892.
21. Mendelssohn Bartholdy, A. The war and German society. New Haven: Yale Univ. Press, 1937.
22. Mole, G. The ignition of explosive gases. *Proc. Phys. Soc. Lond.,* 1936, **48**, 857-864.
23. Nernst, W. Theoretical chemistry. Trans. by Codd, L. W. London: Macmillan, 1923.

24. *Parliamentary debates, official report* (alias *Hansard*). London: H. M. Stationery Office.
25. Pear, T. H. Peace, war and culture patterns. *Bulletin of the John Rylands Library*, Manchester, 1948, 31, 3-30.
26. Plato. Timaeus.
27. Rashevsky, N. Mathematical theory of human relations. Bloomington, Indiana: Principia Press, 1947.
28. Richardson, L. F. Mathematical psychology of war. Oxford: Wm. Hunt, 1919. (A few copies remain with the author.)
29. Richardson, L. F. The behavior of an Osglim lamp. *Proc. Roy. Soc. Lond.* A. 1937, 163, 380-390.
30. Richardson, L. F. Generalized foreign politics. *British J. Psychol. Monog. Supplt.* No. 23, 1939.
31. Richardson, L. F. The number of nations on each side of a war. *J. Roy. Statistical Soc.*, London, 1946, 109, 130-156.
32. Richardson, L. F. Arms and insecurity, a microfilm, 35 mm. Kilmun, Argyll, Britain: from the author, $5, 1947.
33. Semenoff, N. Chemical kinetics and chain reactions. Oxford: Clarendon Press, 1935.
34. Soviet Union, official history of the communist party of the. Moscow: Foreign Languages Publishing House, 1939.
35. Spillane, J. P. in The neuroses in war, edited by Miller, E. London: Macmillan, 1940.
36. Times, The. Daily newspaper, London.
37. Trotter, W. Instincts of the herd in war and peace: London: Fisher Unwin, 1921.
38. *U. D. C.*, a periodical. London: Union of Democratic Control.
39. Vedel-Petersen, see Dumas.
40. Volterra, V. Theory of functionals, edited by Fantippié, trans. by Long, M. Glasgow: Blackie & Son, 1930.
41. Volterra, V. Leçons sur la théorie mathématique de la lutte pour la vie. Paris: Gauthier-Villars, 1931.
42. Wright, M. B. in The neuroses in war, edited by Miller, E. London: Macmillan, 1940.

PSYCHOMETRIKA—VOL. 13, NO. 4
DECEMBER, 1948

WAR-MOODS: II

LEWIS F. RICHARDSON
HILLSIDE HOUSE, KILMUN, ARGYLL, BRITAIN

PART II A

Introduction to Part II A

This is a continuation of a paper in the previous issue of *Psychometrika* (Sept., 1948, pp. 147-174) in which some historical facts about the First World War were expressed in mathematical form. The present introduction is intended to make the sequel readable by a person who has never seen Part I. This can perhaps just be done; although for ease, clarity, and conviction a prior study of Part I is strongly to be recommended. The independent variable is t, the time. There are supposed to be two opposing nations, or alliances, distinguished by suffixes 1 and 2, their populations initially numbered N_1 and N_2. The numbers of war-dead at any time, t, are θ_1^*, and θ_2^*. When these are expressed as fractions of their respective populations, then the asterisk is omitted, thus:

$$\theta_1^*/N_1 = \theta_1, \quad \theta_2^*/N_2 = \theta_2 .$$

Similarly, an asterisk applied to the symbol for any fraction of the population makes it into the symbol for the corresponding number of persons. It is characteristic of the present theory, in contradistinction to those of Volterra (41) and Rashevsky (27), that θ_1 and θ_2 are both small fractions. Defeat is here regarded as coming, not by extermination, but by a feeling of hopelessness. The other dependent variables are the fractions of the population in various war-moods. Each mood is regarded as dual, having an overt part, and a concealed, subconscious, or unconscious part. The names of the two parts of a mood are placed in the same bracket, the overt part standing either above or before the other. The whole course of the moods from peace through war to peace again is summarized thus:

$$\begin{matrix} & \text{Arms-race} & & \text{Outbreak} & & \text{Attrition} & & \text{Armistice} & \\ \left\{\begin{matrix}\text{Friendly}\\ \text{Friendly}\end{matrix}\right\} & \rightarrow & \left\{\begin{matrix}\text{Friendly}\\ \text{Hostile}\end{matrix}\right\} & \rightarrow & \left\{\begin{matrix}\text{Hostile}\\ \text{Friendly}\end{matrix}\right\} & \rightarrow & \left\{\begin{matrix}\text{Hostile}\\ \text{War-weary}\end{matrix}\right\} & \rightarrow & \left\{\begin{matrix}\text{War-weary}\\ \text{Hostile}\end{matrix}\right\} \\ \beta & & \xi & & \eta & & \rho & & \omega \end{matrix}$$

Under each dual mood is placed the Greek letter which represents the fraction of the population in that mood at time t. These letters are to be given suffixes 1 or 2 according to the side to which they belong.

In Part IB the successive changes of mood have been discussed separately. Each process was regarded as a mental infection, after the manner of the simplest of Kermack and McKendrick's (15) theories of epidemics of disease. Each process was provided with an adjustable constant. Of these w is a pure number; but each of A, K, C, E, B, D, F, is the reciprocal of a time. The orders of magnitude of several of them have been estimated with reference to the First World War. If a change of mood is purely internal to a side, then its constant is given a single suffix 1 or 2; if instead the change of mood involves both sides, then its constant is given a double suffix 12 or 21. The meanings of the constants can be gathered from their situation in the following equations.

Because Part II contains several references to Figures 1 and 2, they are reprinted here.

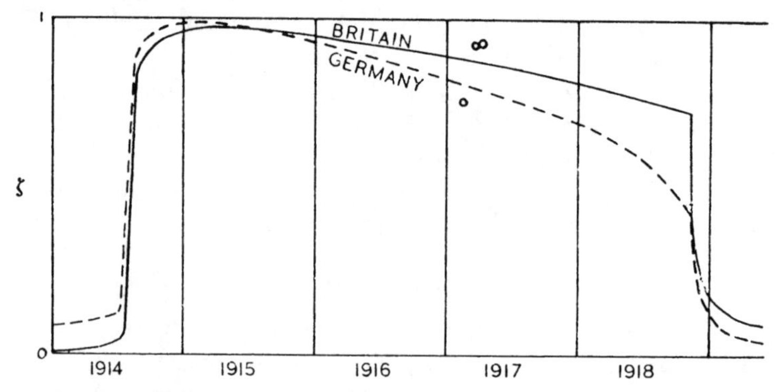

FIG. 1.—The ordinate ζ represents the fraction of the population that was overtly in favor of the war of 1914-18. This diagram is based on historical facts, not on mathematical theory.

Circles mark the British by-elections.

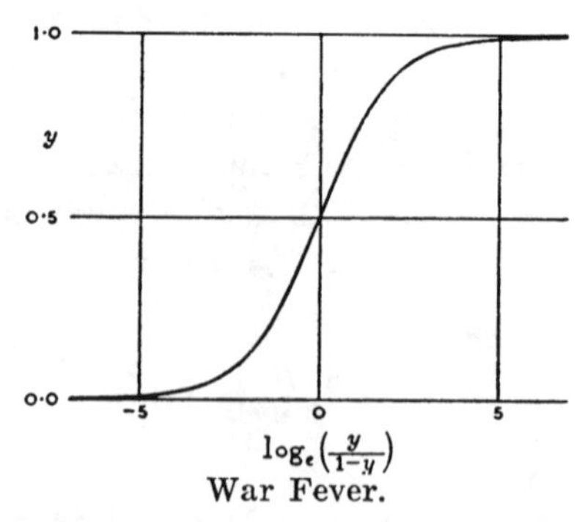

$\log_e\left(\frac{y}{1-y}\right)$
War Fever.

FIG. 2. —Simple approximation to the outbreak of a symmetrical war. The abscissa is proportional to the time. The ordinate is the fraction, of either population, overtly in favor of war.

The complete system of hypotheses

The results of the piecemeal discussion in Part I have now to be collected and fitted together without contradicting one another. Each of the changes is a transfer of persons from one category to another, so that every term must appear twice with opposite signs. Some of the equations of Part I are complete statements and are simply copied here: thus (26) is (7); (28) is (13); (34) is (19); and (38) is (25).* On the contrary, (16) and (23) are partial statements, which are now completed so as to give (30), (32), and (36). The time-rates are printed for the first nation. Those for the second nation can be obtained by interchanging the suffixes, and so are sufficiently represented by merely giving them reference numbers.

$$\beta_1 + \xi_1 + \eta_1 + \theta_1 + \rho_1 + \omega_1 = 1 \qquad (26),(27)$$

$$d\beta_1/dt = \beta_1\{A_1(\xi_1 + \eta_1) - K_{12}(\xi_2 + w_{12}\eta_2)\} \qquad (28),(29)$$

$$d\xi_1/dt = -\beta_1 A_1 \xi_1 + \beta_1 K_{12}(\xi_2 + w_{12}\eta_2) - C_{12}\xi_1(\eta_2 + \rho_2) \qquad (30),(31)$$

$$d\eta_1/dt = -\beta_1 A_1 \eta_1 + C_{12}\xi_1(\eta_2 + \rho_2) - B_{12}\eta_1(\eta_2 + \rho_2) - E_{12}\eta_1(\eta_2 + \rho_2) \qquad (32),(33)$$

$$d\theta_1/dt = E_{12}(\eta_1 + \rho_1)(\eta_2 + \rho_2) \qquad (34),(35)$$

$$d\rho_1/dt = -D_1\rho_1\omega_1 - F_{12}\rho_1\omega_2 + B_{12}\eta_1(\eta_2 + \rho_2) - E_{12}\rho_1(\eta_2 + \rho_2) \qquad (36),(37)$$

$$d\omega_1/dt = D_1\rho_1\omega_1 + F_{12}\rho_1\omega_2 \qquad (38),(39)$$

Here w_{12} is a positive, pure number, regarded as constant. Each capital letter in (26) to (39) is positive, is the reciprocal of a time, and is regarded for simplicity as a constant, except that E_{12} vanishes outside the period of hostilities. This verbal statement will be referred to as - - - - - - - - - - - - - - - - (40), (41). These statements (26) to (41), if found to be true, or even tolerable approximations, will be the most valuable part of the theory; for they express a psychological analysis of observations which first come to our notice in quite other forms. But, before the statements (26) to

*We have so many quantities that some readers may like to have a mnemonic for the symbols. The following is offered on the understanding that it is rather silly and must never be allowed to take the place of the definitions which have already been given: β for beatitude, ξ for excitation, η for eagerness, θ for thanatos, ρ for rueful rumination, and ω because it is the last; A and K because they are respectively related, as will be shown, to the a and k of Richardson's linear theory of arms-races, C because it is like K, E for efficiency in extermination, B for boredom, D for defeatism, F for foreign infection, and w for weight.

(41) can gain credence, they must be subjected to a variety of tests, some simple, some elaborate. Let us begin with the simplest.

Constancy of total population

By adding equations (28), (30), (32), (34), (36), and (38), we obtain

$$\frac{d}{dt}(\beta_1 + \xi_1 + \eta_1 + \theta_1 + \rho_1 + \omega_1) = 0,$$

which agrees with (26).

Buffering

It is necessary to make sure that the hypotheses automatically prevent any one of β_1, ξ_1, η_1, θ_1, ρ_1, ω_1 from over-running its termini 0 and 1. If any one of these six variables is equal to unity we see from (26) that each of the other five must be zero. Several over-runnings are prevented by zero factors arising in this way. To express the results of the test compactly, let x stand for any one of β_1, ξ_1, η_1, θ_1, ρ_1, ω_1. The indications in the body of the following table refer to dx/dt and are deductions from equations (26) to (38).

If x stands for	β_1	ξ_1	η_1	θ_1	ρ_1	ω_1
then at $x = 1$, dx/dt is	$\leqslant 0$	$\leqslant 0$	$\leqslant 0$	$= 0$	$\leqslant 0$	$= 0$
and at $x = 0$, dx/dt is	$= 0$	$\geqslant 0$	$\geqslant 0$	$\geqslant 0$	$\geqslant 0$	$\geqslant 0$

It is seen that when any variable is at one of its termini it is, according to the hypotheses (28) to (39), either stationary or moving into the permitted range. Some, at first sight, harmless changes in the hypotheses would infringe this necessary condition. For example, the author at first supposed that the decrease of β_1 represented by (28) was entirely compensated by an increase of ξ_1. But if so, $d\xi_1/dt$ would contain a term $-\beta_1 A_1 \eta_1$, which could be negative when ξ_1 was zero, contrary to the requirements of buffering. This term had therefore to be placed instead in the equation (32) for $d\eta_1/dt$, where there is no difficulty, as $\beta_1\eta_1$ vanishes when $\eta_1 = 0$ or 1. The remainder of $d\beta_1/dt$ is compensated in $d\xi_1/dt$.

DEDUCTIONS

General remarks on methods of deduction

The hypotheses have now passed the easier tests and are therefore worthy to be tested by more elaborate deductions. The author is not aware of any explicit general solution of the fourteen equations (26) to (39). We are, however, able to gain considerable insight into the form of integrals by assuming, in accordance with observation, that certain variables are negligible at certain phases in the process. For the majority of the population passes in turn through the moods: {friendly, friendly}, {friendly, hostile}, {hostile, friendly}, {hostile, weary}, {weary, hostile}; so that in turn β, ξ, η, ρ, ω exceed $\frac{1}{2}$; and the other variables are left with less than $\frac{1}{2}$ between them. We can study:

(i) an early phase of an uneasy peace, in which β is near unity but θ, ρ, and ω are zero.

(ii) the outbreak of hostilities, during which ξ and η are important but θ, ρ, and ω are negligible.

(iii) a middle phase of persistence and attrition, in which ξ, η, and ρ are important but β and ω can be neglected.

(iv) the cessation of hostilities at a time when ρ and ω are important but β and ξ can be neglected.

Another device for simplifying the mathematics is to halve the number of differential equations by considering a war between two nations that are equal, at each instant, in all the respects which are represented by symbols in the theory. Such a theoretical scheme will be called for short a *symmetrical war*. It involves no victory-and-defeat, but a process of equal attrition until both nations are simultaneously willing to agree. The first three years of the Great War of 1914-'18 were moderately symmetrical as between France and Britain on the one side and Germany and Austria-Hungary on the other; but marked asymmetry developed at the end. Symmetrical wars are particularly worthy of attention because it will be shown in Case V that, if the constants are symmetrical, the variables tend to become so in the earliest phase, though not necessarily in the final phase.

At first sight some of the more general statements in Volterra's (41) theory of competing species may appear to include the present hypotheses as a special case. Volterra was mainly concerned with populations whose total varies. Nevertheless, constancy can be arranged by putting Volterra's $\beta_i = 1$. There remains the difficulty that in Volterra's most general hypothesis (41, p. 127, eqn. 44), which in his notation reads

$$\frac{dN_i}{dt} = f_i(N_1, \cdots, N_n)N_i \qquad (i = 1, 2, \cdots, n),$$

the N_i appears as a factor in the second member, whereas ξ_1 is not a factor of my equation (30) for $d\xi_1/dt$, nor are η_1, θ_1, ρ_1 and ω_1 factors, respectively, of my equations for $d\eta_1/dt$, $d\theta_1/dt$, $d\rho_1/dt$, $d\omega_1/dt$. Of course, this difficulty could be got over by choosing special forms for f_i; but Volterra appears to exclude such forms by the hypothesis that f_i is continuous when $N_r = 0$. Actually no application has been made of Volterra's formulas.

Case I. The earliest phase

Let us suppose that each nation consists entirely of those who are overtly friendly towards the other nation, but that a small fraction of these people harbor a subconscious hostility. The hypotheses (26) to (35) simplify in this case to

$$\beta_1 + \xi_1 = 1, \qquad \beta_2 + \xi_2 = 1 \qquad (42),(43)$$

$$d\beta_1/dt = \beta_1 A_1 \xi_1 - \beta_1 K_{12} \xi_2, \qquad d\beta_2/dt = \beta_2 A_2 \xi_2 - \beta_2 K_{21} \xi_1$$
$$= -d\xi_1/dt \qquad\qquad = -d\xi_2/dt \qquad (44),(45)$$

Wherever β_1, β_2 occur as multipliers, we may replace them approximately by unity, - - - - - - - - - - - - - - (46)

so obtaining linear equations

$$d\xi_1/dt = -A_1\xi_1 + K_{12}\xi_2, \qquad d\xi_2/dt = K_{21}\xi_1 - A_2\xi_2 . \quad (47)\,(48)$$

From these it can be proved, as in Richardson (30, §2-11) that ξ_1 and ξ_2 will tend to zero as $t \to \infty$, provided that

$$A_1 A_2 > K_{12} \cdot K_{21} . \qquad (49)$$

If (49) were satisfied, peace would be stable. We have, however, seen a reason, namely, the mote and beam effect as expressed in inequality (14), to expect that $K_{12} > A_1$ and $K_{21} > A_2$. If so, ξ_1 and ξ_2 would increase until our approximation (46) became invalid.

Comparison with the linear theory of arms races, as published by Richardson (30 and 32).

To make this comparison we require to express x_1, the psychological attitude of the first nation towards the second, as a function of β_1^* and ξ_1^*. In conformity with Ockham's "razor," let us choose the simplest formulas that make sense, and accordingly put

$$x_1 = -a_1\beta_1^* + b_1\xi_1^*, \quad x_2 = -a_2\beta_2^* + b_2\xi_2^*, \qquad (50),(51)$$

where a_1, b_1, a_2, b_2 are constants. The sign of x_1 or x_2 was conventionally taken throughout the linear theory of arms-races to be positive if the feeling was hostile, negative if friendly. Accordingly, each of a_1, b_1, a_2, b_2 is positive. - - - - - - - - - (52)

Ten million persons, all of the same opinion, would naturally be regarded as having a more intense outward attitude than 100 persons of the same opinion. As a more definite assumption, let us for simplicity suppose that the outward attitude is proportional to the number of persons of like mind - - - - - - - - - (53)

Now β_1^*, ξ_1^*, β_2^*, ξ_2^* are numbers of persons. So in (50) and (51) the constants a_1, b_1, a_2, b_2 do not depend on how many people there are. - - - - - - - - - - - - - - - (54)

They may be called personal as distinct from national constants.

On transformation to fractions of the population, (50) and (51) become

$$x_1/N_1 = -a_1\beta_1 + b_1\xi_1\,, \quad x_2/N_2 = -a_2\beta_2 + b_2\xi_2\,. \qquad (55),(56)$$

Let β_1, β_2 be eliminated by means of (42) and (43), so that

$$x_1/N_1 = -a_1 + (a_1+b_1)\,\xi_1\,, \quad x_2/N_2 = -a_2 + (a_2+b_2)\,\xi_2\,. \qquad (57),(58)$$

From now on it will suffice to attend to the first of each pair of equations. The second, which can be obtained by interchanging the suffixes, is given a reference number to remind us that it is available.

The derivative of (57) is

$$\frac{d(x_1/N_1)}{dt} = (a_1 + b_1)\,\frac{d\xi_1}{dt}. \qquad (59),(60)$$

Let $d\xi_1/dt$ be eliminated between (59) and (47), so that

$$\frac{d(x_1/N_1)}{dt} = (a_1 + b_1)\,(-A_1\xi_1 + K_{12}\xi_2)\,. \qquad (61),(62)$$

Finally, in (61) let ξ_1 and ξ_2 be expressed in terms of x_1 and x_2 by way of (57) and (58), giving

$$\frac{d(x_1/N_1)}{dt} = (a_1 + b_1)\left\{-\frac{A_1}{a_1 + b_1}\left(\frac{x_1}{N_1} + a_1\right) + \frac{K_{12}}{a_2 + b_2}\left(\frac{x_2}{N_2} + a_2\right)\right\}. \qquad (63),(64)$$

This is a linear equation comparable with the fundamental assumption of Richardson's (**30**, p. 6) linear theory of arms-races, namely,

$$dx_1/dt = -\alpha_1 x_1 + k_{12}x_2 + g_1\,. \qquad (65),(66)$$

On comparison of the coefficients of the variables in (63) with those in (65), we see that:

The fatigue-and-expense coefficient, α_1, is given by $\alpha_1 = A_1$. (67),(68)

The defense-coefficient k_{12} is given by

$$k_{12} = K_{12}\left(\frac{a_1 + b_1}{a_2 + b_2}\right)\frac{N_1}{N_2}. \qquad (69),(70)$$

The "grievance," g_1, is given by

$$g_1 = N_1\left\{-A_1 a_1 + K_{12}\left(\frac{a_1 + b_1}{a_2 + b_2}\right)a_2\right\}. \qquad (71),(72)$$

These formulas are of considerable interest. A new psychological interpretation is given to α_1 and to A_1 by their equality. Formerly α_1 was supposed to depend on the fatigue of personal service in preparation for war and on a reluctance to incur the expense of armaments; whereas A_1 was supposed to be an expression of persistence in friendliness in spite of threats. It will be better in future to regard α_1 and A_1 as expressing the joint effect of all such restraining influences and to call them *"restraint coefficients."* This statement will be referred to as - - - - - - - - - - - - - - - - - (73).
If we eliminate A_1 and K_{12} between (67), (69), and (71), we obtain

$$g_1 = -N_1 a_1 \alpha_1 + N_2 a_2 k_{12}. \qquad (74),(75)$$

Thus g_1 is an independent constant, not expressible in terms of α_1 and k_{12} alone. Also, g_1 is capable of either sign. Both these verbal results accord with the linear theory of arms-races. But the relation (74) makes also other assertions, which are new, and problematic.
On multiplying (69) by (70), we have

$$k_{12} \cdot k_{21} = K_{12} \cdot K_{21}. \qquad (76)$$

So in view of (67) and (68),

$$k_{12} \cdot k_{21}/\alpha_1\alpha_2 = K_{12} \cdot K_{21}/A_1 A_2, \qquad (77)$$

that is to say, the conditions of stability in the two theories are consistent with one another.

The orders of magnitude of A_1 *and* K_{12} can be estimated by way of (67) and (69).

The numerical value of α was shown by Richardson (32), from the rate of demobilization, to be about one year^{-1}. Therefore:

The constant A_1 is of the order of one year^{-1}. - - - (78)

Of the constants that occur in (69), we may expect from the definition in (50) and (51) that:

a_1 is of the order of a_2, and b_1 of the order of b_2, so that $(a_1 + b_1)/(a_2 + b_2)$ is of the order of unity. - - - - (79)

In picking out a defense coefficient from the observations in Richardson (32), for comparison with K_{12}, we must pay attention to the note at the end of Type IX, which is to the effect that it may be proper to regard the world as composed either of two alliances as was done in one section, or of ten nations as was done in another; but that the defense coefficients appropriate to these two views will be different. So it is perhaps clearest to state simply that: if the nations of the pair are equal, so that $N_1 = N_2$, and if their instability coefficient is λ, then

$$k_{12} = a + \lambda; \text{ and so by (67) and (69) } K_{12} = A + \lambda, \tag{80}$$

where λ has been of the order of $0._3$ year^{-1} during the two great arms-races.

When the nations are very unequal we should attend to K_{21}/K_{12}. By dividing (69) by (70), we obtain

$$\frac{k_{12}}{k_{21}} = \frac{K_{12}}{K_{21}} \left(\frac{a_1 + b_1}{a_2 + b_2} \right)^2 \left(\frac{N_1}{N_2} \right)^2. \tag{81}$$

But reasons were given by Richardson (30, pp. 74, 75, revised in 32) for believing the defense coefficient to be proportional to the "size" of a nation in the sense of its industry and population. The same idea appeared in the microfilm (32) in Type IX and Type XII, and was widely applied. As far as it can be expressed in terms of population alone, it is that

$$k_{12}/k_{21} = N_1/N_2, \tag{82}$$

the suffix of N corresponding to the *first* suffix of k. When this is introduced into the previous equation, there is left

$$\frac{N_2}{N_1} = \frac{K_{12}}{K_{21}} \left(\frac{a_1 + b_1}{a_2 + b_2} \right)^2, \tag{83}$$

in which the suffix of N corresponds to the *second* suffix of K.

In view of (79), we conclude that

$$K_{21}/K_{12} \textit{ is probably of the order of } N_1/N_2. \tag{84}$$

Compare statement (22) on page 171.

Case II. Simple approximation to the outbreak of a symmetrical war

As the two nations are assumed to be equal in respect of each quantity that occurs in the theory, we may omit the suffixes. For simplicity, it will be supposed that there has been a long arms-race in the course of which nearly everybody has changed from the mood {friendly, friendly} to the mood {friendly, hostile}; so that when the present phase begins β is negligible and ξ is near to unity. Near the beginning of a war ω is zero and ρ and θ not far from it. Observation shows that the processes of attrition, represented by $d\theta/dt$ and $d\rho/dt$ are very slow in comparison with the sudden uprising of indignation represented by $d\eta/dt$. We may therefore neglect B and E in comparison with C. The hypotheses (26) to (41) accordingly reduce to

$$\xi + \eta = 1 \tag{85}$$

$$d\xi/dt = -C\xi\eta = -d\eta/dt\,. \tag{86}$$

On eliminating ξ between (85) and (86) and separating the variables, we obtain

$$Ct + \text{Const.} = \int \frac{d\eta}{\eta(1-\eta)} = \log_e\left(\frac{\eta}{1-\eta}\right). \tag{87}$$

A graph of this equation is shown in Fig. 2 (on page 198) for the special case $C = 1$, const. $= 0$, by a graph of y as a function of $\log_e\{y/(1-y)\}$. The graph has a shape suitable for representing the observations insofar as: (i) the straight line $\eta = 0$ is an asymptote, (ii) from which the curve rises continuously, (iii) with the fewest possible changes of sign of its curvature, (iv) to approach the straight line $\eta = 1$ as another asymptote, and (v) with any degree of suddenness, or of its opposite, obtainable by adjustment of the constant C. There are plenty of other functions which are also suitable because they possess the properties (i) to (v). The rather vague observational data are insufficient to decide the choice between these functions. We may accept (87) on this understanding, noting also that, when θ and ρ are taken into account, η will have a maximum near unity instead of tending towards an asymptote. A maximum is shown in Case VI.

In order to determine the value of C, let $\eta = \eta'$ when $t = t'$, and $\eta = \eta''$ when $t = t''$. Then from (87)

$$C = \frac{1}{t'' - t'} \log_e \frac{\eta''(1-\eta')}{\eta'(1-\eta'')}. \tag{88}$$

With reference to Figure 1 and the collection of historical facts in Part IA, we may reasonably assume that between July 23, 1914 and August 4, 1914, there was in Britain a rise from $\eta' = 0.1$ to $\eta'' = 0.9$, which occurred in the course of about 10 days; so that $t'' - t' = 0.028$ year. When these particulars are inserted in (88) they give

$$C = 1.6 \times 10^2 \text{ year}^{-1}. \tag{89}$$

On referring back to equations (78) and (80) we notice that C, which expresses a rate of conscious change, is of the order of 100 times K, which expresses a rate of subconscious change.

Case III. The phase of attrition in the middle of a symmetrical war

Let us suppose that there are no people left in the overtly friendly moods; so that $\beta = 0$ and $\xi = 0$; but that on the other hand no one yet admits a wish to make peace; so that $\omega = 0$. With these assumptions the hypotheses (26) to (41) reduce to

$$\eta + \theta + \rho = 1 \tag{90}$$

$$d\eta/dt = -(B + E)\eta(\eta + \rho) \tag{91}$$

$$d\theta/dt = E(\eta + \rho)^2 \tag{92}$$

$$d\rho/dt = (B\eta - E\rho)(\eta + \rho). \tag{93}$$

Let $\eta + \rho$ be eliminated between (90) and (92), giving

$$d\theta/dt = E(1 - \theta)^2, \tag{94}$$

the integal of which is

$$Et + \text{Const.} = 1/(1 - \theta). \tag{95}$$

Let $\theta = \theta'$ at $t = t'$ and $\theta = \theta''$ at $t = t''$. It follows, on elimination of the arbitrary constant, that

$$E = \frac{\theta'' - \theta'}{(t'' - t')(1 - \theta')(1 - \theta'')}. \tag{96}$$

This agrees with the rougher estimate of equation (20) to within a few per cent.

For example, the Italian war-deaths among both the military and the civil population are given for a period in the middle of the Great War by Vedel-Petersen (39, pp. 144, 165). At the beginning of 1916, $\theta^* = 1.0 \times 10^5$, if one distributes the deaths of prisoners. Two years later, $\theta^* = 4.6 \times 10^5$. Meanwhile $N_1 = 36.7 \times 10^6$; so that $\theta' = 0.0027$, $\theta'' = 0.0125$. When these data are inserted in (96), it gives

$$E = 0.005 \text{ year}^{-1}. \tag{97}$$

The separate casualties for 1916 and 1917 show that E varied in a ratio of 1.3 between those years.

Let $t = 0$ at the beginning of hostilities when both $\theta = 0$ and $\rho = 0$. In view of (90), we have also

$$\eta = 1 \text{ at } t = 0. \tag{98}$$

It may seem unnatural to suppose that $\eta = 1$, for a nation is seldom quite unanimously in favor of war. But we can avoid that difficulty by supposing that the instant $t = 0$, though used as a reference mark, may lie outside the phase for which the approximations of the present case are valid; so that the formulas are to be applied only for values of t that are not too near to zero. With this origin of time, the arbitrary constant in (95) is unity, so that

$$1 - \theta = 1/(Et + 1). \tag{99}$$

By (90) and (99), $\eta + \rho = 1/(Et + 1)$, and when this is substituted in (91), the variables can be separated, giving, with the convention (98),

$$\log \eta = -\frac{B + E}{E} \log (Et + 1). \tag{100}$$

From (99) and (100),

$$\eta = (1 - \theta)^{1+B/E}. \tag{101}$$

This formula appears to be suitable for representing the facts insofar as it shows that η will steadily diminish at a rate which is controlled by the constant B. The difficulty in the way of determining B from observations is that the persons counted in ρ^* differ qualitatively from those counted in η^* only by a change in the subconscious, which they usually conceal, and concerning which therefore no statistics were, or could have been, collected at the time of its occurrence. From admissions made at a later phase, say just after the end of the Great War, it may be gathered, rather vaguely, that a majority of the population had passed through the mood {aggressive, war-weary}, and so ρ had approached unity and η had approached zero. We can make a conditional estimate of B/E from (101) by supposing that when $\theta = 0.02$, a value typical of several nations towards the end of the war of 1914-'18, then η was, say, η_T. On solving (100), we obtain

$$\frac{B}{E} = \frac{\log \eta}{\log (1 - \theta)} - 1 = \frac{\log_{10} \eta_T}{-0.00877} - 1. \tag{102}$$

This relation is shown by pairs of values in the following table:

η_T	0.3	0.1	0.03	0.01
B/E	59	113	173	227

As E, for severely smitten nations, is known from (21) and (97) to be of the order of 10^{-2} year^{-1}, we may conclude that:

$$\text{B } \textit{is of the order of } 1 \text{ year}^{-1}. \tag{103}$$

Case IV. A simple approximation to the final phase of a symmetrical war

Let us suppose that the initial population of N now consists only of $\theta_T{}^*$ dead, of ρ^* in the mood {aggressive, war-wearied} and of ω^* in the mood {war-wearied, aggressive}; so that

$$\theta_T + \rho + \omega = 1. \tag{104}$$

Observation shows that the collapse at the end of the First World War, though not as sudden as the uprush of indignation at the beginning, was more rapid than the attrition in the middle phase; so that we may regard θ_T as constant during the collapse. This is equivalent to neglecting E. Accordingly, the hypotheses (28) to (38) reduce to

$$d\rho/dt = -(D+F)\rho\omega, \tag{105}$$

$$d\omega/dt = (D+F)\rho\omega. \tag{106}$$

On eliminating ρ between (104) and (106), we have

$$d\omega/dt = (D+F)(1-\theta_T-\omega)\omega. \tag{107}$$

By separating the variables, we obtain the integral

$$(D+F)t + \text{Const.} = \frac{1}{1-\theta_T}\log\left\{\frac{\omega}{1-\theta_T-\omega}\right\}. \tag{108}$$

We may reasonably suppose that hostilities will come to an end when about half the survivors are unwilling to continue the struggle. To found a precise definition on this rough notion,

$$\text{let } t = T \text{ when } \omega = \tfrac{1}{2}(1-\theta_T) \tag{109}$$

so that $t = T$ at, or about, the end of hostilities. On substituting (109) into (108), we find that the arbitrary constant is $-(D+F)T$, and so

$$(1-\theta_T)(D+F)(t-T)=\log\left\{\frac{\omega}{1-\theta_T-\omega}\right\}. \tag{110}$$

This equation is closely similar to (87) ; and the same graph, Fig 2, on page 198, can, by suitable interpretation of its scales, be made to represent both. Time runs the same way in either case; but the ordinate represents those openly in favor of war at the outbreak, and those openly in favor of peace at the cessation of hostilities.

Otherwise if $t=t'$ when $\omega=\omega'$ and $t=t''$ when $\omega=\omega''$, then

$$D+F=\frac{1}{(t''-t')(1-\theta_T)}\log_e\left\{\frac{\omega''(1-\theta_T-\omega')}{(1-\theta_T-\omega'')\omega'}\right\}. \tag{111}$$

To determine the order of magnitude of $D+F$ from (111) taken in conjunction with the observational data shown in Fig. 1, we should first draw a curve midway between the curves for Britain and Germany so as to represent an artificially symmetrized war. On this mean curve the ordinate, which is in general $\eta+\rho$, but in this phase simply ρ, falls from 0.67 to 0.15 during the last half of the year 1918. Also, the mean value of θ_T for Britain and Germany was 0.02. These values, when inserted into (104) and (111), yield

$$D+F=5\ \text{year}^{-1}. \tag{112}$$

According to (110), $\omega\to 0$ as $t\to-\infty$. The hypotheses thus entail that ω grows, so to speak, from seed; so that unless there were a few defeatists at the beginning of the war there could not be any at the end.

PATRIOT: "Let us get rid, then, of these accursed pre-war defeatists!"

AUTHOR: "Before you do that, pray let me point out that, if there were no defeatists, the end of this symmetrical war could not come until every person in both nations had been killed.

It may well be a defect of the hypotheses that they have this entailment. The question is not easily decided by observation; for a great variety of small minorities exist. One is reminded of the difficulty that Pasteur had in proving the nonexistence of spontaneous generation of life.

PATRIOT: "May be. But I wasn't thinking of symmetrical wars."

AUTHOR: "When gregariousness is taken into account, as in Case IX, the formulas are different."

Case V. Symmetry in the constants but not necessarily in the variables

It was for mathematical convenience that we considered a perfectly symmetrical system. Actual wars, on the contrary, have mostly ended unsymmetrically in victory and defeat. So let us enquire whether the mathematical model has any corresponding tendency. That is to say, if the two nations are equal in the constants, does an inequality in the variables grow or diminish as time goes on? This question will be answered for two phases separately.

In the *earliest phase*, for which, as in Case I,

$$\beta_1 + \xi_1 = 1, \qquad \beta_2 + \xi_2 = 1, \qquad (113),(114)$$

the hypotheses (28) to (38) become, when we omit the suffixes on the constants,

$$d \log \beta_1 / dt = A\xi_1 - K\xi_2, \qquad d \log \beta_2/dt = A\xi_2 - K\xi_1, \quad (115),(116)$$

whence by subtraction

$$\frac{d \log (\beta_1/\beta_2)}{dt} = (A + K)(\beta_2 - \beta_1). \qquad (117)$$

If at any instant $\beta_2 > \beta_1$, then $\log \beta_2 > \log \beta_1$, and so $\log (\beta_1/\beta_2) < 0$. But $A + K$ is positive. So $\log (\beta_1/\beta_2)$ is increasing and is moving towards zero. But if, on the contrary, $\beta_2 < \beta_1$ at any instant, then $\log (\beta_1/\beta_2)$ is positive and is decreasing. Thus (117) entails that $\log (\beta_1/\beta_2)$ is either zero or is always moving towards zero; so that β_1/β_2 is either unity or is moving towards unity. We may express this result by saying that *in the earliest phase symmetry in the variables is stable.* - - - - - - - - - - - - - - - - - (118)

Common observation supports this deduction.

Solving for A. Because attempts to determine the restraint-constant α from a study of defense-expenditure have so far failed except for the phase of demobilization (see the microfilm, 32), any hope of determining the theoretically equal quantity A is all the more welcome. On eliminating K between (115) and (116), we obtain

$$A = \frac{\xi_1 \, d \log \beta_1/dt - \xi_2 \, d \log \beta_2/dt}{\xi_1^2 - \xi_2^2}. \qquad (119)$$

From this we can remove β_1 and β_2 by means of (113) and (14) for

$$\xi_1 \, d \log \beta_1/dt = \left[\frac{-\xi_1}{1 - \xi_1} \right] \frac{d\xi_1}{dt} = -\frac{1}{2(1-\xi_1)} \frac{d\xi_1^2}{dt}. \qquad (120)$$

In the earliest phase when ξ_1 and ξ_2 are both near to zero, we may replace the factors $1-\xi_1$ and $1-\xi_2$ by unity, so that

$$A = -\frac{1}{2}\frac{d}{dt}\log|\xi_1^2 - \xi_2^2|, \text{ approximately.} \tag{121}$$

It has already been shown that $\xi_1^2 - \xi_2^2$ drifts towards zero. If some observations of this drift towards symmetry are found, then (121) or more accurately (119), can be used to determine A.

In the *final phase*, for which θ is treated as in Case IV as a common constant, θ_T, we have

$$\rho_1 + \omega_1 = 1 - \theta_T, \qquad \rho_2 + \omega_2 = 1 - \theta_T. \qquad (122),(123)$$

Also the hypotheses (28) to (38) become

$$d\log\rho_1/dt = -F\omega_2 - D\omega_1, \tag{124}$$

$$d\log\rho_2/dt = -F\omega_1 - D\omega_2, \tag{125}$$

whence by subtraction

$$\frac{d\log(\rho_1/\rho_2)}{dt} = (F-D)(\omega_1-\omega_2) = (F-D)(\rho_2-\rho_1). \tag{126}$$

If $F > D$ the equation (126) resembles (117); and arguments similar to those stated above show that ρ_1/ρ_2 moves towards unity.

But if $F < D$, then $\log(\rho_1/\rho_2)$ moves away from zero on either side; and ρ_1/ρ_2 moves away from unity.

That is to say: *in the final phase symmetry in the variables is stable if* F > D, *and unstable if* $F < D$. - - - - - - - (127)

When we look back at the psychological meanings of D and F, we see that they both express the infectiousness of an open admission of war-weariness; but that F relates to an infection coming from the enemy, presumably by radio; whereas D relates to an infection passing between persons inside the same nation by radio, newspaper, and conversation. To the author, the above relation (127) seems credible but new.

An equation like (126) occurred in the theory of the *reciprocal inhibition* between a pair of osglim lamps wired in parallel, reference (29). This opportunity is taken to correct, with apologies, two mistakes in that paper. In the last member of its equation (6), for $+u_2$, read $-u_2$. In the second line below equation (7), for "second member," read "last bracket."

Case VI. The outbreak of a symmetrical war and the beginning of suppressed weariness

The treatment of the outbreak in Case II was made as easy as possible; but the approximations failed at the time when war-fever, as represented by η, should have reached its maximum. Let us now improve the method by retaining ρ so that the formulas of Case VI may replace those of Case II for the outbreak, and may extend to the phase in which suppressed weariness begins. We may still, however, neglect ω, for it belongs to the last phase of a war. Also, in view of the orders of magnitude of B and E, as found in Case III, we may neglect E while retaining B. Thus the hypotheses (26) to (39) reduce to

$$\xi + \eta + \rho = 1 \tag{128}$$

$$d\xi/dt = -\xi(\eta + \rho)C \tag{129}$$

$$d\eta/dt = \xi(\eta + \rho)C - \eta(\eta + \rho)B \tag{130}$$

$$d\rho/dt = \eta(\eta + \rho)B\,. \tag{131}$$

By (129) $d\xi/dt$ has always the same sign, thus permitting ξ to be used as a parameter defining the time. On elimination of $\eta + \rho$ between (128) and (129), we have

$$-Ct + \text{Const.} = \int \frac{d\xi}{\xi(1-\xi)} = \log\frac{\xi}{1-\xi}. \tag{132}$$

A graph of y as a function of $\log\{y/(1-y)\}$ appears in Fig. 2 on page 198, and if read from right to left to suit the negative coefficient of t in (132), that graph shows the manner in which ξ falls from 1 to 0 as t increases from $-\infty$ to ∞. According to (132) and (128), $\eta + \rho \to 1$ as $t \to \infty$. This is a fault of the approximation, arising from the neglect of θ and ω. But it merely indicates that the formulas of the present Case VI must not be applied to the end of a war. Until the maximum of η is passed, the formulas may be expected to be useful approximate deductions from the complete hypotheses.

It is convenient to let $t = 0$ when $\xi = \frac{1}{2}$, - - - - - (133)
for with that choice the arbitrary constant in (132) vanishes, leaving simply

$$t = \frac{1}{C}\log\frac{1-\xi}{\xi}, \tag{134}$$

which is shown in Fig. 3.

Elimination of dt between (129) and (130) gives

$$\frac{d\eta}{d\xi} - \frac{\eta}{\xi}\frac{B}{C} = -1. \tag{135}$$

To determine η as a function of t, we can first integrate (135) so as to express η in terms of ξ, and afterwards bring in the time by way of (134). The integrating factor is $\xi^{-B/C}$; and so the solution of (135) is

$$\eta = \xi^{B/C}\left\{\text{Const.} - \frac{C}{C-B}\,\xi^{(C-B)/C}\right\}. \tag{136}$$

To determine the arbitrary constant in (136), let us assume an initial condition in which at a late stage of the arms race

$$\xi = 1, \qquad \eta = 0. \tag{137}$$

It follows that

$$\eta = \frac{C}{C-B}\left(\xi^{B/C} - \xi\right), \tag{138}$$

which is also shown in Fig. 3.
From (138)

$$\frac{d\eta}{dt} = \frac{C}{C-B}\left(\xi^{(B-C)/C}\cdot\frac{B}{C} - 1\right)\frac{d\xi}{dt}, \tag{139}$$

and, as $d\xi/dt$ does not vanish for finite values of t, the maximum η_M of η in time occurs when $\xi = \xi_M$, say, such that

$$\xi_M = \left(\frac{C}{B}\right)^{C/(B-C)}; \tag{140}$$

and from (138) and (140)

$$\eta_M = \frac{C}{C-B}\left\{\left(\frac{C}{B}\right)^{B/(B-C)} - \left(\frac{C}{B}\right)^{C/(B-C)}\right\} = \left(\frac{C}{B}\right)^{B/(B-C)}. \tag{141}$$

As a check we have, directly from (135),

$$\eta_M/\xi_M = C/B. \tag{142}$$

From (141) the following pairs of values have been computed

$$\begin{array}{lllll} C/B = 50 & 100 & 200 & 400 & \\ \eta_M = 0.923 & 0.955 & 0.974 & 0.985. & \end{array} \tag{143}$$

In any voting on the question whether a war should be continued, (see Part IA) it is probably not η but $\eta + \rho$ that can be observed, or, in a secret ballot, perhaps some number between η and $\eta + \rho$. Nevertheless we are not entirely ignorant of the actual historical value of

η_M. For η_M must be less than or equal to the greatest fraction of the population ever observed to be in favor of the war. Also observation indicates that there is a time, soon after the outbreak of a war, when enthusiasm for the war is at its greatest and there is no cause to suspect the existence of much suppressed war-weariness, so that at that time $\rho = \rho_M$, say, must be almost zero and η_M roughly observable. This is an interim approximation, for we shall find out more about ρ_M presently. On looking at the graph (Fig. 1) of the course of $\eta + \rho$ in Germany and Britain during the war of 1914-'18 and taking the average for the two countries in order to compare it with the present theory of a symmetrical war, we see that the maximum of $\eta + \rho$ was 0.98 and so $\eta_M \leqslant 0.98$. On comparison with (143), it follows that

$$C/B \leqslant 300\,. \tag{144}$$

But, from (89), $C = 1.6 \times 10^2$ year^{-1}.
Therefore

$$B \geqslant 0.5\,\text{year}^{-1}. \tag{145}$$

This estimate is of the same order as that recorded in (103) although obtained by an entirely different method.

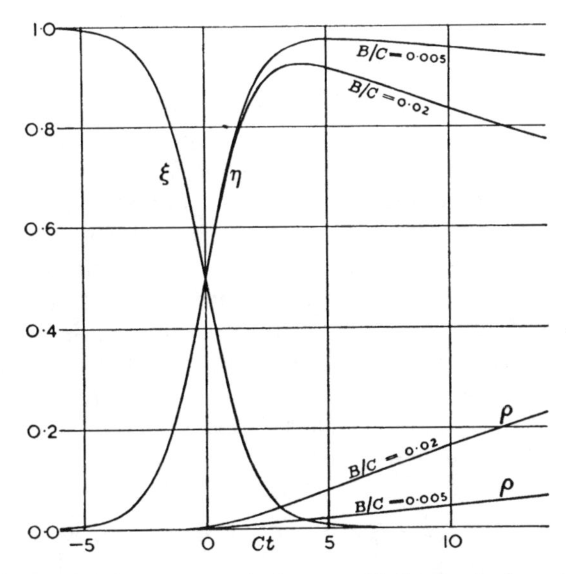

FIG. 3. Outbreak of a symmetrical war and the beginning of weariness, as deduced from hypotheses. Time runs horizontally to the right. The ordinates are fractions of the population. Three moods are shown: ξ marks those who are outwardly friendly but subconsciously hostile: η marks those who are outwardly hostile but subconsciously friendly; ρ marks those who are outwardly hostile but secretly weary. For η and ρ there are two curves, depending on the value of B/C.

For the purpose of drawing graphs to illustrate the formulas let us choose, in conformity with (144),

$$B/C = 0.005\,. \tag{146}$$

Then

$$\eta = \frac{1}{0.995}\,(\xi^{0.005} - \xi)\,. \tag{147}$$

In Fig. 3 the curves show ξ, η, ρ as functions of t. They have been taken from formulas (134), (138), and (128) for $B/C = 0.005$ and again for $B/C = 0.02$. The graphs of η in particular are suitable for comparison with observations of the fraction of a population that is keen on a war. These graphs are seen to be an improvement on that of Case II because now the curve rises to a maximum and falls slowly. The rate of fall relative to the rate of rise depends on B/C. The time-scale depends on C. If C is given the value in (89), namely, $C = 1.6 \times 10^2$ year^{-1}, then the whole extent of the diagram, which is 20 units of Ct, amounts to $\frac{1}{8}$ year $= 6\frac{1}{2}$ weeks.

A war that is extinguished at its beginning

The maximum attained by η is controlled in the mathematics chiefly by the constant B, which regulates the rate at which the number of secretly weary persons increases. If we make B large enough, η_M can be made as small as we please. It may be possible in this way to represent a mutual flaring up of indignation in two countries such as the Fashoda incident between Britain and France in the autumn of 1898, when war seemed likely, but was avoided. It is not possible, however, to explore this suggestion thoroughly without taking account of ω, the number of those who admit their weariness; for if $\eta_M < 1/2$, then ρ becomes important by the time that η attains its maximum and so ω develops early. As much as is worth doing, in the present neglect of ω, is to note, as a first approximation, that formula (141) shows that $\eta_M = \frac{1}{2}$ if $B/C = \frac{1}{2}$. An alternative explanation of how a war may be just avoided is given in Part II B with reference to gregariousness.

Case VII. An outbreak between unequal nations

As in Case II let us neglect β, ρ, θ, ω; but let us restore from (26), (27), (30), and (31) the suffixes which were not needed in the symmetrical case. The hypotheses now are

$$\xi_1 + \eta_1 = 1\,, \qquad \xi_2 + \eta_2 = 1\,, \qquad (148)\,(149)$$

$$d\xi_1/dt = -C_{12}\xi_1\eta_2, \qquad d\xi_2/dt = -C_{21}\xi_2\eta_1. \quad (150)\,(151)$$

On eliminating dt, we have

$$d\xi_1/d\xi_2 = C_{12}\xi_1\eta_2/(C_{21}\xi_2\eta_1). \qquad (152)$$

The variables can be separated, giving

$$C_{21}(1-\xi_1)\, d\log\xi_1 = C_{12}(1-\xi_2)\, d\log\xi_2. \qquad (153)$$

Whence on integration

$$C_{21}(\log\xi_1 - \xi_1) = C_{12}(\log\xi_2 - \xi_2) + \text{constant}. \qquad (154)$$

The constant of integration can be fixed by the *supposition that the two nations began the quarrel at the same time in the sense that there was some early time when*

$$\xi_1 = 1 = \xi_2. \qquad (156)$$

At that instant $-C_{21} = -C_{12} +$ Constant.

Therefore at all times

$$C_{21}(\log\xi_1 - \xi_1 + 1) = C_{12}(\log\xi_2 - \xi_2 + 1). \qquad (157)$$

The course of the function is illustrated in the following table.

ξ	1.0	0.9	0.8	0.7	0.6	0.5	0.4
$\log\xi - \xi + 1$	0	—.0054	—.0231	—.0567	—.1108	—.1931	—.3163
ξ	0.3	0.2	0.1	0.05	0.03	0.01	0
$\log\xi - \xi + 1$	—.5040	—.8094	—1.403	—2.046	—2.537	—3.615	$-\infty$

Let this theory now be compared with historical facts concerning some very unsymmetrical outbreaks. The facts are usually given us as incidents separated by pauses. It is reasonable to suppose that the changes of mood were less jerky, and so more like the smooth curves of the theory.

Paul Kruger, President of the South African Republic, records in his *Memoirs* (17) that in the autumn of 1899 prior to his ultimatum of October 9:

On the 22nd September, the mobilization of an army corps for South Africa was announced in England, and, on the 28th of September, it was announced that the greater part of that army corps would leave for South Africa without delay. The Government thereupon commandeered the greater part of the burghers to take up their position near the frontiers of the Republic, in order to be prepared for a sudden attack on the part of England.

That is to say, the mobilization went on at about the same rate in men on both sides, but very much faster among the Boers when reckoned as a fraction of the population. Mobilization is of course not identical with the change from the mood {friendly, aggressive} to the mood {aggressive, friendly}. Nevertheless we may safely conclude that η_2 among the Boers rose much more rapidly than η_1 among the British.

There is a similar fact about the Russo-Finnish war of 1939-'40. On December 1, 1939 the Russions invaded Finland. But six weeks earlier it was reported that Finland's mobilization was completed (*Glasgow Herald,* October 20, 1939). On November 15th, a fortnight before the invasion, the same paper reported that "One-fifth of Finland's population of 3,000,000 is now under arms." There is no suggestion that a fifth of the Russian population was under arms; and so huge a mobilization could not have escaped notice, had it existed.

It is understandable that in any dispute between very unequal populations, the less numerous people are likely to mobilize first, because they feel more threatened. This is one of the many difficulties in the way of giving any equitable answer to the question: who began it?

Now let us return to equation (157). If at any time $\eta_2 > \eta_1$, then $\xi_2 < \xi_1$ and $|\log \xi_2 - \xi_2 + 1| > |\log \xi_1 - \xi_1 + 1|$; whence $C_{21} > C_{12}$. So the comparison of the Boer-British and Finn-Russian wars with the theory leads to the conclusion that

$$\text{if } N_1 >> N_2 \quad \text{then} \quad C_{21} > C_{12}. \tag{158}$$

To estimate C_{21}/C_{12}, let us consider the time in 1899 when the British people were equally divided between overt friendliness and overt hostility; so that ξ_1 was 0.5 and $\log \xi_1 - \xi_1 + 1$ was -0.1931. According to (22) E_{21}/E_{12} was of the order of $(N_1/N_2)^{0.7}$; and, according to (84), K_{21}/K_{12} was of the order N_1/N_2. So let us now suppose for the sake of argument that $C_{21}/C_{12} = N_1/N_2 = 180$. Then $\log \xi_2 - \xi_2 + 1 = -0.1931 \times 180 = -34.76$ and therefore $\xi_2 = 10^{-15}$. But that is an incredible degree of unanimity; for Botha opposed the Boer ultimation to Britain (7, 3, p. 948). So let us suppose instead that $C_{21}/C_{12} = \sqrt{(N_1/N_2)} = \sqrt{180} = 13.4$. Then $\log \xi_2 - \xi_2 + 1 = -0.1931 \times 13.4 = -2.59$; whence $\xi_2 = 0.029$; which indicates that only three Boers in a hundred were still overtly friendly. This is more credible.

Conclusion.

$$C_{21}/C_{12} \textit{ may be of the order of } \sqrt{(N_1/N_2)}. \tag{159}$$

Summary on the bipersonal theory of war-fever

Practical politics has been called the "art of the possible"; and the same may be said of mathematical politics. This bipersonal theory has been arranged so as to have one constant for each important psychological effect and so that the set of equations is sufficiently tractable. The gist of the theory is contained in the hypotheses (26) to (41). They have been tested by making from them a variety of deductions and by comparing these with the war of 1914-'18. The deductions have been found to be in interesting agreement with the observations. The orders of magnitude of most of the constants have been estimated and are collected in the following table. Support is thus given to the psychological discussion which led up to the formulation of the hypotheses (26) to (41). But only moderate determinism should be expected. For the observations are not definite or consistent enough to confirm or to refute any theory with the precision expected in physics. And the theory can only express what is habitual, traditional, or instinctive; whereas we know, in everyday life, the operation of free choice.

Collection of Constants. Those without suffixes are mostly for a symmetrized description of the war of 1914-'18 as between Britain and Germany.

Constant	A	K	C	B	E
Unit	year^{-1}	year^{-1}	year^{-1}	year^{-1}	year^{-1}
Order of magnitude	1	$A + 0.3$	$1._6 \times 10^2$	1 $\geqslant 0.5$	10^{-2} 0.5×10^{-2}
Reference	(78)	(80)	(89)	(103) and (145)	(21) and (97)
Constant	$D + F$	K_{21}/K_{12}	C_{21}/C_{12}		E_{21}/E_{12}
Unit	year^{-1}	unity	unity		unity
Order of magnitude	5	N_1/N_2	$\sqrt{(N_1/N_2)}$		$(N_1/N_2)^{0.7}$
Reference	(112)	(84)	(159)		(22)

PART II B

GREGARIOUSNESS AND WAR-FEVER*

Introduction

The unity of a nation in war for defense or attack was attributed by Trotter (37), in a book which expresses British war-sentiments of the year 1917, to what he called the "instinct of the herd," a tendency to act together. On the contrary, the hypotheses (26) to (41) of the preceding theory of war-fever explain the observed national unity without making any reference to gregariousness. For example, the maximum value of η as shown in Fig. 3 for Case VI depends on C/B in the manner specified by (141) and (143). Here C is a constant expressing the reaction to hostile foreigners; and B is a constant expressing the wearying effect of war.

This diversity of explanation deserves further consideration.

It may be argued, against Trotter's thesis, that the mere fact of simultaneous and similar action by most of the persons in a nation is in itself no proof that their motive was gregarious. For example, most people sleep at night. But is it not probable that this custom arose primarily because each individual was similarly affected by darkness, and that any persuasion of one another to agree on a conventional bedtime was only a secondary modification? In the foregoing theory of war-fever, the observed national unity is attributed to the threat from the enemy acting simultaneously on all the citizens, just as the darkness of night acts on them all.

However we know from everyday observation how strong can be the control exerted by ideas of fashion, of good form, of doing the done thing, of conformity to the pattern of our culture. Prof. T. H. Pear (25) has recently discussed Dr. Ruth Benedict's statement that war itself is a social theme that may or may not be used in any culture. Pear concludes that "modern warfare is not due to simple instincts, nor is it inevitable," for it depends on culture-patterns which are not congenital, but are learned, and can be altered. The relations between what is congenital and what is learned appear to me personally as very important, but complicated, obscure, and controversial. For example the tendency to learn the national culture-pattern was regarded by Trotter as an instinct. I shall not here attempt to disentangle the inborn from the acquired, but only to describe how people conform to the example of their living fellows. That can be done with differential equations; whereas a discussion of conformity to the be-

*A brief account was read to the British Psychological Society at Brighton on April 11, 1942.

havior of past generations would require the more difficult technique of integral equations.

The effects of gregariousness are likely to show themselves before the outbreak of hostilities as a restraint exercised by the peace-loving majority on a fiery minority; whereas after the outbreak gregariousness instead constrains any who still feel pacific to conform to the war-fevered majority.

Pure fashion

To bring this gregarious effect prominently into consideration let us as a preliminary imagine the extreme and fanciful case in which the only motive is to agree with the majority. As in Case II we may put approximately

$$\xi + \eta = 1 \,. \tag{160}$$

But now, to express the effect of conforming to the majority, $d\eta/dt$ has to be positive if $\eta > \xi$ and negative if $\eta < \xi$. The simplest formula of that type is $d\eta/dt = \eta - \xi$. But that would permit η to stray outside the possible range $0 \leqslant \eta \leqslant 1$. Buffers must therefore be inserted so as to make $d\eta/dt$ vanish at the termini, giving $d\eta/dt = \eta(\eta - \xi)\xi$. We must also insert a constant G in order that there may be the possibility of adjusting the speed of the theoretical changes to agree with observations. Thus we obtain

$$-d\xi/dt = d\eta/dt = G\eta(\eta - \xi)\xi \,, \tag{161}$$

in which G is the reciprocal of a time. That is the simplest formula which makes sense of pure gregariousness, and it is of the third degree in ξ and η jointly. The foregoing hypotheses (28) to (39) of the bipersonal theory are of the second degree in the variables jointly. In accordance with the habits of mathematicians, we have seen what can be done with terms of the second degree before proceeding to those of the third.

Although the present theory is mainly about multitudes, yet valuable hints may be gained by noticing the least number of people for which its concepts make sense. Two men alone on an island could quarrel and make peace; but neither of them could be said to be out of the island-fashion in clothes or speech. Thus the concept of fashion requires at least three people. In this sense the pair of equations (160) and (161) may be called *tripersonal.* The equations almost say so themselves; for, if $N = 2$, the only possibilities are $\eta = 1$, $\eta = \frac{1}{2}$, and $\eta = 0$, and in each of these three cases $\eta(\eta - \xi)\xi$ vanishes; whereas, if $N = 3$, then η can be $2/3$, making $\eta(\eta - \xi)\xi$ positive.

According to (161), the point $\xi = \eta$, at which there is no majority, cannot be crossed. The motion is away from this point on both sides of it. The existence of an uncrossable point is otherwise obvious. For, if the only motive in the population were to be fashionable, the fashion could never change. If peace and war were controlled by gregariousness alone, a nation would be either permanently at war or permanently at peace. Hence it is obvious that, in the real world, gregariousness acts, not as a prime cause of change, but as a modifier of changes caused by other motives. This accords, so far, with what Trotter wrote of effects as being *enhanced* by gregariousness.

The integral of (161) is

$$Gt + \text{Constant} = 2 \log(\eta - \tfrac{1}{2}) - \log \eta - \log(1 - \eta). \qquad (162)$$

A graph of this function is shown in Fig 4. There are two branches, according as η is greater or less than $\frac{1}{2}$. Change is slow when the population is either nearly unanimous or nearly equally divided. Change is most rapid when $\eta = 0.7887$ or 0.2113.

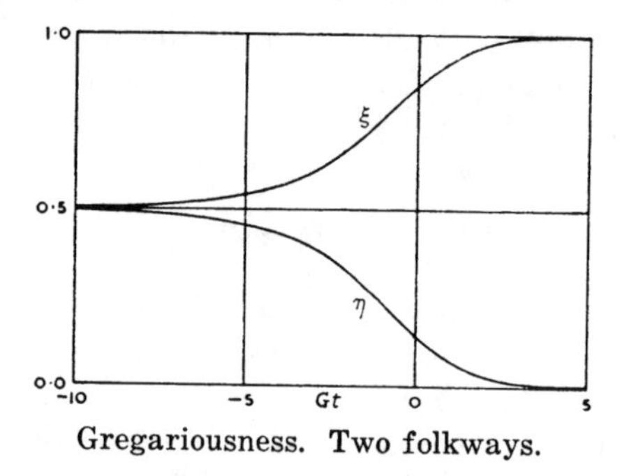

Gregariousness. Two folkways.

FIG. 4. Pure fashion. The abscissa is proportional to the time. The ordinate is the fraction of the population which follows one of two alternative folkways. The relation is theoretical, as specified by equations (160), (161), (162). The constant G is a measure of the intensity of gregariousness.

Case VIII. The outbreak of a symmetrical war, as modified by gregariousness

To examine the interaction between gregariousness and another strong motive, let us consider again the outbreak of a symmetrical war, which has been treated in Case II. But now let the bipersonal effect represented in Case II be combined with the gregarious effect represented in the section on pure fashion. An additive type of interaction will be assumed for mathematical convenience. Additive interaction is also in accordance with a practice widely successful in theoretical physics. So let the new hypothesis be

$$d\eta/dt = C\eta\xi + G\eta(\eta - \xi)\xi . \tag{163}$$

Our other hypothesis is, as before,

$$\xi + \eta = 1 . \tag{164}$$

Let

$$h = G/C , \tag{165}$$

so that h is a positive pure number expressing the strength of gregariousness relative to that of the motive for change. Accordingly (163) may be arranged as

$$\begin{aligned} d\eta/dt &= C\eta\xi\{1 + h(\eta - \xi)\} \\ &= C\eta(1 - \eta)\{1 + h(2\eta - 1)\} . \end{aligned} \tag{166}$$

The chief question concerning (166) is whether $d\eta/dt$ can be negative; that is to say, whether gregariousness can prevent the outbreak of a war. The factors $C\eta(1 - \eta)$ are necessarily positive; and in the last factor

$$-1 \leqslant 2\eta - 1 \leqslant 1 . \tag{167}$$

So $d\eta/dt$ can be negative if, and only if,

$$h > 1 . \tag{168}$$

Let us consider separately the subcases $h > 1$, $h = 1$, and $0 \leqslant h < 1$.

Subcase $h > 1$. The last factor in (166) vanishes when

$$\eta = (h - 1)/2h = \mathcal{E} \text{, say,} \tag{169}$$

and $d\eta/dt \leqslant 0$ whenever $\eta < \mathcal{E}$; so that η, if initially near enough to zero, will, according to this theory, decrease to zero. We have already noted that gregariousness, if in sole control, would permit no deviation from a uniform and constant state of public opinion. We now see further in what circumstances gregariousness, although competing with a motive for change, may yet, according to this theory, convert the minds of a sufficiently small minority so as to make them agree with the majority. To express the same relations differently, let us suppose that gregariousness, as represented by G, is of fixed intensity while the motive for change, represented by C, has one or other of various constant values. If $C = 0$, the point $\eta = 1/2$ cannot be crossed. For larger constant values of C the uncrossable point is lower, being $\mathcal{E}$ as given by (169).

Subcase $h = 1$. Equation (166) becomes

$$d\eta/dt = C\eta(1-\eta)2\eta\,, \tag{170}$$

so that $d\eta/dt$ is positive for all values of η in the open interval $0 < \eta < 1$, but $d\eta/dt$ vanishes doubly at $\eta = 0$.

Subcase $0 \leqslant h < 1$. For a sufficiently large C, the uncrossable point lies outside the positive region of actual η; and then gregariousness modifies, but does not prevent, the conversion of persons from the mood {friendly, hostile} to the mood {hostile, friendly}. This happens, according to (169) when $h < 1$.

Integrals. Let us examine all three subcases more thoroughly by forming the integrals of (166) and (170). From (166), by separation of the variables and integration by partial fractions it may be shown, after a page or two of routine, that, if $h \neq 1$,

$$Ct + \text{Const.} = \frac{1}{1-h^2}\left\{\log\left(\frac{\eta}{1-\eta}\right) + h\log\left(\frac{\eta(1-\eta)}{4h^2(\eta-\mathcal{E})^2}\right)\right\}\,, \tag{171}$$

which can be verified more easily by differentiation. It is convenient to fix the arbitrary constant by the convention that

$$t = 0 \quad \text{when} \quad \eta = 1/2\,, \tag{172}$$

for with this choice the time is reckoned from about the beginning of hostilities. From (171) and (172) it follows that, if $h \neq 1$

$$Ct = \frac{1}{1-h^2}\left[(1+h)\log\eta + (h-1)\log(1-\eta) - 2h\{\log h + \log|\eta-\mathcal{E}|\}\right]. \tag{173}$$

From this formula η has been computed as a function of Ct for various values of h, and is plotted in Figure 5.

The graph for $h = 0$ is of course that already mentioned in Case II. The graph for $h = 1/2$ resembles that for $h = 0$, but is steeper when $\eta > 1/2$ and less steep when $\eta < 1/2$. Leaving aside for the present the subcase $h = 1$, which requires a special formula, let us pass on to the graph for $h = 2$. It has two branches, which may be explained as follows. If by some disturbance, accidental in the sense that it is not taken into account in the present theory, η were momentarily raised to a value exceeding $\frac{1}{4}$ and there released, then the motives specified in the theory would carry η on towards unity in the manner shown by the upper branch of the curve. But if, on the contrary, the accidental disturbance raised η momentarily to a value less than $\frac{1}{4}$ and released it there, then η would sink back towards zero

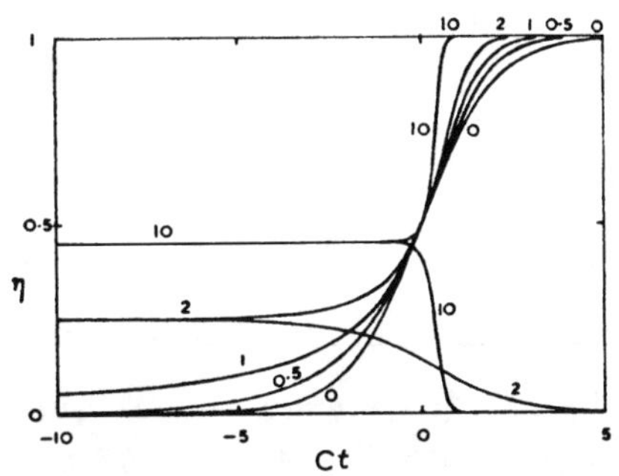

Gregariousness and War Fever.

FIG. 5. Fashion competing with a motive for change.

Different theoretical ways in which a symmetrical war can begin, or can be avoided. The abscissa Ct is proportional to the time. The ordinate η is the fraction of the population that is eager for war. The numbers beside the curves are the values of h, which measures the intensity of gregariousness relative to that of threats.

The same diagram applies to the end of a symmetrical war with time running in the same sense. The ordinate is then ω_8, the fraction of the surviving population which is openly in favor of peace, as explained in Case IX.

in the manner shown by the lower branch of the curve. For $h = 10$ the curve shows a long interval of suspense followed by abrupt alternative decisions. The curves for $h = 2$ and 10 may serve as types for all those for which $h > 1$, on the understanding that the uncrossable value of η, given by (169), increases with h to the limit $\frac{1}{2}$ as $h \to \infty$.

The integral for the subcase $h = 1$ is best found from (170) because of the vanishing denominator in (171). The integral of (170), subject to the convention (172), can be shown to be

$$Ct = 1 + \tfrac{1}{2}\left\{ \log\left(\frac{\eta}{1-\eta}\right) - \frac{1}{\eta} \right\}. \tag{174}$$

Moreover, (174) can be shown to agree with the limit of (171) as $h \to 1$ provided that $\eta \neq 0$. The curve for $h = 1$ separates the one-branched curves from the two-branched curves. When $h = 1$, an infinitesimal accidental disturbance of η from the value zero will start η on a drift towards unity. But for small values of η the drift is much slower for $h = 1$ than for $h = 1/2$.

Comparison of the theory with the outbreak of war in 1914. The diagram of observations (Figure 1) must first be artificially symmetrized by drawing a curve midway between those for Britain and Germany, and then this mid-curve must be compared with the theoretical Figure 5. It is obvious that the uncrossable point $\mathcal{E}$ was less than 1/20; and that therefore, by (169), $h < 10/9$. But it is not possible from those uncertain observations to prove that h exceeeded zero. This is a justification for believing that the bipersonal theory expresses the main effect and that gregariousness comes in only as a modification.

Case IX. The termination of a symmetrical war as modified by gregariousness

Towards the end of a war, gregarious effects may be expected to show themselves when ρ^* people are feeling tired but are concealing their weariness because they notice that others are patriotically doing the same. In a later phase some of these same people may notice others openly admitting that they have had enough of the war and may consequently do likewise.

This case may be regarded as a modification of Case IV. As in Case IV we assume that

$$\rho + \omega = 1 - \theta_T\,. \tag{175}$$

By a process of reasoning similar to that which led up to the hypothesis (163) let us now copy the bipersonal effect from Case IV and add a gregarious term from the theory of pure fashion, and so arrive at

$$d\omega/dt = (D + F)\rho\omega + G\omega(\omega - \rho)\rho\,. \tag{176}$$

If θ_T, which may be about 0.02, could be neglected, then (175) and (176) could be obtained from (164) and (163), respectively, by the simple substitution of ω for η, of ρ for ξ, and of $D + F$ for C. Consequently the same substitution would make the integrals of Case VIII applicable to Case IX, except for the arbitrary constant.

But, if it is desired to take θ_T into account, this can best be done by transforming the variables to fractions of the surviving population by putting

$$\rho/(1 - \theta_T) = \rho_s\,, \qquad \omega/(1 - \theta_T) = \omega_s\,. \tag{177}$$

For then (175) becomes

$$\rho_s + \omega_s = 1\,, \tag{178}$$

and (176) can be arranged as

$$d\omega_s/dt = (1 - \theta_T)\,(D + F)\,\omega_s(1 - \omega_s)\{1 + h_s(2\omega_s - 1)\}\,, \tag{179}$$

in which

$$h_s = G(1 - \theta_T)/(D + F)\,; \tag{180}$$

so that these equations are closely similar to those of Case VIII. It remains to fix the arbitrary constant. As in Case IV, we may reasonably suppose that hostilities end at or about the time when $\omega_s = 1/2$, and so we can conveniently fix the arbitrary constant by the convention that

$$t = T \quad \text{when} \quad \omega_s = 1/2 . \tag{181}$$

This having been done, the integrals and the diagram of Case VIII become applicable to Case IX by the substitution of ω_s for η, of $(1 - \theta_T)(D + F)(t - T)$ for Ct, and of h_s for h. This useful substitution will be referred to as - - - - - - - - - - (182)

The most noteworthy effect of gregariousness, in Case IX, is that a considerable minority, openly in favor of peace, may remain almost constant for a long time prior to the end of the war, provided that $h_s > 1$. This behavior is in contrast with that in Case IV, where ω could be constant only when zero. There were in fact during 1916, 1917, and 1918 minorities opposed to the war in both Britain and Germany, as noted in Part I A. They were persistent, but whether they were constant is not clear.

MENTOR: In all the foregoing theory, you take the effect of the threat from the enemy-nation as acting always in the same sense. But in fact reversals have occurred. One might say that moderate threats weld a nation together; but very severe threats, following on military disaster, crack it in pieces. Witness the revolt of the Fascist Grand Council against Mussolini in July 1943, and the revolt of some German generals against Hitler in July 1944. How do you account for that on your theory?

AUTHOR: I have to leave it out as being a more complicated phenomenon. The last bit of theory is only about the change of mood from {hostile, weary} to {weary, hostile}. That change, to an open longing for peace, may have had a little to do with the revolts which you mention. But their main motive appears to have been discontent with the government for having failed to protect the nation. In the latter aspect they resembled the discontent in Britain with Mr. Neville Chamberlain in May, 1940; except that British parliamentary customs allowed his replacement to be effected in an orderly manner. The mood of "discontent with own government" is not on our present list. Theories, in their beginning, must omit.

Case X. The interaction of gregariousness with indignation in Britain during the Sudeten crisis of September, 1938

The reluctant acceptance in Britain of the Munich agreement of September 30, 1938, may be regarded as an instance in which gregariousness prevailed over indignation. For the political fashion in Britain at that time was appeasement towards Germany and Italy. Yet the aggressive German behavior towards Czechoslovakia from early summer onwards aroused strong indignation in Britain; many Britishers wished somehow peacefully to quell Hitler by threats; and a minority, η_1, of Britishers would have preferred war with Germany to the cession of the Sudetenland, which was agreed to at Munich. These various attitudes can be seen in the Official Report (24) of the debates in the House of Commons on October 3, 4, 5, 6. It will suffice to refer to the speeches of Mr. Duff Cooper (Cols. 29 to 40), of Sir Samuel Hoare and Mr. Dalton (Cols. 150, 151) and to the motion (Cols. 557, 558) "That this House approves the policy of His Majesty's Government by which war was averted in the recent crisis and supports their efforts to secure a lasting peace"; this was carried by 366 to 144. We may conclude that, counting the tellers,

$$\eta_1 < 146/(146 + 368) = 0.28 \tag{183}$$

if the House of Commons represented the nation, which it seldom does precisely.

It is further evident from the history that, between September 1st and 28th,

$$d\eta_1/dt > 0 \, . \tag{184}$$

The order of magnitude of $d\eta_1/dt$ is not easy to estimate; perhaps, at a guess, η_1 may have risen from 0.01 to 0.10 in a month; so that

$d\eta_1/dt$ was of the order of unity during September. (185)

Many speakers in the aforesaid debate, while deploring the situation, expressed relief and satisfaction at the peaceful settlement. Chamberlain had described it as "peace for our time." From this it may be inferred that, just after the Munich agreement of September 30th,

$$d\eta_1/dt < 0 \, . \tag{186}$$

It further seems proper to assume that η_1, regarded as a function of time, was continuous across the crisis. - - - - (187)

Let suffix 1 indicate Britain and suffix 2 indicate Germany. Let us neglect β and ρ, so that, as in Cases II and VIII

$$\xi_1 + \eta_1 = 1 \, . \tag{188}$$

But the situation will not be assumed to be symmetrical.

The gregarious part of $d\eta_1/dt$ will be the same as in Case VIII, equation (163), namely $G\eta(\eta - \xi)\xi$, except that G, η, ξ will now become G_1, η_1, ξ_1. The international part of $d\eta_1/dt$ will be the unsymmetrical term $C_{12}\ \xi_1\ \eta_2$ quoted from (32). On adding these parts, we have

$$d\eta_1/dt = \xi_1\{C_{12}\eta_2 + G_1(2\eta_1 - 1)\eta_1\}\,. \tag{189}$$

Instead of writing an equation of motion for German moods, it will suffice to regard η_2 as a known external disturbance to British moods. The change in German η_2 can then be approximately represented by a sudden jump. Let η_2' be the value of η_2 just before the Munich agreement and η_2'' the value just after. It appeared to the British that

$$\eta_2' > \eta_2''\,. \tag{190}$$

On inserting in turn (184) and (186) in (189), we obtain:

just before the agreement

$$C_{12}\eta_2' + G_1(2\eta_1 - 1)\eta_1 > 0\ ; \tag{191}$$

just after the agreement

$$C_{12}\eta_2'' + G_1(2\eta_1 - 1)\eta_1 < 0\,. \tag{192}$$

From these two inequalities it follows, in view of (187), that $C_{12}(\eta_2' - \eta_2'') > 0$; and therefore, in view of (190), that

$$C_{12} > 0\,, \tag{193}$$

as was expected. Next, (183) implies that

$$1 - 2\eta_1 > 0\,, \tag{194}$$

and so (191) and (192) may be rearranged as

$$\frac{\eta_2''}{\eta_1} < \frac{G_1}{C_{12}}(1 - 2\eta_1) < \frac{\eta_2'}{\eta_1}. \tag{195}$$

We have thus a proof that $G_1 > 0$, and a hint as to how, with a little more information, the value of G_1 could be estimated. To make a guess, for illustration, suppose that $\eta_1 = 0.1$, $\eta_2' = 0.6$, $\eta_2'' = 0.2$; then would $2.5 < G_1/C_{12} < 7.5$. This range corresponds, as we should expect, to values of G/C which, in the symmetrical Case VIII, are large enough to give curves having two branches. The other branch, leading to war, was followed in August and September, 1939.

PART II C

Chemical Analogies

Much of the theory of Parts I B and II A was developed from the simple notion of the probability of a binary encounter, as introduced by Kermack and McKendrick (15) into their 1927 theory of epidemics. Accordingly each of the terms in the second members of equations (26) to (39) is of the second degree in the variables. In chemical language the reactions would be bimolecular. But, as the reagents are persons, we had better call the reactions *bipersonal.* The simple approximation in the Cases II and IV leads to an integral commonly used by chemists; see (23, p. 643). Mole's (22) theory of the ignition of explosive gases has terms resembling, at least in sign, the author's linear theory of the instability of peace, as mentioned by Richardson (30). Those who are more familiar than the present author with the literature of physical chemistry may know where in it to find either helpful analogies or ready-made integrals of the complete set of differential equations (26) to (39), or of substantial parts thereof.

Dr. Ernest B. Ludlam, whose researches on the explosion of phosphorus vapor are mentioned in standard textbooks, has, at various times since 1936, drawn the author's attention to a social analogue to the activated molecules of chemistry and kindly permits it to be mentioned here. Common observation shows us that only a few of the persons who hold an opinion are sufficiently energetic to propagate it; just as only a few of the molecules, which have the formula suited to a reaction, are sufficiently energetic to react. Moreover, the unusual social energy abides at first in a small chain of leaders and their contacts; just as, in a chemical chain, the energy set free by the reaction is not immediately scattered among the whole vesselful of molecules. These suggestions, which may prove to be valuable, are mentioned here lest they be forgotten. But the author does not feel inclined to develop them at present. According to Semenoff (33), after van't Hoff in 1884 and Arrhenius in 1889 had stated the kinetics of bimolecular reactions, chemists pursued their researches for 24 years before Bodenstein in 1913 introduced the conception of chains of reactions. So a psychologist may excuse himself for taking time for deliberation. The author's theory is not yet at the stage of Arrhenius, for it takes no account of the different energy of different persons in the same dual mood.

REFERENCES

1. Bernhardi, F. von. Germany and the next war. Trans. by Powles, A. H. London: Edward Arnold, 1914.
2. British Association for the Advancement of Science. *Report of the Annual Meeting*, 1938, 277-334.
3. Buxton, Mrs. C. R. A weekly survey of the foreign press. *Cambridge Magazine*, Cambridge, England, 1915-1919.
4. Calvert, E. R. Capital punishment in the twentieth century. London: Putnams, 1930.
5. *Commonsense,* a London periodical current during the first world war.

6. Dumas, S. and Vedel-Petersen, K. O. Losses of life caused by war. Oxford: Clarendon Press, 1923.
7. Encyclopaedia Britannica. New York: Encyclopaedia Britannica, Inc., 1929, edition XIV.
8. German, a. J'Accuse, trans. by Gray, A. New York and London: Hodder and Stoughton, 1915.
9. *Glasgow Herald.* Daily Newspaper, Glasgow, Britain.
10. Graham, J. W. Conscription and conscience. London: Allen and Unwin, 1922.
11. Hirst, F. W. The consequences of the war to Great Britain. Oxford: Univ. Press, 1934.
12. Hitler, A. Mein Kampf, translated by Murphy, J. London: Hurst and Blackett, 1939.
13. James, W. The varieties of religious experience. London: Longmans Green, 1902.
14. *Keesing's contemporary archives,* weekly periodical. Bristol, England.
15. Kermack, W. O. and McKendrick, A. G. Mathematical theory of epidemics. *Proc. Roy. Soc. Lond.* A, 1927, 115, 700-722.
16. Kermack, W. O. and McKendrick, A. G. The solution of sets of simultaneous integral equations related to the equation of Volterra. *Proc. Lond. Math. Soc.*, 1936, **41**, 462-482.
17. Kruger, P. Memoirs. London: Fisher Unwin, 1902.
18. Lloyd George, D. War memoirs. London: Odhams, 1938.
19. Matthew, Saint. Gospel, revised version. Oxford: Univ. Press, 1891.
20. McCarthy, J. A history of our own times. London: Chatto and Windus, 1892.
21. Mendelssohn Bartholdy, A. The war and German society. New Haven: Yale Univ. Press, 1937.
22. Mole, G. The ignition of explosive gases. *Proc. Phys. Soc. Lond.*, 1936, **48**, 857-864.
23. Nernst, W. Theoretical chemistry. Trans. by Codd, L. W. London: Macmillan, 1923.
24. *Parliamentary debates, official report* (alias *Hansard*). London: H. M. Stationery Office.
25. Pear, T. H. Peace, war and culture patterns. *Bulletin of the John Rylands Library,* Manchester, 1948, **31**, 3-30.
26. Plato. Timaeus.
27. Rashevsky, N. Mathematical theory of human relations. Bloomington, Indiana: Principia Press, 1947.
28. Richardson, L. F. Mathematical psychology of war. Oxford: Wm. Hunt, 1919. (A few copies remain with the author.)
29. Richardson, L. F. The behavior of an Osglim lamp. *Proc. Roy. Soc. Lond.* A, 1937, **163**, 380-390.
30. Richardson, L. F. Generalized foreign politics. *British Journ. Psychol. Monog. Supplt.* No. 23, 1939.
31. Richardson, L. F. The number of nations on each side of a war. London: *Journ. Roy. Statistical Soc.*, 1946, **109**, 130-156.

32. Richardson, L. F. Arms and insecurity, a microfilm, 35 mm. Kilmun, Argyll, Britain: from the author, $5, 1947.
33. Semenoff, N. Chemical kinetics and chain reactions. Oxford: Clarendon Press, 1935.
34. Soviet Union, official history of the communist party of the. Moscow: Foreign Languages Publishing House, 1939.
35. Spillane, J. P. In The neuroses in war, edited by Miller, E. London: Macmillan, 1940.
36. *Times, The.* Daily newspaper, London.
37. Trotter, W. Instincts of the herd in war and peace. London: Fisher Unwin, 1921.
38. *U. D. C.*, a periodical, London: Union of Democratic Control.
39. Vedel-Petersen, see Dumas.
40. Volterra, V. Theory of functionals, edited by Fantippié, trans. by Long, M. Glasgow: Blackie & Son, 1930.
41. Volterra, V. Leçons sur la théorie mathématique de la lutte pour la vie. Paris: Gauthier-Villars, 1931.
42. Wright, M. B. In The neuroses in war, edited by Miller, E. London: Macmillan, 1940.

Richardson, L. F. (1948)
J. Am. statist. Ass. **43**, 523–46

1948:4

VARIATION OF THE FREQUENCY OF FATAL QUARRELS WITH MAGNITUDE

Journal of the American Statistical Association
Boston, Massachusetts
Vol. 43, December 1948, pp. 523–546

Notes:

1. This paper was described in 1941:1 as 'nearly ready for publication'
2. Nearly all of this paper reappeared unaltered in various places in *Statistics of Deadly Quarrels* (1960:1). Most of the 'Introduction' was incorporated into Chapter I.4., and the three sections on quarrels that caused from 4 to 315 deaths into Chapter II.7. The rest of the paper reappeared unaltered as Chapter IV, 1 and 2. (The 'Introduction' and most of the three sections on small fatal quarrels were not included in the microfilm (1950:4) – for the details see the 'Notes on *Statistics of Deadly Quarrels*', below p. 537.)

Reprinted from the JOURNAL OF THE AMERICAN STATISTICAL ASSOCIATION
December, 1948, Vol. 43, pp. 523-546

VARIATION OF THE FREQUENCY OF FATAL QUARRELS WITH MAGNITUDE

LEWIS F. RICHARDSON
Hillside House, Kilmun, Argyll, Britain

A record of wars during the interval A.D. 1820 to 1945 has been collected from the whole world, and has been classified according to the number of war-dead. The smaller incidents have been the more frequent, according to a fairly regular graph which can be extended to quarrels that caused a single death.

INTRODUCTION

ANYONE WHO TRIES to make a list of "all the wars" (e.g. Wright [18] p. 636) encounters the difficulty that there are so many small incidents, that some rule has to be made to exclude them.

From the psychological point of view a war, a riot, and a murder, though differing in many important aspects, social, legal, and ethical, have at least this in common that they are all manifestations of the instinct of aggressiveness. This Freudian thesis has been developed by Glover [5], by Durbin & Bowlby [3], and by Harding [6]. There is thus a psychological justification for looking to see whether there is any statistical connection between war, riot, and murder. I dealt with this problem in 1941 by forming the inclusive concept of a "fatal quarrel," and then classifying the fatal quarrels according to the number of quarrel-dead [12]. By a *fatal quarrel* is meant any quarrel which caused death to humans. The term thus includes murders, banditries, mutinies, insurrections, and wars small and large; but it excludes accidents, and calamities such as earthquakes and tornadoes. Deaths by famine and disease are included if they were immediate results of the quarrel, but not otherwise. In puzzling cases the legal criterion of "malice aforethought" was taken as a guide.

The record of the number killed in a particular war is often uncertain by a factor of two. The meaningful part of the record can be separated from its uncertainty by taking the logarithm to the base ten and rounding the logarithm off to a whole number, or to the first decimal, according to the quality of the information. For simplicity I have lumped together the deaths on the opposing sides of the quarrel. The *magnitude* of a fatal quarrel is defined to be the logarithm to the base ten of the number of people who died because of that quarrel. The magnitude will be denoted by μ. The range of magnitude extends from 0 for a murder involving only one death, to 7.4 for the Second World War. Other well-known wars had magnitudes as follows: 1899–1902 British versus Boers 4.4; 1939–40 Russians versus Finns 4.83; 1861–65 North American Civil War 5.8. The magnitude of a war is usually known to within ± 0.2; so that a classification by unit ranges of magnitude is meaningful. These ranges have been marked off at 7.5, 6.5, 5.5, 4.5, 3.5, 2.5, 1.5 and perhaps at 0.5 and -0.5. Every fatal quarrel must lie inside one of these cells; it cannot lie on a boundary, because the antilogs of the boundary μ are not integers.

This abstract framework becomes of interest when the facts are sorted into it. As far as I have been able to ascertain, the historians have never sorted their facts by the scale of magnitude. The chief obstacle to any scientific study of wars-and-how-to-avoid-them, is each nation's habit of blaming other nations. National prejudice can however be avoided by taking the whole world as the field of study; and by taking a time-interval longer than personal memory. For this reason I have made a search for the records of fatal quarrels in the whole world since the beginning of A.D. 1820. The task has been long; and the results are here presented in brief summary. For magnitudes greater than 2.5, the facts were mostly obtained from works on history. For magnitudes less than 0.5 they were taken from criminal statistics. For magnitudes between 0.5 and 2.5 the information is scrappy and unorganized; what there is of it suggests that such small fatal quarrels were too numerous and too insignificant to be systematically recorded as history, and yet too large and too political to be recorded as crime.

It is important to know whether the facts have all been gathered. The best evidence is provided by the progress of search. I began in the year 1940. At first my collection grew rapidly, then slowly; then various revisions caused its totals to oscillate slightly. The collection appeared to be sufficiently complete to warrant the publication of a summary, which appeared in *Nature* of 15th November, 1941. Afterwards the publication of Quincy Wright's list of wars ([18] Appendix XX) provided a stimulus to further enquiry, which involved the consultation of some seventy history-books. In the following table three stages of my records are compared. All refer to fatal quarrels which ended from A.D. 1820 to 1929 inclusive.

Date and place of record	Ends of range of magnitude			
	$7\pm\frac{1}{2}$	$6\pm\frac{1}{2}$	$5\pm\frac{1}{2}$	$4\pm\frac{1}{2}$
	Numbers of fatal quarrels recorded			
1941. *Nature* of Nov. 15	1	3	16	62
1944. J.R.S.S. 107, 248	—	—	—	63
1947. Here.	1	3	20	60

It is seen that there has been a sort of convergence, which makes the present numbers worthy to be discussed. That is for $\mu > 3.5$. For smaller incidents in the range $2.5 < \mu < 3.5$ my collection has continued to grow; the sign $\geqq$ is therefore prefixed to its totals.

The world-total for the murders was obtained from the murder-rates of 17 countries. They were weighted by populations. The weighted mean rate for the world was found to be 32 murders per million per year. This is not as definite as the statistics of wars, because the criminal statistics of China, Africa, and South America (except Chile) were not available. But even if that estimate were as much as three times wrong, which is incredible, it still would give significant results in Figures 0 and 1, because they relate to such enormous ranges. The evidence which I had concerning time-changes in the murder-rate was scrappy and conflicting; so that provisionally I took the rate to be constant. The mean population of the world over the 126 years from A.D. 1820 to 1945 was computed from a publication by Carr-Saunders [1A] and was found to be 1.49×10^9. So the world-total of murders was computed as $1.49\times 10^9 \times 32 \times 10^{-6} \times 126 = 6.0\times 10^6$.

A revision based on the murder-rates of 21 countries, gave a mean rate of 37 instead of 32, and a world-total of 7 instead of 6 million. The following diagrams and tables were based on the earlier estimate and have not been altered.

A CONSPECTUS WHICH SUITS THE WARS

The number of fatal quarrels has been counted, or estimated, in unit ranges of magnitude, with the following results for the world as a whole. The date of a quarrel is taken to be that of the termination of hostilities. The interval is A.D. 1820 to 1945 inclusive.

Ends of range of magnitude	$7\pm\frac{1}{2}$	$6\pm\frac{1}{2}$	$5\pm\frac{1}{2}$	$4\pm\frac{1}{2}$	$3\pm\frac{1}{2}\cdots$ Murders
Observed number of fatal quarrels	2	5	24	63	$\geq 188 \cdots 6\times 10^6$

It is desirable to show all these facts together on a single diagram. If the number of fatal quarrels per unit range of magnitude were taken as the ordinate, then either the wars would show, or the murders, but not both. When however the logarithm of the number of fatal quarrels per unit range of magnitude is taken as the ordinate, then all parts show equally well. Each murder presumably involved a small number of deaths, such as 1, 2, 3, or 4, so that the corresponding logarithms were 0.000, 0.301, 0.477, or 0.602. How to group them was not obvious. As a first expedient I extended to the murders the system of classification which had been found to suit the wars, namely by unit ranges of magnitude cut at 7.5, 6.5, 5.5, 4.5, 3.5, and so on by extension to 0.5 and -0.5. Murders involving 4 or more deaths were regarded as negligibly rare. The ordinate f is defined to be

$$f = \log_{10}\left\{\frac{\text{number of fatal quarrels}}{\text{corresponding range of magnitude}}\right\}.$$

This rough and ready scheme gives the conspectus shown in Figure 0. It is the type of diagram with which the author began, and it has now been asked for by a referee. A dotted curve has been drawn across the gap to suggest that there is statistical continuity between the wars and the murders.

THE FAULTS OF FIGURE 0 AND HOW TO AMEND THEM

In order to match the wars, the murders have been enclosed in a unit range of magnitude, as shown at A in Figure 0. But the half of this range, from $\mu = -0.5$ to μ just less than zero, is necessarily empty. Would it be more reasonable to regard the murders as enclosed in a half-unit range? If so the line representing them in Figure 0 would be halved in length. Also the number of murders per unit of magnitude would be doubled. So the representative line would be raised by $\log_{10} 2$ into position marked B. Worse ambiguities occur if we wish to show separately the number of murders each causing one death. Let there be M of them. To what range of magnitude ought they to be attributed? The range must contain the point $\mu = 0$, and must not contain the point $\mu = \log_{10} 2$, but otherwise is unspecified. Yet M has to be divided by this indefinite range in order that the reckoning may be "per unit range of magnitude". Figure 0, though suitable for the wars, is too vague for quarrels that caused only one death.

It may be thought that the root of the trouble is the reckoning "per unit range of" anything, and that if ranges were abandoned then murders and wars could be compared by plotting the logarithm of their numbers against their magnitudes. Such a diagram would consist of

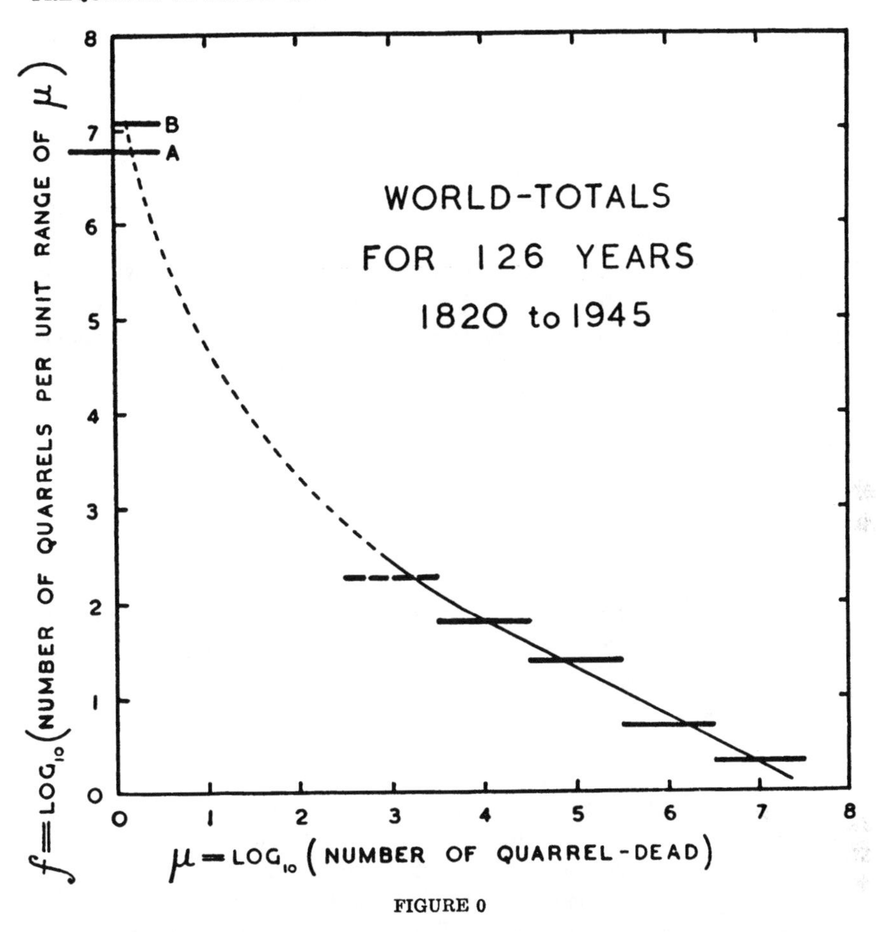

FIGURE 0

points, not segments. There are two objections to this proposal: (i) the magnitudes of wars are imperfectly known and some of them are rounded off to whole numbers; so that there would be many artificially coincident points, (ii) if the magnitudes were better known they would presumably be scattered, so that a point-diagram of wars would present great irregularities, of no interest. For when n exceeds 1000 there is never likely to be any statistically significant distinction between wars involving n and $n+1$ deaths. In brief we must have points for the murders and averages over ranges for the wars, and yet these diverse representations must somehow be compared. The solving idea, which melted all the obstacles, was that the appropriate measure of any range of quarrel-dead is simply the number of integers in that range, and not the corresponding range of magnitude, nor any other artificial conception.

A DIAGRAM WHICH SUITS THE MURDERS

The logically simple idea is that there were, say $q(n)$ quarrels each involving n deaths. (1)

For example in England and Wales during 1935 and 1936 there were known to the police 186 cases of murder of 213 persons aged over one year. In 74 cases (involving 91 persons) the murderer or suspect committed suicide. In 105 cases (115 victims) 111 persons were arrested. In the remaining 7 cases, involving 7 victims, no arrest was made. Of the 111 persons arrested 16 were executed (Whitaker's Almanacks 1938 and 1939, p. 652). These statements can be interpreted, with only slight uncertainty, by the following distribution.

				totals
$n=$	1	2	3	
$q(n)=$	92	71	23	186
$nq(n)=$	92	142	69	303

The number of deaths, n, connected with the quarrel is here taken to include, along with those of the victims, also those of the murderer or suspect, whether by suicide or execution. It seems permissible, for illustration until better information can be found, to regard the 6 million murders in the world from 1820 to 1945 as the number of fatal quarrels for $n=1$, 2 and 3, and to subdivide this total among its three parts in the same ratios as the total for England and Wales is subdivided. Thus one finds for the world, the following enlarged values of $q(n)$. To take one country as a sample of the world is very unsatisfactory. But it should be noted that the uncertainties of these $q(n)$ are insignificant in comparison with the vast range of Figure 1 where $q(n)$ varies in a ratio exceeding 10^{13}.

				totals	
$n=$	1	2	3		
$q(n)=$	3.0	2.3	0.7	6.0	millions
$nq(n)=$	3.0	4.6	2.1	9.7	millions

These give three points in the top left corner of Fig. 1, with the coordinates

$\log_{10} n=$	0	0.301	0.477
$\log_{10} q(n)=$	6.48	6.36	5.85

A GENERAL AND WELL DEFINED CONSPECTUS

This will now be obtained by extending to the wars the method introduced above for the murders.

In the region of wars, say for $n > 1000$, the function $q(n)$ is usually zero and occasionally unity; so that a diagram of it would be bristly and unreadable. Therefore for the wars we should take an average of $q(n)$, say $\bar{q}(n)$. As a preliminary to the definition of $\bar{q}(n)$, it must be understood that the fatal quarrels are arranged in order of n. The definition of $\bar{q}(n)$ is then

$$\bar{q}(n) = \frac{\text{number of fatal quarrels in a range of } n}{\text{number of integers in that range of } n}. \tag{2}$$

Briefly, and somewhat inaccurately, $\bar{q}(n)$ may be called the "average frequency of quarrels per unit range of quarrel-dead." As in all frequency-diagrams so here the choice of the group-range has to be a compromise: if the range is too small, the diagram is spiky; if the range is too large, essential features are flattened out. The question of the best range must be decided by trial.

The diagram of q and $\bar{q}$ as functions of n is the logically simple conspectus of the frequency of wars and murders. This diagram cannot be drawn, because no sheet of paper is large enough to show the facts at both ends of the range. The difficulty is overcome by plotting the logarithms of q, $\bar{q}$ and n. The magnitude μ has already been defined by

$$\mu = \log_{10} n. \tag{3}$$

Another oft-recurring symbol, ϕ, must now be defined.

Let $\log_{10} \bar{q}(n) = \phi(\mu)$, say, for the wars: while for the murders

$$\phi(\mu) = \log_{10} q(n) \text{ without averaging.} \tag{4}$$

The last column of the following table shows $\phi(\mu)$, apart from some corrections for grouping, which will be explained on p. 542. The first

Range of magnitude μ	Number of integers in range of war-dead	Observed number of fatal quarrels Years 1820–1945	$\log_{10}$ (number of quarrels per unit range of war-dead)
2.5 to 3.5	2,846	≧188	≧ −1.180
3.5 to 4.5	28,460	63	−2.655
4.5 to 5.5	284,605	24	−4.074
5.5 to 6.5	2,846,050	5	−5.755
6.5 to 7.5	28,460,499	2	−7.153

and last columns of this table provide the abscissa and ordinate of the accompanying diagram (Figure 1), as far as the wars are concerned. It is evident that the wars can be satisfactorily fitted by a straight sloping line. On account of grouping, the wars are represented, not by points as the murders are, but by horizontal segments. It will be shown later (p. 542) that the $\phi(\mu)$ graph should pass slightly above the mid-points of these segments. That is to say the sloping straight line was actually obtained by a second approximation.

The relation between the ordinates of Figures 0 and 1

The abscissa μ is the same in both diagrams and the ordinates are respectively f and ϕ, defined thus

$$f = \log_{10}\left\{\frac{q(n_1) + q(n_1+1) + q(n_1+2) + \cdots + q(n_1+s)}{\log_{10} n_b - \log_{10} n_a}\right\}$$

where $n_1 - 1 < n_a < n_1$, and $n_1 + s < n_b < n_1 + s + 1$. These extra bits at the ends of the range have to be put in to prevent f becoming infinite at $s = 0$.

$$\phi = \log_{10}\left\{\frac{q(n_1) + q(n_1+1) + q(n_1+2) + \cdots + q(n_1+s)}{s+1}\right\}$$

which is definite at $s = 0$. By subtraction

$$f - \phi = \log_{10}\left\{\frac{s+1}{\log_{10} n_b - \log_{10} n_a}\right\}.$$

This expression involves an ambiguity, because n_b is loose in a range of unity, and so is n_a. The practical question however is whether the ambiguity is greater than the uncertainty of the observations. It is necessary to distinguish different cases.

Case (i). The range here used in treating the wars is $\mu_0 - \frac{1}{2} \leqq \mu \leqq \mu_0 + \frac{1}{2}$ where $\mu = \log_{10} n$ and where $\mu_0 \geqq 3$. In these circumstances arithmetic shows that

$$f - \phi = \mu_0 + 0.454, \tag{5}$$

reliable to the third decimal. Seeing that the μ of a particular war is often uncertain by ± 0.1, equation (5) is admirably accurate for the purpose.

Case (ii). We may conceivably wish to consider a shorter range $\mu_0 - \epsilon \leqq \mu \leqq \mu_0 + \epsilon$ where μ_0 is still in the region of the wars. If ϵ is small, but not too small, there is a quasi-limit such that

$$f - \phi = \mu_0 + 0.362. \tag{5A}$$

This was obtained by approximating to the difference-ratio by a derivative, and was confirmed by arithmetic.

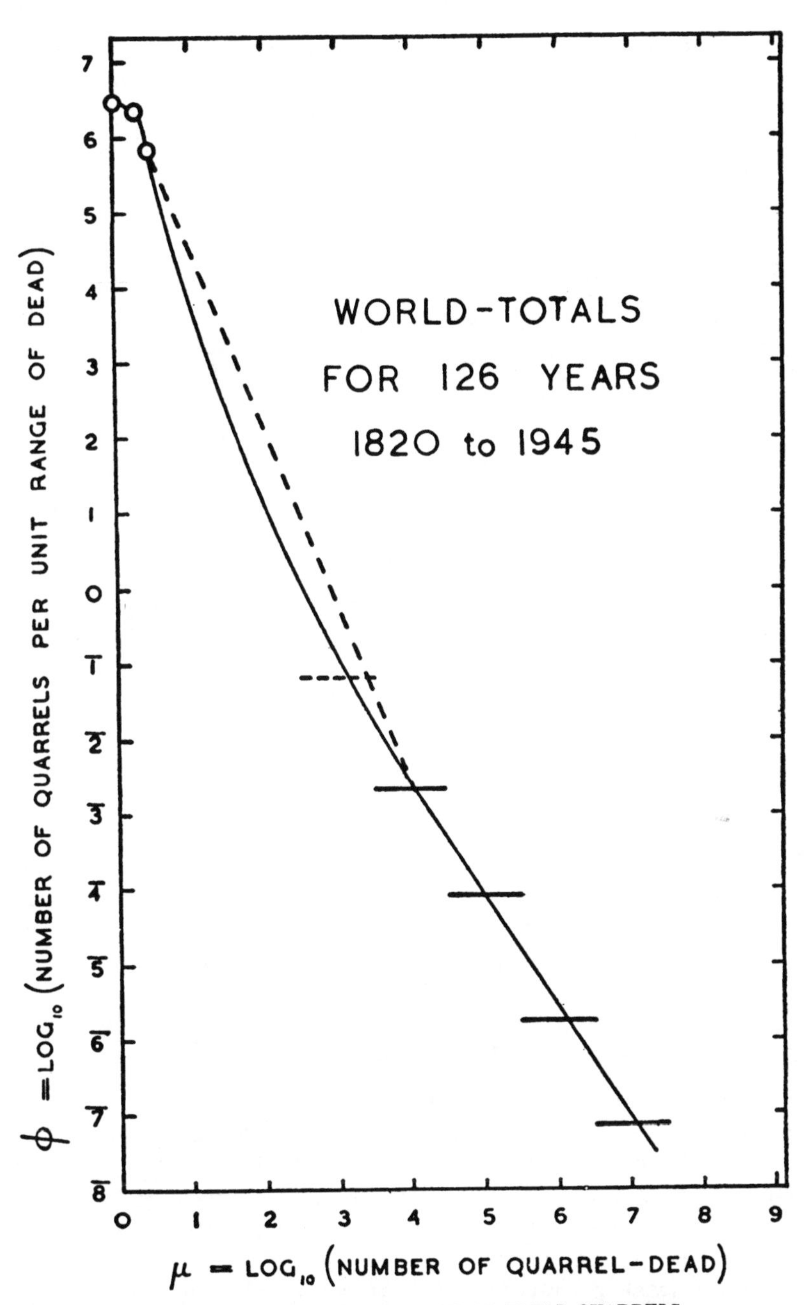

FIGURE 1. THE WHOLE RANGE OF FATAL QUARRELS

THE HORIZONTAL SCALE EXTENDS FROM A SINGLE PERSON ON THE LEFT TO THE POPULATION OF THE WORLD ON THE RIGHT. THE THREE DOTS AT THE LEFT HAND TOP REPRESENT THE MURDERS. THE SHORT HORIZONTAL LINES REPRESENT THE WARS GROUPED BY MAGNITUDES. ONE OF THESE LINES IS DOTTED, BECAUSE IT SHOULD BE RAISED TO AN UNKNOWN HEIGHT.

Case (iii). If there are only a few integers n in the range then the ambiguity becomes troublesome. This is the situation as to the murders. The ambiguity of $f-\phi$ is more than that of some criminal statistics. The fault lies in f, not in ϕ. When $n_1=1$ and $s=0$ it is hardly possible to locate n_a and n_b by any reasonable convention, as already explained on page 526. To revert to Cases (i) and (ii), either of them give the same relation between the slopes of the f- and ϕ-diagrams. As these slopes are in fact negative, it is convenient to write the relation thus

$$-\,d\phi/d\mu_0 = -\,df/d\mu_0 + 1 \tag{5B}$$

which shows that the ϕ-diagram is the steeper.

The slope in the region of the wars

This was found, by the principle of Maximum Likelihood, [4 & 10] to be

$$d\phi/d\mu_0 = -\,1.50. \tag{5C}$$

Interpolation across the gap between the wars and the murders

The dotted horizontal segment for the small wars represents only a lower bound, and otherwise should be ignored. Between the definite end-points of the gap, the mean slope is

$$d\phi/d\mu = -\,2.38. \tag{6}$$

This is shown by a straight line of dashes.

As to the choice of a curve, there are the general clues of simplicity and of continuous slope. For simplicity a circular arc was chosen; and, for continuous slope, the arc was made tangent to the straight line which fits the wars. It was then found, as an independent confirmation that the slope of the circular arc joins on continuously to that for the murders. Some collateral evidence is provided by banditry.

A SEARCH FOR THE FACTS ABOUT QUARRELS THAT CAUSED FROM 4 TO 315 DEATHS $(0.5<\mu<2.5)$

That there were many small fatal quarrels is shown by a statement made by Quincy Wright [18] in the introduction to his long list of "Wars of Modern Civilization." He remarked that "A list of all revolutions, insurrections, interventions, punitive expeditions, pacifications, and explorations involving the use of armed force would probably be more than ten times as long as the present list, . . . " Wright however did not classify fatal quarrels according to the number of quarrel-dead. Similarly whilst reading histories and police-reports and listening to radio-news I have noticed allusions to very many incidents less important than wars, but more important than murders. References [2], [7], [11], [15], contain some remarkable examples. However I see no hope of obtaining world-totals of the numbers of such incidents during the 126 years to which Figure 1 relates, and classified moreover by magnitudes in the range $0.5 < \mu < 2.5$. Any factual test of the interpolated portion of Figure 1 will have to depend on smaller samples. The record of banditry in Manchoukuo is comparable on certain assumptions, and that of gangsters in Chicago on further assumptions. These ordered collections of facts, being rare specimens, are correspondingly valuable.

BANDITRY IN MANCHOUKUO DURING THE YEAR 1935

This is of especial interest, because the facts are given in a form which throws light on aggregation for aggression. The following is quoted from the *Japan and Manchoukuo Year Book*, 1938, pp. 692–95.

"At the time of the founding of Manchoukuo in March of 1932 the total number of bandits exceeded 100,000. By September of the same year the number had increased to 210,000 due principally to the subversive activities of Chang Hsueh-Liang's remnant troops who were thrown out of employment following the downfall of the young marshal. Since then, however the number of such bandits has been on the decrease as a result of their suppression by Manchoukuo and Japanese forces.

"Compared with the condition obtaining in 1932 two factors loom in prominence with regard to the bandit situation. Firstly may be noted the actual reduction of bandits as a whole, and, secondly, the shrinkage in size of bandit groups. In 1932 some bandit groups had an actual fighting force of 30,000 men, but at present the average is below 50 bandits per group. The chief cause for the existence of bandits in Manchoukuo is believed to be an economic one, resulting from unemployment."

STATISTICS OF BANDITRY

Size of bandit groups						
1 to 30	31 to 50	51 to 100	101 to 200	201 to 300	301 to 500	>501
Corresponding number of raids						
28,145	4,784	3,864	1,530	455	240	130

The first step towards making a comparison with larger fatal quarrels is to compensate for the unequal ranges of size in the above quotation, by dividing each number of raids by the corresponding range of size. For example $4784/(50-31+1)=239.2$. Any such ratio may be called the "number of raids per unit range of membership." For the first six groups it runs as follows

938.17 239.2 77.28 15.30 4.55 1.20

and shows plainly that *the smaller incidents were much the more frequent.*

Was a raid like a small battle or like a small war? In other words did the same group of bandits perpetrate many raids? This can be answered by attending to totals. The total number of appearances of bandits can be underestimated from the above table by multiplying the least size in each column by the corresponding number of raids, and comes to 758,371 or more. A moderate estimate is 1.3 million.

But the total number of bandits is stated to have been much less than 210,000. It is evident therefore that on the average the same bandit appeared in several raids. That is to say *a raid resembled a battle rather than a war.*

In the same year and same region the numbers killed in connection with banditry are stated thus: bandits 13,338, suppression troops 1361, civilians 2512. The total dead was therefore 17,211. This is 1.3 per cent of the total number of appearances of bandits.

The facts about banditry are plotted in Figure 2. This is not a $\phi(\mu)$ diagram, yet it is rather like one. For the abscissa, ν say, is $\log_{10}$ of the number of bandits in the group, whereas μ is $\log_{10}$ of the number of quarrel-dead. Again the ordinate, ψ say, is $\log_{10}$ of the average number of raids per unit range of membership, whereas ϕ is $\log_{10}$ of the average number of fatal quarrels per unit range of quarrel-dead. The first group in the data has a membership range of from 1 to 30, which is in a ratio too great for present purposes. Although the first group is shown on Figure 2, it will be ignored in the discussion. The blunted top of Figure 1 has however some resemblance to the horizontal in Figure 2. The other groups are well fitted by a straight line having the slope

$$d\psi(\nu)/d\nu = -2.29. \tag{7}$$

This has a remarkable resemblance to the mean slope, -2.38, across the gap in the world-diagram, as stated in (6). Can this agreement of the slopes be a mere coincidence? Or is it a clue to a general law concerning aggregation for aggression? It is at any rate easily explained by the following two assumptions:—

Firstly let us suppose that the quarrel-dead were on the average proportional to the number of bandits in the group. That is to say that groups continued to make raids until a constant fraction of their members had been killed. The supposition is that

$$n = A \times \text{(number of bandits in group)}$$

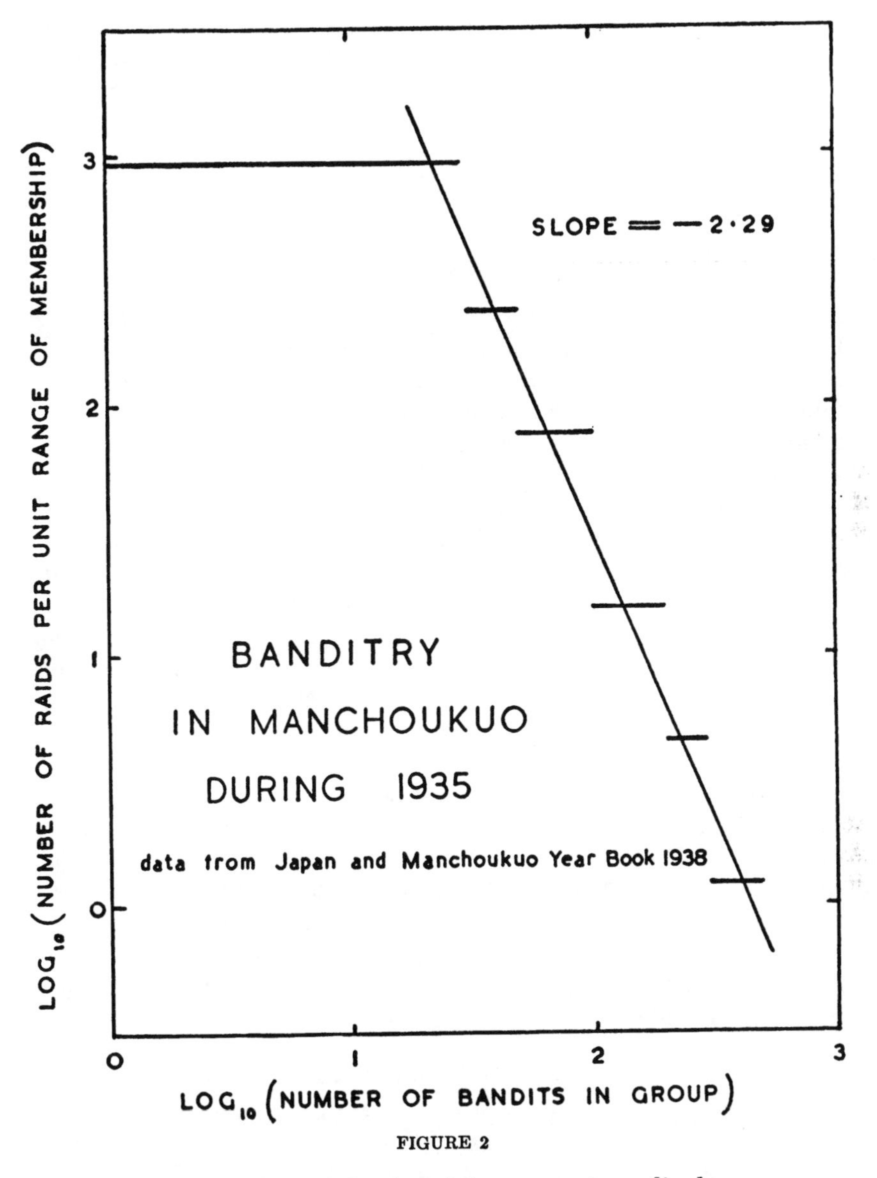

FIGURE 2

where A is independent of the size of the group. Accordingly

$$\mu = \nu + \log_{10} A. \tag{8}$$

It may not be necessary thus to suppose that the percentage of killed was exactly the same for every group, but only to suppose that it had no correlation with the size of the group.

Secondly let us suppose that the number of raids which a group of bandits made, as part of their quarrel with the rest of the community, was not correlated with the size of the group. More precisely the supposition is that:

$$(\text{number of raids made by a group}) = Bq(n)$$

where B is independent of n, so that

$$\psi(\nu) = \phi(\mu) + \log_{10} B. \tag{9}$$

According to these assumptions the $\psi(\nu)$ graph, when slid parallel to itself, horizontally through $\log A$ and vertically through $\log B$, becomes a *local* $\phi(\mu)$ graph. The remarkable agreement of the slopes can therefore be interpreted as meaning that one of the characteristics of aggregation for aggression was the same in Manchoukuo as it was in the world-total.

That broad agreement is interesting; but there remain some doubts or discrepancies in detail. The Manchurian data do not fix A and B, though A can be roughly estimated. For it has been shown that on the average 1.3 per cent of the bandits who appeared in a raid were killed, and that a group on the average perpetrated several raids. The total quarrel-dead, n, may have been about 10 per cent of the number of bandits in the group; that is to say $A=0.1$, so that $\mu=\nu-1$. The mean value of μ on the straight part of the bandit-diagram is accordingly about one. If so, there is a discrepancy with the circular arc on the world-diagram. For at $\mu=1$ the slope of the arc is decidedly steeper, namely $d\phi/d\mu=-3.1$. The Manchurian data therefore suggest that the gap in the world-diagram (Figure 1) should be closed, not by the circular arc, but by the straight segment of slope -2.38 leaving discontinuities of slope at its two ends. I regard that as a suggestion to be remembered, but not to be acted on without further evidence. The shift $\mu=\nu-1$ cuts off most of the horizontal part of Figure 2 and thus increases its resemblance to the top of Figure 1.

GANGING IN CHICAGO

F. M. Thrasher [16] made a study of 1313 gangs in Chicago with a view to their re-direction, or to some other social improvement.

He defined a gang in these words (p. 57): "The gang is an interstitial group originally formed spontaneously, and then integrated through conflict. It is characterized by the following types of behaviour: meeting face to face, milling, movement through space as a unit, conflict, and planning. The result of this collective behavior is the development of tradition, unreflective internal structure, *esprit de corps*, solidarity, morale, group awareness, and attachment to a local territory."

Thrasher gives (p. 319) a table showing the approximate numbers of members in 895 gangs. The first two columns of the following table are copied from Thrasher, the third column is a deduction, designed to smooth out the disparities of the given ranges. The number of members

Number of members in gang	Number of such gangs	Gangs per unit range of membership
inclusively		
3 to 5	37	12.3
6 to 10	198	39.6
11 to 15	191	38.2
16 to 20	149	29.8
21 to 25	79	15.8
26 to 30	46	9.2
31 to 40	55	5.5
41 to 50	51	5.1
51 to 75	26	1.04
76 to 100	25	1.00
101 to 200	25	0.25
201 to 500	11	0.37
501 to 2,000	2	0.0013

in a gang is the same sort of quantity as the number of bandits in a group; so the same symbol ν is here used for $\log_{10}$ of the first column. Log_{10} of the third column is here denoted by $\chi(\nu)$ to distinguish it from $\psi(\nu)$, because the number of gangs is not similar to the number of raids, though related to it. Figure 3 shows a graph of $\chi(\nu)$. Apart from gangs of 15 or less, which show again the blunted top, the $\chi(\nu)$ graph is well fitted by the straight line shown, which has a slope

$$d\chi(\nu)/d\nu = -2.30. \tag{10}$$

It is remarkable that this slope is almost the same as $d\psi(\nu)/d\nu = -2.29$ for banditry in Manchoukuo. This agreement strengthens the suspicion that some fairly general tendency concerning aggregation for aggression is revealed by these otherwise scattered phenomena. Hypothetical explanation is again easy. The distinction between the number of gangs and the number of raids would not affect the slope, if it were true that the number, C say, of raids made by a gang was independent of the size of the gang. For then

$$\psi(\nu) = \chi(\nu) + \log_{10} C \tag{11}$$

and the constant $\log_{10} C$ would disappear on taking $d/d\nu$. Yet several features remain unclear. The Chicago gangs were certainly aggressive.

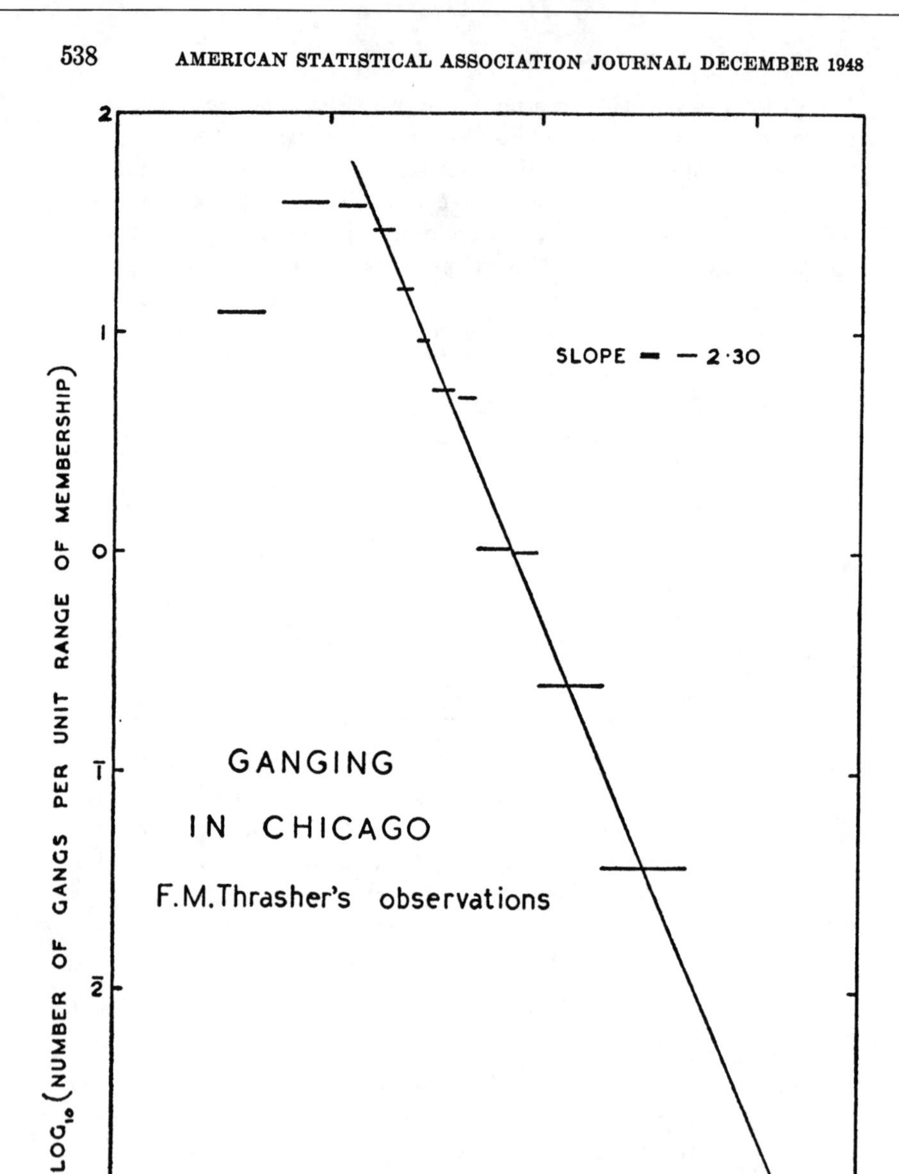

FIGURE 3

Thrasher has a chapter headed "Gang Warfare" and beginning with these words

"The gang is a conflict group. It develops through strife and thrives on warfare. The members of a gang will fight each other. They will even fight for a 'cause.' . . . Gangsters are impelled, in a way, to fight; so much of their activity is outside the law that fighting is the only means of avenging injuries and maintaining the code."

Thrasher mentioned a few gang-fights which went as far as homicide, but he did not give any comprehensive statistics of the quarrel-dead. I am indebted to Prof. Lundberg and Margaret Black Richardson, for the indication of Thrasher's work.

SUMMARY

The $\phi(\mu)$ diagram in Figure 1 is the best conspectus of the facts for the world as a whole. Alternative tracks are shown across the gap where world-totals are lacking. Samples from Manchoukuo and Chicago confirm the slope of the straight alternative, whereas continuity of slope is a consideration in favor of the circular arc.

MORE EXACT CONNECTION BETWEEN THE $\phi(\mu)$ GRAPH OF FIGURE 1 AND THE NUMBERS OF FATAL QUARRELS IN UNIT RANGES OF MAGNITUDE. INTERPOLATION ACROSS THE GAP.

We need to connect the mean over a range with the value at its midpoint. It might be thought that the connection was obvious; for a glance at Figure 1 shows that the graph is fairly straight; and of course the mean ordinate of a straight segment is simply the ordinate at its midpoint. But that would be a delusion based on a doubly wrong type of mean. For here the mean of q, not of the ordinate $\log_{10} q$, has to be taken over the integers in the range of n, and not over the abscissa $\mu = \log_{10} n$. It will be shown that the logarithmic transformation of both coordinates introduces correcting factors which depend on the slope, vanishing when it is either 0 or -2, but otherwise often considerable. They are obtained from the simplest formula that can fit the facts tolerably, namely the formula for a straight line on the $\phi(\mu)$ graph. The straight lines are suitably short segments; for it would be a mistake to smooth away significant detail by the over-wide sweep of any too-simple formula. In particular the sharp bend near the top of Figure 1 comes from British criminal statistics where it appears to be significant; and something rather like it is to be seen in the graphs which represent banditry in Manchoukuo and ganging in Chicago; so this bend should certainly not be smoothed away. But elsewhere a short length of the $\phi(\mu)$ graph can be represented by

$$\phi(\mu) = B - C\mu$$

where B and C are positive constants. On taking antilogarithms this becomes

$$q = 10^B n^{-c}$$

an inverse power law. To find $\bar{q}$ this n^{-c} has to be averaged over a range of n. As n proceeds only by integers, the average should be made by sums, not by integrals. However the integral $\int n^{-c}dn$ gives a useful approximation to the sum.

This programme will now be worked out in proper detail.

The notation is that already defined in connection with the graph. The typical range ends, say, at $\mu=\mu_0 \pm \frac{1}{2}$, where $\mu_0=7, 6, 5, 4, 3, 2, 1$. It is convenient to begin with a rough and ready approximation, R, to the number Q of fatal quarrels in the range, whereby $\bar{q}(n)$ at the midpoint is multiplied by the number of integers in the range of quarrel-dead, which it is convenient for a later purpose, to denote by $\sum_n 1$ where $\sum_n$ sums for all n in the range of μ. Accordingly

$$R = 10^{\phi(\mu_0)} \sum_n 1. \tag{12}$$

For example, as read from the circular arc on Figure 1:

Range of magnitude $\mu=\log_{10} n$	Ordinate at midpoint $\phi(\mu_0)$	$\therefore \bar{q}(n)$ at midpoint	Number of integers in the range of quarrel-dead $\sum_n 1$	First approximation to number of quarrels R
0.5 to 1.5	3.87	7410	28	207,000
1.5 to 2.5	1.23	17.0	285	4,850
2.5 to 3.5	$\bar{1}.17$	0.131	2846	373

The above easy method is however suspect, because $\bar{q}(n)$ varies in the ratio 1740 in one of the ranges thus represented by its midpoint. So we must consider corrections.

Correction for width of range

As the $\phi(\mu)$ graph is nearly or quite straight, let us represent a local portion of it by

$$\phi(\mu) = \phi(\mu_0) - c(\mu - \mu_0) \tag{13}$$

where μ_0 is the midpoint of the range and $c = -d\phi/d\mu$ there. The distinction between $q(n)$ and $\bar{q}(n)$ is one that concerns detailed observations. It does not arise when, as now, deductions from a smooth curve in question. From (13)

$$q(n) = 10^{\phi(\mu)} = 10^{\phi(\mu_0)+c\mu_0}n^{-c}. \tag{14}$$

Let Q denote the total number of fatal quarrels in the range, then

$$Q = \sum_n q(n) = 10^{\phi(\mu_0)+c\mu_0}\sum_n n^{-c} \tag{15}$$

So from (12) and (15)

$$Q = R10^{c\mu_0}\sum_n n^{-c} \Big/ \sum_n 1. \tag{16}$$

The coefficient of R will be called the correcting factor. Two verifications of (16) may be noted: (i) if $c=0$, then $Q=R$. (ii) if the range of quarrel-dead contained only a single integer n, and if it were located at μ_0 so that $10^{\mu_0}=n$, then (16) would become $Q=Rn^c n^{-c}/1=R$, as it should.

The sum in (16) was evaluated by a method connected with the "integral test" for the convergence of series, in the following manner. Let

$$b = 10^{\mu_0+1/2} \quad \text{and} \quad a = 10^{\mu_0-1/2} \text{ for brevity.} \tag{17}$$

In the following chain of equations, the first is an approximation which improves as μ_0 increases, and the other are accurate.

$$\sum_n n^{-c} \simeq \int_a^b t^{-c}dt = \frac{1}{1-c}(b^{1-c} - a^{1-c})$$

$$= \frac{10^{\mu_0(1-c)}}{1-c}\left\{10^{(1-c)/2} - 10^{(c-1)/2}\right\}. \tag{18}$$

In the special case of $c=0$, (18) gives the approximation

$$\sum_n 1 = 10^{\mu_0}(10^{1/2} - 10^{-1/2}) = b - a. \tag{19}$$

Strictly $\sum_n 1$ is the integer next below $b-a$; but it is advisable to use the same type of approximation in both numerator and denominator of (16). Insertion of (18) and (19) into (16) gives the correcting factor

$$Q/R = \frac{1}{1-c}\,\frac{10^{(1-c)/2} - 10^{(c-1)/2}}{10^{1/2} - 10^{-1/2}}. \tag{20}$$

This expression is unity when $c=0$ or $c=2$.
For the wars, $c=1.50$ and $Q/R=0.855$.

The observed quantity is Q in the unit range of magnitude. The $\phi(\mu)$ curve should however be connected to μ at a point, and so to R. We have

$$\log_{10} R = \log_{10} Q - \log_{10} 0.855 = \log_{10} Q + 0.068.$$

In the $\phi(\mu)$ graph (Figure 1) the sloping line for the wars was therefore raised 0.068 above that which best fits the observations grouped in unit ranges of magnitude.

In the unexplored region we have, on taking the first approximation R from the previous table:

World-totals for the 126 years 1820 to 1945

Range of magnitude μ	$c = -\dfrac{d\phi}{d\mu}$	Correcting factor Q/R	Corrected number of fatal quarrels Q	Range of number killed in a quarrel n
0.5 to 1.5	3.14	1.92	397,000	4 to 31
1.5 to 2.5	2.32	1.16	5,630	32 to 316
2.5 to 3.5	1.87	0.95	354	317 to 3162

The fourth column shows the improved estimates for the unexplored region between the wars and the murders according to the circular arc in Figure (1). The difficulties of direct counting have already been described, and they are emphasized by these large numbers. In the range where 354 are expected, I have counted 188. If the straight line in Figure (1) were accepted instead of the circular arc, 354 would be increased to 2530. Although I know that the search is incomplete, I am unable to believe that less than a tenth have been found; and so I prefer the circular arc.

THE TOTAL NUMBER OF PERSONS WHO DIED BECAUSE OF QUARRELS DURING THE 126 YEARS FROM 1820 TO 1945 A.D.

Ends of range of magnitude	Total number of deaths in millions	How computed
$7\pm\frac{1}{2}$	36	By summation over a list of fatal quarrels.
$6\pm\frac{1}{2}$	6.7	
$5\pm\frac{1}{2}$	3.4	
$4\pm\frac{1}{2}$	0.75	
$3\pm\frac{1}{2}$	0.30	From the circular arc in Figure (1) by the method described below.
$2\pm\frac{1}{2}$	0.40	
$1\pm\frac{1}{2}$	2.2	
$0\pm\frac{1}{2}$	9.7	From page 528
	Total 59	

A remarkable feature of the above table is that the heavy loss of life occurred at the two ends of the sequence of magnitudes, namely the World Wars and the murders. The small wars contributed much less to the total. The total deaths because of quarrels should be compared with the *total deaths from all causes.* There are particulars given by de Jastrzebski [9], and in the Statistical Year Books of the League of Nations, which allow the total to be estimated. A mean world population of 1.5×10^9 and a mean death rate of 20 per thousand per year would give during 126 years 3.8×10^9 deaths from all causes. Of these the part caused by quarrels was 1.6 per cent. This is less than one might have guessed from the large amount of attention which quarrels attract. Those who enjoy wars can excuse their taste by saying that wars after all are much less deadly than disease.

The method of computation in the unexplored region was an extension of that explained above. The number D of dead in any range of μ is strictly

$$D = \sum_n nq(n) = \sum_n 10^{\mu+\phi(\mu)}. \tag{21}$$

First a common-sense estimate, E of D was obtained from the graph at the midpoint μ_0 of the range $\mu_0 - \frac{1}{2} \leqq \mu \leqq \mu_0 + \frac{1}{2}$ thus, by (12),

$$E = 10^{\mu_0}R = 10^{\mu_0+\phi(\mu_0)} \sum_n 1. \tag{22}$$

Then E was corrected by multiplication by a factor which was found to be approximately

$$\frac{D}{E} = \frac{1}{2-c}\left\{\frac{10^{(2-c)/2} - 10^{(c-2)/2}}{10^{1/2} - 10^{-1/2}}\right\}. \tag{23}$$

The theory of this correction is based on the approximation (13), whence it follows that

$$\mu + \phi(\mu) = c\mu_0 + \phi_0 + \mu(1-c). \tag{24}$$

From (21) and (24)

$$D = 10^{c\mu_0+\phi(\mu_0)} \sum_n n^{1-c}, \quad \text{and} \tag{25}$$

$$\frac{D}{E} = 10^{\mu_0(c-1)} \frac{\sum_n n^{1-c}}{\sum_n 1}. \tag{26}$$

The sums in (26) were obtained from (18) by suitable alterations of the index and they lead to (22). The correction (22) was verified in particular cases by the "deferred approach to the limit" of Richardson and Gaunt [14].

COMPARISON WITH THE DISTRIBUTION OF SIZES OF TOWNS

Towns and wars are both examples of human aggregation. A referee has asked me to compare them. For simplicity I take the distribution of towns from Lotka's book [10A] in Auerbach's idealized form whereby the town of rank r in a given country has a population n such that

$$nr = A, \tag{1}$$

in which A is a constant, namely the population of the largest town. The desired comparison might conceivably be made by arranging the fatal quarrels in order of rank; but there would be many artificially coincident ranks on account of the rounding off of imperfectly known casualties; moreover the gap in the data prevents the extension of rank from the wars to the murders. It is preferable therefore to leave the fatal quarrels as they are already shown on Figure 1, and to transform Auerbach's law so as to see how towns would appear on a diagram of that type. For the purpose of comparison the same symbols will be used in corresponding meanings. Thus in this section

$$n \text{ is the population of a town, and } \mu \text{ will be } \log_{10} n. \tag{2}$$

Again, to match the fatal quarrels, $\bar{q}$ will here be defined as

$$\bar{q} = \frac{\text{number of towns in a range of population}}{\text{number of integers in that range of population}}. \tag{3}$$

Also, as before, let

$$\phi = \log_{10} \bar{q}. \tag{4}$$

Consider two towns of ranks r', r'', of which r'' is the greater. Let them mark the ends of a range. The number of towns inside the range is strictly $r''-r'-1$. But let us follow the usual statistical practice of regarding half of an end-object as lying on either side of the end. The number of towns in the range is then simply $r''-r'$. Let n' and n'' be the respective populations. With the same convention about ends on the population range, we have from (3)

$$\bar{q} = \frac{r'' - r'}{n' - n''}. \tag{5}$$

Elimination of ranks between (5) and (1) gives

$$\bar{q} = \frac{A}{n'n''}. \tag{6}$$

So that from (4)

$$\phi = \log_{10} A - \log_{10} n' - \log_{10} n''. \tag{7}$$

As for the wars, let μ_0 be the midpoint of the range of μ. It is

$$\mu_0 = \tfrac{1}{2}(\log_{10} n' + \log_{10} n''). \tag{8}$$

From (7) and (8)

$$\phi = \log_{10} A - 2\mu_0, \tag{9}$$

however wide or narrow the range may be. That the constant $\log_{10} A$ should be the same whatever the grouping is a peculiarity of the slope

$$d\phi/d\mu_0 = -2 \tag{10}$$

as shown on page 541, where the correction vanished for $c=2$.

This slope for the towns is quite close to that of the straight line which in Figure 1 might join the murders to the world wars. Such a broad resemblance between two forms of aggregation is certainly interesting, and may suggest theories. Perhaps the most suggestive formula is

$$q = An^{-2} \tag{11}$$

which follows from (6) when n' is almost equal to n''. Rashevsky [11A], who gives formula (11), has tried-out several explanations.

These overall resemblances do not however conduce to an accurate description of fatal quarrels; for they distract attention from the curvature of the $\phi(\mu)$ graph in Figure 1. The slope of the part relating to wars is certainly not -2. If it were so, then the number of fatal quarrels would increase 10 times for each unit decrease of magnitude. The observed ratio is less than four.

REFERENCES

[1] Anderson, J. G. 1939. *China fights for the world.* Kegan Paul, London, p. 106.
[1A] Carr-Saunders, A. M. 1936. *World Population.* Clarendon Press, Oxford.
[2] Coupland, R. 1939. *The Exploitation of East Africa 1856–1890.* Faber & Faber, London. Chh. VII, XI.
[3] Durbin, E. F. M., and Bowlby, J. 1939. *Personal Aggressiveness and War.* Kegan Paul, London.
[4] Fisher, R. A. 1922. *Phil. Trans. Roy. Soc. Lond.* A **222**, 332.
[5] Glover, E. 1933 & 1947. *War, Sadism & Pacifism.* Allen & Unwin, London.
[6] Harding, D. W. 1941. *The Impulse to Dominate.* Allen & Unwin, London.
[7] Hume, M. A. S. 1900. *Modern Spain.* Fisher Unwin, London.
[8] *Japan and Manchoukuo Year Book,* 1938. Tokio.
[9] Jastrzebski, S. de. 1929 article *Death-Rate* in Ency. Brit. XIV edn.
[10] Jeffreys, H. 1939. *Theory of Probability,* p. 148. Clarendon Press, Oxford.
[10A] Lotka, A. J. 1925. *Elements of Physical Biology.* Williams & Wilkins, Baltimore.
[11] Miliukov, P. article *Russia* in *Ency. Brit.* XIV ed., **19**, 720 d.
[11A] Rashevsky, N. 1947. *Mathematical Theory of Human Relations.* The Principia Press, Bloomington, Indiana.
[12] Richardson, L. F. in *Nature* of 15 Nov. 1941.
[13] Richardson, L. F. 1945. *J. Roy. Statistical Soc.* **107**, 242–250.
[14] Richardson, L. F., and Gaunt, J. A. 1927. *Phil. Trans. Roy. Soc. Lond.* A **226**, 299–361.
[15] Sleeman, W. H. 1836. *Ramaseeana.* Military Orphan Press Calcutta, p. 39.
[16] Thrasher, F. M. 1927. *The Gang.* Chicago University Press.
[17] Whitaker's Almanack, annual, London.
[18] Wright, Q. 1942. *A Study of War.* Chicago University Press.

Richardson, L. F. (1949)
Br. J. med. Psychol. **22**, 166–8

1949:4

THE PERSISTENCE OF NATIONAL HATRED AND THE CHANGEABILITY OF ITS OBJECTS

The British Journal of Medical Psychology
Cambridge
Vol. XXII Parts 3 and 4, 1949, pp. 166–168

Note: This paper was read at the Twelfth International Congress of Psychology in Edinburgh on 27 July 1948. An abstract (1948:1, not reproduced here) was published in the *Proceedings and Papers* of the Congress

[166]

THE PERSISTENCE OF NATIONAL HATRED AND THE CHANGEABILITY OF ITS OBJECTS*

BY LEWIS F. RICHARDSON

There are numerous historical examples of quick changes of the object of hatred. They fit, as special cases, into the general Freudian doctrine of the permanence and transferability of affect.

HISTORICAL EXAMPLES

(1) In the First Balkan War Bulgaria was allied with Serbia, Greece and Montenegro against Turkey. Peace was signed on 30 May 1913. One month later Bulgaria attacked Serbia and Greece in the hope of acquiring a larger share of the lands taken from the Turks. Callwell, 1929.

(2) In July 1914 Great Britain was on the verge of civil war about Ulster. When the Germans invaded Belgium the threat of civil war in Britain suddenly ended and British anger was directed against the Germans. By 1915 Sir Edward Carson and F. E. Smith, the leaders of the Ulster revolt, had become colleagues of their former Liberal opponents in a coalition British Government for the prosecution of the war against Germany.

(3) Praise has been bestowed—e.g. by Winston Churchill—on the unity of the British people in time of world war. Such unity and its praise are, however, not peculiar to Britain. In the Johannis Kirche at Stargard in Pomerania on 5 August 1939 I heard the preacher say that when the war broke out 25 years previously the Germans forgot their differences and became a band of brothers.

(4) Nora Waln records that Mrs Sun Yat-Sen, wife of the President of China, preached a boycott of the British, yet was seen in 1926 travelling on a British steamer. An official with long experience of China explained this by remarking that 'British hostility is only a political ruse to create support for the Nationalist Party. They have to have an "enemy" for their cause, or they cannot collect the attention of the populace' (1938, p. 229). Sir Meyrick Hewlett, who was British Consul General at Nanking in 1928, wrote that 'it was a regrettable fact that the only thing which really united China was an anti-foreign outburst' (1943, p. 205).

(5) Chiang Kai-Shek, President of China, had struggled successfully against Russian communists in 1927; but in September 1931, when the Japanese occupied Mukden in Manchuria, Chiang Kai-Shek sent word to Mo Teh-hui at Moscow instructing him to bring Russia over to China's side in the dispute with Japan in Manchuria. (Waln, 1939, p. 283.)

(6) According to Müller-Brandenburg (1938, p. 192) in 1933 when Hitler attained power, 'he was faced with the fact that the German people were divided into two sections, neither of whom—though using the German language—could understand one another. Indeed, they were even prepared to fight one another to the death. The Führer and his movement succeeded in achieving the impossible by putting an end to class hatred.' Thus Müller-Brandenburg jubilated in 1938. He went on to praise the part played by the German Labour Camps in this reconciliation, stating that: 'In our camps, the conception "bourgeoisie", meets with just as much ridicule as the conception "proletariat", for every member looks upon himself as a German, and nothing else' (p. 195). The rest of Hitler's technique is now well known. He organized peace between the German proletariat and the German bourgeoisie by leading both of them into a defensive and offensive alliance against surrounding nations and against Jews. Hitler did not quell hatred, he diverted it.

(7) Hitler denounced the Russian government with great intensity for a number of years up to January 1939, yet in August 1939, when an acute crisis had developed between Germany and Poland, he made a pact with Russia; and simultaneously German propaganda swung its hate away from Russia towards Poland, Britain and France. It may be said that the switch-over was the result of a strategic calculation. But it is remarkable that human emotional disposition *permits* such a sudden displacement of hatred.

* Read at the Twelfth International Congress of Psychology.

(8) An intended railway strike in Britain in August 1939 was cancelled as soon as war with Germany appeared inevitable.

(9) In the early part of 1939 there was three-cornered fighting in Palestine between the Arabs, the Jews and the British. When war broke out in Europe in September 1939 the internal disputes in Palestine were postponed (Lord Samuel's broadcast 27 March 1940, B.B.C.).

(10) At the time of writing (October 1947) the British newspapers and radio have daily been giving reasons why Britain should hate Russia instead of Germany.

(11) In China the Yunnanese and the Szechuanese joined forces in 1916 to expel the northern troops of Yüan Shih-k'ai. Within a year after this co-operation the Yunnanese and the Szechuanese were fighting each other (Hewlett, 1943, pp. 92–3).

(12) In October 1947, the United States of America had been extraordinarily loving and helpful towards non-Soviet Europe, first by the Marshall offer, and then by meatless Tuesdays, eggless and fowlless Thursdays, and the closing of distilleries to save grain. At the same time the debates at the United Nations' Assembly between the representatives of U.S.A. and those of the Soviets had become more acrimonious.

The suggestion is that:

$$(\textit{external hatred}) - (\textit{external love}) = \textit{a constant}.$$

When the U.S.A. was isolationist, the two brackets on the left were both near to zero. Now they are both large, but compensate one another.

Some general impressions

The following are worth quoting:

(1) Durbin & Bowlby (1939, p. 126) wrote that 'Peace within a tribe seems to be bought at the expense of continual warfare without.'

(2) P. J. Noel-Baker (1936, p. 408) wrote that 'It is a law of nature that every General Staff requires an enemy.'

(3) A British Ambassador of wide European experience, Lord Rennell of Rodd, in a letter to *The Times* on 14 November 1933 remarked that: 'The propaganda of confidence is uphill work. But the propaganda of apprehension is easy and dangerous' (Noel-Baker, 1936, p. 99).

(4) Bertram Pickard, who has much experience of international organizations, wrote (1942) that 'There is a type of pacifist who loves his enemies and hates his friends.'

(5) Dr C. G. Jung wrote (1947, pp. xv to xvi) that 'In Switzerland we have built up a so-called "perfect democracy" in which our warlike instincts spend themselves in the form of domestic quarrels called "political life".... Our order would be perfect if people could only take their lust of combat home into themselves.'

Reactions to news

Abundant reasons for hating the new object are usually present. In a love story there are usually reasons why Mr Smith, for example, was jilted by Miss Jones and why he married instead Miss Brown 'on the rebound'. The latter phrase reminds us of his persistent desire to marry somebody. I suggest that the course of a *hate*-story can be similar; and that therefore when we are given reasons for hating a foreign nation we should consider *also*: (*a*) that we, or our compatriots, may harbour a subconscious hatred which is seeking an object; (*b*) that every piece of news is a selection; and that its selector has usually a wide choice among different emotional tones.

The world problem

The foregoing remarks are intended to lead up to the important question: *Is there an irreducible minimum of hatred; and if so, against what objects had it better be directed*?

The psycho-analysts (e.g. Glover, 1947) have plans for a mass reduction of hate by means of a better upbringing of children. It is not clear that they hope thus to abolish hate altogether.

If the whole world were politically united, so that there were no human enemies, would abstractions like dirt, disease, ignorance, poverty, falsehood, ugliness, and cruelty seem sufficiently *exciting* to direct our latent pugnacities all in beneficent directions? Flugel (1940) discusses the difficulty and concludes that: 'Nature in many of her varied aspects, is well qualified to withstand our hate though she will inevitably also, as the universal mother (albeit often a careless or a cruel one) have something of our love. If we can persuade ourselves that she is in truth a none too tender parent, we may be able to unload upon her a larger share of our aggression, and thus hope to deflect some of our hatred from our fellow-men, who can tolerate it so very much less easily, and whom we may then come to look upon as brothers fighting in a common cause....'

The Pope in his Christmas message of 1940 suggested a different sublimation of pugnacity by his reference to: victory over hatred; victory over distrust; victory over the dismal theory that utility is the foundation and aim of law and that might can create right; victory over those potential conflicts arising from the disequilibrium of world economy; and victory over cold egoism (Keesing, 4388).

As a criticism of philosophy it has been said that few people are able to worship an impersonal God. Similarly, few people may be able to hate nature, or to hate the abstract evils mentioned by the Pope, as intensely as they hate a national leader, whose photograph or caricature they see.

Dr Charlotte Banks suggested to me in conversation that the best plan is to let off one's annoyances on to one's immediate associates, who understand and can make allowances. A statistical inquiry into the success of this proposal would be desirable.

Dr C. G. Jung recommends, in effect, that we should hate the evil in ourselves. His broadcast of 1946 contained the following assertions: 'We still labour under the unwholesome conviction that we should be at peace within ourselves....We psychologists have learnt, through long and painful experience, that you deprive a man of his best when you help him to get rid of his complexes. You can only help him to become sufficiently aware of them and to start a conscious conflict with himself (1947, p. xvi).

REFERENCES

Callwell, Sir C. E. *Encyclopaedia Britannica*, 14th ed., **2**, 988–91.

Durbin, E. F. M. & Bowlby, J. (1939). *Personal Aggressiveness and War*, London: Kegan Paul.

Flugel, J. C. (1940). In *New Commonwealth Quarterly*, **6**, 102–14.

Glover, E. (1947). *War, Sadism and Pacifism*. London: Allen and Unwin.

Hewlett, Sir Meyrick (1943). *Forty Years in China*, p. 205. London: Macmillan.

Jung, C. G. (1946). The fight with the shadow. *The Listener*, 7 November 1946. Printed also in *Essays on Contemporary Events* (1947). London: Kegan Paul.

Keesing's Contemporary Archives, Bristol, England.

Müller-Brandenburg (1938). In *Germany speaks*, London: Butterworth.

Noel-Baker, P. J. (1936). *The Private Manufacture of Armaments*. London: Victor Gollancz.

Pickard, B. (1942). *Peacemaker's Dilemma*. Pendle Hill Pamphlets, Wallingford, Pennsylvania, U.S.A.

Waln, N. (1938). *The House of Exile*. Penguin Books, Harmondsworth, Middlesex, England.

1950:2

WAR AND EUGENICS

The Eugenics Review
London
Vol. XLII No. 1, April 1950, pp. 25–36

Note: This is a revision of a paper read before the Eugenics Society in London on 6 December 1949

WAR AND EUGENICS*

By LEWIS F. RICHARDSON, D.Sc. (Physics), F.R.S.

WARS have been regarded from many points of view. There are military historians who write about the technique of winning wars. There are popular romances which emphasize the heroism of their own side and the wickedness of their opponents. The adventure and fun in war have been picked out by Alexandre Dumas and by Ian Hay. There are also literary historians who, long after the event, make carefully impartial summaries of the motives of both sides. Apart from descriptions of particular wars there are investigations of wars in general with a view to their avoidance. The Nye Committee (1935) investigated the influence of arms manufacturers, so did Noel-Baker (1936). One volume of Pitrim Sorokin's *Social and Cultural Dynamics* (1937) is concerned with wars, ancient and modern. Quincy Wright has published a comprehensive *Study of War* (1942), its aim being not to win wars but to prevent them. My own work has the same purpose. The findings may seem emotionally bleak. It is impossible to combine honest research with a gratification of any one nation's patriotic feelings. The *Eugenics Society* is notable for having had the intellectual courage to attend to emotionally embarrassing problems. I shall rely on that courage here. My results are offered as true. Truth, even if at first it seems like tribulation, should in the long run yield peaceable fruit.

Loss of Life Caused by Quarrels

Every accountant is interested in both income and expenditure. Although the main concern of eugenists is with the income of human life, they may on occasion be willing to attend to an item on the other side of the account.

Anyone who tries to make a list of "all the wars" is embarrassed by a multitude of minor incidents which perhaps ought to be included. Thus Quincy Wright (1942), having made a list of 278 wars, remarked that probably ten times as many might have been added. The astronomers count stars by first arranging them according to brightness. I applied this idea to wars by first arranging them according to the number of people killed. Statements of casualties are sometimes uncertain by a factor as large as three. So the subdivisions had to be broad. It seemed statistically convenient to proceed by factors of ten. The first world war caused about 10 million deaths. The next smaller magnitude of wars was taken to have caused each about a million deaths, the next 100,000, the next 10,000. But why stop there? Although smaller incidents would not ordinarily be called wars, it seemed natural to extend the classification to quarrels which had caused respectively about 1,000 or 100 or 10 or 1 deaths. The last class are ordinarily called murders. Some name was then needed for any member of this wide class. At first I called them all fatal quarrels. But a more appropriate adjective would be lethal or deadly.

Let the "magnitude" of any deadly quarrel be defined to be the logarithm, to the base ten, of the number of persons who died because of that quarrel. The middles of our successive classes are then at magnitudes 7, 6, 5, 4, 3, 2, 1, 0. To make a clean cut between adjacent classes we need to know, not the middles of the classes, but their boundaries. Let these boundaries be at 7·5, 6·5, 5·5, 4·5, 3·5, 2·5 . . . on the scale of magnitude. For example, magnitude 3·5 lies between 3,162 and 3,163 deaths, magnitude 4·5 lies between 31,622 and 31,623 deaths, magnitude 5·5 lies between 316,227 and 316,228 deaths, and so on. This concept of magnitude has several functions: (i) It tidies an intellectual muddle; (ii) It helps the research-worker to decide what was important, apart from his national prejudices; (iii) It disguises the distressful facts, so as to allow one to think

* A revision of a paper read before the *Eugenics Society* on December 6th, 1949.

about them undistracted by emotion. Such emotional detachment has, however, its dangers for society.

The whole range of deadly quarrels is shown in Fig. 1, which is reproduced, together with a numerical digest, by permission of the American Statistical Association of Washington, D.C. (Richardson, 1948, *c*). The enormous ranges necessitate logarithmic scales.

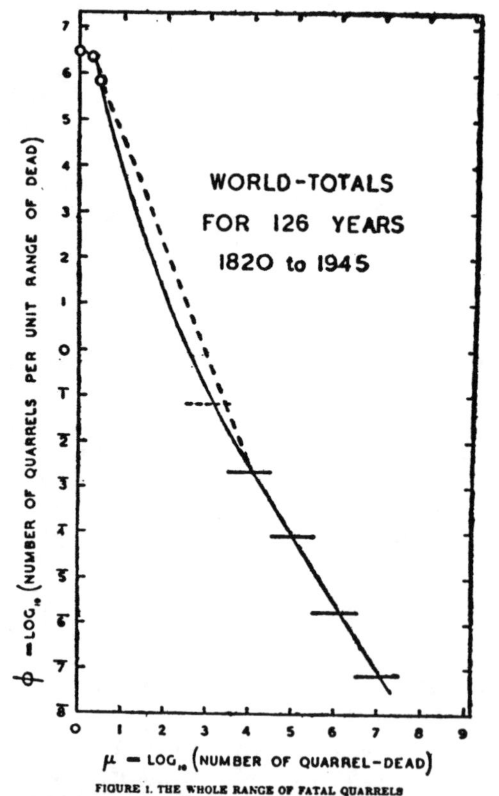

FIGURE 1. THE WHOLE RANGE OF FATAL QUARRELS

THE HORIZONTAL SCALE EXTENDS FROM A SINGLE PERSON ON THE LEFT TO THE POPULATION OF THE WORLD ON THE RIGHT. THE THREE DOTS AT THE LEFT HAND TOP REPRESENT THE MURDERS. THE SHORT HORIZONTAL LINES REPRESENT THE WARS GROUPED BY MAGNITUDES. ONE OF THESE LINES IS DOTTED, BECAUSE IT SHOULD BE RAISED TO AN UNKNOWN HEIGHT.

Fig. 1

There is a gap between the small wars and the murders, because world totals are here lacking. The slope of the graph between magnitudes 2 and 1 is, however, supported by some local samples, one of banditry in Manchukuo and one of ganging in Chicago.

By these and other methods the total number of persons who died because of quarrels has been computed and is summarized in Table I.

TABLE I

THE TOTAL NUMBER OF PERSONS WHO DIED BECAUSE OF QUARRELS DURING THE 126 YEARS FROM A.D. 1820 TO 1945

Ends of range of magnitude	Total number of deaths in millions	How computed
$7\pm\frac{1}{2}$	36	By summation over a list of quarrels
$6\pm\frac{1}{2}$	6·7	
$5\pm\frac{1}{2}$	3·4	
$4\pm\frac{1}{2}$	0·75	
$3\pm\frac{1}{2}$	0·30	From the circular arc in Fig. (1) by the method described in (Richardson, 1948, *c*)
$2\pm\frac{1}{2}$	0·40	
$1\pm\frac{1}{2}$	2·2	
$0\pm\frac{1}{2}$	9·7	From criminal statistics
Total	59	

A remarkable feature of the above table is that the heavy loss of life occurred at the two ends of the sequence of magnitudes, namely the world wars and the murders. The small wars contributed much less to the total. The total deaths because of quarrels should be compared with the *total deaths from all causes*. There are particulars given by de Jastrzebski (1929) and in the *Statistical Year Books* of the League of Nations, which allow the total to be estimated. A mean world population of $1\cdot5 \times 10^9$ and a mean death rate of twenty per thousand per year would give during 126 years $3\cdot8 \times 10^9$ deaths from all causes. Of these the part caused by quarrels was 1·6 per cent. This is less than one might have guessed from the large amount of attention which quarrels attract. Those who enjoy wars can excuse their taste by saying that wars after all are much less deadly than disease.

Intermarriage as a Pacifier

The emotions aroused by quarrels are intense, especially if the quarrel threatens to cause loss of life. What can be strong enough to compete with such an emotional disposition? Modern war affects everybody. In present democratic conditions the drift towards, or away from, war is influenced by whole populations. The formal decision to make, or not to make, war, may be legally reserved for a few statesmen; but in fact leaders are often pushed from behind.

Abstract ideas which might have convinced an intellectual statesman of the eighteenth century may, therefore, be of little avail to-day. As to what might influence the population in general, it would be wiser to take a hint from a request brought by a small boy to a public library: he said "One love story and one murder; for my mother." The librarians, of whom I was one, felt that the child had summarized the popular taste.

The strongest motives that bind ordinary individuals to live harmoniously together are sexual attraction and family affection. Let us, therefore, consider the possibilities of international marriage, on account of the strength of the emotions which marriage involves.

Various objections to international marriage will readily occur to you.

First, the social relations might be strained. A bachelor said to me: "Is it worth while to abolish war at the expense of ruining all marriages?" In Shakespeare's *All's Well that Ends Well*, Bertram remarks that "War is no strife, to the dark house and the detested wife."

Second, the genetic constitution of the offspring may be unsatisfactory for certain crosses. Notable at present is Dr. Malan's policy of racial segregation (*Apartheid*) in South Africa. "All marriages between Europeans and non-Europeans will be prohibited" (1948, p. 8). I shall return to genetic questions later. The difficulty, of course, is to distinguish between a dislike of those manners and customs which could be changed by education, and a reasonable fear of irreversibility. By irreversibility I mean that it may be very easy to let foreign genes into a population, and almost impossible to select those genes out again.

Third, that international marriage on any large and rapid scale is impracticable, and does not meet the demand of those peacemakers who insist that "We must do something now!"

In view of these objections I wish to point out that the problem of pacification goes back as far as history; and that there have been some very long delays between aspiration and satisfaction. Leonardo da Vinci made a drawing of a flying machine. Four centuries of mechanical invention and discovery, much of it at first sight irrelevant to the problem of flight, intervened before any man flew. The results of aviation are partly joyous, partly disastrous. Although the problem of pacification feels urgent, there may be no quick solution. There is a scarcity of reliable pacifying influences; so that peace-makers may have no option but to attend to methods that would be both troublesome and slow. It may be necessary to plan now for our grandchildren or great-grandchildren.

Other Pacificatory Influences

The following other influences have been mentioned or advocated as pacifiers: (i) distraction by sports, (ii) hating a different group of people, (iii) hating the evil in oneself, (iv) armed strength, (v) collective security, (vi) international trade, (vii) plenty of living space, (viii) fewer frontiers, (ix) a common government, (x) a common language, (xi) a common religion. I wish to say something about each of these in turn. The list is perhaps not complete.

Distraction by sports. Alistair Cooke, in a B.B.C. broadcast on October 7th, 1947, from New York, stated that the newspaper headline "Reds have atom bomb," though large, was not so large as a headline announcing the defeat of the New York baseball team. The Communists were temporarily forgotten during an interest in baseball. It seems, however, hardly likely that the release of international tension by such distraction could be maintained permanently.

Hating a different group of people. This is an ancient device, used, according to Shakespeare, by an Archbishop of Canterbury, to protect the Church by setting King Henry V against France. The same device was used by the Nazis when they focused German hate on to the Jews. The present example of this psychological mechanism is perhaps accidental rather than deliberate: during 1945-9 British hate for the Germans has diminished while British hate for the Russians has increased. These are special cases of the Freudian doctrine of the permanence and transferability of affect. Some other political examples have been collected by me in a paper soon to appear (1950*b*). Among psychological processes, the diversion of hatred seems to be unusually mechanistic and reliable. But it does not tend to pacify the world as a whole.

Hating the evil in oneself. C. G. Jung has recently suggested (1947) that people should take their political quarrels home into themselves. Personally I view this recommendation with the greatest respect.

Armed strength. This has been the policy in which many nations have put their trust. Few people feel the need for any statistical test of this policy, because the proposition that "if we are strong, we shall not be attacked," seems to most people quite obvious. We reach that conclusion by merely letting our thoughts flow easily in an instinctive channel. Moreover factual examples of the success of the policy are easy to find. In February, 1948, some Guatemalans proposed to invade British Honduras, to which Guatemala had long laid claim. Then Britain sent warships carrying land forces, and the active threat from Guatemala subsided; although up to September, 1949, there was no agreed settlement (K. 9153, 9636, 10277). Please notice, however, that the powers of Britain and of Guatemala were very unequal. When disputes have occurred between nations of roughly equal power, the effects of armed strength have been remarkably different.

Fig. 2 shows a statistical test of the policy of armed strength for all the nations which took part in the first world war. For comparison with the ancient Roman adage that "If you desire peace, you should prepare for war" the abscissa represents a nation's insurance against war, and the ordinate represents its subsequent sufferings by war. The same theory asserts that the correlation should be negative; that the points should slope from the left down to the right.

About the year 1910 an American professor, whose name I have unfortunately forgotten, published a book in which he said, "What you prepare for, that you will get." On his theory the points should slope up towards the right. The facts in the diagram do not support either of these conflicting theories. The correlation is so small as not to be worth calculating.

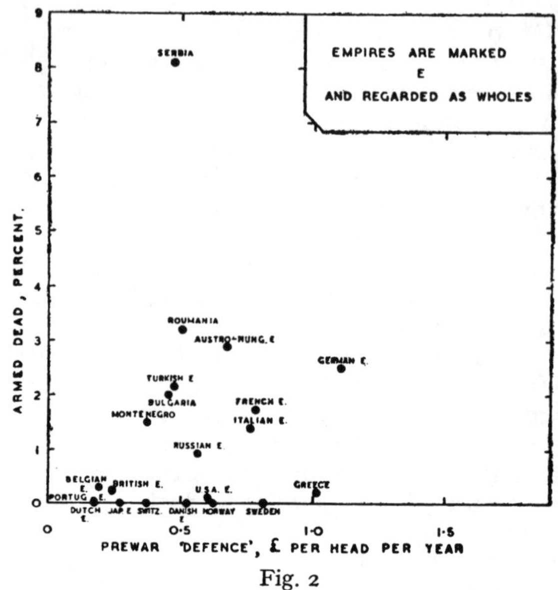

Fig. 2

Perhaps, however, it was injudicious to divide the pre-war defence expenditure by the population of the whole empire. So in Fig. 3 the divisor is the population of the metropolitan core only. But again neither of the opposing theories is confirmed. One

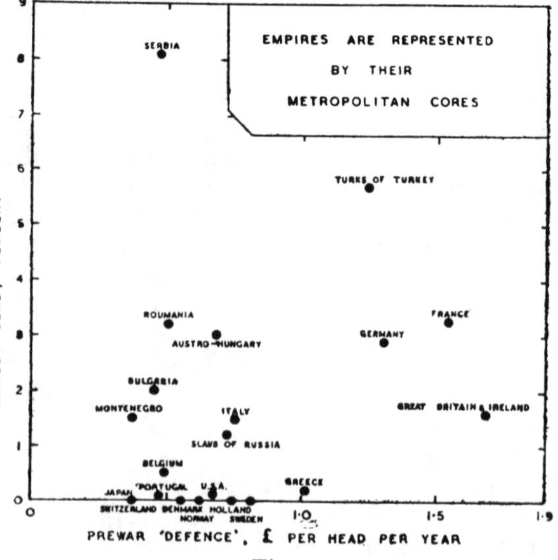

Fig. 3

may conclude, from this sample, that the effects of armed strength were erratic and untrustworthy. Further particulars can be seen in the microfilm *Arms and Insecurity* (Richardson, 1949).

The process whereby armed strength antagonizes Great Powers, instead of pacifying them, has been analysed from the financial point of view by comparing their total "defence" expenditure with the rate at which that total increased. The main cleavage of policy in Europe in the year 1913 was that between France and Russia on one side versus Germany and Austria-Hungary on the other. Britain and Italy were not then so obviously attached. The total of the defence expenditures of the four Powers chiefly concerned is stated as $U + V$ in Table II and is plotted in Fig. 4 against its time rate. These are quotations from a monograph published by the British Psychological Society (Richardson, 1939). A straight line appears. The more those Powers spent on so-called defence, the more rapidly did they increase that expenditure.

TABLE II
THE ARMS RACE OF 1909-14
Defence budgets expressed in millions of £ sterling

	1909	1910	1911	1912	1913
France	48·6	50·9	57·1	63·2	74·7
Russia	66·7	68·5	70·7	81·8	92·0
Germany	63·1	62·0	62·5	68·2	95·4
Austria-Hungary ...	20·8	23·4	24·6	25·5	26·9
Total $= U + V$ ...	199·2	204·8	214·9	238·7	289·0
Time $= \Delta(U+V)/\Delta t$		5·6	10·1	23·8	50·3
$(U+V)$ at same date		202·0	209·8	226·8	263·8

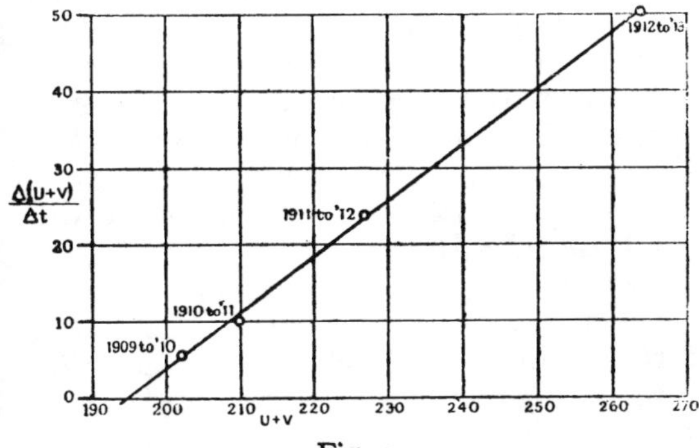

Fig. 4

The same arms race was afterwards re-examined much more thoroughly with the results shown in Fig. 5.

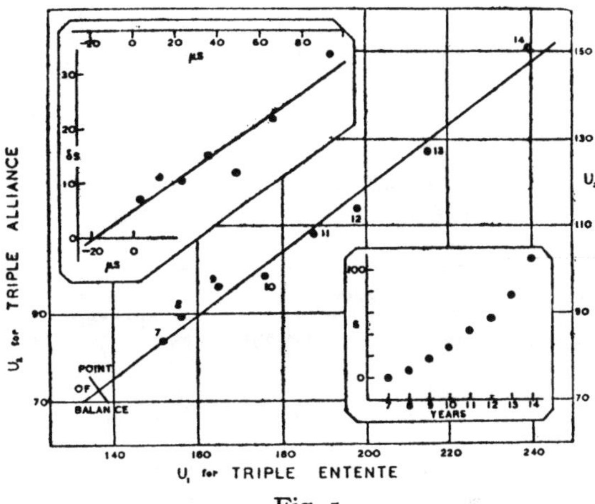

Fig. 5

The Triple Entente was composed of France, Russia and Britain. The Triple Alliance was composed of Germany, Austria-Hungary and Italy. Their "defence" expenditures were expressed as sterling and were denoted respectively by U_1 and U_2. The main diagram shows how the two sides went along together. The top left hand inset is of special interest. Here $S = U_1 + U_2$; μS is the mean of S for successive pairs of years; and δS is the annual increment. A strong positive correlation appears, confirming that the more they spent on so-called defence, the faster did they increase that expenditure. Further particulars are offered in the form of microfilm (Richardson, 1949). The simplest mathematical description of mutual reprisals is the pair of simultaneous differential equations.

$$\frac{dx}{dt} = y \quad , \quad \frac{dy}{dt} = x$$

I started from that idea in 1918. Various other terms were afterwards put in, so as to make the equations describe the financial statistics of the arms races.

This financial analysis was suggested by a psychological analysis, namely that one of the usual reactions to threats is to make counter-threats, if possible of a stronger character. There is mutual stimulation by mutual threats. Gregory Bateson has provided a suitable technical term to describe the process. He calls it "schismogenesis," that is to say the growth of a cleavage.

In his book *Naven* (1936) he described two kinds of schismogenesis in the Iatmul tribe in New Guinea.

In order to explain national moods during the first world war I found it necessary to take a hint from depth-psychology so as to regard each mood as consisting of a conscious part and a subconscious part. The following summary is quoted from a recent paper in *Psychometrika* (Richardson, 1948, *a* and *b*). The name of the overt mood is set above that of the subconscious mood. See the adjoining table, which spreads across both columns.

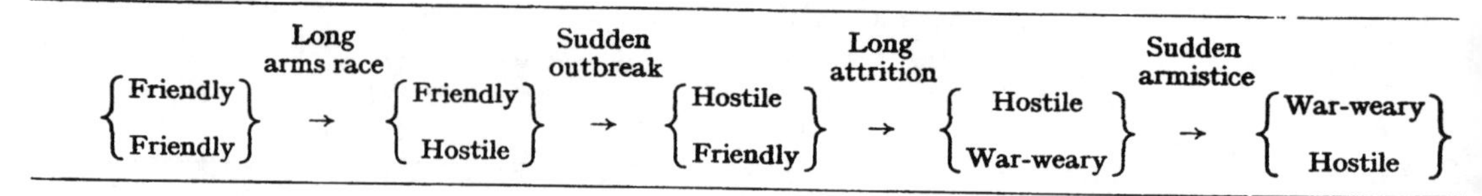

The changes of mood on one side of the conflict are supposed to have been caused partly by the existence of moods on the other side: again a mutual stimulation. The mathematics is rather like that of epidemics of disease. I took the first notion of it from a paper by Kermack and McKendrick on cholera (1927).

From all this mass of evidence I conclude that the popular policy of armed strength is untrustworthy; on some occasions it has been a pacifier; on others a dangerous irritant leading to war. Let us pass on to the next policy, namely:

Collective security. The theory is that the large majority of peace-loving nations should forcibly repress a small minority of aggressive nations just as the police force suppresses a small minority of criminals. Ranyard West (1942) has pointed out that in the course of a few years a nation which we may have been accustomed to regard as criminal may be found to have joined the police, and vice-versa. We may, however, consider the proposition that justice concerning any one international dispute is to be decided by the nations in majority on that occasion, apart from whatever opinions they held previously. If so, then wars of many nations versus one nation may be likened to the police versus a criminal, whereas wars of one nation versus one nation may be likened to ordinary quarrelling. The number of deadly quarrels of each type "r versus s" is shown in Table III, which is quoted by permission of the Royal Statistical Society (Richardson, 1947).

TABLE III

DEADLY QUARRELS OF ALL MAGNITUDES GREATER THAN 3·5 WHICH ENDED FROM A.D. 1820 TO 1939 INCLUSIVE

s \ r	1	2	3	4	5	6
6						0
5					0	0
4				0	0	0
3			0	1	0	0
2		3	2	1	1	0
1	42	24	5	5	2	1

There were also, beyond bounds of this table: 2 wars of 7 versus 1, 1 war of 9 versus 1, 1 war of 16 versus 5. Thus making a total of 91 wars

There were thus forty-two quarrels of the type one versus one, and only eleven quarrels of four or more versus one. It is seen that collective security is a new proposal and is contrary to national habits.

International trade. A more direct opponent to the indignation and suspicion of the arms race is needed. In 1938 I argued thus. The opposite to war is not peace in the sense of tranquil inattention to the doings of foreigners. The opposite to war is active co-operation with foreigners. Of the various forms of co-operation, foreign trade then occupied the greatest number of persons. Therefore, foreign trade seemed the most hopeful antidote to war. This supposition was later examined in connection with the financial statistics of the arms race, and those of the simultaneous foreign trade. The results are available in detail only in the form of microfilm (Richardson, 1949). The rapidly falling foreign trade of the years 1929-33 coincided with a steady high value of the "instability co-efficient" deduced from the expenditures on armaments. It would be reasonable, therefore, to expect that a rapid increase in foreign trade would occur along with a less instability of the political relations. But to quell the subsequent arms race of 1933-9 by reducing the instability co-efficient to zero would have required, if proportional, a boom of foreign trade two and a half times more rapid than the exceptional slump of 1929-33. In brief: foreign trade is a pacifier, but not nearly strong enough.

Plenty of living space. The standard treatise on this subject was written by our President, Sir Alexander Carr-Saunders (1936). I will only remind you that there are two motives: to seek more space and to seek more company. At first glance these motives seem simply opposite. Further consideration shows that their interaction is more complicated. Sir James Jeans (1929) explained how stars would be formed in a hypothetical primordial gas by the interaction between gas pressure and gravitational attraction. A slight modification of Jeans's theory enabled me to explain how towns would be formed by the interaction between population pressure and gregarious attraction (1941).

Fewer frontiers. There is a positive correlation of 0·77 between the number of frontiers which a State has, and the number of its external wars. This is shown by Fig. 6 (Richardson, 1950). For the peace of the world it would be desirable to have fewer frontiers; but, of course, they are not easily altered.

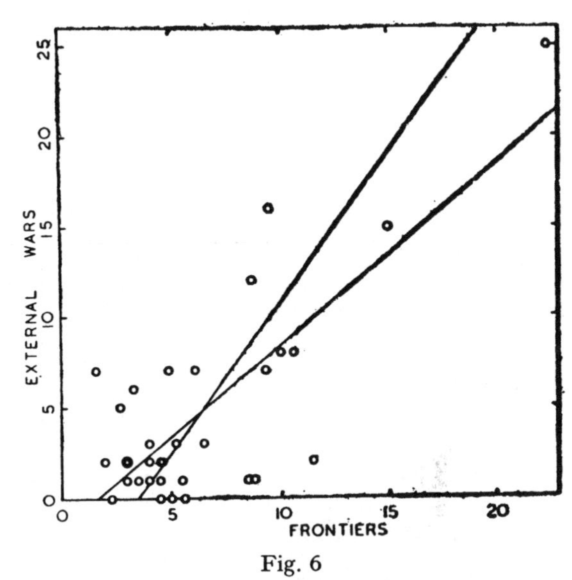

Fig. 6

The geographical opportunities for contiguous fighting have been estimated by attending to maps on which the population is divided among compact cells, each containing the same number of people. Such maps were shown to, and discussed by, the Geography Section of the British Association in September, 1949. A full account is in preparation. The results for the world as a whole are shown in Table IV.

TABLE IV

Persons in each cell	10,000,000	1,000,000	100,000
Cell-edges in A.D. 1910			
between nations	167	592	2,100
within nations	144	3,900	48,000

The actual fightings during the interval A.D. 1820 to 1929 in the whole world have been classified so as to correspond to the geographical opportunities by taking the deaths to be about 2 per cent. of the cell population. This is indeed a very rough correspondence; but it leads to a striking result, shown in Tables V and VI.

TABLE V

Persons killed in the quarrel	From 316,227 to 31,623	From 31,622 to 3,163	From 3,162 to 317
Pairs of opposed belligerent groups, A.D. 1820 to 1929 inclusive			
between nations	52	137	224*
within nations	23	33	179*

* Corrected for incompleteness of collection in the ratio 354 / 188. See Richardson, L. F., 1948 *c*., pp. 523-46.

The smallness of the last two fractions in Table VI shows that fighting within nations has somehow been prevented. National governments usually try to prevent fighting between subordinate groups within their territory. Another possible explanation is that the subordinate groups may have intermarried with one another.

TABLE VI
ACTUAL FIGHTING PER GEOGRAPHICAL OPPORTUNITY

Between nations ...	$\frac{52}{167} = 0{\cdot}31$	$\frac{137}{592} = 0{\cdot}23$	$\frac{224}{2,100} = 0{\cdot}11$
Within nations ...	$\frac{23}{144} = 0{\cdot}16$	$\frac{33}{3,900} = 0{\cdot}008$	$\frac{179}{48,000} = 0{\cdot}004$

The duration of common government prior to an outbreak of civil war is shown in Table VII.

These facts may be explained by saying that obedience to a common authority is easily broken when new, but becomes a habit after a generation or two. Alternatively one may ask whether intermarriage occurred. Two roughly similar time lags have, however, occurred in circumstances where intermarriage was unlikely. Alliances welded by wars have gradually fallen apart. Retaliations against a former enemy have become less frequent as the time interval increased. We may compare the rates at which these effects fade away by naming the time at which they are half gone (as in radio-activity).

Showing disloyalty by civil war ...	23·5 years
Breaking an alliance welded in war	14 years
Retaliation against a former enemy	15 years

It looks as though learning and forgetting were the main causes of these time lags.

The unifying effect of a *common language* and a *common religion* has often been mentioned. To see whether they deserve this reputation I sorted out the number of pairs of opposed belligerent groups who had a common language for political purposes. They were gathered from the whole world over the 110 years from 1820 to 1929. The observed numbers were compared with expectations deduced, according to various theories, from the total number of persons who had that language. Thus P_1 was derived from the theory that wars are made by leaders with a world-wide range of action; P_2 from pairs of populations each of a million with a world-wide range of action; P_3 from pairs of populations each of a million, in compact cells and capable of fighting only with neighbouring cells. Preliminary results were discussed at the British Psychological Society's meeting at Oxford in April, 1943. The argument may soon be available in microfilm (Richardson, 1950). It was found that there was something in the Chinese way of life, up to their revolution of A.D. 1911, which had prevented fightings between Chinese. Contrariwise it was found that there was something which stimulated fightings between groups who both spoke Spanish. For the other chief languages the statistics neither confirmed, nor refuted, Zamenhoff's belief that a common language, by promoting mutual understanding, would have a pacificatory effect. There may have been such an effect; but, if so, it was lost among the various ambiguities.

A similar statistical technique was applied to the effect of a common religion. That of China has already been included as part of its way of life. At first sight it looked as if Christianity incited wars between Christians; also as if Mohammedanism prevented wars between Moslems. These suggestions, startling to the author, were, however, found not to be statistically reliable. There was clear evidence of more fightings of Christians versus Moslems than would be expected from their numbers in the world.

TABLE VII

Preceding years of common government	0 to 20	20 to 40	40 to 60	60 to 80	Over 80	Totals
Observed pairs of opposed belligerents ...	41	20	13	5	15	94
Geometrical progression	37·13	22·46	13·59	8·22	12·59	93·99

Summary of Pacifying influences other than Intermarriage

The old Chinese religion and the habit of obedience to a common authority have both been effective. Less effective is international trade. Fewer frontiers might help. Armed strength is erratic. Collective security would be a new acquisition. These pacifiers are neither so many, nor so convenient, as to make attention to international marriage seem unnecessary.

Examples of Intermarriage

We need some short name for a population resulting from intermarriage in a bounded area during many generations. It used to be called a race; but the Nazis spoilt that word. The anthropologists call it an endogamous group. Let there be no pretence of racial purity. Each endogamous group will be characterized by its statistical distribution of genes.

England. About a century after the Norman conquest, Richard Fitzneal, Treasurer of King Henry II, stated that the English and Normans had so intermarried and the nations had become so intermingled that among the free classes it was impossible to distinguish between them. That is quoted from D. J. Medley (1901). Wars between Normans and English in England had ceased before Fitzneal's time, and did not recur. The French and English languages remained distinct and both in use in England for three or four centuries after the Conquest. (G. M. Trevelyan, 1926.)

England and Scotland. After the union of the Parliaments in A.D. 1707, the Scottish peers journeyed to London. One consequence was a progressive intermarriage between the Scottish and English aristocracy. This is conspicuously shown in the genealogies of Scottish families. For this information I am indebted to the Rev. D. W. P. Strang, M.C., D.D. The last notable fighting of Scots versus English occurred in A.D. 1746, and those Scots were mostly Highlanders.

South Africa. The High Commissioner for the Union of South Africa, Colonel Deneys Reitz, told the Press on January 14th, 1943, that English and Dutch in South Africa were freely intermingling and intermarrying, and within the next thirty years or so even their most fervid racial politicians would find it hard to distinguish an Englishman from a Dutchman. (*The Times*, London, January 15th, 1943, p. 2, col. 5.)

Persia. Alexander the Great, after conquering Persia, held a great marriage feast, in which he and ninety of his generals and friends were married to Persian brides. After Alexander's death there was fighting between his generals, but the division was not the former one of Macedonians versus Persians. (H. G. Wells's *Outline of History*, 1920, pp. 191, 194.)

In *Louisiana* soon after 1769, when the Spanish took it over by treaty from the French, " Many French Creoles were appointed to office, intermarriages of French and Spanish and even English were encouraged by the highest officials, and in general a liberal and conciliatory policy was followed, which made Louisiana under Spanish rule quiet and prosperous." (E. **14**, 430.)

In *New Zealand* there were wars between the Maoris and the British from 1861 to 1871 (W. P. Reeves in E. **16**, 401). By 1929 the situation of the Maoris was described thus: " At present the native people number about 54,000, but the percentage of white blood is rapidly increasing, so that ultimately miscegenation may be expected to absorb the Maori into our New Zealand population." (R. Firth in E. **14**, 834.)

In *Germany* the Nazi discrimination against Jews was moderated for those who had one non-Jewish parent or three non-Jewish grandparents.

The racial mixture in *Brazil* is described by Calogeras (1939, pp. 18-30). The Portuguese discoverers called the natives Indians. " Unions with Indians of both sexes evoked no criticism except when they were illegal or contrary to religious sanctions." Negro slaves from Africa were imported from A.D. 1538 to 1850. The Negroes constituted the lowest social stratum of the colony, and racial crossing with them was looked upon as degrading. Yet, in spite of everything, crossing between the Negroes and Portuguese began very early, and has persisted with intermixture also of Spaniards and Italians. There has been " a progressive and noticeable whitening of the skin of the local population in large sections of Brazil." " Roosevelt rightly pointed out that the future has reserved for us a great boon: the happy solution of a problem fraught with tremendous, even mortal dangers—the problems of a possible conflict between the two races."

In assessing the pros and cons of such a mixture between Europeans, Amerinds and Negroes, it must be noticed that Brazil has not been particularly law-abiding nor peaceful. Numerous armed insurrections are mentioned in Brazilian history; but usually with particulars insufficient to fix the magnitude. From the interval A.D. 1820-1939 I have collected sixteen deadly quarrels, probably of magnitudes between 3.5 and 1.5, Their alleged causes include local autonomy, Conservatives versus Liberals, discontent in the navy, fanaticism, personal ambition, and the world economic depression of 1929-34. Only one of them was a conflict of races. It occurred in 1871 between the Government and Negroes in the province of Minas Geraes. (H. H. **23**, 661.)

A contrast between *Argentina and Peru* during the years A.D. 1810-24 is described by Levene in his *History of Argentina* (1937, p. 316) in these words:

> When the Spanish-American colonies revolted against the motherland in 1810, the viceroyalty of Peru remained apart from that movement. Two chief causes explain that phenomenon: first, the heterogeneous character of its population which was composed of aborigines, Negro slaves, and the mestizos resulting from the admixture of Indians, Africans and Spaniards. The cohesion which the fusion of races had caused in the provinces of La Plata, and which early produced among us the phenomenon of a spontaneous democracy did not, therefore, exist in Peru. A second cause was due to the powerful Spanish resistance in Peru.

Levene's statement is certainly relevant to the relations of intermarriage to co-operation, yet its indications are not clear. For the more thorough race-mixture in La Plata was accompanied not only by " spontaneous democracy " but also by revolt against authority and by numerous faction fights.

Jews in England and U.S.A. Dr. Eliot Slater (1947) collected some evidence that in England and the U.S.A. the marriage of Jews with non-Jews is now occurring so frequently that in two generations there may not be much distinction left for the anti-Semites to talk about.

In the foregoing examples intermarriage was accompanied by common government, so that it would be possible to argue that pacification was caused by either or by both. The following examples are specially interesting because they relate to intermarriage apart from common government.

In *Europe*, Professor Arnold J. Toynbee has expressed the opinion (1937, p. 37) that royal marriages made on the whole more for peaceful change than for new trouble.

Italy and *U.S.A.* A pacifying effect in the midst of war was reported by the British Broadcasting Corporation on September 6th, 1943. Six British soldiers had maintained themselves for a week behind the Axis lines in Italy, having been given bread and cheese by Italians. " Nearly every Italian they met claimed to have relatives in the United States."

Intermarriage may not have a pacifying effect if there are strong religious differences. For example:

(*a*) The Moslems of China have had wars against the Confucians. Edgar Snow (1937, p. 322) noted that " . . . while markedly Turkish features are not infrequent among them, the physiognomy of the majority is hardly distinguishable from the Chinese, with whom they have for centuries intermarried."

(*b*) In 1947 the Government of the Soviet Union forbade Russians to marry foreigners. The British have been complaining about the retention by Russia of the wives of British men who have left Russia.

Action Suggested

Marx and Engels ended their Communist Manifesto of 1848 with the slogan " Working men of all countries, unite," meaning " unite in organization against the owners of property." The experience of the first world war showed that the bond of the socialist international organization was not strong enough to restrain national patriotisms. Lenin complained about it in 1915 (reprinted 1940).

There does not seem to be any immediate practical possibility of whole nations intermarrying. In peace-time it is the wealthier and more enterprising part of the population which travels abroad, and which experiments in new customs. International marriage is already noticeable among those classes of society. So, taking a hint from Marx and Engels, I suggest the slogan :

"Managers of all countries, unite—by matrimony, in order to hold the world together !"

An inner mentor reminds me that it is very rash thus to indulge in a full-throated slogan before the social scientists have thoroughly investigated the proposal, especially its irreversible genetical effects. So I come back to the genetics of the socially desirable qualities such as courage, kindliness, memory and intelligence. Here I feel deplorably ignorant. Moreover, I recently asked an eminent psychologist and an eminent geneticist, and they both said that there was not much sure knowledge on the subject. Strong opinions, however, abound. One is told that half-breeds have all the vices of both parental races. One hears also the counter-argument that the vices of half-breeds are due to the social contempt which they suffer from both parental races. How shall truth be disentangled from amongst such intense prejudices? I suggest two sieves for straining out prejudice and letting truth pass through :

(i) Lay aside, meantime, all general impressions, and attend only to definite tests, designed to be as little dependent on schooling as possible. These at present relate mostly to intelligence and to memory.

(ii) Collect the results of tests on half-breeds only from places where race-prejudice is weak or non-existent, so that there half-breed children grow up in a socially approved environment. According to Romanzo Adams (1934) one such place is Hawaii. According to Levene (1937, p. 316) another is Argentina.

Intelligence and the Race Crossing

S. D. Porteus (1937), Director of the Psychological Clinic of the University of Hawaii, made a study of young persons aged 9 to 18 attending Honolulu schools and found the following average test-quotients on the form and assembling test :

	Boys		Girls	
	Number tested	Average T.Q.	Number tested	Average T.Q.
Chinese ...	114	102·2	170	100·8
Chinese-Hawaiian	145	101·7	121	95·2
Hawaiian ...	94	97·7	77	89·1

That is to say the half-breeds were on the average intermediate between the averages of the parental races.

Russell Munday (1934), a schoolmaster in Buenos Aires, compared the intelligence of 191 children whose grandparents were all from the same nation with the intelligence of 117 children having grandparents from more than one nation. He found no significant difference. The nations considered were European. Negroes and Amerinds were not included.

These quotations from Porteus and from Munday relate to the average intelligence of groups of people. But it should be noted that individuals are usually scattered about the mean of their group to an extent greater than the difference between the means of any two well-known races. For example the Scottish Council for Research in Education (1933) found that the average differences in intelligence quotient between Scottish, Californians and English (at Halifax, Yorkshire) were negligible in comparison with the standard deviation of individuals from their means.

Ruth Byrns (1936) studied intelligence and nationality among 78,560 high school students in Wisconsin, U.S.A. She listed twenty-seven European nationalities, also Negro, Syrian and the large American group. The nationality of the pupil was taken to be that of the parents. The intelligence was tested. The fact that all of these pupils had completed approximately ten to twelve years of schooling minimized the influence of educational differences and tended to eliminate the differences due to language difficulty. The selective influences of migration and of schools are such that it is not profitable for our purposes to compare the mean intelligences of these different national groups; but (p. 459) Byrns states that "... the full range of ability is found within every natio-racial group."

Porteus (1931, p. 363) tested a very primitive people, the Australian aborigines, and found them, on the average, far below the Americans in intelligence quotient, thus: Americans 97, Australian aborigines 80. Yet Porteus remarked that "... there are at present among the aborigines, individuals whose planning capacity is decidedly above the average of whites...."

The managers of all countries, whom it was suggested should unite by matrimony, would, of course, be selected individuals, superior to the averages of their countries.

Conclusion. I submit that a strong *prima-facie* case has been made for the further consideration of international marriage as a hopeful pacifying influence.

REFERENCES

Adams, Romanzo (1934), in *Race and Culture Contacts*, edited by Reuter, E. B. London: McGraw Hill.

Bateson, Gregory (1936), *Naven*. Cambridge University Press.

Byrns, Ruth (1936), *Journ. Social Psych.*, **7**, 455-70, Clark University Press.

Calogeras, J. P., trans. Martin, P.A. (1939), *A History of Brazil*. University of North Carolina Press.

Carr-Saunders, A. M. (1936), *World Population*. Oxford: Clarendon Press.

E. refers to the *Encyclopædia Britannica*, 1929, 14th edn.

H.H. refers to *Historians' History of the World* (1907), edited by Williams, H. S. London: *The Times*.

Jastrzebski, S. de (1929). Article "Death-rate" in E.

Jeans, Sir James (1929), *Astronomy and Cosmogony*. Cambridge University Press.

Jung, C. G. (1947), *Essays on Contemporary Events*. London: Kegan Paul.

K stands for Keesing's Contemporary Archives, Bristol. Page numbers are given.

Kermack, W. O., and McKendrik, A. G. (1927), *Proc. Roy. Soc. London*, A, **115**,, 700-22.

Lenin, V. I. (1940), *Socialism and War*. London: Lawrence & Wishart.

Levene, R., trans. Robertson, W. S. (1937), *A History of Argentina*. University of North Carolina Press.

Malan, D. F. (1948), *What "Apartheid" Means*. Public Relations Office, South Africa House, Trafalgar Square, London, W.C.2.

Medley, D. J. (1901) in Traill & Mann's *Social England*, **1**, 534. London: Cassell & Co.

Munday, Russell (1932), *Brit. J. Educ. Psych.*, **2**, 1-7.

Noel-Baker, P. J. (1936), *The Private Manufacture of Armaments*. London: Victor Gollancz.

Nye Committee (1935) Report No. 944, Part 2. United States Government Printing Office.

Porteus, S. D. (1931), *The Psychology of a Primitive People*. London: Arnold.

—— (1937), *Primitive Intelligence and Environment*. New York: Macmillan.

Richardson, L. F. (1939), "Generalized Foreign Politics," *Brit. J. Psychol. Monog. Supplt.*, No. 23.

—— (1941), *Nature*, of December 27th.

—— (1947), *J. Roy. Statistical Soc.*, **109**, 130-56.

—— (1948, *a* and *b*), *Psychometrika*, **13**, 147-74 and 197-232.

—— (1948, *c*.), *J. Amer. Statistical Assn.* Washington, D.C. **43**, 523-46.

—— (1949), *Arms and Insecurity*, a book in microfilm, from the author, 2nd. edn.

—— (1950, *a*.), *Statistics of Deadly Quarrels*, a book in microfilm, from the author.

—— (1950, *b*.), *Brit. J. Psychol.* (Medical Section).

Scottish Council for Research in Education (1933), *Intelligence of Scottish Children*. University of London Press.

Slater, E. (1947), EUGENICS REVIEW, **39**, 17-21.

Sorokin, Pitrim A. (1937), *Social and Cultural Dynamics*. American Book Co.

Snow, Edgar (1937), *Red Star Over China*. London Victor Gollancz.

Toynbee, A. J. (1937), in *Peaceful Change*. London: Macmillan.

Trevelyan, G. M. (1926), *History of England*. London: Longmans Green.

Wells, H. G. (1920), *Outline of History*. London: Cassell.

West, Ranyard (1942), *Conscience and Society*. London: Methuen.

Wright, Quincy (1942), *A Study of War*. University of Chicago Press.

1950:4

STATISTICS OF DEADLY QUARRELS

Microfilm of reprint, typescript and manuscript, privately produced

(Title-page dated 27 July 1950)
(Photographed 8 August 1950)

[2nd edition] 1960:1

STATISTICS OF DEADLY QUARRELS

Edited by Quincy Wright and Carl C. Lienau
Pittsburgh: Boxwood Press Chicago: Quadrangle Books
London: Stevens and Sons Limited

Notes on *Statistics of Deadly Quarrels*

1. *Statistics of Deadly Quarrels* consists of a comprehensive summary of wars occurring world-wide since AD 1820 and a detailed statistical analysis of their characteristics. The second edition incorporates all Richardson's analyses of these data, some of which had already been published as separate papers, and includes the final version of his list of wars (updated in August 1953).

2. In an introduction dated 27 July 1950 to the first (microfilm) edition Richardson wrote:

> The present microfilm is part one of a comprehensive work on *The Instability of Peace*. Part two deals mainly with arms-races and has been published in microfilm since 1947 with the title *Arms and Insecurity*. Some of part three has appeared as scattered papers on war-moods, on national hatred, on war and eugenics and on voting power....
> The microfilm is intended to ensure the work against fire, battle and sudden death, until such time as it can be printed.

The aim of producing a single comprehensive work in print was never realised. *Statistics of Deadly Quarrels* and *Arms and Insecurity* remained separate, and appeared as books only after Richardson's death (1960:1, 1960:2). The papers referred to above as 'some of part three' are:

1918 International Voting Power

1926:2 Power in the League of Nations

1948:2 War-moods

1949:4 The Persistence of National Hatred and the Changeability of its Objects

1950:2 War and Eugenics

3. *Statistics of Deadly Quarrels* was published by Richardson privately on microfilm (1950:4) because of the difficulty he had experienced in getting the larger book into print (see the 'Notes on *Arms and Insecurity*', above p. 419). Most of the original text of the microfilm had been 'ready in typescript' 'by 1943'. Following advice at that time a few sections had been published separately by 1950, namely:

1944:2 The Distribution of Wars in Time (subsequently revised and extended to become §1/3 of the microfilm [Chapter III of the book])

1946:1 The Number of Nations on each Side of a War (reappeared unaltered as §1/10 [Chapter X])

1948:4 Variation of the Frequency of fatal Quarrels with Magnitude (most of the paper reappeared unaltered in §1/4 [Chapter IV] – see the Notes on 1948:4, above p. 489)

4. At the time of his death in September 1953 Richardson was planning a second edition of *Statistics of Deadly Quarrels*, but there are relatively few differences between the 1950 and 1960 editions, apart from a number of additions to Chapters I and II, and the inclusion unaltered of one later paper, 1952:2 Contiguity and Deadly Quarrels (Chapter XII, introduced by a new Chapter XI). Thus, the rest of the main text of the 1960 book (Chapters III–X) was first published in or before 1950, nearly all of it having apparently been completed about seven years earlier, during the Second World War.

   The following changes and additions were made in the second edition (1960:1). All the page, paragraph and chapter references (here and elsewhere) are to this edition, the book.

   Pp. 2–3: The seven paragraphs 'This arrangement is unusual ... different classes of magnitude.' were added.

   Pp. 6–7: The five paragraphs 'From the psychological ... prefixed to its totals' replaced two short paragraphs, defining the logarithmic scale of magnitude. The new paragraphs were transferred here from 1948:4.

   P. 19: The paragraph 'The present classification ... contrary to expectation' was added.

   P. 21: The two separate paragraphs 'Protection may also ... dangerous to be accepted' and 'Fear, widespread ... Richardson 1953b)' were added.

   P. 26: The section 'On missing the obvious' was added.

   P. 31: The numbers of quarrels 'As revised in 1953 Aug. –' were added.

   Pp. 32–111: No attempt has been made to identify the changes and additions to the list of wars of magnitude 2.5 or more since the microfilm. The earlier list was superseded by that in the book, which was updated to August, 1953.

   P. 112: The paragraph 'That there were many ... important than murders' was added (from 1948:4).

   Pp. 115–19: The paragraphs 'The facts about banditry ... indication of Thrasher's work' (p. 118), including Figures 1 and 2, were added (from 1948:4).

   Pp. 141–2: The three paragraphs 'In a previous table ... the next section does that', including the table, were revised.

   P. 142: The section 'Inquiry concerning improved communications' was added.

   Pp. 288–94: Chapter XI was added as an introduction to Chapter XII.

   Pp. 295–314: Chapter XII (previously published as 1952:2) was added.

5. The table below shows the cross-references between the 'Reader's Guide to the Microfilm', a printed card accompanying the microfilm, and the first ten chapters of the book.

	1950: 4		1960: 1	
Section	Abstract of contents	Page	Chapter	Page
	Preface	3		
	Detailed contents	7		
1/1	Plan for collecting information	13	I	1
1/1/2	Brief guide to the list	15a		2
	Definition of "magnitude"	18		4
	Definition of symbols	48		21
1/2/2	Magnitudes, 7.5 to 6.5 (world wars)	63	II	32
1/2/3	Magnitudes, 6.5 to 5.5	75		40
1/2/4	Magnitudes, 5.5 to 4.5	78		44
1/2/5	Magnitudes, 4.5 to 3.5	90		51
1/2/6	Magnitudes, 3.5 to 2.5	119		73
1/2/7	Magnitudes, 2.5 to 0.5	172		111
1/2/8	Magnitudes, 0.5 to 0 (murders)	178		119
1/2/9	Reliability	187		125
1/3	Distribution of wars in time [1944:2 revised]	190	III	128
1/4	Variation of frequency with magnitude [most of 1948:4]	212	IV	143
1/4/4	The growth of population	238		157
1/5/1	Belligerent groups	250	V	168
1/5/3	Which nations were most involved?	258		173
1/5/4	Frontiers and external wars	264		176
1/6/2	Civil war	275	VI	184
1/6/3	Alliances	290		194
1/6/4	Retaliation and revenge	295		197
1/7/1	Poor versus rich	306	VII	205
1/7/2	Disbanded or unpaid soldiers	309		206
1/7/3	Various economic causes	311		207
1/8	Languages and wars	317	VIII	211
1/9	Religions and wars	349	IX	231
1/10	Number of nations on each side [1946:1]	374	X	247
	References	405		

6. In addition to the main text, the book *Statistics of Deadly Quarrels* (1960:1) contains four other items, not reproduced here, which also appear in the book *Arms and Insecurity* (1960:2). These are:

	1950:4 Page	1960:1 Page	1960:2 Page
The use of mathematics	10	xliii	xvii
The interaction of many causes	402	xlv	xix
Author's preface (dated 1953)	3 (in part)	xxxv	xiii
Bibliography of studies of the causation of wars with a view to their avoidance		xxxi	289

The first two of these items originally appeared as the first and last sections (§0 and §1.11) of the microfilm of *Statistics of Deadly Quarrels* (1950:4). The third consists of the preface to this microfilm followed by the final paragraphs of the preface to the microfilm of *Arms and Insecurity* (1949:2), with some additional acknowledgements. The bibliography was prepared by Richardson, apparently in the summer of 1953 as part of his application for a fellowship at King's College, Cambridge. It was first published as 1957 (not reproduced here).

7. Two footnotes in 1950:4 (on pages 318 and 328 of the microfilm), which were not reproduced in 1960:1, are of interest:

    An earlier version of §1/8 [Languages and Wars, Chapter VIII] was described in abstract and discussed at the British Psychological Society's meeting in April 1943 at Oxford. Since then many particulars have been altered; yet the main conclusions have survived.

    Note added 1950 July 10. The present §1/8 on languages [Chapter VIII] and §1/9 on religions [Chapter IX] were substantially ready in 1942. They stimulated a much fuller enquiry into 'Mapping by Compact Cells of Equal Population' which is now approaching completion. When that geographical study is finished, the author's intention is to use it to refine the rather crude geometry of the present §1/8 and §1/9. As far as he knows today, the main conclusions about languages and religions are likely to be upheld.

    No abstract of the 1943 paper 'Languages and Wars, Religions and Wars' was published. The fuller enquiry referred to in the second footnote eventually became 1951:2 'The Problem of Contiguity', which was published posthumously in 1961. (See the 'Notes on The Problem of Contiguity', below p. 579.) Richardson used the findings in this paper when preparing 1952:2 [also in *Statistics of Deadly Quarrels*, Chapter XII] but never reanalysed the material on languages or religions.

Richardson, L. F. (1950)
Threats and Security; Statistics of Fatal Quarrels
In Pear, T. H. (Ed.) *Psychological Factors of Peace and War*, pp. 219–55. London: Hutchinson

1950:5

THREATS AND SECURITY; STATISTICS OF FATAL QUARRELS

In: *Pear, T. H. (ed.) Psychological Factors of Peace and War*
London: Hutchinson, 1950
Richardson, L. F. *Threats and Security*
Chapter X, pp. 219–235
Richardson, L. F. *Statistics of Fatal Quarrels*
Chapter XI, pp. 237–255

Note: These chapters are described by Richardson as 'mostly brief summaries' of *Arms and Insecurity* and *Statistics of Deadly Quarrels*

CHAPTER X

THREATS AND SECURITY

by L. F. Richardson

THE DIVERSE EFFECTS OF THREATS

THE reader has probably heard a mother say to her child: "Stop that noise, or I'll smack you". Did the child in fact become quiet?

A threat from one person, or group of people, to another person or group has occasionally produced very little immediate effect, being received with contempt. Effects, when conspicuous, may be classified as submission at one extreme, negotiation or avoidance in the middle, and retaliation at the other extreme. The following incidents are classified in that manner; otherwise they are purposely miscellaneous.

Contempt

EXAMPLE 1. About fifty states, organized as the League of Nations, tried in 1935 and 1936, by appeals and by cutting off supplies, to restrain Mussolini's Italy from making war on Abyssinia. At the time Mussolini disregarded the League, and went on with the conquest of Abyssinia. He did not however forget. Four years later in his speech to the Italian people on the occasion of Italy's declaration of war against Britain and France, Mussolini said, "The events of quite recent history can be summarized in these words—half promises, constant threats, blackmail and finally, as the crown of this ignoble edifice, the League siege of the fifty-two States".[1]

Submission

EXAMPLE 2.[2] In 1906 the British Government had a disagreement with the Sultan of Turkey about the exact location of the frontier between Egypt and Turkish Palestine. A British battleship was sent on 3rd May with an ultimatum, and thereupon the Sultan accepted the British view. Some resentment may perhaps have lingered, for after the First World War had been going on for three months Turkey joined the side opposite to Britain.

EXAMPLE 3. In August 1945 the Japanese, having suffered several

[1] *Glasgow Herald*, 11th June, 1940.

[2] *Ency. Brit.* XIV, ed. 14. **2**, 156. Grey, Viscount, 1925, *Twenty-Five Years*. Hodder and Stoughton, London.

years of war, having lost by defeat their Italian and German allies, being newly attacked by Russia, having had two atomic bombs dropped on them, and being threatened with more of the same, surrendered unconditionally.

Negotiation Followed by Submission

EXAMPLE 4.[1] After the Germans annexed Austria on 13th March, 1938, the German minority in Czechoslovakia began to agitate for self-government, and both the German and Czech Governments moved troops towards their common frontier. On 13th August there began German Army manœuvres on an unprecedented scale. On 6th September France called up reservists. In September the cession of the Sudetenland by Czechoslovakia to Germany was discussed, to the accompaniment of threats by Hitler on 12th, the partial mobilization of France on 24th, and the mobilization of Czechoslovakia and of the British Navy on 27th. Finally, at Munich on the 29th the French and British agreed to advise the Czechs to submit to the German demand for the Sudetenland, partly because its population spoke German, and partly because of the German threat to take it by armed force. Intense resentment at this humiliation lingered.

Negotiation Followed by a Bargain

EXAMPLE 5.[2] In the spring of 1911 French troops entered Fez, in Morocco, the German Government protesting. On 1st July, 1911, the German Government notified those of France and Britain that a German gunboat was being dispatched to Agadir on the southern coast of Morocco in order there to protect some German firms from the local tribesmen. The French and British interpreted the movement of the gunboat as a threat against themselves, like a 'thumping of the diplomatic table'. On 21st July Mr. Lloyd George made a speech containing the sentence "I say emphatically that peace at that price would be humiliation intolerable for a great country like ours to endure". Much indignation was expressed in the newspapers of France, Britain and Germany. Negotiations ensued. By 4th November France and Germany had agreed on a rearrangement of their West African territories and rights. The arms race continued.

[1]*Keesing's Contemporary Archives.* Bristol.

[2]*Ency. Brit.*, XIV, ed. 14. **23**, 349, 352. Morel, E. D., 1915, *Ten Years of Secret Diplomacy*, National Labour Press, London. Grey, Viscount, 1925, *Twenty-Five Years*, Hodder and Stoughton, London.

Avoidance

EXAMPLE 6. The normal behaviour of the armed personnel guarding any frontier in time of peace is to avoid crossing the frontier lest they should be attacked.

EXAMPLE 7. During 1941 British shipping mostly went eastward via the Cape of Good Hope avoiding the Mediteranean where the enemy threat was too strong.

EXAMPLE 8. Criminals are usually said to avoid the police: that is, when the criminals are decidedly outnumbered by the police.

Retaliation

EXAMPLE 9. After the Agadir incident had been settled by the Franco-German Agreement of 4th November, 1911, the 'defence' expenditures of both France and Germany nevertheless continued to increase. See the tabular statement on page 227.

EXAMPLE 10. Within six weeks after the Munich Agreement of 29th September, 1938, rearmament was proceeding more rapidly in France,[1] Britain,[2] Germany[3] and U.S.A.[4]

EXAMPLE 11. On 5th November, 1940, the British armed merchantship *Jervis Bay*, being charged with the defence of a convoy, saw a German pocket-battleship threatening it. The *Jervis Bay*, although of obviously lesser power, attacked the battleship, and continued to fight until sunk, thus distracting the battleship's attention from the convoy, and giving the latter a chance to escape.

EXAMPLE 12. On 10th November, 1941, Mr. Winston Churchill warned the Japanese that "should the United States become involved in war with Japan the British declaration will follow within the hour". This formidable threat did not deter the Japanese from attacking Pearl Harbour within a month.

These miscellaneous illustrations may serve to remind the reader of many others. Is there any understandable regularity about the wider phenomena which they represent? The present question is about what happens in fact; the other very important question as to whether the fact is ethically good or bad, is here left aside. Some conclusions emerge:

[1]Athlone broadcast, 5th October and 12th October.
[2]*Parliamentary Debates*, 3rd to 6th October.
[3]*Glasgow Herald*, 10th October.
[4]*Glasgow Herald*, 7th November.

(*a*) People, when threatened, do not always behave with coldly calculated self-interest. They sometimes fight back, taking extreme risks. (Examples 11 and 12.)

(*b*) There is a notable distinction between fresh and tired nations, in the sense that a formidable threat to a fresh nation was followed by retaliation (Example 12) whereas an even more severe threat to the same nation, when tired, produced submission. (Example 3.)

(*c*) A group of people, having a more or less reasonable claim, has sometimes quickly obtained by a threat of violence, more than it otherwise would. But that may not have been the end of the matter. (Examples 2 and 4.)

(*d*) There have often been two contrasted effects, one immediate, the other delayed. An immediate effect of contempt or submission or negotiation or avoidance has been followed by resentful plans for retaliation at some later opportunity. (Example 1, Example 2, Examples 4 and 10, Examples 5 and 9.)

(*e*) What nowadays is euphemistically called national 'defence', in fact always includes preparations for attack, and thus constitutes a threat to some other group of people. This type of 'defence' is based on the assumption that threats directed towards other people will produce in them either submission, or negotiation, or avoidance; and it neglects the possibility that contempt or retaliation may be produced instead. Yet in fact the usual effect between comparable nations is retaliation by counter-preparations, thus leading on by way of an arms race towards another war.

SCHISMOGENESIS

In his study of the Iatmul tribe in New Guinea, Gregory Bateson[1] noticed a custom whereby, at a meeting in the ceremonial hall two men would boast alternately, each provoking the other to make bolder claims, until they reached extravagant extremes.

He also noticed a process whereby a man would have some control over a woman. Then her acceptance of his leadership would encourage him to become domineering. This in turn made her submissive. Then he became more domineering and she became more abject, the process running to abnormal extremes.

Bateson called both these processes 'schismogenesis', which may be translated as 'the manner of formation of cleavages'. When both parties developed the same behaviour, for example, both boasting,

[1] Bateson, G., 1935, in the periodical *Man*, p. 199. Bateson, G., 1936, *Naven*, University Press, Cambridge.

Bateson called the schismogenesis 'symmetrical'. When the parties developed contrasted behaviour, say, one domineering and the other submissive, he called the schismogenesis 'complementary'. In this terminology an arms race between two nations is properly described as a case of symmetrical schismogenesis.

In the year 1912 Germany was allied with Austria-Hungary, while France was allied with Czarist Russia. Britain was loosely attached to the latter group, thus forming the Triple Entente, while Italy was nominally attached to the former group, thus making the Triple Alliance. The warlike preparations of the Alliance and of the Entente were both increasing. The usual explanation was then, and perhaps still is, that the motives of the two sides were quite different, for we were only doing what was right, proper and necessary for our own defence, whilst they were disturbing the peace by indulging in wild schemes and extravagant ambitions. There are several distinct contrasts in that omnibus statement. Firstly that their conduct was morally bad, ours morally good. About so national a dispute it would be difficult to say anything that the world as a whole would accept. But there is another alleged contrast as to which there is some hope of general agreement. It was asserted in the years 1912-14 that their motives were fixed and independent of our behaviour, whereas our motives were a response to their behaviour and were varied accordingly. In 1914 Bertrand Russell[1] (now Earl Russell) put forward the contrary view that the motives of the two sides were essentially the same, for each was afraid of the other; and it was this fear which caused each side to increase its armaments as a defence against the other. Russell's pamphlet came at a time when a common boast in the British newspapers was that the British people 'knew no fear'. Several conspicuous heroes have since explained that they achieved their aims in spite of fear. When we analyse arms races it is, however, unnecessary to mention fear, or any other emotion; for an arms race can be recognized by the characteristic outward behaviour, which is shown in the diagram on page 228. The valuable part of Russell's doctrine was not his emphasis on fear, but his emphasis on mutual stimulation.

This view has been restated by another philosopher, C. E. M. Joad:[2]

"... if, as they maintain, the best way to preserve peace is to prepare war, it is not altogether clear why all nations should regard the armaments of other nations as a menace to peace.

[1]Russell, B. A. W., 1914, "War the Offspring of Fear," Union of Democratic Control, London.

[2]Joad, C. E. M., 1939, *Why War?* Penguin Special, p. 69.

However, they do so regard them and are accordingly stimulated to increase their armaments to overtop the armaments by which they conceive themselves to be threatened. . . . These increased arms being in their turn regarded as a menace by nation A whose allegedly defensive armaments have provoked them, are used by nation A as a pretext for accumulating yet greater armaments wherewith to defend itself against the menace. These yet greater armaments are in their turn interpreted by neighbouring nations as constituting a menace to themselves and so on. . . ."

This statement is, I think, a true and very clear description but needs two amendments. The competition is not usually between every nation and every other nation, but rather between two sides; so that a nation looks with moderate favour on the armaments of other nations on its own side, and with strong dislike on those of the opposite side. Joad's description applies to an arms race which has become noticeable. Motives other than defence may have been important in starting the arms race.

It may be well to translate these ideas into the phraseology of 'operational research' which began to be used during the Second World War. Professor C. H. Waddington[1] explains that "The special characteristic which differentiates operational research from other branches of applied science is that it takes as the phenomenon to be studied the whole executive problem and not the individual technical parts. . . ." Surely the maintenance of world peace is an executive problem large enough to be called an operation and to require an appropriate background of operational research. This book is a contribution thereto. Sir Charles Goodeve,[2] in a survey of operational research, distinguishes between 'self-compensating and self-aggravating systems', and he mentions, as an example of the latter, the system composed of the public and of the store-keepers; a system such that a rumour of scarcity can make a real scarcity. In this phraseology it can be said that a system of two great powers, not in the presence of any common enemy, is a 'self-aggravating system' such that a rumour of war can make a real war.

It will be shown in the next section that arms races are best described in quantitative terms; but, for those who do not like mathematics, Bateson's word 'schismogenesis' may serve as an acceptable summary of a process which otherwise requires a long verbal description such as those given by Russell, Bateson, or Joad.

[1]Waddington, C. H., in *Nature*, Vol. CLXI, p. 404.
[2]Goodeve, Sir Charles in *Nature*, Vol. CLXI, p. 384.

The Quantitative Theory of Arms Races

The facts for the years 1909 to 1914 are interesting. The 'defence' budgets of France Germany and Russia were taken from a digest by Per Jacobsson;[1] those for Austria-Hungary from the *Statesman's Year Books*. To make them comparable, they were all reduced to sterling. In those years the exchange rates between national currencies were held steady by the shipment of gold, so that the conversion to sterling is easy and definite.

Because France was allied to Russia it is reasonable to consider the total of their 'defence' expenditures. Let it be U. For a similar reason let V be the total for Germany and Austria-Hungary. In the accompanying diagram the rate of increase of $(U + V)$ is plotted against $(U + V)$. See page 228.

The accuracy with which the four observed points are fitted by a straight line is remarkable, especially as one of the co-ordinates is a difference. Similar diagrams drawn for other years, for other countries, and from other sources of information, are not so straight; but still they are straight enough to suggest that the explanation of the phenomenon is hardly likely to be found in the caprice of a few national leaders; the financial facts suggest either regular planning, or the regularity which results from the average of many opinions.

The main feature shown by the diagram is that the more these nations spent, the more rapidly did they increase their expenditure. Athletic races are not like that, for in them the speed of the contestants does not increase so markedly with the distance that they have run.

TABLE I. *The arms race of 1909-14*
Defence budgets expressed in millions of £ sterling

	1909		1910		1911		1912		1913
France	48·6		50·9		57·1		63·2		74·7
Russia	66·7		68·5		70·7		81·8		92·0
Germany	63·1		62·0		62·5		68·2		95·4
Austria-Hungary	20·8		23·4		24·6		25·5		26·9
Total = $U + V$	199·2		204·8		214·9		238·7		289·0
Time rate = $\Delta(U + V)/\Delta t$		5·6		10·1		23·8		50·3	
$(U + V)$ at same date		202·0		209·8		226·8		263·8	

Here Δ signifies 'take the annual increase of' whatever symbol follows next.

From Monog. Supplt. No. 23 of *Brit. Journ. Psychol.*, by permission of the British Psychological Society.

[1]Jacobsson, Per (1929?). *Armaments Expenditures of the World*, published by the *Economist*, London.

The sloping line when produced backwards cuts the horizontal, where $\Delta(U+V)/\Delta t$ vanishes, at the point where $U+V=194$ million £. This point may suitably be called a point of equilibrium. To explain how it could be a point of equilibrium we can suppose that the total expenditure of 194 million was regarded as that which would have been so ordinary as not to constitute any special threat.

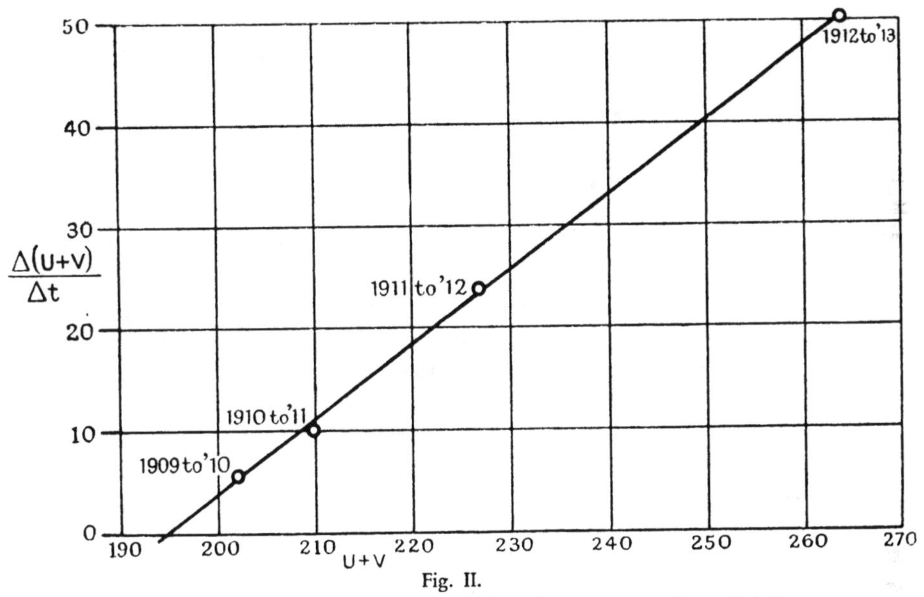

Fig. II.

The financial facts in the table at the foot of page 227 are here plotted.

It was a theory[1] which led L. F. Richardson to make a diagram having those co-ordinates $U+V$ and $\Delta(U+V)/\Delta t$. This theory will now be explained. The opening phase of the First World War afforded a violent illustration of Russell's doctrine of mutuality, for it was evident that warlike activity, and the accompanying hatred, were both growing by tit for tat, alias mutual reprisals. Tit for tat is a jerky alternation; but apart from details, the general drift of

[1]Richardson, L. F., 1919, *Mathematical Psychology of War.* In British copyright libraries. The diagram first appeared in *Nature* of 1938, Vol. CXLII, p. 792.

mutual reprisals was given a smoothed quantitative expression in the statement that the rate of increase of the warlike activity of each side was proportional to the warlike activity of the other side. This statement is equivalent to the following pair of simultaneous differential equations

$$\frac{dx}{dt} = ky, \quad \frac{dy}{dt} = lx \qquad (1)\ (2)$$

where t is the time, x is the warlike activity of one side, y that of the other side, k and l are positive constants, dx/dt is the excellent notation of Leibniz for the time rate of increase of x, and dy/dt is the time rate of increase of y. In accordance with modern custom, the fraction-line is set sloping when it occurs in a line of words. If k were equal to l, then the relation of x to y would be the same as the relation of y to x, so that the system of x and y would be strictly mutual. Strict mutuality is, however, not specially interesting. The essential idea is that k and l, whether equal or not, are both positive. They are called '*defence coefficients*' because they represent a pugnacious response to threats.

The reader may here object that anything so simple as the pair of equations (1) and (2) is hardly likely to be true description of anything so complicated as the politics of an arms race. In reply appeal must be made to a working rule known as Occam's Razor whereby the simplest possible descriptions are to be used until they are proved to be inadequate.

The meaning of (1) and (2) will be further illustrated by deducing from them some simple consequences. If at any time both x and y were zero, it follows according to (1) and (2) that x and y would always remain zero. This is a mathematical expression of the idea of permanent peace by all-round total disarmament. Criticism of that idea will follow, but for the present let us continue to study the meaning of equations (1) and (2). Suppose that x and y being zero, the tranquillity were disturbed by one of the nations making some very slightly threatening gesture, so that y became slightly positive. According to (1) x would then begin to grow. According to (2) as soon as x had become, positive, y would begin to grow further. The larger x and y had become the faster would they increase. Thus the system defined by (1) and (2) represents a possible equilibrium at the point where x and y are both zero, but this equilibrium is unstable, because any slight deviation from it tends to increase. If any historian or politician reads these words, I beg him or her to notice that in the mechanical sense, which is used here, stability is not the same as equilibrium; for on the contrary stable and unstable are adjectives qualifying equilibrium. Thus an equilibrium is said to be stable, or to have stability, if a small disturbance tends to die away; whereas an equilibrium is said to be unstable, or to have instability, if a small disturbance tends to increase. In this mechanical sense the system defined by (1) and (2) has instability. It describes a schismogenesis. "It is an old proverb," wrote William Penn in 1693. "*Maxima bella ex levissimis causis*: The greatest Feuds have had the smallest Beginnings."

One advantage of expressing a concept in mathematics is that deductions can then be made by reliable techniques. Thus in (1) and (2) the nations appear as entangled with one another, for each equation involves both x and y. These variables can, however, be separated by repeating the operation d/dt which signifies 'take the time rate of'.

Thus from (1) it follows that $\frac{d}{dt}\left(\frac{dx}{dt}\right) = k\frac{dy}{dt}$

Simultaneously from (2) $k\frac{dy}{dt} = klx$

On elimination of dy/dt between these two equations there remains an equation which does not involve y, namely

$$\frac{d}{dt}\left(\frac{dx}{dt}\right) = klx. \qquad (3)$$

Similarly

$$\frac{d}{dt}\left(\frac{dy}{dt}\right) = kly \qquad (4)$$

In (3) and (4) each nation appears as if sovereign and independent, managing its own affairs, until we notice that the constant kl is a property of the two nations jointly.

Another advantage of a mathematical statement is that it is so definite that it might be definitely wrong; and if it is found to be wrong, there is a plenteous choice of amendments ready in the mathematicians' stock of formulæ. Some verbal statements have not this merit; they are so vague that they could hardly be wrong, and are correspondingly useless.

The formulae (1) and (2) do indeed require amendment, for they contain no representation of any restraining influences; whereas it is well known that, after a war, the victorious side, no longer feeling threatened by its defeated enemy, proceeds to reduce its armed forces in order to save expenditure, and because the young men are desired at home. The simplest mathematical representation of disarmament by a victor is

$$\frac{dx}{dt} = -\alpha x \qquad (5)$$

where α is a positive constant. For, so long as x is positive, equation (5) asserts that dx/dt is negative, so that x is decreasing. Equation (5) is commonly used in physics to describe fading away. In accountancy, depreciation at a fixed annual percentage is a rule closely similar to (5). As a matter of fact[1] (5) is a good description of the disarmaments of Britain, France, U.S.A., or Italy during the years just after the First World War. In equation (5), which represents disarmament of the victor, there is no mention of y, because the defeated nation no longer threatens. It seems reasonable to suppose that restraining influences of the type represented by (5) are also felt by both of the nations during an arms race, so that equations (1) and (2) should be amended so as to become

$$\frac{dx}{dt} = ky - \alpha x, \qquad \frac{dy}{dt} = lx - \beta y \qquad (6), (7)$$

in which β is another positive constant. At first[2] α and β were called 'fatigue and expense coefficients', but a shorter and equally suitable name is *restraint coefficients*. These restraining influences may, or may not, be sufficient to render the equilibrium stable. The interaction is easily seen in the special, but important, case of similar nations, such that $\alpha = \beta$, and $k = l$. For then the subtraction of (7) from (6) gives

$$\frac{d(x-y)}{dt} = -(k+\alpha)(x-y) \qquad (8)$$

In this $(k+\alpha)$ is always positive. If at any time $(x-y)$ is positive, equation (8) shows that then $(x-y)$ is decreasing; and that moreover $(x-y)$ will continue to decrease until it vanishes, leaving $x = y$. If on the contrary $(x-y)$ is initially negative, (8) shows that $(x-y)$ will increase towards zero. Thus there is a stable drift from either side towards equality of x with y. That is more or less in accord with the historical facts about arms races between nations which can be regarded as similar.

To see the other aspect, let (7) be added to (6) giving

$$\frac{d(x+y)}{dt} = (k-\alpha)(x+y) \qquad (9)$$

The meaning of (9) can be discussed in the same manner as that of (8). The result is that $(x+y)$ will drift towards zero if $(k-\alpha)$ is negative, that is if $\alpha > k$. We may then say that restraint overpowers 'defence', and that the system is thoroughly stable. Unfortunately

[1](*See* 3 on page 234).

[2]Richardson, L. F., in *Nature* of 18th May, 1935, p. 830.

that is not what has happened in Europe in the present century. The other case is that in which $k > \alpha$ so that 'defence' overpowers restraint, and $(x + y)$ drifts away from zero. That is like an arms race. When $k > \alpha$, the system is stable as to $(x - y)$, but unstable as to $(x + y)$. It is the instability which has the disastrous consequences. The owner of a ship which has capsized by rolling over sideways can derive little comfort from the knowledge that it was perfectly stable for pitching fore and aft. People who trust in the balance of power should note this combination of stability with instability.

If at any time x and y were both zero, it would follow from (6) and (7) that x and y would always remain zero. So that the introduction of the restraining terms still leaves the theoretical possibility of permanent peace by universal total disarmament. Small-scale experiments on absence of armament have been tried with success between Norway and Sweden, between Canada and U.S.A., between the early settlers in Pennsylvania and the Red Indians. The experiment of a general world-wide absence of arms has never been tried. Many people doubt if it would result in permanent peace; for, they say, grievances and ambitions would cause various groups to acquire arms in order to assert their rights, or to domineer over their unarmed neighbours. The theory is easily amended to meet this objection. Let two constants g and h be inserted respectively into (6) and (7) thus

$$\frac{dx}{dt} = ky - \alpha x + g; \qquad \frac{dy}{dt} = lx - \beta y + h \qquad (10), (11)$$

If x and y were at any time both zero, then would $dx/dt = g$, and $dy/dt = h$, which do not indicate a permanent condition.

There may be still an equilibrium, but it is not at the point $x = 0$, $y = 0$. To find the new point of equilibrium let $dx/dt = 0$, and $dy/dt = 0$. Then by (10) and (11)

$$0 = ky - \alpha x + g; \quad 0 = lx - \beta y + h \qquad (12), (13).$$

These equations represent two straight lines in the plane of x and y. If these lines are not parallel, their intersection is the point of equilibrium. It may be stable or unstable.

The assertion that the defence-coefficients k and l are positive is equivalent to supposing that the effect of threats is always retaliation. The reader may object that in the opening section of this chapter other effects were also mentioned, namely, contempt, submission, negotiation, or avoidance. The most important of these objections relates to submission, because it is the direct opposite of retaliation. The answer is that the scope of the present theory is restricted to the interaction of groups which style themselves powers, which are proud of their so-called sovereignty and independence, are proud of their armed might, and are not exhausted by combat. This theory is not about victory and defeat. In different circumstances k or l might be negative. A theory of submissiveness showing this has been published.[1] As to contempt, negotiation, or avoidance, they have sometimes gone on concurrently with an arms race, as in Examples (1), (5) and (9), (4) and (10) of the diverse effects of threats.

Let us now return to the 'defence' budgets of France, Russia, Germany and Austria-Hungary. The diagram on page 228 relates to the total $U+V$ of the warlike expenditures of the two opposing sides. The equilibrium point, at $U+V=194$ millions sterling, presumably represents the expenditure which was excused as being customary for the maintenance of internal order, and harmless in view of the treaty situation. In the theory the treaty situation is represented by the constants g and h. Their effects can be regarded as included in the 194, together with the general goodwill between the nations. It appears suitable therefore to compare fact with theory by setting simultaneously $x+y=U+V-194$, together with $g=0$ and $h=0$. (14), (15), (16)
From (14) one derives

$$\frac{d(x+y)}{dt}=\frac{d(U+V)}{dt} \tag{17}$$

The two opposing alliances were about of equal size and civilization, so that it seems permissible to simplify the formulae by setting $\alpha=\beta$ and $k=l$. Addition of the formulae (10) and (11) then gives, as before in (9),

$$\frac{d(x+y)}{dt}=(k-\alpha)\,(x+y) \tag{18}$$

Now $x+y$ can be thoroughly eliminated from (18); from its first member by (17), and from its second member by (14), with the result that

$$\frac{d(U+V)}{dt}=(k-\alpha)\,(U+V-194) \tag{19}$$

This is a statement about the expenditures of the nations in a form comparable with Fig. II on page 228. For $d(U+V)/dt$ is a close approximation to $\Delta(U+V)/\Delta t$. The assumed constancy of $k-\alpha$ agrees with the fact that the sloping line on the diagram is straight. Moreover the

[1]See references 1 and 3 on page 234.

absence from (6) and (7) of squares, reciprocals, or other more complicated functions, which was at first excused on the plea of simplicity, is now seen to be so far justified by comparison with historical fact. The slope of the line on the diagram, when compared with (19) gives

$$k - a = 0{\cdot}7_3 \text{ per year} \qquad (20)$$

Further investigations of this sort have dealt with demobilization,[1,3] with the arms race of 1929–39 between nine nations,[1,3] with war-weariness and its fading,[3] with submissiveness in general,[1,3] and with the submission of the defeated in particular.[2,3]

All those investigations had to do with warlike preparations, an outward or behaviouristic manifestation. The best measure of it was found to be a nation's expenditure on defence divided, in the same currency, by the annual pay of a semi-skilled engineer. This conception may be called 'war-finance per salary" a phrase which can be packed into the new word 'warfinpersal'.

Moods, Friendly or Hostile, Prior to a War

What of the inner thoughts, emotions, and intentions which accompany the growth of warfinpersal? Lloyd George, who was Chancellor of the Exchequer in 1914, describes some of them in his *War Memoirs*, revised in 1938. He relates that although there had been naval rivalry between Britain and Germany during the previous six years, yet as late as 24th July, 1914, only a very small minority of Britons wished for war with Germany. Eleven days later the British nation had changed its mind. Another brilliant description of the moods occurs in H. G. Wells's novel *Mr. Britling Sees It Through*. The contrast between the comparatively slow growth of irritation over years, and the sudden outbreak of war, can be explained by the well-established concept of the subconscious. Suppose, for simplicity, that in a person there are only two mental levels, the overt and the subconscious, and that the moods in these levels are not necessarily the same. In Britain in the year 1906 the prevailing mood towards Germany was friendly openly, and friendly also in the subconscious. The arms race during 1908 to 1913 did not prevent the King from announcing annually to Parliament that "My relations with foreign powers continue to be friendly"; and the majority of British citizens

[1]Richardson, L. F., 1939, *Generalized Foreign Politics*. Monog. Supplt. No. 23 of the *British Journal of Psychology*.

[2]Richardson, L. F., 1944, letter in *Nature* of 19th August, p. 240.

[3]Richardson, L. F., 1947, *Arms and Insecurity* on 35 mm. punched safety microfilm, sold by the author.

continued to speak in friendly terms of their German acquaintances. It is reasonable to suppose, however, that during those same years there was a growing hostility to Germany in the British subconscious mind, caused by the arms race and by diplomatic crises. The hostile mood, having been thus slowly prepared in the subconscious, was ready suddenly to take open control at the beginning of August 1914. A quantitative theory of such changes of mood is offered by L. F. Richardson.[1] Here is a simplified specimen of that theory:

$$d\eta_1/dt = C_{12}\xi_1\eta_2$$

where η_1 is the fraction of the British population that was eager for war at time t, while η_2 is the corresponding fraction for Germany, C_{12} is a constant, and ξ_1 is the fraction of the British population that was in the susceptible mood: overtly friendly, but subconsciously hostile. An equation of this type is used in the theory of epidemics of disease[2] Eagerness for war can be regarded analogously as a mental disease infected into those in a susceptible mood by those who already have the disease in the opposing country. In this theory, as in Russell's *War the Offspring of Fear*, the relations between the two nations are regarded as mutual. Accordingly the same letter, ξ say, is used in relation to either, but is distinguished by suffix 1 for the British, suffix 2 for the Germany quantity. Also there is another equation obtainable from that above by interchanging the suffixes.

Conclusion

This chapter is not about wars and how to win them, but is about attempts to maintain peace by a show of armed strength. Is there any escape from the disastrous mutual stimulation by threat and counter-threat? Jonathan Griffin[3] argued that each nation should confine itself to pure defence which did not include any preparation for attack, while aggressive weapons should be controlled by a supranational authority. The difficulty is that, when once a war has started, attack is more effective than defence. Gandhi's remarkable discipline and strategy of non-violent resistance is explained and discussed by Gregg.[4] The pacifying influence of intermarriage has been considered by Richardson.[5]

[1]Richardson, L. F., 1948, *Psychometrika*, Vol. XIII, pp. 147–74 and 197–232.

[2]Kermack, W. O., and McKendrick, A. G., 1927, *Proc. Roy. Soc. Lond. A* Vol. CXV, pp. 700–22.

[3]Griffin, J., 1936, *Alternative to Rearmament*, Macmillan, London.

[4]Gregg, R. B., 1936, *The Power of Non-Violence*, Routledge, London.

[5]Richardson, L. F., 1950, *The Eugenics Review*, Vol. XLII pp. 25-36.

CHAPTER XI

STATISTICS OF DEADLY QUARRELS

by L. F. Richardson

INTRODUCTION

THERE are many books by military historians dealing in one way or another with the general theme 'wars, and how to win them'. The theme of the present chapter is different, namely, 'wars, and how to take away the occasions for them', as far as this can be done by inquiring into general causes. But is there any scope for such an inquiry? Can there be any general causes that are not well known, and yet of any importance? Almost every individual in a belligerent nation explains the current war quite simply by giving particulars of the abominable wickedness of his enemies. Any further inquiry into general causes appears to a belligerent to be futile, comic, or disloyal. Of course an utterly contradictory explanation is accepted as obviously true by the people on the other side of the war; while the neutrals may express chilly cynicism. This contradiction and variety of explanation does provide a prima-facie case for further investigation. Any such inquiry should be so conducted as to afford a hope that critical individuals belonging to all nations will ultimately come to approve of it. National alliances and enmities vary from generation to generation. One obvious method of beginning a search for general causes is therefore to collect the facts from the whole world over a century or more. Thereby national prejudices are partly eliminated.

COLLECTIONS OF FACTS FROM THE WHOLE WORLD

Professor Quincy Wright[1] has published a collection of *Wars of Modern Civilization*, extending from A.D. 1482 to A.D. 1940, and including 278 wars, together with their dates of beginning and ending, the name of any treaty of peace, the names of the participating states, the number of battles, and a classification into four types of war. This extensive summary of fact is very valuable, for it provides a corrective to those frequent arguments which are based on the few wars which the debater happens to remember, or which happen to support his theory. Wright explains his selection by the statement that his list

[1]Wright, Q., 1942, *A Study of War*, Chicago University Press, Chicago.

"is intended to include all hostilities involving members of the family of nations, whether international, civil, colonial, or imperial, which were recognized as states of war in the legal sense or which involved over 50,000 troops. Some other incidents are included in which hostilities of considerable but lesser magnitude, not recognized at the time as legal states of war, led to important legal results such as the creation or extinction of states, territorial transfers, or changes of government".

Another world-wide collection has been made by L. F. Richardson for a shorter time interval, only A.D. 1820 onwards, but differently selected and classified. No attention was paid to legality or to important legal results, such concepts being regarded as varying too much with opinion. Instead attention was directed to deaths caused by quarrelling, with the idea that these are more objective than the rights and wrongs of the quarrel. The wide class of 'deadly quarrels' includes any quarrel that caused death to humans. This class was subdivided according to the number of deaths. For simplicity the deaths on the opposing sides were added together. The size of the subdivisions had to be suited to the uncertainty of the data. The casualties in some fightings are uncertain by a factor of three. It was found in practice that a scale which proceeded by factors of ten was suitable, in the sense that it was like a sieve which retained the reliable part of the data, but let the uncertainties pass through and away. Accordingly the first notion was to divide deadly quarrels into those which caused about 10,000,000 or 1,000,000 or 100,000 or 10,000 or 1,000 or 100 or 10 or 1 deaths. These numbers are more neatly written respectively as $10,^7$ $10,^6$ $10,^5$ $10,^4$ $10,^3$ $10,^2$ $10,^1$ 10^0 in which the index is the logarithm of the number of deaths. The subsequent discussion is abbreviated by the introduction of a technical term. *Let the 'magnitude' of any deadly quarrel be defined to be the logarithm, to the base ten, of the number of persons who died because of that quarrel.* The middles of the successive classes are then at magnitudes 7, 6, 5, 4, 3, 2, 1, 0. To make a clean cut between adjacent classes it is necessary to specify not the middles of the classes, but their edges. Let these edges be at 7·5, 6·5, 5·5, 4·5, 3·5, 2·5 . . . on the scale of magnitude. For example magnitude 3·5 lies between 3,162 and 3,163 deaths, magnitude 4·5 lies between 31,622 and 31,623 deaths, magnitude 5·5 lies between 316,227 and 316,228 deaths, and so on. Richardson's collection has not yet been published in extenso, but various extracts from it have appeared in print, and it will be available in microfilm. (1950, from the author).

These two world-wide collections provide the raw material for many investigations. Three, which have already been published by learned societies, are summarized below. Others relating to language, religion, and common government will be offered in microfilm.

The Distribution of Wars in Time

This aspect of the collections is taken first, not because it is of the most immediate political interest, but almost for the opposite reason, namely, that it is restfully detached from current controversies.

Before beginning to build, I wish to clear three sorts of rubbish away from the site.

1. There is a saying that "If you take the date of the end of the Boer War and add to it the sum of the digits in the date, you obtain the date of the beginning of the next war, thus 1902 + 1 + 9 + 0 + 2 = 1914". Also 1919 + 1 + 9 + 1 + 9 = 1939. These are merely accidental coincidences. If the Christian calendar were reckoned from the birth of Christ in 4 B.C. then the first sum would be 1906 + 1 + 9 + 0 + 6 = 1922, not 1914 + 4.

2. There is a saying that "Every generation must have its war". This is an expression of a belief, perhaps well founded, in latent pugnacity. As a statistical idea, however, the duration of a generation is too vague to be serviceable.

3. There is an assertion of a fifty-year period in wars which is attributed by Wright (1942, p. 230) to Mewes in 1896. Wright mentions an explanation by Spengler of this supposed period, thus: "The warrior does not wish to fight again himself and prejudices his son against war, but the grandsons are taught to think of war as romantic". This is certainly an interesting suggestion, but it contradicts the other suggestion that "Every generation must have its war". Moreover the genuineness of the fifty-year period is challenged. Since 1896, when Mewes published, the statisticians have developed strict tests for periodicity (*see* for example Kendall's *Advanced Theory of Statistics*, Part II, 1946). These tests have discredited various periods that were formerly believed. In particular the alleged fifty-year period in wars is mentioned by Kendall[1] as an example of a lack of caution.

Having thus cleared the site, let us return to Wright's collection as to a quarry of building material.

The Distribution of Years in Their Relation to War and Peace

A list was made of the calendar years. Against each year was set a mark for every war that began in that year. Thus any year was

[1]Kendall, M. G., 1945, *J. Roy. Statistical Soc.*, **108**, 122.

characterized by the number, x, of wars that began in it. The number, y, of years having the character x was then counted. The results were as follows.[1]

YEARS FROM A.D. 1500, TO A.D. 1931 INCLUSIVE. WRIGHT'S COLLECTION.

Number, x, of outbreaks in a year . .	0	1	2	3	4	>4	Totals
Number, y, of such years	223	142	48	15	4	0	432
Y, as defined below .	216·2	149·7	51·8	12·0	2·1	0·3	432·1

It is seen that there is some regularity about the progression of the numbers y. Moreover they agree roughly with the numbers Y. These are of interest because they are calculated from a well-known formula, called by the name of its discoverer the 'Poisson Distribution' and specified thus

$$Y = \frac{N\lambda^x}{(2{\cdot}7183)^\lambda \, x!}$$

in which N is the whole number of years, λ is the mean number of outbreaks per year, and $x!$ is called 'factorial x' and is equal respectively to 1, 1, 2, 6, 24, when x equals 0, 1, 2, 3, 4. Similar results were obtained from Richardson's collection both for the beginnings and for the ends of fatal quarrels in the range of magnitude extending from 3·5 to 4·5, thus:

Years A.D. 1820 to 1929 inclusive

x outbreaks in a year .	0	1	2	3	4	>4	Total
y for war . . .	65	35	6	4	0	0	110
Poisson . . .	64·3	34·5	9·3	1·7	0·2	0·0	110·0
y for peace . .	63	35	11	1	0	0	110
Poisson . . .	63·8	34·8	9·5	1·7	0·2	0·0	110·0

The numbers in the rows beginning with the word 'Poisson' were calculated from the formula already given, in which N and λ have the same *verbal* definitions as before, and therefore have appropriately altered *numerical* values. Such adjustable constants are called parameters.

If every fatal quarrel had the same duration, then the Poisson distribution for their beginnings would entail a Poisson distribution for their ends; but in fact there is no such rigid connection. The durations are scattered: Spanish America took fourteen years to break free from Spain, but the siege of Bharatpur was over in two months. Therefore the Poisson distributions for war and for peace may reasonably be regarded as separate facts.

Observed numbers hardly ever agree perfectly with the formulae

[1]Richardson, L. F., 1945, *J. Roy. Statistical Soc.*, **107**, 242.

that are accepted as representing them. In the paper cited[1] the disagreement with Wright's collection is examined by the χ^2 test and is shown to be unimportant. It should be noted, however, that the application of this standard χ^2 test involves the tacit assumption that there is such a thing as chance in history.

There is much available information about the Poisson distribution; about the theories from which it can be derived; and about the phenomena which are approximately described by it.[2] The latter include the distribution of equal time intervals classified according to the number of alpha particles emitted during each by a film of radioactive substance.

In order to bring the idea home, an experiment in cookery may be suggested. Take enough flour to make N buns. Add λN currants, where λ is a small number such as 3. Add also the other usual ingredients, and mix all thoroughly. Divide the mass into N equal portions, and bake them. When each bun is eaten, count carefully and record the number of currants which it contains. When the record is complete, count the number y of buns, each of which contains exactly x currants. Theory would suggest that y will be found to be nearly equal to Y, as given by the Poisson formula. I do not know whether the experiment has been tried.

A more abstract, but much more useful, summary of the relations, is to say that the Poisson distribution of years, follows logically from the hypothesis that there is the same very small probability of an outbreak of war, or of peace, somewhere on the globe on every day. In fact there is a seasonal variation, outbreaks of war having been commoner in summer than in winter, as Q. Wright shows. But when years are counted as wholes, this seasonal effect is averaged out; and then λ is such that the probability of a war beginning, or ending, during any short time dt years is λdt.

This explanation of the occurrence of wars is certainly far removed from such explanations as ordinarily appear in newspapers, including the protracted and critical negotiations, the inordinate ambition and the hideous perfidy of the opposing statesmen, and the suspect movements of their armed personnel. The two types of explanation are, however, not necessarily contradictory; they can be reconciled by

[1] *J. Roy. Statistical Soc.*, **107**, 242.

[2] Jeffreys, H., 1939, *Theory of Probability*, Oxford University Press. Kendall, M. G., 1943, *The Advanced Theory of Statistics*, Griffin, London. Shilling, W., 1947, *J. Amer. Statistical Assn.*, **42**, 407–24. Cramér, H., 1946, *Mathematical Methods of Statistics*, Princeton University Press.

saying that each can separately be true as far as it goes, but cannot be the whole truth. A similar diversity of explanation occurs in regard to marriage: on the one hand we have the impersonal and moderately constant marriage rate; on the other hand we have the intense and fluctuating personal emotions of a love-story; yet both types of description can be true.

Those who wish to abolish war need not be discouraged by the persistent recurrence which is described by the Poisson formula. The regularities observed in social phenomena are seldom like the unalterable laws of physical science. The statistics, if we had them, of the sale of snuff or of slaves, would presumably show a persistence during the eighteenth century; yet both habits have now ceased. The existence of a descriptive formula does not necessarily indicate an absence of human control, especially not when the agreement between formula and fact is imperfect. Nevertheless, the Poisson distribution does suggest that the abolition of war is not likely to be easy, and that the League of Nations and its successor the United Nations have taken on a difficult task. In some other fields of human endeavour there have been long lags between aspiration and achievement. For example Leonardo da Vinci drew in detail a flying machine of graceful appearance. But four centuries of mechanical research intervened before flight was achieved. Much of the research that afterwards was applied to aeroplanes was not at first made specifically for that object. So it may be with social science and the abolition of war.

The Poisson distribution is not predictive; it does not answer such questions as 'when will the present war end?' or 'when will the next war begin?' On the contrary the Poisson distribution draws attention to a persistent probability of change from peace to war, or from war to peace. Discontent with present weather has been cynically exaggerated in a comic rhyme:

> As a rule a man's a fool:
> When it's hot he wants it cool,
> When it's cool he wants it hot,
> Always wanting what is not.

A suggestion made by the Poisson law is that discontent with present circumstances underlies even the high purposes of peace and war. There is plenty of psychological evidence in support. This is not the place to attempt a general review of it; but two illustrations may serve as pointers. In 1877 Britain had not been engaged in any considerable war since the end of the conflict with China in 1860.

During the weeks of national excitement in 1877 preluding the dispatch of the British Mediterranean squadron to Gallipoli, in order to frustrate Russian designs on Constantinople, a bellicose music-hall song with the refrain:

'We don't want to fight, but, by Jingo, if we do:
We've got the men, we've got the ships, we've got the money too.'

was produced in London and instantly became very popular.[1]

Contrast this with the behaviour of the governments of Britain, China, USA, and USSR in 1944, after years of severe war, but with victory in sight, who then at Dumbarton Oaks officially described themselves as 'peace-loving'.[2]

Chance in history. The existence of a more or less constant λ, a probability per time of change, plainly directs our attention to chance in history. Thus the question which statisticians are accustomed to ask about any sample of people or things, namely "whether the sample is large enough to justify the conclusions which have been drawn from it" must also be asked about any set of wars.

Have wars become more frequent? In particular the discussion of any alleged trend towards more or fewer wars is a problem in sampling. No definite conclusion about trend can be drawn from the occurrence of two world wars in the present century, because the sample is too small. When, however, the sample was enlarged by the inclusion of all the wars in Wright's collection, and the time was divided into two equal intervals, the following result was obtained.

Dates of beginning	A.D. 1500 to 1715	A.D. 1716 to 1931
Numbers of wars	143	156

The increase from 143 to 156 can be explained away as a chance effect.

This was not so for all subdivisions of the time. When the interval from A.D. 1500 to A.D. 1931 was divided into eight consecutive parts of fifty-four years each, it was found that the fluctuation, from part to part, of the number of outbreaks in Wright's collection was too large to be explained away as chance. The extremes were fifty-four outbreaks from A.D. 1824 to 1877, and sixteen outbreaks from A.D. 1716 to 1769. Other irregular fluctuations of λ were found, although less definitely, for parts of twenty-seven and nine years.[3] All these results may, of course, depend on Wright's selection rules. The problem has been further studied by Moyal.[4]

[1] *Ency. Brit.*, XIV, ed. **13**, 69.
[2] H.M. Stationery Office, London, Cmd. 6666.
[3] *J. Roy. Statistical Soc.*, **107**, 246–7.
[4] Moyal, J. E., 1950. *J. Roy. Statistical Soc.*, 112, 446–9.

The Larger, The Fewer

When the deadly quarrels in Richardson's collection were counted in unit ranges of magnitude, the following distribution was found.[1] The numbers are those of deadly quarrels which ended from A.D. 1820 to 1929 inclusive.

Ends of range of magnitude	$7 \pm \frac{1}{2}$	$6 \pm \frac{1}{2}$	$5 \pm \frac{1}{2}$	$4 \pm \frac{1}{2}$
Quarrel-dead at centre of range	10,000,000	1,000,000	100,000	10,000
Number of deadly quarrels	1	3	16	62

Although Wright's list is not classified by magnitudes, yet some support for the observation that the smaller incidents were the more numerous is provided by his remark (p. 636) that "A list of all revolutions, insurrections, interventions, punitive expeditions, pacifications, and explorations involving the use of armed force would probably be more than ten times as long as the present list". Deadly quarrels that cause few deaths are not in popular language called wars. The usage of the word 'war' is variable and indefinite; but perhaps on the average the customary boundary may be at about 3,000 deaths. From the scientific point of view it would be desirable to extend the above tabular statement to the ranges of magnitude ending at $3 \pm \frac{1}{2}$, $2 \pm \frac{1}{2}$, $1 \pm \frac{1}{2}$, by collecting the corresponding numbers of deadly quarrels from the whole world. There is plenty of evidence that such quarrels, involving about 1,000, or 100, or 10, deaths, have existed in large numbers. They are frequently reported in the radio news. Wright alludes to them in the quotation above. Many are briefly mentioned in history books. But it seems not to have been anyone's professional duty to record them systematically. For the range of magnitude between 3·5 and 2·5 I have made a card index for the years A.D. 1820 to 1929 which recently contained 174 incidents, but was still growing. This number 174, though an underestimate, notably exceeds 62 fatal quarrels in the next unit range of larger magnitude, and is thus in accordance with 'the larger the fewer'.

Between magnitudes 2·5 and 0·5 the world totals are unknown. Beyond this gap in the data are those fatal quarrels which caused 3, 2, or 1 deaths, which are mostly called murders, and which are recorded in criminal statistics. For the murders it is possible to make a rough

[1] Letter in *Nature* of 15th November, 1941.

estimate of the world total in the following manner. Different countries are first compared by expressing the murders per million of population during a year. This 'murder rate' has varied from 610 for Chile[1] in A.D. 1932, to 0·3 for Denmark[2] A.D. 1911–20. The larger countries had middling rates. From various sources, including a governmental report[3] it was estimated the the murder rate for the whole world was of the order of 32 in the interval A.D. 1820 to 1929. As the world population[4] averaged about 1,358 million for the same interval, it follows that the whole number of murders in the world was about

$$110 \times 32 \times 1358 = 5 \text{ million}$$

This far exceeds the number of small wars in the whole world during the same 110 years. Thus 'the larger, the fewer' is a true description of all the known facts about world totals of fatal quarrels.

In the gap where world totals are lacking there are local samples: one of banditry in Manchukuo,[5] and one of ganging in Chicago.[6] Before these can be compared with the world totals it is essential that they should be regrouped according to equal ranges of quarrel-dead or of magnitude; for the maxim 'the larger the fewer' relates to statistics arranged in either of those manners. When thus transformed the statistics of banditry and of ganging fit quite well with the gradation of the world totals, on certain assumptions. A thorough statistical discussion will be found elsewhere.[7]

The suggestion is that deadly quarrels of all magnitudes, from the world wars to the murders, are suitably considered together as forming one wide class, gradated as to magnitude and as to frequency of occurrence. This is a statistical chapter; and for that reason the other very important gradations, legal, social, and ethical, between a world war and a murder are not discussed here. The present conspectus of

[1] *Keesing's Contemporary Archives*, p. 1052, Bristol. Corrected by a factor of ten.

[2] Calvert, E. R., 1930, *Capital Punishment in the Twentieth Century*, Putnam's, London.

[3] *Select Committee on Capital Punishment*, 1931, H.M.S.O., London, for reference to which I am indebted to Mr. John Paton.

[4] Carr-Saunders, A. M., 1936, *World Population*, Clarendon Press, Oxford.

[5] *Japan and Manchukuo Year Book*, 1938, Tokio.

[6] Thrasher, F. M., 1927, *The Gang*, Chicago University Press.

[7] Richardson, L. F., 1948, *Journ. Amer. Statistical Assn.*, Vol. XLIII, pp. 523–46.

all deadly quarrels should be compared with psycho-analytic findings about personal and national aggressiveness which are explained in Chapter VI by Professor J. C. Flugel.

Which Nations Were Most Involved?

This section resembles quinine: it has a bitter taste, but medicinal virtues. The participation of some well-known states in the 278 'wars of modern civilization' as listed by Wright is summarized and discussed by him.[1]

Over the whole time interval from A.D. 1480 to 1941 the numbers of wars in which the several nations participated were as follows: England (Great Britain) 78, France 71, Spain 64, Russia (USSR) 61, Empire (Austria) 52, Turkey 43, Poland 30, Sweden 26, Savoy (Italy) 25, Prussia (Germany) 23, Netherlands 23, Denmark 20, United States 13, China 11, Japan 9.

It may be felt that the year 1480 has not much relevance to present-day affairs. So here are the corresponding numbers for the interval A.D. 1850 to 1941, almost within living memory: Great Britain 20, France 18, Savoy (Italy) 12, Russia (USSR) 11, China 10, Spain 10, Turkey 10, Japan 9, Prussia (Germany) 8, USA 7, Austria 6, Poland 5, Netherlands 2, Denmark 2, Sweden 0.

It would be difficult to reconcile these numbers of wars in which the various nations have participated, with the claim made in 1945 by the Charter of the United Nations[2] to the effect that Britain, France, Russia, China, Turkey, and USA, were 'peace-loving' in contrast with Italy, Japan, and Germany. Some special interpretation of peace-lovingness would be necessary: such as either 'peace-lovingness' at a particular date; or else that 'peace-loving' states participated in many wars in order to preserve world peace.

It would be yet more difficult to reconcile the participations found by Wright with the concentration of Lord Vansittart's invective against Germans, as though he thought that Germans were the chief, and the most persistent, cause of war.[3]

In fact no one nation participated in a majority of the wars in Wright's list. For the greatest participation was that of England (Great Britain) namely in seventy-eight wars; leaving 200 wars in

[1] Wright, Q., 1942, *A Study of War*, Chicago University Press, pp. 220–3 and 650.

[2] H.M. Stationery Office, London, Cmd. 6666, Articles 3 and 4 together with the list of states represented at the San Francisco Conference.

[3] Vansittart, Sir Robert (now Lord), 1941, *Black Record*, Hamish Hamilton, London.

which England did not participate. The distinction between aggression and defence is usually controversial. Nevertheless, it is plain that a nation cannot have been an aggressor in a war in which it did not participate. The conclusion is, therefore, that no one nation was the aggressor in more than 28 per cent of the wars in Wright's list. Aggression was widespread. This result for wars both civil and external agrees broadly with Sorokin's findings after his wide investigation of internal disturbance. He attended to Ancient Greece, Ancient Rome, and to the long interval A.D. 525 to 1925 in Europe. Having compared different nations in regard to internal violence, Sorokin concluded that 'these results are enough to dissipate the legend of "orderly" and "disorderly" peoples'. . . . 'All nations are orderly and disorderly according to the times'.[1] The diversity of the conscious attitudes of individuals is discussed in Chapter III, by Dr. H. J. Eysenck.

There does not appear to be much hope of forming a group of permanently peace-loving nations to keep the permanently aggressive nations in subjection; for the reason that peace-lovingness and aggressiveness are not permanent qualities of nations. Instead the facts support Ranyard West's[2] conception of an international order in which a majority of momentarily peace-loving nations, changing kaleidoscopically in its membership, may hope to restrain a changing minority of momentarily aggressive nations.

THE NUMBER OF GROUPS ON EACH SIDE OF A WAR[3]

Wars can be classified according to the number of organized groups of people on the two sides: for example, 1 government *versus* 1 set of insurgents, or 2 states *versus* 1 state, or 5 nations *versus* 3 nations, or in general r belligerent groups *versus* s belligerent groups. Then the number of wars of the type 'r *versus* s' can be counted, for each r and s, and the results can be written in a table of rows and columns. As there was no good reason for distinguishing 2 *versus* 1 from 1 *versus* 2 the observed number of wars of this type was bisected, and half of it was written in each of the two possible places; and so in general whenever r was not equal to s. This analysis was applied to both Wright's and Richardson's collections with the following results.

[1]Sorokin, Pitrim A., 1937, *Social and Cultural Dynamics*, American Book Co.
[2]West, R., 1942, *Conscience and Society*, Methuen, London.
[3]Being an abstract of L. F. Richardson's paper in the *Journal of the Royal Statistical Society*, **109**, 130–56.

Quarrels of Magnitudes greater than 3·5 which ended from A.D. 1820 to 1939 inclusive. Richardson's Collection

s						
6	0·5	0	0	0	0	0
5	1	0·5	0	0	0	0
4	2·5	0·5	0·5	0	0	0
3	2·5	1	0	0·5	0	0
2	12	3	1	0·5	0·5	0
1	42	12	2·5	2·5	1	0·5
r	1	2	3	4	5	6

There were also beyond the bounds of the above table: 2 wars of 7 *versus* 1; 1 war of 9 *versus* 1; 1 war of 15 *versus* 5; thus making a total of 91 wars.

Wars not Marked Civil in Wright's List from A.D. 1480 to 1941 inclusive

s						
6	1	0·5	0	0	0	0
5	1·5	0·5	0·5	0	0	0
4	6	0	0·5	0	0	0
3	6	3	1	0·5	0·5	0
2	14	4	3	0	0·5	0·5
1	117	14	6	6	1·5	1
r	1	2	3	4	5	6

There were also in Wright's list beyond the bounds of the above table one war of each of the following types: 7 *versus* 1, 8 *versus* 1, 11 *versus* 1, 16 *versus* 1, 20 *versus* 1, 7 *versus* 3, 8 *versus* 5, 20 *versus* 5, 33 *versus* 5, 35 *versus* 7, 9 *versus* 8, thus making a total of 200 wars.

The above two tables, though based on different definitions of war, have a strong resemblance. In both the commonest type is 1 *versus* 1, and the next commonest 2 *versus* 1. Both distributions are tolerably well fitted by the formula

$$(\text{number of wars of type } r \textit{ versus } s) = 5\,\frac{(\text{whole number of wars})}{9\,(rs)^{2\cdot 5}}$$

Professor M. S. Bartlett has pointed out to me that, according to this law, any cell-frequency is equal to the product of the marginal totals for its row and its column, divided by the total for the whole table; so that the variable r would be said to be 'statistically independent' of the variable *s*. Apart from this bit of insight, the formula is empirical: it describes the facts, but does not explain them. Although tables in rows and columns, including especially correlation tables, are a common feature of works on statistics, yet a frequency-distribution of this particular shape was certainly not well known. No ready-made theory, which might have illuminated the causes of wars, was available. So rival theories were made on purpose, under the guidance of the following leading ideas:

(i) International relations have not been so deterministic as to justify any theory which would offer to predict exactly what must happen at any date.

(ii) A theory, however, should indicate what probably would happen. It ought to agree with the historical facts collected from any sufficiently long interval, say, from 100 years or more.

(iii) That because a nation cannot be at war all by itself, therefore the probabilities of war must be attached to pairs of nations, and not to nations singly.

(iv) In the course of a century the same nation may have been peace-loving on several occasions, and aggressive on several others. These characteristics are not sufficiently permanent to form the basis of a long-term theory. Klingberg[1] has published a summary of the opinions of 220 outstanding students of international affairs about the chance of war between pairs of States. Considerable fluctuations of opinion occurred in a few years. It would be an instructive adventure to begin at the opposite extreme by first regarding all nations as of similar pugnacity, and later introducing only such discriminations between nations as are called for by the statistics of *r versus s*.

(v) That any type of war, such, for example, as 2 *versus* 1, comprises many mutually exclusive varieties of conceivable war, which could be specified by naming the two belligerents and the one. The number of such mutually exclusive varieties can be formulated; and so in general for *r versus s*.

(vi) A remarkable feature is that 1 *versus* 1 has been much the commonest type; and that in general the more complicated types have been rarer than the simpler types. This is a characteristic of lack of organization, of chaos. In the molecular chaos of a gas, collisions of molecules two in a bunch are much more frequent than collisions three in a bunch. Mathematically the characteristic of chaos is that the probability of a complicated event contains among its factors the probabilities of simpler events.

[1]Klingberg, F. L. 1941. *Psychometrika* **6**, 335-352.

(vii) These ideas combine to give an expression of the following form for the probability of a war of the type *r versus s*

$$(\text{number of mutually exclusive varieties}) \times p^x \times (1 - p)^y$$

where p is the probability of war between any pair of nations, x is the number of pairs that are in the war, and $(1 - p)^y$ is the probability that the neutrals would keep out. Any such theory may be called, in musical terminology, a variation on the theme by Bernoulli concerning the binomial distribution of probability.

These general ideas were developed in connection with successive special hypotheses until a combination was found which agreed with the facts. For some theories the appropriate facts were numbers of wars, for others durations of wars. It was quite interesting to notice the manner in which some hypotheses failed. The details can be seen in the original paper, to which the Roman numerals refer. In one theory (VIII) which is called 'a simple chaos between sixty nations' it is supposed for simplicity that p, the probability of war inside a pair of nations, is the same for all pairs and at all times. But it was impossible to find any number p that would agree with all parts of the distribution of durations of wars classified as *r versus s*. Most of the misfits could have been avoided if the number of nations in the world had been about six or ten instead of about sixty. The next three theories (IX, X, XI) are devices for reducing the number of nations in effective contact with one another. Thus in theory IX an approximate agreement would be achieved if there were in the world only about ten bellicose nations, all the others being permanently non-belligerent. It is, however, impossible to sort the nations into these two supposed categories; for the total number of names of belligerents, including names of insurgents, but not counting any name twice, was found to be, not ten, but 108, for the ninety-one wars that ended from A.D. 1820 to 1939 according to Richardson's list.

To circumvent this obstacle a different device (X) was next considered for reducing the effective number of belligerents. It was supposed that disputes had occurred in localities scattered at random over the globe so that altogether they concerned numerous possible belligerents; but that each dispute was localized so that it concerned only eight nations or other possible belligerent groups; and that the probability of war about that dispute was 0·35 for every pair that could be formed from the eight groups. These remarkably simple hypotheses led to a good agreement with the facts in Richardson's collection, provided $r + s \leqq 8$. But the theory denied the possibility of any war involving more than eight belligerents, and so in particular that of the First World War.

The failure of these three theories (VIII, IX, X) showed the need for a more inclusive hypothesis designed to explain both the localization of most wars, and the occasional occurrence of long-range or world wars, in the era before aviation became dominant. Accordingly in theory XI the world was supposed for simplicity to consist of only three sorts of nations, namely eleven land-locked, forty-four local-coastal, and five powers capable of reaching any coast by sea. This hypothesis greatly complicated the mathematics; for different sorts of pairs of nations were contemplated, each with its appropriate, but at first unknown, probability of war. The probability of any event is here defined to be the fraction of time during which the event occurred in the course of any very long historical interval. For comparison, the historical data in Richardson's collection had to be rearranged as durations of wars. Moreover the former type 1 *versus* 1 had now to be divided into three subtypes namely:

A long-range power *versus* a long-range power;

A long-range power *versus* a short-range power;

A short-range power *versus* a short-range power;

and so on for the more complicated types. The classification became four-dimensional; the type r *versus* s being analysed into subtypes such as r_1 and r_2 *versus* s_1 and s_2.

The probabilities were deduced from the historical data. For type 1 *versus* 1 the deduction was definite and unique. For type 2 *versus* 1 the deduction was a compromise fitted to redundant data. For type 3 *versus* 1 various difficulties increased. The comparison of theory with fact was carried as far as type 2 *versus* 2. Beyond that the historical facts would be described statistically as 'outliers' in the sense that the classification of them contained many empty compartments with rare observations irregularly dispersed. Similar ragged appearances are usually to be seen in the outer regions of any diagram of observed frequencies. In particular the First World War, which was regarded as 15 *versus* 5, was isolated in the four-dimensional classification, being surrounded on all sides by many empty cells; one cannot easily say how many. A comparison of theory XI with the fact of the First World War would involve a difficult summation of probabilities over the outer regions of the classification, both empty and occupied. Theory XI certainly admits the possibility of such a war; but a quantitative study of outliers was not attempted.

The probability, x, of war between two long-range powers was found to be only of the order of 0·001. The probability, y, of war between a long- and a short-range power was found rather discordantly to be 0·002 or 0·015 or 0·009; on the average one may say that y was of the order 0·01, about ten times as great as x. The probability, z, of war between neighbouring powers was found to vary conspicuously with circumstances, thus:

Number, rs, of pairs of opposed belligerents . . .	1	2	3	4
z	0·008	0·020	0·046	0·119?

On the average z was decidedly greater than y, and all the more so than x. That is to say *propinquity tended to war.*

Theory XII was an amendment to theory XI such that the variation of z with rs was simply accepted and explained as due to the infectiousness of fighting.

Theory XIII was called 'a uniform chaos modified only by infectiousness'; that is to say geographical barriers were ignored, and every nation was supposed (as also in theory VIII) to be in contact with every other nation. Aviation may make it so in future; but for the years before A.D. 1929 theory XIII definitely misfits the history. So one should return to theory XII which is called '*chaos restricted by geography and modified by infectiousness*'; for of all these theories it is the only one which has survived the test of quantitative comparison with historical fact. The possibility remains that someone may invent a different theory which may fit the facts as well or better.

Historians will doubtless be keenly aware of many relevant considerations which have been ignored in the foregoing batch of theories. But a theory is not necessarily to be despised for what it leaves out. This may be gathered from the history of the explanation of the moon's motion.[1] Sir Isaac Newton began the explanation by considering an idealized moon moving uniformly in a circle about the earth as centre. As a description that was crude, for Hipparchus in the second century B.C. had known better. Yet Newton's first simple explanation is so interesting that it is still regularly taught to physics students. It has also been fertile. In 1913 Bohr used it, along with brilliantly novel ideas, to explain the motion of an electron around the nucleus of an atom. Meanwhile a succession of astronomers have laboured to improve lunar theory. In the present century E. W.

[1]John Jackson in *Ency. Brit.*, XIV, ed. **15**, 780–1.

Brown put in all the relevant considerations. Brown's theory is so accurate that it is used in the computation of the *Nautical Almanac*. But, as Brown's theory involves 1,500 terms, it is not teachable to scientists in general.

Richardson, L. F. (1961)
General Systems Yearbook **6**, 140–87

1951:2

THE PROBLEM OF CONTIGUITY:

An appendix to *Statistics of Deadly Quarrels*

General Systems:
Yearbook of the Society for General Systems Research
Ann Arbor, Michigan
Vol. VI, 1961, pp. 140–187

Notes on 'The Problem of Contiguity'

1. The origins of this paper go back to 1942. See Note 7 of the 'Notes on *Statistics of Deadly Quarrels*' (p. 540).
2. On 5 September 1949 Richardson read a paper to the Geography Section of the British Association in Newcastle on 'Mapping by cells of equal population with reference to culture-contact and world statistics of civil wars'. (Br. Ass. Advmt Sci. **6** (1950), 381). At that meeting he showed maps on which the population was divided among compact cells, each containing the same number of people, and these were discussed. In 1950:2 he stated that a full account was 'in preparation', in 1950:4 that the 'enquiry into "Mapping by Compact Cells of Equal Population" is now approaching completion,' and 'is nearly ready', and in 1952:2 that 'the proper account of this long enquiry is ready in typescript, but unfortunately is not yet published'. The year of completion of this paper may therefore be taken as 1951. It is referred to as 1951:2; the figures in the final table are quoted in 1951:3 and 1952:2, replacing the earlier estimates in 1950:2.
3. At the time of his death in 1953 Richardson was preparing a second edition of *Statistics of Deadly Quarrels* and had included the text of this paper as an appendix to the manuscript, entitled 'Appendix on Contiguity: measured by mapping Populations by Cells.' It was however not included in the eventual second edition of *Statistics of Deadly Quarrels* (1960:2) but published separately in 1961 with a new title 'The Problem of Contiguity'.

THE PROBLEM OF CONTIGUITY:
An Appendix to Statistics of Deadly Quarrels

Lewis F. Richardson

1. INTRODUCTION

Various Methods for Mapping Population

A census yields the number of people within a specified boundary. Five methods for mapping these data can be distinguished as follows:

(i) A map of census-areas on each of which is written an Arabic number representing its population. I have not seen such a map in print, but I have had to construct them for many countries as a preliminary to drawing cells of a million. The numbered areas preserve both of the useful sorts of information given by the census. But such a map cannot be read at a glance.

(ii) A map of census-areas each colored or shaded according to its mean density of population. Such have been published by the governments of India and of the Netherlands East Indies. They are pleasant to look at, and quickly give a general impression of the distribution. The mean density is a simple and definite concept. The coloring usually proceeds by large steps in the mean density, which therefore is specified rather vaguely. The densest class is sometimes regrettably left unbounded above. Because they waste some useful information, these maps do not lead on, as system (i) does, to maps in cells of equal population.

(iii) Maps showing isopleths of density-at-a-point. A set of these has been prepared by Professors J. F. Unstead and E. G. R. Taylor, and is published by Messrs. George Philip of London. They have the great merit of covering the whole world. The maps have a pleasing appearance, and give at a glance an impression of where the people live. The coloring mostly proceeds by factors of two in the density. The densest colored class is left unbounded above. Large cities are represented differently. Thus, as in system (ii) some information is wasted. Another example of isopleths of density is a finely detailed map of the population of Great Britain in 1931, drawn by Mr. A. C. O'Dell, and published by the Ordnance Survey. The density-at-a-point, unlike the mean density of a census-area, is a complicated or obscure concept. Although it will be used freely in the theoretical investigations of § 6, it is best avoided in a representation of observed fact.

(iv) Maps in which a dot represents a specified number of persons. There is, for example, such a map of Africa in a book by Fritz Jaeger (1928). It gives an immediate, if not very accurate, impression. But the boundaries which contain the specified numbers of persons are not shown; so that their location is rather indefinite, information being thus wasted.

(v) Maps in compact cells containing equal numbers of persons, are the subject of the present paper. Hitherto, as far as my enquiries go, such maps have not been available. Mr. A. C. O'Dell drew my attention to a set of honeycomb patterns, which are printed in certain copies of the 1851 Census of England and Wales. These patterns, however, were not intended to show the spatial distribution of population, but only the changes of its total in time.

The number of people in a cell is a simple and definite concept, which avoids the doubts that obscure the density-at-a-point or the neighborhood-of-a-dot. These maps are not as accurate as those on system (i) because a little information is wasted in the interpolation by which compact cells of equal population are constructed. But the uncertainties are less than those of systems (ii), (iii), (iv). No coloring or shading is required. Maps of the present type can be read—I submit—more easily than those made on any of the other four systems. There is one defect for which I have not found any neat remedy: what to do about populations that are not enough to form one standard cell.

At the outset of this research I consulted Mr. (now Professor) A. Stevens, head of the geography department at Glasgow University. To him I am indebted for guidance as to what was available and what was lacking.

The Problem Here Discussed

Required to draw, in theoretical imagination, on the surface of the globe a network of cells such that the following five conditions are satisfied simultaneously.

Each cell contains the same number of people. (1/1)
At most three cells meet in a point. (1/2)
No cell completely surrounds another. (1/3)
National frontiers lie in the edges of cells. (1/4)
Each cell is about as broad as it is long, or in a word, each cell is "compact," in the usage of that adjective in common speech. (See below) (1/5)

If condition (5) about compactness were omitted, the other four conditions could easily be satisfied by dividing each country into strips bounded, say, by parallels of latitude. Such strips would however have little sociological interest. Loyalties are not usually confined to strips, for Egypt and Chile are exceptions. It is thus the interaction of the five conditions that makes the problem both difficult and interesting.

When the network has been drawn, or otherwise specified, the further requirement is to count, or to estimate theoretically, for the whole globe:

<u>s</u> the number of cells,
<u>C</u> the number of "civil" cell-edges, namely those that lie within national territories,
<u>F</u> the number of "foreign" cell-edges that separate two distinct national territories,
<u>B</u> the number of "boundary" cell-edges that separate a country from either the sea, or the polar ice. To save repetition, the word "void" will be used inclusively to denote sea, desert, ice, lake, or other uninhabited region; so that B is the number of edges to the external void. Lakes and deserts form the internal void.

Commentary on the Nomenclature, the Assumptions, and on the Accuracy Required

The theory is related to Euler's theorem on the number of faces, edges, and vertices of a closed polyhedron. It is known (J. W. Young, 1929) that this theorem is not restricted to flat faces meeting in straight edges. It is thus convenient to take over from the polyhedral theory the word "vertex" for any point on the globe where more than two cells meet, and the word "edge" for the line on the globe joining two adjacent vertices. But the word "face," having human connotations, has been replaced by "cell."

The simplest possible vertex is one where three cells meet. In the present theory more complicated vertices are excluded. This assumption might be justified by simplicity alone; but there is also abundant factual evidence in support. On political maps there are of course many points where two countries meet the water, and many other points where three countries meet one another; but more complicated meeting points can scarcely be found. The nearest approach to an exception is that Arizona, Utah, Colorado, and New Mexico meet at a common point; but they are all four under the Federal government of U.S.A. In the partition of Palestine, as proposed by the Majority of the United Nations Commission on 31 August 1947, two cross-points were planned. At each cross-point two Arab quadrants were to meet two Jewish quadrants (K 8821). Fighting prevented this extraordinary map from coming into operation. About a year later the United Nations' mediator, Count Bernadotte, proposed a simpler map on which the vertices are of the usual kind (K 9531).

The assumption that no cell completely surrounds another, is a formalization of the usual practice whereby one political country does not completely surround another. The only exceptions at present are that San Marino is surrounded by Italy and that Basutoland is governed from London but is surrounded by the Union of South Africa.

The notion of compactness is also an idealization of the usual and long-established practice whereby most countries, empires excepted, have been more or less compact. Governments presumably have preferred forms which facilitate communications and defense.

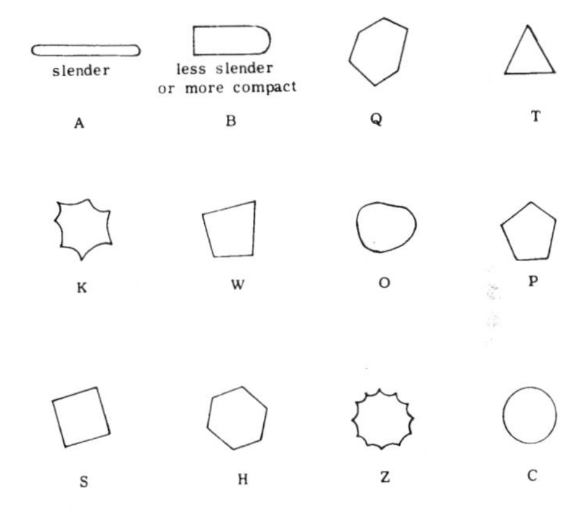

Figure 1

The concept of compactness needs further elucidation. A shape is here called compact when it is the opposite of elongated or slender. A compact shape is (at least for me personally) an immediate visual perception, as immediate as a red color, or a loud sound, or a sweet taste, or hunger. Such qualities are perceived independently of any abstract verbal definition; and are best explained to another person by the demonstration of samples. The twelve shapes reproduced in Figure 1 were drawn separately on cards and were submitted to the judgment of fifty persons, one at a time, with the following instructions.

"Here are twelve shapes, each drawn on a card. One is marked 'slender.' Another is marked 'less slender or more compact' than the first. Please arrange them all in order of slenderness, or of its opposite compactness. There are nearly forty million possible answers. The correct answer is at present unknown. Your personal opinion is requested. The question is about shape not about area." No hints were given. The observer was not hurried. Each card bore on the back a letter. The order of the letters was recorded for each opinion. After all the opinions had been gathered the cards were rearranged in the order of the average opinion, and they appear so in Figure 1 when it is read like a book.

The statistical digest in Table I relates to the eleven shapes for which there was free opinion: shape A serves only to fix the rank one, which was given. It is conceivable that an observer had no opinion, and accordingly rid himself of an impossible task by arranging the cards just anyhow. If every observer had done that, and if they had been very numerous then the mean rank would have been 7 for each shape. On the same hypothesis the variance of rank for each shape would have been

$$\frac{2}{11}\,(1^2 + 2^2 + 3^2 + 4^2 + 5^2) = 10.$$

The actual results are not like that, as may be seen from the last two columns of Table I. The observers agreed much more closely about the regular pentagon than they did about the equilateral triangle. Five of the observers remarked that the triangle was difficult to place. An over-all measure of the agreement between observers has been provided by Kendall and Babington Smith in their "coefficient of concordance," $\underline{W}$, which ranges from $\underline{W} = 1$ for perfect agreement, to $\underline{W} = 0$ for a merely random table. Elaborate studies have also been made of the probability that the computed W might have arisen by chance from the erratic statements of observers who had no genuine opinions. (Kendall, 1943, pp. 410-421.) From the present data,

Table I

PSYCHOLOGICAL EXPERIMENT ON COMPACTNESS

Shape		Rank According to Personal Opinions: 1	2	3	4	5	6	7	8	9	10	11	12	Mean Rank	Variance of Rank
		Frequency Among Fifty Opinions													
Slender 	A	50	0	0	0	0	0	0	0	0	0	0	0		
Less slender or more compact 	B	0	37	1	2	1	0	1	1	2	0	1	4	3.64	10.47
	Q	0	0	13	12	8	4	4	1	1	6	1	0	5.34	5.78
	T	0	7	16	4	4	6	1	0	2	1	2	7	5.50	12.01
	K	0	4	8	7	10	2	5	4	6	1	3	0	5.72	6.72
	W	0	0	0	6	5	9	16	4	5	4	0	1	6.88	3.35
	O	0	1	4	6	6	11	3	2	5	4	7	1	6.90	7.45
	P	0	0	1	0	5	7	8	17	8	3	1	0	7.50	2.45
	S	0	1	2	6	3	3	3	6	5	9	9	3	8.08	7.87
	H	0	0	0	1	3	3	7	7	12	11	6	0	8.52	3.17
	Z	0	0	2	3	2	3	2	7	2	7	9	13	9.18	7.59
	C	0	0	3	3	3	2	0	1	2	4	11	21	9.74	9.19
													Means	7.00	6.91

$\underline{W}$ = 0.3086. The notion that the observers had no genuine opinions is thus proved to be quite incredible, indeed off the range of the available statistical tables.

The observers were miscellaneous, including relatives, visitors, neighbors, both sexes, and ages ranging from 12 to 75 years. From the large body of information about personal differences in various other abilities, one would expect that persons would differ considerably in their ability to perceive compactness. Tentatively I have selected the more able observers in three successive stages, each stage being controlled by a majority of those left in the competition. A majority of the original 50 observers agreed in placing B at rank 2. Those 37 observers were therefore selected. A majority of them, namely 19, agreed in regarding the circle as the most compact of the given shapes; so those 19 observers were selected. A majority of them, namely 10, agreed in placing the triangle at rank 3, and were therefore selected. After this stage the competition came to an end, because there was no further majority of opinion. The eight ranks, 4, 5, . . . 11, were unaffected by the selection. The concordance of the 10 selected observers on the 8 unaffected ranks was measured by W = 0.6305, which could not credibly have arisen by chance. The average ranking by the ten selected observers was

A B T Q W K S P O H Z C,

and may be regarded as the best available opinion. It differs somewhat from the order in Figure 1. No single observer agreed entirely with either order. Mr. John Garnett helped me in the selection and suggested a fairer method.

In psychology the study of intuitive perceptions of shape is entirely respectable, and has developed into the important branch known as Gestalt. In mathematics, on the contrary, to call a process intuitive is to condemn it as unanalyzed. Several strict definitions of compactness will be offered in § 5; but the choice between those various reasonable possibilities cannot be made by logic alone. There are anyway strong reasons why an abstract definition of compactness had better be delayed, namely:

(i) Political maps are so irregular that any useful study of them must be in part inductive. A definition, if too early made precise, might cramp inductive insight; and might prevent any drawing of cells on maps. Later, when many particular examples have been prepared, the ideal of compactness will be reviewed in § 5.

(ii) Some vagueness can be tolerated because there are maximum and minimum properties such that moderate deviations from the ideal cause only slight effects. This will be shown later in § 4.

Another objection to the present popular usage of the word "compact" may be that it has a different meaning in advanced topology. In Lehmer's translation of Pontrjagin (1939, p. 42) the following definition is given: "A subset $\underline{M}$ of a topological space $\underline{R}$ is called compact if every infinite subset $\underline{N} \subset \underline{M}$ has at least one limit point in $\underline{M}$." Such a concept is not necessary in the present study; for an infinite number of human beings cannot cluster in the neighborhood of a point. Although some topological ideas are used in the present study, yet they are only of the most elementary and intuitive kind. Even so the double meaning of compact is regrettable. When this work was nearly finished, I asked the Professor of Greek in Cambridge, D. S. Robertson, F.B.A., for a word to mean "about as broad as it is long." He obligingly invented "homoplatous" to replace "compact," and "homoplaty" to replace "compactness." In the future these new words will probably be the best.[2]

Although this is not the place to discuss history, yet it is necessary to indicate briefly the range and accuracy which will be required elsewhere for historical purposes. The geographical theory has to cover the whole world, and to be applicable at any time since A.D. 1819. The cell-population is a parameter, constant for any particular map, but variable between maps in the range from $10^{2.5}$ to $10^{6.5}$. If no theoretical guidance were available it would still be possible to obtain the required information by drawing cellular maps for 100 or more territories, at three dates, and at each date for 6 different cell-populations. The total number of maps to be drawn would be at least 100 x 3 x 6 = 1800. To draw a map of an average population, say 20 millions in cells of a million people, has been found to amount to at least one day's work (often two or three). For smaller cells the map would take longer. Altogether the operation would exceed 10 person-years of industrious draughtmanship. Such a task being far beyond the powers of the present writer, he sought for theoretical short-cuts, and has effected a great economy of toil in the manner which will here be described. At each stage of the theoretical exploration as much accuracy as possible has been welcomed; but, as will be shown elsewhere, the final contrast between the historical numbers and the geographical numbers is so stark that it would be noticeable even if the latter were in error by a factor of two. It is important to bear this consideration in mind.

2. TOPOLOGICAL THEOREMS

There is a famous theorem by Euler that if a closed polyhedron has $\underline{s}$ faces, $\underline{v}$ vertices, and $\underline{e}$ edges then

2. This sentence was added to the typescript by the author sometime between 1952 and his death in September 1953. [C.C.L.]

$$s + v - e = 2 \qquad (2/1)$$

Four proofs of it are quoted by Sommerville (1929), and one by Lines (1935). J. W. Young (1929) pointed out that Euler's theorem is "topological," in the sense that it remains true when the figure is constructed of India rubber and is stretched and bent, in any manner which excludes tearing or the contact of points previously separated. This topological generality is most valuable here, because no one of the edges which we have to consider is a straight line, and some of them are seacoasts and political frontiers. Present requirements are partly simpler and partly more general than the conditions contemplated in the proofs of Euler's theorem. On the one hand only the simplest type of vertex, that where three cells meet, comes here in question. On the other hand, the concept of closure needs to be replaced by a more general procedure suitable to political territories, or to archipelagoes. So it is convenient to arrange special theorems, although they are intimately related to Euler's. The faces of the polyhedron will here be called cells.

A Theorem on Cell-Division, When There Is No Interior Void

The following theorem has topological generality, and does not involve any assumption about compactness, nor any about population.

Let us begin with one cell bounded by an outline of any sort. On this outline let there be given g vertices where the outline is met by edges coming to it from outside the cell. Let us proceed to divide the cell progressively. At the rth stage of division let there be s_r cells, v_r vertices, and e_r edges. The initial datum has a peculiarity; for it is impossible, under the rules (1/2) and (1/3), that g should be unity. The cases $g = 0$ and $g \geqslant 2$ need separate consideration, and lead to formulae differing as to e_1.

Case $g \geqslant 2$, for application to political territories which are not islands. The initial datum is

$$s_1 = 1, \quad v_1 = g, \quad e_1 = g. \qquad (2/2)$$

Now draw a line dividing the cell into two areas. By assumption (1/3) one of these areas cannot surround the other. Also by assumption (1/2) the dividing line must not meet an existing vertex. Therefore the dividing line must create two new vertices. There are however alternative ways of doing this. Either the two new vertices may both lie on an edge that previously was all one: if so they cut it into three edges. Or the two new vertices may lie on edges that were previously distinct: if so they cut each of them into two edges. Whichever alternative is shown, the number of edges, counting the dividing line as one, is increased by three. After this first division

$$s_2 = 2, \quad v_2 = g + 2, \quad e_2 = g + 3. \qquad (2/3)$$

Precisely the same argument applies at every subsequent division. Consequently s_r, v_r, e_r increase with r in arithmetical progressions having common differences 1, 2, 3, respectively. (2/4)
Therefore

$$s_r = r, \quad v_r = g + 2(r - 1), \quad e_r = g + 3(r - 1). \qquad (2/5)$$

It is now convenient to eliminate r, and afterwards to omit r as a suffix, writing simply

$$v = g + 2(s - 1), \qquad e = g + 3(s - 1). \qquad (2/6)$$

Example. Europe was being mapped in cells of a million persons at mid-1910. Austria-Hungary was the last country to be drawn. For it, $s = 51$. The surroundings gave $g = 33$. Therefore $e + 33 + 3(51 - 1) = 183$. By counting on the map it was found that Austria-Hungary had 4 edges to the sea, 58 edges to foreign countries, and 121 internal edges, total 183 edges, in agreement with the theorem.

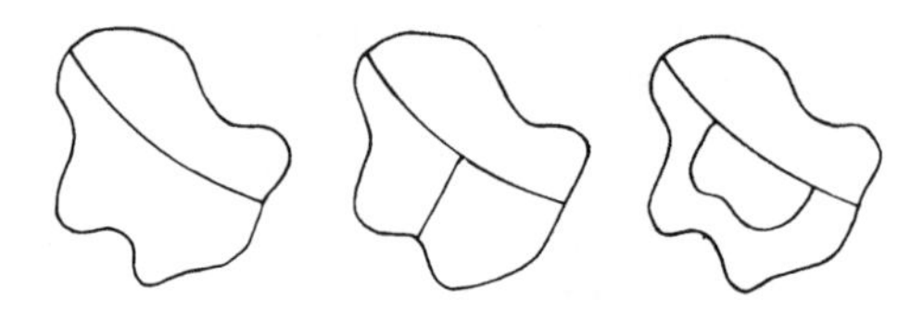

Figure 2. To illustrate the proof about cell-division, when $g = 0$.

Case $g = 0$, for application to islands and continents. The initial datum is

$$s_1 = 1, \quad v_1 = 0, \quad e_1 = 1. \qquad (2/7)$$

This is not obtainable by setting $g = 0$ in (2/2). The initial edge is peculiar in being a closed curve. The first dividing line must create two vertices on it thus cutting it into two edges. After this first division

$$s_2 = 2, \quad v_2 = 2, \quad e_2 = 3. \qquad (2/8)$$

In the subsequent stages every edge joins two vertices, so that the conditions are the same as those under which the arithmetic progressions (2/4) were deduced. But when $g = 0$ the progression for e does not extend back to $r = 1$, but begins at $r = 2$. Accordingly

$$s_r = r, \quad v_r = 2 + 2(r - 2), \quad e_r = 3 + 3(r - 2). \qquad (2/9)$$

On eliminating r and dropping the suffix we have simply

$$v = 2(s - 1), \quad e = 3(s - 1), \text{ provided } s \geqslant 2. \quad (2/10)$$

Although the arguments in the two cases start differently, their results somehow come later into step; for (2/10) can be obtained by setting g = 0 in (2/6). The only exception to the generality of (2/6) is that $e_1 = 1$ when $g = 0$.

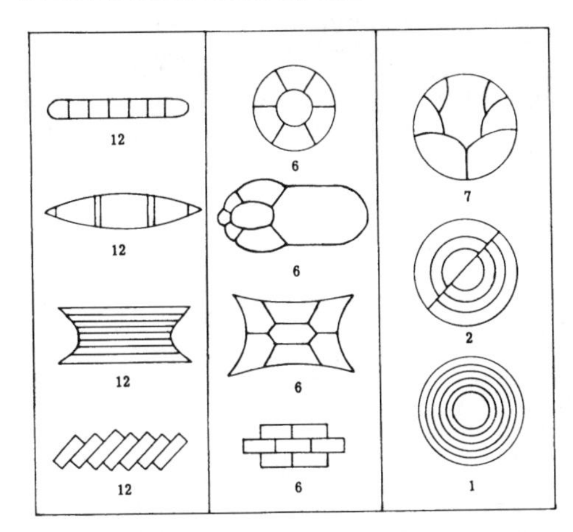

Figure 3. To illustrate the topological theorem (2/13) that C + F + B = 3(s - 1) when g = 0, here are various clusters each of seven cells, and surrounded by the void. Accordingly s = 7, F = 0 and the theorem specializes to C + B = 18. Beneath each cluster is set its number, B, of edges to the void. The bottom right-hand cluster, for which B = 1, contravenes the condition that a cell must not enclose another; but for the remaining ten clusters, C + B = 18. Although the four clusters in the left-hand column look so different, they are topologically alike; that is they can be transformed into one another by continuous deformation. Similarly the four clusters in the middle column are topologically alike.

The number of vertices on a cell is of interest. A corollary tells us something about it. Let there be u vertices on the seacoast. Each of them is on two cells. The other v - u vertices are each on three cells. The average number of vertices on a cell is therefore

$$\frac{2u + 3(v - u)}{s} = \frac{3v - u}{s} = 6\left(1 - \frac{1}{s}\right)\left(1 - \frac{u}{3v}\right), \quad (2/11)$$

provided s ⩾ 2, and g = 0. If the cells were parallel strips, all the vertices could lie on the coast, so that u = v. But, if we bring in the additional assumption that each cell is to be compact, then as $s \to \infty$, $\frac{u}{v} \to 0$; and the average number of vertices on a cell tends to 6. That is why hexagonal cells are a standard idea.

In the sequel three kinds of edges have to be distinguished; civil, foreign, and boundary. Let the numbers of them be C, F, B respectively. The topological theorem does not help in this distinction, which involves the notion of compactness. All that can be done at this stage is to substitute

$$e = C + F + B \quad (2/12)$$

in (2/6) so that

$$C + F + B = 3(s - 1) + g, \text{ provided } s \geqslant 2.$$

The Behavior of Different Varieties of Edges in Totals

Suppose that a number of countries together form a cluster bounded by the void and that the nth country is divided into s_n cells having C_n internal edges, F_n edges to other countries and B_n edges to the void. Let s, C, F, B, without suffixes, refer to the cluster of cells formed by all the countries jointly. Then

$$s = \sum_n s_n, \quad C = \sum_n C_n, \quad B = \sum_n B_n \text{ but } F = \frac{1}{2}\sum_n F_n. \quad (2/14)$$

Effect of political union. Suppose next, that the cells remain as specified in the last paragraph, but that all those countries unite politically. The $\frac{1}{2}\sum_n F_n$ edges then become civil, and there are no longer edges between nations. Thus

$$C + F \text{ is unaltered.} \quad (2/15)$$

Lakes and Deserts, Relevant or Ignorable?

Large internal waters such as the Caspian, Victoria Nyanza, or Lake Superior, obviously claim attention. There are also small lakes and ponds, so numerous that their number cannot be ascertained. The author began with a vague opinion that large lakes were important barriers to culture-contact, but that small ponds, even if very numerous, did not noticeably hinder the percolation of pedestrians. In a search for some definite rule that should distinguish negligible ponds from important lakes, it became apparent that the size of the water surface is not the sole criterion, but that the problem involves the topology of the cells of equal population, which touch the water. This can be explained with the aid of the set of diagrams comprised in Figure 4. In each diagram the outer circle represents a sea shore, and the complete inner circle represents a lake shore. The questions are topological, so that the circles could be deformed into any non-intersecting closed curves without affecting the argument.

In Diagram I the line $\alpha\beta$ has no meaning, for it does not separate two different cells. Let such lines be omitted.

In Diagram II the cells A and B are in contact by land. The lake does not put these cells in communication with any third cell. It is therefore permissible to ignore the lake, and to count both ways past the lake from α to β as together all one edge. This policy makes a great simplification.

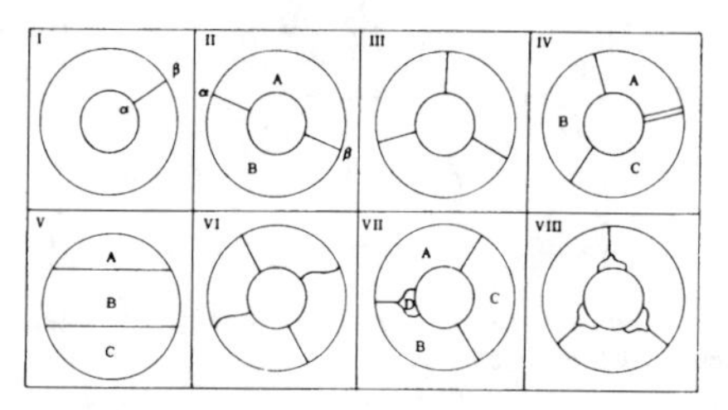

Figure 4. Edges to an internal void.

In Diagram III the lake permits any cell to communicate by water only with cells which it already touches by land. It seems proper therefore not to count as separate edges the three parts of the lake shore.

If only three cells touch a lake and if they have land-contact with one another where they meet the lake, then no amount of complication in the surrounding network of cells and other lakes can abolish that land contact. Similar remarks apply to deserts. So a fairly general principle is:

<u>When counting edges, ignore any lake or desert touched by three or fewer cells.</u> (2/16)

An exception of this rule is shown in Diagram IV where the so-called lake has an outlet to the sea so broad as to form a barrier between the cells A and C. It would be suitable in this case to regard Diagram IV as topologically the same as Diagram V, the cells being lettered to correspond.

In Diagram VI the lake permits each cell to communicate by water with the opposite cell, which it does not touch by land. We should therefore count the four portions of the lake shore as four edges to the internal void.

In Diagram VII, four cells again meet the lake. Two of them, C and D, are not in contact by land; but the other two, A and B, have land-contact. It may seem proper in this network to count only the lake shores of cells C and D as edges to the internal void. To do so would however misfit or complicate a most useful subsequent theorem, (2/20).

It appears to be impossible to draw four cells each touching the lake, and each touching the other three by land.

In Diagram VIII the lake shore of each cell provides it with access by water to some cell which it does not touch by land. Accordingly there are six edges to the internal void.

For simplicity and definiteness exceptions like Diagram VII will be ignored in the rest of this work; and the following principle will be adopted:

<u>If a lake or desert is touched by four or more cells, all the edges to its shore are to be counted.</u> (2/17)

Generalization of the Topological Theorem (2/13) for Those Lakes and Deserts Which Cannot Be Ignored

The general principles are well illustrated by the map of Africa in Figure 7. It includes $117 = \underline{s}$ cells of a million people, $166 = \underline{C}$ civil edges, $90 = \underline{F}$ foreign edges, $42 = \underline{B}$ edges to the sea, $42 = \underline{D}$ edges to deserts, and $22 = \underline{L}$ edges to lakes. The isthmus of Suez is neglected so that $g = 0$. The two members of (2/13) are therefore $\underline{C} + \underline{F} + \underline{B} = 298$ and $3(s - 1) + g = 348$, in disagreement by far more than any ambiguity. The reason why they are so unequal is that theorem (2/13) is applicable only where deserts and lakes can be ignored. We can however relate the real Africa, by easily countable modifications, to an imaginary Africa for which theorem (2/13) is valid. The author hopes that the following vast and impracticable changes can be imagined, for numerical purposes only, without disturbing anyone's political susceptibilities. Let the deserts be moistened and the lakes dried up, so that their sites become habitable. To the site of each desert or lake let $\underline{h}$ immigrants be admitted, so that it becomes one cell. Here $\underline{h}$ is a million.

Let the number of deserts and lakes, which cannot be ignored, be denoted jointly by σ. (2/18) Lake Chad touched only three cells; and so by (2/16) the shores of that lake were not counted in $\underline{L}$; and the lake itself must not be counted in σ. Four lakes are to be counted: Victoria, Tanganyika, Nyasa and Rudolf. The countable deserts include that between Egypt and the Red Sea, the Sahara, but probably not the Kahahari. Therefore $\sigma = 6$.

Foreign and civil edges occur in (2/13) only in the combination $\underline{F} + \underline{C}$; so that the abstract result will be the same whatever the nationality of the immigrants. The counting of the modifications is however at its simplest if the immigrants differ in nationality from any of their new neighbors; for then the edges $\underline{D} + \underline{L}$ to deserts and lakes all become foreign, so that $\underline{F}$ is replaced by $\underline{F} + \underline{D} + \underline{L}$.

The seacoasts of deserts were not counted in $\underline{B}$. But when the deserts become populated cells, their coasts become edges, thereby increasing $\underline{B}$ to $\underline{B} + \beta$, say. (2/19)
The Sahara is shown in Figure 7 with two dotted coasts, the Red Sea with another. The cost of German South West Africa should probably be counted in β. So take $\beta = 4$.

The application of (2/13) to this modified continent gives

$$C + F + L + D + \beta = 3(s + \sigma - 1) + g, \text{ provided } s \geqslant 2. \qquad (2/20)$$

In our African illustration the first and second members of (2/20) are both 366, in exact agreement. There might have been a slight misfit if the Kahahari desert had not been interpreted consistently with German South West Africa.

Theorem (2/20), though illustrated by a particular country, is evidently based on reasoning which is general.

There is an associated inequality. By (2/16)

$$D + L \geqslant 3\sigma \qquad (2/21)$$

the equality occurring when both members are zero. Subtraction of (2/21) from (2/20) leaves

$$C + F + B \leqslant 3(s - 1) + g - \beta. \qquad (2/22)$$

Comparison of (2/22) with (2/13) shows that, for a known total population, the presence of lakes and deserts may decrease, but cannot increase, C + F + B. (2/23)

In theorem (2/13) the population appeared only by its total, via s. But when there are lakes or deserts, the distribution of population is also involved, via σ in (2/20), and the condition (2/16) for ignoring lakes or deserts. Given a map of the lakes and deserts in any country, but only its total population, is it possible to estimate σ? Occasionally this is easy. Finland is said to have a thousand lakes; and its population in 1910 was 3.12 millions. If it were mapped in three cells of 1.04 million each, not more than three cells could touch any lake; and so by (2/16) none of the internal lakes would count. Ladoga was external. If Finland were mapped as 312 cells of 10,000 persons each, we could reason very roughly thus. The diameter of a cell may be about $\frac{1}{\sqrt{312}} = \frac{1}{18}$ of the diameter of Finland. A lake is not likely to touch four or more cells, unless its greatest length is as much as a cell diameter. Inspection of a map suggests that the number of such lakes is less than 100. Therefore $\sigma < 100$. The useful feature of this estimate is that, even in such a lakey country as Finland, σ is a minor correction to s.

3. EMPIRICAL MAPS OF SAMPLE COUNTRIES IN CELLS OF A MILLION PEOPLE

This empirical section is inserted early, in order to prevent theoretical interest from drifting away from actuality. The method of drawing these maps will be explained by taking France as illustration. In the Statesman's Year-Book for 1914 the populations of 87 Departments, as given by the census of March 1911, are listed separately. A map of France in Departments was covered with tracing paper. The frontier was corrected to the date of the census. The number of people in each Department was written to the nearest 10,000 near its center of area in red ink. In soft pencil trial cells were sketched, first along the seacoasts, and land frontiers, and around the largest inland cities, Paris and Lyons. The total number of cells was counted and found to be 39, as required. Considerable rubbing out and amendment went on in attempts to make the cells more compact and more equal in population.

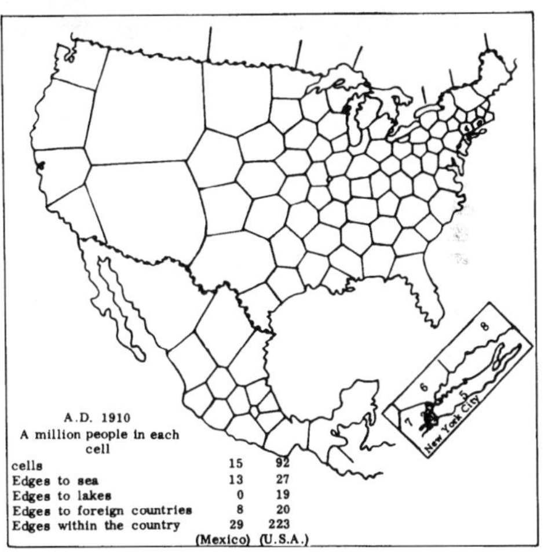

Figure 5

There comes a stage in this process when some success has been attained and when further modifications may, or may not, be improvements. It is thereafter advisable to affirm all proposals in ink, to test them definitely, and to make alterations by tracing on to a blank sheet of paper. A glass-topped table illuminated from below was prepared in order to facilitate some of these many tracings. Precautions were taken to prevent the heat of the lamp from causing the paper to shrink.

Such a process of freehand drawing, with pauses for testing, followed by amendments made on another sheet by tracing, is not to be despised: in solving Laplace's partial differential equation it can give an accuracy of one percent, as I showed long ago. (Richardson, 1908).

Three general questions arise: (i) Has the problem any solution? (Existence). (ii) Has the problem more than one solution? (Uniqueness). (iii) Can the solution be spread out from a portion of the boundary, without reference to other portions? (Marching problems); or is a solution necessarily an agreement between all parts of the boundary acting together? (Jury problems).

Similar questions are answered in treatises of differential equations by deduction from assumed properties of the given functions, which are taken to be analytic, or otherwise simple. Our main datum is the distribution of population, and it is not analytic nor otherwise simple. For example the Ordnance Survey publish a map of the population of Great Britain in 1931 represented by colored contours of density drawn by Mr. A. C. O'Dell. It shows an intricacy of fine streaks and blotches on the scale of 1:1,000,000. There is no prospect that the detail would appear smooth if the map were enlarged; for houses and persons are separate units. Accordingly there does not appear to be much hope of answering questions (i), (ii), (iii), in a precise deductive manner; and so it is all the more desirable to record the experience gathered in the process of drawing maps.

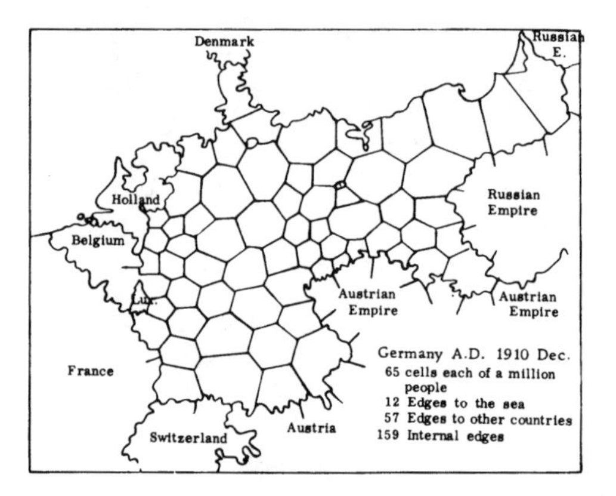

Figure 6. The Austrian edges have since been revised. See the large drawing.

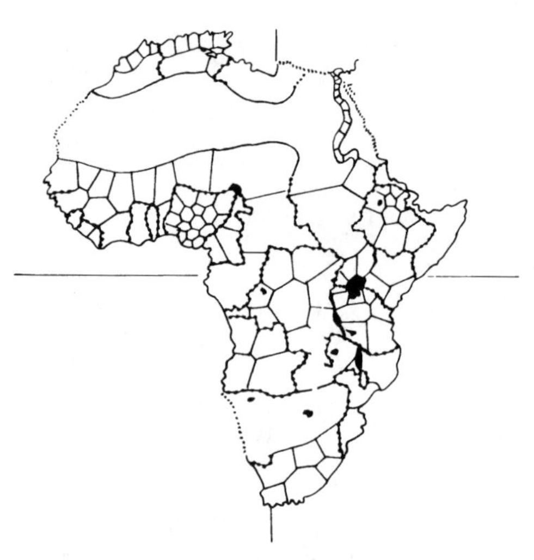

Figure 7. Africa at mid-1910 in cells of a million persons. Lakes are painted black. Any cell-edge which was also a political frontier is emphasized by beads. Coasts where sea meets desert are shown by broken lines.

Although this map gives a correct general impression at a glance, yet it could never be made satisfactory in detail, because: (i) at that date the populations of Abyssinia and of the Belgian Congo were uncertain by several millions; and because (ii) some politically significant territories such as German South West Africa, Spanish Guinea, Portuguese Guinea, Gambia, Rio de Oro, Libya, Eritrea, and the three Somalilands, require for their representation cells of less than a million.

Approximately the same facts have been represented as a dot-map in Jaeger's Afrika (1928, p. 64), and as contours of density in a map published by Messrs. Philips. Thus the reader will be able to compare three styles of presentation.

The author drew his information chiefly from the Statesman's Year Books for 1914 and 1926 and from the maps just mentioned. For Tanganyika and Kenya he was fortunate in having the advice of two district commissioners, respectively D. W..I. Piggott and Colin Parker.

In practice every drawing was treated as a jury problem.

Three of the five conditions were satisfied without any difficulty or approximation in every map: these were (1/2) on simple vertices, (1/3) on non-encirclement, and (1/4) on national frontiers. The condition (1/1) that each cell should contain a million people was satisfied with useful approximation. There are two kinds of errors about the cell-content, the systematic and the random. The systematic error arises because the number of cells is made an integer; for France in A.D. 1911 a population of 39.3 million was divided among 39 cells of 1.01 million each. Luxemburg with a population of only 0.26 million in A.D. 1910 had unfortunately to be rounded off to zero. Its territory was treated as if part of Germany, with which it then had a customs-union. Similarly Andorra was treated as a part of France. The random error arises in the interpolation from irregular census-areas to the overlapping cells. In the present map of France it may amount to a few percent, the data being for Departments. If smaller census-areas had been used the random error would have been less. Neither the systematic nor the random error in the cell content is of much importance.

The questions about existence, uniqueness, a marching or a jury problem are thus all reduced to questions about homoplaty. A cell which touches a frontier can rarely be made perfectly homoplatous. If the uniform cell-content were decreased so as to make more and smaller cells in the same country, then the proportion of them which suffered this imperfection at the frontier would be decreased. There are also some internal situations which make it difficult to draw homoplatous cells. These occur where two contiguous cells have areas in a considerable ratio, say greater than two. This happens in the neighborhood of a city whose population is about equal to the cell-contact. The reader can judge as to whether the cells are compact by looking at Figures 5, 6, 7, 8. The author is not entirely satisfied with the compactness of these cells, although he cannot improve on them.

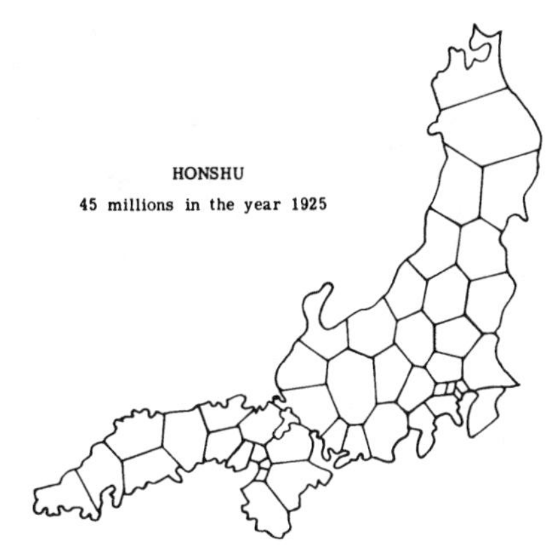

Figure 8

Granted that there is at least one map which satisfies the simultaneous conditions (1/1), (1/2), (1/3), (1/4), (1/5) to the accuracy of Figure 5, the question remains whether a different map could do so equally well? I find that there is not one uniquely best map. Geographers, being accustomed to uniquely determined maps, may feel that any arbitrariness is unsatisfactory. Physicists however are accustomed to equipotentials and stream-lines which represent a field in a manner partly arbitrary, being selections from two infinite families of lines. The important relation, I suggest, is not that the census should uniquely determine the map, but almost conversely that the map should quickly give a good idea of the census. A relation connecting one map of population to many different censuses would be deplorable; but a relation connecting many different maps to one census is valuable. The number of edges seems to be rather invariant. For example after drawing a map of France in 39 cells I waited until I had forgotten most of its appearance, and then drew another independently. The cell-boundaries are considerably different in these two maps; yet both of them were found to have 90 internal edges and 16 edges to the sea. The number $\underline{F}$ of edges between France and other countries had of course to wait for similar mapping of the contiguous portions of Spain, Belgium, Germany, Switzerland, and Italy.

The numbers $\underline{F}$, $\underline{C}$, $\underline{B}$, of edges were not consciously controlled; they were allowed to happen, attention being occupied with making the cells compact and of the prescribed population. Any agreement of $\underline{F}$, $\underline{C}$, or $\underline{B}$ with any formula is therefore an experimental finding.

Approximate Empirical Formulae

The comparison of two adjacent columns in Table II shows that $3s_n - B_n - \frac{F_n}{2}$ is a good approximation to C_n. (3/1) A theory is not far away, if lakes and deserts may be neglected altogether.

Table II

EMPIRICAL RESULTS FOR VARIOUS COUNTRIES

The Countries Are Arranged in Order of Population

Country	Date of Census	s_n	B_n	L_n	F_n	C_n	$3s_n - B_n - \frac{F_n}{2}$	$B_n + F_n$	$2(12s_n - 3)^{\frac{1}{2}}$
Switzerland	1910.XII	4	0	0	14	5	5	14	13.4
Netherlands	1911.XII	6	5	0	12	8	7	17	16.6
Portugal	1911.XII	6	5	0	7	8	9.5	12	16.6
Belgium	1910.XII	7	1	0	16	12	12	17	18.0
Canada*	1911	7	7	6	14	6	7	21	18.0
Mexico	1910	15	13	0	8	29	28	21	26.6
Spain	1910	19	13	0	12	38	38	25	30.0
Italy	1911.VI	30	21	0	14	59	62	35	37.8
France	1911.III	39	16	0	24	90	89	40	43.1
Honshu	1925	45	41	0	0	91	94	41	46.3
Germany	1910.XII	65	12	0	56	159	155	68	55.8
U.S.A.‡	1910	92	27	19	20	223	239	47	66.4

*The five Great Lakes, Alaska, and Labrador being regarded as void.
‡The five Great Lakes being regarded as void.

Europe and Asia together form a cluster bounded by the void; so that by the topological theorem (2/13), as $s > 2$,

$$C + B + F = 3(s - 1)$$

in which the letters are totals for Europe and Asia. There were in A.D. 1911 about 26 countries in this area. The last equation can be rearranged, by the aid of the theorems (2/14) on totals, as

$$\frac{1}{26}\sum_{n=1}^{n=26}\left\{C_n + B_n + \frac{1}{2}F_n - 3s_n + \frac{3}{26}\right\} = 0. \quad (3/2)$$

The $\frac{3}{26}$ is almost negligible. So, on the average over the 26 countries, C_n must be nearly equal to $3s_n - B_n - \frac{1}{2}F_n$. There is no suggestion in this theory that the individual countries should conform to the average, but the empirical studies in Table II show that they nearly do so. Honshu, and Britain being islands, should give

$$C_n + B_n = 3(s_n - 1) \text{ strictly.}$$

The last two columns of Table II show that $B_n + F_n$ is highly correlated with, and roughly equal to, $2(12s_n - 3)^{\frac{1}{2}}$. (3/3)
This must be accepted as an empirical fact. I stumbled across it by accident, for I was expecting $2B_n + F_n$ to be nearly equal to $2(12s_n - 3)^{\frac{1}{2}}$. That would be so if the density of population were uniform, and the outlines were compact, as will be shown later in §4. The fact that $\underline{B_n}$ not $\underline{2B_n}$ is appropriate can be explained if the density of population and seacoasts is suitably more than that on land frontiers.

4. STANDARD CLUSTERS AND THEIR MODIFICATIONS

The political map and the distribution of population are both so irregular that the exact formulae which will here be proved for special clusters are not immediately applicable. Their use is indirect. They serve to test methods which are more adaptable, such as the integrals which follow. With a similar intention electrical technicians are accustomed to keep a Weston standard cell for checking the errors of a portable voltmeter. The first six of these standard clusters are shown in Figure 9. They contain 1, 7, 19, 37, 61, 91 cells. Each is obtained from the next smaller cluster by fitting round it a complete border of one layer of cells. This process can be continued indefinitely. The importance of this set of clusters lies in the fact that they have a maximum property analogous to that of the circle, which has the greatest area for a given perimeter. In consequence these clusters are as significant for the theory of networks, as a circle is for the theory of pipes. I propose to name them "standard." As far as equation (4/15), the standard clusters will be regarded as surrounded by further portions of the same honeycomb pattern; so that $\underline{B} = 0$. Each outer side of a hexagon is accordingly counted as an edge. On this view the cluster is analogous to landlocked states such as Switzerland, Afghanistan, or Paraguay. Later, each standard cluster will be regarded as an island all under one government; so that $\underline{F} = 0$. The number B is then expressed by equation (4/17).

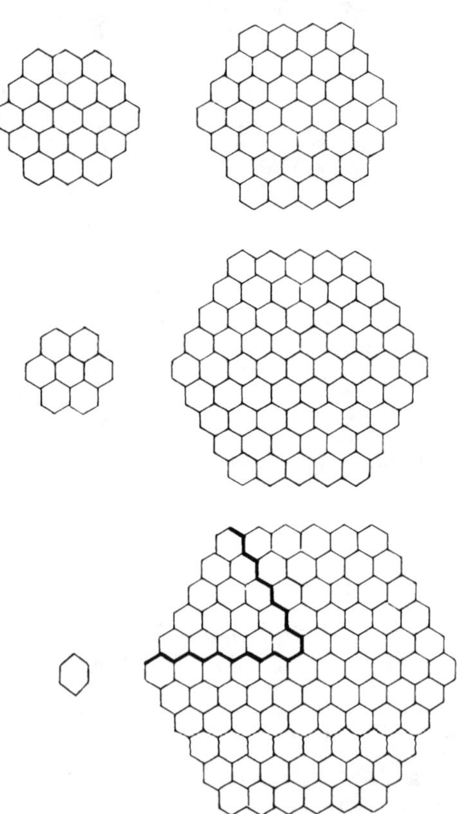

Figure 9. The first six of the standard clusters. For its given number of equal hexagonal cells, each cluster has a maximum number of internal edges, and a minimum number of external edges. The successive numbers of cells are 1, 7, 19, 37, 61, 91.

The Maximum and Minimum Property

A standard cluster has more internal edges and fewer external edges than any slightly modified arrangement of the same cells. The proof is as follows:

Let any cell in the border where it has α external edges be moved to a place where it has β external edges. On counting the occurrences at the two sites it is found that $\underline{F}$ for the cluster is increased by $2(\beta - \alpha)$. But the only choices open are those for $\beta > \alpha$. Therefore $\underline{F}$ was a minimum. The same transfer of a cell increases the number $\underline{C}$ of internal edges of the cluster by $\alpha - \beta$. Therefore $\underline{C}$ was a maximum. This result contrasts with the fact that the circle has the greatest area for a given perimeter. The standard hexagonal cluster of 547 cells can be made to appear, to an unfocused eye, more circular by removing three cells from each corner and placing them in the middle of the next side. But $\frac{C}{F}$ is thereby decreased. There are several contrasts between this property and the analogous one for the circle. The circle can be deformed continuously: the cluster only by moving a whole cell. The area of a circle is almost equal to that of any slightly deformed figure of the same perimeter: the internal edges of the altered cluster become fewer. It is not a "stationary" property.

Historical Note. The Pythagoreans represented numbers by patterns of dots. Hogben (1942, p. 207) when explaining Pythagorean arithmetic, showed hexagonal figures of dots containing severally 1, 7, 19, 37, . . . and called these "the hexagonal numbers." The present Figure 9 can be obtained by replacing each of Hogben's dots by a regular hexagon. Unfortunately the name "hexagonal numbers," which otherwise would have been most appropriate, has for two thousand years denoted the set beginning 1, 6, 15, 28, 45, 66, . . . these being numbers of dots arranged by the Pythagoreans in the irregular patterns of which the first four are shown in Figure 10. (Heath, 1921, pp. 1, 76-79 and 105-106, C. Smith, 1921, p. 407).

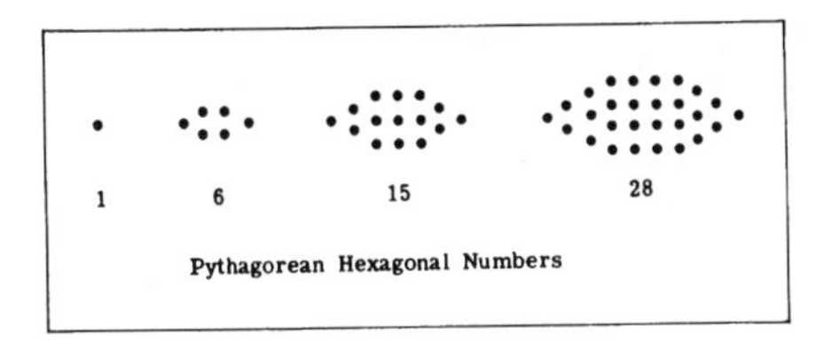

Pythagorean Hexagonal Numbers

Figure 10

It is not possible to fit equal regular hexagons together so that their centers coincide with those of the dots of the Pythagorean hexagonal patterns; and accordingly the numbers 1, 6, 15, 28, 45, 66 . . . are of no interest for the theory of hexagonal cells. They are mentioned only to prevent confusion.

The Number of Edges Belonging to Standard Clusters

Let us start with one hexagon, and fit round it a border of six hexagons. Then fit a border round the resulting cluster of 7 cells; and so on by repetition. The first six of these clusters are shown in Figure 9. Let the successive clusters be given serial numbers $\underline{k} = 1, 2, 3, \ldots$; and let $\underline{k}$ be used as a suffix qualifying $\underline{s}$, $\underline{C}$, and $\underline{F}$; so that

$$s_1 = 1, \quad C_1 = 0, \quad F_1 = 6. \qquad (4/1)$$

Let b_k be the number of cells in the border which already forms part of the cluster $\underline{k}$, not the border which will be put round the cluster $\underline{k}$ at the next operation. Then

$$s_{k+1} - s_k = b_{k+1}. \qquad (4/2)$$

By inspection of the diagrams in Figure 9 it is seen that each successive border contains six more cells than the previous border; and that this relation must persist in all subsequent clusters, so that

$$b_{k+1} - b_k = 6 \quad \text{for} \quad k \geqslant 1. \qquad (4/3)$$

The first cell might be regarded as a border or not; but anyway $b_2 = 6$, whence it follows by summing (4/3) that

$$b_k = 6(k - 1) \quad \text{for} \quad k \geqslant 2. \qquad (4/4)$$

A summation of (4/2), starting from (4/1) and using (4/4) gives

$$s_k = 1 + 6 \sum_{j=2}^{j=k} (j - 1) = 3k^2 - 3k + 1. \qquad (4/5)$$

The numbers $\underline{s}$ begin 1, 7, 19, 37, 61, 91, ...

The external edges, not the void but to further hexagons, are two for every border-cell plus an extra edge at each of the six bends in the border. That is,

$$F_k = 2b_k + 6 = 12k - 6. \qquad (4/6)$$

The internal edges consist of b_k edges between adjacent cells in the border, plus all the edges of the previous cluster. This gives the recurrence formula

$$C_k - C_{k-1} = b_k + F_{k-1} = 18k = 12 \text{ for } k \geqslant 2. \qquad (4/7)$$

On summation, starting from $C_1 = 0$, this gives

$$C_k = (k - 1)(9k - 6). \qquad (4/8)$$

It is now desirable to change the independent variable from $\underline{k}$ to $\underline{s}$. The only permissible solution of (4/5)

$$k = \frac{1}{2}\left\{1 + \sqrt{\left(\frac{4}{3}s_k - \frac{1}{3}\right)}\right\}. \qquad (4/9)$$

The suffixes $\underline{k}$ may now be omitted. Elimination of $\underline{k}$ between (4/9), (4/6), (4/8) gives

$$F = 2(12s - 3)^{\frac{1}{2}}, \quad C = 3s - \frac{1}{2}F, \qquad (4/10), (4/11)$$

$$\frac{2C}{F} = 3s(12s - 3)^{-\frac{1}{2}} - 1. \qquad (4/12)$$

To show how these expressions vary for large $\underline{s}$, the square roots have been expanded by the binomial theorem, which gives

$$F = 6.9282s^{\frac{1}{2}}\left\{1 - \frac{1}{8s} - \frac{1}{128s^2} - \frac{1}{1024s^3}\cdots\right\}, \qquad (4/13)$$

$$\frac{2C}{F} = -1 + 0.86602s^{\frac{1}{2}}\left\{1 + \frac{1}{8s} + \frac{3}{128s^2} + \frac{5}{1024s^3}\cdots\right\}. \qquad (4/14)$$

At $s = 1$, where these truncated series are at their worst, the terms shown above give

$$F = 6.001,3, \quad C = -0.000,6, \quad \frac{2C}{F} = -0.001,3$$

as approximations to 6, 0, 0 respectively. As this accuracy is quite sufficient, the series may be regarded as ending with the terms in s^{-3}.

All so far is on the assumption that $B = 0$. (4/15)

Islands. Alternatively the cluster may be regarded as an island, under a single government, so that $F = 0$. (4/16)
The seaward boundary of any cell is then counted as all one edge, the angles in it being of no significance. Accordingly $\underline{B}$ is equal to the number $\underline{b}_k$ of cells in the border. So by (4/4) and (4/9)

$$B = -3 + (12s - 3)^{\frac{1}{2}} \quad \text{for} \quad s > 7. \qquad (4/17)$$

A relation between the two interpretations is

$$3 + \left(B + \frac{1}{2}F\right)_{\text{island}} = \left(B + \frac{1}{2}F\right)_{\text{landlocked}}. \qquad (4/18)$$

For large $\underline{s}$ and special purposes the 3 may be negligible. Contrast $\underline{B} + \underline{F}$ in Table II.

The number of internal edges remains unchanged at

$$C = 3s - (12s - 3)^{\frac{1}{2}}. \qquad (4/19)$$

THE IDEA OF LOCAL UNIFORMITY

Successful treatment of a physical distribution, alias "field," has often begun with the assumption that any suitably small part of it is simpler than the whole. This hypothesis leads to a differential equation, and so on to its integral. The corresponding assumption about population would be that any suitably local distribution could be mapped by a honeycomb pattern. Jeans (1916, p. 15) discussed the choice of a small element in a gas and alluded to the analogous problem for population. The element of space must be large enough to contain many individuals, and yet must be so small that the density in it is effectively uniform. The theory is thus useful in a range of size bounded at both ends. The smallest permissible size of cell is easily discussed. For if there were $\underline{h}$ persons in it, random variation would probably be represented by a standard deviation of $\sqrt{h}$. So the choice of the least $\underline{h}$ is the same all over the map. In view of towns, villages and open spaces, the greatest permissible $\underline{h}$ is not obvious. Let us proceed in the hope that there may be a considerable range of $\underline{h}$ for which both conditions are tolerably satisfied.

To study the idea of local uniformity let us imagine a continent containing some elongated countries and some conspicuous gradients of population. Suppose that it has been mapped, by the methods of §3, into $\underline{s}_0$ fairly compact cells, each containing $\underline{h}_0$ persons, and having altogether $\underline{C}_0$ civil edges, $\underline{F}_0$ foreign edges, and $\underline{B}_0$ edges to the void. Suppose further that within each cell the density is almost uniform. Required to estimate $\underline{C}$, $\underline{F}$, $\underline{B}$, when the continent has been further subdivided into $\underline{s}$ cells each of population $\underline{h}$.

The simplest mode of subdivision is to replace each of the given cells by a standard cluster of seven, as shown in Figure 11. The given edges cannot be simply subdivided: each has to be replaced by a zigzag of three edges which fit it fairly well on the average. When the given edges are unalterable, as at a seacoast or land frontier, neither pattern can be quite regular. The suggestion made by the diagram is that:

$$\text{When } \underline{s} = 7s_0, \quad \text{then } \underline{F} = 3F_0 \quad \text{and} \quad \underline{B} = 3B_0, \qquad (4/20), (4/21), (4/22)$$

but that there are also twelve new internal edges for each given cell, so that

$$C = 3C_0 + 12s_0. \qquad (4/23)$$

These relations would be strict if the given cells could all be regular hexagons. Actually some of them, though fairly compact, may have other than six edges. There is a strict topological theorem, namely (2/10) which must be satisfied. For the given map it asserts that

$$F_0 + B_0 + C_0 = 3s_0 - 3. \qquad (4/24)$$

When F_0, B_0, C_0, s_0 have been eliminated between the above five equations the result is

$$F + B + C = 3s - 9, \qquad (4/25)$$

However the same topological theorem, when applied directly to the new map, shows that (4/25) is an error by six of the new edges. The suggestion that each given edge can be replaced by a zigzag of three new edges cannot be quite correct. But the error becomes negligible as $\underline{s}$ becomes large, and will be ignored in the rest of this section.

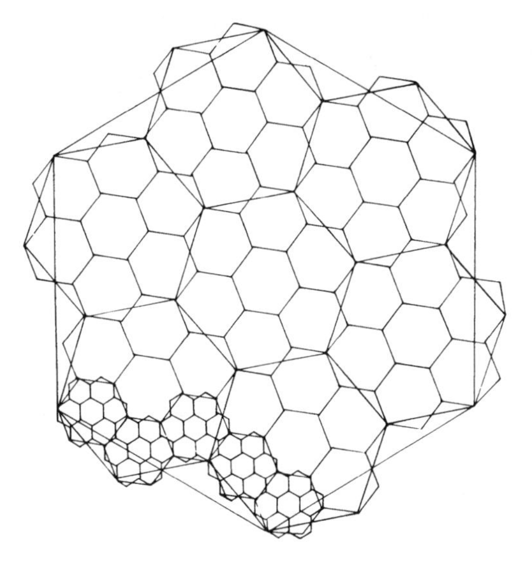

Figure 11. Replacement of one honeycomb pattern by another having cells of $\frac{1}{7}$ the area, or $\frac{1}{7^2}$ or $\frac{1}{7^3}$. Although there are infinitely many other possible arrangements, the one shown is specially convenient for counting edges, because the vertices of the given pattern persist in all its successors. Each pattern is rotated from its predecessor through the angle arctan $\left(\sqrt{\frac{3}{5}}\right)$ $= 19^{\circ}6'.4$.

Because the lengths of edges vary as the square-roots of the areas of similar cells, one might incautiously suppose that $\underline{F}$ and $\underline{B}$ would vary as $\underline{s}^{0.5}$. Such reasoning would be fallacious, because the problem is not to divide a given straight length, or a smooth curve, into parts, but to fit a polygon to a seacoast or a land frontier. It will be shown in §7 that such loci, though usually definitely given, seldom have given lengths.

It is advantageous, however, to take a hint from that imperfect reasoning by expressing $\frac{F}{F_0}$ and $\frac{B}{B_0}$ as powers of $\frac{s}{s_0}$. The power is found to depend on the mode of division. When, as above, each given cell is replaced by 7 cells, and each given edge by 3 edges, then because $3 = 7^{0.564575}$ it follows that

$$\frac{F}{F_0} = \frac{B}{B_0} = \left(\frac{s}{s_0}\right)^{0.564575}. \qquad (4/26)$$

The same formula holds good when each of the new cells is replaced by seven cells, and so on repeatedly as far as local uniformity persists.

Many other modes of subdivision are possible. Each of the given cells can be cut into the standard cluster having in general the serial number k. By (4/5) the cluster contains $3k^2 - 3k + 1$ cells; by (4/8) it has $(k - 1)(9k - 6)$ internal edges; and by (4/6) each given edge is replaced by a zigzag of $2k - 1$ edges. In general let α be defined by

$$\frac{F}{F_0} = \left(\frac{s}{s_0}\right)^{\alpha}. \qquad (4/27)$$

The table shows how α depends on k.

Table III

k	$\frac{s}{s_0}$	$\frac{F}{F_0}$	α
2	7	3	0.5646
3	19	5	0.5466
4	37	7	0.5389
5	61	9	0.5345
6	91	11	0.5316
1387	5,767,147	2773	0.5092
∞	∞	∞	0.5 exactly

Almost the same final number of cells can be obtained by two contrasting methods of division. In the first, k = 2, so that each cell is replaced by seven cells, and the operation is performed 8 times in succession. In the second method k = 1387, and the operation is performed once only. The resulting numbers of cells are given respectively by

$$\frac{s}{s_0} = 7^8 = 5{,}764{,}801 \quad \text{and} \quad \frac{s}{s_0} = 5{,}767{,}147.$$

The numbers of civil edges obtained by these two modes of division were worked out, and were found to agree closely, as one would expect. The numbers of foreign edges, however, are in marked contrast; $\frac{F}{F_0}$ being $3^8 = 6561$ by the first method and 2773 by the second. The explanation is that the zigzag by which a given foreign edge is replaced, is in the first method much more complicated than in the second. The first zigzag cannot be drawn in full detail; but the stages in its construction as far as 7^3 are illustrated in Figure 11. The second zigzag is more like a common "straight" saw, although with blunt teeth.

That is perhaps as far as it is worth while to pursue the theory. For it has appeared that the appropriate index α depends on the shape of the zigzag which will fit the given coasts and land frontiers; and that is an empirical question which will be examined in §7.

The Most Compact Clusters for Numbers of Cells That Are Not Standard

When the number of cells in the cluster was other than s = 1, 7, 19, 37, 61, 91 . . . , the author had no theory; and so he proceeded inductively by drawing clusters which looked to him to be homoplatous, and then counting their cells and edges. Among alternative forms he looked for those that had the least F and the greatest C. Often several alternative forms were found to have the same F and C. For example for 9 cells each of the four clusters shown in Figure 12 has F = 22, C = 16. In this stationary property clusters of 9 cells contrast with clusters of 7 cells, because for the latter there is a unique pattern which has less F than any other. These empirical counts are set in the following table beside the values given by inserting the counted s into the formulae (4/10) and (4/11). The list is complete up to s = 20. The run from s = 140 to s = 147 is intended to be a random sample of larger s. It lies between the adjacent numbers s = 127 and s = 169 which are of the set $3k^2 - 3k + 1$.

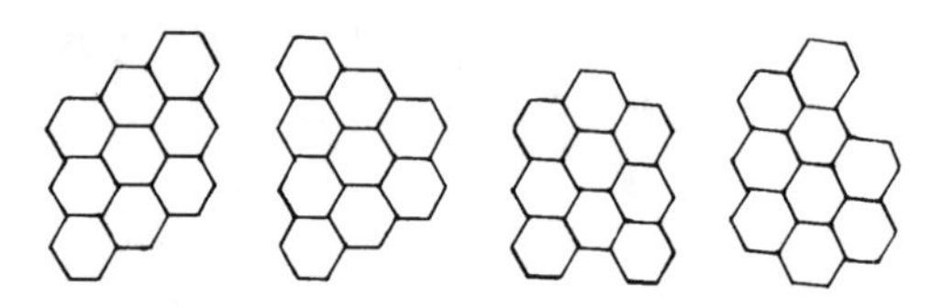

Figure 12. Clusters which, although different, agree as to their numbers of cells, of internal edges, and of external edges to further portions of the hexagonal net.

Table IV

Cells	External Edges* F		Internal Edges C	
s	Counted	Formula (4/10)	Counted	Formula (4/11)
2	10	9.2	1	1.4
3	12	11.5	3	3.3
4	14	13.4	5	5.3
5	16	15.1	7	7.5
6	18	16.6	9	9.7
8	20	19.3	14	14.4
9	22	20.5	16	16.8
10	22	21.6	19	19.2
11	24	22.7	21	21.6
12	24	23.7	24	24.1
13	26	24.7	26	26.6
14	26	25.7	29	29.2
15	28	26.6	31	31.7
16	28	27.5	34	34.3
17	30	28.4	36	36.8
18	30	29.2	39	39.4
20	32	30.8	44	44.6
. . .	. . .	. . .	. . .	. . .
140	82	81.9	379	379.05
141	84	82.2	381	381.9
142	84	82.5	384	384.8
143	84	82.8	387	387.6
144	84	83.1	390	390.5
145	84	83.4	393	393.3
146	84	83.6	396	396.2
147	84	83.9	399	399.04

*Not to the void, but to further portions of the hexagonal net.

Inspection of Table IV shows:

(i) that the deviations from the formulae are in every example towards greater F and less C. In other words the formulae (4/10) and (4/11) set an ideal of homoplaty for a cluster.

(ii) that the deviations from the formulae are slight, at least in comparison with the irregularities of geographical politics. These facts are remarkable, because the formulae were deduced only for values of s which do not occur in the table.

Generalization by Conformal Representation

The numbers F, C, or B of edges and the number s of cells are not altered when the cluster is drawn on India rubber which is then stretched or bent in any manner which does not tear it nor cause separated edges to meet. Not every manner of stretching will preserve the metrical property that each cell is to be compact. The marvelous method of conformal representation (See for example H. and B. Jeffreys, 1946, Chps. 11 and 13) allows the honeycomb pattern to be strained in all those manners which leave the cells almost regular hexagons but no longer all equal. There are exceptions at singular points of the transformation, but these are usually rare. Each such conformal representation corresponds to a particular distribution of the density of population ρ. An endless variety of distributions of ρ can thus be represented by six-edged cells.

Take two Mercator's projections of the globe. On one draw any cluster composed of equal regular hexagons. Let ξ, η be rectangular equi-scale coordinates in this plane. The figure is to be represented on the other plane where the rectangular equi-scale coordinates are x, y. Let the point (x,y) be connected to the point (ξ,η) by the relation

$$\zeta \equiv \xi + i\eta = f(x + iy) = f(z) \qquad (4/28)$$

in which f is an analytic function. A point at which $\frac{d\zeta}{dz}$ is either 0 or ∞ is called a "singular" point,

(H. and B. Jeffreys, 1946, p. 382) and has a widespread influence, as will be illustrated presently. Let us consider first the case when there is no singular point in the region. Then a network of equal regular hexagons in the (ξ,η) plane corresponds to a network in the (x,y) plane having the same topological properties, such as six vertices to every cell. By well-known theorems the angles at the vertices are the same in both planes; and the linear magnification of one figure relative to the other is the same in all directions at a point, being $\left|\frac{d\zeta}{dz}\right|$. The process of conformal representation offers a vast extension of the theory of standard hexagonal clusters; for the formulae (4/10) and (4/11), namely $F = 2(12s - 3)^{\frac{1}{2}}$ and $C = 3s - (12s - 3)^{\frac{1}{2}}$ are equally true for the cluster and its representation in the absence of a singular point.

The density of population then transforms as follows. By hypothesis there are the same number of people in every cell in both planes. Let c denote the uniform density of population in the ζ plane, and ρ the variable density in the z plane. Let dS be the area of a small cell in the z plane, and $d\Omega$ the corresponding area in the ζ plane. Then

$$c\,d\Omega = \rho\,dS \qquad (4/29)$$

also

$$\frac{d\Omega}{dS} = \left|\frac{d\zeta}{dz}\right|^2 \qquad (4/30)$$

whence

$$\rho = c\left|\frac{d\zeta}{dz}\right|^2 = c\left\{\left(\frac{\partial\xi}{\partial x}\right)^2 + \left(\frac{\partial\xi}{\partial y}\right)^2\right\}. \qquad (4/31)$$

Special cases including singular points are illustrated by Figure 13.

For the "central desert" the transformation is $\zeta \propto z^2$. (4/32)

From (4/31) it follows that $\rho \propto |z|^2$. Polar coordinates make this transformation clearer. Let them be such that

$$\zeta = \xi + i\eta = R(\cos\phi + i\sin\phi) \text{ and}$$

$$z = x + iy = r(\cos\theta + i\sin\theta). \text{ Then}$$

$$r \propto R^{\frac{1}{2}}, \quad \theta = \frac{\phi}{2} \text{ and } \rho \propto r^2 \qquad (4/33)$$

The curves are arcs of rectangular hyperbolae. The center of the desert is a singular point; and the cell which contains it has 12 edges. As ϕ increases from 0° to 360°, θ increases from 0° to only 180°. As ϕ continues from 360° to 720°, the other half of the conformal representation is completed. Except for the central cell, each cell in the ζ plane generates two cells in the z plane; but the central cell generates in each turn of ϕ only one of the two halves of its fellow. For this reason

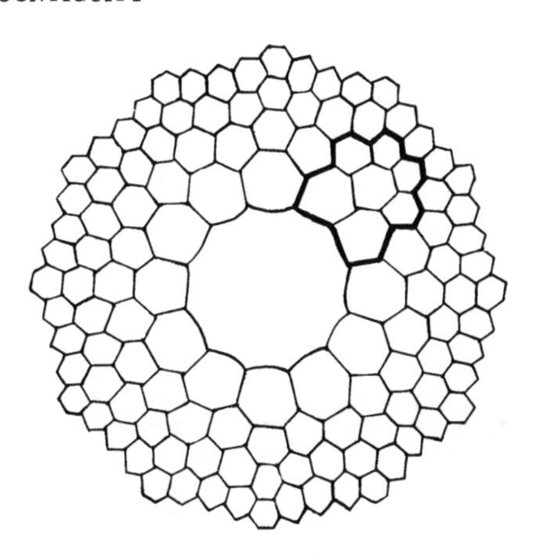

CENTRAL DESERT

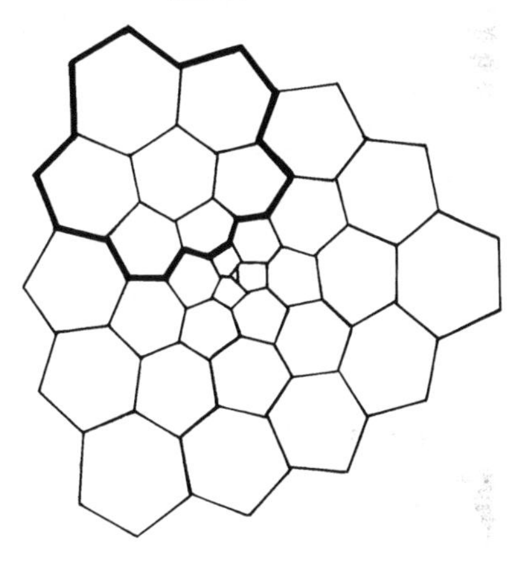

CENTRAL TOWN

Figure 13. Two conformal representations of the standard cluster of 61 cells, which is shown in Figure 9. The number of people in any cell is proportional to its number of sides.

the 12-sided central cell contains twice as many people as does each of its six-sided neighboring cells. Although there are 121 cells in the cluster, the population is therefore 122 units.

For the "central town" the transformation is $\zeta \propto z^{\frac{1}{2}}$ **(4/34)**

$$\text{so that} \quad r \propto R^2, \quad \theta = 2\phi \text{ and } \rho = \frac{k}{r} \qquad (4/35)$$

where k is a constant. The curves are parabolic. The half of the ζ plane bounded by $\phi = 0$ and $\phi = 180°$ corresponds to the whole of the z plane. The portions of the regular hexagons, cut by these terminal radii, join together neatly; so that in effect one cell in the ζ plane generates one cell in the z plane, except the cell which contains the center; for one-half of the central hexagon in the ζ plane generates the whole of a three-edged cell in the z plane, which cell therefore contains only half a unit of population. The number of people within any circle $r = \underline{a}$ is $k\int_0^a \frac{2\pi r dr}{r} = k2\pi a$, and so is finite, although the density is infinite at $r = 0$. To get rid of this monstrous detail let us imagine that at the center of the town there is a round pond of radius $r = b$; so that the central cell contains $\underline{k2\pi b}$ fewer people than it otherwise would. By making $\underline{b}$ small the defect can perhaps be made negligible.

In connection with wars the chief interest of the formal diagrams is that they show the geographical opportunities for the two kinds of wars, civil and external, may or may not be affected by the distribution of population. Much depends on whether the cluster contains a singular point or not. Although the center of the desert and of the town are singular points, we can avoid them by choosing a cluster of cells at one side. For example two clusters of seven cells each are marked by thick outlines. The cells vary in size, but each of them is homoplatous and the marked clusters have exactly the same numbers of internal and external edges as the standard 7-celled cluster shown in Figure 8. The diagram can be imagined as extended to greater radii, and the same property will then hold for the representations of the larger standard clusters selected to one side of the singular point. That is to say: If the cluster does not contain a singular point then the non-uniformity of the density of population has no effect on the geographical opportunities, as here counted. They are given by $F = 2(12s - 3)^{\frac{1}{2}}$, $B = 0$, $C = 3s - \frac{1}{2}F$ if the cluster is a transformation of one of the standards. On the contrary the complete diagrams including the singular points at their centers deviate notably from those formulae as the following particulars show.

Table V

	Central Desert	Central Town
Cells	121	31
Units of population (= s' below)	122	30.5
Internal edges =	312	78
Compare $3s' - (12s' - 3)^{\frac{1}{2}}$ =	327.8	72.4
External edges to further parts of the same net =	108	27
Compare $2(12s' - 3)^{\frac{1}{2}}$ =	76.4	38.1

Proof of Exact Generalized Formulae When the Density Varies As Certain Powers of Distance from a Point

The particular diagrams in Figure 13 open the way to a more general argument, which gives the numbers of foreign and civil edges, $\underline{F}$ and $\underline{C}$. Let us being as before in the ζ plane with a standard cluster of 1 or 7 or 19 or 37 or 61 . . . cells, all being equal regular hexagons. Let the transformation be

$$\zeta = z^n \quad \text{so that} \quad R = r^n, \quad \phi = n\theta. \tag{4/36}$$

Not every $\underline{n}$ is acceptable; for the diagram in the z plane must join on to itself when θ makes a complete turn. The given cluster in the ζ plane has an axis of 6-fold symmetry normal to its plane. Let the corresponding axis in the z plane be β-fold, so that its central cell has β edges. Then

$$\beta = 6n. \tag{4/37}$$

The condition for "joining-on" is that β must be an integer. It has been shown that ρ, the density of population, is proportional to $\left|\frac{d\zeta}{dz}\right|^2$, that is

$$\rho \propto |z^{n-1}|^2 = r^{2n-2} = r^{\frac{1}{3}\beta - 2}. \tag{4/38}$$

The possible cases, when arranged in order of β, begin thus:

β =	3	4	5	6	7	8	9 . . .
n =	$\frac{1}{2}$	$\frac{2}{3}$	$\frac{5}{6}$	1	$\frac{7}{6}$	$\frac{4}{3}$	$\frac{3}{2}$ . . .
$\rho \propto$	r^{-1}	$r^{-\frac{2}{3}}$	$r^{-\frac{1}{3}}$	r^0	$r^{\frac{1}{3}}$	$r^{\frac{2}{3}}$	r . . .
	Crowded centers			Uniform density	Sparse centers		

The complete diagram is made up of β portions each obtainable by transforming the portion marked with the thick outline on the regular cluster in Figure 8. The argument above for $\beta = 6$ can be generalized for other β. We begin with a single cell, named $\underline{k} = 1$, and having β edges, and we form clusters named $\underline{k} = 2, 3, 4, \ldots$ by putting successive borders all round the previous cluster. The number of cells in the border which is just inside cluster $\underline{k}$ amounts to $\beta(\underline{k} - 1)$ for $k \geqslant 2$. They have each two external edges plus an extra edge in β places. Therefore $\underline{F} = \beta(2k - 1)$. (4/39)

It is assumed that $B = 0$. (4/40)

The number of cells is one plus the sum of those in all the added borders. Therefore

$$s = 1 + \sum_{j=2}^{j=k} \beta(j - 1) = 1 + \frac{1}{2}\beta k(k - 1) \tag{4/41}$$

The external edges of cluster $\underline{k}$ number $\beta(2k - 1)$, and therefore those of cluster $k - 1$ number $\beta(2k - 3)$. The latter become internal to cluster k, which acquires also an internal edge between each pair of adjacent cells in its border. There are $\beta(k - 1)$ of these cells. Thus the number of external edges in cluster k exceeds that in cluster $(k - 1)$ by $\beta(2k - 3) + \beta(k - 1) = \beta(3k - 4)$. The original cell had no internal edges. By addition of all the subsequent acquisitions one obtains for C, the number of internal edges of cluster k

$$C = \sum_{j=2}^{j=k} \beta(3j - 4) = \beta(k - 1)\left(\frac{3}{2}k - 1\right). \qquad (4/42)$$

The ideas about geometrical patterns, represented by k and β have been essential to the argument, but they should now be replaced by ideas about population and its distribution. The central cell is peculiar. It contains $\frac{\beta}{6}$ of the standard cell population. Let

$$s' = s - 1 + \frac{\beta}{6}. \qquad (4/43)$$

Then s' is the population of the country measured in that of a standard cell as unit.

To eliminate k we have first from (4/41) and (4/43)

$$s' = \beta\left\{\frac{1}{6} + \frac{1}{2}k(k - 1)\right\}. \qquad (4/44)$$

The solution of this quadratic for k is

$$k = \frac{1}{2} + \frac{1}{2}\left(\frac{8s'}{\beta} - \frac{1}{3}\right)^{\frac{1}{2}}. \qquad (4/45)$$

A negative sign before the square root would not agree with particular cases, and so has been rejected. The substitution of k from (4/45) into (12) and (15) gives

$$F = (8s'\beta)^{\frac{1}{2}}\left(1 - \frac{\beta}{24s'}\right)^{\frac{1}{2}} \qquad (4/46)$$

and

$$C = 3s' - \frac{1}{2}F. \qquad (4/47)$$

These formulae have been checked against the diagrams for $\beta = 3$, 6, and 12. The factor $\left(1 - \frac{\beta}{24s'}\right)^{\frac{1}{2}}$ tends rapidly to unity as the number of cells increases.

Lastly, in the cases here considered, the density of population varies as the mth power of the radius and

$$\beta = 3(m + 2) \qquad (4/48)$$

so

$$F = \left\{24(m + 2)s'\right\}^{\frac{1}{2}}\left\{1 - \frac{(m + 2)}{8s'}\right\}^{\frac{1}{2}}. \qquad (4/49)$$

If $m > 0$ there is a central desert and a crowded frontier; if $m < 0$ there is a central town and a sparse frontier. For a given population the number F of foreign edges increases nearly as $(m + 2)^{\frac{1}{2}}$. The number C of civil edges is much less affected than F by the distribution of the population, because in (4/47) the term $\frac{1}{2}F$ is swamped by $3s'$.

So far it has been assumed that the cluster is bounded by further portions of the same network, making $B = 0$.

Islands. If instead the cluster is an island under a single government then $F = 0$. (4/50) B is then equal to the number of cells touching the sea. That is $B = \beta(k - 1)$ for $k \geqslant 2$. Therefore by (4/45)

$$B = -\frac{\beta}{2} + \left(2s'\beta - \frac{\beta^2}{12}\right)^{\frac{1}{2}} \quad \text{for} \quad s \geqslant 1 + \beta. \qquad (4/51)$$

C remains unchanged, but must now be expressed apart from F, thus

$$C = 3s' - \left(2s'\beta - \frac{\beta^2}{12}\right)^{\frac{1}{2}} = 3s' - B - \frac{\beta}{2}. \qquad (4/52)$$

A relation between the two interpretations is

$$\frac{\beta}{2} + \left(B + \frac{1}{2}F\right)_{\text{island}} = \left(B + \frac{1}{2}F\right)_{\text{landlocked}} \qquad (4/53)$$

Contrast B + F in the empirical Table II.

Restriction on Conformal Representation

The process of conformal representation is of wonderful generality, yet is subject to a peculiar restriction, which was proved by Clerk-Maxwell (1904, Art. 187). When

$$\frac{\partial\xi}{\partial x} = \frac{\partial\eta}{\partial y} \quad \text{and} \quad \frac{\partial\eta}{\partial x} = -\frac{\partial\xi}{\partial y}$$

then ξ and η are said to be "conjugate functions" of x and y. It follows that

$$\frac{\partial^2\xi}{\partial x^2} + \frac{\partial^2\xi}{\partial y^2} = 0$$

and that

$$\xi = i\eta = f(x + iy).$$

Now $\frac{\partial\xi}{\partial y}$ and $\frac{\partial\xi}{\partial x}$ are conjugate functions of x and y. For brevity put $\frac{\partial\xi}{\partial y} = u$, $\frac{\partial\xi}{\partial x} = v$. Then it follows by taking u, v derivatives that $\frac{1}{2}\log(u^2 + v^2)$ and $\arctan\frac{v}{u}$ are conjugate functions of u and v. But the property is transitive, so that $\frac{1}{2}\log(u^2 + v^2)$ and $\arctan\frac{v}{u}$ are conjugate functions of x and y.

Therefore

$$\left(\frac{\partial^2}{\partial x^2} + \frac{\partial^2}{\partial y^2}\right)\log(u^2 + v^2) = 0.$$

That is

$$\left(\frac{\partial^2}{\partial x^2} + \frac{\partial^2}{\partial y^2}\right)\log \rho = 0. \tag{4/54}$$

The density ρ of population must satisfy this equation if the process of conformal representation is to be possible. For example if ρ is any power of the radius r the condition (4/54) is satisfied. Two special cases, $\rho = r^2$ and $\rho = r^{-1}$, have already been illustrated by Figure 13. But a simple gradient of population, namely $\rho = x$, does not satisfy (4/54) and so cannot be mapped by a conformal representation of the honeycomb pattern. Nor can $\rho = e^{-x^2-y^2}$, although that form will be treated easily in the approximate theory in § 6.

The restriction on ρ may also be approached from the theory of strain, as given by A. E. H. Love (1906, p. 50). To preserve all angles the shear e_{xy} is to be zero everywhere, and $e_{xx} = e_{yy}$. The area is increased in such a manner that

$$\rho(1 + e_{xx})(1 + e_{yy}) = \text{constant}.$$

One of the identical relations of strain, namely

$$\frac{\partial^2 e_{xx}}{\partial y^2} + \frac{\partial^2 e_{yy}}{\partial x^2} = \frac{\partial^2 e_{xy}}{\partial x \partial y},$$

then leads to

$$\left(\frac{\partial^2}{\partial x^2} + \frac{\partial^2}{\partial y^2}\right)\rho^{-\frac{1}{2}} = 0.$$

This appears to differ from (4/54); but as the strains are restricted to be small, the variation of ρ must be slight; and when ρ is almost constant the expressions agree. In equation (4/54) the strains may be of any size.

When Conformal Representation Fails

General theorems about the numbers of edges are no longer available; but it is still possible, in examples, to draw compact cells of equal population. The rule that not more than three lines may meet at any point is maintained. Some properties of the figure were deduced from the specified distribution of density. Other properties of the figure are not deducible, they were left to chance, or to the draughtman's feeling for a pleasant appearance. The network so obtained was studied inductively by comparison with the formulae (4/10) and (4/11) for the standard clusters, or with those of the approximate general theory to be given later in § 6.

Example 1. $\rho = x$ (4/55)

This is shown in Figure 14. The diagram was constructed in the following manner. Let the typical cell extend from x_n to x_{n+1}. In order that it may be compact, it was made square. The number of people in it is

$$(x_{n+1} - x_n)\int_{x_n}^{x_{n+1}} \rho\, dx;$$

and this must be the same for every n in the region. As $\rho = x$ it follows that

$$(x_{n+1} - x_n)(x_{n+1}^2 - x_n^2) = (x_1 - x_o)(x_1^2 - x_o^2). \tag{4/56}$$

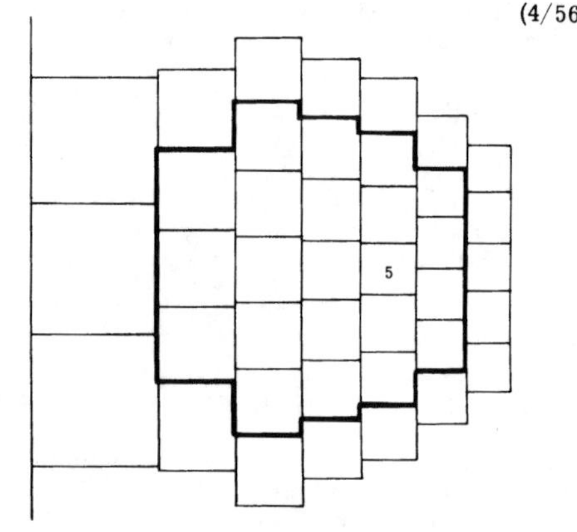

Figure 14. The density of population is proportional to distance from the long line at the western extreme. Each cell contains the same number of people.

Two of the x must be chosen arbitrarily: they were taken to be

$$x_o = 0, \qquad x_1 = 1. \tag{4/57}$$

It follows that

$$(x_{n+1} - x_n)^2 = \frac{1}{(x_{n+1} + x_n)}. \tag{4/58}$$

This cubic equation was solved in turn for $n = 1, 2, 3, 4, 5, 6$. Given x_n the method was to insert a trial value of x_{n+1} in the second member only, and to proceed by successive approximations. The results were

$$x_2 = 1.6180, \quad x_3 = 2.1343, \quad x_4 = 2.5942,$$

$$x_5 = 3.0163, \quad x_6 = 3.4108, \quad x_7 = 3.7836.$$

From them the diagram was drawn. The second differences of $x_n^{\frac{3}{2}}$ become small as n increases. Here ends the deductive portion.

To study the diagram inductively we must first note that an "edge" in our topological sense, being the line, straight or bent, between adjacent vertices, is quite distinct from the side of the square cell. The external edges of a cell are not specified unless it is in contact with external cells. Therefore we must confine attention to the cells within the thick outline. There are 22 of them. One, marked 5, has five edges. Each of the others has 6 edges. This prevalence of 6-edged cells was not prearranged: it just happened. In contrast with Figure 13 the edges of any one cell have very unequal lengths. They could be made less unequal by suitable bending of lines.

The marked cluster of 22 cells has 31 external edges and 50 internal edges. The formulae (4/10) and (4/11) namely $\underline{F} = 2(12s - 3)^{\frac{1}{2}}$ and $\underline{C} = 3\underline{s} - (12\underline{s} - 3)^{\frac{1}{2}}$ are a description of the standard clusters of $\underline{s} = 1, 7, 19, 37, \ldots$ equal regular hexagons. Let us, by inserting $\underline{s} = 22$, see how far the present cluster departs from these standards. The formulae give $\underline{F} = 32.3$ instead of 31, and $\underline{C} = 49.8$ instead of 50. This remarkable agreement probably depends on the fact that the thick outline was selected so as to enclose a cluster which looked compact. The observed $\underline{F}$ is slightly less than the minimum $\underline{F}$ for uniform density. A slight deviation in that sense will be deduced in "Comparison II" of the theory by integrals on page 165 in § 6.

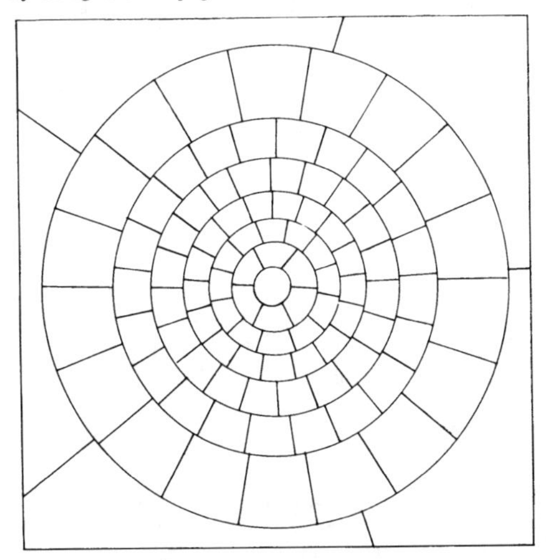

Figure 15. A sparse frontier surrounding an almost uniform middle. The density varies as e^{-r^2}. There are the same number of people in each cell. The chief use of this diagram is to check the errors of the subsequent theory of the two integrals.

Example 2. $\rho = N\pi^{-1}e^{-r^2}$ with cells each containing $\frac{N}{100}$ people. (4/59)

This is of interest because: (i) it shows again that the failure of conformal representation does not preclude mapping, (ii) because it provides another check on the subsequent approximate theory in § 6, and (iii) because it raises some subtle questions regarding compactness. Before we come to these, the reader is requested to examine Figure 15 and to decide whether each cell looks about as broad as it is long.

In order that a cell shall look compact, its radial length must be some sort of a mean between the chords of its two circular arcs. The best sort of mean is not evident. Moreover, if any precise definition of a compact shape were accepted, it would seldom be strictly applicable, because there must be a whole number of cells in a ring. Let γ_n be the number of cells in the ring between r_{n-1} and r_n; and let $\gamma_1 = 1$, so that the central cell is a circle. The geometrical mean may reasonably be preferred to the arithmetical mean, because the former agrees better with $\gamma_2 = 6$ when the density is uniform. The condition for compactness is accordingly

$$2(r_{n-1} \cdot r_n)^{\frac{1}{2}} \sin\left(\frac{\pi}{\gamma_n}\right) = r_n - r_{n-1}. \quad (4/60)$$

This can be solved as a quadratic, giving

$$\left(\frac{r_n}{r_{n-1}}\right)^{\frac{1}{2}} = \sin\left(\frac{\pi}{\gamma_n}\right) + \left\{1 + \sin^2\left(\frac{\pi}{\gamma_n}\right)\right\}^{\frac{1}{2}}. \quad (4/61)$$

It does not involve the distribution of density. Each cell will contain the same population provided

$$\gamma_n\left\{1 - \text{Exp}(-r_1^2)\right\} = \text{Exp}(-r_{n-1}^2) - \text{Exp}(-r_n^2). \quad (4/62)$$

The equations (4/61) and (4/62) are simultaneous, but not quite compatible, as γ_n must be an integer.

The radius $\underline{r}_1$ of the central cell was fixed by the condition that it is to contain $\frac{N}{100}$ people. Then r_1, together with tentative integers for γ_2, were inserted into the two formulae for $\underline{r}_2$; and that value of γ_2 was chosen which made the two values of $\underline{r}_2$ agree best. It was $\gamma_2 = 6$ giving $\underline{r}_2 = 0.2631$ for compactness, or $\underline{r}_2 = 0.2694$ for equal cell-population. The latter was accepted. A similar advance was made from $\underline{r}_2$ with tentative γ_3 to $\underline{r}_3$; and so on from ring to ring.

Table VI

ρ at frontier / ρ at center	$\underline{r}$ Radius of Frontier	$\underline{s}$ Cells Within	$\underline{F}$	$\underline{C}$
0.99	0.1005	1	6	0
0.93	0.2694	7	18	12
0.81	0.4590	19	29	42
0.64	0.6681	36	37	88
0.44	0.9061	56	41	145
0.23	1.2123	77	39	207
0.05	1.7308	95	23	264
0.00		100	--	292

These numbers will be used later in "Comparison IV" of § 6. (4/63)

The cells were made strictly equal in population; but slight random deviations from the ideal of compactness had to be tolerated. These were of the order of one percent of $\underline{r}$. The cells of the outermost row are exceptional, for they extend to infinity and cannot be made compact. The results are shown in the preceding table.

Example 3. The density varies as r^2.

This has already been treated by conformal representation in Figure 13 which was satisfactory with two exceptions: (i) the central cell contained twice the standard population; (ii) the outline, not being circular, was not suitable for testing the approximate, but versatile, theory which will be given later in § 6. Both these difficulties are here removed by taking the cells to be bounded by portions of concentric circles and radii. Let the ring between $\underline{r}_{n-1}$ and $\underline{r}_n$ contain γ_n equal cells, so that

$$\gamma_n = \underline{s}_n - \underline{s}_{n-1}\,. \tag{4/64}$$

To make a start, take $\underline{r}_1 = 1$, $\gamma_1 = 1$, so that the central cell is a circle. Its population will equal that of any other cell provided that

$$\gamma_n = \underline{r}_n^4 - \underline{r}_{n-1}^4\,. \tag{4/65}$$

This condition was satisfied strictly. There is a simultaneous condition to make each compact. As in the previous example it was taken to be that the radial length of the cell should equal the geometrical mean of the chords of its two circular arcs. This condition, expressed in formula (4/60), was satisfied as nearly as it can be with γ_n a whole number. It is not necessary to draw the figure; for when the numbers γ_n have been computed, the other properties follow by a simple argument. Let $\underline{F}_n$ be the number of edges on the circle $\underline{r}_n$; and let $\underline{C}_n$ be the number of edges inside it. Then

$$F_n = \gamma_n + \gamma_{n+1} \text{ provided } \gamma_n \geqslant 2,\quad \gamma_{n+1} \geqslant 2. \tag{4/66}$$

Also

$$C_n = C_{n-1} + F_{n-1} + \gamma_n. \tag{4/67}$$

The results were as follows:

Table VII

$\underline{r}$ Radius of Frontier	$\underline{s}$ Cells Within	$\underline{F}$	$\underline{C}$
1.0000	1	10	0
1.8212	11	32	20
2.3968	33	57	74
2.8716	68	82	166

Table VII (continued)

$\underline{r}$ Radius of Frontier	$\underline{s}$ Cells Within	$\underline{F}$	$\underline{C}$
3.2747	115	107	295
3.6371	175	132	462
3.9643	247	---	666

These numbers will be used later in "Comparison III" of § 6.

5. MATHEMATICAL DEFINITIONS OF COMPACTNESS, ALIAS HOMOPLATY

It is necessary to enquire about the homoplaty of two sorts of things: outlines and meshes. The outlines are those of islands, or of political countries, or of the shapes in Figure 1. The meshes are cells in a network drawn so that each contains the same number of people. The ideal of homoplaty for a mesh may perhaps differ from the ideal for an outline.

There are three main geometrical criteria by which compactness may be judged: one depends on shortening boundaries, another on concentrating elements of area, the third on area and perimeter jointly.

Area and Perimeter Jointly

Let the homoplaty of a closed figure be measured by

$$\frac{2(\pi \times \text{area})^{\frac{1}{2}}}{\text{perimeter}} \tag{5/1}$$

This quantity is independent of the size of similar figures, and is equal to unity for the circle. For regular polygons it runs as follows:

triangle	square	pentagon	hexagon
0.7776	0.8862	0.9299	0.9523

Concentrating Elements of Area

In order to avoid difficult questions about the length of seacoasts one might ignore the perimeter and attend instead to the radial moments of the area about its center. Thus take polar coordinates $(\underline{r}, \theta)$ and let $\underline{r}$ be zero at the intersection of any two straight lines which bisect the area. Let the boundary of the figure be specified by $\underline{r} = \underline{R}$, a given function of θ. In the elementary triangle, $d\theta$, the radial moment of order $\underline{n}$ is

$$d\theta \int_0^R r^{n+1}\,dr = \frac{R^{n+2}}{n+2}\,d\theta.$$

So the moment for the whole figure is

$$\frac{1}{n+2}\int_0^{2\pi} R^{n+2}\, d\theta = M_n, \text{ say.} \qquad (5/2)$$

The area is M_0. A variable figure of fixed area would reasonably be regarded as becoming more slender when its higher radial moments, M_1, M_2, M_3, . . . increased, and more compact when M_1, M_2, M_3, . . . decreased. Slenderness or compactness are regarded as functions of shape but not of size. Therefore each higher moment should be combined with M_0, in such a way as to cancel the size. It is not obvious whether M_1 or M_2 is the more important. Suppose that M_2 is chosen. Then an appropriate measure of compactness is

$$\frac{1}{2\pi}\frac{M_0^2}{M_2} \qquad (5/3)$$

in which the 2π is inserted to make the measure unity for a circle. For regular polygons it runs as follows:

triangle	square	pentagon	hexagon
0.8270	0.9549	0.9833	0.9924

The numbers differ from the previous measure; but the succession of shapes is the same on either, and agrees moreover with the mean visual ranking by ten selected observers (see page 140).

Minimizing the Sum of Squares of Cell-Edges

It would be idle to consider the shortening of edges apart from the distribution of population. Here the distribution will be specified not by a formula, but by a map of cells each containing the same number of people. It will be supposed that the map is accurate as to the population, but may perhaps need improvement as to the compactness of its cells. As usual three edges will meet in each vertex. For simplicity the adjustable edges will be straight. Frontiers and the cells that touch them will not be considered in the theory, but will be left to visual judgment, because of the difficulty of defining either the length or the direction of a seacoast. Let ℓ be the length of a typical edge and let $\Sigma\ell$ sum for all edges except those forming part of, or ending on, frontiers. It is proposed to define a compact map as one which minimizes $\Sigma\ell^2$, while maintaining the equality of the cell-populations. Here $\Sigma\ell^2$ is preferred to $\Sigma\ell$ or to $\Sigma\ell^{\frac{1}{2}}$ merely for mathematical conveniences, because ℓ^2 has a simpler formula than ℓ or $\ell^{\frac{1}{2}}$. Although the proposed condition involves all the interior parts of the map, yet for theoretical discussion it is desirable to attend to the most localized amendment that can alter edge-lengths, but does not alter the number of persons in any cell. Any motion of a single vertex alters the population of some cell. By taking two adjacent vertices, P and Q and arranging for their motions to compensate one another, the number of persons can be kept constant in each cell. Let P be connected to fixed vertices R and S, and similarly Q to fixed vertices T and U, as in Figure 16, where cell-edges are shown by thick lines. The actual distribution of population, in houses, villages, and towns among empty country, is too complicated for theoretical treatment. Some smoothing will be necessary. The simplest type of smoothing is that which permits us to keep constant the area of each cell, instead of the number of persons in it, during a small deformation of the map. Let that principle be adopted. To keep constant the area of the cell below RPS it is necessary that P should move parallel to RS. Similarly Q must move parallel to TU. (5/4)

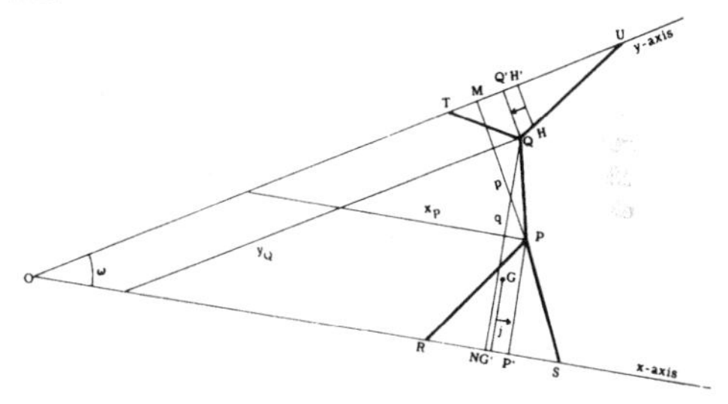

Figure 16. Cell-areas fixed while edge-lengths vary.

Produce SR and UT until they meet in O, and take OU and OS as oblique Cartesian axes of x and y respectively. Let $T\hat{O}R = \omega$. The letters for points will be used as suffixes to their coordinates. The coordinates fixed by the given map will be marked by a bar above; thus $\bar{x}_Q$ and $\bar{y}_P$ are given, but x_P and y_Q are variables in the neighborhood of the configuration which makes $\Sigma\ell^2$ a minimum. Draw QN normal to the x-axis and PM normal to the y-axis. Let the length PM be p, and the length QN be q so that

$$p = x_P \cdot \sin\omega, \qquad q = y_Q \cdot \sin\omega. \qquad (5/5)$$

If x_P varies alone, then the area $ORPQT$ increases by $\frac{1}{2} q \cdot dx_P$. If y_Q varies alone, the area $ORPQT$ increases by $\frac{1}{2} p \cdot dy_Q$. So to keep constant the area of the cell to the left of $RPQT$ it is necessary that

$$q \cdot dx_P + p \cdot dy_Q = 0 \qquad (5/6)$$

Therefore in view of (1)

$$x_P \cdot y_Q = \bar{x}_P \cdot \bar{y}_Q \,. \qquad (5/7)$$

The constancy of the area of the cell to the right of SPQU then follows automatically, because it is the only other cell affected. There is only one degree of freedom; but for symmetry, it is neater to vary $\underline{x}_P$ and $\underline{y}_Q$ at first independently, and afterwards to connect them by (5/6). For each edge there is an expression like

$$(RP)^2 = \ell^2_{RP} = (\bar{x}_R - x_P)^2 + (\bar{y}_R - \bar{y}_P)^2$$
$$+ 2(\bar{x}_R - x_P)(\bar{y}_R - \bar{y}_P) \cos \omega. \quad (5/8)$$

On working out the derivatives, and collecting them from the five variable edges, it is found that if x_P and y_Q were independent, then would

$$\frac{1}{6}\frac{\partial \Sigma \ell^2}{\partial x_P} = \left[x_P + \bar{y}_P \cdot \cos \omega\right]$$
$$- \frac{1}{3}\left[\bar{x}_Q + \bar{x}_R + \bar{x}_S + (y_Q + \bar{y}_R + \bar{y}_S) \cos \omega\right]$$
$$= \underline{j}, \text{ for short;} \quad (5/9)$$

$$\frac{1}{6}\frac{\partial \Sigma \ell^2}{\partial y_Q} = \left[y_Q + \bar{x}_Q \cdot \cos \omega\right]$$
$$- \frac{1}{3}\left[\bar{y}_P + \bar{y}_T + \bar{y}_U + (x_P + \bar{x}_T + \bar{x}_U) \cos \omega\right]$$
$$= \underline{k}, \text{ for short;} \quad (5/10)$$

$$\frac{\partial^2 \Sigma \ell^2}{\partial x_P^2} = 6, \quad \frac{\partial^2 \Sigma \ell^2}{\partial x_P \partial y_Q} = -2 \cos \omega = \frac{\partial^2 \Sigma \ell^2}{\partial y_Q \partial x_P},$$

$$\frac{\partial^2 \Sigma \ell^2}{\partial y_Q^2} = 6. \quad (5/11)$$

The second order terms in Taylor's expansion of $\Sigma\ell^2$ are therefore

$$3(dx_P)^2 - 2 \cos \omega \cdot dx_P \cdot dy_Q + 3(dy_Q)^2. \quad (5/12)$$

This quadratic form is positive for all values of dx_P and dy_Q, and so in particular for those which satisfy the restriction (5/6). Any stationary value of $\Sigma\ell^2$ must therefore be a minimum.

The conditions for stationary $\Sigma\ell^2$ are (5/6) together with

$$\frac{\partial \Sigma \ell^2}{\partial x_P} dx_P + \frac{\partial \Sigma \ell^2}{\partial y_Q} dy_Q = 0. \quad (5/13)$$

The long expressions in (5/9) and (5/10) have simple geometrical meanings. Let $\underline{P}'$ and $\underline{Q}'$ be the feet of the perpendiculars from $\underline{P}$ and $\underline{Q}$ on to the $\underline{x}$ and $\underline{y}$ axes respectively. Then

$$x_P + \bar{y}_P \cdot \cos \omega = OP', \quad y_Q + \bar{x}_Q \cdot \cos \omega = OQ'. \quad (5/14)$$

Let $\underline{G}$ be the centroid of equal particles placed at $\underline{Q}$, $\underline{\bar{R}}$, $\underline{S}$. Then

$$\bar{x}_G = \frac{1}{3}(\bar{x}_Q + \bar{x}_R + \bar{x}_S), \quad \bar{y}_G = \frac{1}{3}(y_Q + \bar{y}_R + \bar{y}_S). \quad (5/15)$$

Let $\underline{G}'$ be the foot of the perpendicular from G on to the x-axis. Then the second number of (5/$\bar{9}$) is the length $\overline{G'P'}$, reckoned positive when $\overline{OP'} > \overline{OG'}$. Denote this length by $\underline{j}$ in general, but by $\bar{\underline{j}}$ when $\underline{P}$ is in the position on the given map. Similarly, on the upper side of the figure, let $\underline{k}$ or $\bar{\underline{k}}$ be the length $\underline{H'Q'}$ reckoned positive when $\overline{OQ'} > \overline{OH'}$. Equation (5/13) may now be written

$$j\, dx_P + k\, dy_Q = 0. \quad (5/16)$$

Elimination of the differentials between (5/6) and (5/16) leaves

$$kq - pj = 0, \quad (5/17)$$

as the condition for stationary $\Sigma\ell^2$. In this $\underline{p}$ and $\underline{q}$ are both positive, so $\underline{j}$ must have the same sign as $\underline{k}$. For example the accompanying diagram shows a non-minimum of $\Sigma\ell^2$, because the arrows marked $\underline{j}$ and $\underline{k}$ point oppositely.

It often happens that ω is small or zero; so that the coordinates x_P and y_Q are too large to be measured. The distances which appear in formula (5/17) remain however easily measurable as $\omega \to 0$. Some special cases are noteworthy.

(i) If the given map is locally a honeycomb pattern, then $\underline{G}$ coincides with $\underline{P}$ and $\underline{H}$ with $\underline{Q}$, so that $\underline{j}$ and $\underline{k}$ both vanish, (5/17) is satisfied, and no amendment is needed.

(ii) If the given map has the sort of symmetry for which $\underline{p} = \underline{q}$, $\underline{j} = \underline{k}$, then no amendment is needed.

(iii) In Figure 14, where the cells are square, the condition (5/17) is not satisfied, except for special choices of the vertices.

In a perfect map the test (5/17) for local homoplaty would be satisfied for every set of five edges connected as in Figure 16.

If some amendment is required, its amount, if slight, can be computed in the following manner. Let barred symbols refer, as before, to quantities on the given map. Let Δ_P, Δ_Q be corrections such that the minimum of $\Sigma\ell^2$ occurs at

$$x_P = \bar{x}_P + \Delta_P, \quad y_Q = \bar{y}_Q + \Delta_Q. \quad (5/18)$$

If Δ_P and Δ_Q are so near to zero that their squares and product can be neglected, then it follows from (5/5, 6, 9, 10, 17), after some algebra, that

$$\frac{\Delta_P}{\bar{p}} = \frac{\bar{k}\bar{q} - \bar{j}\bar{p}}{\bar{p}^2 + \frac{2}{3}\bar{p}\bar{q} \cos \omega + \bar{q}^2 + (\bar{k}\bar{q} + \bar{j}\bar{p}) \sin \omega} = \frac{-\Delta_Q}{\bar{q}}. \quad (5/19)$$

Other Criteria by Shortening Edges

The political map of the world looks as though the makers of peace-treaties had desired to avoid drawing frontiers through dense populations. Frontiers often run along mountain ridges or through the middle of lakes. Frontiers rarely divide great cities: the partition of Berlin in 1945 was a very unusual arrangement, and it has not functioned smoothly. Shortening is a milder form of avoidance. These considerations suggest that, in a theory about shortening edges, $d\ell$ might suitably be weighted, not by 2ℓ as when $\Sigma\ell^2$ is minimized, but by some inverse power of ℓ, so as to emphasize the desirability of shortening edges that are already short. Preliminary experiments suggest that a suitable weight might vary as $\ell^{-\frac{1}{2}}$ or $\ell^{-\frac{1}{4}}$. These correspond to minimizing $\Sigma\ell^{\frac{1}{2}}$ or $\Sigma\ell^{\frac{3}{4}}$. The formulae would be more complicated than those already given for $\Sigma\ell^2$.

THE IDEAL OF HOMOPLATY REVIEWED

Besides the four or more possible definitions which have been offered in the last section, the following materials are also available for discussion:

(V) The opinions of fifty persons during visual inspection of outlines. Figure 1.

(VI) Maps of several countries drawn in cells intended to be homoplatous. Figures 5, 6, 7, 8.

(VII) The standard plane clusters of equal regular hexagons for special numbers of cells, namely s = 1, 7, 19, 37, 61, 91, and in general $3k^2 - 3k + 1$ where k is an integer. Each cluster is marked off as part of a honeycomb pattern extending beyond the cluster in all directions. The numbers of edges are

$$B = 0, \quad F = 2(12s - 3)^{\frac{1}{2}}, \quad C = 3s - (12s - 3)^{\frac{1}{2}}. \qquad (5/20)$$

For the honeycomb pattern $\Sigma\ell^2$ is a minimum for deformations which do not alter the areas of cells. The outline of a standard cluster is not circular and so is not of the most homoplatous form according to the two measures for outlines given in the last section.

(VIII) Plane clusters of equal regular hexagons for numbers of cells other than $3k^2 - 3k + 1$, where k is an integer. It has been shown that the above formulae for F and C set an ideal of compactness for an outline marked off in an extensive honeycomb pattern.

(IX) The standard clusters modified so as to make them bounded by the void. The numbers of edges are then

$$F = 0, \quad B = -3 + (12s - 3)^{\frac{1}{2}} \quad \text{for} \quad s \geqslant 7,$$
$$C = 3s - (12s - 3)^{\frac{1}{2}}. \qquad (5/21)$$

The zigzag outline forming part of the honeycomb pattern could then be smoothed, but it could not be made circular, except for s = 1 or 7. As the cells would be maintained of equal area the density of population would still be uniform.

(X) Plane conformal representations of standard clusters to which the same formulae (5/20) and (5/21) apply precisely in the absence of a singular point. Many non-uniform distributions of population can thus be mapped in hexagonal cells; but not all, for the density ρ must satisfy

$$\left(\frac{\partial^2}{\partial x^2} + \frac{\partial^2}{\partial y^2}\right) \log \rho = 0. \qquad (5/22)$$

Also a cell which contains a singular point may not have six sides. The cell-edges are often slightly curved (see Figure 13) so that there is a slight misfit between this ideal and the criteria of shortening straight edges.

(XI) The cellular mappings shown in Figures 14 and 15 for distributions of population for which (5/22) is not satisfied. The cells in these diagrams were made square merely because such cells were the least difficult to compute and draw, and were about as broad as they were long. The total length of the edges could certainly be decreased, by modifying the diagrams. The attainable decrease in length appears to be slight, perhaps about ten percent. A side of many of these squares is marked off by vertices into more than one edge, so that most of the squares have six edges.

(XII) The topological theorem (2/11) which relates to a cluster bounded by the void, and which shows that the average number of vertices to a cell is less than 6, but tends to 6 as the number of cells become infinite.

(XIII) Euler's theorem that for a closed polyhydron $s - e + v = 2$ where s, e, v are the total numbers of faces, edges and vertices respectively. This shows that if a sphere be covered by a network of cells having three edges meeting in each vertex (as they do for example in the tetrahedron, the cube, and the pentagonal dodecahedron) then the cells cannot all be six-edged. Although we do not actually wish to cover the whole sphere, Euler's theorem is a warning against placing too much emphasis on hexagonal cells.

Conclusion. This mass of evidence includes both agreements and conflicts. When the density of the population is uniform there seems to be little doubt that we should accept the "standard" clusters as the ideal of homoplaty, both for mesh and outline. Their internal cells have each six edges; most of their cells that touch the void have five edges, but some have only four. This acceptance of "standard" clusters involves the rejection of the circle as an ideally homoplatous outline, except when the cluster is bounded by the void and contains either 1 or 7 cells.

When the density of population is not uniform, the choice is not so clear. The theorem about $\Sigma\ell^2$ could be taken as a fairly general guide. But it can be argued that a minimum of $\Sigma\ell^{\frac{3}{4}}$ might be preferable. We may reasonably expect most internal cells to have 6 edges; but there is no reason to forbid other numbers. Coastal cells are likely to have about 5 or 4 edges. The various outstanding disagreements are noticeable in the abstract, yet in practice they are negligible for the historical purpose in view.

6. APPROXIMATIONS BY WAY OF INTEGRALS

In order to apply the integral calculus let it now be assumed that:

the cells are very small and numerous, (6/1)

the density of population is a continuous function of position, (6/2)

the frontiers of the cluster are of simple shapes. (6/3)

The reader may well demur to the artificiality of the second and third assumptions. They are necessary for easy theoretical progress. The errors of the third assumption will be compensated in a later section.

On a land-frontier an edge going outwards cannot start from the same point as an edge going inwards, for that would be contrary to the fundamental rule that vertices must be as simple as possible. Thus the number of vertices on the frontier approximates to the total number of cells which touch the frontier from within and without. The approximation becomes an exact equality when the land-frontier is a closed curve having at least two cells touching it on either side.

In general $\underline{F}$ is approximately equal to the total number of cells which touch the land-frontier within and without. (6/4)

For a seacoast there are no cells outside and $\underline{B}$ is approximately equal to the number of cells which touch the coast from within. (6/5)

Consider a nation of $\underline{N}$ humans divided into a large number, $\underline{s}$, of cells each containing $\underline{h}$ humans, so that $N = hs$. (6/6)

Let ρ be the local density of population.

Then $$s = \frac{N}{h} = h^{-1}\iint \rho\, d\,(\text{area}) \qquad (6/7)$$

The area of a cell is $h\rho^{-1}$. The average diameter of a cell will be $a\left|h^{\frac{1}{2}}\rho^{-\frac{1}{2}}\right|$ where $\underline{a}$ is a constant. (6/8)

The number of cells in a row of unit length would be $a^{-1}\left|h^{-\frac{1}{2}}\rho^{\frac{1}{2}}\right|$ if ρ were constant; but, as ρ may vary, a line-integral of $a^{-1}\left|h^{-\frac{1}{2}}\rho^{\frac{1}{2}}\right|$ must be taken. According to (6/4) we need two such line-integrals, one around the belt of cells lying just inside the frontier, the other around the belt just outside. The two line-integrals are then to be added. As an easy approximation, let the sum of the two line-integrals be replaced by twice the line-integral taken along the frontier itself. This gives

$$F = 2a^{-1}\left|h^{-\frac{1}{2}}\right|\int\left|\rho^{\frac{1}{2}}\right| d\,(\text{land-frontier}), \text{ approximately.} \qquad (6/9)$$

The approximation (6/9) is likely to be satisfactory if the three conditions (6/1, 2, 3) are satisfied. (6/10)

The approximation (6/9) is too small if the density is less on the frontier than on either side of it. (6/11)

This is of interest because there are several frontiers which run along the crests of mountain ranges. A correction will be considered later in special circumstances. See equation (6/25).

It is desired to express $\underline{F}$ as a function of $\underline{s}$ when $\underline{h}$ varies. By taking the square root of (6/7) and then eliminating $\underline{h}$ with (6/9), one obtains, on the understanding that all square roots are to be taken positive,

$$F = \frac{2s^{\frac{1}{2}}\int \rho^{\frac{1}{2}}\, d(\text{land-frontier})}{a\left[\iint \rho\, d\,(\text{area})\right]^{\frac{1}{2}}}, \text{ approximately.} \qquad (6/12)$$

Similarly for a seacoast, except that, as there are no cells covering the sea, 2B replaces F. In general, for a country bounded partly by land and partly by sea,

$$\frac{a(2B+F)}{2s^{\frac{1}{2}}} = \frac{\int \rho^{\frac{1}{2}}\, d\,(\text{perimeter})}{\left[\iint \rho\, d\,(\text{area})\right]^{\frac{1}{2}}}. \qquad (6/13)$$

The numeric $\underline{a}$ will depend on the shape of the cell. In Figure 14 each cell is a square, so that $a = 1$. (6/14)

In Figure 15 the cells are bounded by arcs of concentric circles and by radii as to look squarish. For them $a = 1$ will be an approximation. But in each diagram of Figure 13 the cells, except one, approximate to regular hexagons. For a regular hexagon

$$a = \left(\frac{4}{3}\right)^{\frac{1}{4}} = 1.0746, \quad \text{whence} \quad \frac{2}{a} = 1.8612. \qquad (6/15)$$

The range between $\underline{a}$ for the square and $\underline{a}$ for the hexagon is only 7 per cent.

The integrals in (6/12) and (6/13) are properties of the territory and population as a whole, independent of their division into cells. It appears accordingly that for a given distribution of population on a given territory:

$$\underline{F} \text{ and } \underline{B} \text{ vary as } \underline{s}^{\frac{1}{2}}. \qquad (6/16)$$

A slight amendment of this power $\underline{s}^{\frac{1}{2}}$ will be obtained later in §7 by replacing an integral by a sum.

The formulae (6/12) and (6/13) are fruitful, for they allow the effects of the outline of the territory and of the distribution of population within it to be studied in great variety with comparative ease. For example the effect of an elongated form, or of a gradient of density between a rich plain on one side and a barren mountain range on the other will now be shown by formulae. So also will the effect of town and rural area. Some important outlines cannot be represented by any formulae, for example the west coast of Great Britain. This will later be studied on maps in §7, but again with reference to (6/13). In all these examples the main question in the background is: how much do the various countries differ from a compact country of uniform density? The latter can be regarded as an approximation, if it does not deviate by a factor of more than two. The reason for this wide tolerance has been explained on page 140.

Theoretical countries that are completely surrounded by land, and that are small in comparison with the radius of the earth, so that they can be treated as if they were plane. The proofs in the following examples, being routine, are omitted. The results are expressed when B = 0 in terms of $\frac{aF}{2s^{\frac{1}{2}}}$ because it is a definite consequence of the assumed contour and density; whereas a depends by a few percent on the shape of the cell. For an island F should be replaced by 2B, or for a mixed boundary by 2B + F.

Comparison I. Various Outlines With Uniform Density

Then ρ cancels leaving

$$\frac{aF}{2s^{\frac{1}{2}}} = \frac{\text{perimeter}}{(\text{area})^{\frac{1}{2}}} . \qquad (6/17)$$

The unit of length cancels in the second member, which is the same for countries of the same shape however small they are. The reciprocal, namely $(\text{area})^{\frac{1}{2}}(\text{perimeter})^{-1}$, has been taken, together with a factor of $2\sqrt{\pi}$, as a measure of compactness in (5/1).

For a circular outline

$$\frac{aF}{2s^{\frac{1}{2}}} = 2\sqrt{\pi} = 3.5449. \qquad (6/18)$$

For a regular hexagon

$$\frac{aF}{2s^{\frac{1}{2}}} = 3.7224. \qquad (6/19)$$

The exact theory of networks, which is given above, shows that the outline, which gives the least F, roughly resembles a hexagon rather than a circle. The present theory puts the hexagon and the circle in the wrong order. This is because the shape of the individual cells is slightly blurred or mushed. Nevertheless the author is convinced that the present approximations are well worth pursuing, because (i) the integrals are so much more manageable than the exact theory, and (ii) the discrepancy is only a few percent, and is less than many of the uncertainties of the geographical and historical facts with which the theory will be compared.

For a rectangular outline

The result depends only on the ratio, say, of adjacent sides of the rectangle.

$$\frac{aF}{2s^{\frac{1}{2}}} = 2\left[\mu^{\frac{1}{2}} + \mu^{-\frac{1}{2}}\right] \qquad (6/20)$$

which works out thus

μ	=	1	4	9	16	25
$\frac{aF}{2s^{\frac{1}{2}}}$	=	4	5	6.66	8.5	10.4

The most elongated country in the world is Chile. It is about 20 times as long as it is broad. The population of Chile is not uniform, and Chile has a seacoast. We may however, imagine a hypothetical country of that extreme elongation, but otherwise conforming to the present specification. Then $\frac{aF}{2s^{\frac{1}{2}}}$ is only 2.35 times its value 4 for a square.

Comparison II. A Uniform Gradient of Density in a Square Outline

This may be regarded as a formalized model of U.S.A. Let the density vary as the distance from a line parallel to a pair of sides of the square. Let the distances of these sides from the line of zero density be x_1 and x_2; and $\frac{x_2}{x_1} = \gamma$, where $\gamma > 0$, so that γ is the ratio of the extreme densities in the country. The result is that

$$\frac{aF}{2s^{\frac{1}{2}}} = \frac{\sqrt{2}}{3}\frac{(7\gamma + 10\gamma^{\frac{1}{2}} + 7)}{(\gamma+1)^{\frac{1}{2}}(\gamma^{\frac{1}{2}}+1)} . \qquad (6/21)$$

For example:

γ	=	1	2	9	1000	∞
$\frac{aF}{2s^{\frac{1}{2}}}$	=	4	3.9617	3.7268	3.4584	3.2998

As the line of zero density recedes to a great distance from the country, $\gamma \to 1$ and $\frac{aF}{2s^{\frac{1}{2}}} \to 4$, which is its value for a square of uniform density. The other extreme occurs when the line of zero density coincides with one side of the country; then $\gamma \to \infty$ and $\frac{aF}{2s^{\frac{1}{2}}}$ is less than for a circle of uniform density. It is remarkable that the whole possible variation of $\frac{aF}{2s^{\frac{1}{2}}}$ is so slight. An exact treatment of a uniform gradient is given via (4/55), but for a different contour.

Comparison III. A Circular Outline With Either a Central City, or a Central Desert

Let the density vary as the mth power of the distance from the center of the area. (6/22)

If $m > 0$ the center is uninhabited, rather like that of Australia. If $m < 0$ the center is the most popular part, rather like that of Mexico. Although for $m < 0$ the density is infinite at the center, yet for $-2 < m < 0$ the total population is finite. The present treatment will be confined to the case of a finite total; and it will be supposed that any infinite density at the center is smoothed away by some slight modification such as a central pond, which need not delay us. The result is

$$\frac{aF}{2s^{\frac{1}{2}}} = \left[2\pi(m+2)\right]^{\frac{1}{2}} \qquad (6/23)$$

The radius of the circle does not appear in this formula. When $m = 0$ the density is uniform and the result then agrees with (18). Slight variations of m from zero are swamped by the additive 2.

The same distribution of density was treated exactly by the conformal representation of hexagonal cells in §4. The outline of the net cannot of course be a circle or any other smooth curve, but it sufficiently resembles a circle (see Figure 13) for a comparison to be interesting. The exact formula was (4/49), namely

$$F = \left\{24(m+2)s'\right\}^{\frac{1}{2}}\left\{1 - \frac{m+2}{8s'}\right\}^{\frac{1}{2}}.$$

For large s this becomes $F = \left\{24(m+2)s\right\}^{\frac{1}{2}}$, and then agrees with the present approximation provided that $a = \sqrt{\left(\frac{\pi}{3}\right)} = 1.02$. An independent estimate made for hexagonal cells in (6/15) gave $a = 1.07$.

For $m = 2$ a better test of the present approximation is available because the problem has been solved for a circular outline in Table VII by compact cells bounded by portions of radii and by concentric circles. As these cells are squarish, it is proper to put $a = 1$. The approximate formula (6/23) then gives $F = 4(2\pi s)^{\frac{1}{2}} = 10.026 s^{\frac{1}{2}}$. The two theories are compared thus:

s =	1	11	33	68	115	175
F counted, =	10	32	57	82	107	132
$10.026 s^{\frac{1}{2}}$ =	10.0	32.2	57.6	82.7	107.5	132.6

The agreement is satisfactory.

Comparison IV. A Circular Territory Having a Uniformly Populated Center and a Density Either Increasing or Decreasing Towards the Frontier.

Let the outline have radius r_1 and let the density be

$$\rho = Ae^{kr^2} \qquad (6/24)$$

where A and k are constants. At the center $\rho = A$ for all k; thus the variation of ρ is milder than in the previous case. The result is

$$\frac{aF}{2s^{\frac{1}{2}}} = 2\left[\frac{\pi \log_e\left(\frac{\rho_o}{\rho_1}\right)}{\frac{\rho_o}{\rho_1} - 1}\right]^{\frac{1}{2}} \qquad (6/25)$$

where ρ_1 is the density at the frontier, and ρ_o is that at the center. Computation from formula (25) gives

$\frac{\rho_1}{\rho_o}$ =	0.001	0.01	0.1	1	10	100	1000
$\frac{aF}{2s^{\frac{1}{2}}}$ =	0.2948	0.7646	1.793	[3.545] as limit	5.670	7.646	9.322

It would be obvious without calculation that, for a given population, a dense frontier makes more external edges than does a sparse frontier. The calculation shows this quantitatively. It is noteworthy that F is insensitive to variations of $\frac{\rho_1}{\rho_o}$; for example from end to end of the above range $\frac{\rho_1}{\rho_o}$ varies in a ratio of a million, but $\frac{aF}{2s^{\frac{1}{2}}}$ only in a ratio of about 32. Frontiers, where the density is much less than the average of the countries which they separate, occur between France and Spain, between Argentina and Chile, between Brazil and Colombia, between Egypt and Libya, and elsewhere. But Tibet seems to be the only political state which is entirely surrounded by a frontier less populated than its body.

The special case obtained by putting $k = -1$ in (6/24) has already been treated for the same outline by a more exact method via (4/59). The result is shown in Figure 15. The form of the cell is approximately square; and so for comparison a should be set at unity in $\frac{aF}{2s^{\frac{1}{2}}}$. The two methods are compared below.

Table VIII

s =	1	7	19	36	56	77
F, counted on Figure 15, =	6	18	29	37	41	39
F, by integrals =	7.1	18.4	29.3	37.9	42.6	41.2

The agreement is tolerable.

Comparison V. Alternations Between Towns and Rural Areas

Let the territory be a square of side 4π bounded by $x = \pm 2\pi$, $y = \pm 2\pi$, and let the density be given by

$$\rho = A(\cos nx \cdot \cos ny)^{2k}, \qquad (6/26)$$

where $\underline{A}$ is a positive constant, and $\underline{n}$ and $\underline{k}$ are positive integers. Then ρ varies between maxima of $\underline{A}$, and minima of zero. The lines on which ρ vanishes divide the territory into squares of side $\frac{\pi}{n}$. There are $16n^2$ such squares in the territory, counting in portions near the edges. At the center of each of these squares on side $\frac{\pi}{n}$ there is a maximum of ρ, which may be likened to the center of a town. As $\underline{k}$ increases, the population concentrates more towards the center of each town, leaving a more empty surrounding space. This artificially planned pattern of towns has been chosen because it makes the integrals manageable. Before integration the powers of the cosines are examined in multiple angles. Both $\underline{A}$ and $\underline{n}$ cancel, leaving a result which depends only on $\underline{k}$, thus:

k =	0	1	2	3	4	5	60	∞ even	
$\frac{aF}{2s^{\frac{1}{2}}}$ =	4.000	5.093	5.333	5.432	5.486	5.519	5.645	5.657	(6/27)

Over the whole range, from uniformity to extreme urbanization, $\frac{aF}{2s^{\frac{1}{2}}}$ varies only in the ratio $\sqrt{2}$.

The fitting of $\underline{k}$ to an observed distribution of population can be done by way of the ratio of the greatest density to the mean density; for this ratio depends only on $\underline{k}$, thus

$$\frac{\text{maximum density}}{\text{mean density}} = \frac{2^{4k}[k!]^4}{[(2k)!]^2} = \phi(k), \text{ say.} \qquad (6/28)$$

The run of this function is as follows:

k =	0		1		2		3		4		5
φ(k) =	1		4		7.1111		10.2400		13.3747		16.5120
differences		3		3.1111		3.1289		3.1347		3.1373	

Although the formula for ϕ is complicated, the arithmetic shows that its difference

$\phi(\underline{k} + 1) - \phi(\underline{k})$ is almost constant.

For large $\underline{k}$, Stirling's approximation gives

$$\phi(k) = \pi k, \text{ simply.} \qquad (6/29)$$

In the densest parts of London, namely in Shoreditch and Southwark, the density at the census of 1911 was 176 times the density for England and Wales as a whole. According to (6/29) this would correspond to k = 56. The order of magnitude of k is thus indicated for a very urbanized country.

Summary on the Method of the Two Integrals

In so far as it ignores the detailed structure of the network of cells, this method is an approximation. On the other hand it derives much advantage from the strength of the integral calculus. Comparisons have been made in special cases with more exact methods, and these show that the present approximation is good enough for the purpose in view, namely for applications to history.

For a population uniformly spread over and beyond a regular hexagon, $\frac{aF}{2s^{\frac{1}{2}}} = 3.72$. This may be regarded as the standard. Apart from the great irregularity of some frontiers, which will be studied in §7, it appears that $\frac{aF}{2s^{\frac{1}{2}}}$ is rather insensitive to such variations of outline or of distribution of population as occur in the actual world. The alternation between towns and rural areas may increase $\frac{aF}{2s^{\frac{1}{2}}}$ to 5.65. A form as elongated as that of Chile would, apart from other influences, increase $\frac{aF}{2s^{\frac{1}{2}}}$ to 9.5. A uniform gradient of density may decrease it to 3.3. A dense frontier might raise it to 8, and a sparse frontier might lower it to 0.8. Altogether the range

$$1 < \frac{a(2B + F)}{2s^{\frac{1}{2}}} < 10 \qquad (6/30)$$

seems likely to include every country.

7. LENGTHS OF LAND FRONTIERS OR SEACOASTS

In the previous section integrals were taken around simple geometrical figures, as a preliminary to taking them around frontiers shown on political maps. An embarrassing doubt arose as to whether actual frontiers were so intricate as to invalidate that otherwise promising theory. A special investigation was made to settle this question. Some strange features came to notice; nevertheless an over-all general correction was found possible. The results will now be reported.

To divide the difficulties connected with

$\int \rho^{\frac{1}{2}} d$ (frontier), the effect of alterations of ρ on a straight frontier have already been studied in (6/27), and now, on the contrary, ρ will be supposed constant, but the outline will be studied in all its natural intricacy.

At first I tried to measure frontiers by rolling a wheel of 1.8 centimeters diameter on maps; but there is often fine detail, which the wheel cannot follow; some convention would be needed as to what detail should be ignored and what retained; considerable skill would be needed to guide the wheel in accordance with any such decision; and in practice the results were erratic.

Much more definite measurements have been made by walking a pair of dividers along a map of the frontier so as to count the number of equal sides of a polygon, the corners of which lie on the frontier. In this respect the polygon resembles one "inscribed" to a circle, but some sides may lie outside the frontier, and the polygon need not be closed. Its total length, $\Sigma \underline{\ell}$, has been studied as a function of the length, $\underline{\ell}$, of its side. This process comes down to use from Archimedes, and is standard in pure mathematics. For perfection the dividers should be stepped along a map on a globe; but only plane maps were available. To avoid most of the errors caused by the projection of the globe on to a plane, the investigation has been restricted to moderately small portions of the earth's surface, the largest being Australia. Also, for definiteness, the maps have been specified. Usually $\underline{\ell}$ was fixed in advance, and a fractional side was estimated at the end of the walk. If it was desired to have a whole number of sides, then $\underline{\ell}$ had to be adjusted by successive approximations. The main purpose was to study the broad average variation of $\Sigma \underline{\ell}$ with $\underline{\ell}$. But some of the incidental details are so interesting that they deserve mention; for they are in marked contrast with the properties of the smooth curves which have their lengths integrated in the textbooks. The moving spike of the dividers describes a circle which may intersect the frontier in more than one point; if so, the intersection to be chosen is that one which comes next in order forward along the frontier. This obvious rule has surprising consequences: it sometimes prevents the polygon from having a whole number of sides.

As an explanation of how chance can arise in a world which he regarded as strictly deterministic, Henri Poincaré* (no date) drew attention to insignificant causes which produced very noticeable effects. Seacoasts provide an apt illustration. For the spike of the dividers may just miss, or just catch, a promontory of land or the head of a loch; so that an insignificant change in $\underline{\ell}$ may alter $\Sigma \underline{\ell}$ noticeably.

The west coast of Britain from Land's End to Duncansby Head was chosen as an example of a coast that looks more irregular than most other coasts in an atlas of the world. The quality and scale of the maps of Britain were such that the irregularities of the coast greatly predominated over any errors that are likely to have occurred in the processes of drawing, printing, or reading the maps. For the longer steps the Times Atlas 1900 was used. The 10 kilometer steps were counted on the British Isles Pocket Atlas 1935 by John Bartholemew F.R.G.S. Narrow waters were regarded as barred by any bridges shown in the latter atlas. The Mersey tunnel was also regarded as a barrier, but the Severn tunnel was disregarded. These rules were kept the same for all lengths of step. The over-all length in one step was found to be 971 km. The attempt to make exactly two equal steps is worth stating in detail because it illustrates principles and peculiarities. The map was page 15 of the Times Atlas dated 1900. A circle with center at Duncansby Head and radius 490 km. cuts the coast of Cumberland at a point $\underline{P}$ near Silecroft. There are other intersections, but they are further forward along the coast, and therefore must be ignored. A circle with center $\underline{P}$ and the same radius cuts the coast in about ten points in the north of Scotland; these, being backward along the coast, must be ignored. The first forward intersection is at Land's End. So there are two steps with a total length of 980 km. on the southward journey. The northward journey is quite different. There is a point $\underline{Q}$ on Morecambe Bay which is 498 km. from either Land's End or Duncansby Head. A circle with center at Land's End and radius 498 km. cuts the coast in several points, but $\underline{Q}$ comes first along the coast. A second circle with center $\underline{Q}$ and the same radius cuts the coast in several points, of which the next forward along the coast is near Cape Wrath. So we cannot arrive at Duncansby Head. Moreover a search for some other midpoint, which would allow the northward journey to be made in two equal steps, failed to find any. The northward journey is impossible according to the rule of "next intersection forward along the frontier"; yet this rule seems too reasonable to be abandoned. The strange impossibility arises from the attempt to hit off a whole number of sides; it does not occur, if a fractional side is estimated at the end. The results are summarized below.

The West Coast of Britain

Start	N or S	N	S	S	S	S	S
Length of side, km.	971	490	-	200	100	30	10
Number of sides	1	2	2	5.9	15.4	69.1	293.1
Total length, km.	971	980	-	1180	1540	2073	2931

As to how the total length $\Sigma \underline{\ell}$ may be expected to vary with the length $\underline{\ell}$ of the side, I have no theory. Quite empirically the logarithms of these

*Poincaré, H. Science and Method. Trans. by Maitland, F. London, Thomas Nelson.

variables were plotted against one another in Figure 17; and a straight line was drawn through the points. More evidence would be needed before one could say whether the deviations from the straight line are of any interest. I am inclined to regard them as random. The important feature for present purposes is that the slope of the graph is only moderate even for such a ragged line as the western shore of Great Britain. On the straight line in Figure 17 the total length varies inversely, as the fourth root of the side, that is

$$\Sigma \ell \propto \ell^{-0.25} . \qquad (7/1)$$

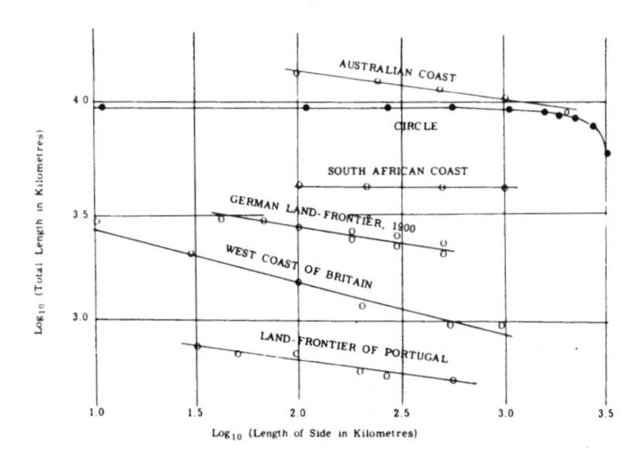

Figure 17. Measurements of curves by way of polygons (not shown) which have equal sides and have their corners on the curve. The slope of a graph shows how the total length of the polygon increases as its side becomes shorter. The convergence for the circle contrasts with the behavior for frontiers.

As an example of a less indented coast, that of Australia was selected, in particular as shown on Map 30 of the Times Atlas dated 1900. The starting point was always at the southernmost point, and the motion was always counterclockwise. These are the results:

The Coast of the Australian Mainland

Length of step, km.	2000	1000	500	250	100
Number of steps	4.7	11.04	23.9	52.8	144.2
Total length, km.	9400	11,040	11,950	13,200	14,420

The logarithms are plotted in Figure 17. The points lie near a straight line of slope−0.13. This is about half the slope for the west coast of Britain. For comparison with the Australian coast, a circle containing a plane area of 7.636×10^6 km.^{2}, which is given as the area of the Australian mainland, was inscribed with regular polygons, and their calculated perimeters are shown on the diagram.

The coast of South Africa was studied because it is exceptionally smooth over a long length. The starting point was Swakopmund, and the end point was Cape Sta. Lucia. The map was page 118 of the Times Atlas, 1900.

Coast of South Africa

Side in km.	1000	500	215.5	100
Number of sides	4.12	8.31	19.78	43.34
Total length, km.	4120	4155	4263	4334

The same type of empirical formula is again found to be suitable but with a smaller parameter so that

$$\Sigma \ell \propto \ell^{-0.02} . \qquad (7/2)$$

The land-frontier of Canada is in part defined by a meridian and a parallel of latitude which are so far definite in length; but many other land-frontiers follow winding rivers. It is not surprising therefore to find disagreement between official statements about the lengths of frontiers. Here are some collected from amongst otherwise concordant data in Armaments Year-Book 1938.

Table IX

Land-frontier Between	Kilometers as Stated by:	
	the Former Country	the Latter Country
Spain and Portugal	987	1214
Netherlands and Belgium	380	449.5
U.S.S.R. and Finland	1590	1566
U.S.S.R. and Roumania	742	812
U.S.S.R. and Latvia	269	351
Estonia and Latvia	356	375
Yugoslavia and Greece	262.1	236.8

There is evidently no official convention which can hamper the intellectual discussion of the lengths of frontiers.

Each of these frontiers is irregular. That between Spain and Portugal was selected for measurement, and was more precisely defined as extending from a railway bridge at the northwest to the tip of an estuary at the south, as shown in the Times Atlas dated 1900, map 61. Some of the results depend on whether the polygon is started at the S. or the N.W. end, as may be seen from the following table. When an attempt is made to construct a polygon of three sides starting at the northwest, a peculiar difficulty occurs. If the side $\leqslant$ 206 km., it is much too short; if the side $>$ 206 km., it is much too long; so that there is no solution. The critical length, 206 km., is that at which the point of the dividers just grazes the northeast shoulder of Portugal. In spite of this peculiarity the general run of the points on the diagram is tolerably straight.

Land-frontier Between Spain and Portugal

Start	S or NW	S or NW	S	NW	S	NW	S	NW	S	NW
Number of sides	1	2	3		7.05	7.07	13.10	13.06	27.2	27.05
Side, km.	543	285	201	---	100	100	56.29	56.36	30	30
Total, km.	543	570	603	---	705	707	737	736	816	812

As an example of a variegated land-frontier bounded partly by rivers, partly by mountains, and elsewhere by neither, I chose that of Germany as shown in the Times Atlas, dated 1900, pp. 39-40. Since then the frontier has been much altered, but previously it had persisted from 1871. Results follow:

Land-frontier of Germany in 1900

Start	E	W	E	W	E	W	E	W	W	W
Side in km.	500	50	300	299	180	180	100	99.6	66.0	40.5
Number of sides	4.02	4.52	7.62	7.85	13.75	14.92	28.05	28.1	44.45	74.9
Total, km.	2010	2260	2286	2353	2475	2686	2805	2799	2934	3053

Again the total length depends on whether the start is at the east or west. Nevertheless, for a conspectus, a straight line on the graph may serve.

Conclusion on Polygons of Equal Sides Inscribed to Frontiers

(1) The fitting of a whole number of sides can only be done by troublesome successive approximations.

(2) It is occasionally impossible to fit a whole number of sides.

(3) The possibility of fitting a whole number of sides occasionally depends on which end of the frontier is taken as the starting point.

(4) For the above three reasons it is preferable to allow the final side to be an estimated fraction of the standard side.

(5) The total polygonal length, including the estimated fraction, usually depends slightly on the point that is taken as the start.

(6) It is doubtful whether the total polygonal length of a seacoast tends to any limit as the side of the polygon tends to zero.

(7) To speak simply of the "length" of a coast is therefore to make an unwarranted assumption. When a man says that he "walked 10 miles along the coast," he usually means that he walked 10 miles near the coast.

(8) Official statements of the lengths of some land-frontiers disagree with one another in ratios such as 1:1 or 1:2.

(9) Although the phenomena mentioned in (2), (3), (5) and (6) above are peculiar and disconcerting, yet coexisting with them are some broad average regularities, which are summarized by the useful empirical formula

$$\Sigma \ell \propto \ell^{-\alpha} \qquad (7/3)$$

where $\Sigma \ell$ is the total polygonal length, ℓ is the length of the side of the polygon, and α is a positive constant, characteristic of the frontier.

(10) The constant α may be expected to have some positive correlation with one's immediate visual perception of the irregularity of the frontier. At one extreme $\alpha = 0.00$ for a frontier that looks straight on the map. For the other extreme the west coast of Britain was selected, because it looks one of the most irregular in the world; for it α was found to be 0.25. Three other frontiers which, judging by their appearance on the map were more like the average of the world in irregularity, gave: $\alpha = 0.15$ for the land-frontier of Germany in about A.D. 1899; $\alpha = 0.14$ for the land-frontier between Spain and Portugal; and $\alpha = 0.13$ for the Australian coast. A coast selected, as looking one of the smoothest in the atlas, was that of South Africa, and for it $\alpha = 0.02$.

(11) The relation $\Sigma \ell \propto \ell^{-\alpha}$ is in marked contrast with the ordinary behavior of smooth curves, for which

$$\Sigma \ell = A + B\ell^2 + C\ell^4 + D\ell^6 + \ldots \qquad (7/4)$$

where A, B, C, D, . . . are constants. This property is used in the "deferred approach to the limit" (Richardson and Gaunt, 1927).

The Effect of the Irregular Form of Frontiers on the Variation of the Numbers F and B of Edges with the Numbers s of Cells in a Fixed Country.

Many interesting results have been obtained from the formulae (6/12) and (6/13). For a fixed country and a fixed distribution of population these assert that

$$\text{F or B varies as } s^{\frac{1}{2}} \int \rho^{\frac{1}{2}} d \text{ (frontier)}. \qquad (7/5)$$

From (7/3) it appears that a frontier may not have a finite length, and so the integral may diverge. Fortunately the integral was put into the formula merely as an easy approximation to a sum. The corrected form of the statement is

$$F \propto s^{\frac{1}{2}} \sum_n \rho_n^{\frac{1}{2}} \ell_n \qquad (7/6)$$

in which ℓ_n is the side of the nth cell on the land-frontier; similarly for B and the seacoast.

Case I. The density of population is independent of position. Then (7/6) simplifies to

$$F \propto s^{\frac{1}{2}} \sum_n \ell_n \qquad (7/7)$$

Moreover the ℓ_n are all equal. The present studies of frontiers have shown that $\Sigma \ell_n \propto \ell_n^{-\alpha}$, so that $F \propto s^{\frac{1}{2}} \ell_n^{-\alpha}$. But $\ell_n \propto s^{-\frac{1}{2}}$. Therefore finally

$$F \text{ varies as } s^{\frac{1+\alpha}{2}} \qquad (7/8)$$

There is a similar formula for B, but with its appropriate α. On the average for the world, α is probably about 0.1, so that $\frac{1}{2}(1 + \alpha)$ is about 0.55. (7/9)

Case II. Density non-uniform. The variation of ρ has already been treated. The present question is whether the factor by which these results should be corrected is $s^{\frac{\alpha}{2}}$, the same as for uniform density. To clarify this question let the terms of $\sum_n \rho_n^{\frac{1}{2}} \ell_n$ be rearranged so as to bracket together those with the same value of ρ. That is to say, the sum, given in the form of Riemann, is to be rearranged in that of Lebesgue. To each bracket separately the results already stated in case I apply; but α may perhaps vary from bracket to bracket. The author has no information as to whether the local density of population is correlated with the local irregularity of the frontier. So, as the factor $s^{0.05}$ is not very important, he proposes to regard the correction furnished by case I as sufficient, pending further investigation.

8. CATALOGUE OF TERRITORIES ARRANGED IN ORDER OF POPULATION

The cellular map becomes more complete as h, the standard number of persons in a cell, becomes less. It is unfortunately necessary to neglect populations that are not sufficient to form a cell. At $h = 10^7$ many of the smaller sovereign states are omitted; so that this is the greatest h that is tolerable for world-totals. At $h = 10^6$, nearly all the important sovereign states are included; so that the representation of the world as a whole is fairly satisfactory. At $h = 10^5$, the mapping would be adequate also for some more local purposes. The first step towards attaining world-totals of cells and of edges is therefore to arrange the parts of the land-area in the order of their populations, so that the summations can be stopped at suitable termini. Incidentally this arrangement reveals an interesting statistical distribution.

The date for the populations has been taken to be the middle of A.D. 1910. An earlier date, say 1875, would have been preferable for comparison with the history of 1820 to 1929, if censuses had been available; but many were then lacking. Even for 1910 it is necessary, for some territories, to take later censuses and to extrapolate from them backwards. One has to do this for the French Congo. After adjustment to mid-1910, unreliable digits have been curtailed, or set below the line.

In order to appreciate whether a new classification is valid, it should be played with a game, strictly in accordance with its own rules. One rule about these maps is that a cell may not spread across a political frontier. There are many degrees of political unity or separation. Here foreign policy and defence have been taken as the criterion. Continuous territories having a unified foreign policy and defence are treated as one block; for example Austria, Hungary, Bosnia, and Herzegovina in 1910. The Union of South Africa was instituted one month before the date for this list; but at first the Union had no foreign minister and British troops continued to be stationed on its territory. Accordingly there was in 1910 a British block consisting of the Union of South Africa, together with Basutoland, Swaziland, Bechuanaland, Nyasaland and the Rhodesias.

Another rule is that a cell must not spread across the void. The population of any archipelago, such as Britain or Japan, has therefore been subdivided among its separate islands. A bridge or causeway, however, has been regarded as a land-connection; so for example, Manhattan, Long Island, Rhode Island, Galveston, Walcheren, Tholen, and Funen were reckoned as parts of their mainlands. A politically continuous block of French Africa (see a map in E 1, 319) is regarded as separated by the Sahara desert into northern and southern portions. The oases are classified among the islands.

As the author did not find any list ready-made according to these rules, he had laboriously to construct that which follows. The preliminary method of search for the larger populations was to turn over consecutively the pages of the Statesman's Year Book for 1914.

Much the worst uncertainty is that about the population of China.

For populations below 50,000 persons, a card-index was constructed having, for each territory, a card bearing an estimate of the population at mid-1910. The cards were arranged in order of population p. The index went down as far as 5,000 persons. When the continuing search, among the sources of information mentioned on page 151, had failed to add any more territories to the collection, the territories were numbered consecutively in order of population, beginning with China which was numbered one, and continuing to the end of the index at about 431. The number thus assigned has been called the serial number and denoted by t.

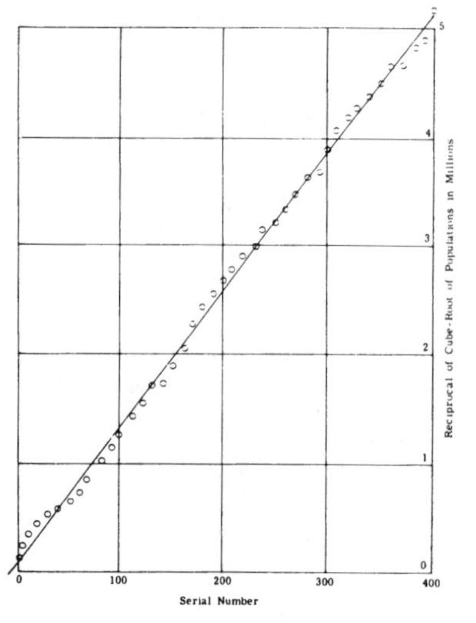

Figure 18

It is not obvious that territories specified thus, partly by pure geography and partly by political organization, could be expected to have any simple statistical distribution; but Figure 18 shows that in fact they do, at least approximately. The straight line in that diagram has the formula

$$p^{-\alpha} = \beta t + \gamma \qquad (8/1)$$

where, the unit of p being a million persons,

$$\alpha = \frac{1}{3}, \quad \beta = 0.0127, \quad \gamma = 0.06. \qquad (8/2)$$

The deviations of the observed points from the straight line are a wavy function of the serial number. For $t > 200$ some of this waviness may be due to the rounding off of uncertain populations to the nearest whole number of thousands; but the departures around $t = 10$ or $t = 70$ are much too large to be attributed to any inaccuracy in the data.

At an earlier stage, before Figure 18 was conceived, a formula of the type (8/1) had been fitted to three points, on or near the observed distribution, with the result that

$$\alpha = 0.28963, \quad \beta = 0.01046, \quad \gamma = 0.17205. \qquad (8/2A)$$

It is not suggested that so many as five digits are significant; two might be enough; but, in the ensuing argument about curtailment, all those digits were retained for the sake of internal consistency.

The frequency of territories on the scale of p is

$$\frac{dt}{dp} = -\frac{\alpha}{\beta} p^{-\alpha-1} = f(p), \text{ say.} \qquad (8/3)$$

Such a truncated inverse power law is called by Cramér (1946, p. 248) a Pareto distribution. A great many empirical distributions, of a general type of which (8/1) is a particular example, have been collected by G. K. Zipf (1949) who offers theoretical, though rather vague, explanations of them.

How to Curtail Digits in This Distribution

For comparison with deadly quarrels that have been classified so that the number of quarrel-dead increases by whole powers of ten, it is necessary to divide the list of territories in a similar manner, namely into sections relevant to cell-contents of $h = 10^7, 10^6, 10^5, 10^4$ persons. The question where to cut the list for this purpose depends on the least population which can be regarded as approximately one cell. Is it to be $\underline{h}$ or $\frac{1}{2}\underline{h}$, or something intermediate between $\underline{h}$ and $\frac{1}{2}\underline{h}$? In ordinary computations the custom is to round off to the nearest integer, because there is reason to believe that the frequency of the neglected decimals is constant; so that the total of any large collection of numbers is almost the same whether individually they have been rounded off or not. That custom would lead one to regard any population between $\frac{1}{2}\underline{h}$ and $\frac{3}{2}\underline{h}$ as approximately one cell; but here it lacks justification because the frequency varies rapidly. The usual statistical practice is to adhere to the rule of rounding off to the nearest integer, but afterwards to correct, in the manner of W. F. Sheppard, 1898, for the joint effect of grouping and of variable frequency. Corrections for grouping have been reviewed by M. G. Kendall (1943, pp. 68-79). Although the present author is addicted to such deferred approaches to the limit (Richardson and Gaunt, 1927, pp. 306-307) he felt that territories, which might have to be drawn on maps, had too strong an individuality to be thus slurred over; and that it would be better to know the names of those which were to be included. Accordingly the cuts have been made at points chosen so as to keep the total population of a large collection of territories the same, whether the populations of the individual territories have been rounded off or not. This will be called the principle of the invariance of totals. The following development may possibly be original.* Let K be an integer and suppose that the populations, p, to be founded off to Kh are all those lying in the range

$$h\theta_{K-\frac{1}{2}} < p < h\theta_{K+\frac{1}{2}} \qquad (8/4)$$

where the θ are constants to be determined.

The customary assumption that $\theta_{K+\frac{1}{2}}$ equals $K + \frac{1}{2}$ will here be avoided, and instead a small deviation $\epsilon_{K+\frac{1}{2}}$ such that, for any K,

$$\theta_{K+\frac{1}{2}} = K + \frac{1}{2} + \epsilon_{K+\frac{1}{2}} \qquad (8/5)$$

will be expected. It will be assumed as usual that the single cut at $\theta_{K+\frac{1}{2}}$ serves both for rounding up to K + 1 and down to K. For then

$$K - 1 < \theta_{K-\frac{1}{2}} < K < \theta_{K+\frac{1}{2}} < K + 1. \qquad (8/6)$$

As the question is about probabilities, it seems fitting to answer it by way of the smoothed, rather than the actual frequencies. Accordingly the frequency of territories is taken to be $\frac{dt}{dp}$ as given by (8/3). So the total population between $h\theta_{K-\frac{1}{2}}$ and $h\theta_{K+\frac{1}{2}}$ will be unaffected by the rounding off of p to Kh if

$$0 = \int_{h\theta_{K-\frac{1}{2}}}^{h\theta_{K+\frac{1}{2}}} (p - Kh)p^{-\alpha-1}\,dp. \qquad (8/7)$$

This condition is simplified by setting $\frac{p}{Kh} = x$, (8/8)

for then

$$0 = \int_{\frac{\theta_{K-\frac{1}{2}}}{K}}^{\frac{\theta_{K+\frac{1}{2}}}{K}} (x - 1)x^{-\alpha-1}\,dx. \qquad (8/9)$$

The indefinite integral in (8/9) is

$$\frac{x^{1-\alpha}}{1-\alpha} + \frac{x^{-\alpha}}{\alpha} = E(x), \text{ say.} \qquad (8/10)$$

The condition connecting any two successive cuts is therefore

$$E\left(\frac{\theta_{K+\frac{1}{2}}}{K}\right) = E\left(\frac{\theta_{K-\frac{1}{2}}}{K}\right). \qquad (8/11)$$

When K becomes large the range of integration shrinks, because by (8/6)

$$\frac{\theta_{K-\frac{1}{2}}}{K} > 1 - \frac{1}{K} \text{ and } \frac{\theta_{K+\frac{1}{2}}}{K} < 1 + \frac{1}{K}. \qquad (8/12)$$

Ultimately $x^{-\alpha-1}$ has not room to vary. The circumstances are then like those in ordinary computations, so that:

<u>The custom of rounding off to the nearest integer is here correct if that integer be large.</u> (8/13)

Starting from $\theta_{K+\frac{1}{2}} = K + \frac{1}{2}$ when K is infinite, the relation (8/11) which decreases the suffix of θ by unity, was applied repeatedly so as finally to yield $\theta_{\frac{1}{2}}$. The reduction was made by two processes, namely from infinite K to K = 5 by infinite series, and onward from 4 to 1 by computing particular values of E(x), and of its derivative. In comparison with the above plain principles, the detailed working out may seem intricate. The complications arose from the fact that the two functions inverse to E(x) were not well known in detail. Readers who wish to pass directly to the result will find it numbered (8/31).

In general let $x = 1 + \xi$. (8/14)

Given $\xi_1 > 0$ it was required to find $\xi_2 < 0$ such that

$$E(1 + \xi_2) = E(1 + \xi_1). \qquad (8/15)$$

The computation of ξ_2 was made by logarithms when $\xi_1 > 0.1$; but when $\xi_1 < 0.1$ seven-figure logarithms did not yield ξ_2 with sufficient accuracy; and instead the following series was obtained by substituting (8/14) in (8/9) and expanding the binomial.

$$E(x) - E(1) = \sum_{m=2}^{\infty} (-1)^m \frac{(\alpha+1)(\alpha+2)\ldots(\alpha+m-2)}{(m-2)!\,m}\,\xi^m. \qquad (8/16)$$

This is convergent when $|\xi| < 1$, and useful when $|\xi| < 0.1$. On equating the series for ξ_2 to that for ξ_1, dividing by $\frac{1}{2}(\xi_1 - \xi_2)$ and separating the first term, it follows that

$$-\xi_2 = \xi_1 + 2\sum_{m=3}^{\infty} (-1)^m \frac{(\alpha+1)(\alpha+2)\ldots(\alpha+m-2)}{(m-2)!\,m}\left(\frac{\xi_1^m - \xi_2^m}{\xi_1 - \xi_2}\right). \qquad (8/17)$$

On insertion of the numerical value of α from (8/2A), formula (8/17) becomes well arranged for iteration. A trial value of ξ_2 was inserted in the second member only. The first member then was

*But it is certainly tedious; so that the hasty reader is advised to skip to the result expressed in (8/31). The result comes at the end of a chain of operations, and the chain cannot be anchored anywhere nearer than infinity.

an improved value of $-\xi_2$. Leibniz' rule showed when the tail was negligible.

A graph of $y = E(x)$, in rectangular coordinates, is rather like a parabola, but steeper on the side where x is less. The length of a chord, at constant y, is $\xi_1 - \xi_2$. Its mid-point is at

$$\xi = \frac{1}{2}(\xi_1 + \xi_2). \qquad (8/18)$$

By reversion of the series (8/16) after the manner of Bromwich (1926, p. 200 Ex. 23) it was proved that

$$\xi_1 + \xi_2 = 0.21494(\xi_1 - \xi_2)^2 - 0.00907(\xi_1 - \xi_2)^4 + \dots \qquad (8/19)$$

provided ξ was near enough to zero. However computed points showed that even at $\xi_1 = 0.27$, which is beyond the required range, the terms stated in (8/19) gave $\xi_1 + \xi_2$ to 1 part in 4500. There is no need to consider the higher terms. This (8/19) is one of the simplest properties of the function E.

These properties of E become entangled with the successive integers K. It is convenient to compute small deviations from custom, namely the ϵ defined by (8/5). For then (8/11) becomes

$$E\left(1 + \frac{1}{2K} + \frac{\epsilon_{K+\frac{1}{2}}}{K}\right) = E\left(1 - \frac{1}{2K} + \frac{\epsilon_{K-\frac{1}{2}}}{K}\right), \qquad (8/20)$$

which is satisfied provided we set

$$\frac{1}{K}\left(\frac{1}{2} + \epsilon_{K+\frac{1}{2}}\right) = \xi_1 \text{ and } \frac{1}{K}\left(-\frac{1}{2} + \epsilon_{K-\frac{1}{2}}\right) = \xi_2. \qquad (8/21)$$

These entail that

$$\xi_1 + \xi_2 = \frac{1}{K}\left(\epsilon_{K+\frac{1}{2}} + \epsilon_{K-\frac{1}{2}}\right), \qquad (8/22)$$

also that

$$\xi_1 - \xi_2 = \frac{1}{K}\left(1 + \epsilon_{K+\frac{1}{2}} - \epsilon_{K-\frac{1}{2}}\right). \qquad (8/23)$$

When the last two equations are substituted in (8/19) they yield

$$\epsilon_{K+\frac{1}{2}} + \epsilon_{K-\frac{1}{2}} = \frac{0.21494}{K}\left(1 + \epsilon_{K+\frac{1}{2}} - \epsilon_{K-\frac{1}{2}}\right)^2 - \frac{0.009071}{K^3}\left(1 + \epsilon_{K+\frac{1}{2}} - \epsilon_{K-\frac{1}{2}}\right)^4. \qquad (8/24)$$

It is known from (8/13) that $\epsilon \to 0$ as $K \to \infty$. To obtain a simple first approximation, which shall be at its best for large K, replace the first member of (8/24) by $2\epsilon_K$ and the brackets in the second member by unity and neglect the term in K^{-3}. This yields, as a first approximation,

$$\epsilon_K \simeq \frac{0.10747}{K} \qquad (8/25)$$

whence,

$$\epsilon_{K+\frac{1}{2}} - \epsilon_{K-\frac{1}{2}} \simeq -\frac{0.10747}{K^2}. \qquad (8/26)$$

To obtain a second approximation let (8/25) and (8/26) be inserted into the second member only of (8/24) and retain terms only as far as K^{-3}. Also, in view of a prospective summation, replace K by the dummy n obtaining

$$\epsilon_{n+\frac{1}{2}} + \epsilon_{n-\frac{1}{2}} = \frac{0.21494}{n} - \frac{0.05527}{n^3}. \qquad (8/27)$$

Multiply (8/27) by $(-1)^{n-K}$ and sum from $n = K$ to $n \to \infty$. There is a progressive cancellation in the first member, which leaves

$$\epsilon_{K-\frac{1}{2}} = 0.21494 \sum_{n=K}^{\infty} \frac{(-1)^{n-K}}{n} - 0.05527 \sum_{n=K}^{\infty} \frac{(-1)^{n-K}}{n^3}. \qquad (8/28)$$

The sums for $K = 1$ are given in Dale's tables as

$$\sum_{n=1}^{\infty} \frac{(-1)^{n-1}}{n} = 0.693{,}147{,}2 \text{ and } \sum_{n=1}^{\infty} \frac{(-1)^{n-1}}{n^3} = 0.901{,}542{,}7. \qquad (8/29)$$

From these, by allowing for the terms in n = 1, 2, 3, 4, it was found that

$$\epsilon_{4.5} = 0.023{,}32 \qquad (8/30)$$

not much different from the first approximation which, by (8/25) was 0.023,88. The above method of approximation is excellent for large K, but fails conspicuously as K decreases from 5 to 1. So instead, the reduction of K from 4 to 1 was made by computing E(x) and its inverse functions by logarithms. The result was that, for the purpose of rounding off the whole numbers of cells, the list of territories should be cut at the following multiples of h: 6.516, 5.519, 4.523, 3.530, 2.540, 1.562, 0.622; (8/31)
of these the last is the most noteworthy because it differs most from the half integer. The lists I, II, III, IV have been separated at 0.622h. Any population between 0.622h and 1.562h should be regarded as approximately one cell. (8/32)

The effects of this slight deviation from the custom of the "nearest integer" are considerable; for, if $\theta_{0.5}$ had been 0.5 instead of 0.622, then about 50 territories would have been classified differently.

In the following lists, the type of boundary is indicated by prefixes to the names of territories, thus:

* indicates partly a seacoast and partly a land frontier.
Absence of any prefix indicates a seacoast only.
‡ indicates desert only.
† indicates a lake shore only.
** indicates a completely surrounding land, part at least of which was inhabited.

A glance at the lists I, II, III, IV will show that most of the more populous territories had both seacoast and land frontier, whereas the least populous territories were mostly islands in the sea.

The columns marked F, B, D, L, in lists I and II logically belong in §9 where they will be explained and discussed. They are set here for economy in printing. Though they appear in list I, they relate to cells of one, not ten, million people.

List I. When the cell-content is ten million persons, then the territories which must receive attention have been taken to be all those, and only those in list I.

	Millions		Method	Edges for Cells of One Million See § 9			
				F Foreign Land	B Sea or Ice	D Desert	L Lake
1	say 379	*China, Tibet and Mongolia	U	47	56	0	17
2	313.3	*India, Burma, Bhutan, & Baluchistan less Cutch	U	52	68	0	0
3	166.3	*Russia, Finland, Bokhara & Khiva	U	63	29	0	15
4	91.97	*U.S.A., mainland	T	20	27	0	19
5	64.51	*Germany, less Femern and Rügen	T	56	12	0	0
6	51.14	*Austria, Hungary, Bosnia & Herzegovina	T	58	4	0	0
7	40.31	Great Britain, mainland	T	0	27	0	0
8	39.26	*France, omitting Corsica	T	24	16	0	0
9	36.5	Honshu, less Sado, Oki & Awaji	T	0	36	0	0
10	29.93	*Italy, less Sicily, Sardinia & Elba	T	15	21	0	0
11	29.8	Java	T	0	27	0	0
12	24.0	*Brazil	T	12	12	0	0
13	18.82	*Spain continental	T	13	13	0	0
14	18	*Turkey in Asia	T	11	20	9	0
15	16.9	*Nigeria	T	15	4	0	0
16	16.8	*French Indo-China	T	20	12	0	0
17	15.16	*Mexico	T	8	13	0	0
18	15	*Egypt and Anglo-Egyptian Sudan	T	2	2	29	0
19	13.6	*French West and Equatorial Africa	T	31	5	9	0
20	13.3	*Korea	T	6	13	0	0
21	10	*Abyssinia	T	12	0	0	0
22	10	*Persia	T	14	4	7	0
23	9	*Belgian Congo	T	18	1	0	2
24	8.7	*Nyasaland, Rhodesia, & South Africa	T	11	5	2	3
25	8.15	*Siam	T	17	5	0	0
26	7.4	*German East Africa	T	9	2	0	8
27	7.4	*Algeria and Tunisia	T	4	6	0	2
28	7.39	*Belgium	T	16	1	0	0
29	7.09	*Argentina, less islands	T	11	4	0	0
30	7.0	Kyushu	T	0	9	0	0
31	6.99	*Rumania	T	21	1	0	0
32	6.91	*Canada & Labrador less Prince of Wales Island	T	14	7	0	6
33	6.7	*Uganda & British East Africa Protectorate	T	13	1	0	6
		TOTALS FOR LIST I					
	1496.33	for 33 territories		613	463	47	78

Opinions may differ as to whether Egypt, the Anglo-Egyptian Sudan, Uganda, and the British East Africa Protectorate were all one block, or not, at mid-1910. In Figure 7 they are shown as one, but here as two blocks in agreement with the map in E 1, 319.

List II. When each cell contains one million persons, then the territories in list I above must be supplemented by those in list II below.

	Millions		Method	F'	F''	B	D	L
				Edges for Cells of One Million See §9				
				Foreign Land		Sea or Ice	Desert	Lake
34	6.1	*Turkey in Europe	T	1	9	8	0	0
35	6	**Afghanistan	T	17	0	0	0	0
36	5.90	*Netherlands less Schouwen & Over-Flakkee	T	12	0	5	0	0
37	5.6	**Nepal	T	25	0	0	0	0
38	5.48	*Portugal, continental	T	8	0	4	0	0
39	5.42	*Sweden less islands	T	1	3	6	0	0
40	5.4	*Columbia	T	1	6	3	0	0
41	5.3	Sumatra	T	0	0	8	0	0
42	4.7	*Morocco	T	3	0	5	1	0
43	4.4	Luzon	T	0	0	6	0	0
44	4.4	*Peru	T	3	5	4	0	0
45	4.40	Ireland	T2	0	0	4	0	0
46	4.32	*Bulgaria	T	6	7	1	0	0
47	4.14	Australia excluding Tasmania	T2	0	0	4	0	0
48	4.1	Ceylon	T2	0	0	4	0	0
49	4.0	*Angola	T	7	0	2	2	0
50	3.72	**Switzerland	T	14	0	0	0	0
51	3.66	Sicily	T2	0	0	5	0	0
52	3.4	Formosa	T2	0	0	3	0	0
53	3.35	*Chile less islands	T	5	3	3	0	0
54	3.04	Madagascar	T2	0	0	4	0	0
55	3.0	*Mozambique	T	6	0	3	2	1
56	3.	Hainan	T2	0	0	3	0	0
57	2.91	**Serbia	T	6	5	0	0	0
58	2.71	*Venezuela	T	1	3	2	0	0
59	2.6	*Kamerun	T	11	0	1	0	0
60	2.6	Shikoku	T	0	0	4	0	0
61	2.55	*Malay Peninsula (British part)	T	1	0	5	0	0
62	2.38	*Norway	T	1	3	2	0	0
63	2.3	Celebes	T	0	0	2	0	0
64	2.22	Cuba	T	0	0	2	0	0
65	2.2	*The Republic of Haiti	T	0	2	2	0	0
66	2.11	*Greece, less islands	T2	0	1	3	0	0
67	2.02	*Guatemala	T	2	1	2	0	0
68	2.	*Bolivia	T	4	3	0	0	0
69	1.8	*Liberia	T	3	1	2	0	0
70	1.78	Madura	T	0	0	2	0	0
71	1.5	*Ecuador	T	0	3	1	0	0
72	1.49	*Gold Coast and Ashanti	T	3	1	1	0	0
73	1.4	*Dutch-Borneo	T	0	1	1	0	0
74	1.4	Yezo	T	0	0	1	0	0
75	1.39	*Sierra Leone, Colony & Protectorate	T	2	1	1	0	0
76	1.19	*Jutland	T	1	0	1	0	0
77	1.17	*Salvador	T	0	1	1	0	0
78	1.12	Puerto Rico	T	0	0	1	0	0
79	1.12	*Uruguay	T	3	0	1	0	0
80	1.	Nejd	T	0	0	0	1	0
81	1.0	*Togoland	T	3	1	1	0	0
82	1.00	Zealand	T	0	0	1	0	0
83	0.93	Bali	T	0	0	1	0	0
84	0.86	Panay	T	0	0	1	0	0
85	0.846	Sardinia	T	0	0	1	0	0

(List II continued)

	Millions		Method	F'	F''	B	D	L
86	0.83	**Paraguay	T	4	1	0	0	0
87	0.82	Jamaica	T	0	0	1	0	0
88	0.73	*Sarawak, British N. Borneo, & Brunei	T	0	1	1	0	0
89	0.69	*Dominican Republic	T	0	2	1	0	0
90	0.68	Cebu	T	0	0	1	0	0
		TOTALS FOR LISTS I & II TOGETHER						
1652.506		for 90 territories		831		595	53	79

List III. When each cell contains 100,000 persons, then the territories in lists I and II above must be supplemented by those in list III below.

91	Nicaragua	0.60
92	Mindanao	0.57
93	*Honduras	0.553
94	New Zealand, North Island	0.553
95	Lombok	0.54
96	*Tripoli & Cyrenaica	0.52
97	*Goa	0.516
98	Cutch	0.51
99	Negros	0.51
100	*Oman	0.50
101	*Japanese Kwantung	0.49
102	New Zealand, South Island	0.44
103	*Panama & Canal Zone	0.418
104	Leyte	0.41
105	Flores in Dutch East Indies	0.40
106	*Italian Somaliland	0.4
107	*Costa Rica	0.39
108	Okinawajima	0.38
109	Mauritius	0.37
110	*British Somaliland	0.35
111	Crete	0.34
112	*Portuguese Timor, main	0.33
113	*Dutch Timor	0.33
114	Trinidad	0.33
115	Bohol	0.32
116	*Etritrea	0.3
117	*British Guiana	0.29
118	Corsica	0.29
119	*British Papua	0.27
120	Cyprus	0.27
121	**Luxemburg	0.26
122	Majorca	0.26
123	Sámar	0.26
124	*Montenegro	0.25
125	*Portuguese Guinea	0.25
126	Newfoundland	0.232
127	*Kowloon & adjuncts	0.23
128	Malta	0.227
129	Hongkong (island only)	0.22
130	Chusan Island	0.2
131	*Dutch New Guinea	0.2
132	*French Somali Coast	0.2
133	*Kaiser Wilhelmsland (main)	0.2
134	Lesbos	0.2
135	*Spanish Guinea (mainland) (=Rio Muni)	0.2
136	Sumbawa	0.2
137	Awaji	0.196
138	Martinique	0.192
139	Guadeloupe	0.19
140	Tasmania	0.190
141	Quelpart	0.183
142	Hawaii	0.18
143	Barbados	0.173
144	*German Kiau-Chau	0.17
145	Reunion	0.170
146	*Pondicherry	0.168
147	Madiera	0.164
148	Teneriffe	0.156
149	Boetoeng	0.15
150	*French Kwang Chau Wan	0.15
151	*British Wei-Hai-Wei	0.146
152	Gran Canary	0.144
153	Cambia	0.13
154	Oahu	0.13
155	Banca (=Bangha)	0.127
156	Nias	0.125
157	Euboea	0.123
158	Amoy	0.12
159	Sumba (=Sandalwood)	0.12
160	Sado	0.115
161	São Miguel	0.115
162	Laaland with Falster	0.11
163	Zanzibar	0.11
164	Corfu	0.103
165	Cape Breton Island	0.10
166	*German S.W. Africa	0.1
167	Penang (Island only)	0.1
168	Prince Edward Island	0.095
169	Isle of Wight	0.087
170	Iceland	0.085
171	*Dutch Guiana	0.084
172	Pemba	0.082
173	Chios	0.08
174	Pantar (=Pandai)	0.08
175	Vancouver Island	0.08
176	Cephalonia	0.076
177	*Macao	0.074
178	Ceram (=Seram)	0.07
179	Grand Comoro	0.07
180	Neu Pommern (=New Britain)	0.07

(List III continued)

181	Rotti	0.07	184	Grenada	0.0664
182	Saleyer (=Saleier)	0.07	185	Mindoro	0.065
183	Samos	0.067	186	*Alaska	0.064

TOTALS FOR LISTS I, II, III TOGETHER

186 territories containing 1674.4774 million persons.

List IV. When each cell contains 10,000 persons, then the territories in the lists I, II and III above must be supplemented by those in list IV below. There is no natural break between these lists but, as some of the smaller places are not well known, their identification is aided by the latitude and longitude of a point in or near the territory.

187	*Karikal	11.0	N,	79.8	E	0.062
188	Masbate	12	N,	123	E	0.061
189	Alor (=Ombai)	8	S,	129.5	E	0.06
190	Billiton (=Belitong)	2.8	S,	108	E	0.06
191	Oesel (=Saare Maa)	58.4	N,	22.6	E	0.06
192	Sao Thiago	14.9	N,	23.5	W	0.06
193	Marinduque	13.4	N,	122.0	E	0.058
194	Gottland	57.5	N,	18.5	E	0.0551
195	Siquijor (=Sikihor)	9.2	N,	123.6	E	0.053
196	Isle of Man	54.2	N,	4.6	W	0.0522
197	Jersey	49.2	N,	2.1	W	0.0520
198	Jolo (=Holo=Sulu)	6.0	N,	121.2	E	0.051
199	New Caledonia	22	S,	166	E	0.0508
200	Bahrein Island	26.1	N,	50.5	E	0.05
201	Chiloé Island (only)	42.5	S,	74	W	0.05
202	*Kabinda district	5.3	S,	12.3	E	0.05
203	Malaita	9	S,	161	E	0.05
204	Rügen	54.2	N,	13.4	E	0.05
205	St. Lucia	13.9	N,	61	W	0.0484
206	Catanduanes	13.9	N,	124.4	E	0.048
207	*French Guiana (=Cayenne)	4	N,	53	W	0.048
208	Palma	28.5	N,	17.8	W	0.047
209	Terceira	38.7	N,	27.2	W	0.047
210	Maui	21	N,	156	W	0.046
211	*Aden	13.3	N,	45	E	0.045
212	Amboina (island only)	3.6	S,	128.1	E	0.045
213	Bornholm	55.1	N,	15	E	0.045
214	Daman (=Damao)	20.4	N,	72.9	E	0.045
215	Sangihe	3.6	N,	6	E	0.045
216	San Thomé	0	N,	6	E	0.044
217	Leucas (=Leukadia Santa Maura)	38.7	N,	20.6	E	0.043
218	St. Vincent	13.2	N,	61.2	W	0.042
219	Zante (=Zacynthos)	37.8	N,	20.8	E	0.042
220	Guernsey	49.45	N,	2.6	W	0.041
221	Minorca	40	N,	4.1	E	0.0406
222	*British Honduras	17	N,	87.7	W	0.0401
223	Bougainville	7	S,	155	E	0.04
224	Florianopolis (=Desterro-Sta Catherina)	27.5	S,	48.4	W	0.04
225	‡Hofuf Oasis	25.3	N,	49.7	E	0.04
226	Lesbos (=Mytilene)	39.2	N,	26.3	E	0.04
227	Malekula (=Mallicolo)	16.3	S,	167.5	E	0.04
228	Siau	2.6	N,	125.5	E	0.04
229	Camiguin	9.2	N,	124.7	E	0.037
230	Iki Shima	33.8	N,	131.7	E	0.037
231	Kauai	22	N	160	W	0.036
232	Tsu Shima	34.6	N,	131.3	E	0.035
233	Masbate	12.1	N,	123.6	E	0.034

(List IV continued)

234	Dominica	15.3	N,	61.3	W	0.0333
235	Salang (=Jukseylon=Puket Island)	7.8	N,	98.4	E	0.033
236	Lewis with Harris	58.2	N,	6.7	W	0.0328
237	Curaçao	12.2	N,	69	W	0.0325
238	Antigua	17.1	N,	61.8	W	0.032
239	Lomblen	8	S,	123.5	E	0.032
240	Mindoro	13	N,	121	E	0.032
241	Miyakojima	24.8	N,	125.3	E	0.032
242	Muharraq	26.3	N,	50.7	E	0.032
243	Basilan	6.6	N,	122.1	E	0.031
244	Rhodes	36.2	N,	28	E	0.031
245	Anjouan (=Johanna)	12.2	S,	44.4	E	0.03
246	Buru	3	S,	127	E	0.03
247	Halmahera (=Jilolo)	1	N,	128	E	0.03
248	Lemnos	39.9	N,	25.2	E	0.03
249	Moena	5.1	S,	122.5	E	0.03
250	Neu Mecklenburg (=New Ireland)	4	S,	153	E	0.03
251	Öland	56.8	N,	16.7	E	0.03
252	Tanegashima	30.6	N,	131	E	0.03
253	Syros (=Syra)	37.4	N,	24.9	E	0.029
254	**Chandernagor	22.8	N,	88.4	E	0.028
255	Goeree & Over-Flakkee	51.75	N,	4.1	E	0.028
256	*Japanese Saghalien (=Karafuto)	48	N,	142.5	E	0.028
257	Tablâs	12.4	N,	122.1	E	0.028
258	Ischia	40.7	N,	13.9	E	0.0275
259	Elba	42.8	N,	10.3	E	0.027
260	St. Kitts (=St. Christopher)	17.4	N,	62.8	W	0.0266
261	Pico	38.4	N,	28.3	W	0.0265
262	Majorca	39.6	N,	3	E	0.026
263	‡Qatif oasis (=Katif)	27	N,	49.5	E	0.026
264	*Gibraltar					0.0255
265	Adunara (=Andonare)	8	S,	123	E	0.025
266	Alsen	55	N,	9.9	E	0.025
267	Peleng	1.4	S,	123.	E	0.025
268	Veglia (=Krk)	45.1	N,	14.6	E	0.025
269	Guimaras	10.6	N,	122.6	E	0.024
270	Santo Antão	17	N,	25.3	W	0.024
271	Schouwen & Duivenland	51.7	N,	3.9	E	0.024
272	Fayal	38.6	N,	28.7	W	0.023
273	Fernando Po	3.5	N,	8.7	E	0.023
274	Iviza	39	N,	1.5	E	0.023
275	Bilaran	11.6	N,	124.5	E	0.022
276	Brazza	43.3	N,	16.7	E	0.022
277	Bantayan	11.2	N,	123.8	E	0.021
278	Gozo	36.1	N,	14.3	E	0.021
279	Tobago	11.2	N,	60.7	W	0.021
280	*Monaco	43.75	N,	7.4	E	0.021
281	Gomera	28	N,	17.3	W	0.0202
282	Bintan	1.0	N,	104.6	E	0.02
283	Espiritu Santo	15	S,	167	E	0.02
284	Laoet	36	S,	116.2	E	0.02
285	Mactan	10.3	N,	124.0	E	0.020
286	Naxos	37.1	N,	25.5	E	0.02
287	*Portuguese enclave in Dutch Timor	9.3	S,	124.3	E	0.02
288	*Rio de Oro and Adrar	25	N,	14	W	0.02
289	Saparua	3.6	S,	128.7	E	0.02
290	Sta. Cruz (=St. Croix)	17.7	N,	64.7	W	0.02
291	Upolu	13.9	S,	171.7	W	0.020
292	Wowoni	4.2	S,	123.2	E	0.02

(List IV continued)

293	Yamdena	7.5	S,	133	E	0.02
294	Lanzarote	29	N,	13.5	W	0.0198
295	Shetland Mainland	60.3	N,	1.3	W	0.0197
296	Andros	37.8	N,	24.8	E	0.019
297	Langeland	54.9	N,	10.8	E	0.019
298	Aland	60	N,	20	E	0.018
299	Mahe in Seychelles	4.7	S,	55.7	E	0.018
300	Dakla Oasis	25.5	N,	29.3	E	0.017
301	Nuhu Yut	5.6	S,	133.0	E	0.017
302	Sao Jorge	38.6	N,	28	W	0.017
303	South Andaman	12	N,	92.6	E	0.017
304	Fogo	14.8	N,	24.5	W	0.0167
305	Oleron	46	N,	1.3	W	0.016
306	Panglao	9.5	N,	123.8	E	0.016
307	Savu	10.7	S,	122	E	0.016
308	Taragan	7	S,	134	E	0.016
309	Möen	55	N,	12.4	E	0.015
310	Qishm (=Kishm-Tawilah)	27	N,	56	E	0.015
311	Truk	6.9	N,	151.9	E	0.015
312	Ukara	1.9	S,	33.0	E	0.015
313	Pomona	59	N,	3	W	0.0147
314	Cythera (=Cerigo)	36.2	N,	23.0	E	0.014
315	Dagö (=Hiiu Maa)	58.8	N,	22.6	E	0.014
316	Nuhu Rowa	5.7	S,	132.7	E	0.014
317	*Russian Sakhalin	52	N,	143	E	0.014
318	Siaoe	2.7	N,	125.4	E	0.014
319	Tidore	0.7	N,	127.4	E	0.014
320	Siassi	5.5	N,	120.9	E	0.0137
321	New Providence	25	N,	77.4	W	0.0136
322	Diu	20.75	N,	71.1	E	0.0134
323	Greenland					0.0134
324	Biak	1	S,	136.3	E	0.013
325	*Ceuta	35.9	N,	5.3	W	0.013
326	Buru	3.5	S,	126.6	E	0.013
327	Simeuloe	2.6	N,	96.1	E	0.013
328	Nevis	17.2	N,	62.6	W	0.0129
329	Savaii	13.7	S,	172.4	W	0.0128
330	Skye	57.3	N,	6.3	W	0.0128
331	Nakanoshima	36.1	N,	133.1	E	0.0126
332	Palawan (=Paragua)	9.7	N,	118.5	E	0.0126
333	Ishigakijima	24.5	N,	124.3	E	0.0123
334	Sibuyán	12.4	N,	122.6	E	0.0123
335	Montserrat	16.8	N,	62.2	W	0.0122
336	Kabaena	5.3	S,	121.9	E	0.012
337	Karakelong	4.3	N,	126.8	E	0.012
338	Mayotte	12.8	S,	45.2	E	0.012
339	St. Thomas	18.4	N,	64.9	W	0.012
340	Santorin (=Thera)	36.4	N,	25.4	E	0.012
341	Siargao	9.9	N,	126.1	E	0.012
342	Tenos (=Tino)	37.6	N,	25.2	E	0.012
343	Tahiti	17.6	S,	149.4	W	0.012
344	Ternate Island (only)	0.8	N,	127.4	E	0.012
345	Ticao (=Tikao)	12.5	N,	123.7	E	0.012
346	Bute	55.8	N,	5.1	W	0.0018
347	Guam	13.4	N,	144.7	E	0.0118
348	**Liechtenstein	47.2	N,	9.55	E	0.011
349	Sao Nicolão	16.5	N,	24.4	W	0.011
350	**San Marino	43.9	N,	12.5	E	0.0110
351	Romblón	12.6	N,	122.4	E	0.0108

(List IV continued)

352	Fuerteventura	28	N,	13.8	W	0.0103
353	*Mahé (in French India)	11.6	N,	75.5	E	0.0102
354	Aruba	12.5	N,	70	W	0.010
355	Batan	1.1	N,	140.0	E	0.01
356	Cherso with Lussin	44.7	N,	14.4	E	0.01
357	Cos (=Kos-Stanko)	36.8	N,	27.2	E	0.01
358	Ghardaia	32.5	N,	4	E	0.01
359	Haruku	3.6	S,	128.5	E	0.01
360	Île de Ré	46.2	N,	1.4	W	0.01
361	**Italian Tientsin	39.1	N,	117.2	E	0.01
362	Ithaca (=Thiaki)	38.4	N,	20.7	E	0.01
363	Japen	1.8	S,	136.3	E	0.01
364	Laguan (=Laoang)	12.6	N,	125.0	E	0.01
365	Lipari	38.5	N,	14.9	E	0.01
366	Lissa	43.1	N,	16.2	E	0.010
367	Moa	8	S,	128	E	0.01
368	Siberoet	1.4	S,	98.9	E	0.01
369	Solor	8	S,	123	E	0.01
370	Streymoy (=Stromo)	62.2	N,	7	W	0.01
371	Thasos	40.6	N,	24.7	E	0.010
372	Poro	10.7	N,	124.4	E	0.0099
373	Fehmern (=Fehmarn=Femern)	54.5	N,	11.1	W	0.0098
374	Panaón	10.1	N,	125.2	E	0.0098
375	Nossi-Bé	13	S,	48	E	0.0095
376	Aegina	37.7	N,	23.5	E	0.009
377	Belle-Île-en-Mer	47.3	N,	3.2	W	0.009
378	Cuyo	10.8	N,	121.0	E	0.009
379	Flores in Azores	39.4	N,	31.3	W	0.009
380	Great Natuna	4.0	N,	108.2	E	0.009
381	Imbros	40.2	N,	25.8	E	0.009
382	Koendoer	1.2	S,	103.4	E	0.009
383	*Melilla	35.3	N,	3.0	W	0.0090
384	Lingga	0.2	S,	104.7	E	0.009
385	Pantelleria	36.8	N,	12.0	E	0.009
386	Singkep	0.5	S,	104.4	E	0.009
387	Taliaboe	1.8	S,	124.7	E	0.009
388	Brava	14.7	N,	24.8	W	0.0086
389	Pasijan	10.7	N,	124.3	E	0.0085
390	San Cristoval	10.5	S,	162.5	E	0.0084
391	Graciosa	39.1	N,	28	W	0.0083
392	Guadaleanar	9.5	S,	160	E	0.0081
393	Buka	5.2	S,	154.7	E	0.008
394	Kharga Oasis	25.4	N,	30.7	E	0.008
395	Paros	37.1	N,	25.2	E	0.0078
396	Nishinoshima	36.1	N,	133.0	E	0.0076
397	São Vicente	16.8	N,	25	W	0.0076
398	Wetar (=Wetter)	7.7	S,	126.5	E	0.0075
399	Hierro	27.7	N,	18	W	0.0074
400	Lubang	13.8	N,	120.2	E	0.0073
401	Capri	40.6	N,	14.2	E	0.007
402	Groot Karimoen	1.1	N,	103.4	E	0.007
403	Hydra	37.3	N,	23.5	E	0.007
404	Labuan	5.3	N,	115.2	E	0.007
405	Mafia	8	S,	39.7	E	0.007
406	Makian	0	N,	27.3	E	0.007
407	Spezzia	37.3	N,	23.2	E	0.007
408	Tangulandang (=Tagulanda)	2.4	N,	125.5	E	0.007
409	Yap	9.4	N,	138.2	E	0.007
410	Samsö	55.9	N,	10.6	E	0.0066

411	Eleuthera	25	N,	76.2	W	0.00657
412	Bachian	0.6	S,	127.7	E	0.0065
413	Texel	53.1	N,	4.8	E	0.0065
414	Islay	55.8	N,	6.2	W	0.0063

TOTALS FOR LISTS I, II, III, IV TOGETHER

414 Territories containing 1679.59347 million persons.

Are These Lists Complete?

The search among the Philippines came easily to a definite end, because the census of 1903 was published by islands separately. Not so the Netherlands East Indies, for which the populations of small islands in 1930 were published by groups, such as the Kei, the Schouten, and the Talaud. The best that the author could do, was to divide the group-population proportionately to the areas of the component islands. Judging by the slowness with which recent enquiries have added further territories to the foregoing lists, the author feels sure that the lists I and II are complete, and that list III is tolerably complete, but he would not be surprised if three percent more territories ought to be added to list IV. There are of course a great many islands having populations less than 6220 persons; it would be easy to make a long list of them; but it would be impossible to make such a list sufficiently complete to be statistically comparable with lists I, II, III, and IV.

The information about little-known places was searched out in the University and Mitchell Libraries of Glasgow, and finally in the University Library at Cambridge. I am particularly indebted to Mr. H. R. Mallett, head of the map-department of the last-named, for his resourceful aid.

The sources of information, from which the lists of territories were compiled, and the cellular maps were drawn, included the following:

Official census publications for: England & Wales 1911; Scotland 1911; India 1911; Philippines 1903; Netherlands East Indies 1930; Siam 1925-26.

Abstracts of census arranged by subordinate areas, in: The Statesman's Year Books 1912-49, but especially 1914; China Year Books 1913 and 1931-32; Japan Year-Book 1931; The Japanese Empire 1932; Far East Year Book 1941; West Indies Year Book 1945; New Zealand Handbook; Australian Year-Book 1945-46; Kolb, A., "Die Philippinen" Leipzig, Koehler; Admiralty Manual of Netherlands India 1921.

Total populations of countries, islands and towns from: League of Nations Statistical Year-Books; The Encyclopaedia Britannica; Grand La Rousse Illustre; Gazetteers of the world by Chambers 1906, by Lippincott 1906, and by Chisholm; Whitaker's Almanacks; The World Almanac 1937, New York; Blanche et Gallois Geographie Universelle 1929; Admiralty Handbook of Arabia; Admiralty Pilots, various. F. Jaeger's Afrika 1928, Leipzig, Bibliographisches Institut.

Personal Messages from Mr. D. W. I. Piggott about Mafia; from Miss Cornelia Schouw and the Burgemeester of Bruinisse about Schouwen and Duivenland; and from Mr. Keith Bovey about Japanese islands.

Maps of subordinate census areas in: The Time Atlas 1900; The Census of India 1911; Netherlands Indies Census 1930; Statistical Year Book, Siam No. 19; Atlas van tropisch Nederland 1938; Grand La Rousse Illustre 1931 for French Indo-China; Ordnance Survey of Britain, map of administrative areas 1944; The Daily Telegraph Atlas 1919-1920 for Colombia. Finding a map to fit a census has often been the crucial difficulty.

Maps of density of population: for the whole world those edited by Prof. J. F. Unstead and Prof. London; for Britain in 1931 prepared by Mr. A. C. O'Dell and published by the Ordnance Survey; for Africa a dot-map in Jaeger's "Afrika."

* * *

The author is much indebted to the compilers of works of reference, but he begs to draw their attention to an incongruity. Censuses have often been quoted, accurately to one person, but without any mention of their dates. It ought to be borne in mind that populations usually change by as much as 1 in 2000 in a month.

9. WORLD TOTALS OF CELLS AND OF FIVE SORTS OF EDGES

In this final section the many threads of argument in this long appendix are woven together.

The World-Total of Population

Carr-Saunders (1936) emphasized the uncertainties of the world-total of population. The chief of them relates to China, which stands first in the present arrangement, so that its uncertainty disturbs all the rest of the running total. Carr-Saunders quoted estimates by Willcox, which when reduced to A.D. 1910 would make the total 1640 millions, but himself preferred 1725. The League of Nations made the total about 1760 when reduced to the same date.

In the present arrangement of territories, the total population comes to 1679.59 millions by direct addition as far as the end of list IV. Beyond that list there were certainly numerous territories each having less than 6220 inhabitants; but how many such inhabited territories there were, or what their total population was, probably no one can say definitely. A likely estimate can however be obtained by assuming that the same Pareto distribution, which has been shown, in Figure 18, to describe the known facts tolerably between serial numbers t = 100 and t = 414, is valid also for the unknown smaller pieces. Let us first find what that formula predicts for the serial number of a territory having a single inhabitant, and therefore by (18/1) $p = 10^{-6}$ millions. The answer is t = 7869, far too many to be summed in a list. On the other hand, so long an extrapolation by an empirical formula is indeed risky. However let us continue. From the same formulae (18/1) and (18/2) we can estimate the total population of all the pieces smaller than those which have been factually totaled.

This "tail" of population is

$$\sum_{t=415}^{t=7869} \frac{1}{(\beta t + \gamma)^3}.$$

Because, in this series, t moves on by a small fraction of itself, (which it would not do near t = 1) the sum is nearly equal to the integral

$$\int_{414.5}^{7869.5} (\beta t + \gamma)^{-3} dt$$

which works out to 1.38 Millions. Thus the present estimate of the world population at mid-1910 is 1679.59 + 1.38 = 1681 million, nearly midway between those of Carr-Saunders and of Willcox.

The Number of Cells

The total number of cells in the world will here be noted by **s**, in heavy type and without any suffix. By definition **s** is a whole number, for we cannot deal with fractions of a cell. As <u>h</u>, the cell-population diminishes, the number of cells increases. The first and roughest idea is that h**s** equals the world population. But this is of course not correct; for numerous bits, not individually amounting to 0.622h, have to be left out. The correct statement is that h**s** is the part of the world-population that can be included in cells of the specified h. This part becomes equal to the whole when h is one person. Otherwise it runs as follows, as mid-1910. The numbers have been taken from the foregoing lists.

h =	10^7	10^6	10^5	10^4
s =	149.6	1652.5	16745	167959
10^{-6} h**s** =	1496	1653	1674.5	1679.59

(9/1)

The large uncertainty in the population of China leaves each value of 10^{-6} h**s** uncertain by ±60. Nevertheless the <u>difference</u> between 1674.5 and 1679.6 may perhaps be correct to the first decimal. In these circumstances the usual conventions for indicating uncertainty are not appropriate. The numbers of cells and edges are necessarily integers. The author has taken the liberty to state them in full, leaving aside meantime the difficult questions about their uncertainty.

In order to obtain a result roughly applicable to dates other than mid-1910, it may be imagined that the population changed everywhere in the same ratio, the geographical pattern remaining unaltered. If so, the various h**s** would all change with time in the same ratio, and their ratios to each other would be a permanent feature of world-geography. Here they are:

<u>h</u>	10^7	10^6	10^5	10^4
Ratio of h**s** to its value at <u>h</u> = 10^4	0.891	0.984	0.997	1.000

(9/2)

The Number of Edges of All Sorts Together

Most of the world-total of edges is conveniently obtainable from the topological theorem (2/20), because this is already in the form of a sum, not indeed over the whole world, but over a considerable region. By taking the region to be bounded by the sea or ice, we get rid of <u>g</u>, which is the number of external edges that meet the outline. The regions are then Europe with Asia with Africa, North and South America, Great Britain, Honshu, Java, and so on down to islands of two cells in the sea.

Let the number in the world of regions, each bounded by sea or ice, and each containing at least two cells, be denoted by λ. (9/3)

Let Σ sum over these λ regions. Then (2/20) gives

$$\Sigma \underline{C} + \Sigma \underline{F} + \Sigma \underline{L} + \Sigma \underline{D} + \Sigma \underline{B} + \Sigma \beta = 3[\Sigma \mathbf{s} + \Sigma \sigma - \lambda]. \quad (9/4)$$

In contrast with (2/14), there is here no double counting of foreign edges, because the void intervenes. There exist also cells of one edge, to which the theorem (2/20) does not apply and which therefore need to be counted separately. To qualify for one edge, the territory must be bounded by the void and must have a population between 1.562<u>h</u> and 0.622<u>h</u>, according to the rule (8/32) for rounding off digits.

Let the number of one-edged cells in the world be denoted by ν. (9/5)

Most of them are islands in the sea; but for suitable <u>h</u>, there are a few oases, or islands in lakes. The existence of San Marino, completely surrounded by Italy, may cause a slight confusion; for there is a risk of contradiction with the fundamental rule (1/3) on which theorem (2/20) depends. The population of San Marino was however only 0.011 million.

The world total, $\Sigma\sigma$, of non-ignorable lakes and deserts was directly counted for cells of a million and was found to be 18, which is 0.0109 of the corresponding number of cells. It became obvious in §2 that $\Sigma\sigma$ would increase as $\underline{h}$ decreased, but would always remain a small fraction of the number of cells. The author does not know how to obtain an exact functional relation between $\Sigma\sigma$ and $\underline{h}$. A plausible assumption seems preferable to mere neglect, and is made thus:

$\Sigma\sigma$ is assumed to be $0.0109\mathbf{s}$ for all $\underline{h}$. (9/6)

Let bold-faced letters denote world totals. (9/7)

In like manner the world total, $\Sigma\beta$, of the seacoasts of non-ignorable deserts was directly counted for cells of a million, and was found to be 4, which is 0.0067 of the corresponding number, 595, of cell edges to the sea. Accordingly:

$\Sigma\beta$ is assumed to be $0.0067\mathbf{B}$ for all $\underline{h}$. (9/8)

The ν one-edged cells, together with the Σs cells in the λ regions, make up the world total; so that

$$\mathbf{s} = \nu + \Sigma s. \qquad (9/9)$$

When the corrections (9/6, 7, 9) have been inserted into (9/4) it becomes

$$\mathbf{C} + \mathbf{F} + 1.0067\mathbf{B} + \mathbf{L} + \mathbf{D} = 3[1.0109\mathbf{s} - \lambda] - 2\nu. \qquad (9/10)$$

The numerical expression of (9/10) is as follows. The numbers ν and λ were obtained by counting along the lists in §8 with attention to the asterisks, and to politically divided islands such as Haiti, Borneo, New Guinea, Timor, and Saghalien.

$\underline{h}$	= 10^7	10^6	10^5	10^4
$1.0109\mathbf{s}$	= 151	1671	16928	169790
λ	= 5	20	63	212
ν	= 1	9	33	99
$\mathbf{C} + \mathbf{F} + 1.0067\mathbf{B} + \mathbf{L} + \mathbf{D}$	= 436	4935	50562	508635

(9/11)

Next **F**, **B**, **L**, **D** will be found by other processes. Finally **C** will be obtained by difference. At the end of the computation the numbers of edges should be curtailed to three significant digits.

The Number of Non-Civil Edges for Cells of One Million

These have been stated for individual territories in §8. It remains here to explain and discuss the numbers. They were obtained by two methods which are distinguished here by the same marks as in the lists I and II of §8. All methods involve correction to a standard date, which here is mid-1910.

The easiest, but roughest, method is lettered Z, and is mentioned first. The others follow as they arose from the author's attempts to improve the accuracy, without too much toil. The most accurate method is lettered T.

(Z) Each political territory was assumed to be of compact shape and of uniform density. The formulae for standard clusters (4/10) and (4/17) were combined so as to give for the 4th territory, after neglecting small constants,

$$F_r + 2B_r = 2(12s_r)^{\frac{1}{2}}. \qquad (9/12)$$

On summation for all the territories, this gives in view of (2/14)

$$2\mathbf{F} + 2\mathbf{B} = 4\sqrt{3}\sum_r s_r^{\frac{1}{2}}. \qquad (9/13)$$

This was applied to the list of territories and populations for the end of A.D. 1930, published in the Statistical Year Book 1931-32 of the League of Nations, omitting territories with populations less than half a million. The result was that, for cells of a million,

$$\mathbf{F} + \mathbf{B} = 1090 \text{ at end of A.D. 1930}. \qquad (9/14)$$

The present accepted value is $\mathbf{F} + \mathbf{B} = 1426$ at mid-1910. (9/15)

It may seem surprising that such obviously incorrect assumptions should give the right order of magnitude. They do so because of the minimum property of standard clusters, which is mentioned in §4.

(Y) A preliminary study of twelve countries, by drawing cells, led to the empirical correlation (3/3). When idealized into an equality it is

$$F_r + B_r = 2(12s_r - 3)^{\frac{1}{2}}. \qquad (9/16)$$

This was summed over the lists I and II of territories in §8, and it gave, for cells of a million,

$$2\mathbf{F} + \mathbf{B} = 2\sum_r(12s_r - 3)^{\frac{1}{2}} = 1810 \text{ at mid-1910}. \qquad (9/17)$$

This easily obtained estimate is $\frac{4}{5}$ of the present accepted value namely

$$2\mathbf{F} + \mathbf{B} = 2257 \text{ at the same date}. \qquad (9/18)$$

(X) The shape of a territory can be taken into account by the formulae of §6. For example, Java was regarded as a rectangle having sides in the ratio 7:1, and uniform density. Formula (6/20) accordingly gave

$$B_r = \frac{1}{2}F_r = \frac{2s^{\frac{1}{2}}}{a}\left\{7^{\frac{1}{2}} + 7^{-\frac{1}{2}}\right\}.$$ For s = 30 and $\underline{a}$ given as 1.0746 by (6/15) this works out to $\underline{B_r} = 31$. Drawing cells gave $\underline{B_r} = 27$.

(W) Both shape and the distribution of density are taken into account in some of the formulae of

§6. But the complications of fact are beyond the reach of formulae.

(V) The maps of density, published by Messrs. Philips were used in the following manner. From the local colors on the coast and on its hinterland the author computed the side of a square which would contain a million people. That side was then marked as a straight edge having its ends on the coastline. The process was then repeated, each square being adjusted to the local density of population. Special allowance had to be made for cities, where Philip's color-scheme is replaced by black dots. This method is very rapid for areas of many cells. The author at first applied it to the coasts of all the continents, with the following results for coastal edges:

Edges to Sea or Ice at Mid-1910

	Method V	Accepted Value
Brazil	14	12
Africa	57	44
Eurasia	307	310
Chile	8	3

The method V goes wrong for Chile, because the square cells overlap the Argentine frontier.

(U) Constructing from a map of census-areas, together with their populations in Arabic numbers, a strip of compact cells of a million persons along the coasts and both sides of the land frontiers. The influence of the interior parts of the population is neglected.

(T) This is like method U, but extended to the entire territory. This is the most accurate method. It is slow and troublesome when there are 40 or more cells in the territory, but rapid and easy when the number of cells is five or less. It seems to be definite to about 2 percent.

If we are given that the total population of any territory forms one cell, no information about its distribution is necessary; for method T then degenerates into the obvious. A similar but moderate relaxation occurs when there are only 2, 3 or 4 cells in the territory; for then some information about the distribution is desirable, but it need seldom be as detailed or as accurate as a census by provinces. In method T2 the cellular map was made from a secondary source, usually from a Philips' map of population-density.

The world-totals at mid-1910 were:

F = 831 to foreign countries, B = 595 to sea or ice, D = 53 to desert, L = 79 to lakes. (9/19)

Variation of F, B, D, and L With h

The author at first supposed that F and B would vary as $h^{-0.5}$. That is indeed the main effect, and it may suffice for some historical arguments. Further investigation however has shown that amendments are required.

(i) The idea of local uniformity of density led in §4 to variations with h ranging from $h^{-0.5}$ to $h^{-0.565}$, according to the shape of the detail on the boundaries. An empirical study of coasts in §7 showed that $h^{-0.55}$ is appropriate to them on the average.

(ii) The contrast between urban and rural areas has an effect which depends on h. The sign of this effect can be made evident, and its order of magnitude can be roughly estimated, from the following considerations. The argument goes roundabout, as in thermodynamics, thus:

contrasts large → uniform large → uniform small → contrasts small.

If the cell population exceeded that of the largest cities, then the cell-boundaries could remain fixed while the population flowed out of the towns and cities and spread itself uniformly over each cell. That is to say; when h is as large as 10^7, the urban-rural contrasts of density have a negligibly small effect on the numbers of edges.

At the other extreme, when h is small in comparison with a town's population, the effect of the alternations have been studied in §6 by way of integrals. It is shown in (6/27) that, for a given total number of cells, there is an increase in the number of external edges as the urban-rural contrasts become more marked. The ratio of increase of F or B is about 1.3 for the actual degree of urbanization.

The last two paragraphs, when compared, show that F and B must increase, as h decreases, more rapidly than they do for uniform density. To see whether the effect is negligible or important, we can express it crudely as a power of h thus:

When $\log_{10}h$ decreases from 7 to 3, then the urban-rural contrast causes an extra increase of $\log_{10}B$ equal to 0.114.

Therefore the urban-rural effect is of the order of $h^{-0.03}$. (9/20)

This is comparable with the effect of irregularities in coasts and land frontiers.

(iii) In the absence of information to the contrary, it will be assumed that the amendments (i) and (ii) may be combined by adding their indices, so that;

the number of external edges of given territories usually vary as $h^{-0.58}$. (9/21)

(iv) The preceding amendments relate to territories that are given; but, as h is decreased, additional territories come into consideration in the manner set out in §8. When h is decreased in the ratio ten, the additional territories have from 1 to 6 cells each. They were thus fairly easy to sketch in cells. This was done; their edges were counted and added to the others.

The application of these principles will be illustrated by their numerical consequences as to B.

When $\underline{h} = 10^7$ only the territories in list I of §8 need attention. They were mapped in cells of a million, and on those maps the total of B was 463. To reduce this to cells of ten million it can be divided, in accordance with (9/21), by $10^{+0.58}$ which equals 3.802. The result is B = 122.

When $\underline{h} = 10^6$ a direct count gave B = 595.

When $\underline{h} = 10^5$ we have, from the territories in lists I and II, 595 x 3.802 = 2262. To this must be added a contribution from list III. It was estimated by sketching cells of 100,000, and was found to be 248, thus making a total of 2262 + 248 = 2510 for the world.

When $\underline{h} = 10^4$ the territories in lists I, II, III give 2510 x 3.802 = 9543. The contribution from list IV was roughly estimated to be 549, so that the world total was B = 10092. In preparing this estimate of 549 for list IV the territories with land frontiers were considered individually as far as the scant information permitted; but the islands were grouped according to the number of their cells, and were credited with the following average numbers of edges, taken from list III, and smoothed.

number of cells in island	1	2	3	4	5	6
number of edges to the sea	1	2	3.4	4.9	6.7	8.5

The world totals F, D and L were computed in a similar manner, except that for F, there is the following complication.

The extra territories, which come into consideration whenever $\underline{h}$ is decreased, have foreign edges of two sorts: some to the territories which had previously received attention, others entirely among themselves.

Let the numbers of these two sorts be denoted respectively by F' and F''. (9/22)

They are functions of two values of $\underline{h}$. The application of these numbers depends on circumstances. In §8 the cell edges in list I are stated, incongruously but conveniently, for cells of one, not ten, millions. Thus when we work backwards to cells of ten million, F is not

$$\frac{613}{3.802}, \text{ but is } \frac{(613 - 154)}{3.802} = 121. \text{ On the contrary}$$

when we work forwards to cells of 10^5, the contribution from list III is $\underline{\Sigma F'} = \frac{1}{2}\underline{\Sigma F''}$ where $\underline{F'}$ and $\underline{F''}$ are for the desired size of cell and Σ sums for list III.

<u>Summary of World-totals of Edges at the Middle of A.D. 1910</u>

The values for C were obtained from equation (9/11).

Table

$\underline{h}$	10^7	10^6	10^5	10^4
F	121	831	3,319	12,685
B	122	595	2,510	10,092
D	12	53	212	820
L	21	79	304	1,157
C	159	3373	44,200	483,813
$\frac{C}{F}$	1.31	4.06	13.32	38.14

The problem set in §1 has thus at last been solved.

It is difficult to state the accuracy concisely. For some purposes everything is spoiled by the uncertainty about China. For other purposes the accuracy may at best be about one percent for individual territories, and better in the total of many such. For small $\underline{h}$ the value of C is controlled mainly by the number of cells, and is but little affected by inaccuracies in F, B, D, L.

REFERENCES FOR THE APPENDIX

(See also a special list in §8)

1. ADMIRALTY (no date) I.D. 1128, <u>Handbook of Arabia</u>, Vol. I, London, H. M. Stationary Office.
2. "BARLOW" <u>Report, 1940, of the Royal Commission on the Distribution of the Industrial Population</u>, London, H. M. Stationary Office Cmd. 6153.
3. BARTHOLOMEW, J., 1935. <u>The British Isles Pocket Atlas</u>, Edinburgh, Bartholomew & Sons, Ltd.
4. BROMWICH, T. J. I'A 1926. <u>An Introduction to the Theory of Infinite Series</u>. Revised by T. M. MacRobert, London, Macmillan.
5. GAUNT, J. A., see Richardson & Gaunt.
6. HEATH, Sir Thomas, 1921. <u>A History of Greek Mathematics</u>, Oxford, Clarendon Press.
7. HOGBEN, L., 1942. <u>Mathematics for the Million</u>. 2 edt. London, Allen & Unwin.
8. JAEGER, Fritz, 1928. <u>Afrika</u> Leipzig, Bibliographisches Institut.

9. JEFFREYS, H. and JEFFREYS, B. S., 1946. Mathematical Physics, Cambridge, University Press.
10. K = Keesing's Contemporary Archives, Bristol, Keesing's Publications, Ltd.
11. KENDALL, M. G., 1943. The Advanced Theory of Statistics, Vol. I, London, Charles Griffin & Co.
12. LEAGUE OF NATIONS, Armaments Year Book. Geneva, International Statistical Year Book. Geneva.
13. LINES, L., 1935. Solid Geometry, p. 135, London, Macmillan.
14. LOVE, A. E. H., 1906. The Mathematical Theory of Elasticity, 2 edit., Cambridge, University Press.
15. MAXWELL, J. C., 1904. A Treatise on Electricity and Magnetism, 3 edit., Oxford, Clarendon Press.
16. O'DELL, A. C., Map of Population of Britain in 1931. Ordnance Survey.
17. POINCARÉ, H. (no date), trans. Maitland F., Science and Method, London, T. Nelson & Sons.
18. PONTRJAGIN, L., trans. Lehmer, E., Topological Groups, Princeton, University Press.
19. RICHARDSON, L. F., 1908. Phil. Mag. S 6, 15, 237-269.
20. RICHARDSON, L. F. and GAUNT, A., 1927. Phil. Trans. A 226, 299-361.
21. RICHARDSON, L. F., 1950a. Eugenics Review, 42, 25-36.
22. RICHARDSON, L. F., 1952. Contiguity and Deadly Quarrels: The Local Pacifying Influence, J. Roy. Statistical Soc., 115, 219-231.
23. SHEPPARD, W. F., 1898. Proc. Lond. Math. Soc., 29, 353.
24. SOMMERVILLE, D. M. Y., 1929. An Introduction to the Geometry of n Dimensions, pp. 141-144, London, Methuen.
25. TAYLOR, E. G. R., see Unstead and Taylor.
26. Times Atlas, 1900, London at the office of "The Times."
27. UNSTEAD, J. F. and TAYLOR, E. G. R., 1921. . . . Maps of Population, London, George Philip & Son.
28. YOUNG, J. W., 1929. Ency. Brit., 14 edit. 10, 177.

Richardson, L. F. (1993)
Collected Papers of Lewis Fry Richardson, **2**, 631–644

1951:3

HINTS CONCERNING INTERNATIONAL ORGANIZATION

(Manuscript dated 19 April 1951,
with a revised 'Summary' dated 6 July 1951)

(Not previously published)

Note: Professor C. C. Lienau of Columbia University invited Richardson to contribute a chapter to a book on *Measures of Organization* which he was planning to edit. This is Richardson's contribution, but the projected book never appeared, 'for failure of funding'.

Hints concerning international organization

By L. F. Richardson

Introduction

In the nineteenth century the world could not be organized as a whole, because of the slowness of messages and of travel. Modern radio and aircraft have greatly speeded interaction, yet the world is still not organized as a whole. The chief remaining obstacles appear to be in human nature.

Every organizer uses information. If his stock of information is copious, accurate, and well-arranged, he may claim that his process of organization is scientific. Every organizer must also make decisions about value, distinguishing better from worse, good from evil. The process of organization is an interaction between knowledge and judgments of value, and therefore it cannot possibly be reduced to social science alone. This puts me in a difficulty. Professor Lienau has asked me to write this chapter because I have published various researches on the causation of wars. In those articles I have tried to be objective and impersonal. But now, when writing about organization, I must inevitably take sides, by saying what I personally think better, what worse. One cannot discuss world-organization without indicating what sort of world one would like. Yet in order to avoid excessive personalism, I will allude to several ideals, as a reminder of how diverse they can be.

Ideal A. A world in which there is plenty of variety, excitement, and disorder.

Ideal B. A world where love, joy, peace, long-suffering, gentleness, goodness, faith, meekness, and temperance prevail everywhere.

Ideal C. A world in which there are plenty of opportunities for the able and energetic to become millionaires.

Ideal D. A world in which no one suffers poverty.

Ideal E. A world designed to make the next generation better than this one, by selective breeding or otherwise.

Ideal F. A world where nearly everyone loyally follows the leader, while dissident minorities are suppressed.

Ideal G. A world where friendly cooperation abounds, where competition and controversy are permitted, but fighting is forbidden.

Having mentioned these various ideals, I go on to discuss the difficulties of attaining only one of them, namely G, the last. Allow me to take, as a working model or microcosm of this last world-ideal, the British House of Commons. It has been said to be 'the best club in London'. To dine occasionally with political opponents is regarded by some members as a duty. Yet the House is also the scene of almost daily strenuous contention. In foreign policy and in defence, the opposition have often supported the government; but on other problems the opposition generally try to modify, to delay, or to frustrate governmental action. A spare team, known as the shadow cabinet, is ready at any time to supplant the government. The Leader of the Opposition, for his services in organizing all this, has been paid from state funds a salary of £2000 a year.

A never-mentioned, but very influential, feature of the parliamentary situation is the fact that the opposing parties do not keep party armies. This appears the more remarkable when it is compared with the declaration by Mr Anthony Eden that, in foreign affairs, diplomacy needs to be supported by armed force. On the rare occasions when a member of Parliament has descended to fisticuffs he has been ordered by the Speaker to leave, and, if he refused to go, he has been forcibly removed by the Serjeant-at-Arms. These are officers of the House of Commons as a whole, not of one party. How different is international behaviour!

On this ideal, the present world-wide controversy between the communists and the rest could be quite in order, if it were not for the fact it has led to slaughter in many places, especially in Greece and Korea.

I write with diffidence about what ought to be done, for I suspect that the connection between social science and practical politics may be very loose. That is why this chapter is called 'hints'. Even in physics, where so many exact quantitative generalizations are well known, a graduate trained in the simple tidy statements of the textbooks, will have much to learn before he can decide what ought to be done in the more diversified situation in a factory. Nevertheless, industry has benefited greatly by applications of pure science. It seems likely that politics will benefit, within the next 50 years, from studies which at present appear to be merely academic social science. When expressed in the language of engineering students, the rest of this chapter is more like 'dynamics' or 'strength of materials' than 'machine design'. They need to know something about dynamics and the strength of materials, before they can design reliable machines.

Arms and insecurity

The present practice of the great powers is based on the maxim that if you wish for peace you should prepare for war. This is the ancient Roman proverb 'Si vis pacem para bellum'. A statistical test of the truth of it, in modern conditions, was made on the facts relating to the First World War (Richardson 1949a). The warlike preparation of each nation was measured by its cost in pounds sterling per head of population for a year shortly before the war began. The subsequent suffering was measured for each nation by its war-dead reckoned as a percentage of its prewar population. The preparation was plotted against the suffering for the 21 nations that were more or less involved in the war. The diagram showed a miscellaneous scatter, but no conspicuous relation of any kind. To examine the same facts more thoroughly, the correlation coefficient between preparation and suffering was computed by the product-moment method. The ancient proverb would lead us to expect that, the more a nation had prepared, the less would it suffer: so that the correlation coefficient would be negative. The computed correlations were found to be positive: namely 0.15 when empires were regarded as wholes, but 0.23 when empires were represented by their metropolitan cores. According to R. A. Fisher's test (1936, p. 212) these coefficients are not significantly different from zero in a sample so small as 21 nations. That test is not quite appropriate, because the distribution of suffering is skew. Nevertheless, I think we must conclude that, although the ancient proverb is certainly not verified, yet neither is the contradictory saying 'what you prepare for, that you will get' established in its place. No simple rule emerges from those facts.

The ancient proverb has also been compared with historical fact by a different method, which extends over a much longer time. Quincy Wright (1950) has done this by an analysis of the wars in his list. (Wright 1942, pp. 636–46). He states his conclusion in the following words:

> Generalizations should be made with caution, but the historical record appears to indicate that, in a large proportion of the balance-of-power wars of the last three centuries, the militarily better-prepared countries initiated the war and that in about half of these wars, that state, though winning the early battles, eventually lost the war largely because its less prepared enemies were able to convince the world that they were the victims of attack and thereby to gain sufficient allies to achieve eventual victory. This seems to have been true of about two-thirds of the fifteen major balance-of-power wars during this period including the Thirty Years War of the 17th century, the wars of Louis XIV, the Napoleonic wars, and World wars I and II.

Some at least of the recent difficulties in foreign politics appear to be caused by a euphemism. The *Concise Oxford Dictionary* defines a euphemism as the 'substitution of a mild or vague expression for a harsh or blunt one'. Although a euphemism is an untruth, yet it is generally supposed to be harmless, because the real meaning is only thinly veiled. The trouble about the euphemism 'defence', when used to denote warlike preparations of all sorts together, is that it conceals the real meaning from one side only, namely from the side which makes those preparations. The other side see quite plainly that battleships, submarines, tanks, and long-range bombing aeroplanes, could be used for aggression. Neither side pays much attention to the verbal assurances of the other side: both rely on the principle that deeds speak louder than words. Each side says: '*Our* preparations are purely defensive. Of course we have the right to defend ourselves! *Their* preparations are obviously for attack'.

Another troublesome mental confusion about foreign politics arises from the neglect to notice that the behaviour of a small nation may be opposite to that of a great nation because of the effects of submissiveness, as formulated below in equations (1) and (2). It is commonly observed that a policeman can overawe a small boy. It is easy to find, in the history books (Camb. Mod. or E), occasions when a great power has overawed a small power by a threat of armed violence, although with little or no actual fighting. Here are some miscellaneous examples: (i) In 1850 Britain collected from Greece money owed to a British subject, Don Pacifico. (ii) In 1853 the US Americans opened Japan to foreign trade. (iii) In 1856 Oudh was misgoverned, and the British annexed it to their Indian possessions. (iv) In 1898 after the murder of two German missionaries in China, Germany occupied the Chinese port of Tsingtao. (v) In 1915 USA occupied Haiti in order to stop murder and misgovernment. (vi) In 1938 Germany annexed

Austria. (vii) In 1939 Russia occupied Estonia, Latvia, and Lithuania. (viii) In 1948 Britain quelled a threat from Guatemala to British Honduras.

Whether the great power was acting like a policeman, or in self-defence, or like a bully, is a very important question; but probably controversial. However for the present discussion it suffices to notice that the small power submitted with little or no resistance, and that the great power did not perpetrate a massacre. One might say that, in such examples, peaceful change was effected by a show of armed force.

But to conclude, from such contests between very unequal powers, that armed strength in general keeps the peace, would be extremely rash. Yet some such too-simple generalization has misled even the learned. An example occurs in volume XII of the *Cambridge Modern History*, which was published in 1910. In an introductory chapter, Stanley Leathes, one of the editors, when describing the condition of Europe at about that date, wrote (p. 7):

> On the whole, the existence of this tremendous military equipment makes for peace. The consequences of war would be felt in every household; and statesmen, as well as nations, shrink from the thought of a conflict so immense.

Presumably the heads of the fighting services did not also shrink; for at the time when Leathes wrote, there had begun the arms-race which led to the First World War. One might have thought that this opinion of Leathes had been sufficiently refuted by the occurrence of two subsequent world-wars. Yet the same idea was revived in 1946 by Moon and Burhop of the Atomic Scientists Association, in these words:

> The advent of the atomic bomb has given us a rare opportunity to make future warfare less likely on account of the horrors of mass destruction it will entail.

Moon and Burhop did however point out in 1946 the risk that an arms-race might develop. Now by 1951 the race is conspicuous.

Thus a show of armed strength, without actual fighting, has had diverse effects according to circumstances, the chief effects having been: (i) submission when the opposing strengths were very unequal; but (ii) an arms-race, leading in 1914 and 1939 to war, when the opposing strengths were of the same order of magnitude. These two contrasted effects have been shown as special cases of an inclusive statement (Richardson 1939 and 1949a). For the explanation of this problem mathematics is indispensable. Consider the pair of simultaneous differential equations

$$dx/dt = ky\{1-\sigma(y-x)\} \tag{1}$$

$$dy/dt = lx\{1-\rho(x-y)\} \tag{2}$$

in which t denotes the time, and x and y denote the variable warlike activities of the opposing sides. The best measures of x and y were found to be the annual 'defence' budgets divided by the wages, for that country and year, of a semiskilled engineer. The name 'warfinpersal' has been proposed for such a measure, as being a contraction of 'war finance per salary'. (Richardson 1949a) In equations (1) and (2) each of k, l, σ, ρ is to be regarded as constant in time and positive, but they depend on national characteristics. Of these, k and l are called 'defence coefficients'. There is some evidence that k/l is equal to the ratio of the industrial capacity of the x-side to that of the y-side. The coefficients σ and ρ measure submissiveness.

Mathematics can be written in various moods of doubt or certainty: see a 'Symposium on stochastic processes' published by the Royal Statistical Society (J. E. Moyal, M. S. Bartlett, and D. G. Kendall 1949). The observed facts about submission and arms-races are not sufficiently simple, numerous, and agreed, to allow anyone 'to prove by mathematics what certainly must happen'. But one can neatly describe by mathematics the connections between various scattered events that have happened. The following is a sketch. In order to emphasize the essentials, I have left out the 'restraint coefficients α and β, also the 'grievances or ambitions' g and h (Richardson, 1939).

Case (i) *Overwhelming armed strength.* Suppose that at any instant, say $t = 0$, we may neglect y in comparison with x in the bracket $(x-y)$. Then at $t = 0$ equations (1) and (2) are approximately

$$dx/dt = ky(1+\sigma x), \quad dy/dt = lx(1-\rho x) \tag{3), (4}$$

Suppose further that x is so large that $1-\rho x$ is negative. Then (4) shows that dy/dt is negative. This represents the y-nation submitting to the overwhelming threat from the x-nation. Incidentally (3) shows that dx/dt is positive. This represents the x-nation being roused by what it probably called the 'impudence' of the y-nation.

Case (ii) *Opposing groups of comparable size* such as Russia and company versus USA and company at the time of writing (April 1951). For mathematical simplicity let the situation be formally represented by the assumption that

$$k = l, \quad \sigma = \rho \tag{5}$$

The courses in time of x and y are then described by

$$dx/dt = ky\{1-\sigma(y-x)\} \tag{6}$$

$$dy/dt = kx\{1-\sigma(x-y)\} \tag{7}$$

It happens to be easier to attend, not to x and y separately, but to $x-y$ and $x+y$. Subtract (7) from (6) and divide by $x-y$ obtaining

$$\frac{d\log|x-y|}{dt} = -k\{1-\sigma(x+y)\} \tag{8}$$

Similarly by addition

$$\frac{d\log|x+y|}{dt} = k\left\{1-\sigma\frac{(x-y)^2}{(x+y)}\right\} \tag{9}$$

Now suppose that at some initial time, such as the years 1907 or 1929 or 1946 the warlike activities x and y were moderately small so that $1-\sigma(x+y)$ was positive. Then according to (8) $d\log|x-y|/dt$ was negative. That is to say x and y were tending to become equal. There was a drift towards an equality of warlike effort. A chance disturbance to x and y, unless it made $\{1-\sigma(x+y)\}$ negative, would not reverse this drift towards equality. The coordinate $x-y$ is in equilibrium at $x-y = 0$; and this equilibrium is stable.

If $x-y$ is near enough to zero, then $\{1-\sigma(x-y)^2/(x+y)\}$ is practically unity, so that the integral of (9) is

$$x+y = \text{constant}\cdot e^{kt} \qquad (10)$$

This represents an arms-race such as occurred during 1907 to 1914, and 1929 to 1939, and is again happening now. The financial statistics of the first two arms-races have been compared with exponential formulae, which were generalizations of (10). The agreement was good enough to be very interesting (Richardson 1949a).

A machine in balance is seldom useful unless its equilibrium is stable for all likely disturbances. It would be small comfort to the owner of a ship which had rolled over sideways, to know that it was perfectly stable for pitching fore and aft. It would be no comfort to a world which had been devastated by a general war, to understand that the difference between the preparations of the opposing sides had been tending stably towards equilibrium before the war broke out.

But how can $x+y$ be unstable, seeing that British statesmen, from Cardinal Wolsey to Viscount Halifax, have put their trust in the balance of power as a practical policy? The explanation may perhaps be that the instability of $x+y$ is a new phenomenon, which first appeared at some time between AD 1877 and 1900, and which may perhaps be connected with the application of science to war-industry. A search for unstable arms-races between 1868 and 1907 was made by Dr Curt Rosenberg. He did not find any unambiguous example of instability. In particular during the years 1868 to 1870 the sum of the warlike expenditures of France and of the North German Confederation was drifting stably towards an equilibrium [*Arms and Insecurity* Chapter VI]. As far as I know, the first clear description of the instability of an arms-race was published by Bertrand Russell in 1914.

Could a modern arms-race end peacefully?

The equations (6) and (7) become linear, if the submissiveness σ is negligible. To describe the course of the war of 1914 to 1918 it was found necessary to use more thoroughly quadratic expressions, a typical term being $d\eta_1/dt = c_{12}\xi_1\eta_2$ (Richardson 1948b). The equations (8) and (9) are therefore probably incapable of describing victory and defeat. Yet they do suggest a rather similar possibility which might occur without fighting. For suppose that an arms-race is going on, that $x = y$, while $(x+y)$ is increasing in the manner shown by (10). After $x+y$ has grown sufficiently, the second member of (8) becomes positive. When that has happened, $(x-y)$ is still at zero, and is in equilibrium at zero; but this equilibrium has become unstable, so that a chance disturbance might cause $|x-y|$ to grow. If $|x-y|$ were thus to grow sufficiently, the second member of (9) would become negative, and then $x+y$ would begin immediately to decrease. The change in $|x-y|$ would be lingering and chancy, the subsequent change in $(x+y)$ would be immediate and definite. The reason for this contrast is seen if we multiply (8) and (9) respectively by $(x-y)$ and $(x+y)$. For then $d(x-y)/dt$ contains the zero factor $(x-y)$, whereas $d(x+y)/dt$ contains the large positive factor $(x+y)$. These consequences of equations (6) and (7) correspond to an imagined situation in which after an arms-race, but no fighting, one side says to the other: 'We admit that you could beat us in war, we are therefore closing our armament factories, and returning our armed personnel to civil life'. To this the other side replies 'Very well, as you are really doing that, we also shall disarm'. Many people seem to expect a show-down like that actually to happen in the future. Yet in 1914 and again in 1939 an arms-race ended, not in agreement but in war; and, so far, those are the only two direct experiments on this question that have been completed.

'*Negotiation from strength*'. On 5 March 1951 a conference began in Paris between the deputy foreign ministers of Russia on the one part and of USA, France, and Britain on the other part. The purpose of the conference was merely to settle the agenda for a subsequent meeting of the four foreign ministers. During the first five weeks the deputies met almost daily, but did not succeed in agreeing on any agenda. Much of their time was occupied by each side objecting to the warlike preparations of the opposite side (K11309; BBC). Is this procedure properly called negotiation from strength? Is it not rather an illustration of how armed strength can prevent negotiation from beginning?

Defence without menace. The problem of separating defence from menace was discussed by Jonathan Griffin (1936). His recommendation was (pp. 207–9) that the more aggressive types of armament should be reserved for an international authority, while its subordinate nations should be allowed to keep only the more defensive types of weapons. The difficulty has been to obtain agreement.

The yearning for a confident belief. I have been trying to modify the almost universal belief in the efficacy of warlike preparations for keeping the peace. My wife holds that it is wrong to disturb anyone's fundamental beliefs unless one has something better to offer instead. On the contrary I think that people will not bother to search for, nor even to attend to, new truth until they have become somewhat dissatisfied with traditional statements. This domestic controversy was illustrated one day when my wife missed a train by relying on a time-table which was placarded in a railway station. She complained about it to the station master. He said he was sorry, he knew that the placard was out-of-date, but he had nothing better to put in its place.

A time may be coming when statesmen will resemble that station master in so far as they have personally ceased to believe in the Roman proverb 'si vis pacem para bellum', and yet do not possess any alternative belief which could be offered to their peoples to guide their conduct and to make them feel secure. Let us therefore look around for any pacifying influences other than armed strength.

Non-violent resistance has been developed by M. K. Gandhi, the Mahatma, as a proceeding in which large masses of people can join (Louis Fischer 1951, R. B. Gregg 1936). It has left in its wake in India far less hate and destruction than war would have done. For the resisters it is not an escape from suffering. Like soldiers, they have to control their fears, and they have to control their anger more than soldiers need do. The discipline of Gandhi's followers was not always as strict as he intended. Non-violent resistance works on the conscience of opponents. It succeeded against the British in India, because the British had, on the whole, a Christian conscience. When the Koreans tried it in 1919 against their overlords the Japanese, the effects were different (E **13**, 490).

The books by Gregg and Fischer contain much detailed information and critical discussion about non-violent resistance. As my personal experience of it is nil, I briefly pay my respects to the religious heroism of Gandhi, and pass on to some statistical investigations of other pacifying influences.

The local pacifying influence

It has been proved statistically that there has been some powerful influence which has prevented small wars. The gist of the argument is that, although small wars have occurred more often than great wars, yet the geographical opportunities for small wars have been so much more numerous than those for great wars, that the existence of some selective influence is evident. The stages, by which this investigation was built up, will now be indicated.

First an interval of history was selected, namely AD 1820 to 1929, or for some purposes to 1945. A search was next made in numerous works on history for all the wars, or other considerable deadly quarrels, which ended during that interval anywhere in the world. Particular attention was directed to the number of people who died because of each quarrel. The logarithm of this number, to the base ten, was named the 'magnitude' of the quarrel, and was made the independent variable by which frequencies were classified. This bare and strict statistical procedure misses some of the connections between events which historians regard as causal; but it has the compensating advantage that the importance of any quarrel tends to be decided objectively, without bias by the nationality or personal feelings of the compiler. The list so prepared with the magnitude, dates, names of the contestants, and a concise summary of the social background of each of 290 quarrels is published in microfilm (Richardson 1950b). A statistical summary of the variation with magnitude has been printed separately (Richardson 1948a). The latest revision of the observed numbers of quarrels is shown in Table I.

Table I. *Whole world, AD 1820 to 1945 inclusive*

Ends of range of magnitude	$7\pm\frac{1}{2}$	$6\pm\frac{1}{2}$	$5\pm\frac{1}{2}$	$4\pm\frac{1}{2}$	$3\pm\frac{1}{2}$	$2\pm\frac{1}{2}$	$1\pm\frac{1}{2}$	$0\pm\frac{1}{2}$
Observed number of deadly quarrels	2	5	24	68	$\geqslant 196$	?	?	7×10^6
Compare	2	6	20	64	202	640	2024	6400

The rapid increase of observed frequency with diminishing magnitude is remarkable. The bottom row of the table contains the integers nearest to the members of a geometrical progression with common ratio $\sqrt{10}$. This progression agrees well with the observed numbers in the first four or five classes, but widely underestimates the eighth class. This geometrical progression is an empirical generalization waiting for explanation. Let μ_0 denote the central magnitude of any unit range. The formula for the progression is then

$$6400\ 10^{-\mu_0/2} \tag{11}$$

The next stage of the investigation was geographical. Let w denote the population of the world. Let the map be marked off in cells each containing h people. There are then w/h cells in the world. It is a fact of political geography that hardly anywhere do more than three countries meet at a point. The number of frontiers possessed by 33 persistent countries was counted, and was found to average 6.3 per country. Euler's theorem on the faces, edges, and vertices of polyhedra would lead one to expect an average of about 6 edges per country, or a little less, between 5 and 6. Few countries are very elongated; Egypt and Chile are exceptional; most countries are about as broad as they are long. Let us suppose that the fictitious cells, each of population h, resemble the actual political countries in those abstract properties; so that however large or small h may be, the average number of frontiers to a cell remains the same. Let E denote the total number of frontiers between cells in all the world. It follows that

$$E \quad \text{varies as} \quad h^{-1}. \tag{12}$$

E is a measure of the geographical opportunities for contiguous conflict.

The next stage in the investigation relates to defeat. If the populations of two adjacent cells began to fight one another, it seems likely that they would go on fighting until one cell was defeated. The percentage loss of life at which defeat has been admitted is assumed to have been of the same order, whether the population was large or small. Let n denote the number of people killed in the war. Roughly we may assume that

$$n \quad \text{would vary as} \quad h. \tag{13}$$

The parts of the investigation relating severally to history, geography, and casualties are next to be connected together. When h is eliminated between (12) and (13), it follows that

$$E \quad \text{varies as} \quad n^{-1}. \tag{14}$$

But the magnitude μ was defined by

$$n = 10^{\mu}, \tag{15}$$

so that

$$E \quad \text{varies as} \quad 10^{-\mu}. \tag{16}$$

Comparison of (16) with the geometrical progression (11) shows that:

$$\frac{\text{number of actual fightings in unit range of } \mu}{\text{geographical opportunities in the middle of that range}} \text{ varied as } 10^{\mu/2}, \text{ that is as } \sqrt{n}. \tag{17}$$

This may be interpreted as meaning that there was some strong influence which prevented wars of magnitude 3 or 4 relative to those of magnitudes 5 or 6.

The above sketch of a theory is intended to show that there are interesting facts, and relations worthy of investigation. It is offered as an appetizer, not as a proof. Elsewhere (Richardson 1952) many other features have received attention. They include seacoasts, airpower, contiguous and non-contiguous belligerents, the number of nations on each side, the duration of fighting, and the probability that the neutrals would stay out. These complicate the argument, but they do not alter the main conclusion.

Doubts as to what motives operate in the local pacifying influence. It is obvious that if you take at random a compact area containing one million people you are more likely to find them of one language, of similar religion, all under one government, and intermarried, than if your randomly chosen area had contained ten million, or a hundred million people. So, at first sight, any or all of those bonds of unity might explain the prevention of small wars relative to large wars, relative also to the different geographical opportunities which exist for small or large. There is another psychological mechanism which becomes a bond of local unity in a roundabout way. We are free to love our neighbours, if our hate is directed against people at a distance. The Freudians have drawn attention to this process under the name of projection. There are specially relevant discussions of it by Glover (1935) and by Durbin and Bowlby (1939). Some historical examples of quick shifts of hatred have been collected by Richardson (1949 b).

The connection with common government has been particularly investigated (Richardson 1952). Any pair of opposed belligerents can be classified as civil or foreign, according as its two members had, or had not, a common government just before the outbreak of fighting. For comparison with the geographical notion of contiguous cells, the foreign pairs were classified as contiguous or non-contiguous, and only the contiguous were retained. For the same purpose the wars involving more than two belligerents were omitted. This selection is necessary for a definite test, but it is regrettable in so far as it cuts out the two world-wars.

Table II. *The whole world, AD 1820 to 1945. Contiguous quarrels of the type one versus one*

Ends of range of magnitude	$6\pm\frac{1}{2}$	$5\pm\frac{1}{2}$	$4\pm\frac{1}{2}$	$3\pm\frac{1}{2}$	$2\pm\frac{1}{2}$	$1\pm\frac{1}{2}$	$0\pm\frac{1}{2}$
Civil quarrels	3	6	10	65	?	?	7×10^6
Foreign quarrels	0	3	15	64	?	?	Few
Ratio of foreign to civil	0	0.5	1.5	0.99	?	?	< 0.01?

The geographical opportunities for contiguous fighting have also been classified as civil or foreign. This was begun by mapping the population of the world in compact cells each containing one million persons. No cell was allowed to spread across an international frontier. Not more than two cells and the sea were allowed to meet at a point; or, away from the sea, not more than three cells at any point. These maps having been drawn, a count was made of the number F of pairs of contiguous cells, one cell being on either side of any international frontier. The number C of pairs that lay entirely within nations was either counted, or deduced by the aid of Euler's theorem on the

Table III. *Geographical opportunities for contiguous fighting at AD 1910*

Cell population,	h	10^7	10^6	10^5	10^4
Foreign,	F	121	831	3,319	12,680
Civil,	C	159	3,373	44,200	483,800
	F/C	0.763	0.246	0.075	0.026
Corresponding magnitude about		5	4	3	2

polyhedra. This was done for cells of a million, and was extended to other cell-populations by devices partly argumentative and partly cartographic. The chief result is shown in Table III.

We have now the ratio of foreign to civil for two sorts of things: in Table II for actual fightings, in Table III for geographical opportunities. In order to compare them, they are brought together in Table IV at roughly corresponding magnitudes. At magnitude 5 the two ratios are of the same order. But at magnitude 3 they differ widely, in the sense that some influence has either stimulated foreign quarrels, or suppressed civil quarrels, or done both.

Table IV. *Ratios of foreign to civil*

Magnitude	5	4	3
Ratio for numbers of quarrels	0.5	1.5	0.99
Ratio for geographical opportunities	0.763	0.246	0.076

This sketch is again an appetizer rather than a proof; but a proper investigation (Richardson 1952) leads to much the same result.

The habit of loyalty to a common government would explain the effect that has just been noticed. There is some statistical evidence [*Statistics of Deadly Quarrels*, Chapter VI.2] as to how long it takes for loyalty to become established. Half of the revolts and other civil wars broke out after less than 23.5 years of common government. In more detail the distribution in time is shown in Table V.

Table V. *The growth of loyalty. Whole world,* AD *1820 to 1945. All magnitudes from 3.5 to 7.5*

Preceding years of common government	0 to 20	20 to 40	40 to 60	60 to 80	over 80	Total
Observed pairs of opposed civil belligerents	41	20	13	5	15	94
Geometrical progression	37.13	22.46	13.59	8.22	12.59	93.99

The geometrical progression, which is shown for comparison, decreases in the ratio 0.778 per decade.

In connexion with national habits and memories, it should be mentioned that alliance in war and retaliation have both faded away in geometrical progressions like the above, but going somewhat faster [*Statistics of Deadly Quarrels*, Chapter VI.2, 3].

The alleged pacifying influences of common language, and of common religion, have also been investigated statistically; although by rather crude techniques, such that strong effects would show, but slight effects may have escaped notice [*Statistics of Deadly Quarrels*, Chapters VIII, IX]. It was found that there was something pacificatory associated with the Chinese language and religion prior to the revolution of 1911. It was found that there was something bellicose associated with the Spanish language.

Statistics confirmed the historians' observation that the religious difference had instigated wars between Christians and Moslems. No definite conclusion was reached as to whether common language was in general a pacifying influence, as Zamenhof, the founder of Esperanto, assumed that it would be. There were rather more wars between Christians than one would expect if their religion had no influence; but the deviation was slight and not conclusive. The argument is too long to be given here.

A strong prima-facie case has been made out for the further consideration of *international marriage* as a hopeful pacifying influence (Richardson 1950a). It is hopeful because the binding emotions are strong.

Foreign trade. The great trade-depression of 1930 to 1935 embittered social relations both within and between nations. Vice versa it may be expected that abundant foreign trade would soothe international relations. This idea was tested statistically [*Arms and Insecurity*, p. 225]. The result disappointed the hopes earlier expressed by the author (1939, p. 14–16 and 34–45). The increase in foreign trade which seemed likely to be enough to compete with the fears and hates of an arms-race was found to be more than one could expect to occur.

Other forms of international cooperation. Some sorts of legal peace are described by the journalists as 'cold war'; but active cooperation between two nations appears to be the direct opposite of war between them. It seems likely therefore that any form of cooperation between nations has some tendency to prevent war between the same. The existing varieties of cooperation are very numerous. The League of Nations published particulars of about 500 associations, bureaux, or committees in its *Handbook of International Organisations* (1926). Unfortunately I cannot think of any technique for testing the importance of their joint influence.

Which of the local or regional pacifying influences are capable of extension to the whole world? The extraordinary unity and friendliness within a nation while it is engaged in foreign war have often been noticed, and have been explained by the re-direction of hatred away from neighbours and on to the common enemy. This type of unity is obviously not capable of expansion over the whole world.

There is the ordinary type of alliance between nations, whereby the allies do not quarrel with each other, because they are united in opposition to some other group of people. Well known examples have been: the Triple Alliance versus the Triple Entente in 1913; the Axis versus the Franco-British in 1938; and the Soviet Bloc versus the Atlantic Treaty Powers of today. Such alliances are not capable of extension to the whole world; because, if there was no threat from outside, the chief binding motive would disappear. The habit of working together might persist awhile.

C. G. Jung, the eminent psychotherapist at Zürich, has suggested (1947, p. xv–xvi) that hate might be harmlessly disposed of by shortening, instead of lengthening, the distance of its object. Switzerland is remarkable for having kept out of foreign wars, although it is surrounded by nations who have often been at war with each other. Jung emphasizes that there is much political strife within Switzerland, 'We fight each other within the limits of law and constitution, and we are inclined to think of democracy as a chronic state of mitigated civil war'. Then follows Jung's recommendation: 'Our order would be perfect if people could only take their lust of combat home into themselves'.

There are other pacifying influences, which although at present mainly local or regional, look as though they could be extended to the whole world. These include intermarriage and the habit of obedience to a common government.

Dr Ranyard West has suggested (1942) that the mistake is to expect that the great alliance should consist always of the same nations; for when justice is the controlling ideal, the just nations ought to oppose the unjust, whoever they may be. This fits with Sorokin's finding (1937) after a study extending from AD 525 to 1925, that, as regards internal disturbance: 'All nations are orderly and disorderly according to the times'.

Summary

So far this chapter has been a brief sketch of social science relating to pacification. It is partly an analysis of motives, and partly a statistical record of objective events. The emphasis has been on motives that are ordinary, and on events that are totalled over the whole world since AD 1819. In such a treatment, individuals tend to be lost from sight in the crowd. This neglect of the individual can be partially excused by saying that a statesman is called great, when he embodies the motives of a great many people. After the manner of physical science, moral judgment has so far been subdued. This chapter is therefore in marked contrast with newspapers, where it is customary to praise or to blame named individuals. The author suggests that a combination of these contrasted methods is more powerful than either separately.

Interim precepts for pacification. Friends ask me for definite practical recommendations. I am reluctant for the following reasons. Social science, however much it might be developed, could never tell us exactly what we ought to do: it could warn us off some actions, and suggest that others are harmless; but a wide range of free choice would always remain open. Similarly dynamics does not specify the machines that ought to be made. Moreover dynamics has persisted as enduring truth while machines have been scrapped and replaced by new designs. Social science should persist, and in order that it may do so, it should be kept distinct from ephemeral practical policy. However, with that caution, here follows an attempt to be practical:

(1) If you have the affection and the courage which go to make saints and heroes, try Gandhi's method.

(2) If you find it impossible to love your enemies, try, as next best, to understand them. One good way to get under their skin is to read the novels and plays which they enjoy.

(3) Don't object if your relatives wish to marry foreigners. Such bonds may help to hold the world together.

(4) Don't ban goods merely because they are foreign. Trade is a mild pacifier.

(5) Develop loyalty towards world government.

(6) Always remember that some of the 'defence' preparations, which your nations might intend to be purely for defence, are certain to appear to some other

nation as a dangerous threat against which they must make counter-preparations. Nevertheless, some purely defensive preparations do exist, for example air-raid precautions do not alarm unaggressive foreigners.

(7) Because research on the science of pacification is now in progress, look out for new and better techniques.

In this connection let me warn the reader that the present chapter is very far from being a general review of all the research work that has been published on pacification. Much more could be traced by the aid of the following list of references. In particular I wish to allude to my eight co-authors in the collection edited by Pear (1950). The international organization UNESCO pursues a 'group tensions project'.

References

Camb. Mod. stands for the *Cambridge Modern History*. Cambridge University Press.

Durbin, E. F. M. and Bowlby, J. 1939. *Personal aggressiveness and war*. London, Kegan Paul.

E. stands for the *Encyclopaedia Britannica*, 1929 edition.

Fischer, L. 1951. *The Life of Mahatma Gandhi*. London, Jonathan Cape.

Fisher, R. A. 1936. *Statistical Methods for Research Workers*. Edinburgh, Oliver and Boyd.

Glover, E. 1935. *War, Sadism and Pacifism*. London, Allen and Unwin.

Gregg, R. B. 1936. *The Power of Non-Violence*. London: Routledge.

Griffin, J. 1936. *Alternative to Rearmament*. London, Macmillan.

Jung, C. G. 1947. *Essays on Contemporary Events*. London, Kegan Paul.

K stands for *Keesing's Contemporary Archives*, Bristol.

Moon, P. B. and Burhop, E. H. S. 1946. *Atomic Survey*. Atomic Scientists' Association.

Moyal, J. E., Bartlett, M. S. and Kendall, D. G. 1949. *Journ. Roy. Statistical Soc.* B **12**, 150–282.

Pear, T. H. 1950. *Psychological Factors of Peace and War*. London, Hutchinson.

Richardson, L. F. 1939. *Generalized Foreign Politics*. Brit. J. Psychol. Monog. Supplt. No. 23.

— 1948a. *J. Amer. Statistical Assn* **43**, 523–46.

— 1948b. *Psychometrika* **13**, 147–174 & 197–232.

— 1949a. *Arms & Insecurity*, a book in 35 mm microfilm, published by the author.

— 1949b. *Brit. J. Medical Psych.* **22**, 166–8.

— 1950a. *Eugenics Review* (London) **42**, 25–36.

— 1950b. *Statistics of Deadly Quarrels*, a book in 35 mm microfilm, published by the author.

— 1952. *Contiguity and Deadly Quarrels*. Accepted for publication. *Journ. Roy. Statistical Soc.* A.

Russell, B. (now Earl) 1914. *War the Offspring of Fear*. London, Union of Democratic Control.

Sorokin, P. A. 1937. *Social and Cultural Dynamics*. American Book Co.

UNESCO 1950. *International Social Science Bulletin*. UNESCO **2**, 90–103.

West, R. 1942. *Conscience and Society*. London, Methuen.

Wright, Q. 1942. *A Study of War*. University of Chicago Press.

Richardson, L. F. (1951)
Nature, Lond. **168**, 567–8

1951:4

COULD AN ARMS-RACE END WITHOUT FIGHTING?

Nature
London
Vol. 168, No. 4274, 29 September 1951, pp. 567–8
(Letter dated 29 August)

Note: The 'preliminary note...sent to Prof. C. C. Lienau' on this topic is part of 1951:3

(*Reprinted from Nature, Vol.* 168, *p.* 567, *September* 29, 1951)

Could an Arms-Race End Without Fighting?

THERE have been only three great arms-races. The first two of them ended in wars in 1914 and 1939; the third is still going on. From so few events we cannot hope to draw any reliable conclusions by statistics. Let us instead see whether an analysis of motives throws any light on the problem. Arms, in peace-time, are intended to overawe possible enemies; therefore the motive of submission must come into the picture. Arms, in peace-time, in fact provoke possible enemies to prepare to defend themselves; so defensiveness must be taken into account. There is also considerable grumbling about the cost of rearmament. The behaviour of large groups of people is more regular and less capricious than the behaviour of individuals. It is instructive to regard large groups as deterministic, and to represent their behaviour by differential equations, provided that we remember that such a treatment is a caricature. The following simultaneous pair of equations were published[1] for that purpose in 1939.

$$dx/dt = ky\,\{1 - \sigma(y - x)\} - \alpha x + g, \qquad (1)$$

$$dy/dt = lx\,\{1 - \rho(x - y)\} - \beta y + h. \qquad (2)$$

Here t is time, x and y are the war-like preparations of the opposing sides, and the other letters are constants. Of these, k and l are positive 'defence coefficients', σ and ρ are positive measures of 'submissiveness', α and β are positive measures of the objection to the cost of rearmament; but g and h, which represent feelings, not about arms but about the treaty-situation, may have either sign. The names of the opposing nations do not appear, because the motives are supposed to be common to mankind; but the numerical values of the constants may differ from nation to nation.

The validity of these equations was reconsidered in more detail, and by comparison with more historical evidence, in a microfilm[2] published in 1947. The equations were found to be a tolerably good representation of what had happened. Better measures of x and y were introduced under the name of 'warfinpersal', which means the annual 'defence' expenditure divided by the annual earnings of a semi-skilled engineer. By the time that an arms-race had grown to alarming proportions, it was found to be difficult to distinguish the effects of α, β, g or h, for they were swamped by the effects of k and l. The ratio of k to l was reasonably regarded as equal to the ratio of the 'sizes' of their two respective groups as measured by population and industrial development.

The coefficients σ and ρ were originally introduced to represent the well-known fact that great Powers have sometimes suppressed very much smaller Powers without any fighting. The present note is about a comparatively subtle question, namely, the effects of submissiveness between two Powers that are in some sense equal. It will now be shown that the above equations (1) and (2) indicate the possibility that an arms-race might end without any fighting. I first noticed this theoretical possibility in April 1951. A preliminary note about it has been sent to Prof. C. C. Lienau of Columbia University for his book on "Measures of Organization".

The present world-tension between the devotees of communism and of private enterprise is between two groups sufficiently nearly equal to be alarmed about each other. For the sake of mathematical simplicity, let us represent this approximate equality by setting $k = l$, $\alpha = \beta$, $\sigma = \rho$; and let us neglect g and h. Equations (1) and (2) then become

$$\mathrm{d}x/\mathrm{d}t = ky - k\sigma y^2 + k\sigma xy - \alpha x, \tag{3}$$

$$\mathrm{d}y/\mathrm{d}t = kx - k\sigma x^2 + k\sigma xy - \alpha y. \tag{4}$$

It happens to be easier to discuss $(x + y)$ and $(x - y)$ rather than x and y separately. From (3) and (4) by addition and subtraction it follows that:

$$\mathrm{d}(x + y)/\mathrm{d}t = (k - \alpha)(x + y) - k\sigma(x - y)^2, \tag{5}$$

$$\mathrm{d}(x - y)/\mathrm{d}t = -(k+\alpha)(x-y) + k\sigma(x^2-y^2). \tag{6}$$

The relations are further clarified by dividing (5) and (6) respectively by $(x + y)$ and $(x - y)$; for then

$$\mathrm{d}\log(x + y)/\mathrm{d}t = (k - \alpha) - k\sigma(x - y)^2/(x + y), \tag{7}$$

$$\mathrm{d}\log|x-y|/\mathrm{d}t = -(k + \alpha) + k\sigma(x + y). \tag{8}$$

There was a time, during 1945 or 1946, before the present arms-race began to be mentioned. It may, nevertheless, have existed in embryo unnoticed, on one hand in Russia's tardy demobilization, and on the other in the United States' secrecy about the atomic bomb.

For mathematical simplicity it is convenient to contemplate an origin of time, say t_0, at which x and y were positive but both so near to zero that the last terms in (7) and (8) were negligible, leaving approximately:

$$\mathrm{d}\log(x + y)/\mathrm{d}t = k - \alpha, \tag{9}$$

$$\mathrm{d}\log|x - y|/\mathrm{d}t = -(k + \alpha). \tag{10}$$

The second member of (10) is negative, so that $\log|x - y|$ was tending towards minus infinity, and therefore x was tending to equal y. This is the equality of power, which many people rashly regard as a security. The second member of (9) is positive, so that log $(x + y)$ was increasing. This is the arms-race beginning slowly and inconspicuously.

Next let us contemplate various later times at which $(x - y)$ is practically zero, but $(x + y)$ has grown enough to make the arms-race noticeable. In this stage the last term in (7) is more negligible than it was at t_0, because after t_0 its numerator has decreased, while its denominator has increased. So, in accordance with (9), the sum $(x + y)$ continues to increase exponentially. Therefore, a critical time, t_1 say, arrives after which $k\sigma(x + y)$ becomes greater than $k + \alpha$, thus making $d \log | x - y | /dt$ become positive. But because $d | x - y | /dt = | x - y | . d \log | x - y | /dt$, and because $(x - y)$ is practically zero at t_1, there may be a long lingering after t_1 before $(x - y)$ deviates notably from zero. In other words, $(x - y)$ has been in equilibrium at zero, and at t_1 this equilibrium becomes unstable, so that a chance disturbance may cause either x to exceed y, or y to exceed x. The national politicians would be chiefly interested in which of these two ways the change began; but to that question the answer given by the equations would be untrustworthy, because of chance. The cosmopolitans and the non-politicals might have more cause to remember the subsequent changes described in the next paragraph.

Equation (7) shows that immediately after t_1 the sum $(x + y)$ goes on increasing. Yet, when $(x - y)$ has deviated sufficiently from zero, there will come a second critical time, t_2 say, after which $k\sigma(x - y)^2/(x + y)$ will become greater than $(k - \alpha)$, thus making $d \log (x + y)/dt$ negative. But because $d(x + y)/dt = (x + y).d \log (x + y)/dt$, and because $(x + y)$ is large and positive, the change in $(x + y)$ after t_2 will come on without delay. The contrast between these two reversals may be expressed by saying that $(x - y)$ was in equilibrium at t_1, but $(x + y)$ was far from equilibrium at t_2. The decrease of $(x + y)$ would be obvious as the end of the arms-race.

Both these theoretical reversals are consequences of submissiveness, for they would not occur if σ were zero in (7) and (8).

So say the equations (3) and (4). But could events really happen thus? As far as I know, they never yet have done so. Nevertheless, as the more general theory (1) and (2) has been verified in several other cases, it seems desirable that the topical case (3) and (4) should be published for discussion before the actual event becomes known.

The foregoing description is likely to be received with general incredulity, because it asserts, after the manner of Bertrand Russell's 1914 pamphlet called "War the Offspring of Fear", that the motives in the present arms-race are mutual; whereas the reader, whether he believes the capitalist or the communist propaganda, has been told that the arms-race is all due to the aggressiveness of the other side. I can scarcely hope to dispel a prejudice that is so natural anyway, and is daily reinforced by broadcasts and newspapers. Some evidence about present mutuality has been brought back from the U.S.S.R. by Prof. Kathleen Lonsdale.

There may be among the learned another quite minor cause of incredulity, namely, that the above simple argument is mainly about maxima and minima, and does not give a fully quantitative solution of the differential equations (3) and (4). These do not belong to any of the simple types treated in the text-books. Indeed, it seems remarkable that one can obtain considerable insight by a discussion of maxima and minima. The equations have, however, been integrated numerically in a particular case, and some approximate formulæ have been found. The constants have been roughly determined thus: $\alpha = 2$ year^{-1}, $k = 4$ year^{-1}, and perhaps $\sigma > 0{\cdot}4$ per million people. But an account of these doings might easily double the length of this article.

L. F. RICHARDSON

Hillside House, Kilmun,
Argyll.
Aug. 29.

[1] Richardson, L. F., *Brit. J. Psychol., Mon. Supp.* No. 23, p. 23, equations (31) and (32) (1939).

[2] Richardson, L. F., "Arms and Insecurity", a book in 35-mm. microfilm (published by the author, 1947 and 1949).

Printed in Great Britain by Fisher, Knight & Co., Ltd., St. Albans.

1951:5

COULD AN ARMS-RACE END WITHOUT FIGHTING?

Nature
London
Vol. 168, No. 4282, 24 November 1951, p. 920
(Letter dated 15 October)

Note: This is a reply to a letter from M. R. Horne, printed on the same page of *Nature*. Horne had argued that an alternative pair of equations of the type

$$\frac{dx}{dt} = ky - \sigma(y-x) - \alpha x$$

was 'at least as reasonable' as Richardson's 'submissiveness' equations (in 1951:4) of the type

$$\frac{dx}{dt} = ky\{1-\sigma(y-x)\} - \alpha x$$

920 NATURE November 24, 1951 VOL. 168

Could an Arms-Race end Without Fighting?

MR. M. R. HORNE's letter should be welcomed as raising questions that will not be understood until they have been argued. The question at issue may be stated thus.

Let us name $\frac{\partial}{\partial y}\left(\frac{dx}{dt}\right)$ the 'effective' defence-coefficient. Now, on Mr. Horne's hypothesis,

$$\frac{\partial}{\partial y}\left(\frac{dx}{dt}\right) = k - \sigma,$$

which can have either sign, but cannot change sign. But on Richardson's (1939) hypothesis,

$$\frac{\partial}{\partial y}\left(\frac{dx}{dt}\right) = k\,\{1 + \sigma\,(x - 2y)\},$$

which can change sign according as $2y - x \gtrless 1/\sigma$. The common saying that 'all bullies are cowards' suggests that a change of sign occurs somewhere. However, the present equations are intended to represent the behaviour, not of individuals, but of nations. The effective defence-coefficient of Germany has been estimated from statistics of warlike expenditure, and was found to be positive during 1908–14, negative during 1920–30, and positive again during 1930–39. The discussion of the submissive interval is at present published only in microfilm[1].

LEWIS F. RICHARDSON

Hillside House,
Kilmun, Argyll.

October 15.

[1] Richardson, L. F., "Arms and Insecurity", published by the author (1949).

Richardson, L. F. (1952)
Br. J. Sociol. **3**, 77–84

1952:1

IS IT POSSIBLE TO PROVE ANY GENERAL STATEMENTS ABOUT HISTORICAL FACT?

The British Journal of Sociology
London
Vol. III, No. 1, March 1952, pp. 77–84

Note: This paper was described by Richardson as 'statistics for historians'

Is it Possible to Prove any General Statements about Historical Fact?

L. F. RICHARDSON

Introduction

DURING THE past eleven years it has been my habit to read alternately books on history and books on statistical method. The contrast between these two disciplines has provoked the following remarks, which were also stimulated by a broadcast discussion on 4 January, 1948, between Professor P. Geyl of Utrecht and Professor A. J. Toynbee.

Examples of generalizations concerning historical fact will be taken chiefly from four works: Pitrim Sorokin's *Social and Cultural Dynamics*; Quincy Wright's *A Study of War*; my own microfilm *Statistics of Deadly Quarrels*; and A. J. Toynbee's *A Study of History*, as abridged by D. C. Somervell. On the literary grace, poetry, philosophy and religion which pervade Toynbee's book I shall not presume to comment, except that they make it delightful to read, and that the reader may find it difficult to maintain his critical judgment while thus enchanted. As to Toynbee's particular historical illustrations, so impressively numerous and gathered from all over the world, I shall respectfully assume them all to be accurately ascertained facts. The question at issue is whether his endeavours to prove generalizations have succeeded.

The statisticians have agreed on certain proprieties of method which they regard as important. Some of the simplest of these will be mentioned in turn, and the question will be raised whether they have been, or can be, attained when the subject matter is history. In this paper, intended for historians, it would be inappropriate to go into the mathematical elaborations in which professional statisticians delight. Suffice it to say that by "statistical methods" I am alluding to techniques more fully described in such books as Yule and Kendall's *An Introduction to the Theory of Statistics*, R. A. Fisher's *Statistical Methods for Research Workers*, Harold Jeffreys' *Theory of Probability*, and M. G. Kendall's *Advanced Theory of Statistics*. Detailed references are collected in a list at the end of this article.

Statisticians, like other specialists, while showing a united front to laymen, have their internal controversies, the most notable of which relates to the philosophical foundations of statistics. R. A. Fisher founds on a hypothetically infinite collection, from which the observed facts are a random sample; H. Jeffreys founds on reasonable degrees of belief, obeying logical rules. Nevertheless when the question is whether a given set of facts justify a proposed generalization, Fisher and Jeffreys almost always arrive at the same conclusion.

Exceptions Necessitate Counting

To any proposed historical generalization it will usually be possible to find some exceptions. So the question is whether the exceptions are sufficiently numerous and important to invalidate the proposed general statement? Problems of this sort have been considered by statisticians for centuries. They have by now developed an agreed technique for deciding such questions. It all depends on counting. Even in Jeffreys' treatment, which begins and ends with degrees of reasonable belief, the middle portion is concerned with counting instances. If this valuable technique is to be applied to history, it will first be necessary to specify a type of historical event, and to mark off a region of space and time, so that the number of such events in that region can be counted.

The Choice of a Field of Time and Space in which the Events are to be Counted.

Toynbee goes back to 4000 B.C., Sorokin to ancient Rome, Wright to A.D. 1480, Richardson only to A.D. 1820.

All four of those investigators take the whole world, as far as it was known. The known part shrinks as we go backwards in time.

A long time-interval contains a more convincingly numerous assortment of events; but, if they are all to be counted together, any trend is concealed. Whenever the purpose is to study the changing frequency of a phenomenon, it is necessary to mark off a succession of sub-intervals. Suppose for example that the question is whether wars have become more frequent or less frequent. A year is too short a sub-interval, because the number of wars that began in any year varied irregularly from 0 to 5 (Richardson, 1945). A thousand years is too long a sub-interval, because we should be interested in changes of shorter duration. Thus a compromise between long and short is necessary. Moyal (1950) took fifty years.

The Specification of Historical Events so that they can Suitably be Counted

The absurdity of counting incongruous things is brought out in the old story of the army contractor who offered to supply pies composed in equal proportions of rabbit and horse. On examination they were admitted to be "fifty-fifty; one horse, one rabbit".

On the contrary a valuable method of counting unequal objects, namely stars, was introduced by the ancient astronomers, Ptolemy and Al Sufi. They first graded the stars into six classes according to their conspicuousness, and afterwards counted the number of stars in each class. Modern astronomers still use refinements of Ptolemy's method.

Let us take the question: how should wars be counted? Different principles have been followed severally by Sorokin, Wright and Richardson.

Sorokin's measure of an internal disturbance, such as a civil war, is a geometric mean of four estimates of its importance depending severally on: (i) its social area, (ii) its duration, (iii) its intensity as indicated by the amount of violence and the number of sociopolitical changes, (iv) the numbers of people actively engaged. Sorokin vigorously defended these estimates, against the criticism that they are not accurate, by pointing out that they are certainly more quantitative than the vague phrases customarily used by historians.

Wright has published a valuable list of 278 "wars of modern civilization" ranging from A.D. 1480 to 1941. He mentions that "A list of all revolutions, insurrections, interventions, punitive expeditions, pacifications, and explorations involving the use of armed force would probably be more than ten times as long as the present list", which thus appears to be a narrowly selected class. Yet from another point of view Wright's class of "wars" appears very wide, for it includes both the Crimean War of 1853–6 and the 1st Transvaal War of 1880–1, as though they could be counted one each. In judging importance Wright, as a professor of international law, naturally placed much emphasis on the legal aspect. For example, the American Civil War of 1861–6 is included in his list, but is stated to have involved only one participant, because the Confederate South had no legal status either before or after the fighting.

While recognizing, as Sorokin and Wright did, the necessity for somehow judging the importance of any fighting, Richardson did not feel any confidence in his own personal judgment, but yearned after some criterion which should be acceptable to students of all nations. The most objective criterion that he could find was the total number of persons who died because of the war. Unfortunately the statistics of casualties are notoriously inaccurate, and therefore the classes must be broad. They may suitably be taken to be such that the number of deaths was severally of the order of 10,000,000 or 1,000,000 or 100,000 or 10,000 or 1,000 or 100 or 10 or 1. These numbers are more neatly written respectively as 10^7, 10^6, 10^5, 10^4, 10^3, 10^2, 10^1, 10^0. As a further degree of conciseness we may with advantage omit the 10 and write only its index 7, 6, 5, 4, 3, 2, 1, 0, which is the logarithm of the number of dead. Accordingly the "magnitude" of any deadly quarrel was defined by Richardson to be the logarithm to the base ten of the number of persons who died because of that quarrel.

The Desirability of Counting All

To collect all the events in the specified class is certainly the ideal, and is occasionally attainable. Thus Toynbee asserts that there have been only 21 civilizations, and he gives a list of them all. Wright indicates that the

278 wars of modern civilization, according to his definition, form a complete list. Richardson (1948) arrived at the following summary.

NUMBER OF DEADLY QUARRELS THAT ENDED BETWEEN A.D. 1919 AND A.D. 1945 ANYWHERE IN THE WORLD

Ends of range of magnitude	7·5 to 6·5	6·5 to 5·5	5·5 to 4·5	4·5 to 3·5	3·5 to 2·5	2·5 to 1·5	1·5 to 0·5	0·5 to − 0·5
Number of deadly quarrels	2	5	24	63	$\geq$ 188	?	?	6×10^6

Slight revisions are still in progress. The author hopes that if the attention of historians could be brought to bear on the matter, the counts might be extended into years prior to A.D. 1820 and into classes where there is at present an inequality sign or only a question-mark.

FAIR SAMPLES

Because the labour of examining every individual in the specified class would often be prohibitive, the statisticians have developed a technique to avoid bias in the selection of a sample. A convenient method is to make use of numbers printed in a sequence which is chaotic. Tables of such "random sampling numbers" have been published by Tippett (1927), by Kendall and Babington Smith (1939), and by Fisher and Yates (1938). Suppose for illustration that a historian were to require a random sample of letters printed in *The Times* newspaper during the 1,000 months which began with 1868. The first three digits of Tippett's numbers begin thus: 295, 664, 399, 979, 797, 591, 317, 562, 416, 952, 154, 139. Let these be the numbers of the months; January, 1868, being number one. By reading a microfilm of *The Times* for only those twelve months he could obtain a random sample of 1,000 months.

This method of selection can be applied to any phenomena which can be numbered consecutively.

Toynbee offers a wonderful collection of interesting facts which support his generalizations; but he seems to ignore the question whether his selection was a fair sample.

PROBABILITY IN HISTORY

When a student, unfamiliar with the detailed events and discussions, has read an authoritative summary of why any particular war occurred, he is likely to feel that events could not have happened otherwise. On the contrary, when one takes a comprehensive list of wars, either Wright's or Richardson's, and analyses it by a suitable technique, a familiar statistical distribution appears and thereby calls attention to the random and chance aspect of events.

The method of analysis can be explained by its application to Wright's list. The beginning and end were omitted as possibly liable to uncertainty, and the long stretch from A.D. 1500 to 1931 inclusive was retained. Wright lists four of the largest wars both as wholes and as parts. The parts were preferred, because that choice made the wars less unequal in size. A list of calendar years was prepared. Each war was taken in turn and a mark was placed against the year in which it began. When all the wars in the comprehensive list had thus been marked, it was found that there were 223 years with no mark, 142 years with one mark, 48 years with two marks, 15 years with three marks, 4 years with four marks, and no years with more than four marks. This distribution of years can be fitted, at least tolerably, with a formula named after S. D. Poisson, a French mathematician who published it in A.D. 1837. The interest of this connection is that the Poisson law can be deduced from the assumption that the probability that some war would break out somewhere on a day is very small and is the same for each day.

The philosophical basis of probability, according to the dominant school of thought to which R. A. Fisher belongs, is a hypothetical infinite collection from which one takes samples. That seems very suitable when the sample is, say, a netfull of fishes out of the sea. But can one imagine an infinite collection of wars, from which the actual wars could be fished out as a sample ? Instead it is more suitable to contemplate the probability that a war might break out somewhere on any day. The facts are consistent with the assumption that this probability is very small and almost constant, although with a seasonal variation, and perhaps also a variation over centuries (see Richardson, 1945, and Moyal, 1950).

To Prove that Two Attributes are Associated

In an enquiry as to whether two attributes, say A and B, are associated or independent, the standard statistical procedure is to count the number of events which occurred in *all four* of the following classes : (A and B), (A but not B), (B but not A), (not A and not B). Afterwards the " chi-squared test " is applied. A common mistake in popular discussions is to attend to less than four of these possibilities, and to ignore the chi-squared test. Such arguments are inconclusive.

Statisticians call the four compartments the " 2 × 2 bivariate table ". It is the simplest that can be used in a discussion of association.

The reader may wish to see this well-known method illustrated by an application to history. Let the question be : whether alliance in war is persistent ? To make a neat illustration rather than a comprehensive investigation, let it be restricted to a comparison of the First with the Second World War. There are, of course, several problems of classification on which individual judgments might differ ; the following numbers of pairs of belligerents are quoted from Richardson (1950) (see page 82).

That is to say there were 51 pairs of named nations who were allies on both occasions ; 25 pairs who were enemies on both occasions ; 8 pairs who

		At 1st July, 1917		Totals of Rows
		Allies	*Enemies*	
At 1 July, 1942	Allies	51	8	59
	Enemies . . .	24	25	49
Totals of columns		75	33	108

were enemies on the first occasion, but allies on the second; and 24 pairs who changed in the opposite sense. These four observed numbers form what is called the body of the 2 × 2 table. The totals follow from them, and are regarded as extras. The standard statistical argument proceeds in the following manner. Accept all the marginal totals of rows and columns as given; but consider what the four numbers in the body of the table would have been if previous alliance or enmity had been quite irrelevant. On that supposition the row-total of 59 would have been divided between its two columns in the ratio of the given totals of those columns, namely as 75 to 33, thus giving in the upper left-hand compartment $59 \times 75/108 = 41{\cdot}0$ in place of the actual 51. Then, in order to keep all the marginal totals fixed, the 8 must become 18, the 24 must become 34, and the 25 must become 15. So, on the hypothesis of irrelevance, which Fisher calls the "null" hypothesis, we should have had,

$$\begin{matrix} 41 & 18 \\ 34 & 15 \end{matrix} \quad \text{instead of the observed} \quad \begin{matrix} 51 & 8 \\ 24 & 25 \end{matrix}$$

The deviation of the observed numbers from the null hypothesis is in the direction of persistence of both alliance and enmity.

But before drawing any definite conclusion, we must pause to think about chance. We have been led to suppose that historical events are partly controlled by chance, and therefore we must admit the possibility that, even if the null hypothesis were true, some deviations from the null hypothesis might happen accidentally, and therefore be no guide to occurrences on other occasions.

The usual overall measure of the deviations from the null hypothesis is χ^2 defined thus

$$\chi^2 = \left\{ \begin{matrix} \text{Sum for all} \\ \text{compartments} \end{matrix} \right\} \text{of} \left\{ \frac{[(\text{observed number}) - (\text{hypothetical number})]^2}{\text{hypothetical number}} \right\}$$

In the present example

$$\chi^2 = \frac{100}{41} + \frac{100}{18} + \frac{100}{34} + \frac{100}{15} = 17{\cdot}6$$

Historians can safely read the interpretation of χ^2 in Fisher's tables or in Yule's table, which is reprinted in Kendall (1946), Appendix 6. Yule's table gives the probability P for the accidental occurrence of all the χ^2 as great as, or greater than, the observed χ^2. The table stops at $\chi^2 = 10$ for which $P = 0{\cdot}00157$. For our $\chi^2 = 17{\cdot}6$, the P would be still smaller. So the conclusion is that, in spite of the observed exceptions, the general tendency for alliance and enmity to recur after the named 25 years cannot be explained away by chance.

Toynbee makes many assertions about the association of attributes, for example that civilizations have arisen in hard rather than in easy environments. This would require a 2×2 table. He afterwards qualifies this by noticing that the challenge of the environment can be too severe: proof would accordingly involve a 3×2 table of six compartments.

While reading Toynbee's book I kept watch for any problem of association of attributes in which the minimum four compartments were all mentioned, but I did not find any. In that book events are described, but seldom counted; nor does the chi-squared test appear. The statistical comment must be that although Toynbee has noticed, illustrated, and explained many associations of attributes, he has not gone so far as to provide a satisfactory statistical proof of the existence of any of them. To complain about lack of proof at this stage would be sheer ingratitude; for, in the development of any study, relations have usually been noticed or suspected some time before they were proved; and the relations which Toynbee has noticed are such that, if they were proved, they would be of great importance.

The Time-order of Discoveries

Problems have not always been solved in the order of their seeming importance. At a time when the important practical problem seemed to be the transmutation of base metals into gold, the alchemists, by heating things in crucibles, noticed incidentally facts about arsenic, antimony, bismuth, zinc and platinum, which, though unsatisfying to them, have since proved useful, because they were definitely ascertained. At a time when the important question, according to A. J. Toynbee, is whether our civilization has already broken down, the sort of relations which one can prove by statistics may be exemplified by the distribution of wars in time or by the variation in the frequency of deadly quarrels with magnitude. These relations may now seem uninspiring. But because they have been carefully ascertained, they are likely to be fitted later into combinations at present unforeseen.

Conclusion

It seems most desirable that undergraduate students of history should acquire a few simple notions about statistical method. The subject of statistics leads on into mathematical depths; but students sometimes choose history, because they are not mathematically inclined. It would therefore be necessary to protect students of history from being choked with an overdose of mathematical statistics by a severe restriction of their syllabus to such elementals as: the importance of counting, the art of taking fair samples, the need of four classes in a proof of association of attributes, and how to use the chi-squared test. The full theory of the chi-squared test should be excluded as far too difficult; but one can use the test as one can use a clock without knowing how it was made.

To guard against any unwarranted exaggeration of the power of statistical method, let me say plainly that, when applied to history, it can only be a secondary analysis. First must come the sorting of documentary and other evidence by historians. The merit of the statistical conclusions depends jointly on a good statistical technique and on the judicious accuracy with which historians have previously summarized the evidence relating to particular events.

REFERENCES

AL SUFI, (trans. by Schjellerup, 1874). *Description des Étoiles Fixes*, St. Petersburg.

FISHER, R. A. (1950). *Statistical Methods for Research Workers*, 11th edition, Edinburgh, Oliver and Boyd.

FISHER, R. A., and YATES, F. (1938). *Statistical Tables*, Edinburgh, Oliver and Boyd.

JEFFREYS, H. (1950). *Theory of Probability*, 2nd edition, Cambridge University Press.

KENDALL, M. G. (1946). *The Advanced Theory of Statistics*, London, Charles Griffin & Co.

KENDALL, M. G., and BABINGTON SMITH, B. (1939). *Tables of Random Sampling Numbers.* Tracts for Computers, No. 24, Cambridge University Press.

MOYAL, J. E. (1950). *J. Roy. Statist. Soc.*, A **92**, 446–9.

RICHARDSON, L. F. (1945). *J. Roy. Statist. Soc.*, **107**, 242–50.

—— (1948). *J. Amer. Statist. Assn.*, **43**, 523–46.

—— (1950). *Statistics of Deadly Quarrels*, a book on 35-mm. microfilm, published by the author.

SOROKIN, P. A. (1937). *Social and Cultural Dynamics*, American Book Co.

TIPPETT, L. H. C. (1927). *Random Sampling Numbers.* Tracts for Computers, No. 15, Cambridge University Press.

TOYNBEE, A. J., abridged by SOMERVELL, D. C. (1949). *A Study of History*, Oxford University Press.

WRIGHT, Q. (1942). *A Study of War*, University of Chicago Press.

YULE, G. U., and KENDALL, M. G. (1946). *An Introduction to the Theory of Statistics*, 13th edition, London, Charles Griffin & Co.

Richardson, L. F. (1952)
Jl. R. statist. Soc. A **115**, 219–31

1952:2

CONTIGUITY AND DEADLY QUARRELS:
The local pacifying influence

Journal of the Royal Statistical Society, Series A
London
Vol. CXV, Part II, 1952, pp. 219–231

Note: This paper reappeared unaltered in *Statistics of Deadly Quarrels* (Chapter XII)

Contiguity and Deadly Quarrels: The Local Pacifying Influence

By Lewis F. Richardson

1. *Summary and Outlook*

The design of a new machine is not entirely free, for it has to conform to the laws of dynamics. Similarly in politics, any new proposal for decreasing the frequency of wars should be judged in view of what usually happened in the past. The present treatment does not go so far as to discuss "what ought to be done" (i.e., political planning), but is restricted to the preliminary question "what has usually happened" (i.e., social dynamics).

It is shown in section 7 that some strong pacifying influence has prevented small-scale fighting, and in section 8 that civil fighting has been prevented more than foreign fighting. What is offered here is not a surmize based on general impressions, but statistical proofs. A statistical technique which the author devised in 1943 for the study of the alleged pacifying effects of common language or common religion (1950*b*, pp. 317–373), was afterwards applied by him to the simpler question about common government (1950*a*, pp. 31–32). Although he still believes that those previous attempts were crudely correct, he has since improved them, and now wishes to explain.

Since 1914 there has been a slow, diffident and fluctuating drift towards world-government. If, in the future, world-government were to become the norm, then any subsequent wars would have to be called civil. It is well to be prepared for this situation by studying now the causes of civil wars.

The existence of a pacifier is here proved, but its nature is not entirely clear. It may well be the habit of obedience to a common government. But there are several other social features which have positive correlations with common government, so that their pacifying effects could easily be confused with those of government. Such are: intermarriage, common language, common religion and the tendency to direct one's hatred on to foreigners. Some of these have been examined elsewhere. By statistical methods no general pacifying effect was found for either common language or common religion; but particular effects, either pacificatory or bellicose, were found to be connected with particular languages or religions, notably pacificatory with the religion of China before the revolution of A.D. 1911, and bellicose with the Spanish language (Richardson, 1950*b*). It is more difficult to distinguish the effects of intermarriage from those of common government (Richardson, 1950*a*).

In the sequel the word "pacifier" is used merely as an abbreviation for "pacifying influence".

2. *Preliminaries*

The essence of statistics is counting. Wars, having been of such different sizes, cannot suitably be counted until they have somehow been arranged according to their importance. The measure of importance which is here adopted is the number of war-dead on both sides jointly. This concept extends readily to the smaller deadly quarrels, which would not ordinarily be called wars. The "magnitude", μ, of any deadly quarrel has been defined to be the logarithm to the base ten of the number of those who died because of that quarrel (Richardson, 1941 and 1948*a*). For examples: the Second World War was of magnitude $7 \cdot_4$, the North American Civil War was of magnitude 5·8, the Seven Weeks' War was of magnitude 4·6, the Boxer Rising was of magnitude $4 \cdot_2$, the annexation of Hyderabad by India was of magnitude 3·3, Louis Riel's rebellion in Saskatchewan in 1885 was of magnitude 2·3, and examples might be continued down to a murder of magnitude $\log_{10} 1 = 0$.

The historical facts must be numerous enough to yield a statistically significant summary. The necessity for a large sample competes with the desire to be up-to-date. One has to rely on the belief that, although human nature may change, yet it changes slowly, so that an interval of recent history may contain lessons for to-day. The interval actually chosen for study here began with A.D. 1820.

The sample must also be unbiased. Because the chief sources of bias are national, the facts have been collected from the whole world (Richardson, 1950*b*). The particulars of any war were selected after consulting sources of information which were, if possible, either neutral, or of opposite sympathies.

A deadly quarrel is here called civil when the contestants owed allegiance to a common government immediately before the outbreak.

3. *Pairs of Opposed Belligerents*

Some wars have been partly civil and partly international. To obtain a clean-cut statistical classification it is preferable to count, not wars, but pairs of opposed belligerents. A belligerent is usually a group of people, such as a nation, or a group of insurgents. By exception the victim of a murder may be a single person.

Table 1 is partly a quotation from a microfilm (Richardson, 1950*b*, p. 287), where much fuller explanations and particulars can be seen; but the numbers have been revised and subdivided.

TABLE 1

For Deadly Quarrels that Ended Anywhere in the World from A.D. 1820 *to* 1945 *inclusive*

Ends of Range of Magnitude, μ, of the Whole Quarrel	*Pairs of Opposed Belligerents*			*Ratio of Civil Pairs to Total Pairs*
	Civil	*Foreign Contiguous*	*Foreign Non-contiguous*	
7±½ (world-wars)*	12	40	71	0·10
6±½	3	8	14	0·12
5±½	23	59	7	0·26
4±½	49	100	42	0·26
3±½ (incomplete collection)	129	114	39	0·46
2±½ {systematic collection	—	—	—	—
1±½ wanting}	—	—	—	—
0±½ (murders)	—	—	—	1·0

* Neglecting some minor incidents, but counting everything in the so-called "matrices".

World-wars have been mostly international; murders have been nearly all civil; the intermediate deadly quarrels show a fairly regular gradation between these extremes, as may be seen in the last column of Table 1.

4. *The Importance of Contiguity*

The obvious reason why the murderer and his victim were usually subjects of a common government is their localization. Presumably an extension of the same notion, namely the more extensive geographical contacts of the more populous belligerents, would also explain the ratio of total to civil for the other magnitudes. For it has been shown that the number of a State's external wars has a positive correlation of 0·77 with the number of its frontiers (Richardson, 1950*a*, p. 31, or more fully 1950*b*, pp. 263–8). The importance of contiguity also became conspicuous in a research on the number of nations on the two sides of a war (Richardson, 1947), for the statistics were explained by a chaos, restricted by geography, and modified by the local infectiousness of fighting.

These three lines of evidence show that, as a preliminary to estimating the pacificatory effects of common government, or common language, or common religion, it is necessary to prepare a suitable measure of the geographical opportunities for fighting. This measure must apply to the whole world, and must somehow be related to the number of those who died because of the various quarrels. In previous publications the author has offered different crude approximations suited to different purposes.

In order to study the effects of contiguity they must be separated from those of long range power, whether air power or sea power. This is done by alternative methods. In section 7 the sea power receives attention by particulars of navies and by maps; whereas the air power is mostly eliminated by ending the historical interval with A.D. 1929, as in Table 5. The same purpose is attained differently in sections 6 and 8 by specifically excluding foreign non-contiguous quarrels; and when that is done there is no objection to extending the time interval, as in Tables 3 and 9.

5. *A Fundamental Assumption about Those who Did Not Fight*

Any search for pacifying influences must take account both of those who fought and of those who did not. The deadly quarrels have been classified by the number of quarrel-dead or by its logarithm, the "magnitude." Those who did not fight are specified by the number alive. So a statistical link between "magnitude" and population is required. An alternative plan would be to ignore the quarrel-dead, and to classify the actual quarrels instead by the populations of the opposing sides. However, if that were done, the number and the varieties of actual quarrels would become vague and uncountable, because the newspapers of any one nation are usually complaining bitterly about the conduct of some other nation. The author regrets that he sees no escape from the need to make an assumption to connect populations with casualties in quarrels that are hypothetical. The assumption must of course be founded on fact.

The amount of sufferings at the time of defeat can be crudely expressed by reckoning the war-dead as a percentage of the pre-war population. This is done in Table 2. The column marked "Ref." indicates the source of the information about losses of life. The percentages have been rounded off, because the number of war-dead is usually very uncertain. Some of the dates also may not be quite suitable. Belgium is suitably taken at the end of 1914 about two months after it was overrun; but Serbia had to be taken at the general armistice of 1918, because the record of its casualties up to its defeat in 1915 was not found separately. Another uncertainty is whether casualties should be allocated to an empire or to its metropolitan core. Nevertheless, when Table 2 is viewed as a whole, a statistical distribution appears, having its median at $1\cdot_4$ per cent., and its quartiles at $0\cdot_3$ and $2\cdot_9$. During the interval A.D. 1820 to 1945 there were far more defeats than the 22 shown in Table 2. It does not seem possible to collect them all, nor to assert that Table 2 is a fair sample. The author, however, has not wittingly introduced any bias.

Defeat may of course have depended on many considerations other than past or present sufferings. The Russians in 1905 made peace partly because they had only one railway line to the site of their conflict with Japan. Austria in 1859 listened to reasonable terms of peace. Reasonable offers of peace, however important they were on some occasions, cannot be taken into account in the present treatment, because it is restricted to be simply numerical. Moreover it is noticeable that nations which had been thoroughly roused by fighting, but were not yet overcome by exhaustion, have sometimes been remarkably unwilling to attend to terms of peace; so that the war came to an end by sufferings, and not by intelligent compromise (Richardson, 1948*b*, pp. 158–9).

TABLE 2

War-dead as a Percentage of the Whole Population, at or after Over-running or Defeat

War-dead per cent.	*Nation*	*Date*	*Ref.*
$\leqslant$ 83	Paraguay	1870	E, **17**, 259.
$\leqslant$ 22	Serbia	1918	Yovanovitch.
4	Germany	1945	K 7508.
3·1	Roumania	1918	Huber.
3·0	Austr.-Hung. Emp.	1918	Huber.
2·9	Southern U.S.A.	1865	Rhodes.
2·8	Germany	1918	Brockhaus, **20**, 193.
2·35	Holland	1945	K 9579.
2	Turkish Emp.	1918	Emin.
2·0	Bulgaria	1918	Danaillow.
1·6	Boers	1902	E, **21**, 66.
1·2	Japan	1945	K, 7837.
0·93	Russian Emp.	1917	Kohn.
0·9	Italy	1943	K, 7784.
0·5	France	1871	Dumas.
0·5	Finland	1939	K, 4089
0·3	Peru	1879	Dumas.
0·12	Belgium	1914	Otlet.
0·11	Denmark	1864	Dumas.
0·09	Russia	1854	Dumas.
0·04	Russia	1905	Dumas.
0·03	Austro-Hung.	1859	Myrdacz.

With Table 2 in view it seems reasonable to make the following assumption about hypothetical wars:

If two contiguous groups of people began to fight one another, they would, if not restrained by external authority, go on until one side was defeated; and defeat would usually occur when the less populous side had lost in dead some number between 0·05 and 5 per cent. of its population, while the larger population having sustained about equal casualties, would therefore have lost a smaller percentage . . . . (1)

Accordingly the total war-dead of both sides together would be from 0·1 to 10 per cent. of the smaller population. To put this assumption into symbols, let h be the lesser of the two opposing populations, and let μ be the magnitude of the hypothetical deadly quarrel, then

$$\mu = \log_{10} h - j, \quad \text{where } 1.0 < j < 3{\cdot}0. \qquad (2)$$

The median of $1{\cdot}_4$ per cent. in Table 2 corresponds to $j = 1{\cdot}6$. . . . . . (2*a*)

In subsequent abstract models the groups of people who might have fought, but did not, are imagined to occupy cells of equal population, h. The qualification "lesser" is then not needed.

There may be a feeling that the assumption about hypothetical quarrels should agree more closely with fact. Instead of assuming, as is done in sections 7 and 8, that j is independent of μ, would it not be more correct to plot the observed $\log_{10}h$ against the observed μ, to fit the plotted points with a line for predicting h when μ is given, and to accept this prediction-line, instead of the relation (2), as the fundamental assumption about hypothetical quarrels?

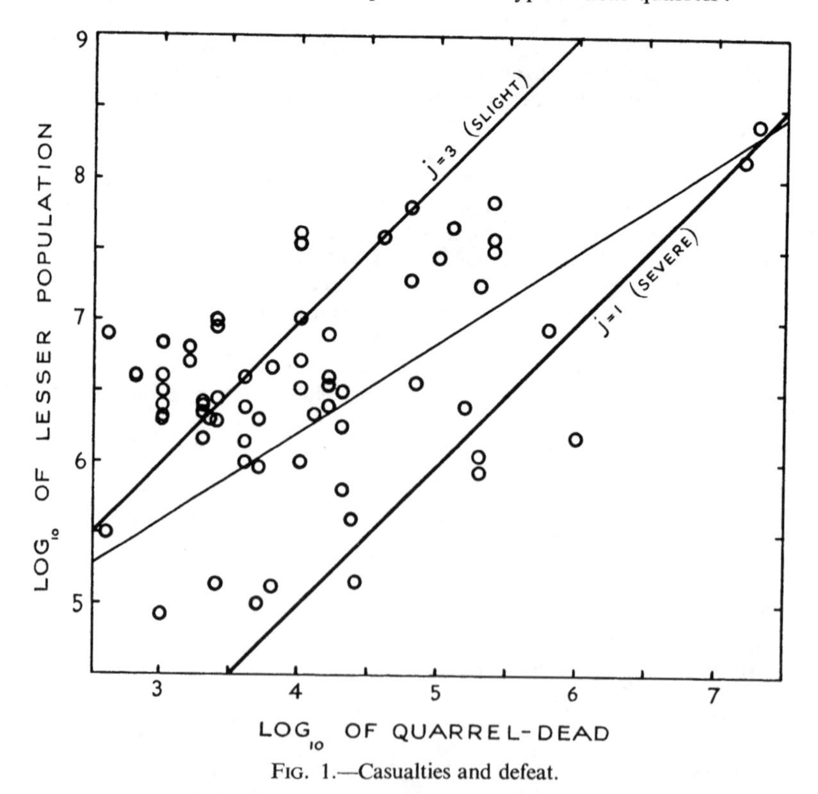

FIG. 1.—Casualties and defeat.

Fig. 1 shows an attempt to do this. The deadly quarrels were those in the author's list, which begins with A.D. 1820 (Richardson, 1950*b*). In order to gather enough facts, wars involving any number of belligerents were included, provided that they were, in the main, arrangeable as between two sides. Thus the two world wars appear on the diagram in the right-hand upper corner. Quarrels were excluded if the uncertainty either of μ or of $\log_{10}h$ appeared to exceed 0.3. This is only a small fraction of the range of scatter shown on the diagram. One might hesitate as to whether to take the population of the defeated side, or of the less populous side, for occasionally the larger population was defeated. I have chosen the less populous side. If this be an error of judgment, it will make a subsequent argument all the stronger. Moreover the lesser population was definite even when there was no definite victory and defeat.

Assumption (2) is shown on the diagram by way of the lines sloping at 45° for $j = 1$ (severe sufferings) and for $j = 3$ (slight sufferings). The band between these lines does indeed include a majority of the observed points for $\mu > 3{\cdot}5$; but for $\mu < 3{\cdot}5$ the band is too low. Some or all of this misfit near $\mu = 3$ is probably not a fact of history, but is merely caused by lack of information. The populations of sovereign states can usually be found in such books as the *Almanach de Gotha* or the *Statesman's Year Book*; but the populations of rebellious fractions of states are usually much more difficult to find. This may cause a bias; because for given μ, the known populations are likely to be larger than the unknown. Moreover the lack of information was most selective around $\mu = 3$, as the following table shows:

Ends of range of magnitude, μ	$3\pm\frac{1}{2}$	$4\pm\frac{1}{2}$	$5\pm\frac{1}{2}$	$6\pm\frac{1}{2}$	$7\pm\frac{1}{2}$
Quarrels ended A.D. 1820 to 1949	198	68	25	6	2
Number of quarrels on diagram	23	25	13	2	2

Consequently the lack of information is likely to have tilted the prediction line, making it slope less steeply than it would have done had the information been complete.

Suppose, however, for the sake of argument, that some future team of investigators were to succeed in finding the populations of all the insurgent groups; and that in consequence the line for predicting $\log_{10}h$ from given μ were found to slope less than 45°, as suggested in the figure. *The slope of such a line would by itself provide a proof of the existence of a local pacifying influence.* For the great wars cluster about the line $j = 1$, which corresponds to severe suffering, whereas the very small wars cluster about the line $j = 3$, which corresponds to much less suffering. Some influence, other than suffering alone, would then appear to have caused the less populous belligerents to cease from fighting. That is an interesting possibility for future investigation; but, because of the incomplete information, it is not yet proved.

For the purpose of investigating, by other arguments, whether a local pacifying influence exists, we should set out from the assumption that it does *not* exist (the "null hypothesis"). This is done in sections 7 and 8, by assuming j to be independent of μ.

6. *The Type One Versus One*

For simplicity the assumption (1) relates to a hypothetical deadly quarrel which involves only a single pair of belligerents. Among actual wars, that type has been fairly frequent (Richardson, 1947). The consequences of assumption (1) may suitably be compared with the historical facts relating to that type alone. When Table 1 has thus been pruned, it becomes Table 3.

TABLE 3

Deadly Quarrels, of the Type "One Versus One", that Ended Anywhere in the World from A.D. 1820 *to* 1945 *inclusive*

	Number of Quarrels, alias number of Pairs of Opposed Belligerents			
Ends of Range of Magnitude	*Civil*	*Foreign Contiguous*	*Foreign Non-contiguous*	*Ratio of Civil to Total*
$7\pm\frac{1}{2}$	0	0	0	—
$6\pm\frac{1}{2}$	3	0	0	$1{\cdot}_0$
$5\pm\frac{1}{2}$	6	3	0	0·7
$4\pm\frac{1}{2}$	10	15	2	$0{\cdot}3_7$
$3\pm\frac{1}{2}$	65	64	10	0·47
. . .	. . .	. . .	. . .	. . .
$0\pm\frac{1}{2}$	—	—	—	1·0

By comparison of Table 3 with Table 1 it is seen that the restriction to a single pair of belligerents has raised the ratio of civil to total, only in the larger magnitudes, where, however, the observed numbers are small. The middle portion of Table 3 will be quoted in Table 9.

7. *The Existence of a Local Pacifying Influence*

A proof of this has been developed from the study of the number of nations on each side of a war (Richardson, 1947, Theory XI) in the following manner. The historical facts were there compared with a formalized political geography consisting of:

α, world-wide sea-Powers which were in contact with so many other Powers by sea that their contacts by land alone could be ignored in comparison;

β, local coastal countries each touching the sea, also touching five different states by land;

γ, land-locked countries each touching five different States; where $\alpha = 5$, $\beta = 44$, $\gamma = 11$, total 60 countries . . . . (3)

The probability p of war between any two named nations was there defined to be the fraction of any very long historical interval during which such war occurred. Therefore $0 \leqslant p \leqslant 1$. According to the geographical situation of the two named nations, p was given the different symbols specified in Table 4.

TABLE 4

Scheme

Members of the Pair	*Symbol for the Probability* p
(i) Two world-wide sea-powers	x.
(ii) A world-wide sea-power and a local coastal power	y.
(iii) Two local coastal powers	z if in contact, or if a local belligerent touched both, otherwise zero.
(iv) A local coastal power and a land-locked power	
(v) Two land-locked powers	
(vi) A world-wide sea-power and a land-locked power	zero, unless a local coastal belligerent put them in contact, when p becomes y.

The historical wars were sorted into types "r nations versus s nations", and these were further sorted into subtypes according to the number of long-range or local powers severally among r and s. It was found in fact that the most frequent subtype was a war between only two powers, both of them local. This subtype was briefly indicated by the symbol "0,1 versus 0,1". The "expectation" of a war of any subtype was defined to be the fraction of the historical interval during which such a subtype would probably occur. This fraction may conceivably exceed unity. Let the expectation here be denoted by ψ. The argument was modelled on Bernoulli's concerning the binomial. For the subtype 0,1 versus 0,1 it gave the following formula for the expectation:

$$\psi = \tfrac{5}{2}(\beta + \gamma)\, z\, (1 - y)^{2\alpha\beta(\beta+2\gamma)/(\beta+\gamma)^2}\, (1 - z)^{12} \quad . \quad . \quad (4)$$

The author still regards that as a valid rough approximation for its original purpose, namely the discussion of wars between known principal countries, all magnitudes from 7·5 to 3·5 being lumped together.

Now however the purpose is quite different, and the treatment must be appropriately modified. The magnitudes are to be separated into unit ranges, and are to extend down as far as possible. In these circumstances the conceivable belligerents are not only or mainly the sovereign states. It is more appropriate to think of cells of equal population h. Each cell may still be conceived as having contact with about five neighbouring cells. As the magnitude diminishes, β and γ must increase. Let W denote the population of the world. Then W/h is the number of cells in the world. So let

$$\beta + \gamma = W/h. \quad . \quad . \quad . \quad . \quad . \quad . \quad . \quad . \quad (5)$$

It will suffice to take for W a mean value over the 110 years in question. The mean was computed from the data published by Carr-Saunders (1936), and was found to be

$$W = 1{\cdot}4 \times 10^9 \text{ persons.} \quad . \quad . \quad . \quad . \quad . \quad . \quad (6)$$

The fundamental assumption (2) as to the relation between population and casualties gives

$$h = 10^{\mu+j} \qquad \text{where } 1\cdot0 < j < 3\cdot0. \quad . \quad . \quad . \quad . \quad . \qquad (7)$$

Elimination of h and of W leaves

$$\beta + \gamma = 1\cdot4 \times 10^{9-\mu-j}. \quad . \quad . \quad . \quad . \quad . \quad . \qquad (8)$$

This is ready for insertion in the formula (4) for ψ.

It will next be shown that the long-range powers can be dropped out of consideration. None of them fought in these purely local wars; nevertheless the probability that they remained neutral appears correctly in the formula for ψ as a bracket $(1 - y)$ having an index involving α. In the former study this probability was

$$(1 - 0\cdot00166)^{9\cdot6} = 0\cdot984.$$

In the smaller magnitudes it seems likely that y may be less than $0\cdot00166$; also that α may remain at 5. In the index the coefficient of 2α is

$$\beta\,(\beta + 2\gamma)/(\beta + \gamma)^2 = 1 - \gamma^2/(\beta + \gamma)^2,$$

and so lies between 0 and 1. When the cells become small, γ predominates over β; and so the index tends towards zero. The probability that the long-range powers remained neutral is therefore near to unity and can be omitted from (4). With these modifications, formula (4) becomes

$$\psi = \tfrac{7}{2}\, 10^{9-\mu-j}\, z(1 - z)^{12} \quad . \quad . \quad . \quad . \quad . \quad . \qquad (9)$$

For comparison with this formula, the historical facts previously published in "Collection J" are now slightly revised and sorted by magnitudes in Table 5.

The sorting by magnitude introduces a new problem; for in order to correspond with assumption (1), the war-dead should be those of the single pair of belligerents only. In the former publication (Richardson, 1947) they were not so entirely. For example in 1875 a war began between the Turks and the Christians in Herzegovina and Bosnia. This may suitably be regarded as of the subtype 0,1 versus 0,1; and it continued like that for 0·8 year. It would be difficult or impossible to find the record of casualties for that 0·8 year by itself. Later the Montenegrins, the Serbs and the Russians joined in the war, the magnitude of which grew to $5\cdot_4$. But the portion which was of the subtype 0,1 versus 0,1 must have had a lesser magnitude.

This difficulty will be surmounted by an over-proof, alias an *a fortiori* argument. The durations in the column of Table 5 headed "in parts of wars" may perhaps properly belong to a range of magnitude lower than that to which they are there assigned. The author does not know whether this is so, nor how much lower. Let these loose pieces therefore be either left where they are, or moved to a lower row, in such a manner as most to oppose the conclusion at which the subsequent argument arrives. This is done by making the aggregate duration increase as rapidly as possible as the magnitude decreases. The result is the column 0, 17, 44, 77, in Table 5.

TABLE 5

Observed Aggregate Durations of Fighting, A.D. 1820 *to* 1929 *inclusive*

Ends of Range of Magnitude of Quarrel as a Whole	*Aggregate Durations of the Subtype* 0,1 *versus* 0,1			
			Reclassification most Adverse to the Conclusion	
	Total years	*In Parts of Wars, years*	*Years*	*Fractions of* 110 *years*
$7\pm\frac{1}{2}$	1 ?	1 ?	0	0
$6\pm\frac{1}{2}$	17	0	17	0·155
$5\pm\frac{1}{2}$	44	30	44	0·400
$4\pm\frac{1}{2}$	47	16	77	0·700

The history is now ready to be connected with the geography. In making the connection, we need to notice that the definition of the probability z tacitly involves the two ends of a range of magnitude. In the original paper this range corresponded roughly to "wars" in the popular sense; but here it is a unit range.

The numbers in the last column of Table 5 were equated to ψ in equation (9), and the equation was then solved for z. The computation is summarized in Table 6.

TABLE 6

Deduced Probabilities

Ends of Range of Magnitude	*Number of Cells in the World*			*Therefore the Probability* z		
	$j=1$	$j=2$	$j=3$	$j=1$	$j=2$	$j=3$
$6\pm\frac{1}{2}$	140	14	—	0·000,45	0·004,7	—
$5\pm\frac{1}{2}$	1,400	140	14	0·000,11	0·001,2	0·013
$4\pm\frac{1}{2}$	14,000	1,400	140	0·000,02	0·000,2	0·002

The last three columns of Table 6 show that whichever value of j is chosen, z decreases rapidly as μ decreases. That is to say, *the fraction of the total time, during which any particular pair of contiguous cells fought one another, was remarkably less as the cells were drawn smaller. This proves the existence of a local pacifier.* Its nature is not revealed by this argument. In particular no distinction has been made between civil and foreign fighting.

The only geographical feature which is involved in the conclusion is the number 5, which is assumed to be the number of neighbouring cells that touch each cell. This 5 appears in formula (4) in the coefficient 5/2; also the index 12 is connected with it. The existence of the local pacifying influence would still be evident if in those connections 5 were replaced by 6.

Some reader may perhaps object that magnitude μ is such an artificial concept that the probability z in unit range of μ is of slight sociological interest. Therefore nothing of any importance can be immediately inferred from the gradation of z shown in Table 6. The primary fact is not the magnitude μ, but the number of war-dead, $n = 10^{\mu}$. It would be more rational to reckon the probability, not per unit range of μ, but for each of the integers n. The author himself in a previous paper about the variation of frequency with magnitude came round to that point of view, although it considerably complicated the presentation of the data. But if n, not μ, is taken as the basis for probability, the gradation changes its direction. For the number of integers n in the range from $\mu_0 - \frac{1}{2}$ to $\mu_0 + \frac{1}{2}$ is about $2{\cdot}846 \times 10^{\mu_0}$. Let great Z denote the probability per integer of n. The probabilities of the $2{\cdot}846 \times 10^{\mu_0}$ separate integers n add up to make the probability of the unit range of μ centered at μ_0. So that $z = 2{\cdot}846 \times 10^{\mu_0} Z$. Then instead of

$$z = 0{\cdot}0047,\ 0{\cdot}0012,\ 0{\cdot}0002,$$
$$\text{at } \mu_0 = \quad 6, \quad 5, \quad 4,$$

as in Table 6, column 6, we have respectively

$$10^9 Z = 1{\cdot}65,\quad 4{\cdot}2,\quad 7,$$

gradated the other way.

In reply to this objection it may be said that there is an awkward question like that in statistical mechanics. It relates to the importance (alias "weight", alias "*a priori* probability") to be assigned to different states of the system. R. H. Fowler in his great treatise (1936, pp. 9–10) confessed that the generally accepted measure of importance was intuitive. My answer will also have to be intuitive. Which is the more meaningful probability, z or Z? Allow me to illustrate the question by an extreme and abstractly numerical contrast, in which all the important social and geographical features are ignored, except only the number of quarrel-dead. If we could thus abstract, then two quarrels involving severally one and two deaths would seem remarkably different; yet two other quarrels involving severally 1,000,001 and 1,000,002 deaths would seem practically indistinguishable. This belongs with the Weber-Fechner doctrine in psychology. The scale of μ corresponds with our intuitive feelings better than does the scale of n. One of the best justifications of intuitive practice is the fact that the intuitive scale of magnitudes for stars was in use for a thousand years before it was explained and improved by photometers and logarithms.

In the next section, however, the range of μ or n is eliminated by attending to the *ratio* of civil to foreign for the same range of μ or n.

8. *A Geographically More Thorough Investigation of the Pacifying Effect Connected with Common Government*

Let a historical interval be chosen between the dates T_1 and T_2; also let the range of magnitude be chosen from μ_1 to μ_2. A technique usual in studies of contingency would then be to ascertain, in that interval and range, the number of pairs a, b, c, d as specified in Table 7.

TABLE 7

Definition of Symbols

	Prior Governments Common	*Prior Governments Separate*	*Totals*
Fought one another . .	a	b	$a + b$
Did not fight one another .	c	d	$c + d$
Totals . . .	$a + c$	$b + d$	$a + b + c + d$

The numbers $a + b$ and a in Table 7 are already available in Table 3; indeed for five separate ranges of magnitude μ. But how are c and d to be obtained? For comparison, c must be the number of pairs of conceivable belligerents who might have fought one another in civil wars, but in fact did not do so. This concept is at first sight bafflingly elusive. Yet without c and d we could not proceed to draw any statistical conclusion about the alleged pacifying effect of common government; so an attempt must be made to catch and tame these elusive concepts. The number c cannot be ascertained from works on history, because insurgents were often not recognized as a group until they had declared themselves to be such by revolting. We are thus compelled to think about groups of people who have never had a collective name, and whose frontiers have never been drawn on any map. In so far as contiguity is important, the problem is geographical; but, in comparison with geography as ordinarily understood, the present problem is much more abstract and statistical. How it can be treated will now be explained.

The skin and body effect for sovereign states.—If we took a political map of, say, Europe, and laid over it a piece of fishing net, we could count the number of strings which separated all those pairs of contiguous meshes which were entirely within the body of sovereign states, also we could count separately the remaining number of strings, which therefore pertain to the frontiers, or skins, of states. The ratio of skin to body could thus be estimated. Next let the experiment be repeated with mosquito net. Its meshes are also hexagons, but much smaller than those of the fishing net. The ratio of skin to body would be found to be much smaller than for the fishing net. The notion, thus crudely illustrated, has been the subject of an elaborate treatment entitled "Mapping by compact cells of equal population". The problem was first re-defined as follows:

The map is required to satisfy the following conditions:

(i) Each cell is to contain the same number, h, of persons.
(ii) At most three cells are to meet at a point.
(iii) No cell is to surround another completely.
(iv) National frontiers are to lie in the edges of cells.
(v) Each cell is to be "homóplatous", that is, about as broad as it is long.

The concept of "homóplaty" was discussed from various points of view, including a psychological experiment, and a theory about minimizing the sum of the squares of cell-edges. For these appropriate new words, the adjective "homóplatous" and the abstract noun "homóplaty", I am indebted to the Professor of Greek in Cambridge, D. S. Robertson, F.B.A., who kindly invented them.

For such a map, covering the whole globe, it is required to find, as functions of h,

- s, the number of cells;
- C, the number of cell-edges within nations;
- F, the number of cell-edges between nations;
- B, the number of cell-edges to the sea, polar ice, or desert.

The problem was attacked by a variety of methods, graphical and argumentative. Euler's theorem on the faces, edges and vertices of polyhedra was adapted to political geography, and it then played a leading part. So did the honeycomb pattern and its conformal representations. The effects of the shape of countries, and of the distribution of population in them, were treated approximately by way of integrals; in particular the alternation between town and country was thus formalized. The lengths of coast-lines had to be studied, and were found to be extremely peculiar. A catalogue was constructed of about 390 territories, arranged in order of population. Maps of homóplatous cells, containing a million persons each, were drawn along all the world's sea coasts and land frontiers. The proper account of this long inquiry is ready in typescript, but unfortunately is not yet published. Its principal outcome however is summarized in Table 8.

TABLE 8

Geographical Facts

Summary of World-total of Cells and Edges at the Middle of A.D. 1910

h	=	10^7	10^6	10^5	10^4
s	=	150	1,652	16,745	168,000
C	=	159	3,373	44,200	483,800
F	=	121	831	3,319	12,680
B	=	122	595	2,510	10,090
C/F	=	1·31	4·06	13·32	38·14

The accuracy of these numbers is a difficult question. They are, however, certainly much better than the author's previous estimates (Richardson, 1950*a*, p. 31).

The contiguous cells of every pair meet along one, and only one, edge. Every edge counted in C or F separates one, and only one, pair of cells. The easiest way to count pairs of contiguous cells is therefore to count the intervening edges. Thus C and F resemble the numbers a, b, c, d in Table 7 in so far as they are all numbers of pairs of contiguous groups of persons. Any cell-edge, between or within nations, is the frontier between the two cells which it separates. Like other frontiers, the edge provides a geographical opportunity for land-fighting. The world-totals C and F measure the total geographical opportunities for civil, and respectively foreign, fighting between contiguous cells each of population h. Thus C/F in Table 8 has a resemblance to $(a + c)/(b + d)$ in the notation of Table 7.

To make this resemblance more like equality, the population h of the cells in Table 8 must somehow be connected with the mean magnitude, $\bar{\mu} = \frac{1}{2}(\mu_1 + \mu_2)$, chosen for Table 7. It is here that the assumption (2) comes in.

Let it be accepted that $\bar{\mu}$ in the historical Table 1 is to be related to h in the geographical Table 8 by the assumption (2) . . . . (10)

Then C/F will be roughly equal to $(a + c)/(b + d)$. There is, however, a correction for date which needs attention. The year 1910 was chosen because before that date censuses were lacking in too many countries. Better for our purpose than C and F at 1910 would be the time-means $\bar{C}$ and $\bar{F}$ over the 110 years A.D. 1820 to 1945. There is no hope of obtaining $\bar{C}$ and $\bar{F}$ accurately, but some approximations are possible. The two main alterations were the growth of world population W, and the changes of frontiers.

If the density of population had varied everywhere in the same ratio, while all political frontiers had remained fixed, then C would have varied roughly as W, and F roughly as $W^{0.5}$, if the distribution had been smooth. The irregularities of frontiers and of the population-density raise the latter index to about $0{\cdot}5_8$. So C/F varied as $W^{0.42}$. From data-collected by Carr-Saunders

(1936) by the League of Nations and by Dr. Swaroop (K 11676) I have computed that the mean of $W^{0\cdot 42}$ was 0·93 of $W^{0\cdot 42}$ at 1910.

On this account $\bar{C}/\bar{F} = 0\cdot 93(C/F)$ at mid 1910. . . . . (11)

The other chief alteration was in the frontiers. Old atlases (Bartholomew, 1860) or historical atlases (*Cambridge Modern History Atlas*; Muir & Philip, 1929) give one some impression of what happened. They are not adequate, because the populations are not all known. Also the boundaries of less civilized kingdoms in Africa and elsewhere are omitted; that might not so much matter if all the wars between such kingdoms were also omitted from the historical statistics. Both amalgamations and secessions occurred. In Africa many frontiers were formed between new colonies. In Asia old principalities were absorbed into empires. It is difficult to say which tendency predominated. The author abandons the attempt, accepts (11) meantime, but at the end of the argument considers the possibility of error.

The position so far is that, provided the connection (10) is made,

$$(a + c)/(b + d) = \bar{C}/\bar{F}. \qquad (12)$$

For the purpose of a χ^2-test we need to think about $(a + c)$ separately from $(b + d)$. This can be arranged by replacing the ratio by a pair of simultaneous equations connected by a common parameter τ, thus

$$a + c = \tau\bar{C}, \qquad b + d = \tau\bar{F}. \qquad (13)$$

The contingency-table 7 is intended to represent some feature of world-society, a feature beyond the control of the investigator. Yet the investigator chooses the historical interval and the range of magnitude, both of which affect a and b. There must be some considerations, akin to "dimensional analysis" in physics, which restrict τ so as to compensate for any irrelevancies introduced by the choice of T_1, T_2, μ_1, μ_2. For definiteness let $T_1 < T_2$ and $\mu_1 < \mu_2$.

The distribution of wars in time has been expressed by the probability λdt that some unspecified war would come to an end in the time-element dt; and λ was found to vary slowly (Richardson, 1945; Moyal, 1949). The geographical opportunities for fighting may suitably be compared with this moderately constant λ. Or, to express the same idea differently, suppose that the investigator, while keeping the mid-date $\frac{1}{2}(T_1 + T_2)$ fixed, were to reduce the interval $(T_2 - T_1)$ from 110 years to 55. Then a and b would be roughly halved. In order that c and d might continue to be comparable with a and b, it is necessary that τ should be proportional to the historical interval.

The variation of the frequency of deadly quarrels with magnitude has also been studied (Richardson, 1941 and 1948*a*). It is more complicated than the distribution in time, because a steep frequency-curve is involved. Nevertheless a similar argument about τ can be made, provided that $\mu_2 - \mu_1$ is small, but not too small.

Therefore τ may be written

$$\tau = \theta(T_2 - T_1)\ (\mu_2 - \mu_1), \qquad (14)$$

provided $\mu_2 - \mu_1$ is suitably small, where θ does not involve $T_2 - T_1$ nor $\mu_2 - \mu_1$, but is otherwise an unknown coefficient. The investigator makes two other choices, namely the unit of time and the base of the logarithms. Neither of these choices is relevant to the social relations. Neither of them alter a, b, $\bar{C}$, nor $\bar{F}$ in (13); therefore they do not affect τ. Consequently the compensation for their arbitrariness must be made by corresponding alterations in the numerical value of θ, in the usual manner. As in physics, so here, dimensional analysis tells us how coefficients depend on arbitrary choices, but does not fix the numerical coefficients absolutely.

There is, however, some other evidence about θ. The specially drawn maps on which C and F were counted were not unique; they were only samples of all the maps satisfying the conditions (i), (ii), (iii), (iv), (v) for the given cell population h. There was evidence that C and F would be almost the same for any such map. Yet the positions of the cells and the names of their inhabitants could vary a great deal. Therefore:

there is likely to be a numerically large factor in θ . . . . (15)

The above discussion of τ and θ has cleared away various obscurities, but unfortunately has not yielded a definite numerical value for τ. The problem will be resumed on a subsequent page.

The historical facts about fighting in Table 3 can now be compared with the geographical opportunities for fighting in Table 8 by way of the logical scheme in Table 7. The correction (11) is introduced; and j is the uncertain number in equation (2). The comparison is displayed in Table 9.

TABLE 9

Comparison of History with Geography, Years A.D. 1820 *to* 1945 *inclusive*

Ends of Range of Magnitude of the Quarrel as a Whole	*Pairs who Actually Fought*		*Ratio of Civil to Foreign*				*Least of the* χ^2 *from* (19)	$P(\chi^2)$
			Actual	*For Geographical Opportunities* $\bar{C}/\bar{F}$				
	a	b	a/b	$j = 1$	$j = 2$	$j = 3$		
$5 \pm \frac{1}{2}$	6	3	2	3·78	1·22	0·4	—	—
$4 \pm \frac{1}{2}$	10	15	0·67	12·39	3·78	1·22	2·23	0·14
$3 \pm \frac{1}{2}$	65	64	1·02	35·47	12·39	3·78	64·0	0·0000

The interim interpretation of Table 9 is made by comparing a/b with $\bar{C}/\bar{F}$ for all the values of j. For magnitudes around 5 it is seen that a/b lies between the extreme values of $\bar{C}/\bar{F}$: that is to say actual fighting was distributed between civil and foreign about in the ratio of their respective geographical opportunities. On the contrary for magnitudes less than 4·5 the actual a/b was less than the geographical $\bar{C}/\bar{F}$. Table 9 suggests therefore that *for magnitudes less than* 4·5 *there must have been some influence which repressed civil fighting relatively to foreign fighting.*

Before we accept this conclusion, let us try to suppose that the apparent contingency has arisen merely by chance, and let us test this supposition in the usual manner by forming, in the notation of Table 7,

$$\chi^2 = \frac{(ad - bc)^2 \ (a + b + c + d);}{(a + b)(c + d)(a + c)(b + d)} \quad . \quad . \quad . \quad . \quad . \quad (16)$$

and by comparing χ^2 with Yule's table of its distribution for one degree of freedom (quoted by Kendall, 1943, pp. 444–5). This standard procedure is here impeded by an unusual difficulty, because c and d have to be found from the two equations (13) which involve an unknown parameter τ. It follows that

$$\chi^2 = \frac{\tau(\bar{C} + \bar{F}) \ (a\bar{F} - b\bar{C})^2}{(\tau\bar{C} - a + \tau\bar{F} - b)\ \bar{C}\,\bar{F}(a + b)} \quad . \quad . \quad . \quad . \quad . \quad . \quad (17)$$

To make further progress we must know more about τ. The following considerations are decisive. In Table 6 the fraction of the whole time during which any two specified contiguous cells fought one another was denoted by z, and was found to be less than 0·005, when no other cells joined in. This fraction has some resemblance to $(a + b)/(a + b + c + d)$. They are not strictly comparable, because one is a fraction of time, and the other of the number of deadly quarrels in a time-interval. These would be comparable if each fighting had the same duration; but actually the durations have varied from a few days to thirty years. Nevertheless it is conspicuous that sovereign states have usually approached the "dread arbitrament of war" with caution. The fightings that went on at any time were much less numerous than those which were geographically possible. Prospective insurgents have also had reason for caution. It will therefore be assumed that:

In Table 7 the fractions $a/(a + c)$ and $b/(b + d)$ were much nearer to zero than to unity. This assumption will be referred to as . . . . (18)

In view of (18), the quantity a may be neglected in comparison with $\tau\bar{C}$ and b in comparison with $\tau\bar{F}$. Accordingly (17) simplifies to

$$\chi^2 = \frac{b^2}{(a+b)} \frac{\bar{C}}{\bar{F}} \left\{ \frac{a}{b} \frac{\bar{F}}{\bar{C}} - 1 \right\}^2, \qquad (19)$$

from which τ is absent, while $\bar{C}$ and $\bar{F}$ occur only as their ratio. This limiting form of χ^2 may have other applications. I have not noticed it in any textbook.

The values of χ^2 were computed from formula (19) for the three values of j, and the least χ^2 is shown in Table 9, together with the probability $P(\chi^2)$ that a greater χ^2 might have arisen merely by chance. At magnitudes around 4, and for $j = 3$, the conclusion is not significant. But it would be if $j = 2$ were taken, for that gives $\chi^2 = 22$, $P(\chi^2) = 0\ 000{,}00$. Here it should be remembered that, according to (2*a*), the uncertain j is more likely to be 2 than 1 or 3; these outliers having been inserted to guard against rash conclusions.

For magnitudes around 3 the attempted explanation by mere chance is utterly incredible. The connection with common government is extremely significant. Now is the time to remember that alterations of frontiers were neglected, and to notice that there is room for a considerable "factor of safety".

The conclusion has already been stated in the initial "Summary and Outlook".

References

BARTHOLOMEW, J. (1860), *Black's General Atlas of the World.* Edinburgh: A. & C. Black.
BROCKHAUS (1935), *Der grosse Brockhaus.* Leipzig.
Cambridge Modern History Atlas (1910). Cambridge: University Press.
CARR-SAUNDERS, A. M. (1936), *World Population.* Oxford: Clarendon Press.
DANAILLOW, G. T. (1932), *Les Effets de la Guerre en Bulgarie.* Paris: Presses Universitaires; Yale University Press.
DUMAS, S. (1923), *Losses of Life caused by War.* Oxford: Clarendon Press.
EMIN, A. (1930), *Turkey in the World War.* Yale: University Press.
E refers to the *Encyclopaedia Britannica*, 1929.
HUBER, M. (1931), *La population de la France pendant la Guerre.* Paris: Presses Universitaires.
K refers to *Keesing's Contemporary Archives.* Bristol: Keesing's publications. The number following K is that of the page.
KENDALL, M. G. (1943), *Advanced Theory of Statistics, Vol. I.* London: Griffin.
KOHN, S. (1932), *The Cost of the War to Russia.* Yale: University Press.
MOYAL, J. E. (1949), *J. Roy. Statist. Soc.*, **112**, 446.
MUIR, R. & PHILIP, G. (1929), *New School Atlas of Universal History.* London: G Philip & Son, Ltd.
MYRDACZ, as quoted by Dumas, S.
OTLET, P. Photoprint from the Mundaneum, Brussels.
RHODES, J. F. (1919), *History of the United States.* New York: Macmillan.
RICHARDSON, L. F. (1941), Letter in *Nature* of November 15th.
—— (1945), *J. Roy. Statist. Soc.*, **107**, 242.
—— (1947), *J. Roy. Statist. Soc.*, **109**, 130.
—— (1948*a*), *J. Amer. Stat. Assn.*, **43**, 523.
—— (1948*b*), *Psychometrika*, **13**, 147–174 and 197–232.
—— (1950*a*), *Eugenics Review*, **42**, 25.
—— (1950*b*), *Statistics of Deadly Quarrels.* A book in microfilm published by the author.
YOVANOVITCH, D. (1930), *Les effets économiques et sociaux de la guerre en Serbie.* Paris: Presses Universitaires.

Richardson, L. F. (1952)
Br. J. Psychol. **43**, 169–76; 192–4

1952:4

DR S. J. F. PHILPOTT'S WAVE-THEORY

The British Journal of Psychology – General Section
Cambridge
Vol. XLIII, Part 3, July 1952

Notes:

1. Dr Philpott's 'Reply to Dr Richardson's criticisms' is on pp. 177–91 of the same issue, and his 'Reply to Dr Richardson's further comments' on pp. 195–9
2. D. W. Harding, Editor of *The British Journal of Psychology*, had invited Richardson to make a thorough assessment of Dr Philpott's work on fluctuations of human output when performing routine tasks, as a critical review of this work was long overdue. Richardson, who had formerly studied under Philpott and was on good terms with him, accepted with some reluctance. See Ashford (1985) pp. 231–3

VOLUME XLIII JULY 1952 PART 3

DR S. J. F. PHILPOTT'S WAVE-THEORY

BY LEWIS F. RICHARDSON

I. INTRODUCTION

For the past twenty-two years Dr Philpott has conducted researches on the fluctuations in human output when routine tasks are performed. Together with his students he has accumulated a great store of experimental results, which must surely be valuable for various purposes, whatever one may think of a particular interpretation.

In various conversations I have heard and expressed general vague incredulity about Dr Philpott's wave-theory. However, remembering that the most unfair type of criticism is that which is so vague that it cannot be answered, I shall now try to dissect my incredulity, in order to emphasize precise objections, and to blow away what was misunderstanding or mere chaff.

The most startling of Dr Philpott's assertions is his claim to have measured $4{\cdot}076 \times 10^{-23}$ sec. by experiments on the time variations of the rate at which people perform routine mental tasks such as arithmetic. This is not a new nor a passing claim. It appears in his Monograph of 1932 (13) and again in his paper dated 1950 (15). In that interval of 18 years there has been no revision of the significant digits, such as usually follows the first measurement of a physical constant. He calls this remarkable time T_0; and it will be so denoted here. In the 1950 paper T_0 is compared with various constants in atomic physics. There are two surprising features about T_0: the alleged determination of a possibly electronic constant in a psychological laboratory, and the enormous ratio of 10^{-23} sec. to the least time which one can estimate mentally, say 10^{-1} sec. Let us consider these features in turn.

II. PROFESSIONAL DEPARTMENTS

We are accustomed to regard psychology and physics as separate disciplines. Their separation is a matter of social convenience, an affair of salaries and appointments. Below the level of social organization there is no reason to expect psychology to be detached from physics. It is generally believed that arithmetic somehow depends on the brain, and that the brain contains electrons. So it would be rash to declare that the speed of working arithmetic could not possibly depend on electronic constants. The objection to this part of Dr Philpott's claim seems to me to resemble a gardener's objection to a potato plant coming up among his cabbages.

III. EXPERIMENTAL PRECISION

Dr Philpott's apparatus was quite ordinary. It consisted of paper, print, pencils and a chronograph. The determination of any ratio to a large number of significant digits usually requires elaborately refined apparatus and technique. Dr Philpott states his remarkable number as $4{\cdot}076 \times 10^{-23}$ sec. I assume that he would be willing to jettison the ·076 if he could convince psychologists about the 4; and that what we mainly have to discuss is the 10^{-23}.

IV. The estimation of large ratios

Let us next consider whether very large ratios such as 10^{-23} have ever been measured, at least roughly, by means of crude apparatus.

Dr G. F. C. Searle, F.R.S., used to show his students how to measure the wave-length of light, about 6×10^{-5} cm., with a candle and a pocket handkerchief. The result was convincing, because a century of research with diverse and better apparatus had established the wave-theory of light.

In order to notice pioneer measurements made with crude apparatus one should search in out-of-date publications. A notable example is Loschmidt's estimate of the diameter of molecules which he published in 1865. The measurements were of the diffusivity of gases and of the ratio of the volume of a gas to that of its liquid, both of them simple by modern standards of experimentation. The theory was the kinetic theory of gases (see Winkelmann (31)).

Another early estimate of the diameter of a molecule, about 10^{-7} cm., was published by Lord Rayleigh in 1890 (see Poynting & Thomson (20)). Rayleigh dropped shavings of camphor on to a water surface which had been contaminated by a small and measured amount of oil. Then by merely cubing 10^{-7} cm. and multiplying by the density of oil, one can estimate the mass of a molecule of oil to be of the order of 10^{-21} g. The modern value for oleic acid is $0{\cdot}469 \times 10^{-21}$ g.

From these examples we may conclude that to make a rough estimate of a very extraordinary number, such as Dr Philpott's 10^{-23} sec., it is not always necessary to have elaborate apparatus; the indispensable necessity is trustworthy theory.

V. The notion of strict periodicity on the scale of log time

What is the theory in which Dr Philpott places so much trust? It is his belief in the existence of waves of constant period on the scale of $\log_{10}$ time. And how did this belief arise? He first noticed such waves in the observations themselves. Afterwards he sought for collateral evidence, and claims to have found some in the rate of evolution of hydrogen from zinc in acid. That might be distantly relevant, for there are electrochemical reactions in nerves. He also claims affinity with the log time in E. A. Milne's astronomy; but that analogy is, I suggest, altogether too far-fetched. Dr Philpott does not mention any other human periodicities in the scale of log time. But, if we may seek guidance from periodicity in ordinary time, there are many human alternations, such as sleeping and waking, breathing, the heart-beat, and the alpha-rhythm shown by the encephalogram. Each of these has a frequency which wanders a bit. In B. van der Pol's (18) nomenclature they are probably 'relaxation-oscillations', which differ from pendular oscillations in being less strictly periodic. For emphasis my chief objections are numbered and restated.

Objection I. As all other human behaviour is somewhat irregular, ought we to imagine that mental waves in log time would be periodic, quite strictly?

VI. EXTRAPOLATION

There is a statistical maxim that, when the theory is derived only from the observations themselves, interpolation is usually safe, but extrapolation is usually risky. This maxim was strikingly illustrated, in its application to regression formulae, by Emily Perrin (11) in a paper called 'The dangers of extrapolation'. It is desirable therefore to be clear as to whether T_0 lies between Dr Philpott's observations or outside their range. Nominally the observational range was from 0 to 1000 sec., and 10^{-23} sec. lies within that nominal range. But actually an experiment could not be started with an accuracy better than 0·1 sec., and the first observational datum was the amount of the task accomplished during the first 5 sec., which may suitably be concentrated at 2·5 sec. Therefore T_0 was obtained by extrapolation. To see whether the extrapolation was long or short we must remember that it was done on the scale of log time.

The times in seconds were	4×10^{-23}		2·5		1000
Their logarithms are respectively	−22·4		0·4		3
The differences of these logarithms are		22·8		2·6	

This very long overhang is another of my chief objections.

Objection II. Extrapolation over a range as long as 22·8 from observations made in a range as short as 2·6, by way of a law inferred only from those observations themselves, is far too risky.

VII. A TEST FOR RANDOMNESS

A grand-total work curve, actually a polygon, which appeared in Dr Philpott's Monograph ((13), p. 97), has been selected by him recently ((14), p. 124) as representative. It runs for 300 sec. and therefore presumably is constructed from 60 successive intervals of 5 sec. each. M. G. Kendall ((9), p. 124, art. 21·43) gives a very simple test to decide whether such a polygon is drawn from a mere random assortment of disconnected ordinates. Let there be N ordinates, so that here $N=60$. Let p be the number of crests and troughs, counting both. If there is nothing systematic about the ordinates, p may, for finite N, be taken as normally distributed about the mean $\frac{2}{3}(N-2)$ with variance $(16N-29)/90$. Insertion of $N=60$ gives mean $=38·7$, variance $=10·34$; therefore standard deviation $=3·216$. On the diagram $p=33$. This differs from the mean by 5·7, which is 1·77 times the standard deviation.

Objection III. This overall test leaves one with the impression that the spikiness of the 'grand-total work-curve' may be mostly random, and that to search among it for periodicity may be fruitless.

VIII. THE ASCERTAINMENT OF OBSERVED PERIODS

Let us next consider the alleged waves in the observed, as distinct from the extrapolated, range of log time. A great store of fluctuations exists in the meteorological records of pressure, temperature, wind, rainfall and humidity. Some of these fluctuations are conspicuous and understandable, being daily and annual variations. Other meteorological fluctuations are as obscure, complicated and mysterious as those in individual work-curves. The history of meteorological research may therefore provide some useful hints.

Sir Gilbert Walker, F.R.S. (29), who paid special attention to periodicity and was looking back over his long experience of meteorology, wrote in 1936 that probably 'ninety-five per cent of the periods announced are non-existent'. This experience suggests that there must be something seductive about the idea of periodicity, which leads even careful researchers to see periods which are not present. When the fluctuations are as complicated as work-curves, the method of spotting troughs by eye is quite discredited. Yet Dr Philpott, as recently as 1949, was alluding to it as though it were respectable ((14), Fig. 1). Personally I find his visually spotted periods quite unconvincing; but I cannot number that as a serious objection, for my visual judgement may be as untrustworthy as his. M. G. Kendall ((9), p. 402) remarked that 'Experience seems to indicate that few things are more likely to mislead in the theory of oscillatory series than attempts to determine the nature of the oscillatory movement by mere contemplation of the series itself'.

The respectable methods for this purpose are the Schuster periodogram and the method of serial correlation. Both are laborious computations. Strictly periodic waves suit the periodogram. Dr Philpott has long since toiled at both these methods and has published considerable results in his Monograph ((13), pp. 87–96).

The interpretation of the periodogram requires a test of statistical significance. Such tests are available (Walker (28), Kendall (9)). But I cannot find any in Philpott's work. Nor are his periodograms published in such a manner that the reader can apply the test himself.

Objection IV. The strictly periodic waves which Dr Philpott claims to have observed were first noticed by a discredited method and were afterwards investigated by a reliable method which, however, stopped just short of a decisive conclusion.

IX. The method of extrapolation

Dr Philpott habitually writes about waves without specifying the wave-form. This will not do for precise argument. For in popular speech there are sea-waves, hair-waves, heat-waves; almost anything that moves to and fro continuously is called a wave. Among precisely defined curves the Bessel function $J_0(x)$ would be called a wave, yet it is not a periodic function. From the context it may be gathered that Dr Philpott is probably thinking about sines or cosines; so for definiteness I shall translate his argument into cosines.

For this purpose it is necessary to accept, at least temporarily, Dr Philpott's idea that the periodicities on the scale of log time are strict. Accordingly, let my objections I, II, III, IV now be put away into cold storage while we look at the matter from another point of view. Let t be the time in seconds, and let the experiment begin at $t=0$. He relies chiefly on 'grand-total curves' formed by adding the ordinates of the individual work-curves of large numbers of persons. He avers that the grand-total curve is the sum of waves, that the wave-lengths are integral multiples of 0·0016 on the scale of $\log_{10} t$. (I have not found any convincing reason for that number 0·0016; but let it pass for the present.) He begins by assuming (e.g. (14), p. 129) that there is a time T_0 at which all waves have a common trough point. From the context it appears that a 'trough point' is another name for a minimum. The assumption that T_0 exists is permissible, because, at this stage, the numerical value of T_0 is left undetermined.

To argue by easy steps, we may note that $-\cos(\log_{10} t)$ has minima at $\log_{10} t = 0, 2\pi, 4\pi,$

and in general at $2s\pi$, where s is an integer positive, zero, or negative. Then $-\cos(2\pi \log_{10} t)$ has minima at $\log_{10} t = 0, 1, 2, 3, \ldots, s$. And $-\cos\left\{\frac{2\pi \log_{10} t}{0{\cdot}0016n}\right\}$ has minima at $\log_{10} t = 0{\cdot}0016n$, $0{\cdot}0032n$, $0{\cdot}0048n$, and in general at $0{\cdot}0016sn$.

It will be assumed, on Dr Philpott's authority, that n is a positive integer. Further it is plain that

$$-\cos\left\{\frac{2\pi(\log_{10} t - \log_{10} T_0)}{0{\cdot}0016n}\right\} \text{ has minima at } (\log_{10} t - \log_{10} T_0) = 0{\cdot}0016n,\ 0{\cdot}0032n,\ 0{\cdot}0048n, \text{ and in general at } 0{\cdot}0016sn. \tag{1}$$

Dr Philpott assumes that such waves are occurring simultaneously for various integers n, and that the work-curve is found by adding their ordinates. He does not specify their amplitudes. For definiteness I insert an amplitude A_n which is constant in time, and, to prevent troughs changing into crests, must be positive. The sum of the waves is accordingly

$$A_0 + \sum_{n=1}^{\infty} -A_n \cos\left\{\frac{2\pi(\log_{10} t - \log_{10} T_0)}{0{\cdot}0016n}\right\} = \phi(t) \quad \text{say.} \tag{2}$$

This function $\phi(t)$ is supposed to represent the ordinate of the 'grand-total work-curve'. As the ordinate of the grand-total work-curve is finite, the sequence $A_1, A_2, A_3, \ldots, A_n$, must be such that the series (2) converges as $n \to \infty$. This will happen provided ΣA_n converges. There are evidently abundant possibilities of convergence, so that we need not stop to consider them in detail.

In a Fourier series n is in the numerator of the angle, so that the wave-lengths of the terms become shorter as n increases. In (2), which may suitably be called Philpott's series, n is in the denominator of the angle, so that, as n increases, the wave-length of the term becomes longer.

Dr Philpott is much impressed by an observed minimum of his grand-total work-curve at $t = 900$ sec. (Personally I do not find it conspicuous; but let that also pass for the present.) He wished therefore to choose the numerical value of T_0 so that plenty of the individual waves in his series (2) should have minima at $t = 900$ sec. By trial he found that $T_0 = 4{\cdot}076 \times 10^{-23}$ sec. gave a remarkable crop of minima at $t = 900$ sec. Let us verify the arithmetic. The coefficient of A_n in (2) will be a minimum if, and only if,

$$\frac{\log_{10} t - \log_{10} T_0}{0{\cdot}0016n}$$

is an integer, positive, zero, or negative. (3)

Now $$\frac{\log_{10} 900 - \log_{10}(4{\cdot}076 \times 10^{-23})}{0{\cdot}0016n} = \frac{25{\cdot}3440083}{0{\cdot}0016n} = 15840{\cdot}0_0/n.$$

When expressed in its prime factors, $15840 = 2^5 \times 3^2 \times 5 \times 11$. Therefore $15840/n$ is an integer for $n = 1, 2, 3, 4, 5, 6, 8, 9, 10, 11, 12$ and for many larger n. But $15840/n$ is not an integer for $n = 7, 13, 14, 17, 19, 21, 23, 25, 26, 27, 28, 29, 31$, and many larger n. The lowest exception, namely, that for $n = 7$, especially spoils the beauty of the scheme. But we can easily remedy that. Let T_0 be determined so that

$$\frac{\log_{10} 900 - \log_{10} T_0}{0{\cdot}0016n} = \frac{7 \times 15840}{n}.$$

This gives $$T_0 = 10^{-175} \times 3{\cdot}517 \text{ sec.} \tag{4}$$

It is easy to discover marvellous numbers by such methods!

It is reasonable to ask, however, whether a more ordinary T_0 could satisfy Dr Philpott's requirements? Or whether he has obtained plenty of coincident minima in the neatest possible manner? Let us suppose that it would suffice if the first n of the elementary waves in (2) had minima at $t = 900$ sec.; and let us determine T_0 for successively increasing n by means of the condition (3). Let L_n denote the least common multiple of the first n whole numbers. Then the condition (3) becomes

$$(\log_{10} 900 - \log_{10} T_0)/0{\cdot}0016 = L_n.$$

When solved for T_0 this gives

$$T_0 = 900 \times 10^{-0{\cdot}0016 L_n}. \tag{5}$$

By obvious arithmetic I find from (5) that T_0 runs thus as a function of n:

n	1	5 or 6	7	8	9 or 10	11 or 12	13 or 14 or 15	
L_n	1	60	420	840	2520	27720	360360	(6)
T_0 (sec.)	896·7	721·5	191·5	40·76	0·0836	$10^{-42} \times 4{\cdot}0017$	$10^{-574} \times 2{\cdot}389$	

Any observational uncertainty about 0·0016, when multiplied by 27720, or by Dr Philpott's 15840, would cause a considerable alteration in T_0. This is quite a minor objection, aimed merely at so many as four significant digits in his 4·076.

As thus determined, T_0 is the *greatest* time that will give coincident minima at $t = 900$ sec. for the first n of the waves. For example, Dr Philpott's $4{\cdot}076 \times 10^{-23}$ sec. does not give a minimum for the seventh wave, and so, from that restricted point of view, is no better than 721·5 sec.

Objection V. There are many widely different numerical values of T_0 that seem reasonably likely to give a minimum of 'Philpott's series' (2) at any particular time, such, for example, as $t = 900$ sec.

It may be said that this objection is unfair, because Dr Philpott was considering not merely a single trough, but several sequences of troughs; and because he has published a comprehensive discussion of sequences in Appendix B of his Monograph (13). I do not understand that Appendix. Anyway the theory of T_0 is very complicated and unclear. Moreover, as soon as sequences of observed troughs are mentioned, my suppressed objections III and IV rush out of their prison, clamorously protesting.

X. Conclusions about T_0

The above objections numbered I, II, III, IV, V compel me to reject the finding $4{\cdot}076 \times 10^{-23}$ sec. It has been derived from the observations by methods of computation that are not at all trustworthy. Nor is any moderate amendment in sight that could rescue either the theory by which T_0 was obtained, or the numerical value of T_0. To look for any physical explanation of T_0 is therefore a waste of intellect. I must plead guilty to having stimulated the search for physical interpretations some ten or fifteen years ago by suggesting, in a letter to Dr Philpott, that T_0 was the time taken by light to go round the equator of an electron. That remark was made in jest, but unfortunately it has been taken seriously. The proper treatment of T_0 would be to consign it to oblivion; but after so much publication that will now be difficult. I am sorry to find myself in strong disagreement, about T_0, with one of my former teachers, for whom otherwise I share in the general high respect.

XI. SINGLE CURVES

It would indeed be tragic if no conclusions were to emerge from a study of 700 work-curves. Although, as I find, the grand totals contain little of interest, it may well be that the individual curves are most instructive. From the titles of theses deposited in the University of London Library by Dr Philpott's students F. W. Warburton, N. F. H. Butcher, F. Akil, R. Bath and A. Z. Saleh, it looks as though the research which he directs has turned in that more promising direction.

Suppose that someone, with remarkable persistence, were to collect the records of 700 experiments on breathing, each curve starting at the word 'Go' and continuing for 900 sec. Would the grand-total breathing curve, formed by adding the ordinates of the separate curves, be likely to tell us anything about breathing? Probably very little, because the different rhythms would blur each other out. It would probably be far more instructive to study the breathing curves one at a time.

If that be accepted, another question follows. Would it be suitable to apply, to a single breathing curve, Schuster's method of periodogram analysis? I think not, because the periodogram is related to the ideal of a strictly periodic function y, which oscillates with a fixed amplitude according to

$$\frac{d^2y}{dt^2}+\omega^2y=0, \qquad (7)$$

where ω^2 is a positive constant; whereas the period of an individual's breathing wanders irregularly, so that the salient facts would be blurred in the periodogram.

G. U. Yule (32) in 1927 had become discontented with simple harmonic motion. He took instead as his ideal the well-known equation for damped harmonic motion, namely,

$$\frac{d^2y}{dt^2}+2k\frac{dy}{dt}+\omega^2y=0, \qquad (8)$$

where k and ω^2 are positive constants. Yule developed a statistical method, intended to replace the Schuster periodogram, and to determine k as well as ω^2. Yule thus, I suggest, took several steps in the right direction, but did not go far enough. For in 1926 B. van der Pol (18) published an ideal, which he called a 'relaxation-oscillation', which is much more appropriate to mental waves than is damped harmonic motion. In a relaxation-oscillation most of the cycle is describable deterministically by a differential equation, yet there may be a pause which comes to an end when a chance disturbance triggers-off the next motion. The record consists of a succession of similar pieces spaced at unequal intervals. Anson & Pearson (2) in 1922 had studied the flashing of a neon lamp when connected in parallel with a condenser, fed from a battery through a resistance. This is typically a relaxation-oscillation. B. van der Pol and J. van der Mark (19) made an electrical apparatus, comprising neon lamps, which imitated the heart-beat remarkably well. W. McDougall (10), in a paper 'On the seat of the psycho-physical processes', had argued in 1901 that the passage of the nerve current through a synapse is a discontinuous process comparable with sparking between the knobs of a Wimshurst machine, and that each spark generates a pulse of sensation. L. F. Richardson (22) in 1930 developed McDougall's idea further by arranging sixteen analogies between mental images and sparks. The most relevant kind of spark seemed to be that inside a neon lamp. Richardson therefore went on to investigate the behaviour of a neon, or more strictly an 'Osglim' lamp, and found mathematical descriptions of its behaviour both in the Anson-Pearson circuit where a single lamp flashes intermittently, and in Richardson's distraction circuit, where either of two lamps inhibits the other (23, 24).

B. van der Pol's fertile ideas were originally about oscillations described by

$$\frac{d^2y}{dt^2}-\epsilon\,(1-y^2)\,\frac{dy}{dt}+y=0, \tag{9}$$

where ϵ is constant and $\epsilon \gg 1$ ((18), equations (6) and (7 *a*)). He seems to have been under the impression that this described oscillations involving the neon lamp. Richardson, however, found by experiment on an Anson-Pearson circuit that its oscillations were described by

$$\frac{d^2 \log y}{dt^2}+(\alpha+\beta y)\,\frac{d \log y}{dt}+y-1=0, \tag{10}$$

where α and β are positive constants ((23), pp. 319 and 329).

In the analysis of a single work-curve it would be advisable, as I suggested fourteen years ago (25), to think in terms of relaxational, rather than harmonic phenomena.

References

(2) Anson, H. St G. & Pearson, S. O. (1922). *Proc. Phys. Soc.* xxxiv, 175–6 and 204–12.
(9) Kendall, M. G. (1946). *The Advanced Theory of Statistics*, ii. London: Charles Griffin and Co.
(10) McDougall, W. (1901). *Brain*, xxiv, 577.
(11) Perrin, Emily (1904). *Biometrika*, iii, 99–103.
(13) Philpott, S. J. F. (1932). Fluctuations in human output. *Brit. J. Psychol. Monogr. Suppl.* no 17. Cambridge University Press.
(14) Philpott, S. J. F. (1949). *Brit. J. Psychol.* xxxix, 123–41.
(15) Philpott, S. J. F. (1950). *Brit. J. Psychol.* xl, 137–48.
(18) Pol, B. van der (1926). *Phil. Mag.* ii, 978–92.
(19) Pol, B. van der and Mark, J. van der (1928). *Phil. Mag.* vi, 763–75.
(20) Poynting, J. H. and Thomson, Sir Joseph (1920). *Properties of Matter*, London: Charles Griffin and Co.
(22) Richardson, L. F. (1930). *Psychol. Rev.* xxxvii, 214–27.
(23) Richardson, L. F. (1937). *Proc. Roy. Soc.* A, clxii, 293–335.
(24) Richardson, L. F. (1937). *Proc. Roy. Soc.* A, clxiii, 380–90.
(25) Richardson, L. F. (1937). *Brit. J. Psychol.* xxviii, 212–15.
(28) Walker, Sir Gilbert (1930). *Mem. R. Met. Soc.* no. 25.
(29) Walker, Sir Gilbert (1936). *Quart. J. R. Met. Soc.* lxii, 1–2.
(31) Winkelmann, A. (1906). *Handbuch der Physik*, ii, Aufl. iii, 763–4. Leipzig: J. A. Barth.
(32) Yule, G. U. (1927). *Philos. Trans.* A, ccxxvi, 267–98.

192

COMMENT ON DR PHILPOTT'S REPLY

By LEWIS F. RICHARDSON

I have read Dr Philpott's reply and have considered whether I ought to modify any of my criticisms.

The same test for randomness ((9), p. 124, art. 21·43) has been applied to each of two curves, alias polygons, in Dr Philpott's 1949 paper (14). One of them (Fig. 1A) is a grand-total work-curve; in the other (Fig. 2) the ordinate, reckoned downwards, represents the 'total number of factors in ten consecutive natural numbers'. The results are rather similar. Both curves have rather more turning points, alias spikes, than one would expect them to have, if they had been produced by purely random processes. The deviations from expectation are not extreme, being respectively 1·77 and about 1·4 of the standard deviation, as given by Kendall's formulae. That is to say, by this test, both curves could credibly be regarded as random. Dr Philpott rejects the test because it has failed to tell us that the second curve was not chance determined; for in fact every point of it was calculated by a stated rule from the natural numbers. He has thus raised a most interesting question concerning the philosophy of chance and randomness. Henri Poincaré (17), in his book *Science and Method*, asked how the concept of chance can arise in a world that he regarded as deterministic. One of Poincaré's answers (p. 75) is that 'we attribute them to chance because their causes are too complicated and too numerous'. The causes of the aforesaid Fig. 2 were moderately numerous and complicated, namely: a long run of the natural numbers, counting how many factors less than 361 each has, and totalling over ten of the original numbers. I suggest: (i) that these causes may well have been sufficient to make Fig. 2 random in Poincaré's sense, (ii) so that it is understandable why Kendall's turning-point test has given us a warning about it, (iii) that therefore Dr Philpott's objection to my application of the same test to his Fig. 1A is washed out.

The curve derived from the natural numbers ((14), Fig. 2) is offered by Dr Philpott as a theoretical curve of fluctuation; and, strange to say, it can hardly be analyzed. For, in its construction, he took totals over ten consecutive elements, and such totalling is an irreversible operation. Even when we know the rule by which the curve was constructed, we cannot, from its ordinates, deduce the natural numbers. One of the obstacles to understanding Dr Philpott may be that, where most research workers would attempt an analysis, he is content with a synthesis. However that may be, the following discussion, via Bayes's theorem, applies to analysis and synthesis indiscriminately.

As a philosophical by-product of this controversy we are left with the question whether irreversibility of calculation can ever be an ingredient of the concepts of chance and randomness? I do not know.

The general pattern of this controversy is illuminated by Prof. H. Jeffreys's theory of probability. Jeffreys (6) regarded probability as a numerical expression of a degree of reasonable belief, 0 denoting certainty that the proposition in question is false, and 1 denoting certainty that it is true.

There is a theorem, derived from Bayes, whereby Jeffreys compares the probabilities of rival hypotheses. It may be stated thus:

(the probability of each hypothesis after the experiments are known) varies as (the probability of the same hypothesis before the experiments are known) multiplied by (the probability of the experimental results if that hypothesis were true).

In all three brackets the same general information is supposed to be available in the background.

The three phrases in brackets are too long for repetitive use. Jeffreys states the theorem briefly thus (p. 46):

Posterior probability $\propto$ prior probability $\times$ likelihood, in which each short expression is a technical term defined by the *respective* longer phrase above. Let our rival hypotheses be: (i) the theory that the observed effects in grand-total curves are merely random, (ii) Dr Philpott's wave-theory. The degrees of belief which I shall assign are merely personal guesses, yet I think that they will illustrate, although perhaps in caricature, the perplexities which many people have experienced.

To estimate the prior probabilities we need to find people who are reasonable and learned, except that they never heard of Dr Philpott's experiments. There were plenty of such in the year 1920. We ask them in imagination two questions. Is it credible that the average of 700 work-curves would be merely random? Answer: Not improbable, except for initial and final spurts; probability say 0·5. Is it credible that the average of 700 work-curves would be explicable in terms of a time of $4{\cdot}076 \times 10^{-23}$ sec.? Answer: Quite incredible; probability zero.

To estimate the 'likelihood', remembering that it is a technical term defined as above, we have to put two questions to persons who have read Dr Philpott's publications. I take myself as a specimen of such. Is it credible that the fluctuations shown in the grand-total work-curve ((14), Fig. 1 A) are merely random? Answer: Yes, by one of Kendall's tests, they may be. I reckon the likelihood, rather diffidently, as 0·08. That anyway will serve as a token value in the sequel. Is it credible that Dr Philpott's wave-theory can explain the fluctuations in the same curve? Answer: Yes, he certainly does explain some of the larger fluctuations. A lot of minor fluctuations he seems to leave undiscussed. He makes a number of small adjustments, and deprecates the tendency to call them cookery. I do not know what the overall likelihood is, but for the sake of argument, I give him the benefit of various doubts, and put it in at the token value of 0·7.

Next the above degrees of belief are to be inserted into Bayes's theorem, so as to give the posterior probabilities, which purport to be degrees of reasonable belief in the rival hypotheses, when all the evidence is considered together:

$$\text{Posterior probability of the random hypothesis} = 0{\cdot}5 \times 0{\cdot}08 = 0{\cdot}04,$$
$$\text{Posterior probability of Dr Philpott's wave-theory} = 0 \times 0{\cdot}7 = 0.$$

Thus the random hypothesis wins, because Dr Philpott contends against a prior probability of zero. Those who seek to prove the existence of precognition by experiments on card-guessing are in a similar difficulty. Let us consider what moves Dr Philpott could make in order to prevail over the random hypothesis. It would be of little avail for him to show more conclusively that his wave-theory explains the observed facts; for that likelihood is already guessed at 0·7, and it cannot rise above 1·0. He complains that I have ignored his analysis of variance ((14), p. 124). All he says about it there is confined to three lines. It would be more likely to receive attention if a systematic account were published in one of the statistical journals. But, anyway, what role could this analysis of variance play in Bayes's theorem? It might, conceivably, reduce the 'likelihood' for the random hypothesis from the 0·08, at which I have tentatively set it, to some number much nearer to zero. Even if the likelihood were actually zero for the random hypothesis, that would not be a victory for Dr Philpott, but only a drawn game. He cannot win unless he can somehow remove the prior probability of his wave-theory a little bit away from zero. Such is the strength of prejudice!

Another by-product of controversy here presents itself. Granted that Bayes's theorem gives an approximate account of the way in which people arrive at beliefs, is not multiplication by zero too strong an expression of prejudice? For how did such great innovators as Copernicus, Darwin and Einstein ever come to be believed?

Dr Philpott asks me emphatically why I have never alluded to his statement that grand-total curves resemble one another. The answer is that I have never seen the proof of it set out with proper statistical clarity, such as one would expect to find in a statistical journal.

I must accept Dr Philpott's assertion, although without his emphasis, that my § IX headed 'The method of extrapolation' misrepresents the argument by which he calculated T_0. He rejects my § IX with vehemence; on the contrary, I now admit plainly what I had previously indicated, namely that it is not *quite* fair, it is 'variations on a theme by Philpott', it presents clearly ideas that would be certain to arise if his theory were ever to become generally accepted, and it has performed the useful service of provoking him to explain himself.

As to my other comments on his work, Dr Philpott's reply does not move me to withdraw or to modify any of them.

There are numerous minor disagreements not worthy of detailed comment. I should be content now for the arguments to 'go to the jury', which consists of readers of this journal.

Additional references

(6) Jeffreys, H. (1939). *Theory of Probability*. Oxford, Clarendon Press.

(17) Poincaré, H. Trans. by Maitland, F., no date. *Science and Method*. London: Nelson.

Richardson, L. F. (1933)
Collected Papers of Lewis Fry Richardson **2**, 697–713

1953:1

VOTING IN AN INTERNATIONAL ORGANIZATION

(Typescript, not dated)

(Not previously published)

Note: The folder containing this typescript was dated 24 August 1953. The 'recent number' of the *International Social Science Bulletin*, referred to in the text, is Vol. V, No. 2, for the second quarter of 1953

Voting in an international organization

by Lewis F. Richardson
Hillside House, Kilmun, Dunoon, Argyll, Britain

Introduction

It is too much to expect everyone to agree. The most that we can hope for is a method that most people will think tolerable for deciding what, if anything, ought to be done when there are obstinate disagreements. Vote or fight sometimes appears to be the alternative; and nowadays fighting is so disastrous.

In 1918, when the possibility of forming a league of nations was under discussion, I published a paper on 'Voting strength in an international assembly'. Its leading notion was that the assembly would probably deal only with affairs arising between nations, and would be prohibited by its constitution from interfering in affairs that are purely internal to a nation; and that therefore voting strength should be a measure of internationality. An index of internationality was suggested, having foreign trade as one of its ingredients. This proposal was, I believe, considered by a committee of the British Foreign Office; but they preferred the principle of the equality of sovereign states; and this was eventually incorporated in the Covenant of the League of Nations. During the Second World War the statesmen decided that the League of Nations had failed hopelessly and must be liquidated, so after the war they set up an organization with a different name and a different headquarters, but much the same purpose, and based, like its predecessor, on the fictitious equality of sovereign states. The United Nations may in its turn destroy itself, and forthwith arise like the mythical phoenix from its ashes. Proposals for reform, like the present paper, should be published in advance of any such crisis.

Present practice

The United Nations Organization 'is based on the principle of the sovereign equality of all its members'. These words are quoted from the Charter, Article 2. The members include USA, USSR, Costa Rica and Iceland. The equality of these is evidently a legal fiction. On any reasonable reckoning USA might be many times more important than Iceland. This large present anomaly sets the very low standard of accuracy which will suffice to make an improvement. An error of the order of fifty times is due to be corrected. It would be a remarkable improvement if the error could be reduced to one of five times. An error of twicc or half hardly matters. To strive after one per cent accuracy may be mere fussiness; yet it is usually advisable to aim at rather more accuracy than is strictly indispensable.

This fictitious equality is moderated in practice by rules of procedure and by commonsense. The rules are briefly as follows. In the United Nations Assembly all member-states have equal votes; but certain questions of importance require a two-thirds majority. The Security Council consists of five permanent members, namely China, France, the Soviet Union, Britain and USA, together with six non-permanent members elected by the General Assembly for a term of two years. Each member of the Security Council has one vote. Decisions on procedural matters are made by an affirmative vote of seven members. Decisions on all other matters are made by an affirmative vote of seven members including the concurring votes of the permanent members. This provides the opportunity for any single permanent member to veto a decision favoured by the majority. A nation that is a party to what has been technically recognized as a 'dispute' abstains from voting on that issue.

Commonsense also presumably moderates fiction. For when statesmen consider whether there is any prospect of giving effect in the world to a decision of one of these voting bodies, they presumably think about the real importance of states, and not about their fictitious equality.

The Specialized Agencies of the United Nations mostly have conferences at which the system of voting is rather like that of the General Assembly; but for some of them (ILO and WMO) a delegate does not necessarily represent a national government. The details can be found in the United Nations Yearbook. Two of the Specialized Agencies, namely the Fund and the Bank, have entirely different systems which depend on money contributed. See below.

The recently formed Consultative Assembly of the Council of Europe and the Coal and Steel Assembly both have graduated votes. See K.12636.

It seems likely that the distinction between internal and external affairs, which is already rather difficult to maintain (e.g. as to Indians in South Africa) will, a hundred years hence, have become considerably blurred.

A recent number of the *International Social Science Bulletin* (Vol. V, No. 2, 1953) contains many articles on 'The technique of international conferences'. Implicit throughout these articles are the possibility of unanimity, or of situations in which it is wiser not to vote, or the technique of voting when unavoidable; and voting is explicitly reviewed by G. Scelle (pp. 250–253), by C. Chaumont (pp. 260–261), by W. R. Sharp (pp. 357–358), and by L. Kopelmanas (pp. 352–354). Several of these writings express dissatisfaction with the present technique of voting, but none of them goes so far as to discuss in detail a quantitative measure of voting strength.

Keeping up to date

It is futile to yearn after a proper, a pre-war, or a normal state of the world; for change has always been going on somewhere. Our 'good old days' may have been some other nation's 'hard times'. The best that can be done in connection with voting strength is to specify the state of the world at a recent specified date. The 'present' state of the world can hardly be specified, for time must be allowed for collecting the information. The mere uncontroversial arithmetic would need to be revised periodically according to the latest information, perhaps once a year. The principles on which the arithmetic was to be performed would be so controversial that, if ever agreed, they had better be such as not to need rediscussion for a long time, say 40 years.

Advantage of having several ingredients

It is certainly beyond the power of the present writer to suggest any ingredient that is quite obviously a proper measure of voting strength; and he doubts whether anyone can be so far-sighted. But there are several possible ingredients for which a plausible case can be stated, and against which there is no conclusive objection.

An index, such as the well-known index of cost-of-living, is usually compounded of several ingredients. It is a merit of the index that its ingredients should differ among themselves. The debate, as to what are worthy to be accepted as ingredients, is likely to be prolonged, so that if ever agreement were reached there might be a reluctance to re-open the question for many years. Compare for instance the rather long intervals during which the qualification for voting within a state has remained unchanged. In such an interval unforeseen circumstances would presumably arise. If the ingredients are several in number and different in quality there is more chance that the index compounded from them will remain reasonable, than if voting strength were made to depend on any one ingredient alone.

Degree of definiteness required

Anything that can be considered as an ingredient of voting strength must be ascertainable, to the convincement of the most suspicious opponent. It must not be such that a state could exaggerate it, in order dishonestly to claim more votes. This requirement rules out various vague characteristics of nations, which might otherwise be judged important, such as degree of civilization or literacy; it rules them out independently of the fact that they are internal not international characteristics.

Pretended fission

Another device against which the suspicious are sure to be on their guard is pretended fission. There were questions about it at the time when the League of Nations was formed; for that was also about the time at which Canada, South Africa, Australia and New Zealand acquired the right to conduct their foreign policies independently of the United Kingdom. Recently there have been similar questions about the Ukraine, Byelorussia, and USSR. Any state which allows its parts to vote separately, takes of course the risk that someday they may vote differently.

Possible ingredients

A wealth of information about the recent condition of the world is now conveniently available in the Statistical Yearbook of the United Nations. Most of it however specifies the internal affairs of states, leaving only a small amount that in any way measures internationality.

R. B. Cattell has published a run of remarkable studies of the quantitative characteristics of nations, using the method of correlation analysis. He relates the results to the wider purpose of understanding national behaviour in terms of culture patterns, and not at all specially to voting strength. In a recent paper by Cattell, Breuer and Hartman (1952) there are the names of some of the 72 variables which they set out to correlate. Most of these relate to characteristics internal to a nation, such as 'many miles of railroad per person', 'large number of riots', 'large gross area', 'high real income per head', 'polygamous marriage'. The only variables that specifically allude to international relations are 'frequent involvement in war', 'large number of clashes with other countries', 'high expenditure of tourists abroad', 'large number of secret treaties concluded', 'great number of treaties contracted', 'many Nobel prizes in science, literature and peace'.

We need ingredients ascertainable in a specified year. The following will be discussed in turn: (A) Money contributed to the international organization, (B) Number of geographical contacts, (C) Number of inhabitants, (D) Square root of number of inhabitants, (E) External trade, (F) Armed strength, (G) Number of men contributed to the international army.

About any proposed ingredient the following four questions usually arise:

(a) Is it a measure of internationality?

(b) How would it be affected by the splitting or coalescence of states?

(c) Is it ascertainable in favourable circumstances?

(d) How have the numbers in the tables at the end of this paper been obtained?

A Money contributed to the international organization

Of the objections to other proposed ingredients, two cannot be raised against money contributions, for they are definitely ascertained when they are paid, and there is no objectionable pretence about splitting a state into parts which contribute separately.

In the *United Nations Year Book* for 1951 (pp. 156–159) there is a summary of a debate on contributions. No very definite principles emerge; but there are references to 'capacity to pay', and to 'per capita income', and to a previous recommendation that no single member state should contribute more than one third of the total. After several protests, the contributions for 1952 were agreed. They may be seen here in Table II, as Table I is, for this ingredient alone, unnecessary. The United States was assessed at 36.90%, although this is more than one third of the total. The least assessment was 0.04%. The inequality of these contributions had not the slightest formal influence on the right of voting in the General Assembly.

On the contrary, the International Monetary Fund and the International Bank for Reconstruction and Development each have systems of numerically unequal voting strength. These institutions are closely related. They have the same address in Washington DC. A government, to become a member of the Bank, must first be a member of the Fund. The Fund's main purpose is to regulate rates of exchange between currencies, whereas the Bank lends money for large constructions. The membership of the Fund and Bank are not identical with that of the United Nations; for in particular USSR does not belong to the Bank or Fund. Each member of the Bank has 250 votes plus one additional vote for each $100,000 of stock held. At the end of 1951 this worked out to percentage votes as follows. The largest were USA 33.03; UK of GB & NI 13.68; China 6.45; France 5.68; India 4.39; Canada 3.61; while the smallest were El Salvador, Honduras, Iceland, Nicaragua, Paraguay with 0.27 per cent each, and finally Panama 0.26. These are similar to, but not identical with, the percentage contributions to the United Nations.

One may ask why the General Assembly prefers to vote by fictitious equality? The answer may be that the Assembly has to deal with wider problems; or it may be that bankers are accustomed to arithmetic, whereas statesmen try to settle matters by oratory, which is in this connection akin to fiction.

B Geographical contact with other states

General explanation. The larger, or longer established, contacts are to be seen on political maps; and the smaller or more recent features can be searched out in verbal records. Some principles of interpretation are however required. For example: does the British government share a land frontier with the Netherlands government? Certainly, because British Guiana shares a land frontier with Dutch Guiana. Similarly Norway shares a frontier with Australia in the Antarctic. Does Canada have more than one land frontier with the United States? Spatially several, one with Alaska and the other interrupted by the Great Lakes; yet, as a relation between governments, it seems proper to count these as only one. Accordingly let the count be, for any state, of the number of other states which it meets by land, either directly or indirectly via colonies, occupied, mandated or trust territories. A seacoast, or several seacoasts, were counted as one and added to the number of land frontiers. I had some doubt as to whether a seacoast ought to count as more than one.

Is the number of contacts a measure of internationality? This method of counting has considerable relevance for world politics, for it was shown by Richardson [*Statistics of Deadly Quarrels*, Chapter V.4] that there was a correlation coefficient of 0.77 between a state's number of frontiers and its number of external wars. That generalization was based on 33 states which had each persisted for at least 109 out of the 126 years AD 1820–1945, which controlled their own foreign policy, and had populations exceeding 0.2 million.

Fission. If a state, having n contacts with other states, splits into two parts, one new frontier is formed between the parts; and there may also be other new frontiers with the surrounding states. Exactly how many one cannot say without a map of the situation.

Ascertainability. This is more definite and simple than for any other of the four accepted ingredients.

Tabular data. The numbers of contacts were counted at the single date 1951 Dec. 31; and the contacts were those between the 60 then members of the United Nations, all other states being ignored. The maps used were those in *Chambers Encyclopaedia* 1950, in the *UN Year Book* 1952 and, for the occupation zones in Germany, *Keesing* p. 10161. Recent changes of sovereignty were sought in the *Statesman's Year Book* 1952 and in *Keesing's Contemporary Achives.*

By the end of 1951 the three western zones of occupied Germany had fused (K.9966 and 10239); so that any member of UN that touched the Western Zone of Germany was counted as having geographical contact with the governments of France, Britain, and USA. Similarly any member of UN that touched the Anglo-Egyptian Sudan was counted as touching both British and Egyptian territory. Similarly for other condominiums.

The contacts of USA have increased remarkably during the present century. A glance at old maps suggested that USA touched only Canada and Mexico. By the end of 1951 it touched also: Panama by way of the Canal Zone; New Zealand in the Antarctic; Britain, Czechoslovakia, France, Denmark, Belgium, Luxembourg, Netherlands and USSR in occupied Germany; Yugoslavia at Trieste; and various countries at its widely scattered armed bases. Of the latter only those need be listed which were UN members and not already mentioned; these are Cuba, the Philippines, Iceland, and Saudi Arabia. Altogether at the end of 1951 USA had land contact with seventeen UN members.

Consistently ignored were merely diplomatic contacts at embassies, consulates, or Tangier.

The results of this count are set out in Table I, column 2.

C The number of inhabitants, briefly called the *population*

(a) *Is population a measure of internationality*? No, it is an internal characteristic. This becomes evident when we compare Norway with Tibet. Both states have about three million inhabitants; the former are given to seafaring, the latter stay at home. Nevertheless the consideration of population leads on to its square root which is, on the average, more appropriate, though no more so in the comparison of Norway with Tibet.

(b) *Fission*. If voting strength were directly proportional to population, the splitting of a state into any assortment of parts would not alter their aggregate voting strength; so that it would be impossible to allege that the fission was a pretence to gain more votes.

(c) *Is it ascertainable*? The number of people can at the best be ascertained with an accuracy of 1 in 10,000 or better; which is far more accuracy than is necessary in an ingredient of voting strength. According to the comments on method in the *UN Statistical Year Book* for 1952, about 46 member states have provided information that is quite sufficient for present purposes; whereas the remaining 14 members leave something to be desired.

If voting strength increased with population there would arise a temptation to exaggerate population; so that it might be necessary to have a census verified by foreign observers.

(d) *Tabular data*. The table of populations of members of UN, as printed on pp. 968–9 of the *UN Year Book* for 1951, was evidently not prepared with any thought of voting; for on the one hand the population of USSR is stated as including the populations of Byelorussia and Ukraine, which have the right to vote independently of USSR; and on the other hand the population of France is stated as 42 millions, thus ignoring the overseas populations of the rest of the French Union, which has, jointly with France, a single vote in the UN Assembly. In Tables I and II the data have been rearranged in accordance with the principle that the population of any area should be credited to the government which directs its foreign policy. Penrose (1952, p. 74) follows some different principle, for he lists Congo separately from Belgium.

The detached portions were found in the *UN Statistical Year Book* and were added. Condominiums were attributed in equal shares to the governments responsible for them, for example Egypt and the United Kingdom were each allocated half the population of the Anglo-Egyptian Sudan. At the date, mid 1951, to which the population statistics refer, Western Germany was regarded as a condominium of Britain, France and USA, while USSR had the whole of Eastern Germany. At the same date, the foreign policy of Indochina was regarded as directed by France.

The populations are not stated directly, but via their square roots in Table I.

D The square root of the number of inhabitants

Where's the joke? The mention of the square root has been observed to cause amusement. Some literary persons may remember square roots as a mystery which they encountered at school but could not 'do'. Scientific people seldom bother to 'do' square roots, preferring to look them up ready-made in Barlow's tables.

Let me further warn any literary reader off the run of words: square root – irrationality – nonsense; it involves a pun; for although some square roots are irrational in the sense that they go to an infinite number of decimal places, yet this elaboration cannot arise when the accuracy is restricted, as it is in all practical affairs; and anyway irrationality in this technical sense has nothing whatever to do with nonsense.

Another pun to be avoided is that on the two meanings of the word population, exemplified in the phrases: the population is a million; the population have black skins and curly hair. Only the former numerical sense is relevant to the square root.

May we now ascend from the ridiculous to the sublime?

The square root of the number of people has been proposed as a measure of voting strength by L. S. Penrose (1946, 1952). It has often been noticed that some voters find it difficult to make up their minds as to the choices that are offered to them. Penrose brought this fact into relation with the theory of the uncertainty of observations. Actually national representatives to the United Nations are appointed by governments, which may or may not have been elected. However, to simplify the argument, Penrose imagines (1950, p. 45) that the national representative is directly elected by the nation as a whole. (Incidentally such election, if ever adopted, would settle such disputes as: who represents China?) Now, Penrose says, this representative really does not represent the whole nation, but only the excess of the number of persons who voted for him, over the number who voted against, the rest neutralizing in pairs. The individual voters are supposed to be unable to foresee the future questions which their representative will have to decide; so their individual votes may be regarded as cast at random. Here I would modify Penrose's statement by pointing out that it is always possible to imagine some other type of candidate which the nation would have emphatically rejected; so that it is better to regard the individual votes as having random deviations *from a mean*. The proposition is not that individuals vote purely at random, for if so democracy would be nonsense. The proposition is that the rival candidates are both so suitable, having been previously selected, that it is difficult to choose between them. Thus the authority of a representative of a nation of n individuals is brought into relation with the uncertainty of the mean of n uncertain observations. The latter problem has been before scientists for 150 years and has been studied both experimentally and theoretically, with the result that the uncertainty of the mean is found to vary inversely as the square root

Table I. *Ingredients for the 60 nations which were members of UN at the end of 1951*

	Geogr. contacts at end of 1951	Square root of population at mid 1951 100 ×	Value of exports during 1950 10^7\$ ×
Afghanistan	4	35	5
Argentina	6	42	189
Australia	5	30	176
Belgium	7	49	170
Bolivia	5	18	10
Brazil	11	73	135
Burma	5	43	14
Byelorussia	3	24	with USSR
Canada	2	37	291
Chile	4	24	28
China	7	215	7
Colombia	6	34	40
Costa Rica	3	9	3
Cuba	2	23	64
Czechoslovakia	6	35	72
Denmark	3	21	66
Dominican Rep.	2	15	7
Ecuador	3	18	6
Egypt	6	50	54
El Salvador	3	14	7
Ethiopia	3	39	3
France	16	119	231
Greece	3	28	9
Guatemala	5	17	7
Haiti	2	18	4
Honduras	4	13	2
Iceland	2	4	3
India	7	189	131
Indonesia	2	88	78
Iran	6	44	80
Iraq	5	23	6
Israel	4	12	4
Lebanon	3	11	3
Liberia	3	13	3
Luxembourg	4	6	with Belgium
Mexico	4	51	46
Netherlands	6	34	196
New Zealand	3	14	51
Nicaragua	3	10	26

Table I cont.

	Geogr. contacts at end of 1951	Square root of population at mid 1951 100 ×	Value of exports during 1950 10^7\$ ×
Norway	5	18	40
Pakistan	5	87	61
Panama	4	9	1
Paraguay	3	12	3
Peru	6	29	19
Philippines	2	45	33
Poland	5	50	53
Saudi Arabia	5	25	10?
Sweden	2	27	110
Syria	5	18	13
Thailand	4	43	11
Turkey	6	47	26
Ukraine	5	56	with USSR
South Africa	2	36	86
USSR	13	132	64
United Kingdom	22	122	767
USA	18	132	1028
Uruguay	3	15	25
Venezuela	4	22	124
Yemen	3	21	0?
Yugoslavia	4	40	16
TOTALS	304	2528	4717

of n. (See for example C.**13**, 153). Therefore, Penrose concludes, the authority of a representative ought to vary directly as the square root of the number of individual electors.

Penrose seems to me to have introduced into the discussion of voting an idea which is likely to prove fertile; but his assumptions are not altogether clear; and I do not see how to clarify them. A principal difficulty is that personality is multidimensional: a candidate may be acceptable in some dimensions but not in others; yet voting must be for, or against, him as a whole. What the day to day news from the United Nations reveals is the formation of alliances against possible enemies. This process is not at all like the random voting on which Penrose bases his theory. The story of Condorcet and the Judges is a warning against rash assumptions about randomness. (Eddington 1935, pp. 123–125). Penrose's conclusion, as distinct from his reasons, can however be supported by the following two quite independent arguments.

A geographical argument in favour of the square root, but not mentioned by Penrose, is the following. Like his theory it is full of doubtful assumptions. The perimeter of similar figures, say regular hexagons of various sizes, is proportional to the square root of their area. The population is proportional to the area, if the density of population is constant. The foreign contacts are proportional to the perimeter. Therefore internationality is more likely to be proportional to the square root of the population than to the population directly.

A third and independent argument in favour of the square root is one of pure compromise, or as Penrose says (1952, p. 45) of expediency. I state it differently. There is a wise political maxim that changes, even seeming improvements, ought not to be made too suddenly, because of the unforeseen ramifications of social effects. Hitherto statesmen have accepted the principle that all sovereign states are equal. Another popular principle is that all men are equal; and in some places 'men' includes women. Those who systematically compare individuals by tests find that they differ widely from one another, but also that they cannot be arranged in a single linear order ranging from best at one end to worst at the other. The differences between individuals are multidimensional. (Gordon W. Allport 1945, Gardner Murphy 1947, L. L. Thurstone 1935, P. E. Vernon 1952.) The false doctrine of the equality of individuals leads directly to the notion that voting ought to be proportional to population. The converse does not follow. It is possible that voting ought to be proportional to population although individuals certainly differ widely. It is logically possible that a million individuals, being all the inhabitants of any geographical area, might be found to have the same average test-score whatever the area chosen. S. D. Porteus's findings (1931, 1937) are to the contrary. But the issue, because it involves racial pride, is naturally very controversial.

When politicans wish to persuade the smaller states to join an international organization, they propose the principle that all sovereign states are equal. When politicians wish to persuade the less competent individuals to vote for their party, they tell them that all men are equal. Both of these setting-forths are pretences, designed to flatter different assortments of people. A merit of the square root of the population is that it is a compromise halfway between two rival pretences. Because it is not designed to flatter anybody, it may meet with resistance.

(a) *Internationality*. How can the square root of the population be a measure of internationality although the population itself is not? Norway and Tibet have about the same arithmetical value for the square root. The answer is that a relation, which fails in a particular instance, may yet hold statistically on the average. The argument, given above about similar figures, indicates that this relation does.

(b) *Fission*. An objection to the square root is that it allows a considerable increase of voting strength to be obtained by pretended fission. Consider for example, a hypothetical state having a population of 144 millions, and voting strength 12. If it were to split into 9 states each of 16 millions, and so each of voting strength 4, the aggregate voting strength after fission would become $9 \times 4 = 36$, three times greater than the original 12. On the present system of sovereign equality, the same hypothetical state of 144 millions would have one vote before fission and 9 votes after fission. So to bring in the square root of the population would make the voting strength less liable to pretended fission than it is at present.

(c) *Ascertainability.* It is well known in physics that taking the square root tends to diminish the importance of errors of observation.

(d) *Tabular data.* The square root of the population-in-millions is stated in Table I for the 60 members of the United Nations.

E External trade

Internationality. External trade is obviously connected with internationality. It distinguishes for instance between two states that are about equal in population, Norway and Tibet, for the trade of Tibet, as far as it is known, appears to have been of the order of one hundredth of that of Norway in value.

Fission. If a state were to split into two parts, the trade between those parts, which was formerly internal, would become external. Thus there would by fission be a gain in aggregate voting strength.

Ascertainability. External trade is ascertainable in so far as there are always two witnesses; the exporting and the importing country. The Statistical Office of the UN has succeeded, with the aid of the International Monetary Fund, in reducing the economic value of the external trade to a common monetary unit, namely US dollars of the gold content fixed on 31st January 1934.

Choice between exports, imports, and their subclasses. Either imports, or exports, or their sum would be a suitable ingredient. Actually I would prefer to choose domestic exports, including gold and silver, but the data are not always so arranged. The available publications distinguish between 'general' and 'special' trade, the distinction being that general trade includes, but special trade excludes, goods that are merely unloaded from a ship into a warehouse to be re-exported to another destination. The distinction may be illustrated by Aden, for which the domestic exports were recently negligible in comparison with the re-exports. It is also customary to distinguish 'merchandise' from gold and silver. That may be a relic of the mercantilist theory; anyway it seems to have no relevance to voting.

Tabular data. Altogether I gather the impression that the various national statistical offices probably have the information, and could arrange it in a relevant form if there were an important motive, such as gain of voting strength, for disclosure. In the meantime it seems worthwhile to state in Table I such imperfect information as I have found, in order to provoke discussion and amendment.

The *UN Statistical Yearbook* for 1952 states the dollar value of the exports from 152 separate geographical areas. Some of these areas are combined politically under the 60 governments which in 1952 had one vote each in the UN General Assembly. There is a snag about totals: for the arithmetical sum of the exports of the parts in general exceeds the exports of the group jointly to the rest of the world, because the parts may export to each other. For this reason it is not in all cases possible to compute, from the data in the *UN Statistical Yearbook*, what was the value of the exports of the group under one government.

The most complicated area that had one vote in the United Nations Assembly was that governed, as to its foreign affairs, from London; it comprised in 1950 the United Kingdom of Great Britain and Northern Ireland together with its Crown Colonies, Protectorates, and Southern Rhodesia. The desired information can be found in the *Statistical Abstract for the Commonwealth* (Trade Statistics) 1948–51 (London: HMSO 1952). This publication shows that total exports of the above group of countries to the rest of the world in 1950 amounted to £2,740 million, re-exports being included. As the Anglo-Egyptian Sudan was a joint U. Kingdom–Egyptian condominium it was, for the purpose of the above calculations, included in the rest of the world. The extraction and computation of the £2,740 [million] was, most obligingly, performed for me by the Statistics Division, Board of Trade.

Personally I think that, if trade were to be an ingredient of voting strength, there would need to be agreement as to whether re-exports should be included or not. The practice of different countries has been various. Also some generally acceptable treatment of condominiums would need to be devised. One might for instance attribute half to Egypt and half to Britain the value of the exports from the Anglo-Egyptian Sudan to countries that belonged neither to Egypt, nor to the above-specified British group.

The next group in order of decreasing complication is the Union Français, for which the desired information is clearly stated in the *Bulletin Mensuel de Statistique d'Outre-Mer* Nov.–Dec. 1951. As to the exports of the Netherlands group, I gratefully acknowledge help from the central Bureau of Statistics at The Hague. The group formed by Belgium and its Congo has been so far regularized by the data in the *Statesman's Year Book* of 1952. However it is not possible to clear this entirely, for Belgium and Luxembourg vote separately, but their exports are published jointly. A similar difficulty may arise about Byelorussia and the Ukraine, which vote separately from USSR but, in another sense, are part of it. However I found no statement of exports for Byelorussia or Ukraine, and none for USSR since 1938. The USSR exports for 1938 were extrapolated to 1950 by multiplying them by the ratio in which world trade has altered. A similar type of extrapolation, though over a shorter run of years, was applied to Argentine, Bolivia, Afghanistan, China, Thailand, Poland.

F Armed strength

'Surely', it may be said, 'all objectively-minded people know that a nation's importance in the world depends chiefly on its armed strength. The eight powers which were generally recognized as great from AD 1900 to 1914, and the present big three, big four, or big five, are not called great merely because they have a great population, or have been greatly beneficent, but because they could be greatly dangerous if annoyed'.

In reply to that common feeling I can only state my personal conviction that to admit 'annual expenditure on defence' as an ingredient of voting strength would be to foster the very nuisance, namely a threat of violence, which a reasonable system of voting should be designed to avoid.

It would be more practical to regard armed threats, not as an ingredient of voting strength, but as the other and obsolescent way of doing things by the antithesis of voting. The commonsense of statesmen at present moderates the monstrous fiction that all sovereign member states are equal; and under any improved system of voting, statesmen would presumably know what was practicable in a world of unequally armed nations; although that deference to armed power would have in it a touch of appeasement, in the recent derogatory sense of the word.

The word appeasement changed its emotional tone between 1936 and 1940 in consequence of Mr Neville Chamberlain's activities. In 1936 appeasement meant the desirable process of pacification; by 1940 the word had become so emotional that one could hardly obtain an intellectual definition of it, but only exclamations of 'Munich!' and 'Danegeld!' I am discussing the meaning of a word, not the history of the crisis at Munich in 1938. One of the shameful connotations of appeasement was buying peace for oneself by giving away what belonged to a third party (Czechoslovakia). But there is a quite separate difficulty in attaching any clear meaning to appeasement. To grant a favour to one's opponent is honourable if it is done (a) because he has a justifiable claim; indeed in that case it is hardly a favour; or (b) in knightly courtesy to allow him fairplay, or (c) in Christian charity. But the very same outward act is appeasement, in the derogatory sense, if it is done in fear and hate. Thus, apart from the property of third parties, appeasement is not recognizable by any objective criterion alone, but also depends on the emotional mood of one of the principals.

G Number of men contributed to the international army

This ingredient was suggested to me by J. Erskine Harper MA, barrister at law. He said that it would make votes proportional to risk taken, in rather the same way as at a shareholders' meeting of a limited liability company, where votes are proportional to capital risked. The reader will notice that this proposed ingredient is quite different from the armed-strength-for-any-purpose which was discussed in the last section, because the present proposal is definitely connected with international cooperation. It would be ascertainable. There could be no complaint about pretended fission.

Yet we must remember that the international army may be fighting somebody; so that it is not really a world-army, but only a majority's army; and that, as during 1952 in Korea, fighting is associated with intense, and wide-spread, hate. Therefore to take the contributions in men as an ingredient of voting strength would tend to make hate an ingredient of voting, a situation which it is desirable to avoid.

How to form an index of voting strength from various ingredients

Let us suppose that each ingredient, by itself, has been judged to be a more or less reasonable measure of voting strength, better anyway than fictitious equality, but still imperfect; and that it is desired to combine them so that they may support each other's merits, and neutralize each other's defects. A problem like this is familiar to economists when they combine various prices into an index of the cost of living (A. L. Bowley 1929). A similar problem also presents itself to the inventors of mental tests when they choose diverse puzzles, each intended to measure general intelligence, and decide how many marks to allot for a correct answer to each. The simplest method, which is not obviously wrong, will be the best; for it would be unreasonable to expect statesmen to regulate their own voting by mathematical elaborations with which they were not familiar. We are not troubled here by any negative, zero, or infinite numbers among the ingredients. Accordingly the method proposed is a simple average of percentages, as shown in Table II. That is to say any nation's number on any ingredient is first expressed as a percentage of the total of that ingredient for all the member-nations. For example the number of geographical contacts ranges from 2 for Sweden to 22 for the United Kingdom. The total of all such numbers is 304 for the 60 nations on the list. Therefore the percentage of Sweden is $200/304 = 0.66$, and that for the United Kingdom is $2200/304 = 7.24$. Similarly for the other ingredients. For any nation its proposed index of voting strength is the arithmetical mean of its percentages for the several ingredients. This is shown in the last column of Table II.

Table II. *Percentages and index for the sixty nations which were members of UN at the end of 1951*

	Scale of assessments for 1952 UN budget	Geogr. Contacts	Square root of population	Exports	Index
Afghanistan	0.08	1.32	1.38	0.11	0.72
Argentina	1.62	1.97	1.66	4.01	2.32
Australia	1.77	1.64	1.19	3.73	2.08
Belgium	1.35	2.30	1.94	3.60	2.30
Bolivia	0.06	1.64	0.71	0.21	0.66
Brazil	1.62	3.62	2.89	2.86	2.75
Burma	0.15	1.64	1.70	0.30	0.95
Byelorussia	0.34	0.99	0.95	with USSR	0.57
Canada	3.35	0.66	1.46	6.17	2.91
Chile	0.35	1.32	0.95	0.59	0.80
China	5.75	2.30	8.50	0.15	4.18
Colombia	0.37	1.97	1.34	0.85	1.13
Costa Rica	0.04	0.99	0.36	0.06	0.36
Cuba	0.33	0.66	0.91	1.36	0.82
Czechoslovakia	1.05	1.97	1.38	1.53	1.48
Denmark	0.79	0.99	0.83	1.40	1.00
Dominican Rep.	0.05	0.66	0.59	0.15	0.36
Ecuador	0.05	0.99	0.71	0.13	0.47
Egypt	0.60	1.97	1.98	1.14	1.42
El Salvador	0.05	0.99	0.55	0.15	0.44
Ethiopia	0.10	0.99	1.54	0.06	0.67
France	5.75	5.26	4.71	4.90	5.16
Greece	0.18	0.99	1.11	0.19	0.62
Guatemala	0.06	1.64	0.67	0.15	0.63
Haiti	0.04	0.66	0.71	0.08	0.37
Honduras	0.04	1.32	0.51	0.04	0.48
Iceland	0.04	0.66	0.16	0.06	0.23
India	3.53	2.30	7.48	2.78	4.02

Table II cont.

	Scale of assessments for 1952 UN budget	Geogr. Contacts	Square root of population	Exports	Index
Indonesia	0.60	0.66	3.48	1.65	1.60
Iran	0.40	1.97	1.74	1.70	1.45
Iraq	0.14	1.64	0.91	0.13	0.70
Israel	0.17	1.32	0.47	0.08	0.51
Lebanon	0.06	0.99	0.44	0.06	0.39
Liberia	0.04	0.99	0.51	0.06	0.40
Luxembourg	0.05	1.32	0.24	with Belgium	0.40
Mexico	0.65	1.32	2.02	0.97	1.24
Netherlands	1.27	1.97	1.34	4.16	2.18
New Zealand	0.50	0.99	0.55	1.08	0.78
Nicaragua	0.04	0.99	0.40	0.55	0.50
Norway	0.50	1.64	0.71	0.85	0.92
Pakistan	0.79	1.64	3.44	1.29	1.79
Panama	0.05	1.32	0.36	0.02	0.44
Paraguay	0.04	0.99	0.47	0.06	0.39
Peru	0.20	1.97	1.15	0.40	0.93
Philippines	0.29	0.66	1.78	0.70	0.86
Poland	1.36	1.64	1.98	1.12	1.52
Saudi Arabia	0.08	1.64	0.99	0.21	0.73
Sweden	1.73	0.66	1.07	2.33	1.45
Syria	0.09	1.64	0.71	0.28	0.68
Thailand	0.21	1.32	1.70	0.23	0.86
Turkey	0.75	1.97	1.86	0.55	1.28
Ukraine	1.30	1.64	2.22	with USSR	1.29
South Africa	0.90	0.66	1.42	1.82	1.20
USSR	9.85	4.28	5.22	1.36	5.18
United Kingdom	10.56	7.24	4.83	16.26	9.72
USA	36.90	5.92	5.22	21.79	17.46
Uruguay	0.18	0.99	0.59	0.53	0.57
Venezuela	0.32	1.32	0.87	2.63	1.28
Yemen	0.04	0.99	0.83	0.00	0.46
Yugoslavia	0.43	1.32	1.58	0.34	0.92
TOTALS	100.00	100.02	99.97	99.97	99.98

Note: For Byelorussia and Ukraine with USSR, also for Luxembourg with Belgium, there is no satisfactory way of computing the index. The exports of Byelorussia, Ukraine and Luxembourg have been treated as if they were zero. This brings the total of the index to nearly 100.

The greatest voting strengths

On looking down the column headed 'index' in Table II, we see that the greatest is USA at 17.46. Then follow in order of rank United Kingdom 9.72; USSR 5.18; France 5.16; China 4.18. The above-named coincide with what journalists call 'the big five'. The next six in rank are India 4.02; Canada 2.91; Brazil 2.75; Argentine 2.32; Belgium 2.30; Netherlands 2.18.

The average voting strength

This is, by definition, $100/60 = 1.66\dot{6}$. The pretence at the United Nations is that they all have $1.66\dot{6}$.

The least voting strengths

The least is Iceland 0.23; then follow, in reversed order of rank, Costa Rica 0.36; Dominician Republic 0.36; Haiti 0.37; Lebanon 0.39; Paraguay 0.39.

The consistency of the different percentage ingredients for any one state

A very poor consistency is tolerable because the aim is to improve on the present practice whereby sovereign states are nominally equal; and because the ingredients are averaged to form the index. Nevertheless if, for any state, one of the percentage ingredients were found to be more than a hundred times another, that would raise the question whether the ingredients were ill chosen. An inspection of Table II yielded the following result:

Ratio of the greatest to the least percentage ingredient for the same state

81	81 to 27	27 to 9	9 to 3	3 to 1	unknown	
Number of such states						Total
1	4	19	23	10	3	60

On the whole I regard the discrepancies as tolerable. The one state for which the ratio exceeded 81 was Yemen, and that happened because the value of Yemen's exports was so small and uncertain that I rounded it off to zero. The four states for which the ratio lay between 81 and 27 were Bolivia, China, Honduras, and Panama.

Conclusion

The author submits that the 'index' in Table II is a systematization of commonsense about voting strength.

References

Allport, G. W. 1945. *Personality*. New York: Henry Holt.

Bowley, A. L. 1929 article 'Cost of Living' in *Encyclopaedia Britannica*. 14 edn.

C. stands for *Chamber's Encyclopaedia* 1950 edition. London: Newnes.

Cattell, R. B., Breul, H. and Hartman, H. P. 1952. *Amer. Sociol Rev.* Vol. 17, pp. 408–419.

Eddington, Sir Arthur. 1935. *New Pathways in Science*. Cambridge University Press.

K. refers to *Keesing's Contemporary Archives*. Bristol: Keesing's Publications, Ltd.

Murphy, G. 1947. *Personality*. New York: Harper.
Penrose, L. S. 1946. *J. Roy. Statistical Soc.* **109**, 53.
Penrose, L. S. 1952. *On the Objective Study of Crowd Behaviour*. London: H. K. Lewis.
Porteous, S. D. 1931. *The Psychology of a Primitive People*. London: Arnold.
Porteous, S. D. 1937. *Primitive Intelligence and Environment*. New York: Macmillan.
Richardson, L. F. 1918. *War and Peace 1918*. A periodical now defunct.
Richardson, L. F. 1950. *Statistics of Deadly Quarrels*. A book in microfilm, published by the author.
Thurstone, L. L. 1935. *The Vectors of Mind*. Chicago: University Press.
Vernon, P. E. 1952. *The Advancement of Science*, Vol. 9, pp. 207–218.

Richardson, L. F. (1953)
Br. J. statist. Psychol. **6**. 77–90

1953:2

THE SUBMISSIVENESS OF NATIONS

The British Journal of Statistical Psychology
Cambridge
Vol. VI, Part II, November 1953, pp. 77–90

Vol. VI Part II The British Journal of Statistical Psychology November, 1953

THE SUBMISSIVENESS OF NATIONS

By LEWIS F. RICHARDSON
Hillside House, Kilmun, Dunoon, Argyll, Britain

I. *Introduction.* II. *The Linear Theory of Arms-races.* III. *First Hypothesis about Submissiveness.* IV. *Second Hypothesis about Submissiveness.* V. *Third Hypothesis (for Sides Equal in their Constants).* VI. *Conclusion.*

I. INTRODUCTION

The problem to be discussed in this paper can be stated quite simply as follows. It is well known that a thorough military defeat leaves the vanquished nation for some years in a mood of submission to its victors (examples are the submissive behaviour of Vichy-France to the Germans during 1941 to 1943, or of West Germany to the Western Powers and of East Germany to the Soviet Union during 1945 to 1952). As this effect seems universal, it is not necessary to name the nations ; they may be distinguished as A and B. Now suppose that A and B are both rested from war and are roughly equal in size and in industrial power ; and suppose further that A makes a show of force against B. Does such behaviour induce in B a submissive mood or a retaliatory mood ?

Whatever may be thought of the following treatment, there can be no doubt that some of the questions connected with submissiveness are important. For example, in September, 1952, the North Atlantic Treaty Organization held combined naval manœuvres (exercise " Mainbrace ") off Norway and in the Baltic simultaneously with land manœuvres (exercise " Holdfast ") involving nearly 200,000 troops in Western Europe. The declared intention of N.A.T.O. was to deter any tendency in the Soviet Union towards aggression. On the contrary the Russian newspaper *Pravda* said that exercise " Mainbrace " was itself aggressive (B.B.C., September 13th). Personally I accept N.A.T.O.'s description of its intention ; but I appreciate that *Pravda* may find this intention very difficult to believe, because of the universal tendency to suspect foreign forces unless those of allies— a tendency represented in the following formulæ by the defence coefficients k or l. To be deterred from any tendency towards aggression is here regarded as a mild form of submission.

Professor C. A. Mace asked whether my paper contained any suggestion for research on small groups of people, a type of enquiry now being pursued by many psychologists. Probably it does ; for submission has been occasionally observed wherever two or more individuals meet. For examples : even between wolves (Lorenz (14), 1952, ch. 12) ; or the pecking order in the hen yard (references are given by Landau (11), 1951) ; or tests on individual persons in their relation to the community (Allport (1), 1945, pp. 410–14). W. McDougall (1917 (15), pp. 62–6) took self-subjection to be one of the primary human instincts. It might be rash, however, to expect that by research on a group of, say, 50 people one could predict the behaviour of a nation : for, at least in physics, large aggregates of molecules show some conspicuous forms of behaviour that are not noticeable in small aggregates. For example, I doubt whether anybody, by intensive observation of a drop of water, could have discovered the peculiar law of eddy-diffusion which operates in the sea (Richardson and Stommel (27), 1948).

II. THE LINEAR THEORY OF ARMS-RACES

This theory must first be mentioned, because submissiveness was formulated by a modification of it. From a pamphlet by Bertrand Russell (28) in 1914 I took the idea that much of what is nowadays called defence does not in fact defend, because between

The Submissiveness of Nations

equal groups in peacetime it provokes retaliation. Gregory Bateson ((2), 1936, p. 266) observed competitive boasting in New Guinea and generalized the concept as 'symmetrical schismogenesis,' of which an arms-race is another example. Russell's concept of mutual stimulation was formulated by me as a pair of simultaneous differential equations (1919, revised 1935). In 1938 I hit on a way of testing the formulæ quantitatively in terms of expenditure. The testing was continued in a monograph published a few months before the war (*Generalized Foreign Politics, Brit. J. Psychol. Mon. Sup.*, 1939 ; a better title would have been *The Theory of Arms-races*).

The financial facts have been re-examined in much greater detail in a microfilm called *Arms and Insecurity* ((23), 1949), where the technical term 'warfinpersal' is introduced as a contraction for 'war finance per salary.' The 'warfinpersal' of any nation is its governmental warlike (alias 'defence') expenditure divided by the annual earnings of an average sort of citizen, usually those of a semi-skilled engineer. The latest summary, including a glance at the present arms-race, is being printed for *Sankya* (26). The sceptical reader is referred to the aforesaid publications for the facts ; here there is room only for the equations which summarize them. As we pass from the study of individual-differences to that of nations, the vagaries of individuals merge into the quasi mechanical regularities of mass-behaviour.

Let x and y denote the warfinpersals of the opposing sides, t the time, and g, h, α, β, k, l be constants of which α, β, k, l are positive. The hypothesis is that

$$dx/dt = g - \alpha x + ky, \qquad (2/1)$$

$$dy/dt = h - \beta y + lx. \qquad (2/2)$$

In particular, if $g = 0$, $h = 0$, $k = l$, and $\alpha = \beta$, then the solution is

$$x + y = Ae^{(k-\alpha)t},\ x - y = Be^{-(k+\alpha)t}, \qquad (2/3, (2/4)$$

where A and B are arbitrary constants. Functions like $x + y$ and $x - y$, which are proportional to a single exponential of the time, are called 'normal co-ordinates.' They lead on well to modifications intended to represent submissiveness : that is why, in the subsequent diagrams, the axes are turned at 45° to those of x and y.

When we are trying to interpret historical facts, we are not given that $\alpha = \beta$ and that $k = l$ exactly, although we may have reasons based on population, on industrialization, and on an assumption about common human nature, for believing that α/β and k/l are not far from unity. In these circumstances the normal co-ordinates would not be simply $x + y$ and $x - y$, but would be $l_1x + m_1y$ and $l_2x - m_2y$, where the l and m are positive weights. Now in an arms-race x is roughly proportional to y, and therefore $x + y$ is roughly proportional to $l_1x + m_1y$ for any not too different weights l_1, m_1. So we can safely study the historical statistics by replacing the unknown $l_1x + m_1y$ by simply $x + y$. But there is no such automatic immunity from error in a replacement of $l_2x - m_2y$ by $x - y$; for they may even have different signs. This difficulty crops up in Sect. V below, where $x - y$ suits the First but not the Second World War.

All the varieties of motion that can be described by linear equations like (2/1) and (2/2), but without any restriction on the signs of α, β, k, l, have long been known (Lagrange, Weierstrass (31), 1875). They were considered, in the context of arms-races, by Richardson ((23), 1949, § 3/2/4) ; but none of them was found to resemble the phenomena of submissiveness, which must therefore require either variable coefficients, a higher order differential equation, an integral equation, or a non-linear description. That was my objection to a linear hypothesis with constant coefficients proposed by Mr. M. R. Horne as a representation of submissiveness ((9), 1951).

III. FIRST HYPOTHESIS ABOUT SUBMISSIVENESS

The statesmen of U.S.A., Britain, France, and U.S.S.R. still appear to pay no attention to mutual stimulation, for they take a pride in accumulating the type of weapons, such as bombing aircraft, or rockets, or submarines, which are most likely to provoke counter-measures. Perhaps they are relying on some unexpressed theory about submissiveness, thus : much force, much submission ; therefore a show of force, some submission. That I call "thinking proportionately." The behaviour of living things is not always so simple. I remember an impertinent schoolboy who was excited and made more rebellious by various mild reproofs and punishments from a schoolmaster in the presence of a class of amused boys, but who was finally suppressed by a much more severe punishment. The behaviour of nations seems to involve a similar reversal.

In 1939 Richardson (21) modified (2/1) and (2/2) by non-linear terms intended to represent submissiveness, thus:

$$dx/dt = g - \alpha x + ky\,\{1 - \sigma(y - x)\}, \tag{3/1}$$

$$dy/dt = h - \beta y + lx\,\{1 - \rho(x - y)\}, \tag{3/2}$$

where σ and ρ were positive constants representing the tendency to submit.

Application of the First Hypothesis (a) *To Very Unequal Nations.*—Eqn. (3/1) and (3/2) were originally intended to represent a large x-nation overawing a much smaller y-nation. If we leave out g and h, as not of the essence of the matter, and if we neglect y in $x - y$, these equations become

$$dx/dt = -\alpha x + ky\{1 + \sigma x\}, \tag{3/3}$$

$$dy/dt = -\beta y + lx\,\{1 - \rho x\}. \tag{3/4}$$

It is dy/dt that will become negative. The smaller nation is suppressed.

For example in 1948 claims by Guatemala to British Honduras were suppressed by a show of armed force by Britain, without any fighting (K9153, K11511). I am not suggesting that the suppression of a smaller nation is necessarily either just or unjust; but am merely asserting that it is what usually happens in a world of power politics.

(b) *To Nations Equal as to their Constants.*—In 1951 I noticed that the terms representing submissiveness would have remarkable consequences even if the two opponents were equal in their constants. How this might happen was explained in a letter to *Nature* (1951, ix, 29) headed "Could an arms-race end without fighting?", in which arguments about increase or decrease were applied only in circumstances not far removed from the actual. Wishing, however, to see a general integral formula over the whole field, and not finding any, I applied to Dr. M. L. Cartwright, F.R.S., who has made a special study of non-linear differential equations ((4), 1949, (6), 1952). As an apology for the lack of general integral formulæ in the first two theories, I may quote her remark ((5), 1952, p. 88) that "what the pure mathematicians have done shows that the non-linear phenomena are genuinely complicated, and no easily applicable general theory can be expected." As the subject matter is economic or political, 3 per cent. accuracy coupled with a plainly intelligible meaning is to be preferred to infinite accuracy coupled with a complicated formula.

A theory of the submissiveness of equal groups must not be too deterministic: it must not say *which* of them will submit, for that is obviously a chancy affair. Such "stochastic processes" have been the subject of a symposium at the Royal Statistical Society (Moyal, J. E., Bartlett, M. S., and Kendall, D. G. (16), 1949).

By the time an arms-race has gone so far as to be alarming, former grievances and ambitions seem relatively unimportant; so that we may neglect g and h. On setting also $\alpha = \beta$, $k = l$, $\sigma = \rho$ eqn. (3/1) and (3/2) simplify to:

$$dx/dt = -\alpha x + ky + k\sigma(xy - y^2), \tag{3/5}$$

$$dy/dt = kx - \alpha y + k\sigma(xy - x^2). \tag{3/6}$$

From a study of the submission of Germany in the 1920s I estimated that:

$$\sigma > 0{\cdot}34 \times 10^{-6} \text{ per person}, \tag{3/7}$$

(Richardson (23), 1949). As this information is meagre, let us change to new variables which involve σ in such a way that we can discuss its general effects without knowing its numerical value. Thus define U, V by

$$x = U/\sigma, \quad y = V/\sigma, \qquad (3/8), (3/9)$$

so that U and V are pure numbers. Eqn. (3/5) and (3/6) then become

$$dU/dt = -\alpha U + kV + k(UV - V^2) \tag{3/10}$$

$$dV/dt = kU - \alpha V + k(UV - U^2). \tag{3/11}$$

Further let

$$t = \tau/k, \tag{3/12}$$

so that τ is a pure number. For thus

$$\frac{dU}{d\tau} = -\frac{\alpha}{k}U + V + UV - V^2, \tag{3/13}$$

$$\frac{dV}{d\tau} = U - \frac{\alpha}{k}V + UV - U^2. \tag{3/14}$$

Only one constant remains explicit, namely, α/k, and it is a pure number. Before 1877 arms-races did not occur, and statesmen believed in a stable balance of power, such as would occur if $\alpha/k \geqslant 1$. From a study of the arms-race which led to the First World War it was concluded that $k - \alpha =$

groups. In another place ((22, 1948) I have given a theory of war-moods in which the people in German groups are represented by five variables ; but here only one variable is available. On December 12th, 1916, the German Chancellor suggested a peace conference. The opposing Powers, after consultation, said they " refuse to consider a proposal which is empty and insincere." (Lloyd George, p. 661.) There were various other tentative and secret peace feelers in 1917 and 1918, but mostly Austrian. According to a Reichstag Committee, quoted by Lloyd George (p. 1932), " Up to 15th July, 1918, the Supreme Army Command rejected the view that victory was no longer possible of attainment by force of arms, and gave no support to peace negotiations upon the basis of a military stalemate . . ." During the next three weeks the fresh Americans came to the aid of the tired French and British. A battle on August 8th, 1918, in which the Germans had no reserves left, changed Ludendorff's opinion, and on August 14th the Kaiser, presiding over a Crown Council at Spa, admitted that Germany would have to find a suitable moment in which to come to an understanding with the enemy.

(iii) After the First World War, the sides were not quite the same. There was the Russian revolution and its consequences. U.S.A. withdrew into isolation. A new diagram was drawn showing the warfinpersal of Germany plotted against that of its entourage, consisting of France, Britain, Poland, Czechoslovakia, and Russia. This appears in a microfilm (Richardson (23), 1949) and is modified here into part of Fig. 3. It shows motion, along a nearly straight track and slowing down as a point of balance is approached. There was a delay of eleven years in the neighbourhood of this point, and then a departure, with increasing speed along another almost straight track. This was the second great arms-race beginning.

(iv) It led to the Second World War during which the warfinpersals approached the state of total mobilization, but were difficult to ascertain. The sides were rather different from those of the first war, Japan and Italy in particular having changed over previously, and Italy changing back again. In October, 1939, Hitler made a peace offer : but Britain and France did not regard it as at all submissive. After that the demand for the unconditional surrender of Germany delayed any sign of submissiveness until defeat.

(v) The Occupation Statute for Germany, an expression of submission, has now been in operation for seven years. Disarmament after the Second World War was barely over when enmity between U.S.A. and the Soviet Union began to dominate the scene.

Fitting Formulæ to the Facts.—It seems permissible to ignore the changes in the membership of the sides, so that x and y may be regarded as having the same meaning throughout. This has already been assumed in Fig. 3.

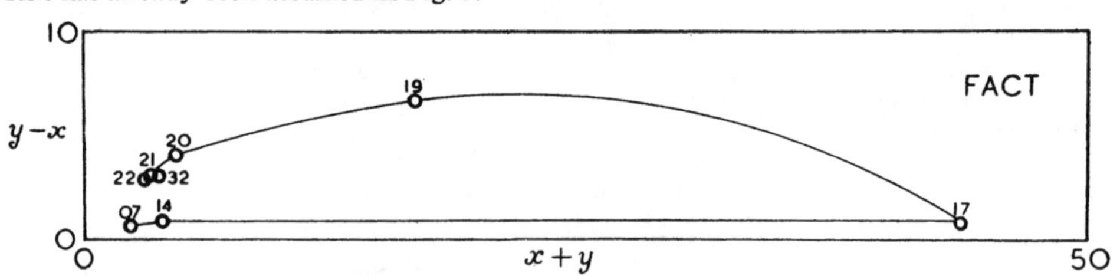

Fig. 3.—Pro- and anti-German warfinpersals in Europe, 1907 to 1932, simplified by ignoring other changes in the sides ; and in particular by ignoring the brief, but balance-altering, intervention of U.S.A. The co-ordinates are the sum and difference. The numbers set beside the plotted points are the dates.

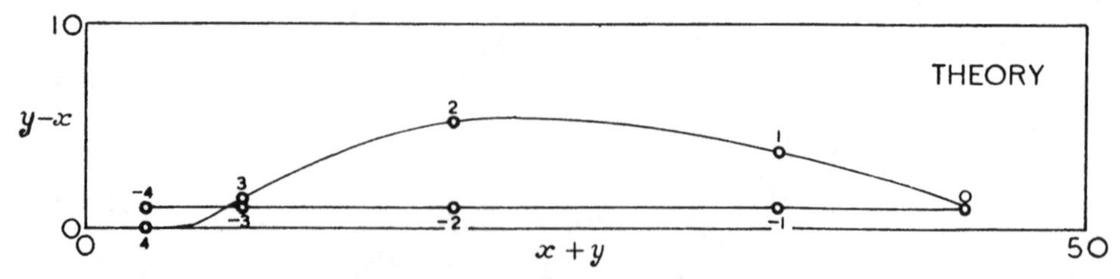

Fig. 4.—A mathematical representation, plotted from formulæ (6/17), (6/19), and (6/20) of the warfinpersals before, during, and after the First World War. For comparison with Fig. 3. Numbers set beside the plotted points are values of t. A unit of one year makes Fig. 4 agree roughly with Fig. 3 as to time.

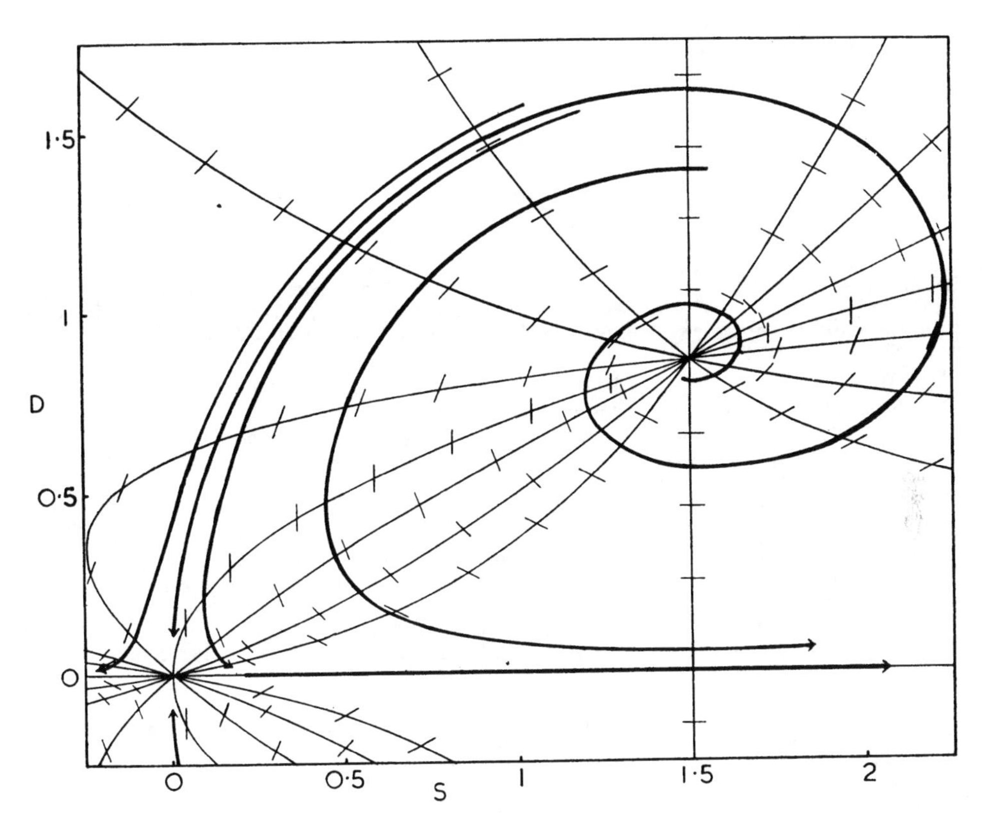

FIG. 1.—On the first hypothesis this shows the field of two opposing sides that are equal in their constants. S is proportional to the sum and D to the difference of their warfinpersals. The thin curves are the isoclines along each of which dD/dS is constant. They are drawn for $dD/dS = 0, \frac{1}{2}, 1, \frac{7}{2}, \infty, -\frac{7}{2}, -1, -\frac{1}{2}$. The directions of motion are indicated by short segments crossing each isocline. The thick lines are trajectories drawn so as to fit the prescribed directions.

To investigate in more detail the neighbourhood of the point of balance at $S = \frac{3}{2}$, $D = \frac{\sqrt{3}}{2}$, define ε and ζ by

$$S = \frac{3}{2} + \varepsilon, \quad D = \frac{\sqrt{3}}{2} + \zeta, \qquad (3/29), (3/30)$$

and substitute the expressions in the eqn. (3/20) and (3/21) obtaining

$$\frac{d\varepsilon}{d\tau} = \frac{1}{2}\varepsilon - \sqrt{3}\zeta - \zeta^2, \quad \frac{d\zeta}{d\tau} = \frac{\sqrt{3}}{2}\varepsilon + \varepsilon\zeta. \qquad (3/31), (3/32)$$

The terms of zero degree have cancelled. On the circle $\varepsilon^2 + \zeta^2 = 1$ the terms of the second degree are of the same order of magnitude as those of the first degree. Much within that, say, inside the circle, $\varepsilon^2 + \zeta^2 = (1/10)^2$, the terms of the second degree become almost negligible. This circle of radius 1/10, to the interior of which the following approximation is restricted, is quite a small part of Fig. 1. In the standard manner try the assumption that, where A, B are constants,

$$\varepsilon = Ae^{\lambda\tau}, \quad \zeta = Be^{\lambda\tau}, \qquad (3/33), (3/34)$$

then

$$A\lambda = \frac{1}{2}A - \sqrt{3}\,B, \qquad (3/35)$$

and

$$B\lambda = \frac{\sqrt{3}}{2}A. \qquad (3/36)$$

The Submissiveness of Nations

These simultaneous equations for A and B are consistent if the determinant

$$\begin{vmatrix} \frac{1}{2}-\lambda \;, & -\sqrt{3} \\ \frac{\sqrt{3}}{2} \;, & -\lambda \end{vmatrix} = 0\,; \qquad (3/37)$$

that is, if

$$\lambda^2 - \frac{1}{2}\lambda + \frac{3}{2} = 0, \qquad (3/38)$$

the roots of which are

$$\lambda_1, \lambda_2 = 0{\cdot}25 \pm i\,1{\cdot}199. \qquad (3{\cdot}39)$$

Because these are complex with positive real part therefore the particle moves spirally outwards. The period of ε or ζ is $2\pi/1{\cdot}199$ on the scale of τ. But $t = \tau/k$; and k has been found from the data of 1949, 1950, 1951 very doubtfully to be $4{\cdot}_1$ years^{-1}. Therefore the period is :

$$\frac{2\pi}{1{\cdot}199 \times 4{\cdot}1} \text{ years} = 1{\cdot}_3 \text{ years}. \qquad (3/40)$$

For a political double ' swing of the pendulum ' that seems incredibly rapid. Maybe the warfinpersals for 1952 and 1953, when available, will show that k is less and the period greater. A method for making an accurate drawing of the trajectories in the neighbourhood of any point of balance is worked out in the author's *Arms and Insecurity* ((23), 1949, § 3/2/4). When the roots λ are complex, the trajectory can be obtained from an auxiliary equiangular spiral by stretching or squeezing it at right angles to a characteristic direction E. Application of those general formulæ to the present case shows that the radius vector of the equiangular spiral revolves counter-clockwise and that its length increases in the ratio 3·707 per turn, or in the ratio 1·388 per quarter turn. The characteristic direction is at $E = 105°$ from the ε-axis. The trajectory is obtained by stretching the equiangular spiral in the ratio 1·5005.

Dr. Chike Obi (17), of University College, Ibadan, Nigeria, who was formerly a pupil of Dr. Cartwright, has made an extensive study, from the point of view of pure mathematics, of the equation

$$\ddot{x} + (\alpha + \beta x)\,\dot{x} + x + \gamma x^2 = 0, \qquad (3/41)$$

where α, β, γ are constants not too far removed from zero. It is possible by a chain of transformations to connect the present hypothesis exactly to Obi's form. For this purpose we first introduce a new variable $\omega = D^2$, then shift the origin to the vortex point, and finally eliminate that deviation which is not in the S direction.

It is sometimes asserted that ' history repeats itself.' If so, we should see a closed orbit. Actually there are only spirals in Fig. 1.

The origin is what I have elsewhere (1949) named an ' in-and-out ' point. Lefschetz ((12), 1948) names it a ' col.' Cartwright calls it a ' saddle point ' or ' col.' To the right of it the drift is towards rearmament ; to the left of it the drift is towards negative armament, which I formerly suggested ((21), 1939, p. 7) was another name for co-operation. The practical difficulty was : how could the nations ever get across to the co-operative side, seeing that the flow diagram does not permit a direct crossing? Fig. 1, however, shows that by going round over the top of the vortex there is a regular way into the co-operative region. These are strange ideas. They all purport to refer to times when there was no fighting.

There is a boundary-spiral which runs right into the ' in-and-out ' point. Any slightly lesser spiral passes first through the arms-race. Any slightly greater spiral goes beyond the ' in-and-out ' point into the co-operative region.

This vortex, which certainly follows from hypothesis I, is very difficult to accept as a description of what nations in fact did. Herein lies a good example of the usefulness of mathematics in social science. The consequences of the hypothesis were not all immediately obvious. In the familiar region they seemed plausible. It was only by deducing consequences in an unfamiliar region that incredible entailments were noticed.

Let us therefore pass on to a different hypothesis.

IV. SECOND HYPOTHESIS ABOUT SUBMISSIVENESS

As in the first hypothesis, the defence-coefficient must become negative when the threat becomes overwhelming; but this change of sign will now be contrived differently. One of the reasons given ((21) 1939, p. 23) for the theory which has just been abandoned was that " the overwhelmingness of the threat depends not on y absolutely, but on y relative to x." The reconsideration is that it is not relative to the x present at that time, but to the x that could be mobilized. This 'x that could be mobilized' is almost a constant for any one nation, but varies very much from one nation to another.

Thus let the new hypothesis be

$$dx/dt = g - \alpha x + ky\{1 - y/m\}, \qquad (4/1)$$

$$dy/dt = h - \beta y + lx\,\{1 - x/n\}, \qquad (4/2)$$

where m and n are constants for those nations, and are such that m is proportional to the total warfinpersal that the x-nation could mobilize, and n is proportional to the corresponding quantity for the y-nation. Therefore

$$x \leqslant m, \qquad y \leqslant n. \qquad (4/3), (4/4)$$

As before, g and h will be neglected, because they are not essential to the phenomena of submissiveness.

Consequences of the Second Hypothesis (a) *For Very Unequal Nations.*—Suppose that $m \gg n$, then n/m, and *a fortiori* y/m, is near zero : so that $1 - y/m$ is almost unity. There is, on the contrary, no reason why x/n should not much exceed unity. In these circumstances we have approximately from (4/1) and (4/2)

$$dx/dt = -\alpha x + ky, \qquad (4/5)$$

$$dy/dt = -\beta y - lx^2/n. \qquad (4/6)$$

Therefore y is decreasing. The smaller nation is suppressed. For this case of very unequal nations there is nothing much to choose between the two hypotheses. Either of them describes what usually happens in a world of power politics.

(b) *For Nations Equal in their Constants.*—From (4/1) and (4/2) we have

$$dx/dt = -\alpha x + ky\,(1 - y/m), \qquad (4/7)$$

$$dy/dt = -\alpha y + kx\,(1 - x/m). \qquad (4/8)$$

The constant m can be made to disappear by a change of variables. Thus define u, v by

$$x = um, \qquad y = vm\,; \qquad (4/9), (4/10)$$

so that

$$u \leqslant 1, \qquad v \leqslant 1. \qquad (4/11), (4/12)$$

For then

$$du/dt = -\alpha u + kv - kv^2, \qquad (4/13)$$

$$dv/dt = -\alpha v + ku - ku^2. \qquad (4/14)$$

The two remaining constants, α and k, can be reduced to one, in the following way :

As in the former theory, define τ by

$$t = \tau/k\,; \qquad (4/15)$$

so that

$$\frac{du}{d\tau} = -\frac{\alpha}{k}u + v - v^2, \qquad (4/16)$$

$$\frac{dv}{d\tau} = -\frac{\alpha}{k}v + u - u^2. \qquad (4/17)$$

Finally, define S and D by

$$S = u + v, \qquad D = v - u, \qquad (4/18), (4/19)$$

obtaining

$$\frac{dS}{d\tau} = \left(1 - \frac{\alpha}{k}\right)S - \frac{1}{2}(S^2 + D^2), \qquad (4/20)$$

$$\frac{dD}{d\tau} = -\left(1 + \frac{\alpha}{k}\right)D + SD. \qquad (4/21)$$

Or, with $\alpha/k = \frac{1}{2}$ as in the former theory, to represent modern conditions

$$dS/d\tau = \frac{1}{2}S - \frac{1}{2}(S^2 + D^2), \qquad (4/22)$$

$$dD/d\tau = -\frac{3}{2}D + SD. \qquad (4/23)$$

The equation for $dD/d\tau$ is exactly the same as on the former hypothesis ; but that for $dS/d\tau$ has $\frac{1}{2}(S^2 + D^2)$ in place of simply D^2. The points of balance where $dS/d\tau$ and $dD/d\tau$ vanish simultaneously are given by

$$0 = S - S^2 - D^2, \qquad 0 = D\left(S - \frac{3}{2}\right). \qquad (4/24), (4/25)$$

The Submissiveness of Nations

If from (4/25) we accept $D = 0$ and substitute in (4/24) we get $0 = S - S^2$ which has solutions $S = 0$ or $S = 1$. Alternatively, if $S = 3/2$, the second condition is satisfied and the first becomes $D^2 = -3/4$, so that D is imaginary. There are therefore only two points of balance, namely,

$$S = 0,\ D = 0 \ \text{and}\ S = 1,\ D = 0. \qquad (4/26), (4/27)$$

We seem well rid of the puzzlingly unsymmetrical point of balance of the former theory. The point $S = 0$, $D = 0$ is of course of the same character as before, namely, an in-and-out point.

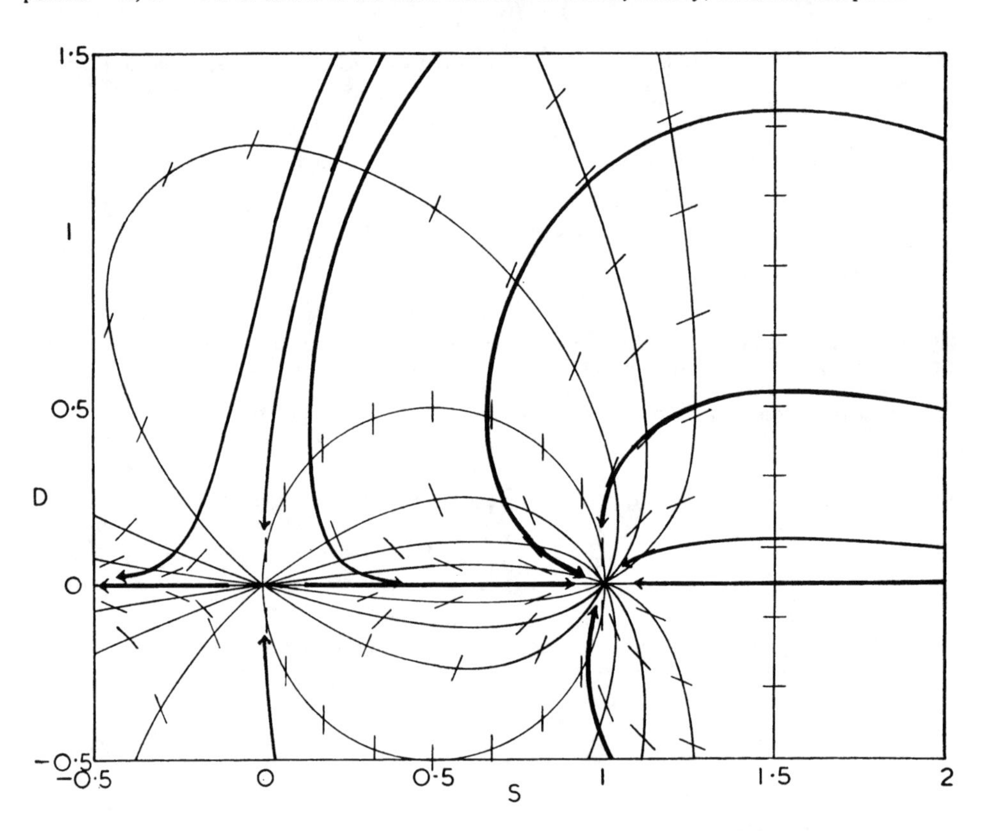

Fig. 2.—The field of two opposing sides on the second hypothesis. As in Fig. 1, S is proportional to the sum and D to the difference of their warfinpersals. The sides are equal as to their constants. The thin curves are the isoclines, along each of which dD/dS is constant. The direction of motion is indicated by a short segment. The thick curves are trajectories, along any one of which the point which represents the two warfinpersals could move.

The general field is shown in Fig. 2 by way of isoclines and trajectories. The isocline

$$\frac{dD}{dS} = j\,, \qquad (4/28)$$

where j is any constant, works out to

$$S - S^2 - D^2 = \frac{D}{j}(2S - 3)\,; \qquad (4/29)$$

so that changing the sign of j has the same effect as changing the sign of D.

In the region of the arms-race, where D tends to zero while S increases, we may neglect D in (4/22) which then simplifies to

$$\frac{dS}{d\tau} = \frac{1}{2}S - \frac{1}{2}S^2\,, \qquad (4/30)$$

in which the variables are separable. Its integra lcan be verified to be

$$\frac{1}{S} = 1 + Ce^{-\tau/2}, \tag{4/31}$$

where C is an arbitrary constant. So as $\tau \to \infty$, $S \to 1$. Let us investigate the neighbourhood of the point $S = 1$, $D = 0$ on the second hypothesis. Define ε here differently from the ε in (3/29), and thus

$$S = 1 + \varepsilon. \tag{4/32}$$

Insert this in (4/22), (4/23). Then

$$d\varepsilon/d\tau = -\frac{1}{2}\varepsilon - \frac{1}{2}(\varepsilon^2 + D^2), \tag{4/33}$$

$$dD/d\tau = -\frac{1}{2}D + \varepsilon D. \tag{4/34}$$

If we neglect the quadratic terms in ε and D, the solutions are

$$\varepsilon = Ae^{-\tau/2}, \qquad D = Be^{-\tau/2}, \qquad (4/35), (4/36)$$

where A, B are arbitrary constants. This point is therefore one of thoroughly stable balance.

The theoretical state of affairs at which we have arrived as a deduction from the second hypothesis might please some people: large armed forces, stable balance, and therefore no fighting. One of the editors of the *Cambridge Modern History* (3), Stanley Leathes, wrote in his preface to Vol. XII: "On the whole, the existence of this tremendous military equipment makes for peace. The consequences of war would be felt in every household; and statesmen, as well as nations, shrink from the thought of a conflict so immense." This was published in 1910. Unfortunately for Leathes' theory, the First World War broke out four years later. Yet this theory persists. Winston Churchill, speaking at the Pilgrims' dinner on October 14th, 1952, said that "a third world war is unlikely because both sides know that it would begin with horrors never dreamt of before."

So my second hypothesis about submissiveness appears to agree with Leathes and with Churchill, but to have the outbreak of the First World War as evidence against it. Personally I have no confidence in it, regarding it as wishful thinking.

V. THIRD HYPOTHESIS (FOR SIDES EQUAL IN THEIR CONSTANTS)

When formulating the two former hypotheses, I had hoped to have some insight into what the motives might be. Now I ignore motives, and try to make a mathematical description of the courses in time of the opposing warfinpersals before, during, and after the First World War. The facts must first be reviewed.

(i) From A.D. 1907 to 1914, the two sides being the Triple Entente and the Triple Alliance, the representative point was moving nearly in a straight line with a speed proportional to its distance from a "point of balance" at $x = £135$ million, $y = £73$ million. The graph is given by Richardson ((26), 1953). This is called an arms-race. The sterling has been converted to warfinpersal by division by £97·3 per year; that being a British wage the conversion is very rough for the other nations. Before these numbers were plotted in Fig. 3, x and y were interchanged.

(ii) There followed the First World War. Italy changed sides. The warfinpersal of Britain attained a maximum of 12·2 million equivalent persons in 1917 (Richardson (26), 1953). I have not succeeded in directly ascertaining the warfinpersals of the other belligerents, but a rough indirect estimate can be made as follows. Lloyd George ((13), *War Memoirs*, 1938, p. 1575) states the composition of the British Army in March, 1918 (excluding Colonial and Indian troops). From this statement I compute that there were 2·96 million British combatants. Now 12·2/2·96 = 4·13. It is understandable that the warfinpersal might considerably exceed the combatant strength, because the warfinpersal accounts for the non-combatants and the munition makers. Lloyd George (p. 1,577) further states that at the end of 1917 the total combatant strength of the Allies, without including any Russians or Roumanians, was 5·4 million against 5·2 million for the Central Powers. My rough estimate is made by assuming that the British factor, 4·13, applied also to the other nations concerned, so that the maximum warfinpersals were: Allies $22{\cdot}3 \times 10^6$, Central Powers $21{\cdot}5 \times 10^6$, total $43{\cdot}8 \times 10^6$ equivalent persons.

Because Germany ultimately submitted, it is relevant to enquire when the first hint of submissiveness appeared in the German state as a whole. The question is not about individuals or

The Submissiveness of Nations

groups. In another place ((22, 1948) I have given a theory of war-moods in which the people in German groups are represented by five variables; but here only one variable is available. On December 12th, 1916, the German Chancellor suggested a peace conference. The opposing Powers, after consultation, said they " refuse to consider a proposal which is empty and insincere." (Lloyd George, p. 661.) There were various other tentative and secret peace feelers in 1917 and 1918, but mostly Austrian. According to a Reichstag Committee, quoted by Lloyd George (p. 1932), " Up to 15th July, 1918, the Supreme Army Command rejected the view that victory was no longer possible of attainment by force of arms, and gave no support to peace negotiations upon the basis of a military stalemate . . ." During the next three weeks the fresh Americans came to the aid of the tired French and British. A battle on August 8th, 1918, in which the Germans had no reserves left, changed Ludendorff's opinion, and on August 14th the Kaiser, presiding over a Crown Council at Spa, admitted that Germany would have to find a suitable moment in which to come to an understanding with the enemy.

(iii) After the First World War, the sides were not quite the same. There was the Russian revolution and its consequences. U.S.A. withdrew into isolation. A new diagram was drawn showing the warfinpersal of Germany plotted against that of its entourage, consisting of France, Britain, Poland, Czechoslovakia, and Russia. This appears in a microfilm (Richardson (23), 1949) and is modified here into part of Fig. 3. It shows motion, along a nearly straight track and slowing down as a point of balance is approached. There was a delay of eleven years in the neighbourhood of this point, and then a departure, with increasing speed along another almost straight track. This was the second great arms-race beginning.

(iv) It led to the Second World War during which the warfinpersals approached the state of total mobilization, but were difficult to ascertain. The sides were rather different from those of the first war, Japan and Italy in particular having changed over previously, and Italy changing back again. In October, 1939, Hitler made a peace offer: but Britain and France did not regard it as at all submissive. After that the demand for the unconditional surrender of Germany delayed any sign of submissiveness until defeat.

(v) The Occupation Statute for Germany, an expression of submission, has now been in operation for seven years. Disarmament after the Second World War was barely over when enmity between U.S.A. and the Soviet Union began to dominate the scene.

Fitting Formulæ to the Facts.—It seems permissible to ignore the changes in the membership of the sides, so that x and y may be regarded as having the same meaning throughout. This has already been assumed in Fig. 3.

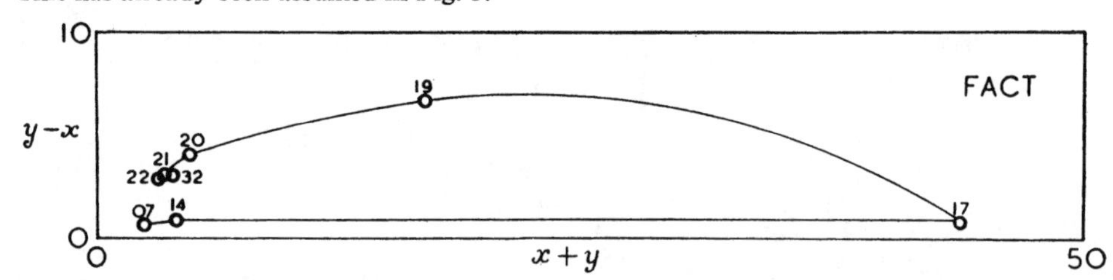

Fig. 3.—Pro- and anti-German warfinpersals in Europe, 1907 to 1932, simplified by ignoring other changes in the sides; and in particular by ignoring the brief, but balance-altering, intervention of U.S.A. The co-ordinates are the sum and difference. The numbers set beside the plotted points are the dates.

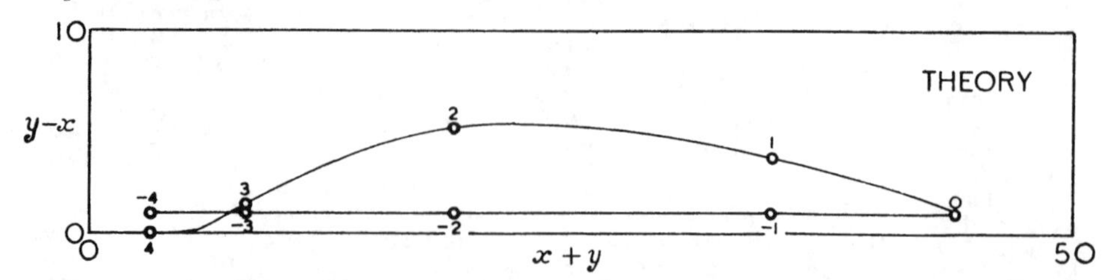

Fig. 4.—A mathematical representation, plotted from formulæ (6/17), (6/19), and (6/20) of the warfinpersals before, during, and after the First World War. For comparison with Fig. 3. Numbers set beside the plotted points are values of t. A unit of one year makes Fig. 4 agree roughly with Fig. 3 as to time.

As the mathematical description is to be of that which usually happened, we are led to beg the question "Could an arms-race end without fighting?" by assuming, on the meagre evidence of two out of three trials, that it usually does not. This assumption is in strong contrast with the first and second hypotheses, which leave that important question open at the outset.

Remembering that an arms-race has to do mainly with the sum $x + y$ and submission mainly with the difference $y - x$ of the warfinpersals, let ξ and η be defined thus:

$$\xi = x + y, \qquad \eta = y - x, \tag{6/1), (6/2}$$

so that conversely

$$x = \frac{1}{2}(\xi - \eta), \qquad y = \frac{1}{2}(\xi + \eta). \tag{6/3), (6/4}$$

Accordingly ξ and η are respectively constant multiples of the S and D of former theories. We can now describe the changes in ξ without mentioning η, thus:

(i) After wars there have been disarmaments. Therefore $d\xi/dt$ must be capable of either sign.

(ii) For small ξ it is observed that $d\xi/dt$ varied as $\pm(\xi - \xi_a)$ where ξ_a is a constant, called a point of balance.

(iii) There was an upper bound to ξ when both the opposing sides were fully mobilized for war. Let γ denote this upper bound.

The above three requirements are satisfied by the assumption that

$$\left(\frac{d\xi}{dt}\right)^2 = a^2(\xi - \xi_a)^2(\gamma - \xi), \tag{6/5}$$

where a^2 is some constant. A simpler expression could not be found; η does not appear; ξ cannot exceed γ, for if it did then $d\xi/dt$ would become imaginary; when ξ is near zero then $d\xi/dt \simeq \pm a(\xi - \xi_a)\sqrt{\gamma}$.

The variables are separable, so that (6/5) can be integrated. The result, which can be verified by differentiation, is

$$\xi - \xi_a = (\gamma - \xi_a)\operatorname{sech}^2 Bt, \tag{6/6}$$

where

$$B = \frac{a\sqrt{(\gamma - \xi_a)}}{2}. \tag{6/7}$$

At $t = 0$, $\xi = \gamma$; that is to say, the constant of integration has been chosen so that $t = 0$ at the maximum of ξ, which occurred during the war. As $t \to -\infty$, $\xi \to \xi_a$: there is not, on this theory, any sudden or definite beginning of the increase of warfinpersal. This vaguely agrees with the author's general impression of a collection of deadly quarrels that ended since A.D. 1820 (Richardson (24), 1950). The question "Who began it?" seldom had any non-controversial answer. He abandoned any attempt in that collection to name the aggressors.

To determine B from observation, let t' be the time at which ξ was halfway between γ and ξ_a. It follows from (6/6) that

$$B = \frac{0{\cdot}8814}{t'}. \tag{6/8}$$

The theoretical diagram, Fig. 4, has been drawn with

$$B = 0{\cdot}5 \text{ year}^{-1}. \tag{6/9}$$

Next as to η, the facts represented in Fig. 3 suggest that η did not alter much while ξ was increasing, but that, after the maximum of ξ, changes in η were conspicuous. These were of two kinds: at first, while ξ was large, $|\eta|$ increased; later when ξ was near ξ_b there was a settling of η to about the same value ξ_b, so that finally

$$x = \frac{1}{2}(\xi - \eta) = 0, \tag{6/10}$$

$$y = \frac{1}{2}(\xi + \eta) = \xi_b. \tag{6/11}$$

This expresses the final submission of the x-side. It would have been a mistake to have made η tend to zero, unless ξ also did. To express both kinds of change in η we need something like a disjunction or "either-or" formula. One can be concocted on the model of $\xi + 1/\xi$ which behaves like ξ when ξ is large, but like $1/\xi$ when ξ is near zero. When ξ was large, η departed further from zero without changing sign, as if $d\eta/dt$ were proportional to η. When ξ was small, η tended to ξ_b, as if $d\eta/dt$ were proportional to $\xi_b - \eta$. These diverse tendencies could be combined by weighting them respectively by ξ and $1/\xi$. That, however, would make the two terms of different

The Submissiveness of Nations

dimensions, for ξ is a number of equivalent persons. So also is γ. It is found to be satisfactory to use the dimensionless weights $2\xi/\gamma$ and $\gamma/2\xi$. Thus one arrives at the assumption that

$$\frac{d\eta}{dt} = \sigma_3 \left\{ \eta\left(\frac{2\xi}{\gamma}\right) + (\xi_b - \eta)\left(\frac{\gamma}{2\xi}\right) \right\}, \qquad (6/12)$$

in which σ_3 is some constant measure of submissiveness, and is the reciprocal of a time.

A most important fact about the submissiveness of Germany as a whole was that it came on suddenly. Sudden commencements have been much studied mathematically in connexion with seismographs, and with the switching on of electric circuits. Heaviside introduced a function $H(t)$ defined to be zero for $t < 0$, but unity for $t > 0$ (H. and B. S. Jeffreys (30), 1946, p. 219). If the onset is required to be not quite so sudden, four approximations to $H(t)$ are available (Van der Pol and Bremmer (19), 1950, pp. 56–7). I shall write

$$\frac{d\eta}{dt} = H(t)\,\sigma_3 \left\{ \eta\frac{2\xi}{\gamma} + (\xi_b - \eta)\frac{\gamma}{2\xi} \right\}. \qquad (6/13)$$

So that $d\eta/dt = 0$ for $t < 0$.

The inevitable non-linearity has been introduced, but in a manner quite different from that of hypotheses I and II.

As a discontinuity at $t = 0$ in $d\eta/dt$ has been admitted, a discontinuity also in $d^2\xi/dt^2$ does not seem too fanciful. It allows us to joint together at $t = 0$ two formulæ for ξ, both like (6/6), but the later one having b for a and C for B, so that the final point of balance ξ_b need not agree with the initial ξ_a. Observed constants, in million equivalent persons, were about

$$\gamma = 44, \qquad \xi_b = 3. \qquad (6/14), (6/15)$$

The integration of (6/13) is, however, much simplified if we take $\xi_b = 0$, for then η is a factor of the second member, so that for $t > 0$

$$\frac{d \log|\eta|}{dt} = \sigma_3 \left\{ \frac{2\xi}{\gamma} - \frac{\gamma}{2\xi} \right\}. \qquad (6/16)$$

This simplified problem is solved as a preliminary. Substitute, in view of (6/6) and (6/9),

$$\xi = \gamma \operatorname{sech}^2 \frac{t}{2}, \qquad (6/17)$$

obtaining from (6/16)

$$\frac{d \log|\eta|}{dt} = \sigma_3 \left\{ 2 \operatorname{sech}^2 \frac{t}{2} - \frac{1}{2}\cosh^2 \frac{t}{2} \right\}. \qquad (6/18)$$

The integral is

$$\frac{\eta}{\eta_0} = \operatorname{Exp}\left[\sigma_3 \left\{ 4 \tanh \frac{t}{2} - \frac{1}{4}\sinh t - \frac{1}{4} t \right\} \right], \qquad (6/19)$$

which can be verified by differentiation. The measure σ_3 of submissiveness was, of course, unknown; but Fig. 4 was plotted with

$$\sigma_3 = 1 \qquad (6/20)$$

and comparison of Fig. 4 with Fig. 3 shows that $\sigma_3 = 1$ is of the right order of magnitude. The complete eqn. (6/13), with ξ_b not zero, has for an integrating factor the expression (6/19) (Forsyth (7), 1929, p. 20) and is thus reducible to quadratures. But these look difficult and are perhaps not worth pursuing at this stage.

VI. CONCLUSION

The memory of recent overwhelming military defeat, say, within five years, certainly leaves a tendency to submit. A large state has been known to overawe a very much smaller state, without any fighting. Apart from those two types, I have looked for submissive effects, but have not noticed any. Two hypotheses as to how submissive effects might arise between rested and approximately equal states, called A and B, are considered. Mathematical hypotheses like these are too complicated to be expressed, both briefly and accurately, in words. However, one can say that the description of an arms-race, as caused by mutual stimulation, is modified by supposing either (i) that the overwhelming nature of the threat against A is proportional to the excess of B's armed force over A's at the same time, or (ii) that it is not A's armed force at the same time that matters, but what A could mobilize. Not all the consequences of these hypotheses are immediately obvious. In historically known circumstances the hypotheses seemed plausible. It was only by deducing their consequences in unusual circumstances that incredible entailments were noticed. This is a good example of the usefulness of mathematics in social science.

Most governments seem to expect that a show of armed force, to a rested and equal group of nations, will have an effect of the same sign as an overwhelming military defeat. This expectation appears to be a mistake : the actual effects go opposite ways.

The Quakers, Gandhi, and others have tried to persuade everybody to be as non-violent as the exceptional Christian saints. Most people feel the suggestion to be too utterly contrary to their defensive impulses. Could not a middle way be found, with the motto " defence without provocation." This motto might well be applied to words as well as deeds. The existence of the various arms of nation A presumably have different psychological effects in peacetime on nation B ; for example, one may guess that B is much more provoked to retaliate by the existence of A's long-range bombers than by A's land-based fighters. Jonathan Griffin (8) discussed ' defence without menace ' in 1936, and now again the problem of the classification of armaments into provocative and non-provocative requires much more attention than it receives.

Submission, except in the sense of defeat, is not a spectacular subject for newspaper headlines ; it is not like fighting for freedom. Yet we need to remember that in any organization, whether it be a social club, a scientific society, a business, a national parliament, or the United Nations, a suitable amount and distribution of submissiveness is essential for smooth working. If each member struggled all the time for his own advantage there would be chaos.

The present study has perhaps over-emphasized the bitter submission of the defeated, threatened, suppressed, or snubbed, and has neglected the joyous devotion to an admired leader. To correct the balance let me instance a classical story illustrating the latter tendency, that of Naomi and Ruth. It is described as an affair of tribal loyalties.

In the United Nations, activities of the follow-my-leader type are similarly noticeable on both sides of the major schism.

REFERENCES

1. Allport, G. W. (1945). *Personality*. New York : Henry Holt.
2. Bateson, G. (1936). *Naven*. Cambridge : University Press.
3. *Cambridge Modern History*, XII (1910). Cambridge : University Press.
4. Cartwright, M. L. (1949). Report on Non-Linear Vibrations. *The Advancement of Science*, **6**.
5. Cartwright, M. L. (1952). *Non-Linear Vibrations. Math. Gazette*, **36**, 81–8.
6. Cartwright, M. L. (*no date*). *Forced oscillations in non-linear systems*. (Lectures at Professor Lefschetz's seminar.)
7. Forsyth, A. R. (1929). *A Treatise on Differential Equations*. London : Macmillan.
8. Griffin, J. (1936). *Alternative to Rearmament*. London : Macmillan.
9. Horne, M. R. (1951). Letter in *Nature*, November 24th.
10. *Keesing's Contemporary Archives*. Bristol : Keesing's Publications : cited as ' K.'
11. Landau, H. G. (1951). *Bull. Math. Biophysics*, 13, 1–19.
12. Lefschetz, S. (1948). *Lectures on Differential Equations*. Princeton : University Press.
13. Lloyd George, D. (1938). *War Memoirs*. London : Odhams.
14. Lorenz, K. Z. (1952). *King Solomon's Ring*. London : Methuen.
15. McDougall, W. (1917). *Social Psychology*. London : Methuen.
16. Moyal, J. E., Bartlett, M. S., and Kendall, D. G. (1949). ' Symposium on Stochastic Processes.' *J. Roy. Stat. Soc.*, B., II, 150–282.
17. Obi, C. (1950). (Unpublished.) Thesis for Ph.D. : Cambridge University.
18. Pol, B. van der (1926). *Phil. Mag.*, II, 978–92.

The Submissiveness of Nations

19. Pol, B. van der, and Bremmer, H. (1950). *Operational Calculus based on the Two-sided Laplace Integral.* Cambridge: University Press.
20. Richardson, L. F. (1935). *Nature,* **135**, 830.
21. Richardson, L. F. (1939). 'Generalized Foreign Politics.' *Brit. J. Psychol. Mon. Sup.*, No. 23.
22. Richardson, L. F. (1948). *Psychometrika,* **13**, 147–74 and 197–232.
23. Richardson, L. F. (1949). *Arms and Insecurity* (microfilm, published by the author).
24. Richardson, L. F. (1950). *Statistics of Deadly Quarrels* (microfilm, published by the author).
25. Richardson, L. F. (1951). *Nature,* **168**, 567.
26. Richardson, L. F. (1953). *The Indian Journal of Statistics,* **12,** 205–28.
27. Richardson, L. F., and Stommel, H. (1948). *Journ. Meteor.*, 5, 238–40.
28. Russell, Bertrand (1914). *War the Offspring of Fear.* London: Union of Democratic Control.
29. Volterra, V. (1930). *Theory of Functionals.* London: Blackie.
30. Jeffreys, H. and B. S. (1946). *Methods of Mathematical Physics.* Cambridge: University Press.
31. Weierstrass, K. (1875). *K. Akad. der Wiss.* (October 28th). Also *Werke,* II, 75–6.

Richardson, L. F. (1953)
Sankhyā **12**, 205–28

1953:3

THREE ARMS-RACES AND TWO DISARMAMENTS

Sankhyā: The Indian Journal of Statistics
Calcutta

Vol. 12, Part 3, 1953, pp. 205–228

Note: This paper was described by Richardson as 'for economists'

Reprinted from Sankhyā : The Indian Journal of Statistics, Vol. 12, Part 3, 1953.

THREE ARMS-RACES AND TWO DISARMAMENTS

By LEWIS F. RICHARDSON
Hillside House, Kilmun, Argyll, Britain

CONTENTS

1. Introduction

The warlike expenditures for the first two great arms-races and for the intervening disarmament have already been analysed by the present author (Richardson 1939 and 1947); early results were published in 1939 under psychological patronage, where economists would not be likely to notice them; in 1947 and 1949 a revised account was offered in a microfilm which few people of any sort have seen. The present paper is intended for professional economists, and is long overdue.

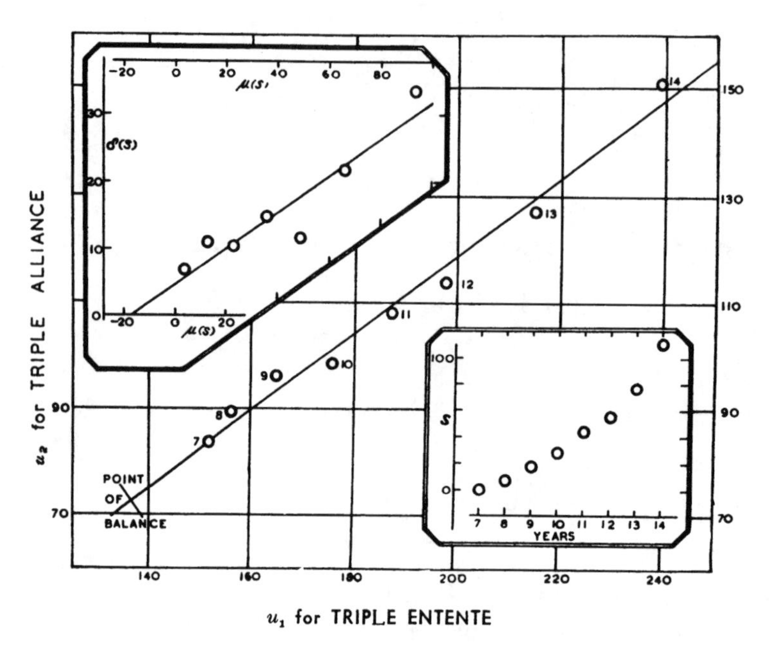

Fig. 1. The European arms-race that preceded the First World War.

Figure 1, with its main diagram and two insets, summarizes and analyses the warlike expenditures of the first great arms-race. The competition was between the Triple Entente consisting of France, Britain and Russia, and the Triple Alliance consisting of Germany, Austro-Hungary and Italy. Of these six countries, France, Britain, Germany, and Italy had important dependencies overseas. The warlike expenditures of the two sides were expressed in million sterling and were denoted by u_1 and u_2. In the main diagram u_2 is plotted against u_1. The numbers set besides the dots are the date of the year minus 1900; more precisely the dot is for the twelve months centered at the beginning of that year. It is seen that the dots are tolerably well fitted by the sloping straight line. Distance along this line from the point marked 7 was measured and is denoted by s. In the lower inset, s is plotted against the date; and the resulting row of dots is conspicuously curved: the more the nations spent, the faster did they increase their expenditure. The upper inset shows a more sensitive analysis of this acceleration: the abscissa $\mu(s)$ is the mean of two consecutive s, while the ordinate $\delta(s)$ is the difference between the same two arithmetical values of s; in in this upper inset it is the slope, not the curvature, of the diagram which is of chief interest; and the sloping straight line when produced backwards until $\delta(s)=0$, arrives there at $\mu(s)=-18$. This point is called the "point of balance" and has been marked by a cross on the main diagram. The suggestion is that if the national warlike expenditures had ever been at the point of balance they might have stayed there.

THREE ARMS-RACES AND TWO DISARMAMENTS

Figure 1 represents a factual analysis; yet it was a theory which moved the author to collect the facts and to analyse them in that manner.

The study of a phenomenon, so inclusive as an arms-race, spreads across professional boundaries; for arms-races are ordinarily discussed by chiefs of armed forces, journalists, statesmen, financiers, economists, philosophers, psychologists, and historians. It was a philosopher and mathematician, Bertrand Russell, who in 1914 asserted that the main motives of the arms-race of 1908 to 1914 had been mutual stimulation by fear. My own contribution to the discussion began in 1919 with the translation of Russell's thesis into a pair of simultaneous differential equations. In a revised notation they can be stated as

$$dX/dt = kY, \qquad dY/dt = lX, \qquad \ldots \quad (1.1), (1.2)$$

in which t is the time, k and l, are positive "defence-constants" and X and Y are variables, which were at first rather vaguely described as the "warlike activity" of the opposing sides. That was a crude beginning; many theoretical elaborations were introduced in 1919 and later in 1935, 1939, and 1947. Nevertheless this simple pair of differential equations suggested a method for the analysis of economic facts, a method which is still useful. For if the opposing sides are equal in the sense that $k=l$, we have by adding equations (1.1) and (1.2)

$$d(X+Y)/dt = k(X+Y) \qquad \ldots \quad (1.3)$$

the integral of which is

$$X+Y = A\, e^{kt} \qquad \ldots \quad (1.4)$$

where A is a constant of integration. The above formulae are only preliminary suggestions; they will be amended in §6.

At the end of a war, it may be supposed that the victorious nations, no longer feeling threatened by the enemies whom they have recently defeated, would be moved to cut down the cost of purposeless armed forces. The simplest representation of this motive is

$$dX/dt = -\alpha X, \qquad \ldots \quad (1.5)$$

where α is a positive "restraint coefficient", formerly named a "fatigue and expense coefficient". The integral of (1.5) is

$$X = B\, e^{-\alpha t} \qquad \ldots \quad (1.6)$$

where B is a constant of integration. The formulae (1.4) and (1.6) resemble one another by each involving a single exponential time-factor.

The study of any very complicated system is likely to be clarified by attending to temporary deviations from its more usual condition. For example the temperature of a fevered patient is advantageously reckoned, not from the absolute zero of the thermodynamic scale, but from the temperature of the same person in health.

Similarly an arms-race is not likely to be the only reason for maintaining armed forces, the others include custom, internal order, and miscellaneous external risks. There is also a background of international trade and other forms of co-operation which modifies the amount that any nation decides to spend on warlike preparations. Again a nation may have a grievance against the treaty-situation, or may be eager for expansion by conquest. All these may affect the expenditure on warlike preparations, and they are too complicated to be analysed in detail. The simplifying assumption made throughout this article is that arms-races and disarmaments have lasted about 3 to 10 years, and that, during such an interval, all the other complicating relationships may be regarded as passably constant. To put this assumption into symbols let the small letters x and y denote what we can estimate from "defence" expenditure and wages in the manner to be described in § 2. For simplicity it will be supposed that

$$x = c_1+X, \quad y = c_2+Y, \qquad \dots \quad (1.7), (1.8)$$

where c_1, c_2 do not vary during any single arms-race or disarmament. The numerical values of c_1 and c_2, whether positive or negative, can be inferred from the observed points of balance.

The reader is certainly not expected to believe, on theoretical grounds, that the above formulae *must* describe arms-races and disarmaments; but he is asked to admit that it is reasonable to enquire whether the various runs of warlike expenditure say $f(t)$ can be fitted with the generalized formula

$$f(t) = Q+Ae^{\lambda t} \qquad \dots \quad (1.9)$$

where Q, A, λ are constants to be determined. The derivative of (1.9) is

$$df/dt = \lambda(f-Q) \qquad \dots \quad (1.10)$$

So, if we could make a diagram having f and df/dt as rectangular co-ordinates, the graph would be expected to be a straight line of slope λ cutting the axis on which $df/dt = 0$ at the point

$$f = Q \qquad \dots \quad (1.11)$$

This point will be called the "point of balance".

The interpretation of λ as a motive will be quite different during an arms-race, from its interpretation during a disarmament.

The formula (1.9) is known to actuaries as Makeham's law for the "force of mortality" (Spurgeon 1929); but the actuarial context is so unlike the present application that an independent treatment seems desirable.

Before arms-races and disarmaments can be properly discussed, two general preliminaries need attention; in §2 the appropriate standard of economic value; and in §3 the effect of being given only annual totals of expenditure, although daily rates would be more comparable with differential equations.

THREE ARMS-RACES AND TWO DISARMAMENTS

2. Standards of economic value for the discussion of arms-races

What is required is some way of comparing the warlike efforts of different nations at different times. The difficulty of finding any absolute standard is mitigated by the low accuracy required: one per cent may suffice from year to year, and ten per cent from country to country.

During the first great arms-race of A.D. 1906 to 1914 the gold standard was in operation, and wages and prices in gold were fairly steady in time; so that a discussion in terms of gold was meaningful.

During the second great arms-race of A.D. 1930 to 1939 the problem of a standard of value became acute, because of the many devaluations of currency. Expecially confusing was the devaluation of the rouble in the ratio 4.4 during the winter of 1935 to 36. The warlike expenditure of U.S.S.R. when expressed in gold, fell at that time in a manner which had no obvious connection with any event in the warlike environment of that country. The expenditure was therefore smoothed by dividing it by wages in roubles (Richardson 1939 p.79). The same procedure, namely the division of national expenditure by wages expressed in the same money, was afterwards applied to a private-enterprize country (U.S.A.) and to a gold-mining country (South Africa), and for both it made the defence-expenditure slightly more understandable when compared with the political environment. Diagrams and details may be seen in a microfilm (Richardson 1947 and 1949 § 3.11). A similar procedure had previously been recommended for more general theoretical purposes by a famous economist who called it the "labour unit" (J.M. Keynes 1936 p.41) Various names have been used. In the aforesaid microfilm the national warlike expenditure, after division by wages, was at first called the "warlike worktime" and later the "warfinpersal" a contraction for war-finance-per-salary. *In the Economic Survey of Europe in 1950* the figures of defence expenditure have, in table 67, been divided by the average wage of industrial workers, and then stated as "man-years".

There has been some unnecessary complication about units of time. My "warlike worktime per year" could be simplified. Similarly, in the Economic Survey's "man-years per year" the unit of time cancels, leaving simply "men", or more precisely persons of the specified type, alias "equivalent" persons.

None of these usages, neither Keynes' labour-unit, nor my warfinpersal, nor the Economic Survey's man-years, involves the false doctrine that all men are equal; for one field-marshal would be expressed as many typical persons. The standard type of person should, I think, be taken in the middle of the range of ability, by avoiding unskilled labourers at one end, and professional or managerial people at the other. When particulars were available, I took adult male semi-skilled engineers. But one has to make-do with such information as can be found. Indices of wage-rates can be used, but then the ratio is not in persons.

The warfinpersal of any country usually far exceeds the number of people in its fighting forces, because warfinpersal includes also munition makers. The Economic Commission has introduced a refinement in so far as by them "an attempt has been made to include in wage-costs social benefits such as family allowances, paid holidays and social insurance the cost of which is borne by the employer...". They intend, presumably, to allude only to such social benefits as are recovered by the employer from the warlike government departments by increasing the price of his contracts. For example in Britain it would not improve warfinpersal to include family allowances paid by the Post Office.

3. CONTINUOUS CHANGE, BUT ANNUAL MEANS

In the psychological analysis of § 1 the motives are regarded as changing continuously; but the warlike expenditure is usually published only as totals over consecutive years. This contrast brings in some adjustments, which are negligible when the changes are slow, but become important when they are rapid. All theories should, if possible, be rather more definite than the facts which they are intended to interpret.

The mean of a ratio

Warfinpersal is a ratio; and in general the annual mean of a ratio is not equal to the ratio of the annual means of its terms. An approximate correction can be made by the formula

$$\overline{(u/v)}=\bar{u}/\bar{v}\left\{1+\frac{1}{12}\frac{dv}{vdt}\left(\frac{dv}{vdt}-\frac{du}{udt}\right)\right\} \qquad \ldots \quad (3.1)$$

where u and v are any quantities such that du/udt and dv/vdt are small constants, where a bar denotes an annual mean, and the unit of t is one year. Usually du/udt has to be approximated as $(u_1-u_{-1})/2u_0$ where u_{-1}, u_0, u_1 are consecutive annual totals. Application of formula (3.1) showed that the distinction between $\overline{(u/v)}$ and $\bar{u}/\bar{v})$ is usually almost negligible. A specimen is shown in Table 3.

The interpretation of a straight line when successive differences are plotted against successive means

The peculiarities of this diagram appear not to be well-known.

The preliminary discussion of motives in §1 suggested that it would be reasonable, if we were given the warfinpersal f as a function of time t, to make a diagram having f and df/dt as rectangular co-ordinates, and to compare it with equation (1.10) The upper inset of Figure 1 was crudely of this type, but not accurately so, because the data were not f but the annual means of f. To keep what physicists call the "dimensions" in good order, it is best to have a symbol for the time interval over which the average is taken. Let it be τ. The data are then of $\phi(t)$ defined by

$$\phi(t)=\frac{1}{\tau}\int_{t-\frac{1}{2}\tau}^{t+\frac{1}{2}\tau} f dt \qquad \ldots \quad (3.2)$$

in which t is the time at the *middle* of the interval τ. On insertion of the formula (1.9) for f, and integration, (3.2) gives

$$\phi(t)=Q+Ae^{\lambda t}\frac{2}{\lambda\tau}\sinh\frac{\lambda\tau}{2}. \qquad \ldots \quad (3.3)$$

THREE ARMS-RACES AND TWO DISARMAMENTS

As we cannot plot df/dt against f, the next best operation is to plot $\delta(\phi)$ against $\mu(\phi)$, where δ and μ are W. F. Sheppard's central difference operators, defined by

$$\delta(\phi(t)) = \phi(t+\tfrac{1}{2}\tau)-\phi(t-\tfrac{1}{2}\tau), \qquad \ldots \quad (3.4)$$

$$\mu(\phi(t)) = \tfrac{1}{2}\{\phi(t+\tfrac{1}{2}\tau)+\phi(t-\tfrac{1}{2}\tau)\} \qquad \ldots \quad (3.5)$$

But when ϕ has the form (3.3) the expressions work out to

$$\delta(\phi) = A\, e^{\lambda t}\left(\frac{2}{\lambda\tau}\sinh\frac{\lambda\tau}{2}\right)\left(2\sinh\frac{\lambda\tau}{2}\right), \qquad \ldots \quad (3.6)$$

$$\mu(\phi) = Q+Ae^{\lambda t}\left(\frac{2}{\lambda\tau}\sinh\frac{\lambda\tau}{2}\right)\cosh\frac{\lambda\tau}{2}. \qquad \ldots \quad (3.7)$$

The graph having $\mu(\phi)$ for abscissa and $\delta(\phi)$ for ordinate has already appeared in Figure 1, upper inset, and will be used repeatedly in the sequel. For brevity it will be named the (μ,δ) diagram. From equations (3.6) and (3.7) it follows that

$$\delta(\phi)=\{\mu(\phi)-Q\}2\tanh\frac{\lambda\tau}{2} \qquad \ldots \quad (3.8)$$

That is to say, if f is of the form $Q+Ae^{\lambda t}$, then (μ,δ) diagram is a straight line, and its slope b is given by

$$b=2\tanh\frac{\lambda\tau}{2}. \qquad \ldots \quad (3.9)$$

from which λ can be computed.

The least number of consecutive values of ϕ that suffice to determine b, is three; and if we have only this minimum information, say ϕ_0, ϕ_1, ϕ_2, then the temporary slope is

$$b = \frac{2(\phi_2-2\phi_1+\phi_0)}{\phi_2-\phi_0} = 2\tanh\frac{\lambda\tau}{2}, \qquad \ldots \quad (3.10)$$

from which λ is so easily computed, that the (μ,δ) diagram need not be drawn.

Conversely, if the (μ, δ) locus is observed to be straight, then the temporary slope is the same at all those times, and therefore λ is so far constant.

Equation (3.9) shows that if b is nearly equal to zero, then b is almost equal to $\lambda\tau$. At other extremes there is a startling discrepancy, for when $\lambda\tau\to\pm\infty$, $b\to\pm 2$. Slopes outside the range $-2\leqslant b\leqslant 2$ are therefore impossible, if the facts agree with equation (1.10). During rapid disarmament b and $\lambda\tau$ have been far enough below zero for the distinction between them to be important.

The straight line represented by (3.8) cuts the axis where $\delta(\phi)=0$ at the point

$$\mu(\phi) = Q. \qquad \ldots \quad (3.11)$$

This coincides exactly with the point of balance (1.11), which would be obtained if we could plot df/dt against f.

A peculiarity of the continuous diagram of df/dt plotted against f may be mentioned here, although it has nothing to do with annual means. The curve cannot cross the axis where $df/dt=0$ except at right angles. At any other angle there is an infinite delay on that axis.

VOL. 12] SANKHYĀ : THE INDIAN JOURNAL OF STATISTICS [PART 3

4. THE ARMS-RACE OF A.D. 1907 TO 1914

This has already been summarized in Figure 1.

The sources of information

It is of course desirable to compare the information published by the opposing sides to this, or any, quarrel. The warlike budgets were therefore extracted both from the Statesman's Year Book (S.Y.B.), published in Britain, and from the Almanack de Gotha (A. de G.), published in Germany. They were compared also with a digest published by Per Jacobsson, an official of the International Bank which might have been used exclusively, had it covered all the years and countries required. There were various minor discrepancies or doubts. In a method which depends on the annual increment $\delta(\phi)$, it does not much matter whether pensions are included or excluded, provided that the same decision is made for all the years. The numbers from which Figure 1 was drawn were actually obtained in the following manner. For Italy A. de G. agreed with S.Y.B. and they were accepted. Similarly for Britain, except for an obvious misprint. For France and Russia the mean of A. de G. and S.Y.B. was accepted. For Germany and Austro-Hungary the budgets for the last year are not to be found in S.Y.B.; and so, to avoid a break in the method, it was necessary to take the budgets from A. de G. alone. In order to make the interval end in the crisis that arose from the murder of the Austrian Archduke, the budgets were interpolated to years ending with June 30th. Linear interpolation was considered to be good enough. The conversion to sterling was made by the, then steady, exchange-rates: £ 1=25.225 francs=25.225 lire=20.43 marks=24 krone=9.46 roubles.

In the (μ, δ) diagram (Figure 1 upper inset) the sloping line is such as to minimize the sum of the squares of its distances to the seven observed points, when those distances are measured parallel to the $\delta(s)$ axis. The equation to that line is

$$\delta(s) = 0.2704\,\{\mu(s)+17.86\}. \qquad \ldots \quad (4.1)$$

The question as to the reliability of the numbers in (4.1) leads right into the contoversy as to whether probability is a frequency or a degree of belief. Let me say that I find sometimes one and sometimes the other view appropriate; and that Jeffreys' theory of reasonable belief (1939), including Bayes's theorem, seems to me reasonable and useful. The events summarized in equation (4.1) happened only once, so that they had no frequency. On the other hand the interpretation of those events is somewhat a matter of opinion: as to whether 6 nations or 4 were involved, and as to what items of expenditure should be included; so that there could be a frequency distribution of interpretative opinion. That some of the events might easily have happened rather differently is suggested in the statement by the British Foreign Secretary, Sir Edward Grey (1925 p. 199) that "The most acute crisis in the Liberal Government came over naval expenditure in 1909. Were we to be committed to the construction of eight new battleships, or would six, or even four, be enough for national safety? For some days there was a Cabinet crisis." In that year the actual British warlike budget, including pensions, was £ 63 millions. The cost of battleships in 1913 is mentioned by Winston Churchill (1923 p. 132) as $2\frac{1}{4}$ millions each. The extreme range of the dispute was therefore one seventh of the warlike budget.

Although the events happened only once, the decisions to make them so, and the interpretation of the record, were both somewhat uncertain. These are justifications for applying a frequency theory. In such a theory, described by R. A. Fisher (1936 pp.137 to 140) the regression coefficient, $b=0.2704$ year^{-1}, is regarded as an estimate, derived from a sample of 7 points of a "true" regression coefficient, β, derived from infinitely many points distributed in the Gaussian normal manner as regards the ordinate $\delta(s)$. This procedure leads, via Gossett's theory of small samples, to the statement that β would lie in the range $0.15<\beta<0.39$ with a probability, in Jeffreys' sense, of 0.95.

The correction for finite differences is in this case negligible; for if $b=0.270$ then (3.9) gives $\lambda\tau=0.272$. In further calculations let us take

$$\lambda = 0.2_7 \text{ year}^{-1} \qquad \dots \quad (4.2)$$

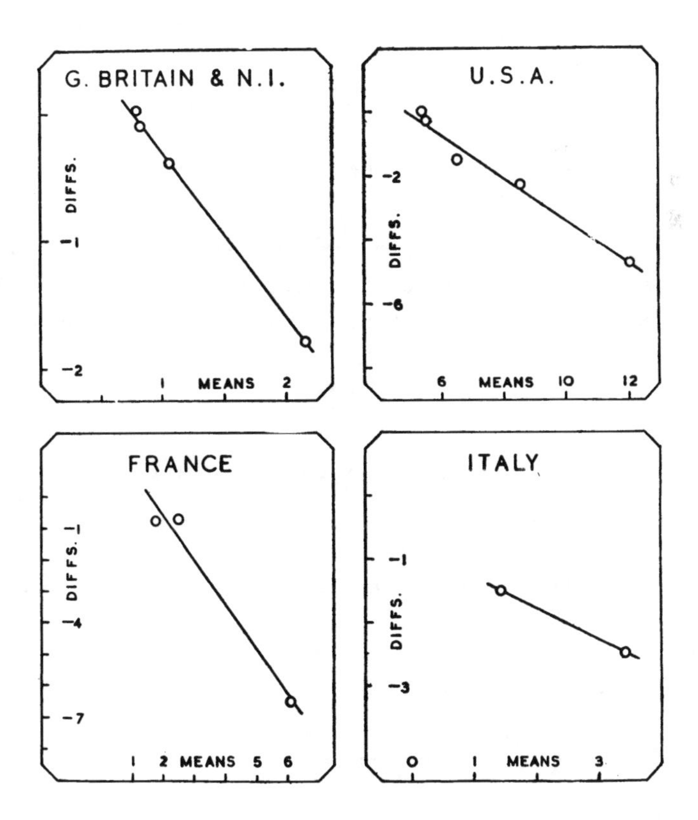

Fig. 2. Disarmament following after the First World War. The motion was from lower right to upper left.

5. Disarmament of victors, A.D. 1919 to 1952

Figure 2 shows the (μ, δ) diagrams. They relate to warfinpersal, except for Italy where an index of wholesale prices had to be used instead of earnings. The sources of information, and the slopes of the diagrams, are summarized in Table 1.

TABLE 1.

	reference for warlike expenditure	reference for annual earnings or the like	b year^{-1}	λ year^{-1}
Great Britain and N. Ireland	R. G. Hawtry (E 10, 695)	E 23, 273	-1.29_2	-1.54
U. S. A.	A. W. Mellon (E 22, 743) Statesmen Year Book	J. R. Commons (E 23, 274)	-0.667	-0.69
France	R. M. Haig (1929)	Lucien March (1925) J. R. Cahill (E 9, 610)	-1.41	-1.8
Italy	C. E. McGuire (1927)	Wholesale price index (McGuire)	-0.49	-0.50

In selecting the data a number of minor decisions had to be made. The question was how much warlike effort the nation was making; the source of the money was irrelevant. In the French budgets the distinction of expenditure as ordinary, extraordinary, or recoverable, was therefore ignored and the total was taken. The wages, if given for years different from the governmental financial year, were interpolated thereto linearly.

Because the threat from the Germanic alliance had ended, each nation was disarming to save expense, and therefore $-\lambda$ can be regarded as the positive "restraint coefficient" α, already mentioned in equations (1.5) and (1.6). The mean for these four victors was

$$\alpha = 1.1_3 \text{ year}^{-1}. \qquad \ldots \quad (5.1)$$

6. The interaction of conflicting motives

In many parts of physics the principle of the "super-position of effects" has been found to agree with observation. It is a consequence of the linearity of the differential equations which describe those phenomena. In the present context the principle would suggest that the restraint-coefficient α, which is conspicuous during disarmament, coexists during the arms-race with the defence-coefficient k. If for simplicity we consider two equal sides, the formulation of these coexisting motives would be

$$dX/dt = kY - \alpha X, \quad dY/dt = kX - \alpha Y, \qquad \ldots \quad (6.1), (6.2)$$

and the reason why k did not effect the disarmament after the First World War was that the threat Y from the defeated side was zero, so that equation (6.1) simplified to $dX/dt = -\alpha X$, which is (1.5).

THREE ARMS-RACES AND TWO DISARMAMENTS

On several occasions I have had passing qualms about this coexistence of α with k because in human beings opposing actions often cannot coexist, but either of them can inhibit the other. For example breathing and swallowing cannot both go on at the same instant. It is conceivable that a nation might be either parsimonious or defensive, but not both at the same time. Formulae for the description of such alternative inhibition can be constructed. We might suppose that α and k were variable in time, that they had a fixed total $\alpha+k=c$, but that they fled rapidly from the middle position $\alpha=\frac{1}{2}c=k$, and would not stay put, except at the extreme $\alpha=c$, $k=0$, or at the opposite extreme $\alpha=0$, $k=c$. A very simple formulation of such alternative inhibition is

$$\frac{d\alpha}{dt}=\frac{\alpha k}{\alpha-k}, \quad \frac{dk}{dt}=\frac{k\alpha}{k-\alpha} \qquad \text{... (6.3) (6.4)}$$

Different formulations are also available (Richardson 1948 Part II pp. 211-212). But anyway parliamentary debates show that the above hypothesis is too extreme, for in Britain parsimony does to some extent coexist with defensiveness. The North Atlantic Treaty Council on 1952. ii.23 proclaimed both motives as its own. The hypothesis should therefore be moderated; and this can easily be done by a modification of (6.3) and (6.4), whereby α is replaced by $\alpha-\alpha_1$, and k by $k-k_1$, in which α_1 and k_1 are constants. Any inhibitory relations would need to be linked with the main relations (6.1) and (6.2); and the details of the linkage are not obvious. The resulting system of equations would be non-linear, and might be troublesome.

A cruder but simpler hypothesis would be to suppose that α had a smaller numerical value during an arms-race than during a disarmament; and I think that this possibility should be borne in mind. On the evidence so far available I reject it as an unnecessary complication; because the explanation by constant k and α is quantatively fairly satisfactory and will be accepted here.

The sum and difference of (6.1) and (6.2) give

$$d(X+Y)/dt=(k-\alpha)(X+Y), \qquad \text{... (6.5)}$$

$$d(X-Y)/dt=-(k+\alpha)(X-Y), \qquad \text{... (6.6)}$$

in which $X+Y$ and $X-Y$ are called 'normal co-ordinates'. The (μ, δ) diagram in Figure 1 corresponds, nearly, to a study of $X+Y$; and the $\lambda=0.27$ year^{-1}, obtained in (4.2) from this study is to be interpreted by comparison with (6.5) to indicate that

$$k-\alpha=0.27 \text{ year}^{-1}. \qquad \text{... (6.7)}$$

If we could make a similar study of $X-Y$ in connection with equation (6.6) we could determine $k+\alpha$ during the arms-race. The difficulty of such an analysis is that $X-Y$ tends to zero so rapidly that the process escapes notice. On taking α from (5.1) and $k-\alpha$ from (6.7) it follows that

$$k=1.4 \text{ year}^{-1} \qquad \text{... (6.8)}$$

$$k+\alpha=2.5_3 \text{ year}^{-1}. \qquad \text{... (6.9)}$$

An evanescence as rapid as that expressed by the substitution of (6.9) into (6.6) would certainly be difficult to observe, the data being annual means. Thus the linear equations (6.1) and (6.2) appear to be a respectable summary of the observations. There were *two* balances of power: that for $X-Y$ was rapidly stable, that for $X+Y$ was slowly unstable.

7. One nation during war and peace

All the other sections of this paper relate to times when ther was no major fighting. Naturally the reader will wish to see how the straight (μ,δ) diagram for the arms-race joins on to the straight, but differently sloping, (μ, δ) diagram for disarmament. Unfortunately it is difficult to collect the information; I have suceeded in doing so only for one nation, namely the United Kingdom. A nation when engaged in war is not, in fact, sovereign and independent; so that to disuss it alone is hardly excusable. However the United Kingdom was peculiar in having connections scattered over the world, which made it sensitive to events in many places. It had also a strict system of keeping and publishing acounts.

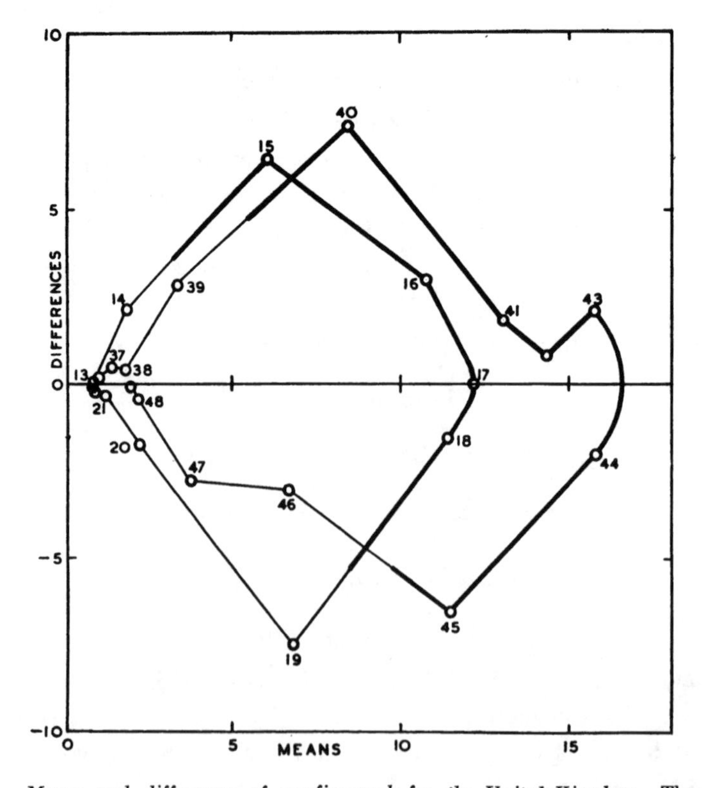

Fig. 3. *Means and differences of warfinpersal for the United Kingdom.* The numbers beside the points are the dates less 1900 : actually the date is that common to th two financial years in question; so that the points are at March 31st, apart from errors of finite differences. The line is drawn thicker when there was fighting in a world war.

THREE ARMS-RACES AND TWO DISARMAMENTS

In Figure 3 we see:—

(i) Two turns of a peculiar spiral, one for each world war. The earlier turn is remarkably regular and approximates to a rhomboid. I have not seen a spiral like it in any other connexion. In comparison with the first turn, the second shows distortions about a year after the beginning and end of Lend-lease from U.S.A.

(ii) The speed with which the point moved around the spiral can be seen from the separation of the annual points. It is fastest at the top and bottom, rather slower on the right, and much slower on the left, where between the first and second world wars there was a pause of ten years. All the rest of the diagram looks deterministic, but the long pause looks indeterminate, like the unstable phase of a relaxation oscillation (Pol, Balth van der. 1926).

The linear theories in other sections of this paper relate to the arms-race and the disarmament, which in Figure 3 are the almost straight portions on the left. Obviously those theories do not apply to the right of the sharp bends at the top and bottom. For the spiral as a whole I have not at present any formula to offer. The difficulties of fitting one are the sharp bends; although no doubt they appear a little sharper than they would do, if data were available at closer intervals. Suggestions, which might be considered, include (a) Volterra's (1931) theory of fish eating fish, (b) Richardson's theory of submissiveness (1939, 1949, 1951), which was altered in 1953, (c) Richardson's theory of war-moods (1948), (d) Rashevsky's theory (1947).

Sources and details for Figure 3. The United Kindom included Ireland until March 1923, after that only Northern Ireland.

Although opinions, as to what expenditure should be included, may differ by as much as ten per cent; yet, if a consistent method is followed from year to year, the diagram should present much the same general features. For the first world war R. G. Hawtry's digest (E **10**, 695) was accepted, after the omission of loans to allies and dominions, exchange account, premoratorium bills and pensions. For the second world war the expenditures were taken from the "Statistical Digest of the War" (1951 H. M. Stationery Office p. 173), expenditures under 'defence' and under the Defence Loan Acts 1937 and 1939 being included. Lease-lend and loans to allies were ignored. At the end of the second war 'terminal charges' were included. The intention was to omit pensions throughout, but during the second world war this detail was obscure.

For the conversion of the sterling expenditure into warfinpersal, the average weekly earnings of men in engineering and the like were extracted from government publications. The other factor of the annual earnings was the number of paid weeks in the year, which was taken to be 50 up to 1937, and afterwards 51 because of holidays with pay.

VOL. 12] SANKHYĀ: THE INDIAN JOURNAL OF STATISTICS [PART 3

8. The arms-race of A.D. 1929 to 1939

A valuable guide to the interpretation of the national accounts of sixty-two countries is the survey of *Public Finance* 1928 *to* 1935 published by the League of Nations (Geneva 1936 to 1938); but it relates to a time when warlike expenditure was near to a minimum. *The Armaments Year Books of the League of Nations* also contain warlike expenditure, stated on a uniform plan, for the years 1922 to 1940; but after 1933 that for Germany had to be sought elsewhere. The earnings were published in the *I.L.O. Year Books of Labour Statistics.*

The author began to study this race in the year 1935 when the nations China, Czechoslovakia, France, Germany, Britain, Italy, Japan, Poland, U.S.A., and U.S.-S.R. were interested, but had not yet all clearly taken sides. At this time the phenomenon appeared to be an interaction between n nations, rather than between two sides. By an obvious generalization of (6.1) and (6.2), the relations between the warfinpersals $x_1, x_2, x_3, \ldots, x_n$ of the n nations were assumed to be expressed by the linear differential equations

$$\frac{dx_i}{dt} = g_i + \sum_{j=1}^{j=n} K_{ij}\, x_j, \qquad (i=1, 2, 3, \ldots, n) \qquad \ldots \quad (8.1)$$

in which the g_i and the K_{ij} were independent of time. By a shift of origin, the g_i were removed, leaving

$$\frac{dX_i}{dt} = \sum_{j=1}^{j=n} K_{ij}\, X_j \qquad (i = 1, 2, 3, \ldots, n). \qquad \ldots \quad (8.2)$$

Each X_i differs from the warfinpersal x_i by a constant. This treatment was almost too ambitious, because it was applied to 9 or later to 10 nations; so that there were 81 or 100 unknown coefficients K_{ij} requiring to be estimated. However some features of the 10×10 matrix $[K_{ij}]$ were obvious. The diagonal elements K_{ij} were restraint-coefficients, certainly negative, and probably of the order of magnitude, namely one year^{-1}, found from the previous disarmament of victors (§ 5). The non-diagonal elements K_{ij}, where i differs from j, were defence-coefficients. If the i-nation was in alliance with the j-nation, K_{ij} was assumed to be zero; zero also if the nations were geographically inaccessible to one another. Otherwise the non-diagonal K_{ij} were taken to be positive, and more so if the nations concerned were known to be at enmity with one another. At first it was supposed that enmity was exactly mutual so that $K_{ij}=K_{ji}$. When any matrix of numerical elements had been proposed for trial, the enquiry proceeded by the computation of some of the latent roots, latent rows, and latent columns of $[K_{ij}]$. A method of successive approximation, called "purification", was designed for this purpose (Richardson 1950). Let $\lambda_1, \lambda_2, \ldots \lambda_n$ denote the n latent roots of the matrix $[K_{ij}]$. The solution of the system of equations (8.2) is often made to look easy by leaving out the awkward special cases. These were dealt with by Weierstrass (1875) in connection with his "elementary divisors." The purification method is capable of finding complex latent roots, or non-linear elementary divisors; these peculiarities, although not so far encountered, may lurk in the uninvestigated intermediate roots $\lambda_2, \ldots, \lambda_9$ of Table 2.

The simplest pattern of thought about the solution of the n equations (8.2) is attained by attending to their "normal co-ordinates", which are generalizations of the $X+Y$ and $X-Y$ of equations (6.5) and (6.6), and, like them, each involve only one exponential time-factor, $e^{\lambda t}$, although in general that may be complex, and or multiplied by a polynominal in t. There are n linearly independent normal co-ordinates. Let S_r denote any one of them. Then if the associated elementary divisor is linear

$$S_r = A_r e^{\lambda_r t} \quad \ldots \quad (8.3)$$

where A_r is independent of time, and where

$$S_r = \sum_{i-1}^{i=n} l_{ri} X_i, \quad \ldots \quad (8.4)$$

the l_{ri} being constant weights proportional to that latent column of the matrix $[K_{ij}]$ which belongs to its latent root λ_r. It is convenient to fix the proportionality by making

$$\sum_{i-1}^{i=n} l_{ri}^2 = 1, \quad \ldots \quad (8.5)$$

for with this convention the weights l_{ri} may be called, allegorically, the "direction cosines of the perpendicular to the plane $S_r=0$".

The search for a matrix $[K_{ij}]$, which, when inserted into (8.1), would provide a credible description of the observed movements of the ten warfinpersals, can now be explained. It proceeded by trial and amendment. For a trial matrix $[K_{ij}]$ the purification method revealed the latent root λ_1 of greatest real part, and its associated l_{1i}. The warfinpersals x_i were known. The following expression was computed for each successive year

$$\sum_{i-1}^{i=n} l_{1i}\, x_i = s_i'. \quad \ldots \quad (8.6)$$

It is called a "hybrid" because it is obtained partly from the observed x_i, and partly from the weights l_{1i} of a trial theory. The course in time of s_1' was analysed by the (μ,δ) diagram of § 3, being fitted approximately with the formula

$$s_1' = Q + A e^{\lambda_1' t} \quad \ldots \quad (8.7)$$

Here λ'_1 is another hybrid, which ought to agree with λ_1, found directly from the assumed matrix, but at the first trial did not. Therefore the maxtrix was amended until its latent root of greatest real part agreed with the hybrid thereof.

A similar procedure was applied to the latent root λ_{10} of least real part. It is negative, so that the hybrid s'_{10} should drift towards, or loiter near, a constant. At first it did not; and the amendment, to make it to do so, involved a departure from the previous assumption that $K_{ij}=K_{ji}$.

The enquiry can be seen in detail in a microfilm (Richardson 1947, second edition 1949). It was not entirely conclusive, because the intermediate latent roots $\lambda_2, \lambda_3, \ldots, \lambda_9$ were not investigated. Nevertheless two results of remarkable interest emerged. The following quotations refer to the matrix, called Type XVIII in the microfilm and copied in Table 2 below. The latent root of greatest real part was found to be $\lambda_1=0.27$ year^{-1}, nearly the same as the hybrid

$$\lambda'_1 = 0.25 \text{ year}^{-1}. \qquad \ldots \quad (8.8)$$

Being positive, it corresponded to instability. The direction cosines l_{1i} associated with λ_1, were all positive; so that its normal co-ordinate S_1 was a generalization of the $X+Y$ of the simple theory of two equal sides. Moreover $\lambda'_1=0.25$ agrees fairly with $k-\alpha=0.27$ stated in (6.7) as appropriate to $X+Y$ for the previous arms-race.

The latent root of least real part was found to be

$$\lambda_{10} = -2.25 \text{ year}^{-1}. \qquad \ldots \quad (8.9)$$

It corresponds to stability. Some of the direction cosines associated with λ_{10} were positive, some negative; so that its normal co-ordinate S_{10} was a generalization of the $X-Y$ of the simple theory of two equal sides. Moreover $\lambda_{10}=-2.25$ year^{-1} is of the same order as $-(k+\alpha)=-2.53$ year^{-1} stated in (6.8) as appropriate to $X-Y$ in the previous arms-race. For $X-Y$ the direction cosines for the different nations were found to be in A.D. 1935:—

France	U.S.S.R.	Czechoslovakia	G. Britain & N.I.	U.S.A.
0.73	0.56	0.22	0.17	0.06
Poland	**China**	**Japan**	**Italy**	**Germany**
−0.00	−0.01	−0.07	−0.10	−0.23

This polarity seemed tolerably credible when compared with political information.

The gist of all this analysis of the arms-race of A.D. 1930 to 1939 is of course the numerical matrix $[K_{ij}]$. As far as I am aware, no one else has published it. So I offer in Table 2 my much amended result, with the warning that it probably needs further amendments related to $\lambda_2, \lambda_3, \ldots, \lambda_9$.

THREE ARMS-RACES AND TWO DISARMAMENTS

To find, in reciprocal years, the elements of $[K_{ij}]$, divide each of the numbers in the body of the following table by thirty.

TABLE 2. INTERACTION OF TEN NATIONS IN A.D. 1935

first suffix	second suffix	1	2	3	4	5	6	7	8	9	10
1	Czechoslovakia	−30	0	0	2	0	0	0	1	0	0
2	China	0	−30	0	0	0	0	12	0	0	18
3	France	0	0	−54	4	0	4	0	0	0	0
4	Germany	36	0	36	−30	18	0	0	3	0	72
5	G. Britain & N. I.	0	0	0	4	−45	6	2	0	0	0
6	Italy	0	0	18	0	36	−15	0	0	0	18
7	Japan	0	12	0	0	0	0	−30	0	36	36
8	Poland	9	0	0	3	0	0	0	−30	0	9
9	U. S. A.	0	0	0	2	6	2	4	0	−21	6
10	U. S. S. R.	0	2	0	8	6	2	4	1	0	−30

9. DISARMAMENT OF VICTORS A.D. 1945 TO 1948

In these circumstances it is reasonable to attend to nations separately, and to compute for each the restraint coefficient α.

The United States of America. The actual expenditure on "military services" was taken from a summary (K 11409). The "hourly earnings" and "hours worked per week" were taken from the *Statistical Year Book of the United Nations* 1949-50 and were multiplied together, and interpolated linearly to the financial years which ended on June 30th. It was assumed that there were fifty working weeks in the year. The results are shown in Table 3.

TABLE 3. DISARMAMENT OF U. S. A.

year ended June 30 of	military services, million \$	earnings in financial year, \$	correcting factor from (3.1) $\overline{(u/v)}/(\bar{u}/\bar{v})$	mean warfinpersal, million typical persons
1945	—	2262	—	—
1946	45 134	2207	1.0016	20.483
1947	14 314	2346	1.0089	6.156
1948	10 961	2603	1.0011	4.216
1949	11 914	2726	—	—

The restraint coefficient was computed by formula (3.10) and found to be

$$\alpha = -\lambda = 2.00 \text{ year}^{-1}. \qquad \dots \quad (9.1)$$

The U. K. of Great Britain and Northern Ireland. The financial years ended on March 31st. For this country I was able to collect the information directly from official publications, a process which entangles one in puzzling questions. The struggle between rival wishes to spend the governmental income in different ways is centralized by the Treasury and is finally decided by the House of Commons. It seems likely therefore that we shall obtain the simplest psychological picture if we attend to expenditure authorized by Parliament under the heading of defence in any year. This is a reason for accepting the "net" amount provided by Parliament instead of the "gross" total, which includes receipts by the departments of rents, services, and sales, appropriated in aid. The departments included were the navy, army, air-force, Ministry of Supply, also, when they existed, the Ministry of Aircraft Production, and the Ministry of Defence. Little difference to the restraint coefficient, is made by the treatment of "non-effective services" mostly pensions, because they varied little: actually they are included in Table 4. On the contrary the treatment of "terminal changes" is important, because they diminished rapidly from year to year. These are described in the *Statement relating to Defence* of February 1946 (Cmd. 6743) as including: war gratuities and other payments to be made at the end of service, termination of contracts, winding up of closed contracts, compensation for damage on the termination of requisitions of land and property. These "terminal changes" might be excluded as being delayed payments for a war that had ended; but on the other hand the same description would cover the pay of armed personnel who were idly waiting to be transported home, or interest on wartime loans. Personally I am in favour of including terminal charges; but Table 4 is arranged both ways. Interest on loans, being almost constant, would not affect the term $e^{\lambda t}$, and is excluded.

As to the distinction between intentions and actualities, the expenditures of the navy, army, and air-force were those reported after the event in *House of Commons papers*, whereas the expenditures of the Ministries of Supply and Aircraft Production, together with the deductions for terminal charges were estimates published in *Command papers* (Cmd. 6743, 7042, 7327), a month or two before the financial year began. The result in sterling appears in the second column of Table 4.

The conversion to warfinpersal was done by statistics published in the *Ministry of Labour Gazette* of the weekly earnings of men of 21 years and over in metal, engineering, and shipbuilding. Unfortunately a change in the classification occurred in October 1948 to metal manufacture as one class, and jointly engineering, shipbuilding and electrical engineering as another class; so that the last half-year had to be extrapolated. To allow for holidays with pay, each financial year in Table 4 was taken to include 51 paid weeks. The warfinpersal was calculated by (3.1).

Table 4. Disarmament of Britain

financial year	warlike expenditure in million £		typical annual earnings £	mean warfinpersal in millions of typical men	
	with terminal	without terminal		with terminal	without terminal
1946 to 47	1734.1	1158.1	330.2	5.28	3.52
1947 to 48	880.7	765.2	355.0	2.49	2.16
1948 to 49	769.8	709.3	375.6	2.05	1.89

From the three financial years in Table 4 the coefficient $\alpha = -\lambda$ was computed by formula (3.10) and found to be

$$\alpha = 1.85 \text{ year}^{-1}, \text{ if terminal charges are included,} \quad \dots \quad (9.2)$$

or

$$\alpha = 1.62 \text{ year}^{-1}, \text{ if terminal charges are omitted.} \quad \dots \quad (9.3)$$

The Soviet Union, U.S.S.R. The financial year coincides with the calendar year. The warlike expenditure was extracted from the *Statesman's Year Books* for 1947, 1950, and 1951 and is entered in Table 5 column 2. The average annual earnings of workers and employees in the Soviet economy was taken from a book by Harry Schwartz (1951 page 460). Gaps in Schwartz's statement have been bridged on the assumption that earnings increased in a constant ratio for equal intervals of time. These numbers are entered in column 3. The warfinpersal in column 4 was computed as usual, by formula (3.1).

TABLE 5. DISARMAMENT OF U.S.S.R.

year	warlike expenditure 10^9 roubles	average annual earnings roubles	mean warfinpersal millions of people
1943	124.7	[4960]	—
1944	137.9	[5286]	26.10
1945	128.2	[5633]	22.77
1946	72.2	6000	12.05
1947	67.0	7100	9.31
1948	66.0	7400	8.92
1949	79.2	—	—

Whereas the demobilization of U.S.A. and U.K. persisted conspicuously, into 1946, that of U.S.S.R. was mostly over by the end of 1945. There are obvious reasons for this difference of dates. The war for U.S.S.R. practically ended with the defeat of Germany on 1945.V.8; whereas U.S.A. and U.K. continued to fight Japan for 3 months longer. Moreover, when the fighing ceased, the forces of U.S.A. and U.K. were at the end of longer lines of communication than those of U.S.S.R. had been. For the purpose of computing the restraint coefficient α, the changes in Russian warfinpersal during the years 1946, '47, '48 are rather too small to give an accurate result. They give in fact, via equation (3.10)

$$\alpha = 1.95 \text{ year}^{-1} \quad \dots \quad (9.4)$$

It seemed however, desirable to bring in the Russian warfinpersal for 1945, although this involved some intellectual effort. The restraint coefficient α operates by itself only when the nation is free from threats. Therefore the early part of 1945 must be accounted separately. The year was divided in the ratio 2 : 3 by a cut at May 26th. This allows some days after victory for celebrations, confirmation of the absence of threat, and cancellation of orders. In the first 2/5 of the year the warfinpersal was assumed to be at its mean for 1944, namely 26.10. Accordingly the mean warfinpersal, ϕ_1, say, during the remainder of the year was given by $\frac{2}{5}\times 26.10+\frac{3}{5}\phi_1=22.77$, whence

$$\phi_1 = 20.55 \qquad \ldots \quad (9.5)$$

To fit the formula (1.10), namely $f(t)=Q+Ae^{\lambda t}$, to the observed warfinpersal given thus for a part of a year, it is best to use integrals, in the following manner. Take $t=0$ at the beginning of the year 1946. For the last 3/5 of the previous year, the mean warfinpersal is

$$\phi_1 = 20.55 = \frac{5}{3}\int_{-3/5}^{0} f dt = Q+\frac{5}{3}\,\frac{A}{\lambda}\{1-e^{-3\lambda/5}\}. \qquad \ldots \quad (9.6)$$

For the whole of 1946 we have in like manner

$$12.05 = \int_0^1 f dt = Q+\frac{A}{\lambda}\{e^{\lambda}-1\}. \qquad \ldots \quad (9.7)$$

Similarly for 1947

$$9.31 = \int_1^2 f dt = Q+\frac{A}{\lambda}\{e^{2\lambda}-e^{\lambda}\}. \qquad \ldots \quad (9.8)$$

Elimination of Q between the last three equations in successive pairs leaves

$$8.50 = \frac{A}{\lambda}\left\{\frac{8}{3}-\frac{5}{3}e^{-3\lambda/5}-e^{\lambda}\right\} \qquad \ldots \quad (9.9)$$

$$2.74 = \frac{A}{\lambda}\{-1+2e^{\lambda}-e^{2\lambda}\}. \qquad \ldots \quad (9.10)$$

Elimination of A then leaves

$$9.307 = \frac{3e^{\lambda}+5e^{-3\lambda/5}-8}{e^{2\lambda}-2e^{\lambda}+1}. \qquad \ldots \quad (9.11)$$

A graph of the function in (9.11) was plotted, and it showed that

$$\alpha = -\lambda = 1.66 \text{ year}^{-1}. \qquad \ldots \quad (9.12)$$

In a previous publication (Richardson 1951) I rashly alluded to the Russian demobilization as "tardy", that word being intended as a summary of adverse British comment. The present investigation shows on the contrary that the Russian demobilization began early and proceeded rapidly. Adverse comment would be more likely to be aimed at the minimum levels of warfinpersal. These can be seen in Tables 3, 4, 5, 6.

Summary on restraint coefficients during A.D. 1945 *to* 1948

The values of α in reciprocal years are :—

U.S.A. 2.00, U.K. 1.85 or 1.62, U.S.S.R. 1.66 or 1.95

Personally I accept the first alternatives, the mean of which is

$$\alpha = 1.84 \text{ year}^{-1}. \qquad \ldots \quad (9.13)$$

After the First World War the mean value of α was less, namely 1.1 year^{-1} (5.1). Speed may have been increased by practice.

10. The arms-race that began in A.D. 1948

The two previous great arms-races began so inconspicuously that it is impossible, by a study of warfinpersal, to fix the precise dates of their beginnings. In this respect a suitable mathematical description is $X+Y = e^{\lambda t}$+random effects. For the same reason it is difficult to say which side took the first moves in those previous races. We need not be surprised if the beginning of the present race is hidden in a similar obscurity.

The Cominform was organized in September 1947 by the communist parties of the Soviet Union, Bulgaria, Czechoslovakia, France, Hungary, Italy, Poland, Rumania, and Yugoslavia; but in France and Italy the communists were not in power (K 8864). The warfinpersals of U.S.A. and U.S.S.R. were both minimum in the year 1948; see Tables 3 and 5. The Brussels Treaty between Britain, France, Belgium, Holland and Luxemburg was signed on 1948.iii.17 (K 9157). Preliminary discussions about a North Atlantic alliance began in 1948.vii (K 9766) and by 1949. vii.21 U.S.A., Canada, United Kingdom, France, Belgium, Holland, Luxemburg, Denmark, Iceland, Italy, Norway, and Portugal had ratified the North Atlantic Treaty (K 10159 and index). Communist Yugoslavia was expelled from the Cominform on 1948.vi.28 (K 9381). Because the partisanship of the nations was almost definite by the beginning of 1948 the economic facts may suitably be compared with the simple theory of two sides, instead of with the more elabortate theory (8.1) of n nations.

The judicious labour of collecting the warlike expenditure and the earnings, computing their ratio, and publishing the result as "man-years" has been performed by the U.N. Economic Commission for Europe. In Table 6 their results are rearranged by sides, the neutrals Spain, Sweden and Switzerland are omitted, man-years is renamed warfinpersal, and is expressed in million typical persons.

Table 6 does not agree accurately with Tables 3, 4, 5. I do not know the exact reason; but there always are differences of opinion about warlike affairs; and

I think it proper to leave the discordance obvious, in order to promote future clarification.

TABLE 6. THE ARMS-RACE A.D. 1949 TO 1951

calendar years*	1949		1950		1951
	warfinpersal in millions of typical persons				
western					
Belgium	0.120		0.130		0.210
Denmark	0.050		0.060b		0.075b
France	1.600		1.700		2.200
Italy	0.950		1.350		1.350
Netherlands	0.250		0.375		0.475b
Norway	0.055		0.055		0.070
United Kingdom	2.300		2.400		4.200
Yugoslavia	0.475		0.550		0.575
United States	4.700		4.600		11.400
totals	10.500		11.220		20.555
eastern					
Czechoslovakia	0.160a		0.160		0.170
Poland	0.275		0.325		0.450
U. S. S. R.	8.400		8.800		10.200
Hungary, Rumania, Bulgaria, China — information lacking					
totals	8.835		9.285		10.820
totals for both sides	19.335		20.505		31.375
successive differences		1.170		10.870	
means of pairs		19.920		25.940	

Note : a Budget Forecast, b Preliminary estimate.

* "Except in the case of the United States and Norway, the figures relate to the fiscal years beginning in the years indicated in the column headings." (p. 138).

It would be extremely unwise, on such incomplete information, to make a study of the difference $X-Y$; but the sum $X+Y$ is comparatively meaningful because the nations are all going the same way. Taking therefore the totals for both sides together, the slope b for the (μ, δ) diagram is found by (3.10) as $b = 1.611$. Therefore

$$\lambda = 2.23 \text{ year}^{-1} \qquad \dots \quad (10.1)$$

This is very much larger than the $\lambda = 0.27$ year^{-1} found for both the two previous arms-races. The analysis of the earlier races was more reliable in so far as the data extended over a longer run of years : about six instead of three. Inspection of Table 6 shows that the present sudden increase of $X+Y$ arises chiefly from the contribution of U.S.A. A more trustworthy estimate of λ may perhaps be obtainable in future years.

Already, by the time of proof-correction in May 1953, there is some imperfect evidence which suggests that, while $X+Y$ has continued to increase, $d(X+Y)/dt$ has attained a maximum at about the beginning of the year 1952. If so, that would be quite unlike the two previous great arms-races; the concept of a constant λ would not apply; and the (μ, δ) diagram would lose its simple interpretation.

THREE ARMS-RACES AND TWO DISARMAMENTS

11. SUMMARY

I submit, for the consideration of economists, the proposition that the financial facts about arms-races, and about the disarmament of victors, fit tolerably well with the psychological theory of motives which is summarized, in its simplest form, for the arms-races of 1906-1914 and 1929-39, by equations (6.1) and (6.2) namely

$$dX/dt = kY - \alpha X, \qquad dY/dt = kX - \alpha Y$$

and that in the years 1945 to 1948:

$$\alpha \text{ was about } 1.8 \text{ year}^{-1},$$

also, although with considerably more observational uncertainty, and for the year 1950

$$k-\alpha \text{ was about } 2.2 \text{ year}^{-1},$$

but has since diminished.

This coefficient $k-\alpha$, which measures the instability of the arms-race, was formerly much smaller, having been 0.2_7 year^{-1} during A.D. 1907 to 1914. If we may extrapolate backwards, we should expect to come to an earlier time when $k-\alpha$ was zero, or negative; so that arms-races of the unstable modern type did not occur. There is direct evidence that this was so; for Dr. Curt Rosenberg made a search into the warlike expenditures preceding the Russo-Turkish war of 1877 and did not find any unstable arms-race. Still earlier the expenditures of France and of the North German Confederation during A.D. 1868, 1869, 1870 were found by him to have been drifting towards a stable balance of power (Richardson 1947 and 1949 §3.5). The belief of statesmen in the Balance of Power, as a safe practical policy, presumably arose in those earlier times when $k-\alpha$ was negative; but unfortunately the belief became a tradition, which survived long after $k-\alpha$ had become positive, with disastrous results.

The cause of the increase of $k-\alpha$ during the last seventy years is not obvious. Personally I guess that it is a consequence of the application of science to war-industry.

From the ethical point of view the chief peculiarity of this theory is that the three arms-races are described, not as struggles of heroes against criminals, but as affairs of mutual stimulation by way of the defence-constants k. It was Bertrand Russell in 1914 who, I believe, first emphasized mutuality. It is proper now to ask whether the economic facts prove the existence of mutual stimulation? I am not clear that they prove it, but certainly they do not contradict it. The evidence for mutuality appears to be partly psychological. Moral judgments as to what acts are righteous, or what are criminal, should, I suggest, be made in the presence of the belief that, between equal groups, mutual stimulation is very important. Beaucoup condamner c'est peu comprendre.

In A.D. 1913 and in 1938 the opposing sides were roughly equal to one another, and again now they are of the same order of size. This has allowed the main features of the theory to be illustrated very simply by the special cases of exact equality. The motives formulated in § 6 are parsimony and defensiveness, which depend on the warfinpersals, together with other motives independent of the warfinpersals. The equations of §6 are linear, so that they generalize, as in §8, for $2, 3, \ldots, n$ unequal nations. The generalization has been published in detail in microfilm (Richardson 1947 and 1949).

When war has gone on so long that one side is beginning to feel exhausted, quite different motives come into operation, and the aforesaid linear equations no longer apply. Non-linear equations, intended to represent submission, have been discussed by Richardson (1953).

REFERENCES

CHURCHILL, WINSTON (1923): *The World Crisis* 1911-1914. London, Thornton Butterworth.

E is short for *Encyclopaedia Britannica.* 14th edition 1929.

FISHER, R. A. (1936): *Statistical Methods for Research Workers.* Edinburgh, Oliver and Boyd.

GREY, VISCOUNT (1925): *Twenty Five Years.* London, Hodder & Stoughton.

HAIG, R. M. (1929): *The Public Finances of Post-War France.* New York, Columbia University Press.

HUBBARD, L. E. (1938): *Soviet Trade and Distribution.* London, Macmillan.

JEFFREYS, H. (1939): *Theory of Probability.* Oxford, Clarendon Press

K is short for *Keesing's Contemporary Archives.* Bristol, England. Keesing's Publications Ltd. The number following K is that of the page.

KEYNES, J. M. (1936): *The General Theory of Employment, Interest and Money.* London, Macmillan.

MCGUIRE, C. E. (1927): *Italy's International Economic Position.* London, Allen & Unwin.

Pol, Balth van der. (1926): *Phil Mag.* **2**, 978-992.

RASHEVSKY, N. (1947): *Mathematical Theory of Human Relations.*Principia Press, Bloomington, Indiana.

RICHARDSON, L. F. (1919): *Mathematical Psychology of War.* In British copyright libraries.

——— (1939): *Generalized Foreign Politics.* Brit. Journ. Psychol. Mong. Supplt. No. 23.

——— (1948): *Psychometrika.* **13**, 197-232.

——— (1947): 2nd edn. 1949. *Arms and Insecurity.* A book on microfilm, 35 mm. wide, published by the author.

——— *Phil. Trans. Roy. Soc. London,* **242**, 439-491.

——— (1951): *Nature.* **168**, 567.

——— (1953): *Brit. J. Psychol.* (Statistical Section), November.

RUSSELL, B. A. W. (1914): *War the Offspring of Fear.* London, Union of Democratic Control.

SCHWARTZ, H. (1951): *Russia's Soviet Economy.* London, Jonathan Cape.

SHEPPARD, W. F. (1899): *London Math. Soc. Proc.* **31**, 449-488.

SPURGEON, E. F. (1929): *Life Contingencies.* Cambridge, University Press.

UNITED NATIONS (1951): *Economic Survey of Europe in* 1950. Geneva, Palais des Nations.

VOLTERRA, V. (1931): *Leçons sur la Théorie Mathématique de la Lutte pour la Vie.* Paris: Gauthier-Villars.

Weierstrass, K. (1875): *K. Akad. der Wiss.* also 1895 *Werke,* **2**, 75 Berlin: Mayer und Müller.

Contents of Volume 1

Note on the patents

The patents

Check-list of publications not reproduced in either volume

1905 (with C. H. Carpenter). Note on the structure of steel plates. *Minut. Proc. Instn civ. Engrs* **155**, 411.

1906 Index to Volumes I–V. *Biometrika.*

1907:1 (*K. Pearson & A. F. C. Pollard, assisted by C. H. Wheen & L. F. Richardson*). *An experimental study of the stresses in masonry dams. Drap. Co. Res. Mem. tech. Ser.* 5.

1907:2 On a freehand potential method. *Rep. Br. Ass. Advmt Sci.* 457.

1913:2 Review of magnetic disturbances at Eskdalemuir in the year 1913. *Br. met. magn. Yb. 1913*, Pt IV, 78–81.

1914:1 Magnetic disturbances 1913. *Nature, Lond.* **94**, 450.

1914:2 Annual review of magnetic disturbances at Eskdalemuir 1914. *Br. met. magn. Yb. 1914*, Pt IV, 75–8.

1915:2 The detection of distant thunderstorms by clicks in a telephone. *Advisory Committee for Aeronautics*, Report T 623.

1921:1 Tables of 0.288th powers (by T. N. Doerr). Introduction by L. F. Richardson. *Q. J. R. met. Soc.* **47**, 196.

1921:2 'Memorandum on the upper air works by V. Bjerknes, in collaboration with L. F. Richardson. *International Commission for the Investigation of the Upper Air, Proc. Seventh Meeting, Bergen July 1921*, Appendix II.

1922:1 *Weather Prediction by Numerical Process.* London: Cambridge University Press.

1922:2 *Computing forms for numerical calculations described in 'Weather Prediction by Numerical Process'*. London: Cambridge University Press.

1922:4 Review of *On the Dynamics of the Circular Vortex* by J. Bjerknes. *Q. J. R. met. Soc.* **48**, 375–6.

924:3 Review of *The Calculus of Observations* by E. T. Whittaker & G. Robinson. *Q. J. R. met. Soc.* **50**, 163.

1924:5 A holiday resort for geophysicists? *Q. J. R. met. Soc.* **50**, 381.

1924:6 *Hints to Meteorological Observers.* 8th edn by W. Marriot. London: Stanford. Section on turbulence, pp. 29–30, by L. F. Richardson.

1932:1 Review of *Manual of Meteorology* by Sir Napier Shaw. *Nature, Lond.* **129**, 220–1.

1932:1 Contribution and addendum to the discussion on a paper by W. A. Leyshon 'On periodic movements of the negative glow in discharge tubes'. *Proc. Phys. Soc. Lond.* **44**, 188–9.

1938:1 A psychology class at an evening institute. *Education for Commerce* Jan. 1938, 103–24.

1939:2 A visit to Danzig and neighbourhood. *Paisley Daily Express* 21 and 23 Aug. 1939.

1939:3 A visit to Danzig. *Northern Echo* 28 Aug. 1939.

1940 Quantitative estimates of sensory events. Final report of committee: Contribution E to Appendix I: On the sone scale. *Advmt Sci.* **1**, 339–40.

1942 Comments on M.R.P. 43. *Met. Res. Paper* **55**.

1943:1 Reviews of Mechanical Physics by H. Dingle, *Sub-atomic Physics* by H. Dingle, and *University Physics* by F. C. Champion. *Nature, Lond.* **152**, 7–8.

1943:2 Review of *Analytical Experimental Physics* by H. B. Lemon & M. Ference. *Nature, Lond.* **152**, 432–3.

1943:3 Review of *Practical Physics* by M. W. White, K. V. Manning, R. L. Weber & R. O. Cornett. *Nature, Lond.* **152**, 491–2.

1947:2 *Arms and Insecurity.* (A book on microfilm, privately produced.)

1947:3 Successive approximations to latent vectors and latent roots of a numerical matrix. MS published as §3/17 of 1947:2, pp. 433–44.

1948:1 The persistence of national hatred and the changeability of its objects. (Abstract of 1919:4.) *Proceedings and Papers Twelfth International Congress of Psychology.* Edinburgh: Oliver & Boyd, 1950, pp. 109–10.

1949:2 *Arms and Insecurity.* (A book on microfilm, privated produced; a slightly revised version of 1947:2.)

1949:3 Successive approximations to latent vectors and latent roots of a numerical matrix. MS published as §3/17 of 1949:2, pp. 433–44.

1950:4 *Statistics of Deadly Quarrels.* (A book on microfilm, privately produced.)

EI 1950:6 Gilbert Hancock Richardson (Obituary). *The Friend* **108**, 583.

1956 *The World of Mathematics* (ed. J. R. Newman). Volume 2, Part VI, Chapter 6, *Mathematics of War and Foreign Politics* and Chapter 7, *Statistics of Deadly Quarrels.* New York: Simon & Schuster. (Reprinted, omitting the final section of the second chapter, from Richardson 1950:5.)

1957 A bibliography of Lewis Fry Richardson's studies of the causation of wars with a view to their avoidance. *J. Conflict Resolution* **1**, 305–7. Also included in 1960:1, 2.

1960:1 *Statistics of Deadly Quarrels* (ed. Quincy Wright and C. C. Lienau). Pittsburgh: Boxwood, Chicago: Quadrangle, London: Stevens and Sons

1960:2 *Arms and Insecurity: a Mathematical Study of the Causes and Origins of War* (ed. N. Rashevsky and E. Trucco). Pittsburgh: Boxwood, Chicago: Quadrangle, London: Stevens and Sons.

1965 *Weather Prediction by Numerical Process.* Reprint of Richardson (1922:1) with a new introduction by Sydney Chapman. New York: Dover.